Quantum Cosmology

Quantum Cosmology

Editor

Paulo Vargas Moniz

MDPI • Basel • Beijing • Wuhan • Barcelona • Belgrade • Manchester • Tokyo • Cluj • Tianjin

Editor
Paulo Vargas Moniz
CMA-UBI and Physics
Department
Universidade da Beira
Interior
Covilhã
Portugal

Editorial Office
MDPI
St. Alban-Anlage 66
4052 Basel, Switzerland

This is a reprint of articles from the Special Issue published online in the open access journal *Universe* (ISSN 2218-1997) (available at: www.mdpi.com/journal/universe/special_issues/quantum_cosmology).

For citation purposes, cite each article independently as indicated on the article page online and as indicated below:

LastName, A.A.; LastName, B.B.; LastName, C.C. Article Title. *Journal Name* **Year**, *Volume Number*, Page Range.

ISBN 978-3-0365-4726-8 (Hbk)
ISBN 978-3-0365-4725-1 (PDF)

Cover image courtesy of Paulo Vargas Moniz

© 2022 by the authors. Articles in this book are Open Access and distributed under the Creative Commons Attribution (CC BY) license, which allows users to download, copy and build upon published articles, as long as the author and publisher are properly credited, which ensures maximum dissemination and a wider impact of our publications.

The book as a whole is distributed by MDPI under the terms and conditions of the Creative Commons license CC BY-NC-ND.

Contents

About the Editor vii

Preface to "Quantum Cosmology" ix

Paulo Vargas Moniz
Editorial to the Special Issue "Quantum Cosmology"
Reprinted from: *Universe* **2022**, *8*, 336, doi:10.3390/universe8060336 1

Shahram Jalalzadeh, Seyed Meraj M. Rasouli and Paulo Moniz
Shape Invariant Potentials in Supersymmetric Quantum Cosmology
Reprinted from: *Universe* **2022**, *8*, 316, doi:10.3390/universe8060316 5

Salvatore Capozziello and Francesco Bajardi
Minisuperspace Quantum Cosmology in Metric and Affine Theories of Gravity
Reprinted from: *Universe* **2022**, *8*, 177, doi:10.3390/universe8030177 25

Przemysław Małkiewicz, Patrick Peter and Sandro Dias Pinto Vitenti
Clocks and Trajectories in Quantum Cosmology
Reprinted from: *Universe* **2022**, *8*, 71, doi:10.3390/universe8020071 43

Hugo García-Compeán, Octavio Obregón and Cupatitzio Ramírez
Topics in Supersymmetric and Noncommutative Quantum Cosmology
Reprinted from: *Universe* **2021**, *7*, 434, doi:10.3390/universe7110434 55

Bei-Lok Hu
Weyl Curvature Hypothesis in Light of Quantum Backreaction at Cosmological Singularities or
Bounces
Reprinted from: *Universe* **2021**, *7*, 424, doi:10.3390/universe7110424 71

Salvador J. Robles-Pérez
Quantum Cosmology with Third Quantisation
Reprinted from: *Universe* **2021**, *7*, 404, doi:10.3390/universe7110404 99

Dong-han Yeom
Fuzzy Instantons in Landscape and Swampland: Review of the Hartle–Hawking Wave
Function and Several Applications
Reprinted from: *Universe* **2021**, *7*, 367, doi:10.3390/universe7100367 161

Gabriele Barca, Eleonora Giovannetti and Giovanni Montani
An Overview on the Nature of the Bounce in LQC and PQM
Reprinted from: *Universe* **2021**, *7*, 327, doi:10.3390/universe7090327 181

Jerónimo Cortez, Guillermo A. Mena Marugán and José M. Velhinho
A Brief Overview of Results about Uniqueness of the Quantization in Cosmology
Reprinted from: *Universe* **2021**, *7*, 299, doi:10.3390/universe7080299 237

João Marto
Hawking Radiation and Black Hole GravitationalBack Reaction—A Quantum
Geometrodynamical Simplified Model
Reprinted from: *Universe* **2021**, *7*, 297, doi:10.3390/universe7080297 257

Teodor Borislavov Vasilev, Mariam Bouhmadi-López and Prado Martín-Moruno
Classical and Quantum $f(R)$ Cosmology: The Big Rip, the Little Rip and the Little Sibling of the Big Rip
Reprinted from: *Universe* **2021**, 7, 288, doi:10.3390/universe7080288 . . . . **279**

Carla R. Almeida, Olesya Galkina and Julio César Fabris
Quantum and Classical Cosmology in the Brans–Dicke Theory
Reprinted from: *Universe* **2021**, 7, 286, doi:10.3390/universe7080286 . . . . **309**

Yaser Tavakoli
Cosmological Particle Production in Quantum Gravity
Reprinted from: *Universe* **2021**, 7, 258, doi:10.3390/universe7080258 . . . . **327**

Alexander Yu Kamenshchik, Jeinny Nallely Pérez Rodríguez and Tereza Vardanyan
Time and Evolution in Quantum and Classical Cosmology
Reprinted from: *Universe* **2021**, 7, 219, doi:10.3390/universe7070219 . . . . **343**

Martin Bojowald
Cosmic Tangle: Loop Quantum Cosmology and CMB Anomalies
Reprinted from: *Universe* **2021**, 7, 186, doi:10.3390/universe7060186 . . . . **367**

Maurizio Gasperini
Quantum String Cosmology
Reprinted from: *Universe* **2021**, 7, 14, doi:10.3390/universe7010014 . . . . **377**

About the Editor

Paulo Vargas Moniz

Paulo Moniz is a full professor at UBI (Portugal). He graduated from Lisbon and then moved to DAMTP, Cambridge, for some years. He has occasionally returned to Cambridge and is also a life member at Clare Hall (college). He has been on the Editorial Board and the Advisory Board at CQG. The author of several books and several (much) more research papers, he has supervised students and post-docs, and has been a research visitor at many institutions, multiple times. Prof. Moniz has served on conference committees, notably the MG series. He was once a vice-rector, and as such has been a representative at several EU (and affiliated) agencies, too. He has received several scientific prizes. His research interest is in SUSY quantum cosmology (mostly the DAMTP 'eigen' line). More about quantum cosmology is also shared at his website, with somewhat of an updated version at his Linkedin webpage.

Preface to "Quantum Cosmology"

Within the second half of the last century, quantum cosmology concretely became one of the main research lines within gravitational theory and cosmology. Substantial progress has been made. Furthermore, quantum cosmology can become a domain that will gradually develop further over the next handful of decades, perhaps assisted by technological developments. Indications for new physics (i.e., beyond the standard model of particle physics or general relativity) could emerge and then the observable universe would surely be seen from quite a new perspective. This motivates bringing quantum cosmology to more research groups and individuals.

This Special Issue (SI) aims to provide a wide set of reviews, ranging from foundational issues to (very) recent advancing discussions. Concretely, we want to inspire new work proposing observational tests, providing an aggregated set of contributions, covering several lines, some of which are thoroughly explored, some allowing progress, and others much unexplored. The aim of this SI is motivate new researchers to employ and further develop quantum cosmology over the forthcoming decades. Textbooks and reviews exist on the present subject, and this SI will complementarily assist in offering open access to a set of wide-ranging reviews. Hopefully, this will assist new interested researchers, in having a single open access online volume, with reviews that can help. In particular, this will help in selecting what to explore, what to read in more detail, where to proceed, and what to investigate further within quantum cosmology.

Paulo Vargas Moniz
Editor

Editorial

Editorial to the Special Issue "Quantum Cosmology"

Paulo Vargas Moniz

Departamento de Física, Centro de Matemática e Aplicações (CMA-UBI), Universidade da Beira Interior, Rua Marquês d'Avila e Bolama, 6200-001 Covilha, Portugal; pmoniz@ubi.pt or prlvmoniz@gmail.com

Citation: Moniz, P.V. Editorial to the Special Issue "Quantum Cosmology". *Universe* **2022**, *8*, 336. https://doi.org/10.3390/universe8060336

Received: 15 June 2022
Accepted: 16 June 2022
Published: 20 June 2022

Publisher's Note: MDPI stays neutral with regard to jurisdictional claims in published maps and institutional affiliations.

Copyright: © 2022 by the author. Licensee MDPI, Basel, Switzerland. This article is an open access article distributed under the terms and conditions of the Creative Commons Attribution (CC BY) license (https://creativecommons.org/licenses/by/4.0/).

Some time ago, when I first inquired as to 'what quantum cosmology is about', I did approach the hall with a combination of caution as well as eagerness. At my earnest, I was proceeding within what one could label as festina lente[1], i.e., albeit going broad and wide, also proceeding serenely as much and as well I could. I wanted eventually to contribute somehow, if possible in a worthy manner. Plus, I had to find and properly study the footsteps previously carved by grand researchers in order to follow them, learn from them and then venture myself.

This Special Issue (SI) also took a while to emerge as we come across several difficulties during the pandemic years ranging from 2020 to 2022. However, we managed to gather a set of diverse and focused contributions, surveying a considerably wide range of topics that represent the main current avenues, as well as the potential for either new and/or still open routes to further explore. Of course, the SI is not exhaustive and plenty other directions remain to be mentioned and further addressed. The main point is that we (i.e., myself and those that added to the SI) (co-)wrote review papers promoting those lines such that eager (young) minds would feel challenged. Then, they can explore 'seas ruled by uncharted trade winds' (cf. https://en.wikipedia.org/wiki/Trade_winds#History, accessed on 15 June 2022). If we were successful, someone will have been influenced and will make a significant addition, after having read selected reviews from this SI. This will mean an unequivocal positive output. This will mean our effort was worthy and, furthermore, it will mean we did convey some kind of legacy from our own work (or part of it), herewith presented.

This SI contains sixteen review papers, each of about 20 pages, with 50 to 100 references each. Some overlapping exists and this is most welcome: quantum cosmology is very much multi-(sub)disciplinary; in particular, progress has been made when immersing oneself within methods from other domains. However, some routes do cross or border each other, so that some references or ingredients emerge 'here, there, somewhere', in one or two (or even more) reviews alike, as expected in a young domain of exploration. In spite of the challenges, quantum cosmology may well be the area that makes progress in this century, in the next decades. To this, we bequeath the content of this SI.

One of the themes is as follows. Reviews herewith did consider several settings for quantum cosmology probing, explanations and promotion, mostly building from General Relativity (GR) as the classical onset. However, then, wider scopes were appraised and discussed. Concretely, in [1] a (super)string setup was presented. Related to it, non-commutative properties were described in [2]. Within frameworks that encompass GR as the specific limit, the review in [3] brought Brans-Dicke context and content, whereas [4] took f(R) and [5] stretched modified gravity up to affine geometries. In a different use of observables, refs. [6,7] plus [8] as well as [9], each differently in either focus or tools, imported from loop quantum gravity. Within a broader coordinatization of space, supersymmetry was employed in [2,10].

However, this was not all. Besides the prospecting line above mentioned, we have other reviews that instead probed the fitting and realism (i.e., if quantum cosmology can be cast as to match either (i) real observed physics in some (semi)classical domain or (ii) known methodologies of well-established field theories): in [11], the issue of third quantization was imported and appraised; likewise present in [1]. Recently, the 'landscape' feature from string and alike settings has been a frequent 'participant' in discussions. In [12], this is taken in a sincere appraisal concerning a specific (now iconic) solution for the universe in quantum cosmology. The fulcrum of 'time' was amply debated through [13,14]. Finally, a majestic review was provided in [15] (if I can put it that way) bringing us into a grand perspective of quantum cosmology. It also conveys our attention to the issue of back-reaction (which is something [4,16] also captivate into).

It may be also of relevance to associate reviews with additional thematic issues. More specifically, in [3,13] the de Broglie–Bohm interpretation of quantum mechanics was employed whereas in [14] the Montevideo interpretation was taken for discussion. Furthermore, the issue of a 'bounce' (and somewhat of an alternative for the very early universe paradigm) at the creation event was used in [9,13]. Although a feature not yet fully elaborated in quantum cosmology, (quantum) entanglement may become primordial and [16] elaborated upon it, besides on back-reaction. Though not yet something, either methodically or thoroughly incorporated in quantum cosmology, chaos was mentioned in [9,15]. Within this domain, providing a contribution for effects created and then eventually causing change, particle creation is important to be considered; see [7]. Another feature not often widely discussed, are symmetries (within the Noether framework) in quantum cosmology; see [5]. Moreover, the fundamental crucible of a singularity is unavoidable and in reference [4] an opportunity was provided to discuss on a particular case. Still elusive but a most important 'grail' in quantum cosmology is to retrieve direct, unequivocal observational evidence of quantum gravity; Furthermore, in reference [6] a discussion about the topic is brought to the literature.

In summary, the above-cited review articles reflect the present state of the art, regarding either already resolved or still unresolved problems. The Guest Editor of this Special Issue does hope that it will be useful to stimulate subsequent new research routes in quantum cosmology. The primordial baseline to embrace is that, beyond the mere 'longitudinal' effort, adding just an incremental borderline result, it is 'transversal' questioning that is now much required (meaning, taking from some quantum cosmology mainstream issues and then formulating them in a new unexpected format and/or move one or two steps away from the current known footsteps).

Acknowledgments: The Guest Editor (P.V.M.) acknowledges the FCT grants UID-B-MAT/00212/2020 and UID-P-MAT/00212/2020 at CMA-UBI plus the COST Action CA18108 (Quantum gravity phenomenology in the multi-messenger approach). Last and not least, it is fundamental to thank all contributors for their outstanding reviews, without which this SI would have not been possible.

Conflicts of Interest: The author declares no conflict of interest.

Note

1. cf. Festina lente—Wikipedia: https://en.wikipedia.org/wiki/Festina_lente, accessed on 15 June 2022.

References

1. Gasperini, M. Quantum String Cosmology. *Universe* **2021**, *7*, 14. [CrossRef]
2. García-Compeán, H.; Obregón, O.; Ramírez, C. Topics in Supersymmetric and Noncommutative Quantum Cosmology. *Universe* **2021**, *7*, 434. [CrossRef]
3. Almeida, C.R.; Galkina, O.; Fabris, J.C. Quantum and Classical Cosmology in the Brans–Dicke Theory. *Universe* **2021**, *7*, 286. [CrossRef]
4. Borislavov Vasilev, T.; Bouhmadi-López, M.; Martín-Moruno, P. Classical and Quantum f(R) Cosmology: The Big Rip, the Little Rip and the Little Sibling of the Big Rip. *Universe* **2021**, *7*, 288. [CrossRef]
5. Capozziello, S.; Bajardi, F. Minisuperspace Quantum Cosmology in Metric and Affine Theories of Gravity. *Universe* **2022**, *8*, 177. [CrossRef]
6. Bojowald, M. Cosmic Tangle: Loop Quantum Cosmology and CMB Anomalies. *Universe* **2021**, *7*, 186. [CrossRef]
7. Tavakoli, Y. Cosmological Particle Production in Quantum Gravity. *Universe* **2021**, *7*, 258. [CrossRef]
8. Cortez, J.; Mena Marugán, G.A.; Velhinho, J.M. A Brief Overview of Results about Uniqueness of the Quantization in Cosmology. *Universe* **2021**, *7*, 299. [CrossRef]

9. Barca, G.; Giovannetti, E.; Montani, G. An Overview on the Nature of the Bounce in LQC and PQM. *Universe* **2021**, *7*, 327. [CrossRef]
10. Jalalzadeh, S.; Rasouli, S.M.M.; Moniz, P. Shape Invariant Potentials in Supersymmetric Quantum Cosmology. *Universe* **2022**, *8*, 316. [CrossRef]
11. Robles-Pérez, S.J. Quantum Cosmology with Third Quantisation. *Universe* **2021**, *7*, 404. [CrossRef]
12. Yeom, D.-H. Fuzzy Instantons in Landscape and Swampland: Review of the Hartle–Hawking Wave Function and Several Applications. *Universe* **2021**, *7*, 367. [CrossRef]
13. Małkiewicz, P.; Peter, P.; Vitenti, S.D.P. Clocks and Trajectories in Quantum Cosmology. *Universe* **2022**, *8*, 71. [CrossRef]
14. Kamenshchik, A.Y.; Pérez Rodríguez, J.N.; Vardanyan, T. Time and Evolution in Quantum and Classical Cosmology. *Universe* **2021**, *7*, 219. [CrossRef]
15. Hu, B.L. Weyl Curvature Hypothesis in Light of Quantum Backreaction at Cosmological Singularities or Bounces. *Universe* **2021**, *7*, 424. [CrossRef]
16. Marto, J. Hawking Radiation and Black Hole Gravitational Back Reaction—A Quantum Geometrodynamical Simplified Model. *Universe* **2021**, *7*, 297. [CrossRef]

Short Biography of Author

Paulo Vargas Moniz is a full professor at UBI (Portugal). He graduated from Lisbon and then moved to DAMTP, Cambridge for some years. He has been occasionally returning ever since and is also a life member at Clare Hall (college). He has been at the Editorial Board and then Advisory Board at CQG. The author of several books and several (much) more research papers, supervised students and post-docs, research visitor at many places, several times. Prof. Moniz has been serving at conferences committees, notably the MG series. Once a vice-rector, as such has been a representative at several EU (and affiliated) agencies, too. He received several science prizes. His research interest is on SUSY quantum cosmology (mostly the DAMTP 'eigen'line). More about quantum cosmology is also shared, idiosyncratically and at times, in http://www.dfis.ubi.pt/~pmoniz/ and somewhat update in https://www.linkedin.com/in/pvmoniz/.

Review

Shape Invariant Potentials in Supersymmetric Quantum Cosmology

Shahram Jalalzadeh [1], Seyed Meraj M. Rasouli [2,3] and Paulo Moniz [2,*]

1. Departamento de Física, Universidade Federal de Pernambuco, Recife 52171-900, PE, Brazil; shahram.jalalzadeh@ufpe.br
2. Departamento de Física, Centro de Matemática e Aplicações (CMA-UBI), Universidade da Beira Interior, Rua Marquês d'Avila e Bolama, 6200-001 Covilhã, Portugal; mrasouli@ubi.pt
3. Department of Physics, Qazvin Branch, Islamic Azad University, Qazvin 341851416, Iran
* Correspondence: pmoniz@ubi.pt

Abstract: In this brief review, we comment on the concept of shape invariant potentials, which is an essential feature in many settings of $N = 2$ supersymmetric quantum mechanics. To motivate its application within supersymmetric quantum cosmology, we present a case study to illustrate the value of this promising tool. Concretely, we take a spatially flat FRW model in the presence of a single scalar field, minimally coupled to gravity. Then, we extract the associated Schrödinger–Wheeler–DeWitt equation, allowing for a particular scope of factor ordering. Subsequently, we compute the corresponding supersymmetric partner Hamiltonians, H_1 and H_2. Moreover, we point out how the shape invariance property can be employed to bring a relation among several factor orderings choices for our Schrödinger–Wheeler–DeWitt equation. The ground state is retrieved, and the excited states easily written. Finally, the Hamiltonians, H_1 and H_2, are explicitly presented within a $N = 2$ supersymmetric quantum mechanics framework.

Keywords: supersymmetric quantum mechanics; shape invariant potentials; supersymmetric quantum cosmology

1. Introduction

Shape Invariant Potentials (SIP) constitute one of the hallmarks of supersymmetric quantum mechanics (SQM), in the sense that it enables a prolific framework to be elaborated. Being more specific, the presence of SIP allows us to easily obtain the set of states for a class of quantum systems, suitably based on an elegant algebraic construction. Hence, let us begin by mentioning that there is an algebraic structure associated with the SIP framework. It has gradually been acquiring a twofold relevance and within most of the exactly solvable problems in quantum mechanics [1–7].

On the one hand, such a structure has provided a method to determine eigenvalues and eigenfunctions, by means of which a spectrum is generated. More specifically, a broad set of those exactly solvable cases can be assembled and assigned within concrete classes; very few exceptions are known [1–7]. The distinguishing feature of any of such classes is that any exactly solvable case bears a *shape invariant potential: supersymmetric partners* are of the same shape, and their spectra can be determined entirely by an algebraic procedure comparable to that of the harmonic oscillator. In other words, operators can be defined, namely $A := \frac{d}{dx} + W(x)$, and its Hermitian conjugate $A^\dagger := -\frac{d}{dx} + W(x)$, Hamiltonians H_1 and its superpartner H_2 being expressed as $A^\dagger A$ and $AA^\dagger$, respectively. From this, we can produce and operate with other (more adequate) ladder operators for correspondingly appropriate quantum numbers. These can be maneuvered within a $J_\pm, J_3$ algebra, with comparable features to textbook ladder operators of angular momentum within either $SU(2)$ or $SO(3)$; please see [1–7] for relevant details.

On the other hand, several of these exactly solvable systems also possess a *potential algebra*: the corresponding Hamiltonian can be written as a Casimir operator of an underlying algebra, which in particular cases is of a $SO(2,1)$ nature [2]. Remarkably, there is a close correspondence: shape invariance can be expressed as constraint, which assists in establishing the spectrum; moreover, this shape invariance constraint can be written as an algebraic condition. For a specific set of SIP, the algebraic condition corresponds to the mentioned $SO(2,1)$ potential algebra, where unitary representations become of crucial use. Interestingly, these can be related to that of $SO(3)$; several SIPs are as such [1–7]. On the whole, a connection between SIP and *potential algebra* was attained. Nevertheless, it is also clear that, in spite of the structural similarity between $SO(2,1)$ and $SO(3)$ algebras, there are caveats to be aware of, related to the differences between those unitary representations.

There are also a couple of additional points that we would like to emphasize. To start with, some quantum states can be retrieved by group theoretical methods. This is further endorsed from the connections between shape invariance and potential algebra, wherein the former is translated into a concrete formulation within the latter [2]. As a result, the scope of the class of potentials where that could be applied was made more prominent [1]. Furthermore, other classes have been explored, related to harmonic oscillator induced second order differential equations, bearing group and algebra features, which subsequently allowed more SIP to be found [1–12]. In particular, this was further extended toward graded algebras in [8].

Secondly, these algebraic/group theory procedures (within the concrete use of algebras such as $SO(2,1)$ or $SO(3)$) have a striking resemblance to the approach and descriptive language used in [13–16]; a review of this idea is found in [17]. Therein, it was pointed out that intertwining boundary conditions, the algebra of constraints and hidden symmetries in quantum cosmology could be quite fruitful. Specifically, group/algebraic properties within ladder operators, either from angular momentum or from within the explicit presence of specific matter fields (and their properties), determined a partition of wave functions and boundary conditions, according to the Bargmann index [13–16]. Moreover, we proposed in [17] to extend this framework towards SIP, which could include well known analytically solvable cosmological cases. Being more clear, provided we identify integrability in terms of the shape invariance conditions, we could eventually import those specific features of SQM towards quantum cosmology [18–21]. That was the challenge we laid out in [17], which is still to be addressed: we hope our review paper herein can further enthuse someone to pick this up. A somewhat related and interesting direction to explore is to also consider an elaboration following [22,23]; specifically, accommodating the lines in [13–17] plus supersymmetry (SUSY) [18,19]. In brief, this paragraph conveys our central motivation to produce this review, building from a suggestion advanced in [17].

Thirdly, the interest in the above elements notwithstanding, there are still obstacles that ought to be mentioned, namely, about the scope of the usefulness of SIP in quantum cosmology. In fact, the list of SIP is quite restrictive, and most potentials therein do not emerge naturally within a minisuperspace. A few do, but for very particular case studies [24]. The classification of spatial geometries upon the Bianchi method implies that the potentials extracted from the gravitational degrees of freedom are very specific. Any 'broadness' can be introduced by inserting: (i) very specific matter fields into the minisuperspace (and therein we ought to be using realistic potentials (as indicated by particle physics)) or instead, (ii) try more SIP fitted choices but at the price of being very much *ad hoc*, i.e., an artificial selection. Nevertheless, the list of SIP and similar cases, where the algebraic tools could be adopted, has been extended. Although not a strong positive endorsement, there is work [25–34] that allows us to consider that eventually an extended notion of SIP may be soundly established, such that more cosmological minisuperspaces can be discussed within (see e.g., [24]). For the moment, this is a purpose set in construction and is what this review paper aims to to enthuse about and promote.

Upon this introductory section, this review is structured as follows. In Section 2, we summarize the features of SQM that we will be employing. In particular, a few technicalities about SIP will be presented in Section 2.2. Then, in Section 3, we take a case study, typically a toy model, by means of which we aim to promote work in SUSY quantum cosmology with the novel perspective of SIP. We emphasize that this is a line of investigation that has not yet been attempted before (see Section 3.3 for details). Section 4 conveys the Discussion and suggestions for the outlook for future research work.

2. Supersymmetric Quantum Mechanics

In this section, let us present a brief review of some of the pillars that characterize SQM. Then, we proceed to add a summary of the shape invariance concept. This section contains neither new results nor any innovated procedure, but only a very short overview of the results presented within seminal papers, e.g., [35–37].

2.1. Hamiltonian Formulation of Supersymmetric Quantum Mechanics

In order to describe SQM, let us start with the Schrödinger equation:

$$H\Psi_n(x) = \left[-\frac{\hbar^2}{2m}\frac{d^2}{dx^2} + V(x)\right]\Psi_n(x) = E_n\Psi_n(x). \tag{1}$$

We assume that the ground state wave function $\Psi_0(x)$ (that has no nodes[1]) associated with a potential $V_1(x)$ is known. Then, by assuming that the ground state energy E_0 to be zero, the Schrödinger equation for this ground state reduces to:

$$H_1\Psi_0(x) = \left[-\frac{\hbar^2}{2m}\frac{d^2}{dx^2} + V_1(x)\right]\Psi_0(x) = 0, \tag{2}$$

which leads to construct the potential $V_1(x)$:

$$V_1(x) = \frac{\hbar^2}{2m}\frac{\Psi_0''(x)}{\Psi_0(x)}, \tag{3}$$

where a prime denotes differentiation with respect to x. It is straightforward to factorize the Hamiltonian from the operators as follows:

$$A = \frac{\hbar}{\sqrt{2m}}\left[\frac{d}{dx} - \frac{\Psi_0'(x)}{\Psi_0(x)}\right], \quad A^\dagger = -\frac{\hbar}{\sqrt{2m}}\left[\frac{d}{dx} + \frac{\Psi_0'(x)}{\Psi_0(x)}\right], \tag{4}$$

through the equation

$$H_1 = A^\dagger A. \tag{5}$$

In order to construct the SUSY theory related to the original Hamiltonian H_1, the next step is to define another operator by reversing the order of A and $A^\dagger$, i.e., $H_2 \equiv AA^\dagger$, by which we indeed get the Hamiltonian corresponding to a new potential $V_2(x)$:

$$H_2 = -\frac{\hbar^2}{2m}\frac{d^2}{dx^2} + V_2(x), \tag{6}$$

where

$$V_2(x) = -V_1(x) + \frac{\hbar}{m}\left[\frac{\Psi_0'(x)}{\Psi_0(x)}\right]^2. \tag{7}$$

The potentials $V_1(x)$ and $V_2(x)$ have been referred to as the supersymmetric partner potentials. It should be noted that H_2, as the partner Hamiltonian corresponding to H_1 is in general not unique, but there is a class of Hamiltonians $H^{(m)}$ that can be partner Hamiltonians[2]. This point has been specified [35,38,39], conveying a better understanding of the relationship between SQM and the inverse scattering method, established by Gelfand and Levitan [40,41].

In SQM, instead of the ground state wave function $\Psi_0(x)$ associated with H_1, the superpotential $W(x)$ is introduced, being related to $\Psi_0(x)$ and its first derivative (with respect to x) by means of

$$W(x) = -\frac{\hbar}{\sqrt{2m}}\left(\frac{\Psi_0'(x)}{\Psi_0(x)}\right). \tag{8}$$

At this stage, it is worth expressing the operators A and $A^\dagger$ and the supersymmetric partner potentials $V_1(x)$ and $V_2(x)$ in terms of the superpotential. Therefore, using (8), Equations (3), (4) and (7) are rewritten as:

$$A = \frac{\hbar}{\sqrt{2m}}\frac{d}{dx} + W(x), \qquad A^\dagger = -\frac{\hbar}{\sqrt{2m}}\frac{d}{dx} + W(x), \tag{9}$$

$$V_1(x) = -\frac{\hbar}{\sqrt{2m}}W'(x) + W^2(x), \qquad V_2(x) = \frac{\hbar}{\sqrt{2m}}W'(x) + W^2(x), \tag{10}$$

where the expressions for V_1 and V_2 in (10) constitute Riccati equations. Equations (9) and (10) imply that $W'(x)$ is proportional to the commutator of the operators A and $A^\dagger$, and $W^2(x)$ is the average of the partner potentials.

It is easy to show that the wave functions, the energy eigenvalues and the S-matrices of both the Hamiltonians H_1 and H_2 are related. Let us merely outline the results and abstain from proving them. In this regard, we take $\Psi_n^{(1)}$ and $\Psi_n^{(2)}$ as the eigenfunctions of H_1 and H_2, respectively. Moreover, we denote their corresponding energy eigenvalues with $E_n^{(1)} \geq 0$ and $E_n^{(2)} \geq 0$ where $n = 0, 1, 2, 3, \ldots$ is the number of the nodes in the wave function. It is straightforward to show that supersymmetric partner potentials $V_1(x)$ and $V_2(x)$ possess the same energy spectrum. However, we should note that, for the ground state energy $E_0^{(1)} = 0$ associated with the potential $V_1(x)$, there is no corresponding level for its partner $V_2(x)$. Concretely, it has been shown that:

$$E_n^{(2)} = E_{n+1}^{(1)}, \qquad E_0^{(1)} = 0, \tag{11}$$

$$\Psi_n^{(2)} = \left[E_{n+1}^{(1)}\right]^{-\frac{1}{2}} A \Psi_{n+1}^{(1)}, \tag{12}$$

$$\Psi_{n+1}^{(1)} = \left[E_n^{(2)}\right]^{-\frac{1}{2}} A^\dagger \Psi_n^{(2)}. \tag{13}$$

In what follows, let us express some facts. (i) If the ground-state wave function $\Psi_0^{(1)}$, which is given by $(A\Psi_0^{(1)} = 0)$

$$\Psi_0^{(1)} = N_0 \exp\left[-\int^x W(x')dx'\right], \tag{14}$$

is square integrable, then the ground state of H_1 has zero energy ($E_0 = 0$) [35]. For this case, it can be shown that the SUSY is unbroken; (ii) if the eigenfunction $\Psi_{n+1}^{(1)}$ of H_1 ($\Psi_n^{(2)}$ of H_2) is normalized, then the $\Psi_n^{(2)}$ ($\Psi_{n+1}^{(1)}$), will be also normalized; (iii) Assuming the eigenfunction $\Psi_n^{(1)}$ with eigenvalue $E_n^{(1)}$ ($\Psi_n^{(2)}$ with eigenvalue $E_n^{(2)}$) corresponds to the Hamiltonian H_1 (H_2), it is easy to show that $A\Psi_n^{(1)}$ ($A^\dagger \Psi_n^{(2)}$) will be an eigenfunction of H_2 (H_1) with the same eigenvalue; (iv) In order to destroy (create) an extra node in the eigenfunction as well as convert an eigenfunction of H_1 (H_2) into an eigenfunction of H_2 (H_1) with the same energy, we apply the operator A ($A^\dagger$); (v) The ground state wave function of H_1 has no SUSY partner; (vi) Applying the operator A ($A^\dagger$), all the eigenfunctions of H_2 (H_1, except for the ground state) can be reconstructed from those of H_1 (H_2); please see Figure 1.

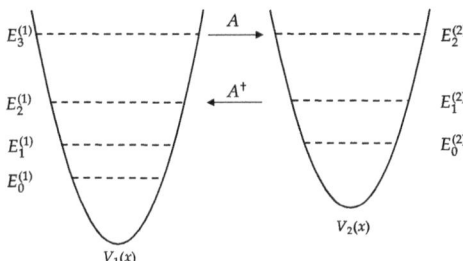

Figure 1. The energy levels of $V_1(x)$ and $V_2(x)$ as two supersymmetric partner potentials. The figure is associated with unbroken SUSY. It is seen that, except an extra state $E_0^{(1)} = 0$, the other energy levels are degenerate. Moreover, in this figure, it is shown that how the operators A and $A^\dagger$ connect eigenfunctions.

It has been believed that this fascinating procedure, which leads to an understanding of the degeneracy of the spectra of H_1 and H_2, can be provided by applying the properties of the SUSY algebra. Therefore, let us consider a matrix SUSY Hamiltonian (which is part of a closed algebra including both bosonic and fermionic operators with commutation and anti-commutation relations) containing both the Hamiltonians H_1 and H_2 [37]:

$$H = \begin{bmatrix} H_1 & 0 \\ 0 & H_2 \end{bmatrix}. \tag{15}$$

Supersymmetric quantum mechanics begins with a set of two matrix operators, Q and $Q^\dagger$, known as supercharges:

$$Q = \begin{bmatrix} 0 & 0 \\ A & 0 \end{bmatrix}, \tag{16}$$

$$Q^\dagger = \begin{bmatrix} 0 & A^\dagger \\ 0 & 0 \end{bmatrix}. \tag{17}$$

The matrix H is part of a closed algebra in which both bosonic and fermionic operators with commutation and anti-commutation relations are included, such that the bosonic degrees of freedom are changed into the fermionic ones and vice versa by the supercharges.

It is straightforward to show that:

$$[H, Q] = [H, Q^\dagger] = 0, \tag{18}$$
$$\{Q, Q^\dagger\} = H, \tag{19}$$
$$\{Q, Q\} = 2Q^2 = \{Q^\dagger, Q^\dagger\} = 2\left(Q^\dagger\right)^2 = 0, \tag{20}$$

by which the closed superalgebra $sl(1,1)$ is described [42] (see also Section 3.3). Note that the relations (18) are responsible for the degeneracy.

In SQM, when the two partner potentials have continuum spectra, it is possible to relate the reflection and transmission coefficients. A necessary condition for providing scattering in both of the partner potentials is that they must be finite when $x \to -\infty$ or $x \to \infty$.

2.2. Shape Invariance and Solvable Potentials

In the context of the non-relativistic quantum mechanics, there are a number of known potentials (e.g., Coulomb, harmonic oscillator, Eckart, Morse, and Pöschl–Teller) for which we can solve the corresponding Schrödinger equation analytically and determine all the energy eigenvalues and eigenfunctions explicitly. In this regard, the following questions naturally arise: Why are just some potentials solvable? Is there any underlying symmetry property? What is this symmetry?

Gendenshtein was the first to answer these questions by introducing the *shape invariance* concept [43]. In fact, for such potentials, all the bound state energy eigenvalues, eigenfunctions and the scattering matrix can be retrieved by applying the generalized operator method, which is essentially equivalent to the Schrödinger's method of factorization [44,45].

In [43], the relationship between SUSY, the hierarchy of Hamiltonians, and solvable potentials has been investigated from an interesting perspective (for detailed discussions see, for instance, [35,37]). In what follows, let us describe briefly the shape invariance concept. "*If the pair of SUSY partner potentials $V_{1,2}(x,b)$ are similar in shape and differ only in the parameters that appear in them, then they are said to be shape invariant*" [46]. Let us be more precise. Consider a pair of SUSY partner potentials, $V_{1,2}(x)$, as defined in (10). If the profiles of these potentials are such that they satisfy the relationship:

$$V_2(x,b) = V_1(x,b_1) + R(b_1), \tag{21}$$

where the parameter b_1 is some function of b, say given by $b_1 = f(b)$, the potentials $V_{1,2}(x)$ are said to bear shape invariance. In other words, to be associated within shape invariance the potentials $V_{1,2}$, while sharing a similar coordinate dependence, can at most differ in the presence of some parameters. To make the definition of shape invariance clear, consider, for example,

$$W = b \tanh\left(\frac{\sqrt{2m}}{\hbar}x\right). \tag{22}$$

Then, inserting this superpotential into (10) gives us:

$$V_1(x,b) = -\frac{b(b+1)}{\cosh^2\left(\frac{\sqrt{2m}}{\hbar}x\right)} + b^2,$$
$$V_2(x,b) = -\frac{b(b-1)}{\cosh^2\left(\frac{\sqrt{2m}}{\hbar}x\right)} + b^2. \tag{23}$$

The above expressions show that one can rewrite V_2 in terms of V_1, as expressed in (10) where, in this example, $b_1 = b - 1$, and $R(b_1) := 2b_1 + 1$. Thus, the potentials $V_{1,2}$ bear shape invariance in accordance with the definition (10). Then, to use the shape invariance condition, let us assume that (10) holds for a sequence of parameters, $\{b_k\}_{k=0,1,2,...}$, where,

$$b_k = \underbrace{f \circ f \circ f \circ \ldots \circ f}_{k \text{ times}}(b) = f^k(b), \quad k = 0,1,2,\ldots, \quad b_0 := b. \tag{24}$$

Consequently,

$$H_2(x, b_k) = H_1(x, b_{k+1}) + R(b_k). \tag{25}$$

Now, we write $H^{(0)} = H_1(x,b)$, $H^{(1)} = H_2(x,b)$, and we define $H^{(m)}$ as:

$$H^{(m)} := -\frac{\hbar^2}{2m}\frac{d^2}{dx^2} + V_1(x,b_m) + \sum_{k=1}^{m} R(b_k) = H_1(x,b_m) + \sum_{k=1}^{m} R(b_k). \tag{26}$$

Using (25), we can extract $H^{(m+1)}$ as:

$$H^{(m+1)} = H_2(x,b_m) + \sum_{k=1}^{m} R(b_k). \tag{27}$$

Therefore, in this way, we are able to set up a hierarchy of Hamiltonians $H^{(k)}$ for various k values.

Employing condition (21) and the hierarchy of Hamiltonians [37], the energy eigenvalues and eigenfunctions have been obtained for any shape invariant potential when SUSY is unbroken. It should be noted that there is a correspondence between the condition (21) (associated with SQM) and the required mathematical condition applied in the method of the factorization of the Hamiltonian [47]. Although the terminology and ideas associated with these methods are different, they can be considered as the special cases of the procedure employed to handle second-order linear differential equations [48,49]. Notwithstanding the above, it has been believed that a better understanding of analytically solvable potentials could be achieved by SUSY and shape invariance. Let us elaborate more on this aspect.

H_2 contains the lowest state with a zero energy eigenvalue, according to the SQM concepts discussed in Section 2. As a result of (11), the lowest energy level of $H^{(m)}$ has the value of:

$$E_0^{(m)} = \sum_{k=1}^{m} R(b_k). \tag{28}$$

Therefore, it is simple to realize that because of the chain $H^{(m)} \to H^{(m-1)} \ldots \to H^{(1)}(:= H_2) \to H^{(0)}(:= H_1)$, the nth member in this sequence carries the nth level of the energy spectra of $H^{(0)}$ (or H_1), namely [35]:

$$E_n^{(0)} = \sum_{k=1}^{n} R(b_k), \quad E_0^{(0)} = 0. \tag{29}$$

Let us now return to the example (22). We rewrite (21) as:

$$V_1(x,b) = V_2(x, b-1) + b^2 - (b-1)^2. \tag{30}$$

We can generate b_k from $b_0 = b$ as $b_k = b - k$. Hence, the energy spectrum from $V_1(x,b)$ yields:

$$E_n^{(0)} = \sum_{k=1}^{n} R(b_k) = \sum_{k=1}^{n} (b^2 - b_k^2) = b^2 - b_n^2 = b^2 - (b-n)^2. \tag{31}$$

It is worth noting that, according to the requirement (21), the well-known solvable potentials (such as those were listed in the first paragraph of this subsection) are all shape invariant and, therefore, their energy eigenvalue spectra are given by (29). "In [43], Gendenshtein then conjectured that shape invariance is not only sufficient but may even be necessary for a potential to be solvable" [35]; In addition, let us mention that new developments have been achieved on this domain, where new approaches (see [50–54]) have either challenged or broaden this assertion. Moreover, by applying SUSY, it is also possible to retrieve the bound-state energy eigenfunctions of H_1 for shape invariant potentials [36]. In particular, in the same paper, by taking $\Psi_0^1(x,b)$ as the ground-state wave function of H_1 (which is given by (14)), and employing relation (13), a relation is obtained for nth-state eigenfunction $\Psi_n^1(x,b)$ as:

$$\Psi_n^1(x,b) = A^\dagger(x,b) A^\dagger(x,b_1) \ldots A^\dagger(x,b_{n-1}) \Psi_0^1(x,b). \tag{32}$$

For later convenience, let us concentrate on a specific shape invariance that only involving translation of the parameter b_0 with a translation step η [27] (for other kind of relations between the parameters, see, for instance, [37]):

$$b_1 = b_0 + \eta. \tag{33}$$

It is feasible to introduce a translation operator as:

$$T(b_0) = \exp\left(\eta \frac{\partial}{\partial b_0}\right), \quad T^{-1}(b_0) = T^\dagger(b_0) = \exp\left(-\eta \frac{\partial}{\partial b_0}\right), \tag{34}$$

which act merely on objects defined on the parameter space.

By composing the translation and bosonic operators, we can construct the following generalized creation and annihilation operators:

$$B_1(b_0) = A^\dagger(b_0)T(b_0), \tag{35}$$

$$B_2(b_0) = T^\dagger(b_0)A(b_0). \tag{36}$$

Applying the shape invariant potentials to solve the Schrödinger equation is similar to the factorization method employed to the case of the harmonic oscillator potential [27]. Therefore, we have:

$$B_2(b_0)A(b_0)\Psi_0(x;b_0) = A(b_0)\Psi_0(x;b_0) = 0. \tag{37}$$

In order to obtain the excited states, the creation operator should repeatedly act on $\Psi_0(x;b_0)$:

$$\Psi_n(x;b_0) = [B_1(b_0)]^n \Psi_0(x;b_0). \tag{38}$$

We should note that the translation operators (T and $T^\dagger$) and ladder operators, A and $A^\dagger$ do not commute with any b_k-dependent and any x-dependent objects, respectively. Therefore, the generalized creation and annihilation operators (B_1 and B_2) act on the objects defined on the dynamical variable space and the objects defined on parameter space via the bosonic operators and the translation operators, respectively. According to (37), we have:

$$\Psi_0(x;b_0) \propto \exp\left(-\int^x W(\tilde{x};b_0)d\tilde{x}\right), \tag{39}$$

which is transformed by a normalization constant (that should, in general, depend on parameters b) into a relation of equality. Concretely, the action of the generalized operators affects in determining such a normalization constant; for more details, see, [55].

Now let us obtain the relations of the energy eigenvalues and energy spectrum. From using (34), we can write:

$$R(b_n) = T(b_0)R(b_{n-1})T^\dagger(b_0), \tag{40}$$

where

$$b_n = b_0 + n\eta \tag{41}$$

is a generalized version of (33). Employing (40), we get:

$$R(b_n)B_1(b_0) = B_1(b_0)R(b_{n-1}). \tag{42}$$

Equations (40) and (42) yield a commutation relation as:

$$[H_1, (B_1)^n] = \left(\sum_{k=1}^n R(b_k)\right)(B_1)^n. \tag{43}$$

Applying (43) on the ground state of H_1, i.e., $\Psi_0(x;b_0)$, it is seen that $[B_1(b_0)]^n \Psi_0(x;b_0)$ is also an eigenfunction of H_1 with eigenvalue $E_n^{(1)}$ given by (29). Therefore, the energy spectrum is:

$$E_n = E_0 + E_n^{(1)}, \tag{44}$$

where the ground state energy E_0 is obtained from either:

$$H_1 = H - E_0, \tag{45}$$

or, equivalently,

$$W(x;b) - W'(x;b) = V(x) - E_0 = V_1(x). \tag{46}$$

Finally, it should be noted that the above established algebraic approach is self-consistent. More concretely, by considering supersymmetric and shape invariance properties of the system, it can be applied as an appropriate method for obtaining not only the eigenvalues and eigenfunctions of the bound state of a Schrödinger equation, but also exact resolutions for this equation [27].

3. SUSY Quantum Cosmology

In order to apply the formalism presented in the previous section, let us investigate a homogeneous and isotropic cosmology, in the context of General Relativity (GR) together with a single scalar field, ϕ, minimally coupled to gravity.

3.1. A Case Study: Classical Setting

By considering the Friedmann–Lemaître–Robertson–Walker (FLRW) line element[3]

$$ds^2 = N(t)dt^2 + a(t)^2 \left\{ \frac{dr^2}{1 - kr^2} + r^2 d\Omega^2 \right\}, \tag{47}$$

the ADM[4] Lagrangian will be[5]

$$L_{\text{ADM}} = -\frac{3}{N} a \dot{a}^2 + 3kNa + a^3 \left(\frac{\dot{\phi}^2}{2N} - NV(\phi) \right), \tag{48}$$

where an over-dot denotes a differentiation with respect to the cosmic time t; $N(t)$ is a lapse function, $a(t)$ is the scale factor, $V(\phi)$ is a scalar potential and $k = \{-1, 0, 1\}$ is the spatial curvature constant associated with open, flat and closed universes, respectively.

In this work, let us consider the scalar potential $V(\phi)$ to be in the form [65,66]

$$V(\phi) = \lambda + \frac{m^2}{2\alpha^2} \sinh^2(\alpha \phi) + \frac{\vartheta}{2\alpha^2} \sinh(2\alpha \phi), \tag{49}$$

where λ may be related to the cosmological constant; $m^2 = \partial^2 V / \partial \phi^2|_{\phi=0}$ is a mass squared parameter; $\alpha^2 = 3/8$ and ϑ is a coupling parameter. Moreover, we will now investigate only the spatially flat FLRW universe. For this case, it has been shown that an oscillator–ghost–oscillator system is produced [66–71]. More concretely, by applying the following transformations [72–75],

$$X = \frac{a^{\frac{3}{2}}}{\alpha} \cosh(\alpha \phi), \qquad Y = \frac{a^{\frac{3}{2}}}{\alpha} \sinh(\alpha \phi), \tag{50}$$

the Lagrangian (48) transform into

$$L_{\text{ADM}} = -\frac{1}{2N} \dot{\xi}^\top J \dot{\xi} + \frac{N}{2} \xi^\top M J \xi, \tag{51}$$

where

$$\xi := \begin{pmatrix} X \\ Y \end{pmatrix}, \quad M := \begin{pmatrix} 2\lambda \alpha^2 & -\vartheta \\ \vartheta & 2\lambda \alpha^2 - m^2 \end{pmatrix}, \quad J := \begin{pmatrix} 1 & 0 \\ 0 & -1 \end{pmatrix}. \tag{52}$$

It is straightforward to decouple (51) into normal modes $\gamma := \Sigma^{-1} \xi$ by means of:

$$\gamma := \begin{pmatrix} u \\ v \end{pmatrix}, \quad \Sigma := \begin{pmatrix} \frac{-m - \sqrt{m^4 - 4\vartheta^2}}{2\vartheta} & \frac{-m + \sqrt{m^4 - 4\vartheta^2}}{2\vartheta} \\ 1 & 1 \end{pmatrix}, \tag{53}$$

which diagonalize the matrix M as follows:

$$\Sigma^{-1}M\Sigma = \begin{pmatrix} \omega_1 & 0 \\ 0 & \omega_2 \end{pmatrix}, \quad \omega_{1,2}^2 = \frac{3\lambda}{4} + \frac{m^2}{2} \mp \frac{\sqrt{m^4 - 4\vartheta^2}}{2}. \tag{54}$$

Thus, we retrieve the Lagrangian associated with a 2D oscillator–ghost–oscillator:

$$\begin{aligned} L_{\mathrm{ADM}}(u,v) &= -\frac{1}{2N}\dot{\gamma}^{\top}\mathcal{I}\dot{\gamma} + \frac{N}{2}\gamma^{\top}\mathcal{J}\gamma \\ &= -\frac{1}{2}\left\{\left(\frac{1}{N}\dot{u}^2 - \omega_1^2 N u^2\right) - \left(\frac{1}{N}\dot{v}^2 - \omega_2^2 N v^2\right)\right\}, \end{aligned} \tag{55}$$

where $\mathcal{I} := \Sigma^{\top} J \Sigma$, and $\mathcal{J} := \Sigma^{\top} M J \Sigma$. The conjugate momenta corresponding to u and v are:

$$p_u = \frac{\dot{u}}{N}, \qquad p_v = -\frac{\dot{v}}{N}. \tag{56}$$

Moreover, the classical Euler–Lagrange equations are given by:

$$\frac{d}{dt}\left(\frac{\dot{u}}{N}\right) + N\omega_1^2 u = 0, \quad \frac{d}{dt}\left(\frac{\dot{v}}{N}\right) + N\omega_2^2 v = 0. \tag{57}$$

It is straightforward to show that the Hamiltonian corresponding to the ADM Lagrangian (55) is:

$$H_{\mathrm{ADM}} = -\frac{N}{2}\left\{\left(p_u^2 + \omega_1^2 u^2\right) - \left(p_v^2 + \omega_2^2 v^2\right)\right\}, \tag{58}$$

which, for the gauge $N = 1$, yields

$$u(t) = u_0 \sin(\omega_1 t - \theta), \qquad v(t) = v_0 \sin(\omega_2 t). \tag{59}$$

In (59), θ is an arbitrary phase factor. From using the Hamiltonian constraint, we obtain $\omega_1 u_0 = \omega_2 v_0$. It is also seen that the classical paths corresponding to solutions (59), in the configuration space (u, v), are the generalized Lissajous ellipsis.

3.2. Quantization

In order to establish a quantum cosmological model corresponding to our model, let us proceed with the Wheeler–DeWitt equation. The canonical quantization of (58) gives

$$\mathcal{H}\Psi(u,v) = \left(-\frac{\partial^2}{\partial u^2} + \frac{\partial^2}{\partial v^2} + \omega_1^2 u^2 - \omega_2^2 v^2\right)\Psi(u,v) = 0. \tag{60}$$

Equation (60) is separable and we can obtain a solution as:

$$\Psi_{n_1,n_2}(u,v) = \alpha_{n_1}(u)\beta_{n_2}(v), \tag{61}$$

where

$$\alpha_n(u) = \left(\frac{\omega_1}{\pi}\right)^{1/4} \frac{H_n(\sqrt{\omega_1}u)}{\sqrt{2^n n!}} e^{-\omega_1 u^2/2}, \tag{62}$$

$$\beta_n(v) = \left(\frac{\omega_2}{\pi}\right)^{1/4} \frac{H_n(\sqrt{\omega_2}v)}{\sqrt{2^n n!}} e^{-\omega_2 v^2/2}. \tag{63}$$

In relations (62) and (63), $H_n(x)$ stands for the Hermite polynomials. Moreover, we should note that the Hamiltonian constrain relates the parameters of the model as:

$$\left(n_1 + \frac{1}{2}\right)\omega_1 = \left(n_2 + \frac{1}{2}\right)\omega_2, \qquad n_1, n_2 = 0, 1, 2, \ldots. \tag{64}$$

The recovery of classical solutions from the corresponding quantum model is one of the essential elements of quantum cosmology. For this aim, a coherent wave packet with reasonable asymptotic behavior in the minisuperspace is often constructed, peaking near the classical trajectory. We can herewith produce a widespread wave packet solution,

$$\Psi(u,v) = \sum_{n_1, n_2} C_{n_1 n_2} \alpha_{n_1}(u) \beta_{n_2}(v), \tag{65}$$

where the summing is restricted to overall values of n_1 and n_2 satisfies the relation (64). Let us consider the simplest case which is when $\omega_1 = \omega_2 = \omega$, which means $m^2 = 2\vartheta$ in the definition of scalar field potential (49). Then, the wave packet will be:

$$\Psi(u,v) = \sqrt{\frac{\omega}{\pi}} \sum_{n=0}^{\infty} \frac{C_n}{n! 2^n} \exp\left(-\frac{\omega}{2}(u^2 + v^2)\right) H_n(\sqrt{\omega}u) H_n(\sqrt{\omega}v), \tag{66}$$

where C_n is a complex constant. We apply the following identity to create a coherent wave packet with suitable asymptotic behavior in the minisuperspace, peaking around the classical trajectory:

$$\sum_{n=0}^{\infty} \frac{t^n}{n!} H_n(x) H_n(y) = \frac{1}{\sqrt{1-t^2}} \exp\left(\frac{2txy - t^2(x^2 + y^2)}{2(1-t^2)}\right). \tag{67}$$

Using this identity and choosing the coefficients C_n in (66) to be $C_n = B 2^n \tanh \xi$, with B and ξ are arbitrary complex constants, we obtain:

$$\Psi(u,v) = C \exp\left(-\frac{\omega}{4} \cos(2\beta_2) \cosh(2\beta_1)(u^2 + v^2 - 2\eta \tanh(2\beta_1) uv)\right)$$
$$\times \exp\left(-\frac{i\omega}{4} \sinh(2\beta_1) \sin(2\beta - 2)(u^2 + v^2 - 2\eta \coth(2\beta_1) uv)\right), \tag{68}$$

where β_1 and β_2 are the real and imaginary parts of $\xi = \beta_1 + i\beta_2$, respectively, $\eta = \pm 1$, and N is a normalization factor. Figure 2a shows the density plot, and Figure 2b illustrates the contour plot of the wave function for typical values of β_1, β_2, and $\eta = 1$ for the following combination of the solutions:

$$\Psi(u,v) = \Psi_{\beta_1, \beta_2}(u,v) - \Psi_{\beta_1 + \delta\beta_1, \beta_2 + \delta\beta_2}(u,v). \tag{69}$$

The classical solutions (59) can easily be represented as the following trajectories (for $\omega_1 = \omega_2 = \omega$):

$$u^2 + v^2 - 2\eta uv \cos(\theta) - u_0^2 \sin^2(\theta) = 0. \tag{70}$$

This equation describes ellipses whose major axes make angle $\pi/4$ with the positive/negative u axis according to the choices ± 1 for η. Additionally, each trajectory's eccentricity and size are determined by θ and u_0, respectively. It can be seen that the quantum pattern in Figure 2 and the classical paths (70) in configuration space, (u,v), have a high correlation.

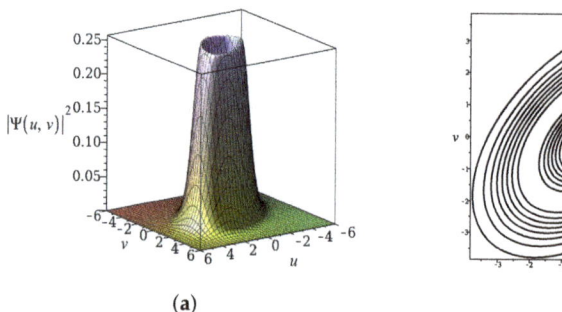

Figure 2. Density plot- (**a**), and contour plot- (**b**), of a wave packet. These figures are plotted for numerical values $\omega = 1$, $\beta_1 = 1$, $\beta_2 = \pi/6$, $\eta = 1$, $\delta\beta_1 = 0.1$ and $\delta\beta_2 = 3\pi/50$.

Let us point out how we can introduce a time-evolving wave-function. By employing a canonical transformation on the (v, p_v) sector of the Hamiltonian (58), we observe that in the total Hamiltonian, the momentum associated with the the new canonical variable appears linearly. Let us be more precise. Consider the following canonical transformation $(v, p_v) \rightarrow (T, p_T)$ given by:

$$v = \sqrt{\frac{2p_T}{\omega_2^2}} \sin(\omega_2 T), \quad p_v = \sqrt{2p_T} \cos(\omega_2 T), \quad \{T, p_T\} = 1. \tag{71}$$

It is easy to check out that the inverse map is given by the following relations:

$$p_T = \frac{1}{2} p_v^2 + \frac{1}{2} \omega_1^2 v^2, \quad T = \frac{1}{\omega_2} \tan^{-1}\left(\omega_2 \frac{v}{p_v}\right). \tag{72}$$

In fact, the new set of phase space coordinates (T, p_T) is related to the harmonic oscillator's action-angle variables, (φ, p_φ), by [76,77]:

$$\varphi = \omega_2 T, \quad p_\varphi = \frac{p_T}{\omega_2}. \tag{73}$$

The ADM Hamiltonian (58) simply takes the form:

$$H_{ADM} = N\left(\frac{1}{2} p_u^2 + \frac{1}{2} \omega_1^2 u^2 - p_T\right). \tag{74}$$

The classical field equations corresponding to (74) are:

$$\begin{cases} \dot{u} = N p_u, & \dot{p}_u = -N\omega_1^2 u, \\ \dot{T} = -N, & \dot{p}_T = 0. \end{cases} \tag{75}$$

For $N = 1$, we find

$$T = -t, \quad p_T = \text{const.} \tag{76}$$

Thus, the motion in 2D phase space (T, p_T) becomes trivial, i.e., flow paths are straight lines with constant p_T. As seen, the second set of solutions for (75) implies that T plays the role of the time parameter. Consequently, the Poisson bracket of the time parameter and super-Hamiltonian does not vanish but instead we have $\{T, \mathcal{H}\} = 1 = \{T, p_T\}$, which implies that T is not a Dirac observable, and therefore, we may consider it as a time variable; see, for instance, [76] and references therein.

3.3. Supersymmetric Quantization

Employing the Hamiltonian constraint upon (74), and then substituting $p_u = -i\frac{d}{du}$ and $p_T = -i\frac{\partial}{\partial T} = i\frac{\partial}{\partial t}$, we get a Schrödinger–Wheeler–WeDitt equation:

$$i\frac{\partial}{\partial t}\Psi(u,t) = \left[-\frac{1}{2}\frac{d^2}{du^2} + \frac{l(l+1)}{2u^2} + \frac{1}{2}\omega_1^2 u^2\right]\Psi(u,t). \tag{77}$$

In the process of obtaining Equation (77), we have further used the following factor ordering procedure:

$$p_u^2 = -\frac{1}{3}\left(u^\alpha \frac{d}{du} u^\beta \frac{d}{du} u^\gamma + u^\gamma \frac{d}{du} u^\alpha \frac{d}{du} x^\beta + u^\beta \frac{d}{du} u^\gamma \frac{d}{du} u^\alpha\right), \tag{78}$$

where the parameters α, β, and γ satisfy the requirement $\alpha + \beta + \gamma = 0$, and we have set $\frac{1}{3}(\beta^2 + \gamma^2 + \beta\gamma) := l(l+1)$.

The time independent sector of Equation (77) reads:

$$H_l \Psi_n^l(u) = E_n^l \Psi_n^l(u), \tag{79}$$

where

$$H_l := -\frac{1}{2}\frac{d^2}{du^2} + \frac{l(l+1)}{2u^2} + \frac{1}{2}\omega_1^2 u^2. \tag{80}$$

According to the Equation (10), we now introduce first-order differential operators:

$$\begin{cases} A_l := \frac{1}{\sqrt{2\omega_1}}\frac{d}{du} + \sqrt{\frac{\omega_1}{2}}u - \frac{l+1}{\sqrt{2\omega_1}u}, \\ A_l^\dagger := -\frac{1}{\sqrt{2\omega_1}}\frac{d}{du} + \sqrt{\frac{\omega_1}{2}}u - \frac{l+1}{\sqrt{2\omega_1}u}. \end{cases} \tag{81}$$

For $l \in \mathbb{N}$, we correspondingly obtain the following supersymmetric partner Hamiltonians:

$$\begin{cases} H_1 = \omega_1 A_l^\dagger A_l = -\frac{1}{2}\frac{d^2}{du^2} + \frac{l(l+1)}{2u^2} + \frac{1}{2}\omega_1^2 u^2 + \omega_1(l - \frac{1}{2}) = H_l - \omega_1(l + \frac{3}{2}), \\ H_2 = \omega_1 A_l A_l^\dagger = -\frac{1}{2}\frac{d^2}{du^2} + \frac{(l+1)(l+2)}{2u^2} + \frac{1}{2}\omega_1^2 u^2 - \omega_1(l + \frac{1}{2}) = H_{l+1} - \omega_1(l + \frac{1}{2}). \end{cases} \tag{82}$$

These two Hamiltonians have the same energy spectrum except the ground state of H_2:

$$\begin{cases} H_1 \Psi_n^{(l)} = \left[E_n^{(l)} + \omega_1 - (l + \frac{3}{2})\right]\Psi_n^{(l)}, \\ H_2 \Psi_{n+1}^{(l+1)} = \left[E_{n+1}^{(l+1)} - \omega_1(l + \frac{1}{2})\right]\Psi_{n+1}^{(l+1)} = \left[E_n^{(l)} - \omega_1(l + \frac{1}{2})\right]\Psi_{n+1}^{(l+1)}. \end{cases} \tag{83}$$

It is seen that these equations refer to the shape-invariance condition, by which we, equivalently, can write:

$$A_{l-1}^\dagger A_{l-1} - A_l A_l^\dagger = \frac{2}{\omega_1}. \tag{84}$$

Thus, altering the sequence of operators A_l and $A_l^\dagger$ causes the value of l to change. This demonstrates how shape-invariance properties link the various factor orderings of the Schrödinger–Wheeler–DeWitt Equation (79). Shape-invariant potentials are well recognized for being simple to deal with when using lowering and raising operators, similar to the harmonic oscillator. However, we should note that the commutator of A_l and $A_l^\dagger$ does not provide a constant value. Namely,

$$[A_l, A_l^\dagger] = 1 + \frac{l+1}{\omega_1 u^2}, \qquad (85)$$

which implies that these operators are not suitable to proceed with. As the eigenvalue relation (82) shows, the potentials $V_1(u; b_0)$ and $V_2(u; b_1)$ introduced in (9) are given by:

$$\begin{aligned} V_1(u; b_0) &= \frac{b_0(b_0+1)}{2u^2} + \frac{1}{2}\omega_1^2 u^2, \\ V_2(u; b_1) &= \frac{b_1(b_1+1)}{2u^2} + \frac{1}{2}\omega_1^2 u^2, \end{aligned} \qquad (86)$$

where $b_1 = l + 1$ and $b_0 = l$. Therefore, in relation (33) this corresponds to $\eta = 1$. According to the Section 2.2, we presume that replacing $l + 1$ with l in a given operator can be accomplished via a similarity transformation, (34), and so we build an appropriate algebraic structure via translation operator:

$$T(l) = \exp\left(\frac{\partial}{\partial l}\right), \qquad T^{-1}(l) = T^\dagger(l) = \exp\left(-\frac{\partial}{\partial l}\right). \qquad (87)$$

Therefore, we introduce the operators:

$$B_l := \frac{1}{\sqrt{2}} T^\dagger(l) A_l, \quad B_l^\dagger := \frac{1}{\sqrt{2}} A_l^\dagger T(l), \quad N_B^l := B_l^\dagger B_l, \qquad (88)$$

which lead us to the simple harmonic oscillator (Heisenberg–Weyl) algebra

$$[B_l, B_l^\dagger] = 1, \quad [N_B^l, B_l] = -B_l, \quad [N_B^l, B_l^\dagger] = B_l^\dagger. \qquad (89)$$

These commutation relations show that $B_l^\dagger$ and B_l are the appropriate creation and annihilation operators for the spectra of our shape-invariant potentials. The action of these operators on normalized eigenfunctions yields:

$$B_l \Psi_n^l = \sqrt{n}\, \Psi_{n-1}^l, \quad B_l^\dagger \Psi_n^l = \sqrt{n+1}\, \Psi_{n+1}^l, \quad N_B^l \Psi_n^l = n \Psi_n^l, \quad n = 0, 1, 2, \ldots. \qquad (90)$$

Equation (79) and the last equation of the above set give:

$$E_n^l = \omega_1 \left(2n + l + \frac{3}{2}\right). \qquad (91)$$

In addition, the condition $B_l \Psi_0^l = 0$ gives us the ground state of the model universe for factor ordering l by:

$$\Psi_0^l = C_l \exp\left(-\frac{\omega_1}{2} u^2 - \frac{l+1}{u^2}\right), \qquad (92)$$

where C_l is a normalization constant. The excited states can be easily determined by applying (37).

In what follows, let us complete our procedure by including the Grassmannian variables ψ and $\bar{\psi}$, which satisfy

$$\psi^2 = 0, \qquad \bar{\psi}^2 = 0, \qquad \psi\bar{\psi} + \bar{\psi}\psi = 1, \qquad (93)$$

involving them in the Hamiltonian (80). By means of such a procedure, we can subsequently construct a supersymmetric extension of our Hamiltonian:

$$H_{\text{SUSY}} = -\frac{1}{2} \frac{d^2}{du^2} + \frac{l(l+1)}{2u^2} + \frac{1}{2}\omega_1^2 u^2 + \omega_1 \bar{\psi}\psi. \qquad (94)$$

In our work herein, the convention of the left derivative for these variables has been adapted. Up to now, we specified the bosonic creation and annihilation operators B and $B^\dagger$ in terms of the dynamical variable u and its conjugate momenta. Here, we can also introduce fermionic creation and annihilation operators $C^\dagger = \bar{\psi}$ and $C = \psi$. Therefore, the Hamiltonian (94) can simply be written as:

$$H_{\text{SUSY}} = 2\omega_1(B^\dagger B + C^\dagger C). \tag{95}$$

Adapting the basic commutator and anticommutator brackets:

$$[B, B^\dagger] = 1, \qquad \{C, C^\dagger\} = 1, \tag{96}$$

and considering all the others to be zero, it is easy to show that the operators:

$$N_B = B^\dagger B, \qquad N_F = C^\dagger C, \qquad Q = B^\dagger C, \qquad Q^\dagger = C^\dagger B, \tag{97}$$

(where the indices B and F refer to the bosonic and fermionic quantities, respectively) will be conserved quantities. Namely,

$$[Q, H_{\text{SUSY}}] = [Q^\dagger, H_{\text{SUSY}}] = 0, \qquad [N_B, H_{\text{SUSY}}] = [N_F, H_{\text{SUSY}}] = 0. \tag{98}$$

Moreover, we have:

$$\begin{aligned}
{} [Q, N_B] &= -Q, & [Q, N_F] &= Q \\
[Q^\dagger, N_B] &= Q^\dagger, & [Q^\dagger, N_F] &= -Q^\dagger, \\
\omega_1\{Q, Q^\dagger\} &= H_{\text{SUSY}}, & Q^2 &= \tfrac{1}{2}\{Q, Q\} = 0, \quad (Q^\dagger)^2 = \tfrac{1}{2}\{Q^\dagger, Q^\dagger\} = 0.
\end{aligned} \tag{99}$$

From (98) and (99) an explicit algebra is therefore produced, where the Hamiltonian H_{SUSY} is a Casimir operator for the whole algebra [78]. If we use a matrix representation, then we can write, alternatively,

$$\psi = C = \begin{pmatrix} 0 & 0 \\ 1 & 0 \end{pmatrix}, \quad \bar{\psi} = C^\dagger = \begin{pmatrix} 0 & 1 \\ 0 & 0 \end{pmatrix}, \tag{100}$$

$$H_{\text{SUSY}} = 2\omega_1 \begin{pmatrix} B_l^\dagger B_l + 1 & 0 \\ 0 & B_l B_l^\dagger \end{pmatrix} = \begin{pmatrix} H_1 & 0 \\ 0 & H_2 \end{pmatrix}. \tag{101}$$

4. Discussion

This paper is a review that embraces a twofold endorsement. On the one hand, it imports SUSY features (in a quantum mechanical setting). However, the current fact is that SUSY has not (yet ...) been found in nature; searches prevail for any evidence, being it directly or indirectly. On the other hand, this review refers to quantum cosmology as a phenomenological domain regarding the full quantization of gravity. Likewise, there is as yet no clear-cut observational evidence of such a stage in the very early universe. Whereas the latter is fairly expected as the cosmos is further probed, proceeding gradually to prior times, the former, although alluringly elegant, may just be a formal framework. So, producing a review on a topic involving these two ideas may seem twice likely to raise discomfort. However, maybe not; perhaps SUSY quantum cosmology deserves to be kept nearby, just over an arm's length, so as to say[6], if the occasion (or data) emerges to either support it or at least enthuse more research about it. There are still open aspects to appraise and the one brought up in this review is among them [18,19].

In what concerns SUSY quantum cosmology, there have been a few books in the past 30 years or so [18,19,79] plus selected reviews on the different procedures that were constructed and subsequently promoted [80–90]. Likewise, there are chapters (and sections) about SUSY quantum cosmology in well known textbooks concerning quantum gravity [91–94]. In particular, the direction and extension of $N = 2$ SQM to SUSY quantum cosmology was led by [95–98] and subsequently by [99,100], referring to conformal issues.

Therefore, the opportunity to produce this review enthused us to refer to and explore a specific particular aspect that was indicated as a concrete open problem in SUSY quantum cosmology (Nb. we emphasize many more still remain; cf. in [18,19]); concretely, investigating the setting of SIP, fairly present in $N = 2$ SQM. This framework has been previously developed and independently from SUSY quantum cosmology, to explore issues in supersymmetric quantum field theory, namely SUSY breaking (which the seminal paper [101] has made possible).

Hence, a necessary and yet to be performed analysis remains to be elaborated: explicitly considering SIP as we just mentioned within quantum cosmology, i.e., bringing up this possibility and using it intertwined within SUSY quantum cosmology. This is a new idea for other researchers to pick up and evolve forward, producing their own assertions. The content of our review is thus very open, conveying a direction for further exploration. It involves algebraic quantum-mechanical aspects that are present when SIP characterizes particular models. It also deals with integrability, which SUSY seems to bring so elegantly.

In this paper, besides contributing a topical review towards this Special Issue, we also provided a constructive example to illustrate how promising the framework can be. Concretely, we provided a case study, consisting of a spatially flat FRW model in the presence of a single scalar field, minimally coupled to gravity. We extracted the Schrödinger–Wheeler–DeWitt equation containing a particular set of possible factor ordering. Next, we computed the corresponding supersymmetric partner Hamiltonians. Intriguingly, the shape invariance properties can be related to the several factor orderings of our Schrödinger–Wheeler–DeWitt equation. The ground state was computed and the excited states as well. Consistently, the partner Hamiltonians, were explicitly presented within an $N = 2$ SQM framework.

We implicitly made another suggestion in Section 1 (Introduction). In more detail, we suggested building a twofold framework; on the one hand, importing the ideas employed in references [13–17], where the presence of constraints, their algebra plus a natural integrability induces separability in the Hilbert space of solutions for the Wheeler–DeWitt equation. On the other hand, exploring, at least to begin with on formal terms, whether any such algebra of the constraints generators for a minisuperspace would bear any similarity to an algebra of supersymmetry generators. In other words, perhaps producing a sequence of *new* operators A_l and $A_l^\dagger$, assisting SIP properties but also related to SUSY constraints of a Schrödinger–Wheeler–DeWitt equation similar to (79). SIP are well recognized for being simple to deal with. In essence, our suggestion is to explore (*i*) if there is any relation between SIP and the descriptive report in [13–17] and, if positive, (*ii*) apply it within SUSY quantum cosmology.

Author Contributions: Conceptualization, S.J., S.M.M.R. and P.M.; Formal analysis, S.J., S.M.M.R. and P.M.; Methodology, S.J., S.M.M.R. and P.M.; Writing–original draft, S.J., S.M.M.R. and P.M.; Writing–review & editing, S.J., S.M.M.R. and P.M. All authors have read and agreed to the published version of the manuscript.

Funding: This research received no external funding.

Institutional Review Board Statement: Not applicable.

Informed Consent Statement: Not applicable.

Data Availability Statement: Not applicable.

Acknowledgments: PVM and SMMR acknowledge the FCT grants UID-B-MAT/00212/2020 and UID-P-MAT/00212/2020 at CMA-UBI plus the COST Action CA18108 (Quantum gravity phenomenology in the multi-messenger approach).

Conflicts of Interest: The authors declare no conflict of interest.

Notes

1. In [37], the excited wave functions have also been studied. More concretely, instead of the choice of a nonsingular superpotential that is based on the ground state wave function $\Psi_0(x)$, a generalized procedure was presented to construct all possible superpotentials.
2. Cf. next subsection, concretely about Equation (28).
3. Throughout this paper we work in natural units where $\hbar = c = k_B = 1$.
4. Adler–Deser–Misner (ADM); see [56] for more details.
5. In this work, we consider a framework in which the scalar field is minimally coupled to gravity, see also [57–59]. Instead, one can choose other interesting gravitational models where ϕ is non-minimally coupled, see for instance [60–64].
6. "Ah, but a man's reach should exceed his grasp, Or what's a heaven for?", Robert Browning (in 'Andrea del Sarto' l. 97 (1855)).

References

1. Cooper, F.; Ginocchio, J.N.; Wipf, A. Supersymmetry, operator transformations and exactly solvable potentials. *J. Phys. A* **1989**, *22*, 3707–3716. [CrossRef]
2. Gangopadhyaya, A.; Mallow, J.V.; Sukhatme, U.P. Shape invariance and its connection to potential algebra. In *Supersymmetry and Integrable Models*; Aratyn, H., Imbo, T.D., Keung, W.Y., Sukhatme, U., Eds.; Springer: Berlin/Heidelberg, Germany, 1998; Volume 502, pp. 341–350. [CrossRef]
3. Balantekin, A.B. Algebraic approach to shape invariance. *Phys. Rev. A* **1998**, *57*, 4188–4191. [CrossRef]
4. Gangopadhyaya, A.; Mallow, J.V.; Sukhatme, U.P. Broken supersymmetric shape invariant systems and their potential algebras. *Phys. Lett. A* **2001**, *283*, 279–284. [CrossRef]
5. Chen, G.; Chen, Z.D.; Xuan, P.C. Exactly solvable potentials of the Klein Gordon equation with the supersymmetry method. *Phys. Lett. A* **2006**, *352*, 317–320. [CrossRef]
6. Khare, A.; Bhaduri, R.K. Supersymmetry, shape invariance and exactly solvable noncentral potentials. *Am. J. Phys.* **1994**, *62*, 1008–1014. [CrossRef]
7. Quesne, C. Deformed Shape Invariant Superpotentials in Quantum Mechanics and Expansions in Powers of $\hbar$. *Symmetry* **2020**, *12*, 1853. [CrossRef]
8. Oikonomou, V.K. A relation between Z_3-graded symmetry and shape invariant supersymmetric systems. *J. Phys. A* **2014**, *47*, 435304. [CrossRef]
9. Bazeia, D.; Das, A. Supersymmetry, shape invariance and the Legendre equations. *Phys. Lett. B* **2012**, *715*, 256–259. [CrossRef]
10. Stahlhofen, A. Remarks on the equivalence between the shape-invariance condition and the factorisation condition. *J. Phys. A* **1989**, *22*, 1053–1058. [CrossRef]
11. Jafarizadeh, M.A.; Fakhri, H. Supersymmetry and shape invariance in differential equations of mathematical physics. *Phys. Lett. A* **1997**, *230*, 164–170. [CrossRef]
12. Amani, A.; Ghorbanpour, H. Supersymmetry Approach and Shape Invariance for Pseudo-harmonic Potential. *Acta Phys. Pol. B* **2012**, *43*, 1795–1803. [CrossRef]
13. Fathi, M.; Jalalzadeh, S.; Moniz, P.V. Classical Universe emerging from quantum cosmology without horizon and flatness problems. *Eur. Phys. J. C* **2016**, *76*, 527. [CrossRef]
14. Jalalzadeh, S.; Rostami, T.; Moniz, P.V. On the relation between boundary proposals and hidden symmetries of the extended pre-big bang quantum cosmology. *Eur. Phys. J. C* **2015**, *75*, 38. [CrossRef]
15. Rostami, T.; Jalalzadeh, S.; Moniz, P.V. Quantum cosmological intertwining: Factor ordering and boundary conditions from hidden symmetries. *Phys. Rev. D* **2015**, *92*, 023526. [CrossRef]
16. Jalalzadeh, S.; Moniz, P.V. Dirac observables and boundary proposals in quantum cosmology. *Phys. Rev. D* **2014**, *89*, 083504. [CrossRef]
17. Jalalzadeh, S.; Rostami, T.; Moniz, P.V. Quantum cosmology: From hidden symmetries towards a new (supersymmetric) perspective. *Int. J. Mod. Phys. D* **2016**, *25*, 1630009. [CrossRef]
18. Moniz, P.V. *Quantum Cosmology—The Supersymmetric Perspective—Vol. 1*; Lecture Notes in Physics; Springer: Berlin/Heidelberg, Germany, 2010; Volume 803. [CrossRef]
19. Moniz, P.V. *Quantum Cosmology—The Supersymmetric Perspective—Vol. 2*; Lecture Notes in Physics; Springer: Berlin/Heidelberg, Germany, 2010; Volume 804. [CrossRef]
20. Moniz, P.V. Origin of structure in supersymmetric quantum cosmology. *Phys. Rev. D* **1998**, *57*, R7071. [CrossRef]
21. Kiefer, C.; Lück, T.; Moniz, P.V. Semiclassical approximation to supersymmetric quantum gravity. *Phys. Rev. D* **2005**, *72*, 045006. [CrossRef]
22. Cordero, R.; Granados, V.D.; Mota, R.D. Novel Complete Non-Compact Symmetries for the Wheeler–DeWitt Equation in a Wormhole Scalar Model and Axion-Dilaton String Cosmology. *Class. Quant. Grav.* **2011**, *28*, 185002. [CrossRef]
23. Cordero, R.; Mota, R.D. New exact supersymmetric wave functions for a massless scalar field and axion–dilaton string cosmology in a FRWL metric. *Eur. Phys. J. Plus* **2020**, *135*, 78. [CrossRef]

24. Díaz, J.S.G.; Reyes, M.A.; Mora, C.V.; Pozo, E.C. Supersymmetric Quantum Mechanics: Two Factorization Schemes and Quasi-Exactly Solvable Potentials. In *Panorama of Contemporary Quantum Mechanics-Concepts and Applications*; IntechOpen: London, UK 2019. [CrossRef]
25. Bhaduri, R.K.; Sakhr, J.; Sprung, D.W.L.; Dutt, R.; Suzuki, A. Shape invariant potentials in SUSY quantum mechanics and periodic orbit theory. *J. Phys. A* **2005**, *38*, L183–L189. [CrossRef]
26. Bougie, J.; Gangopadhyaya, A.; Mallow, J.V. Method for generating additive shape-invariant potentials from an Euler equation. *J. Phys. A* **2011**, *44*, 275307. [CrossRef]
27. Filho, E.D.; Ribeiro, M.A.C. Generalized Ladder Operators for Shape-invariant Potentials. *Phys. Scripta* **2001**, *64*, 548–552. [CrossRef]
28. Gangopadhyaya, A.; Mallow, J.V.; Rasinariu, C.; Bougie, J. Exactness of SWKB for shape invariant potentials. *Phys. Lett. A* **2020**, *384*, 126722. [CrossRef]
29. Bougie, J.; Gangopadhyaya, A.; Mallow, J.; Rasinariu, C. Supersymmetric Quantum Mechanics and Solvable Models. *Symmetry* **2012**, *4*, 452–473. [CrossRef]
30. Cariñena, J.F.; Ramos, A. Shape-invariant potentials depending on n parameters transformed by translation. *J. Phys. A Math. Gen.* **2000**, *33*, 3467–3481. [CrossRef]
31. Su, W.C. Shape invariant potentials in second-order supersymmetric quantum mechanics. *J. Phys. Math.* **2011**, *3*, 1–12. [CrossRef]
32. Dong, S.H. *Factorization Method in Quantum Mechanics*; Springer: Berlin/Heidelberg, Germany, 2007. [CrossRef]
33. Nasuda, Y.; Sawado, N. SWKB Quantization Condition for Conditionally Exactly Solvable Systems and the Residual Corrections. *arXiv* 2021, arXiv:2108.12567.
34. Gangopadhyaya, A.; Mallow, J.V.; Rasinariu, C. *Supersymmetric Quantum Mechanics: An Introduction*; World Scientific: Singapore, 2017. [CrossRef]
35. Cooper, F.; Ginocchio, J.N.; Khare, A. Relationship Between Supersymmetry and Solvable Potentials. *Phys. Rev. D* **1987**, *36*, 2458–2473. [CrossRef] [PubMed]
36. Dutt, R.; Khare, A.; Sukhatme, U.P. Supersymmetry, shape invariance, and exactly solvable potentials. *Am. J. Phys.* **1988**, *56*, 163–168. [CrossRef]
37. Cooper, F.; Khare, A.; Sukhatme, U. Supersymmetry and quantum mechanics. *Phys. Rept.* **1995**, *251*, 267–385. [CrossRef]
38. Nieto, M.M. Relationship Between Supersymmetry and the Inverse Method in Quantum Mechanics. *Phys. Lett. B* **1984**, *145*, 208–210. [CrossRef]
39. Pursey, D.L. Isometric operators, isospectral Hamiltonians, and supersymmetric quantum mechanics. *Phys. Rev. D* **1986**, *33*, 2267–2279. [CrossRef] [PubMed]
40. Gel'fand, I.M.; Levitan, B.M. On the determination of a differential equation from its spectral function. *Izv. Akad. Nauk SSSR Ser. Mat.* **1951**, *15*, 309–360.
41. Abraham, P.B.; Moses, H.E. Changes in potentials due to changes in the point spectrum: Anharmonic oscillators with exact solutions. *Phys. Rev. A* **1980**, *22*, 1333–1340. [CrossRef]
42. Kac, V.G. A sketch of Lie superalgebra theory. *Comm. Math. Phys.* **1977**, *53*, 31–64. [CrossRef]
43. Gendenshtein, L.E. Derivation of Exact Spectra of the Schrodinger Equation by Means of Supersymmetry. *JETP Lett.* **1983**, *38*, 356–359.
44. Schrödinger, E. A Method of Determining Quantum-Mechanical Eigenvalues and Eigenfunctions. *Proc. R. Ir. Acad. A Math. Phys. Sci.* **1940**, *46*, 9–16.
45. Infeld, L.; Hull, T.E. The Factorization Method. *Rev. Mod. Phys.* **1951**, *23*, 21–68. [CrossRef]
46. Khare, A. Supersymmetry in quantum mechanics. *Pramana-J. Phys.* **1997**, *49*, 41–64. [CrossRef]
47. Schrödinger, E. Further Studies on Solving Eigenvalue Problems by Factorization. *Proc. R. Ir. Acad., A Math. phys. sci.* **1940**, *46*, 183–206.
48. Darboux, G. On a proposition relative to linear equations. *C. R. Acad. Sci. Paris* **1882**, *94*, 1456–1459.
49. Luban, M.; Pursey, D.L. New Schrödinger equations for old: Inequivalence of the Darboux and Abraham-Moses constructions. *Phys. Rev. D* **1986**, *33*, 431–436. [CrossRef] [PubMed]
50. Arancibia, A.; Plyushchay, M.S.; Nieto, L.-M. Exotic supersymmetry of the kink-antikink crystal, and the infinite period limit. *Phys. Rev. D* **2016**, *83*, 065025. [CrossRef]
51. Sonnenschein, J.; Tsulaia, M. A Note on Shape Invariant Potentials for Discretized Hamiltonians *arXiv* 2022, arXiv:2205.10100.
52. Cariñena, J.F.; Plyushchay, M.S. Ground-state isolation and discrete flows in a rationally extended quantum harmonic oscillator. *Phys. Rev. D* **2016**, *94*, 105022. [CrossRef]
53. Cariñena, J.F.; Inzunza, L.; Plyushchay, M.S. Rational deformations of conformal mechanics. *Phys. Rev. D* **2018**, *98*, 026017. [CrossRef]
54. Arancibia, A.; Plyushchay, M.S. Chiral asymmetry in propagation of soliton defects in crystalline backgrounds. *Phys. Rev. D* **2015**, *92*, 105009. [CrossRef]
55. Fukui, T.; Aizawa, N. Shape-invariant potentials and an associated coherent state. *Phys. Lett. A* **1993**, *180*, 308–313. [CrossRef]
56. Jalalzadeh, S.; Moniz, P.V. *Challenging Routes in Quantum Cosmology*; World Scientific: Singapore, 2022. [CrossRef]
57. Rasouli, S.; Saba, N.; Farhoudi, M.; Marto, J.; Moniz, P. Inflationary universe in deformed phase space scenario. *Ann. Phys.* **2018**, *393*, 288–307. [CrossRef]

58. Rasouli, S.; Pacheco, R.; Sakellariadou, M.; Moniz, P. Late time cosmic acceleration in modified Sáez–Ballester theory. *Phys. Dark Univ.* **2020**, *27*, 100446. [CrossRef]
59. Rasouli, S.M.M. Noncommutativity, Saez-Ballester Theory and Kinetic Inflation. *Universe* **2022**, *8*, 165. [CrossRef]
60. Rasouli, S.; Farhoudi, M.; Khosravi, N. Horizon problem remediation via deformed phase space. *Gen. Rel. Grav.* **2011**, *43*, 2895–2910. [CrossRef]
61. Rasouli, S.M.M.; Moniz, P.V. Noncommutative minisuperspace, gravity-driven acceleration, and kinetic inflation. *Phys. Rev. D* **2014**, *90*, 083533. [CrossRef]
62. Rasouli, S.; Ziaie, A.; Jalalzadeh, S.; Moniz, P. Non-singular Brans–Dicke collapse in deformed phase space. *Ann. Phys.* **2016**, *375*, 154–178. [CrossRef]
63. Rasouli, S.M.M.; Vargas Moniz, P. Gravity-Driven Acceleration and Kinetic Inflation in Noncommutative Brans-Dicke Setting. *Odessa Astron. Pub.* **2016**, *29*, 19. [CrossRef]
64. Rasouli, S.; Marto, J.; Moniz, P. Kinetic inflation in deformed phase space Brans–Dicke cosmology. *Phys. Dark Univ.* **2019**, *24*, 100269. [CrossRef]
65. Dereli, T.; Tucker, R.W. Signature dynamics in general relativity. *Class. Quant. Grav.* **1993**, *10*, 365–374. [CrossRef]
66. Dereli, T.; Onder, M.; Tucker, R.W. Signature transitions in quantum cosmology. *Class. Quant. Grav.* **1993**, *10*, 1425–1434. [CrossRef]
67. Jalalzadeh, S.; Ahmadi, F.; Sepangi, H.R. Multidimensional classical and quantum cosmology: Exact solutions, signature transition and stabilization. *J. High Energy Phys.* **2003**, *2003*, 012. [CrossRef]
68. Khosravi, N.; Jalalzadeh, S.; Sepangi, H.R. Quantum noncommutative multidimensional cosmology. *Gen. Rel. Grav.* **2007**, *39*, 899–911. [CrossRef]
69. Pedram, P.; Jalalzadeh, S. Quantum cosmology with varying speed of light: Canonical approach. *Phys. Lett. B* **2008**, *660*, 1–6. [CrossRef]
70. Bina, A.; Atazadeh, K.; Jalalzadeh, S. Noncommutativity, generalized uncertainty principle and FRW cosmology. *Int. J. Theor. Phys.* **2008**, *47*, 1354–1362. [CrossRef]
71. Jalalzadeh, S.; Vakili, B. Quantization of the interior Schwarzschild black hole. *Int. J. Theor. Phys.* **2012**, *51*, 263–275. [CrossRef]
72. Khosravi, N.; Jalalzadeh, S.; Sepangi, H.R. Non-commutative multi-dimensional cosmology. *J. High Energy Phys.* **2006**, *2006*, 134. [CrossRef]
73. Vakili, B.; Pedram, P.; Jalalzadeh, S. Late time acceleration in a deformed phase space model of dilaton cosmology. *Phys. Lett. B* **2010**, *687*, 119–123. [CrossRef]
74. Darabi, F. Large scale - small scale duality and cosmological constant. *Phys. Lett. A* **1999**, *259*, 97–103. [CrossRef]
75. Darabi, F.; Rastkar, A. A quantum cosmology and discontinuous signature changing classical solutions. *Gen. Rel. Grav.* **2006**, *38*, 1355–1366. [CrossRef]
76. Jalalzadeh, S.; Rashki, M.; Abarghouei Nejad, S. Classical universe arising from quantum cosmology. *Phys. Dark Univ.* **2020**, *30*, 100741. [CrossRef]
77. Kastrup, H.A. A new look at the quantum mechanics of the harmonic oscillator. *Ann. Phys.* **2007**, *519*, 439–528. [CrossRef]
78. Kumar, R.; Malik, R.P. Supersymmetric Oscillator: Novel Symmetries. *EPL* **2012**, *98*, 11002. [CrossRef]
79. D'Eath, P.D. *Supersymmetric Quantum Cosmology*; Cambridge Monographs on Mathematical Physics; Cambridge University Press: Cambridge, UK, 1996. [CrossRef]
80. Moniz, P.V. Supersymmetric quantum cosmology: Shaken, not stirred. *Int. J. Mod. Phys. A* **1996**, *11*, 4321–4382. [CrossRef]
81. Moniz, P.V. Quantum Cosmology: Meeting SUSY. In *Progress in Mathematical Relativity, Gravitation and Cosmology*; García-Parrado, A., Mena, F.C., Moura, F., Vaz, E., Eds.; Springer: Berlin/Heidelberg, Germany, 2014; pp. 117–125. [CrossRef]
82. García-Compeán, H.; Obregón, O.; Ramírez, C. Topics in Supersymmetric and Noncommutative Quantum Cosmology. *Universe* **2021**, *7*, 434. [CrossRef]
83. Moniz, P.V. Supersymmetric quantum cosmology: A 'Socratic' guide. *Gen. Rel. Grav.* **2014**, *46*, 1618. [CrossRef]
84. López, J.; Obregón, O. Supersymmetric quantum matrix cosmology. *Class. quant. grav.* **2015**, *32*, 235014. [CrossRef]
85. Obregon, O.; Ramirez, C. Dirac-like formulation of quantum supersymmetric cosmology. *Phys. Rev. D* **1998**, *57*, 1015. [CrossRef]
86. Bene, J.; Graham, R. Supersymmetric homogeneous quantum cosmologies coupled to a scalar field. *Phys. Rev. D* **1994**, *49*, 799. [CrossRef] [PubMed]
87. Csordas, A.; Graham, R. Supersymmetric minisuperspace with nonvanishing fermion number. *Phys. Rev. Lett.* **1995**, *74*, 4129. [CrossRef]
88. Kleinschmidt, A.; Koehn, M.; Nicolai, H. Supersymmetric quantum cosmological billiards. *Phys. Rev. D* **2009**, *80*, 061701. [CrossRef]
89. Macías, A.; Mielke, E.W.; Socorro, J. Supersymmetric quantum cosmology for Bianchi class A models. *Int. J. Mod. Phys. D* **1998**, *7*, 701–712. [CrossRef]
90. Damour, T.; Spindel, P. Quantum supersymmetric cosmology and its hidden Kac–Moody structure. *Class. Quant. Grav.* **2013**, *30*, 162001. [CrossRef]
91. Kiefer, C. *Quantum Gravity*; International Series of Monographs on Physics; Clarendon: Oxford, UK, 2004; Volume 124. [CrossRef]
92. Esposito, G. *Quantum Gravity, Quantum Cosmology and Lorentzian Geometries*; Lecture Notes in Physics Monographs; Springer: Berlin/Heidelberg, Germany, 2009; Volume 12. [CrossRef]

93. Calcagni, G. *Classical and Quantum Cosmology*; Springer: Cham, Switzerland, 2017. [CrossRef]
94. Bojowald, M. *Quantum Cosmology: A Fundamental Description of the Universe*; Lecture Notes in Physics; Springer: New York, NY, USA, 2011; Volume 835. [CrossRef]
95. Lidsey, J.E. Quantum cosmology of generalized two-dimensional dilaton-gravity models. *Phys. Rev. D* **1995**, *51*, 6829. [CrossRef] [PubMed]
96. Lidsey, J.E.; Moniz, P.V. Supersymmetric quantization of anisotropic scalar-tensor cosmologies. *Class. Quant. Grav.* **2000**, *17*, 4823. [CrossRef]
97. Graham, R. Supersymmetric Bianchi type IX cosmology. *Phys. Rev. Lett.* **1991**, *67*, 1381. [CrossRef]
98. Lidsey, J.E. Scale factor duality and hidden supersymmetry in scalar-tensor cosmology. *Phys. Rev. D* **1995**, *52*, R5407. [CrossRef] [PubMed]
99. Tkach, V.; Rosales, J.; Obregón, O. Supersymmetric action for Bianchi type models. *Class. Quant. Grav.* **1996**, *13*, 2349. [CrossRef]
100. Obregón, O.; Rosales, J.; Tkach, V. Superfield description of the FRW universe. *Phys. Rev. D* **1996**, *53*, R1750. [CrossRef]
101. Witten, E. Dynamical breaking of supersymmetry. *Nucl. Phys. B* **1981**, *188*, 513–554. [CrossRef]

Article

Minisuperspace Quantum Cosmology in Metric and Affine Theories of Gravity

Salvatore Capozziello [1,2,3,*] **and Francesco Bajardi** [1,2,3]

1. Scuola Superiore Meridionale, Largo San Marcellino 10, I-80138 Naples, Italy; francesco.bajardi@unina.it
2. Department of Physics "E. Pancini", University of Naples "Federico II", Via Cinthia 36, I-80126 Naples, Italy
3. INFN Sezione di Napoli, Complesso Universitario di Monte Sant'Angelo, Edificio G, Via Cinthia, I-80126 Naples, Italy
* Correspondence: capozziello@na.infn.it

Abstract: Minisuperspace Quantum Cosmology is an approach by which it is possible to infer initial conditions for dynamical systems which can suitably represent *observable* and *non-observable* universes. Here we discuss theories of gravity which, from various points of view, extend Einstein's General Relativity. Specifically, the Hamiltonian formalism for $f(R)$, $f(T)$, and $f(\mathcal{G})$ gravity, with R, T, and $\mathcal{G}$ being the curvature, torsion and Gauss–Bonnet scalars, respectively, is developed starting from the Arnowitt–Deser–Misner approach. The Minisuperspace Quantum Cosmology is derived for all these models and cosmological solutions are obtained thanks to the existence of Noether symmetries. The Hartle criterion allows the interpretation of solutions in view of observable universes.

Keywords: quantum cosmology; noether symmetries; ADM formalism; exact solutions

1. Introduction

The Arnowitt–Deser–Misner (ADM) formalism was developed in 1962 with the purpose of solving issues occurring in the attempt to merge the formalism of General Relativity (GR) with Quantum Mechanics [1]. By means of a 3 + 1 decomposition of the metric, it is possible to get a gravitational Hamiltonian and find canonical quantization rules leading to a Schrödinger-like equation, dubbed the *Wheeler–De Witt* (WDW) equation, first obtained by J. A. Wheeler and B. DeWitt in 1967 [2–4]. Nowadays, the ADM formalism is not considered as the ultimate candidate to solve the quantization problem of GR, both because it does not account for a full theory of Quantum Gravity and because it implies an infinite-dimensional superspace which cannot be easily handled. However, the restriction of the problem to cosmology turns out to be useful for several reasons. On the one hand, the configuration superspace can be reduced to a finite-dimensional minisuperspace, where the WDW equation can be analytically solved. On the other hand, Quantum Cosmology can provide information regarding the very early stages of the universe evolution by means of the so-called *Wave Function of the Universe*, which is the solution of the WDW equation. Clearly, this wave function cannot be interpreted as a straightforward probability amplitude like in Quantum Mechanics, due to the lack of a Hilbert space and a definite-positive inner product in gravitational theory. Moreover, the probabilistic meaning, based on many copies of the same system, cannot be applied in the standard Copenhagen interpretation of Quantum Mechanics.

For these reasons, its meaning is still unclear, though different interpretations have been proposed. For instance, it can be thought as an indication of the probability for the quantum system to evolve towards our classical universe [5,6]. Another interpretation, related to the enucleation from nothing, was provided by Vilenkin in [7,8]. Furthermore, according to the so-called *Many World Interpretation* [9], the wave function is supposed to come from quantum measurements that are simultaneously realized in different universes, without showing any collapse, as in standard Quantum Mechanics [10]. In 1983, J. B. Hartle proposed to consider the wave function trend: when it is oscillating, variables are correlated

Citation: Capozziello, S.; Bajardi, F. Minisuperspace Quantum Cosmology in Metric and Affine Theories of Gravity. *Universe* 2022, 8, 177. https://doi.org/10.3390/universe8030177

Academic Editor: Jaime Haro Cases

Received: 7 February 2022
Accepted: 9 March 2022
Published: 10 March 2022

Publisher's Note: MDPI stays neutral with regard to jurisdictional claims in published maps and institutional affiliations.

Copyright: © 2022 by the authors. Licensee MDPI, Basel, Switzerland. This article is an open access article distributed under the terms and conditions of the Creative Commons Attribution (CC BY) license (https://creativecommons.org/licenses/by/4.0/).

and then it is possible to recover observable universes. The analogy comes from the wave function interpretation of non-relativistic Quantum Mechanics. This interpretation is the so-called *Hartle criterion* [11]. Interestingly, using this criterion, in the semiclassical limit one can recast the wave function as $\psi = e^{im_P^2 S}$, with S being the action and m_P the Planck mass. As a consequence, classical trajectories can be straightforwardly found by means of the Hamilton–Jacobi equation.

The formalism of Quantum Cosmology has been successfully applied to GR in the attempt to solve open issues related to the first phases of the Universe. However, in the high-energy regime, GR exhibits several other shortcomings. For this reason, many alternative models have been proposed over the years, such as *Kaluza–Klein Theory* [12–16], *String Theory* [17–21], *Non-Local Gravity* [22–26], *Loop Quantum Gravity* [27–31], etc.

All these theories try to address small-scale shortcoming suffered by GR, and most of them recover Quantum Cosmology as a limit [32].

However, matching the quantum formalism is not the only reason to consider GR alternatives, since many problems also occur at cosmological and astrophysical scales. More precisely, though GR predictions have been successfully confirmed by many experiments and observations, the theory also exhibits incompatibilities at several scales of energy [33,34]. Two of the most controversial problems are related to the existence of Dark Energy and Dark Matter. The former should represent most of the universe content and should be made by a never detected fluid with negative pressure, introduced to address the today observed accelerating expansion; the latter was first introduced to fit the galaxy rotation curves. For a detailed discussion on puzzles and shortcomings in Einstein's gravity see e.g., [35].

Due to the need to address incompatibilities between GR and observations, several modified/extended theories of gravity arose in the last decades. Within the landscape of modified models, some theories extend the Hilbert–Einstein action by including functions of second-order curvature invariants [36,37], some relax GR assumptions such as the Lorentz Invariance [38–40], the Equivalence Principle [41,42], or the metric-compatible connection [43].

One of the most straightforward extensions is the so-called $f(R)$ gravity, whose action contains a function of the scalar curvature. For detailed reviews, see [44,45]. By extending the gravitational action, it is possible to find out explanations for open questions at astrophysical and cosmological level. For instance, $f(R)$ gravity, in the post-Newtonian limit, is capable of fitting the galaxy rotation curve without introducing dark matter [46,47], the cosmological accelerating expansion without any dark energy [48,49], or the mass-radius diagram of neutron stars without exotic equations of state [50,51]. However, to date, no $f(R)$ model can account for the ultimate candidate capable of solving all the problems exhibited by GR simultaneously at any scale.

Furthermore, the Hilbert–Einstein action can be extended by considering also other second-order curvature invariants. An example is $f(R, R^{\mu\nu}R_{\mu\nu}, R^{\mu\nu\rho\sigma}R_{\mu\nu\rho\sigma})$ gravity, with $R_{\mu\nu}$ and $R_{\mu\nu\rho\sigma}$ being the Ricci and the Riemann tensors, respectively. These additional terms naturally arise from the one-loop effective action of GR, with the consequence that the renormalization procedure can be pursued only when higher-order invariants are included, so that UV divergences can be avoided [52–55]. Among all possible choices, there is only a particular combination of curvature invariants giving a topological surface in four dimensions. It is $\mathcal{G} = R^2 - 4R^{\mu\nu}R_{\mu\nu} + R^{\mu\nu\rho\sigma}R_{\mu\nu\rho\sigma}$, the *Gauss–Bonnet invariant*. As a consequence, the integration of $\mathcal{G}$ over the given manifold provides the Euler characteristic, namely, a topological invariant. In four dimensions or less, $\mathcal{G}$ vanishes identically, while in more than four dimensions it provides non-trivial contributions to the field equations. Due to the impossibility of dealing with the linear term $\mathcal{G}$ in four dimensions, usually a function of the Gauss–Bonnet term, that is $f(\mathcal{G})$, is considered into the action. In this way, the corresponding field equations can contribute to the dynamics, since $f(\mathcal{G})$ starts being trivial in three dimensions or less. Therefore, the gravitational action can be extended by considering either the function $f(\mathcal{G})$ or, in general, $f(R, \mathcal{G})$. Both $f(R, \mathcal{G})$ and $f(\mathcal{G})$ models exhibit several interesting features and provide some explanations for unsolved problems of GR at large scales [56–59]. Being a topological surface, the Gauss–Bonnet term can reduce dynamics and provide analytic solutions for the field equations; moreover, it naturally emerges in gauge theories of gravity such as Lovelock, Born–Infeld or Chern–Simons gravity [60–63].

Another class of alternatives to GR is represented by those models relaxing the assumption of torsionless and metric-compatible connections. In particular, it is possible to show that the most general affine connection is made of three different contributions, respectively related to curvature, torsion, and non-metricity. Torsion in the space-time occurs when the connection exhibits an anti-symmetric part, that is when $\Gamma^\alpha_{\mu\nu} \neq \Gamma^\alpha_{\nu\mu}$. Non-metricity occurs when $\nabla_\alpha g_{\mu\nu} \neq 0$, with ∇ being the covariant derivative. Therefore, gravity can be described by means of torsion, curvature and/or non-metricity, giving rise to a three equivalent formalisms [64].

When curvature and non-metricity vanish, the resulting theory is the so-called *Teleparallel Equivalent of General Relativity* (TEGR) (see e.g., [65] for further details); when curvature and torsion vanish we have the so-called *Symmetric Teleparallel Equivalent of General Relativity* (STEGR) [66]. More precisely, in the former theory, it is possible to define two rank-three tensors of the form $T^\alpha{}_{\mu\nu} \equiv 2\Gamma^\alpha{}_{[\mu\nu]}$ and $K^\rho{}_{\mu\nu} \equiv \frac{1}{2}g^{\rho\lambda}(T_{\mu\lambda\nu} + T_{\nu\lambda\mu} + T_{\lambda\mu\nu})$, respectively called *torsion tensor* and *contorsion tensor*. Both of them identically vanish in GR. In this way, by means of the *superpotential* $S^{\rho\mu\nu} \equiv K^{\mu\nu\rho} - g^{\rho\nu}T^{\sigma\mu}{}_\sigma + g^{\rho\mu}T^{\sigma\nu}{}_\sigma$, one can define the torsion scalar as $T \equiv T^{\rho\mu\nu}S_{\rho\mu\nu}$, so that the TEGR action can be chosen to be linearly proportional to T. Interestingly, the action thus constructed turns out to be equivalent to the Hilbert–Einstein one, up to a boundary term [67]. Moreover, TEGR can be recast as a gauge theory with respect to the translation group in the locally flat space-time [67].

Similar considerations apply for STEGR, where the definition of the *non-metricity tensor* $Q_{\rho\mu\nu} \equiv \nabla_\rho g_{\mu\nu} \neq 0$ and the *disformation tensor* $L^\rho{}_{\mu\nu} \equiv \frac{1}{2}g^{\rho\lambda}(-Q_{\mu\nu\lambda} - Q_{\nu\mu\lambda} + Q_{\lambda\mu\nu})$, together with $Q_\mu \equiv Q_\mu{}^\lambda{}_\lambda$ and $\tilde{Q}_\mu \equiv Q_{\alpha\mu}{}^\alpha$, allow to define the non-metricity scalar as: $Q \equiv -\frac{1}{4}Q_{\alpha\mu\nu}\left[-2L^{\alpha\mu\nu} + g^{\mu\nu}(Q^\alpha - \tilde{Q}^\alpha) - \frac{1}{2}(g^{\alpha\mu}Q^\nu + g^{\alpha\nu}Q^\mu)\right]$ [66]. In addition, the STEGR action, chosen to be linearly dependent on the non-metricity scalar, is equivalent to GR and TEGR, up to a boundary term. Modifications of the STEGR action will not be considered in this work; for details on the fundamental structure of the theory and possible applications, see, e.g., Refs. [68,69].

Here we consider an extension of the TEGR action, containing a function of the torsion. This can allow to address the problems suffered by GR at large scales. Notice that, although GR and TEGR are dynamically equivalent, $f(R)$ gravity differs from $f(T)$ gravity. For instance, as the former leads to fourth-order field equations, the latter provides second-order equations with respect to the metric. This makes $f(T)$ gravity easy to handle from a mathematical point of view.

In this paper, we want to discuss the problem of Minisuperspace Quantum Cosmology in metric and affine formulations of gravity theories. In particular, we will consider some metric and affine extensions. The aim is to show that, if the related cosmological models exhibit Noether symmetries, it is possible to interpret solutions under the standard of the Hartle criterion and then achieve observable universes.

The paper is organized as follows: Section 2 is devoted to the main features of ADM formalism, Quantum Cosmology, and the relation between the latter and the Noether symmetries. In Sections 3–5 we apply the Minisuperspace Quantum Cosmology formalism to $f(R)$, $f(T)$ and $f(\mathcal{G})$ models, respectively. In Section 6, we discuss the results and draw conclusions.

2. Quantum Cosmology and Noether Symmetries

As mentioned in the Introduction, the ADM formalism can represent a useful tool in the context of Quantum Cosmology. Here we briefly summarize the foundations of the approach and provide a link between the latter and the Noether theorem. Generally, the most general form of the Hilbert–Einstein action includes the extrinsic scalar curvature K [2,70], defined as the contraction of the three-dimensional spatial curvature tensor K_{ij} with the three-dimensional metric h^{ij}. Here, middle indexes label the three-dimensional space. By defining a set of coordinates X^α, the *deformation vector* $N^\alpha = \dot{X}^\alpha$ can be decomposed in terms of the lapse function N and the shift vector N^i as:

$$N^\alpha = N n^\alpha + N^i X_i^\alpha, \qquad (1)$$

where X_i^α is a tangent vector basis characterizing each point of the three-dimensional surface and n^α a unitary vector satisfying the relations

$$g_{\mu\nu} X_i^\mu n^\nu = 0, \qquad g_{\mu\nu} n^\mu n^\nu = -1. \qquad (2)$$

Therefore, the metric can be recast as:

$$g_{\mu\nu} = \begin{pmatrix} (N^2 - N_i N^i) & N_j \\ N_j & -h_{ij} \end{pmatrix}, \qquad (3)$$

by means of which the Lagrangian density reads:

$$\mathscr{L} = \frac{\kappa}{2} \sqrt{|h|} N \left(K^{ij} K_{ij} - K^2 + {}^{(3)}R \right) + \text{t.d.}, \qquad (4)$$

with h being the determinant of h_{ij} and ${}^{(3)}R$ the intrinsic three-dimensional curvature. From Equation (4), three conjugate momenta follow, respectively related to the shift vector, the lapse function, and the three-dimensional metric:

$$\pi \equiv \frac{\delta \mathscr{L}}{\delta \dot{N}} = 0, \qquad \pi^i \equiv \frac{\delta \mathscr{L}}{\delta \dot{N}_i} = 0,$$

$$\pi^{ij} \equiv \frac{\delta \mathscr{L}}{\delta \dot{h}_{ij}} = \frac{\kappa \sqrt{|h|}}{2} \left(K h^{ij} - K^{ij} \right). \qquad (5)$$

As a consequence, the Hamiltonian can be obtained by using Equation (5) and performing the Legendre transformation of the Lagrangian (4), so that we have:

$$\mathscr{H} = \pi^{ij} \dot{h}_{ij} - \mathscr{L}, \qquad (6)$$

constrained by the relations

$$\begin{cases} \dot{\pi} = -\{\mathcal{H}, \pi\} = \frac{\delta \mathcal{H}}{\delta N} = 0, \\ \dot{\pi}^i = -\{\mathcal{H}, \pi^i\} = \frac{\delta \mathcal{H}}{\delta N_i} = 0. \end{cases} \qquad (7)$$

In the above equations, it is $\mathcal{H} = \int \mathscr{H} d^3x$. In the canonical quantization procedure, the momenta (5) turn into the operators

$$\hat{\pi} = -i \frac{\delta}{\delta N}, \qquad \hat{\pi}^i = -i \frac{\delta}{\delta N_i}, \qquad \hat{\pi}^{ij} = -i \frac{\delta}{\delta h_{ij}}, \qquad (8)$$

while the Poisson brackets (7) turn into commutators. Finally, we obtain

$$\begin{cases} [\hat{h}_{ij}(x), \hat{\pi}^{kl}(x')] = i \, \delta^{kl}_{ij} \, \delta^3(x - x'), \\ \delta^{kl}_{ij} = \frac{1}{2}(\delta_i^k \delta_j^l + \delta_i^l \delta_j^k), \\ [\hat{h}_{ij}, \hat{h}_{kl}] = 0, \\ [\hat{\pi}^{ij}, \hat{\pi}^{kl}] = 0. \end{cases} \qquad (9)$$

Most importantly, from Equation (7), it is

$$\hat{\mathcal{H}}|\psi> = 0,\qquad(10)$$

with ψ being the wave function. Equation (10) can be recast in terms of dynamical variables and momenta, leading to the WDW equation [71], which, in the case of GR, yields

$$\left(\nabla^2 - \frac{\kappa^2}{4}\sqrt{|h|}\,^{(3)}R\right)|\psi> = 0,\qquad(11)$$

with the operator ∇^2 being defined as

$$\nabla^2 = \frac{1}{\sqrt{|h|}}\left(h_{ik}h_{jl} + h_{il}h_{jk} - h_{ij}h_{kl}\right)\frac{\delta}{\delta h_{ij}}\frac{\delta}{\delta h_{kl}}.\qquad(12)$$

Equation (11) shows that the ADM formalism relies on an infinite-dimensional superspace made of all possible 3-metrics, due to which any predictive power is inevitably lost. Moreover, as discussed above, the probabilistic interpretation of the wave function does not apply in this case, mainly because the scalar product $\int \psi^*\psi\,dx^3$ is not positive-definite. As a consequence, an infinite-dimensional Hilbert space cannot be considered. In what follows, we show that the ADM formalism can be suitably applied to cosmology, where the superspace can be reduced to a minisuperspace of configurations where the WDW equation, under some constraints, can be exactly solved. More precisely, the existence of Noether symmetries allows to introduce cyclic variables into the system, thanks to which the Hamiltonians assume handy expressions and classical trajectories can be recovered.

Let us then introduce the main features of the Noether Theorem and how it can be used as a method to select viable cosmological models which, eventually, result in observable universes. To this purpose, we consider a transformation, involving coordinates and fields, which is a symmetry for the Lagrangian density, namely,

$$\begin{cases} \tilde{x}^a = x^a - \xi^a \\ \tilde{\phi}^i = \phi^i - \eta^i, \end{cases}\qquad(13)$$

whose related first prolongation of Noether's vector can be written as:

$$X^{[1]} = \xi^a \partial_a + \eta^i \frac{\partial}{\partial \phi^i} + \left(\partial_a \eta^i - \partial_a \phi^i \partial_b \xi^b\right)\frac{\partial}{\partial(\partial_a \phi^i)}.\qquad(14)$$

The Noether Theorem states that, if the transformation (13) is a symmetry for the Lagrangian, then there exists a gauge function $g^a = g^a(x^a, \phi^i)$ satisfying the condition [72]

$$X^{[1]}\mathscr{L} + \partial_a \xi^a \mathscr{L} = \partial_a g^a\qquad(15)$$

and the quantity

$$j^a = \frac{\partial \mathscr{L}}{\partial(\partial_a \phi^i)}\eta^i - \frac{\partial \mathscr{L}}{\partial(\partial_a \phi^i)}\partial_b \phi^i\,\xi^b + \mathscr{L}\xi^a - g^a,\qquad(16)$$

is an integral of motion. For internal symmetries, Equations (14) and (16) reduce to:

$$X = \eta^i \frac{\partial}{\partial \phi^i} + \partial_a \eta^i \frac{\partial}{\partial(\partial_a \phi^i)},\qquad(17)$$

$$j^a = \frac{\partial \mathscr{L}}{\partial(\partial_a \phi^i)}\eta^i.\qquad(18)$$

Consequently, setting $g^a = 0$, Equation (15) becomes

$$X\mathscr{L} = 0\qquad(19)$$

and can be recast in terms of the Lie derivative along the flux of the vector X as

$$L_X \mathscr{L} = 0. \tag{20}$$

For further details about Noether's theorem and related applications see, e.g., [72–74]. It is worth pointing out that the conserved quantity (18) can be used to properly change the minisuperspace variables if the related point-like Lagrangian is cyclic. Thanks to Noether's Theorem, it is possible to find a methodical procedure aimed at finding suitable new coordinates. To this purpose, let us assume that there exists a transformation which allows to introduce a variable ψ^1, whose related conjugate momentum is an integral of motion, that is

$$\frac{\partial \mathscr{L}}{\partial(\partial_a \psi^1)} = \pi^a_{\psi^1} = \text{const.} \tag{21}$$

From Equation (17), we notice that ψ^1 is a constant of motion only if the related infinitesimal generator is equal to 1. Indeed, considering the general change of variables $\phi^i \to \psi^i(\phi^j)$, such that ψ^1 is cyclic, the infinitesimal generators can be recast as:

$$\eta^i \frac{\partial}{\partial \phi^i} = \eta^i \frac{\partial \psi^j}{\partial \phi^i} \frac{\partial}{\partial \psi^j} = i_X d\psi^j \frac{\partial}{\partial \psi^j}. \tag{22}$$

In this way, the Noether vector X and the conserved quantity j^a, written in terms of the new variables, read

$$X' = \eta^i \frac{\partial}{\partial \phi^i} + \partial_a \eta^i \frac{\partial}{\partial(\partial_a \phi^i)} = (i_X d\psi^k) \frac{\partial}{\partial \psi^k} + \partial_a (i_X d\psi^k) \frac{\partial}{\partial(\partial_a \psi^k)},$$

$$j^a = \eta^i \frac{\partial \mathscr{L}}{\partial(\partial_a \psi^i)} = i_X d\psi^i \frac{\partial \mathscr{L}}{\partial(\partial_a \psi^i)}, \tag{23}$$

where $(i_X d\psi^k)$ is the inner derivative. Requiring the conserved quantity to be equal to the conjugate momentum of ψ^1, the following conditions must hold:

$$i_X d\psi^1 = \eta^j \frac{\partial \psi^1}{\partial \phi^j} = 1, \quad i_X d\psi^i = \eta^j \frac{\partial \psi^i}{\partial \phi^j} = 0, \quad i \neq 1, \tag{24}$$

so that, from Equation (23), one gets

$$j^a = \eta^i \frac{\partial \mathscr{L}}{\partial(\partial_a \phi^i)} = \frac{\partial \mathscr{L}}{\partial(\partial_a \psi^1)} \quad \to \quad \pi^a_{\psi^1} = \text{constant.} \tag{25}$$

Therefore, the conjugate momentum $\pi^a_{\psi^1}$ of the new cyclic variable ψ^1 turns exactly into the Noether current, as expected by construction.

Notice that the condition $X' \mathscr{L}' = X \mathscr{L} = 0$ holds independently of the variables considered, so that any Noether symmetry is preserved under the change of variables. Nonetheless, it is worth remarking that the change of variables in Equation (24) is not unique, but infinite possible field transformations can occur. Therefore, in order to reduce dynamics, new variables must be chosen carefully.

If the variables in the minisuperspace depend only on the general parameter t, like in cosmology, Equations (14) and (16)–(18) provide:

$$X^{[1]} = \dot{\xi}\partial_t + \eta^i \frac{\partial}{\partial q^i} + (\dot{\eta}^i - \dot{q}^i\dot{\xi})\frac{\partial}{\partial \dot{q}^i}, \tag{26}$$

$$X = \eta^i \frac{\partial}{\partial q^i} + \dot{\eta}^i \frac{\partial}{\partial \dot{q}^i}, \tag{27}$$

$$j = \frac{\partial \mathcal{L}}{\partial \dot{q}^i}\eta^i - \frac{\partial \mathcal{L}}{\partial \dot{q}^i}\dot{q}^i \xi + \mathcal{L}\xi - g, \tag{28}$$

$$j = \frac{\partial \mathcal{L}}{\partial \dot{q}^i}\eta^i, \tag{29}$$

with q^i being the variables in the cosmological minisuperspace and $\mathcal{L}$ the Lagrangian. Therefore, the change of variables $q^i \to Q^i(q^j)$, which allows to introduce a constant of motion, can be recast as:

$$i_X dQ^1 = \eta^j \frac{\partial Q^1}{\partial q^j} = 1, \quad i_X dQ^i = \eta^j \frac{\partial Q^i}{\partial q^j} = 0, \quad i \neq 1. \tag{30}$$

After summarizing the main features of ADM formalism and Noether Theorem, it is worth investigating the deep connection between them, introducing the so-called *Hartle criterion*. To this purpose, let us notice that Equation (30) permits to recast the conjugate momenta of the cyclic variables as

$$j_i \equiv \pi^i_{Q^i} = \frac{\partial \mathcal{L}}{\partial \dot{Q}^i}. \tag{31}$$

As a consequence, if the cosmological point-like Lagrangian enjoys m symmetries with related conserved quantities $j_1, j_2 \ldots j_m$, Equation (10) together with (8) yield the following system of $m+1$ differential equations:

$$\begin{cases} \mathcal{H}|\psi> = 0 \\ -i\partial_1|\psi> = j_1|\psi> \\ -i\partial_2|\psi> = j_2|\psi> \\ \vdots \\ -i\partial_m|\psi> = j_m|\psi>. \end{cases} \tag{32}$$

Notice that, thanks to the change of variables (30), suggested by the Noether symmetries, the above system can be suitably integrated. Indeed, the integrals of motion j_i allow to reduce dynamics and generally provide a wave function of the form [75]:

$$|\psi> = e^{ij_k Q^k}|\chi(Q^\ell)>, \quad m < \ell < n, \tag{33}$$

where m are the variables with symmetries (integrals of motion), ℓ are the variables with no symmetries and n the minisuperspace dimension. Interestingly, in the semiclassical limit, the existence of symmetries leads to oscillatory solutions of the WDW equation. The latter, according to the Hartle criterion, are related to observable universes. Specifically, Hartle proposal relies on the analogy with non-relativistic Quantum Mechanics, where the oscillating wave function generally describes a classically permitted region, unlike the exponential wave function which labels the quantum region. Similar arguments can be also applied within the context of Quantum Cosmology, where the existence of Noether symmetries assures that the Hartle criterion can be related to classical trajectories. In other words, the oscillations of some components of the wave function mean correlations among variables while the exponential behavior means no correlation. Moreover, in the *Wentzel–*

Kramers–Brillouin (WKB) limit, the wave function can be linked to the gravitational action S by means of the relation $\psi(h_{ij}, \phi) \sim e^{iS}$. Therefore, using the Hamilton–Jacobi equations $\frac{\partial S}{\partial q^a} = \pi_a$, it is possible to get the dynamics for the generic variable q^a.

In other words, the importance of Noether symmetries in Quantum Cosmology is twofold: on the one hand, symmetries allow to reduce the dynamical system of differential equations arising from the ADM formalism, to find analytic solutions and to link such solutions to observable universes. On the other hand, the wave function can be recast in terms of the cosmological action to recover the Euler–Lagrange equations with respect to the new variables. This permits to suitably find exact solutions. The physical interpretation of such solutions is related to the Hartle criterion. For further readings on Quantum Cosmology and applications to theories of gravity, see, e.g., [35,76,77].

The results discussed above make Quantum Cosmology an important connection between classical and quantum aspects of gravity; while waiting for a complete and self-consistent theory of Quantum Gravity, applications to cosmology give interpretative approach capable of reducing the infinite-dimensional superspace coming from the ADM formalism to minisuperspaces where the equations of motion can be integrated and, eventually, interpreted.

3. Minisuperspace Quantum Cosmology in $f(R)$ Gravity

A first application of the above considerations can be developed for $f(R)$ gravity described by the action

$$S = \int \sqrt{-g} f(R) \, d^4x. \tag{34}$$

We select the functional form of the above action by the Noether Symmetry Approach and get the explicit expression of the point-like Lagrangian, which can be rendered cyclic by applying conditions in Equation (30). Finally, we find the Hamiltonian in terms of the new variables and solve the cosmological WDW equation for this modified model. This procedure leads to the Wave Function of the Universe, by means of which classical trajectories can be suitably obtained. We also show that the Hartle criterion is recovered, in agreement with the above discussion.

By varying Equation (34) with respect to the metric, one gets:

$$G_{\mu\nu} = \frac{1}{f_R(R)} \left\{ \frac{1}{2} g_{\mu\nu} [f(R) - R f_R(R)] + f_R(R)_{;\mu;\nu} - g_{\mu\nu} \Box f_R(R) \right\}, \tag{35}$$

being $G_{\mu\nu}$ the Einstein tensor $G_{\mu\nu} = R_{\mu\nu} - \frac{1}{2} g_{\mu\nu} R$ and $f_R(R)$ the first derivative of $f(R)$ with respect to R. Notice that when $f_R = 1$, Einstein field equations are recovered. In addition, the RHS can be understood as an effective energy–momentum tensor provided by the geometry, capable of reproducing the Dark Energy behavior without any further material ingredient [49].

Before applying the Noether Symmetry Approach, let us find the related point-like Lagrangian in a cosmological spatially flat background of the form $ds^2 = dt^2 - a(t) dx^2$, where $a(t)$ is the scale factor. Using a Lagrange multiplier λ, and considering the cosmological expression of the scalar curvature, the action can be written as

$$S = \int \left[a^3 f(R) - \lambda \left(R + 6 \frac{\ddot{a}}{a} + 6 \frac{\dot{a}^2}{a^2} \right) \right] dt, \tag{36}$$

where we integrated the three-dimensional hypersurface. By varying the action with respect to the curvature scalar, it is possible to find the Lagrange multiplier λ, that is:

$$\frac{\delta S}{\delta R} = a^3 f_R(R) - \lambda = 0, \qquad \lambda = a^3 f_R(R). \tag{37}$$

Replacing the result in Equation (37) into the action (36) and integrating out second derivatives, the canonical point-like Lagrangian turns out to be [78]

$$\mathcal{L}(a, \dot{a}, R, \dot{R}) = a^3[f(R) - Rf_R(R)] + 6a\dot{a}^2 f_R(R) + 6a^2\dot{a}\dot{R}f_{RR}(R). \tag{38}$$

The corresponding Euler–Lagrange equations are

$$\begin{cases} 6a^2\dot{a}\dot{R}f_{RR}(R) + 6a\dot{a}^2 f_R(R) - a^3[f(R) - Rf_R(R)] = 0 \\ \dot{R}^2 f_{RRR}(R) + \ddot{R}f_{RR}(R) + \dfrac{\dot{a}^2}{a^2}f_R(R) + 2\dfrac{\ddot{a}}{a}f_R(R) - \dfrac{1}{2}[f(R) - Rf_R(R)] + 2\dfrac{\dot{a}}{a}\dot{R}f_{RR}(R) = 0 \\ R = -6\left(\dfrac{\ddot{a}}{a} + \dfrac{\dot{a}^2}{a^2}\right). \end{cases} \tag{39}$$

The first equation is the energy condition $E_\mathcal{L} = 0$, corresponding to the (0,0) component of the field equations, namely the modified first Friedmann equation. The second is the equation with respect to the scale factor; the third is the Euler–Lagrange equation with respect to the scalar curvature, which provides again the cosmological expression of R derived from the Lagrange multiplier. In the considered minisuperspace, the Noether vector assumes the form

$$\begin{aligned} X^{[1]} &= \xi(a, R, t)\partial_t + \alpha(a, R, t)\partial_a + \beta(a, R, t)\partial_R \\ &+ (\dot{\alpha}(a, R, t) - \dot{\xi}(a, R, t)\dot{a})\partial_{\dot{a}} + (\dot{\beta}(a, R, t) - \dot{\xi}(a, R, t)\dot{R})\partial_{\dot{R}}, \end{aligned} \tag{40}$$

so that the application of the Noether identity (15), yields the following possible solution

$$\begin{cases} \mathcal{X} = \dfrac{\alpha_0}{a}\partial_a - 2\alpha_0 \dfrac{R}{a^2}\partial_R \\ j = 9\alpha_0 f_0 (2R\dot{a} + a\dot{R})R^{-\frac{1}{2}}, \quad f(R) = f_0 R^{\frac{3}{2}}, \end{cases} \tag{41}$$

where $\mathcal{X}$ is the symmetry generator. For the entire set of solutions see [78,79]. Notice that Equation (41) describes an internal symmetry, thus the infinitesimal generator ξ vanishes identically. This means that Equation (41) is also a solution of the vanishing Lie derivative condition and, consequently, the change of variables in Equation (30) can be adopted.

The WDW Equation and the Wave Function of the Universe

A suitable Minisuperspace Quantum Cosmology can be constructed for the function $f(R) = f_0 R^{\frac{3}{2}}$, selected by the Noether symmetries. Before considering the ADM formalism and finding the related Hamiltonian, we use the Noether Approach to reduce dynamics, by introducing a cyclic variable in the minisuperspace. To this purpose, Equation (24) permits to pass from the minisuperspace $\mathcal{S}\{a, R\}$ to $\mathcal{S}'\{z, w\}$, by means of the following system of differential equations:

$$\begin{cases} \alpha \partial_a z(a, R) + \beta \partial_R z(a, R) = 1 \\ \alpha \partial_a w(a, R) + \beta \partial_R w(a, R) = 0, \end{cases} \tag{42}$$

with z being the cyclic variable. A possible solution is

$$\begin{cases} w = w_0 (a\sqrt{R})^\ell, & z = \dfrac{a^2}{2\alpha_0} \\ a = \sqrt{2\alpha_0 z}, & R = \dfrac{1}{2\alpha_0}z\left(\dfrac{w}{w_0}\right)^{\frac{2}{\ell}}, \end{cases} \tag{43}$$

being ℓ and w_0 integration constants. Replacing Equation (43) into Equation (38), the Lagrangian turns out to be

$$\mathcal{L} = \dfrac{f_0}{\ell w}\left(\dfrac{w}{w_0}\right)^{\frac{1}{\ell}}\left[18\alpha_0 \dot{w}\dot{z} - w\ell\left(\dfrac{w}{w_0}\right)^{\frac{2}{\ell}}\right]. \tag{44}$$

Notice that the new equations of motion are simpler than Equation (39), written in terms of the old variables. They read:

$$\begin{cases} (\ell - 1)\dot{w}^2 - \ell w \ddot{w} = 0 \\ \left(\dfrac{w}{w_0}\right)^{\frac{2}{\ell}} + 6\alpha_0 \ddot{z} = 0, \end{cases} \quad (45)$$

where clearly z is the cyclic variable because there is no potential term depending on it. The Legendre transformation $\mathcal{H} = \pi_i \dot{q}^i - \mathcal{L}$ provides the Hamiltonian

$$\mathcal{H} = \dfrac{\ell}{18\alpha_0 f_0} w \left(\dfrac{w}{w_0}\right)^{-\frac{1}{\ell}} \pi_w \pi_z + f_0 \left(\dfrac{w}{w_0}\right)^{\frac{3}{\ell}}. \quad (46)$$

By promoting the momenta to operators, i.e., $\pi_i \to -i\partial_i$, the primary and secondary constraints in Equations (10) and (8) read as:

$$\begin{cases} \hat{\mathcal{H}}\psi = \left[\dfrac{\ell}{18\alpha_0 f_0} w \left(\dfrac{w}{w_0}\right)^{-\frac{1}{\ell}} \partial_w \partial_z + f_0 \left(\dfrac{w}{w_0}\right)^{\frac{3}{\ell}} \right] \psi = 0 \\ \hat{\pi}_z \psi = -i\partial_z \psi = j_0 \psi, \end{cases} \quad (47)$$

where the former is the WDW equation. The solution of the above system yields the following wave function [80]

$$\psi(z, w) = \psi_0 \exp\left\{ i \left[j_0 z - \dfrac{9\alpha_0 f_0^2}{2 j_0} \left(\dfrac{w}{w_0}\right)^{\frac{4}{\ell}} \right] \right\}. \quad (48)$$

In the semiclassical limit, the wave function can be recast in terms of the action S as

$$\psi \sim e^{iS}, \quad (49)$$

so that comparing Equation (49) with Equation (48), the action turns out to be

$$S = j_0 z - \dfrac{9\alpha_0 f_0^2}{2 j_0} \left(\dfrac{w}{w_0}\right)^{\frac{4}{\ell}}. \quad (50)$$

It can be easily proven that the Hamilton–Jacobi equations $\dfrac{\partial S}{\partial q^i} = \pi_i$ exactly provide the same system as Equation (45). The related solution, after coming back to the old variables, reads [80]

$$a(t) = a_0 \left[c_4 t^4 + c_3 t^3 + c_2 t^2 + c_1 t + c_0 \right]^{1/2}, \quad (51)$$

with c_i integration constants. It is worth noticing that, due to the oscillatory behavior of the wave function (48), the Hartle criterion is recovered by the existence of the Noether symmetry. The cosmological solution, emerging after this process, is an *observable universe* (51). Other solutions of this type can be found, see e.g., [81].

4. Minisuperspace Quantum Cosmology in $f(T)$ Gravity

Let us now consider an action containing a function of the torsion scalar T, namely,

$$S = \int e f(T) d^4 x, \quad (52)$$

defined as an extension of TEGR. As discussed in Section 1, T is the contraction of the superpotential with the torsion tensor. We assume the torsion scalar to be written in terms of the Weitzenböck connection $\Gamma^{\alpha}{}_{\mu\nu} = e_a^{\rho}\partial_{\mu}e_{\nu}^a$, with e_{μ}^a being the tetrad fields. By this choice, the spin connection vanishes identically, but the Lorentz Invariance is formally preserved and the dynamics results unchanged [82].

By varying the action (52) with respect to the tetrad fields, one gets the following field equations:

$$\frac{1}{e}\partial_{\mu}(h\, e_a^{\rho} S_{\rho}{}^{\mu\nu})f_T(T) - e_a^{\lambda} T^{\rho}{}_{\mu\lambda} S_{\rho}{}^{\nu\mu} f_T(T) + e_a^{\rho} S_{\rho}{}^{\mu\nu}(\partial_{\mu}T)f_{TT}(T) + \frac{1}{4}e_a^{\nu}f(T) = 0, \quad (53)$$

with $f_T(T)$ being the first derivative of $f(T)$ with respect to T and e the tetrad fields determinant. Unlike the $f(R)$ model, assigning a cosmological spatially flat space-time is not sufficient to uniquely determine the form of the tetrad fields. To overcome this issue, some criterions of "good" and "bad" tetrads can be used, as shown for example in [83]. In the applications to cosmology, we can use the simplest choice among all possible tetrads leading to a spatially flat cosmological universe, namely, $e_{\mu}^a = diag(1, a(t), a(t), a(t))$. Consequently, the field Equation (53) takes the form

$$Tf_T(T) - \frac{1}{2}f(T) = 0, \quad (54)$$

$$2Tf_{TT}(T) + f_T(T) = 0. \quad (55)$$

The same system of differential equations arises from the Euler–Lagrange equations coming from the point-like Lagrangian, which, in turn, can be obtained by means of the Lagrange Multipliers Method. More precisely, using the cosmological expression of the torsion scalar

$$T = -6\frac{\dot{a}^2}{a^2}, \quad (56)$$

the action (52) can be recast as:

$$S = \int \left[a^3 f(T) - \lambda \left(T + 6\frac{\dot{a}^2}{a^2}\right)\right] dt. \quad (57)$$

Moreover, the Lagrange multiplier λ is straightforwardly provided by the variation of the action with respect to the torsion scalar, that is,

$$\frac{\delta S}{\delta T} = 0 \;\rightarrow\; \lambda = a^3 f_T(T), \quad (58)$$

so that we have

$$\mathcal{L}(a, \dot{a}, T) = a^3[f(T) - Tf_T(T)] - 6a\dot{a}^2 f_T(T). \quad (59)$$

As mentioned above, the dynamical system of Euler–Lagrange equations turns out to be exactly equivalent to that provided by the field Equations (54) and (55). This system, however, can be analytically solved after selecting the form of the function $f(T)$. To this purpose, as in the previous section, we consider the Noether Symmetry Approach and develop the corresponding quantum cosmological model.

Let us start from Lagrangian (59) and select the unknown function by Noether's approach. In the considered two-dimensional minisuperspace, that is $\mathcal{S} = \{a, T\}$, the first prolongation of the Noether vector takes the form:

$$X^{[1]} = \xi \partial_t + \alpha \partial_a + \beta \partial_T + (\dot{\alpha} - \dot{\xi}\dot{a})\partial_{\dot{a}}, \quad (60)$$

where $\alpha = \alpha(a, T)$ and $\beta = \beta(a, T)$ are the components of the above η^i defined in (14). By applying the Noether symmetry existence condition (15) and equating to zero terms with the same time derivatives, we get a system of partial differential equations. A possible solution is

$$\begin{cases} \mathcal{X} = \left[\dfrac{\xi_0}{2k-1}\right]t\partial_t + \left[\dfrac{\xi_0}{3}a + \alpha_0 a^{1-\frac{3}{2k}}\right]\partial_a + \left[\dfrac{1}{k}\left(\beta_1 - 3\dfrac{\alpha_0}{a}^{-\frac{3}{2k}}\right) + \dfrac{\xi_0}{2k-1} + \beta_2\right]T\partial_T, & k \neq \dfrac{3}{2},\dfrac{1}{2} \\ j_0 = -12f_0 k\left(\xi_0 a^2 + \alpha_0 a^{2-\frac{3}{2k}}\right)T^{k-1}\dot{a}. \end{cases} \quad (61)$$

See [84] for other solutions. To introduce a cyclic coordinate by the procedure in Equation (30), we set $\xi_0 = \beta_1 = \beta_2 = 0$, so that internal symmetries are selected. In this way, the components of the function η^i read:

$$\alpha = \alpha_0 a^{1-\frac{3}{2k}}, \qquad \beta = \dfrac{-3\alpha_0}{k}Ta^{-\frac{3}{2k}}. \qquad (62)$$

By means of Equation (30), it is possible to introduce a cyclic variable in the minisperspace $\mathcal{S} = \{a, T\}$. Specifically, Equation (30) provides:

$$\begin{cases} \alpha\partial_a z + \beta\partial_T z = 1 \\ \alpha\partial_a w + \beta\partial_T w = 0, \end{cases} \qquad (63)$$

which, using the solution (62), reduces to

$$\begin{cases} z = \dfrac{2k}{3\alpha_0}a^{\frac{3}{2k}}, & \rightarrow \quad a = \left(\dfrac{3\alpha_0 z}{2k}\right)^{\frac{2k}{3}} \\ w = a^3 T^k, & \rightarrow \quad T = w^{\frac{1}{k}}\left(\dfrac{3\alpha_0 z}{2k}\right)^{-2}, \end{cases} \qquad (64)$$

so the Lagrangian with cyclic variable reads

$$\mathcal{L} = w(1-k) - 6k\alpha_0^2 \dot{z}^2 w^{\frac{k-1}{k}}, \qquad (65)$$

and z is cyclic as expected. The set of Euler–Lagrange equations coming from the above Lagrangian is

$$kw\ddot{z} + \dot{w}\dot{z}(k-1) = 0, \qquad (66)$$
$$w = -6^k \alpha_0^{2k}\dot{z}^{2k}. \qquad (67)$$

Following the same steps as $f(R)$ gravity, we perform a Legendre transformation for Lagrangian (65), that is:

$$\mathcal{H} = w(k-1) + \dfrac{3\pi_z^2}{24k\alpha_0^2 w^{\frac{k-1}{k}}}. \qquad (68)$$

In the canonical quantization scheme, where the conjugate momenta and the Hamiltonian are recast in terms of differential operators, the primary constraints (8) and (10) yield the system:

$$\begin{cases} \left[w(k-1) - \dfrac{3\partial_z^2}{24k\alpha_0^2 w^{\frac{k-1}{k}}}\right]\psi = 0 \\ i\partial_z \psi = j_0\psi, \end{cases} \qquad (69)$$

where the first equation is the WDW equation and the second is the momentum conservation equation. A solution is:

$$\psi \sim \exp\left\{i\left[2\alpha_0 w^{\frac{2k-1}{2k}}\sqrt{2k(k-1)}\right]z\right\}. \qquad (70)$$

Notice that, also here, the wave function is oscillating, confirming that the Hartle criterion holds for this model and observable universes are naturally provided. Moreover, in the WKB approximation, the wave function can be related to the gravitational action, so that the latter can be explicitly recast as:

$$S = \left[2\alpha_0 w^{\frac{2k-1}{2k}} \sqrt{2k(k-1)} \right] z.$$

Using the Hamilton–Jacobi equations

$$\begin{cases} \dfrac{\partial S}{\partial z} = \pi_z = j_0 \\ \dfrac{\partial S}{\partial w} = \pi_w = 0 \, , \end{cases} \tag{71}$$

we obtain the following analytic solution for z and w

$$z(t) = z_0 t, \quad w = -6^k \alpha_0^{2k} z_0^{2k}, \tag{72}$$

which, in terms of the old variables, becomes

$$a(t) = a_0 t^{\frac{2k}{3}}, \quad T(t) = -\frac{8k^2}{3} \frac{1}{t^2}. \tag{73}$$

The free parameter k can be constrained by the energy condition, which is the first Friedmann equation. It is possible to show that the only admissible solution occurs for $k = 1/2$. This means that the only cosmological solution which is compatible with Noether symmetries describes a stiff matter dominated epoch.

5. Minisuperspace Quantum Cosmology in $f(\mathcal{G})$ Gravity

As mentioned in the Introduction, Gauss–Bonnet cosmology has been recently taken into account because it allows to reduce the complexity of the field equations and, at the same time, to solve some high-energy issues exhibited by GR. Moreover, once considering an action proportional to $R + f(\mathcal{G})$, the function can be understood as an effective cosmological constant, with negligible contributions at the level of Solar System. Therefore, in the limit $f(\mathcal{G}) \to 0$, GR is safely recovered. Nevertheless, it is possible to show that Einstein's theory can be obtained even without imposing the GR limit as a requirement, namely, when the action is only proportional to the function $f(\mathcal{G})$. Specifically, in cosmological contexts, the model $f(\mathcal{G}) \sim \sqrt{\mathcal{G}}$ turns out to be dynamically equivalent to the scalar curvature [58]. Similar considerations also apply in a spherically symmetric background [85]. Let us start with the action

$$S = \int \sqrt{-g} f(\mathcal{G}) \, d^4 x, \tag{74}$$

whose variation with respect to the metric tensor provides [36,86]:

$$\frac{1}{2} g_{\mu\nu} f(\mathcal{G}) - \left(2R R_{\mu\nu} - 4R_{\mu p} R^p{}_\nu + 2R_\mu{}^{\rho\sigma\tau} R_{\nu\rho\sigma\tau} - 4R^{\alpha\beta} R_{\mu\alpha\nu\beta} \right) f_\mathcal{G}(\mathcal{G}) + \\ + \left(2R \nabla_\mu \nabla_\nu + 4G_{\mu\nu}\Box - 4R^p_{\{\nu} \nabla_{\mu\}} \nabla_p + 4g_{\mu\nu} R^{\rho\sigma} \nabla_\rho \nabla_\sigma - 4R_{\mu\alpha\nu\beta} \nabla^\alpha \nabla^\beta \right) f_\mathcal{G}(\mathcal{G}) = 0 \, . \tag{75}$$

In a cosmological spatially flat space-time, the Gauss–Bonnet invariant takes the form

$$\mathcal{G} = 24 \frac{\dot{a}^2 \ddot{a}}{a^3}. \tag{76}$$

It is easy to see that, when multiplied by $\sqrt{-g}$, Equation (76) turns out to be a boundary term. In order to apply the Noether Symmetry Approach and find the cosmological

Hamiltonian, we adopt a Lagrangian description as in the previous sections. Therefore, we use the constraint (76) to recast the action (74) in terms of Lagrange multipliers as

$$S = \int \left[a^3 f(\mathcal{G}) - \lambda \left\{ \mathcal{G} - 24 \frac{\dot{a}^2 \ddot{a}}{a^3} \right\} \right] d^4x. \qquad (77)$$

Using the variational principle and integrating out higher derivatives, the canonical point-like Lagrangian turns out to be:

$$\mathcal{L}(a, \dot{a}, \mathcal{G}, \dot{\mathcal{G}}) = a^3 [f(\mathcal{G}) - \mathcal{G} f_\mathcal{G}(\mathcal{G})] - 8 \dot{a}^3 \dot{\mathcal{G}} f_{\mathcal{G}\mathcal{G}}(\mathcal{G}). \qquad (78)$$

The field Equation (75) is thus equivalent to the system of equations of motion coming from Equation (78), which reads

$$\begin{cases} a^3 [f(\mathcal{G}) - \mathcal{G} f_\mathcal{G}(\mathcal{G})] + 24 \dot{a}^3 \dot{\mathcal{G}} f_{\mathcal{G}\mathcal{G}}(\mathcal{G}) = 0 \\ a^2 [f(\mathcal{G}) - \mathcal{G} f_\mathcal{G}(\mathcal{G})] + 8 \dot{a} [2 \dot{\mathcal{G}} \ddot{a} f_{\mathcal{G}\mathcal{G}}(\mathcal{G}) + \dot{a} \ddot{\mathcal{G}} f_{\mathcal{G}\mathcal{G}}(\mathcal{G}) + \dot{a} \dot{\mathcal{G}}^2 f_{\mathcal{G}\mathcal{G}\mathcal{G}}(\mathcal{G})] \\ \mathcal{G} = 24 \frac{\dot{a}^2 \ddot{a}}{a^3}. \end{cases} \qquad (79)$$

The first equation is the energy condition. The second is the equation for the scale factor evolution and the third is the equation for the Gauss–Bonnet term coinciding with the Lagrange multiplier (76).

In the two-dimensional minisuperspace $\mathcal{S} = \{a, \mathcal{G}\}$, the generator of transformations can be

$$\mathcal{X} = \xi(a, \mathcal{G}, t) \partial_t + \alpha(a, \mathcal{G}, t) \partial_a + \beta(a, \mathcal{G}, t) \partial_\mathcal{G}. \qquad (80)$$

The transformation is a symmetry, if the condition (15) holds. Pursuing the same procedure as in Sections 3 and 4, the application of the Noether identity to Lagrangian (78) provides a system of two differential equations:

$$\begin{cases} 3\alpha a^2 [f(\mathcal{G}) - \mathcal{G} f'(\mathcal{G})] - \beta a^3 \mathcal{G} f''(\mathcal{G}) + \partial_t \xi a^3 [f(\mathcal{G}) - \mathcal{G} f'(\mathcal{G})] = 0 \\ 3 \partial_a \alpha f''(\mathcal{G}) + \beta f'''(\mathcal{G}) - 3 \partial_t \xi f''(\mathcal{G}) + \partial_\mathcal{G} \beta f''(\mathcal{G}) = 0 \\ \xi = \xi(t), \quad \alpha = \alpha(a), \quad \beta = \beta(\mathcal{G}) \quad g = g_0, \end{cases} \qquad (81)$$

whose solution is [58]

$$\alpha = \alpha_0 a, \quad \beta = -4 \xi_0 \mathcal{G}, \quad \xi = \xi_0 t + \xi_1, \quad f(\mathcal{G}) = f_0 \mathcal{G}^k, \quad j_0 = \frac{\dot{a}^3}{\mathcal{G}^{3k}}, \qquad (82)$$

with the definitions

$$k \equiv \frac{3\alpha_0 + \xi_0}{4 \xi_0}, \quad f_0 \equiv \frac{4 f_0 \xi_0}{3\alpha_0 + \xi_0}. \qquad (83)$$

Replacing the selected function into the Lagrangian (78), the latter takes the form:

$$\mathcal{L} = -\frac{1}{3} \mathcal{G}^{k-2} \left[3(k-1) a^3 \mathcal{G}^2 + 24 k(k-1) \dot{a}^3 \dot{\mathcal{G}} \right]. \qquad (84)$$

Starting from this Lagrangian, it is possible to find the related Hamiltonian and the Wave Function of the Universe.

The WDW Equation and Wave Function of the Universe

As above, applying Equation (30), the minisuperspace can be reduced. Here, setting $\xi = 0$, leads to trivial solutions. Therefore, only symmetries involving space-time translations can be selected and this fact does not allow to use the procedure, based on

Equation (30), as in previous sections. Nevertheless, starting from Lagrangian (84), it is still possible to write the Hamiltonian as:

$$\mathcal{H} = \frac{f_0}{k}\mathcal{G}^k a^3 + \pi_a \left(-\frac{\pi_\mathcal{G}}{8f_0}\mathcal{G}^{2-k}\right)^{\frac{1}{3}}, \tag{85}$$

with $\pi_a = \frac{\partial \mathcal{L}}{\partial \dot{a}}$ and $\pi_\mathcal{G} = \frac{\partial \mathcal{L}}{\partial \dot{\mathcal{G}}}$. In this form, the system cannot be immediately quantized due to the presence of the fractional exponent. However, we can still use the conserved quantity (82) to reduce the minisuperspace and to get a suitable Hamiltonian. More precisely, the momentum $\pi_\mathcal{G}$ can be rewritten in terms of j_0 as:

$$\pi_\mathcal{G} = -8f_0 j_0 \mathcal{G}^{4k-2}, \tag{86}$$

so that the Hamiltonian becomes

$$\mathcal{H} = \frac{f_0}{k}\mathcal{G}^k a^3 + \pi_a \left(j_0 \mathcal{G}^{3k}\right)^{\frac{1}{3}}. \tag{87}$$

The canonical quantization rules, together with the WDW equation $\mathcal{H}\psi = 0$, provide a system of differential equations of the form

$$\begin{cases} \pi_\mathcal{G}\psi = -i\dfrac{\partial}{\partial \mathcal{G}}\psi & \rightarrow \quad \psi(a,\mathcal{G}) = A(a)\,\exp\left\{i\,\dfrac{8f_0 j_0 \mathcal{G}^{4k-1}}{1-4k}\right\} \\ \mathcal{H}\psi = 0 & \rightarrow \quad \dfrac{f_0}{k}(j_0)^{-\frac{1}{3}}a^3 A(a) - i\dfrac{\partial A(a)}{\partial a} = 0, \end{cases} \tag{88}$$

which can be solved with respect to $A(a)$ to provide the Wave Function of the Universe. The solution is:

$$\psi(a,\mathcal{G}) = \psi_0 \exp\left\{i\left[-\frac{f_0}{4k}(j_0)^{-\frac{1}{3}}a^4 + \frac{8f_0 j_0 \mathcal{G}^{4k-1}}{1-4k}\right]\right\}. \tag{89}$$

Even in this case, the Hartle criterion is preserved by the existence of symmetries, as the wave function is oscillating in the minisuperspace considered. Moreover, in the WKB approximation, the gravitational action can be addressed to the quantity:

$$S = -\frac{f_0}{4k}(j_0)^{-\frac{1}{3}}a^4 + \frac{8f_0 j_0}{1-4k}\mathcal{G}^{4k-1}, \tag{90}$$

so that Hamilton–Jacobi equation with respect to $a(t)$ yields

$$\frac{\partial S}{\partial a} = \pi_a \quad \rightarrow \quad \mathcal{G}^k a^3 = 24\mathcal{G}^{k-2}\dot{a}^3 \dot{\mathcal{G}}, \tag{91}$$

which is exactly the first Euler–Lagrange equation. The second Hamilton–Jacobi equation $\left(\dfrac{\partial S}{\partial \mathcal{G}} = \pi_\mathcal{G}\right)$ instead, provides the identity $\pi_\mathcal{G} = j_0$. The system can be implemented with the energy condition, so that the entire system of three differential equations yields the exact solution:

$$a(t) = a_0 t^{1-4k} \qquad \mathcal{G}(t) = -96k(1-4k)^3 t^{-4} \equiv \mathcal{G}_0 t^{-4}. \tag{92}$$

As shown in [58], the epochs crossed by the universe evolution can be obtained by varying the value of the constant k. To conclude, also in this case observable universes are recovered in the semiclassical limit.

6. Discussion and Conclusions

Minisuperspace Quantum Cosmology is an approach that ultimately allows to select initial conditions for observable universes. Starting from the Hamiltonian formulation of gravity theories, it is possible, by a quantization procedure, to obtain the functional WDW equation whose solution is the Wave Function of the Universe. Selecting particular configuration spaces (minisuperspaces), it is possible to reduce the infinite dimensional problem of superspace, and then the WDW equation to a partial differential equation eventually solvable. According to the Hartle criterion, we can determine if dynamical variables of such a wave function are either correlated or not and then apply the Hamilton–Jacobi equations for achieving classical trajectories. The Noether Symmetry Approach, in its Hamiltonian formulation, gives a straightforward interpretation of the Hartle criterion [75,80]: correlations occur for the oscillating components of the Wave Function of the Universe and they are related to the existence of first integrals of motion. The possibility to select classical trajectories (i.e., observable universes) relies on the existence of Noether symmetries. In other words, Noether symmetries constitute a selection rule in Quantum Cosmology.

In this paper, we considered some classes of gravity theories and selected their functional forms taking into account the existence of Noether symmetries. Specifically, we studied cosmological models related to $f(R)$ gravity, $f(T)$ gravity and Gauss–Bonnet gravity. The application of Noether Symmetry Approach provides (i) the transformation generator (which is a symmetry for the starting Lagrangian), (ii) the conserved quantity and (iii) the form of the action functional. When searching for internal symmetries, it is possible to follow the procedure in Equation (30) by which the cyclic variables for the dynamical system are derived.

After a Legendre transformation, we obtain the Hamiltonian function related to the Noether symmetry. After a canonical quantization, it is possible to derive the corresponding WDW equations and the Wave Function of the Universe. The Hartle criterion is always recovered thanks to presence of first integrals which give rise to oscillating behaviors independently of the considered representation of gravity. This means that classical trajectories, and therefore observable universes, can be always recovered if symmetries exist. As reported also in [79,87], the presence of Noether symmetries seems a criterion to recover physically viable models. In a forthcoming paper, this approach will be developed also in comparison with observational data.

Author Contributions: Conceptualization, F.B. and S.C.; methodology, F.B. and S.C.; formal analysis, F.B. and S.C.; investigation, F.B. and S.C.; writing—original draft preparation, F.B.; writing—review and editing, S.C. All authors have read and agreed to the published version of the manuscript.

Funding: This research received no external funding.

Acknowledgments: The Authors acknowledge the support of *Istituto Nazionale di Fisica Nucleare* (INFN) *(iniziative specifiche* GINGER, MOONLIGHT2, QGSKY, and TEONGRAV). This paper is based upon work from COST action CA15117 (CANTATA), COST Action CA16104 (GWverse), and COST action CA18108 (QG-MM), supported by COST (European Cooperation in Science and Technology).

Conflicts of Interest: The authors declare no conflict of interest.

References

1. Arnowitt, R.L.; Deser, S.; Misner, C.W. The Dynamics of general relativity. *Gravitation: Introd. Curr. Res.* **2008**, *40*, 1997–2027. [CrossRef]
2. DeWitt, B.S. Quantum Theory of Gravity. 1. The Canonical Theory. *Phys. Rev.* **1967**, *160*, 1113–1148. [CrossRef]
3. DeWitt, B.S. Quantum Theory of Gravity. 2. The Manifestly Covariant Theory. *Phys. Rev.* **1967**, *162*, 1195–1239. [CrossRef]
4. Wheeler, J.A.; On the Nature of quantum geometrodynamics. *Ann. Phys.* **1957**, *2*, 604–614. [CrossRef]
5. Vilenkin, A. The Interpretation of the Wave Function of the Universe. *Phys. Rev. D* **1989**, *39*, 1116. [CrossRef]
6. Hawking, S.W. The Quantum State of the Universe. *Nucl. Phys. B* **1984**, *239*, 257. [CrossRef]
7. Vilenkin, A. Creation of Universes from Nothing. *Phys. Lett. B* **1982**, *117*, 25–28. [CrossRef]
8. Vilenkin, A. Quantum Creation of Universes. *Phys. Rev. D* **1984**, *30*, 509–511. [CrossRef]
9. DeWitt, B.S.; Graham, N. (Eds.) *The Many-Worlds Interpretation of Quantum Mechanics*; Princeton University Press: Princeton, NJ, USA, 1973; ISBN 0-691-08131-X.

10. Bousso, R.; Susskind, L. The Multiverse Interpretation of Quantum Mechanics. *Phys. Rev. D* **2012**, *85*, 045007. [CrossRef]
11. Hartle, J.B.; Hawking, S.W. Wave Function of the Universe. *Phys. Rev. D* **1983**, *28*, 2960–2975. [CrossRef]
12. Klein, O. Quantum Theory and Five-Dimensional Theory of Relativity. *Z. Phys.* **1926**, *37*, 895–906. [CrossRef] (In German and English)
13. Kaluza, T. Zum Unitätsproblem der Physik. *Sitzungsberichte Der KöNiglich Preußlschen Akad. Wiss.* **1921**, *1921*, 966–972.
14. Han, T.; Lykken, J.D.; Zhang, R.J. On Kaluza-Klein states from large extra dimensions. *Phys. Rev. D* **1999**, *59*, 105006. [CrossRef]
15. Servant, G.; Tait, T.M.P. Is the lightest Kaluza-Klein particle a viable dark matter candidate? *Nucl. Phys. B* **2003**, *650*, 391–419. [CrossRef]
16. Duff, M.J.; Nilsson, B.E.W.; Pope, C.N. Kaluza-Klein Supergravity. *Phys. Rep.* **1986**, *130*, 1–142. [CrossRef]
17. Green, M.B.; Schwarz, J.H.; Witten, E. *Superstring Theory*; Volume 1: Introduction; Cambridge University Press: Cambridge, UK, 2008.
18. Polchinski, J. *String Theory*; Volume 1: An Introduction to the Bosonic String; Cambridge Monograph on Mathematica Physics: Cambridge, UK, 1998.
19. Seiberg, N.; Witten, E. String theory and noncommutative geometry. *J. High Energy Phys.* **1999**, *9*, 032. [CrossRef]
20. Witten, E. String theory dynamics in various dimensions. *Nucl. Phys. B* **1995**, *443*, 85–126. [CrossRef]
21. Friedan, D.; Martinec, E.J.; Shenker, S.H. Conformal Invariance, Supersymmetry and String Theory. *Nucl. Phys. B* **1986**, *271*, 93–165. [CrossRef]
22. Simon, J.Z. Higher Derivative Lagrangians, Nonlocality, Problems and Solutions. *Phys. Rev. D* **1990**, *41*, 3720. [CrossRef]
23. Modesto, L.; Tsujikawa, S. Non-local massive gravity. *Phys. Lett. B* **2013**, *727*, 48–56. [CrossRef]
24. Koshelev, A.S.; Modesto, L.; Rachwal, L.; Starobinsky, A.A. Occurrence of exact R^2 inflation in non-local UV-complete gravity. *J. High Energy Phys.* **2016**, *11*, 067. [CrossRef]
25. Calcagni, G.; Modesto, L.; Nicolini, P. Super-accelerating bouncing cosmology in asymptotically-free non-local gravity. *Eur. Phys. J. C* **2014**, *74*, 2999. [CrossRef]
26. Capozziello, S.; Bajardi, F. Non-Local Gravity Cosmology: An Overview. *Int. J. Mod. Phys. D* 2230009. [CrossRef]
27. Ashtekar, A.; Singh, P. Loop Quantum Cosmology: A Status Report. *Class. Quantum Gravity* **2011**, *28*, 213001. [CrossRef]
28. Rovelli, C. Loop quantum gravity. *Living Rev. Relativ.* **1998**, *11*, 1–69. [CrossRef]
29. Rovelli, C.; Smolin, L. Loop Space Representation of Quantum General Relativity. *Nucl. Phys. B* **1990**, *331*, 80–152. [CrossRef]
30. Meissner, K.A. Black hole entropy in loop quantum gravity. *Class. Quantum Gravity* **2004**, *21*, 5245–5252. [CrossRef]
31. Bojowald, M. Loop quantum cosmology. *Living Rev. Relativ.* **2005**, *8*, 11. [CrossRef]
32. Bojowald, M. Quantum cosmology: A review. *Rep. Prog. Phys.* **2015**, *78*, 023901. [CrossRef]
33. Bull, P.; Akrami, Y.; Adamek, J.; Baker, T.; Bellini, E.; Jimenez, J.B.; Bentivegna, E.; Camera, S.; Clesse, S.; Davis, J.H.; et al. Beyond ΛCDM: Problems, solutions, and the road ahead. *Phys. Dark Univ.* **2016**, *12*, 56–99. [CrossRef]
34. Clifton, T.; Ferreira, P.G.; Padilla, A.; Skordis, C. Modified Gravity and Cosmology. *Phys. Rep.* **2012**, *513*, 1–189. [CrossRef]
35. Capozziello, S.; Faraoni, V. *Beyond Einstein Gravity: A Survey of Gravitational Theories for Cosmology and Astrophysics*; Springer: Dordrecht, The Netherlands, 2010; p. 170. [CrossRef]
36. Nojiri, S.; Odintsov, S.D. Modified Gauss-Bonnet theory as gravitational alternative for dark energy. *Phys. Lett. B* **2005**, *631*, 1–6. [CrossRef]
37. Bueno, P.; Cano, P.A.; Min, V.S.; Visser, M.R. Aspects of general higher-order gravities. *Phys. Rev. D* **2017**, *95*, 044010. [CrossRef]
38. Colladay, D.; Kostelecky, V.A. Lorentz violating extension of the standard model. *Phys. Rev. D* **1998**, *58*, 116002. [CrossRef]
39. Collins, J.; Perez, A.; Sudarsky, D.; Urrutia, L.; Vucetich, H. Lorentz invariance and quantum gravity: An additional fine-tuning problem. *Phys. Rev. Lett.* **2004**, *93*, 191301. [CrossRef]
40. Horava, P. Quantum Gravity at a Lifshitz Point. *Phys. Rev. D* **2009**, *79*, 084008. [CrossRef]
41. Hui, L.; Nicolis, A.; Stubbs, C. Equivalence Principle Implications of Modified Gravity Models. *Phys. Rev. D* **2009**, *80*, 104002. [CrossRef]
42. Damour, T.; Donoghue, J.F. Equivalence Principle Violations and Couplings of a Light Dilaton. *Phys. Rev. D* **2010**, *82*, 084033. [CrossRef]
43. Olmo, G.J. Palatini Approach to Modified Gravity: $f(R)$ Theories and Beyond. *Int. J. Mod. Phys. D* **2011**, *20*, 413–462. [CrossRef]
44. Capozziello, S.; Laurentis, M.D. Extended Theories of Gravity. *Phys. Rep.* **2011**, *509*, 167–321. [CrossRef]
45. Nojiri, S.; Odintsov, S.D.; Oikonomou, V.K. Modified Gravity Theories on a Nutshell: Inflation, Bounce and Late-time Evolution. *Phys. Rep.* **2017**, *692*, 1–104. [CrossRef]
46. Sanders, R.H. Modified gravity without dark matter. *Lect. Notes Phys.* **2007**, *720*, 375–402.
47. Capozziello, S.; Cardone, V.F.; Troisi, A. Low surface brightness galaxies rotation curves in the low energy limit of R^n gravity: No need for dark matter? *Mon. Not. R. Astron. Soc.* **2007**, *375*, 1423–1440. [CrossRef]
48. Copeland, E.J.; Sami, M.; Tsujikawa, S. Dynamics of dark energy. *Int. J. Mod. Phys. D* **2006**, *15*, 1753–1936. [CrossRef]
49. Capozziello, S. Curvature quintessence. *Int. J. Mod. Phys. D* **2002**, *11*, 483–492. [CrossRef]
50. Astashenok, A.V.; Capozziello, S.; Odintsov, S.D. Extreme neutron stars from Extended Theories of Gravity. *J. Cosmol. Astropart. Phys.* **2015**, *2015*, 001. [CrossRef]
51. Astashenok, A.V.; Capozziello, S.; Odintsov, S.D.; Oikonomou, V.K. Extended Gravity Description for the GW190814 Supermassive Neutron Star. *Phys. Lett. B* **2020**, *811*, 135910. [CrossRef]
52. Stelle, K.S. Renormalization of Higher Derivative Quantum Gravity. *Phys. Rev. D* **1977**, *16*, 953–969. [CrossRef]
53. Adams, F.C.; Freese, K.; Guth, A.H. Constraints on the scalar field potential in inflationary models. *Phys. Rev. D* **1991**, *43*, 965–976. [CrossRef]

54. Cotsakis, S.; Tsokaros, A. Flat, radiation universes with quadratic corrections and asymptotic analysis. *arXiv* **2006**, arXiv:gr-qc/0612190.
55. Amendola, L.; Mayer, A.B.; Capozziello, S.; Occhionero, F.; Gottlober, S.; Muller, V.; Schmidt, H.J. Generalized sixth order gravity and inflation. *Class. Quantum Gravity* **1993**, *10*, L43–L47. [CrossRef]
56. Nojiri, S.; Odintsov, S.D.; Sasaki, M. Gauss-Bonnet dark energy. *Phys. Rev. D* **2005**, *71*, 123509. [CrossRef]
57. Cvetic, M.; Nojiri, S.; Odintsov, S.D. Black hole thermodynamics and negative entropy in de Sitter and anti-de Sitter Einstein-Gauss-Bonnet gravity. *Nucl. Phys. B* **2002**, *628*, 295–330. [CrossRef]
58. Bajardi, F.; Capozziello, S. $f(\mathcal{G})$ Noether cosmology. *Eur. Phys. J. C* **2020**, *80*, 704. [CrossRef]
59. Glavan, D.; Lin, C. Einstein-Gauss-Bonnet Gravity in Four-Dimensional Spacetime. *Phys. Rev. Lett.* **2020**, *124*, 081301. [CrossRef]
60. Bajardi, F.; Vernieri, D.; Capozziello, S. Exact solutions in higher-dimensional Lovelock and AdS $_5$ Chern-Simons gravity. *J. Cosmol. Astropart. Phys.* **2021**, *2021*, 057. [CrossRef]
61. Zanelli, J. Lecture notes on Chern-Simons (super-)gravities. Second edition (February 2008). *arXiv* **2008**, arXiv:hep-th/0502193.
62. Mardones, A.; Zanelli, J. Lovelock-Cartan theory of gravity. *Class. Quantum Gravity* **1991**, *8*, 1545–1558. [CrossRef]
63. Ferraro, R.; Fiorini, F. On Born-Infeld Gravity in Weitzenbock spacetime. *Phys. Rev. D* **2008**, *78*, 124019. [CrossRef]
64. Jiménez, J.B.; Heisenberg, L.; Koivisto, T.S. The Geometrical Trinity of Gravity. *Universe* **2019**, *5*, 173.
65. Maluf, J.W. The teleparallel equivalent of general relativity. *Ann. Phys.* **2013**, *525*, 339–357. [CrossRef]
66. Jiménez, J.B.; Heisenberg, L.; Koivisto, T. Coincident General Relativity. *Phys. Rev. D* **2018**, *98*, 044048. [CrossRef]
67. Aldrovandi, R.; Pereira, J.G. Teleparallel Gravity: An Introduction. *Fundam. Theor. Phys.* **2013**, *173*. [CrossRef]
68. Jiménez, J.B.; Heisenberg, L.; Koivisto, T.S.; Pekar, S. Cosmology in $f(Q)$ geometry. *Phys. Rev. D* **2020**, *101*, 103507. [CrossRef]
69. D'Ambrosio, F.; Fell, S.D.B.; Heisenberg, L.; Kuhn, S. Black holes in $f(Q)$ gravity. *Phys. Rev. D* **2022**, *105*, 024042. [CrossRef]
70. Thiemann, T. *Modern Canonical Quantum General Relativity*; Cambridge Monographs on Mathematical Physics: Cambridge, UK, 2007. [CrossRef]
71. Cianfrani, F.; Lecian, O.M.; Lulli, M.; Montani, G. *Canonical Quantum Gravity*; World Scientific: Singapore, 2014.
72. Dialektopoulos, K.F.; Capozziello, S. Noether Symmetries as a geometric criterion to select theories of gravity. *Int. J. Geom. Methods Mod. Phys.* **2018**, *15*, 1840007. [CrossRef]
73. Acunzo, A.; Bajardi, F.; Capozziello, S. Non-local curvature gravity cosmology via Noether symmetries. *Phys. Lett. B* **2022**, *826*, 136907. [CrossRef]
74. Bajardi, F.; Capozziello, S. Equivalence of nonminimally coupled cosmologies by Noether symmetries. *Int. J. Mod. Phys. D* **2020**, *29*, 2030015. [CrossRef]
75. Capozziello, S.; Lambiase, G. Selection rules in minisuperspace quantum cosmology. *Gen. Relativ. Gravit.* **2000**, *32*, 673–696. [CrossRef]
76. Unruh, W.G.; Wald, R.M. Time and the Interpretation of Canonical Quantum Gravity. *Phys. Rev. D* **1989**, *40*, 2598. [CrossRef]
77. Schwinger, J.S. Quantized gravitational field. *Phys. Rev.* **1963**, *130*, 1253–1258. [CrossRef]
78. Capozziello, S.; Felice, A.D. $f(R)$ cosmology by Noether's symmetry. *J. Cosmol. Astropart. Phys.* **2008**, *8*, 016. [CrossRef]
79. Benetti, M.; Capozziello, S.; Graef, L.L. Swampland conjecture in $f(R)$ gravity by the Noether Symmetry Approach. *Phys. Rev. D* **2019**, *100*, 084013. [CrossRef]
80. Capozziello, S.; Laurentis, M.D.; Odintsov, S.D. Hamiltonian dynamics and Noether symmetries in Extended Gravity Cosmology. *Eur. Phys. J. C* **2012**, *72*, 2068. [CrossRef]
81. Jamil, M.; Mahomed, F.M.; Momeni, D. Noether symmetry approach in $f(R)$–tachyon model. *Phys. Lett. B* **2011**, *702*, 315–319. [CrossRef]
82. Krššák, M.; Saridakis, E.N. The covariant formulation of $f(T)$ gravity. *Class. Quantum Gravity* **2016**, *33*, 115009. [CrossRef]
83. Tamanini, N.; Boehmer, C.G. Good and bad tetrads in $f(T)$ gravity. *Phys. Rev. D* **2012**, *86*, 044009. [CrossRef]
84. Bajardi, F.; Capozziello, S. Noether symmetries and quantum cosmology in extended teleparallel gravity. *Int. J. Geom. Methods Mod. Phys.* **2021**, *18*, 2140002. [CrossRef]
85. Bajardi, F.; Dialektopoulos, K.F.; Capozziello, S. Higher Dimensional Static and Spherically Symmetric Solutions in Extended Gauss–Bonnet Gravity. *Symmetry* **2020**, *12*, 372. [CrossRef]
86. Bamba, K.; Ilyas, M.; Bhatti, M.Z.; Yousaf, Z. Energy Conditions in Modified $f(G)$ Gravity. *Gen. Relativ. Gravit.* **2017**, *49*, 112. [CrossRef]
87. Capozziello, S.; Nesseris, S.; Perivolaropoulos, L. Reconstruction of the Scalar-Tensor Lagrangian from a LCDM Background and Noether Symmetry. *J. Cosmol. Astropart. Phys.* **2007**, *12*, 009. [CrossRef]

Article

Clocks and Trajectories in Quantum Cosmology

Przemysław Małkiewicz [1], Patrick Peter [2,*] and Sandro Dias Pinto Vitenti [3,4]

1. National Centre for Nuclear Research, Pasteura 7, 02-093 Warszawa, Poland; Przemyslaw.Malkiewicz@ncbj.gov.pl
2. $\mathcal{GR}\varepsilon\mathcal{CO}$—Institut d'Astrophysique de Paris, CNRS & Sorbonne Université, UMR 7095 98 bis Boulevard Arago, 75014 Paris, France
3. Departamento de Física, Universidade Estadual de Londrina, Rod. Celso Garcia Cid, Km 380, Londrina 86057-970, PR, Brazil; vitenti@uel.br
4. Instituto de Física, Universidade de Brasília—UnB, Campus Universitário Darcy Ribeiro-Asa Norte Sala BT 297-ICC-Centro, Brasília 70919-970, DF, Brazil

* Correspondence: peter@iap.fr

Simple Summary: We clarify the question of clock transformations and trajectories in quantum cosmology in a vacuum Bianchi I minisuperspace.

Abstract: We consider a simple cosmological model consisting of an empty Bianchi I Universe, whose Hamiltonian we deparametrise to provide a natural clock variable. The model thus effectively describes an isotropic universe with an induced clock given by the shear. By quantising this model, we obtain various different possible bouncing trajectories (semiquantum expectation values on coherent states or obtained by the de Broglie–Bohm formulation) and explicit their clock dependence, specifically emphasising the question of symmetry across the bounce.

Keywords: quantum cosmology; canonical quantum gravity; time; clocks

1. Introduction

The problem of time in quantum cosmology [1,2] is well-known and, as of now, unsolved. It rests on the fact that general relativity (GR) is a totally constrained theory, and its canonically quantised counterpart can be reduced to the Wheeler–DeWitt (WDW) equation $\mathcal{H}\Psi = 0$, which is a timeless Schrödinger equation. Hence, dynamics is absent and, in a sense, meaningless in this framework.

A simple way to reintroduce dynamical properties into the theory consists in deparametrisation, namely by making use of the fact that there exists a constraint and using a variable to serve as clock. Indeed, let us denote the relevant canonical variables $\{q^k\}$ and their associated momenta $\{p_k\}$, one has $\mathcal{H}\left(\{q^k\},\{p_k\}\right) \approx 0$ in the Dirac weak sense. Performing a canonical transformation $\left(\{q^k\},\{p_k\}\right) \mapsto (\{Q^a\},\{P_a\})$ and assuming that there exists a new variable Q^α such that the Poisson bracket $\{Q^\alpha,\mathcal{H}\}_{\text{P.B.}}$ is unity, one obtains $\mathrm{d}Q^\alpha/\mathrm{d}t = 1$, so that the variable Q^α itself can be used as time; this is a classical internal clock.

A simple and illustrative example consists in the Hamiltonian $H_{xy} = H_x + H_y$ with arbitrary H_x for a set of variables x but independent of the variable y, and $H_y = -\frac{1}{2}(\dot{y}^2 + y^2)$ represents a harmonic oscillator with negative sign. The (local) canonical transformation $T = 2\arctan(p_y/y)$ and $p_T = -\frac{1}{2}(p_y^2 + y^2)$ produces $H_y = p_T$, leading to $\dot{p}_T = 0$ and $\dot{T} = 1$, showing that T is a perfectly acceptable (local) clock variable for the Hamiltonian H_x.

Denoting $Q^\alpha \to t$ and its canonically conjugate momentum $P_\alpha \to P_t$, one notes that since $\{Q^\alpha, \mathcal{H}\}_{\text{P.B.}} = 1$, the total Hamiltonian can be split into $\mathcal{H} = P_t + H$, where H may depend on t but not on P_t. At the quantum level, it then suffices to apply the Dirac operator prescription $p_t \mapsto \hat{p}_t = -i\hbar \partial/\partial t$ to the original time WDW equation without

time to transform it into $i\hbar\partial\Psi/\partial t = H\Psi$ and thus recover a time-dependent Schrödinger equation. Although this procedure is not always applicable for configurations in superspace, restriction to a cosmological minisuperspace often permits it.

The question that naturally comes to mind is whether a clock thus defined is unique and what the effect of changing it is. In what follows, we first discuss a simple cosmological model based on a homogeneous but anisotropic Bianchi I metric in Section 2 in which we obtain a clock provided by the shear; this yields a simple free-particle Hamiltonian in which we introduce an affine quantisation procedure (Section 3) to account for the restriction that the scale factor is positive definite. Section 4 is dedicated to exploring in detail the clock transformations relevant to our quantised model, and we discuss the associated trajectories in Section 5 before wrapping up our findings and concluding.

2. Classical Bianchi I Model

We begin by assuming a homogeneous and anisotropic Bianchi type I metric

$$ds^2 = -N^2 d\tau^2 + \underbrace{e^{2(\beta_0+\beta_+ +\sqrt{3}\beta_-)}}_{a_1^2}\left(dx^1\right)^2 + \underbrace{e^{2(\beta_0+\beta_+ -\sqrt{3}\beta_-)}}_{a_2^2}\left(dx^2\right)^2 + \underbrace{e^{2(\beta_0-2\beta_+)}}_{a_3^2}\left(dx^3\right)^2, \quad (1)$$

thereby defining the scale factors a_i and the lapse N. Classically, in order to ensure the required symmetries, all these functions are assumed to depend on time τ only. For the metric (1), the usual Einstein–Hilbert action then reduces to

$$\mathcal{S}_{\text{EH}} = \frac{1}{16\pi G_N}\int\sqrt{-g}R d^4x = \frac{3}{8\pi G_N}\underbrace{\int\sqrt{\gamma}d^3x}_{\mathcal{V}_0}\int\frac{e^{3\beta_0}}{N}\left(\dot{\beta}_+^2 + \dot{\beta}_-^2 + 2\dot{\beta}_0^2 + \ddot{\beta}_0 - \frac{\dot{N}}{N}\dot{\beta}_0\right)d\tau, \quad (2)$$

in which we assume the comoving volume of 3-space to be finite (compact space ensuring the extrinsic curvature surface term to be absent) and set to $\mathcal{V}_0$. Noting that

$$\frac{e^{3\beta_0}}{N}\left(\ddot{\beta}_0 + 2\dot{\beta}_0^2 - \frac{\dot{N}}{N}\dot{\beta}_0\right) = \frac{d}{d\tau}\left(\frac{e^{3\beta_0}}{N}\dot{\beta}_0\right) - \frac{e^{3\beta_0}}{N}\dot{\beta}_0^2,$$

one integrates (2) by each part and discards the boundary term to obtain the reduced action

$$\mathcal{S}_{\text{EH}} = \frac{3\mathcal{V}_0}{8\pi G_N}\int\frac{e^{3\beta_0}}{N}\left(-\dot{\beta}_0^2 + \dot{\beta}_+^2 + \dot{\beta}_-^2\right)d\tau = \int L\left(\beta_i,\dot{\beta}_i\right)d\tau, \quad (3)$$

from which the momenta are found to be

$$p_0 = \frac{\partial L}{\partial\dot{\beta}_0} = -\frac{3\mathcal{V}_0}{4\pi G_N}\frac{e^{3\beta_0}}{N}\dot{\beta}_0 \quad \text{and} \quad p_\pm = \frac{\partial L}{\partial\dot{\beta}_\pm} = \frac{3\mathcal{V}_0}{4\pi G_N}\frac{e^{3\beta_0}}{N}\dot{\beta}_\pm, \quad (4)$$

leading to the Hamiltonian

$$\mathcal{S}_{\text{EH}} = \int\left[p_0\dot{\beta}_0 + p_+\dot{\beta}_+ + p_-\dot{\beta}_- - \underbrace{\frac{2\pi G_N}{3\mathcal{V}_0}e^{-3\beta_0}\left(-p_0^2 + p_+^2 + p_-^2\right)N}_{H=CN}\right], \quad (5)$$

where we emphasise the constraint C, which classically vanishes, with the lapse function $N(\tau)$ always being nonvanishing.

For later convenience, we consider instead of β_0 the volume variable $V = \exp(3\beta_0)$, with momentum $p_V = p_0 \exp(-3\beta_0)/3$, transforming the Hamiltonian into

$$H = \frac{3V}{8}\left(-p_V^2 + \frac{p_+^2 + p_-^2}{9V^2}\right)N = CN. \quad (6)$$

In (6) and in what follows, we assume units such that $16\pi G_N = \mathcal{V}_0$.

As H in (6) depends on neither $\beta_\pm$, these cyclic coordinates have conserved associated momenta $p_\pm$, which we write as

$$p_+ = k\cos\varphi \quad \text{and} \quad p_+ = k\sin\varphi,$$

in which we assume $k > 0$. Correspondingly, we find that the corresponding momenta can be written as $p_k = -\beta_+ \cos\varphi - \beta_- \sin\varphi$ and $p_\varphi = k(\beta_+ \sin\varphi - \beta_- \cos\varphi)$. Plugging these relations into the Hamiltonian, it turns out that the new variable φ can be altogether ignored as neither φ nor its momentum p_φ appears in H. We thus end with

$$S_{\text{EH}} = \int \left[p_V \dot{V} + p_k \dot{k} - \frac{3V}{8}\left(-p_V^2 + \frac{k^2}{9V^2}\right)N\right] d\tau. \tag{7}$$

As the volume is positive definite, solving the constraint $C = 0$ translates into setting $k^2 = 9V^2 p_V^2$, so that $dk/d\tau = [1/(2k)][d(k^2)/d\tau] = [1/(2k)][d(9V^2 p_V^2)/d\tau]$, and finally

$$p_V \dot{V} + p_k \dot{k} = \frac{d}{d\tau}\left[V p_V \ln V + \frac{1}{2}V^2 p_V^2\left(\frac{9p_k}{k} - \frac{\ln V}{V p_V}\right)\right] - \frac{1}{2}V^2 p_V^2 \frac{dY}{d\tau},$$

where

$$Y = \frac{9p_k}{k} - \frac{\ln V}{V p_V}. \tag{8}$$

The variable Y now serves as an integrating measure in the action, and it has therefore turned into a clock variable.

As now the constraint is satisfied; setting aside the boundary term above, one finally obtains the action in the form

$$S_{\text{EH}} = -\frac{1}{2}\int V^2 p_V^2 dY = \int \left[V p_V d(V p_V T) - \frac{1}{2}V^2 p_V^2 d(Y + T) - \frac{1}{2}d(V^2 p_V^2 T)\right], \tag{9}$$

where we have introduced an arbitrary function $T(V, p_V, p_k)$ of the original relevant variables. Discarding the last, integrated term and setting $q = V p_V T$ and $p = V p_V$, the action is expressed in the canonical form

$$S_{\text{EH}} = \int \left[p\frac{dq}{dt} - H(q,p)\right] dt = \int \left(p\frac{dq}{dt} - \frac{1}{2}p^2\right) dt, \tag{10}$$

provided we set $t = Y + T$ as the new time variable.

A thorough discussion of this issue together with that of choosing the otherwise arbitrary function T is given in Ref. [3], where in particular it was shown that there exist two categories of possible choices, namely the so-called fast- and slow-time gauges. In the former case, the singularity is somehow not removed upon quantisation, in the sense that the wavefunction asymptotically shrinks towards a δ-function around the vanishing scale factor (hence a singularity) after an infinite amount of time. In the latter case of slow-time gauge, the singularity is resolved into a bouncing universe.

We shall restrict out attention in what follows to the slow-time gauge only and therefore assume the arbitrary function to take the simple form $T = V^{-1}$, leading the relevant variable q to be identified with the volume V. The classical Hamiltonian is now reduced to that of a free particle confined to the semi-infinite half line $\mathbb{R}^+$. We now turn to the quantisation of this problem.

3. Affine Quantisation

Quantising a Hamiltonian system in principle follows a well-defined procedure, referred to as "canonical quantization" and proposed by Dirac. It consists of replacing the relevant dynamical variables by corresponding operators and the Poisson brackets by

i times the commutators between these operators. In the position representation with wavefunction $\Psi(q,t)$, the operator $\hat{Q}$ becomes the multiplication by q and the momentum yields $\hat{P}\Psi = -i\hbar\partial\Psi/\partial q$.

Canonical quantisation is based on the unitary and irreducible representation of the group of translations in the (q,p) plane, the Weyl–Heisenberg group. For a particle living in a smaller space, it therefore might not apply in a straightforward manner, as one has to reduce the Hilbert space of available states to ensure the mathematical properties of the observables to be satisfied. Instead of adopting this potentially problematic approach, we propose that the so-called covariant integral be considered. This is based on a minimal group of canonical transformations with a nontrivial unitary representation.

For the half-plane that arises in the Bianchi I case of the previous section, the natural choice is the 2-parameter affine group of a real line with elements $(q,p) \in \mathbb{R}^+ \times \mathbb{R}$, transforming $s \in \mathbb{R}$ into $(q,p) \cdot s = s/q + p$ and with composition law

$$\{(q_0, p_0), (q, p)\} \mapsto (q', p') = (q_0, p_0) \circ (q, p) = \left(q_0 q, \frac{p}{q_0} + p_0\right) \tag{11}$$

and left-invariant measure $dq' \wedge dp' = dq \wedge dp$. It is clear that q represents a change of scale, which is what one would expect for a scale factor (dimensionless in our conventions), while the momentum is rescaled and translated as the scale is modified. For the 2-parameter affine group, one can find a unitary, irreducible and square-integrable representation in the Hilbert space $\mathcal{H} = L^2(\mathbb{R}^+, dx)$. It reads

$$\langle x|U(q,p)|\zeta\rangle = \langle x|q,p\rangle_\zeta = \frac{e^{ipx/\hbar}}{\sqrt{q}}\zeta\left(\frac{x}{q}\right), \tag{12}$$

where $\zeta(x) = \langle x|\zeta\rangle \in \mathcal{H}$ and $|\zeta\rangle \in \mathcal{H}$ is an (almost) arbitrary fiducial state vector belonging to the Hilbert space (see Ref. [4] and references therein). As for the unitary operator $U(q,p)$ implementing an affine transformation, it reads

$$U(q,p) = e^{ip\hat{Q}/\hbar}e^{-i\ln q\hat{D}/\hbar}, \tag{13}$$

with $\hat{D} := \frac{1}{2}(\hat{Q}\hat{P} + \hat{P}\hat{Q})$ as the dilation operator, forming with $\hat{Q}$ the algebra $\left[\hat{Q},\hat{D}\right] = i\hbar\hat{Q}$.

Let us define the series of integrals

$$\rho_\zeta(s) := \int \frac{\langle\zeta|x\rangle\langle x|\zeta\rangle}{x^{s+1}} dx = \int \frac{|\zeta(x)|^2}{x^{s+1}} dx < \infty \quad \text{and} \quad \sigma_\zeta(s) := \int \left|\frac{d\zeta(x)}{dx}\right|^2 \frac{dx}{x^{s+1}} \tag{14}$$

assumed convergent, and the quantisation rule

$$f(q,p) \mapsto A_\zeta[f] := \mathcal{N}_\zeta \int_{\mathbb{R}^+\times\mathbb{R}} \frac{dqdp}{2\pi\hbar} |q,p\rangle_\zeta f(q,p)\,_\zeta\langle q,p| \quad \text{with} \quad \mathcal{N}_\zeta = \frac{1}{\rho_\zeta(0)} =: \frac{1}{\rho_0}, \tag{15}$$

associating to each function $f(q,p)$ of the classical dynamical variables a unique operator $A_\zeta[f]$ in the Hilbert space $\mathcal{H}$. The normalisation $\mathcal{N}_\zeta$ comes from the resolution of unity

$$\int \frac{dqdp}{2\pi\hbar\rho_0} |q,p\rangle_\zeta\,_\zeta\langle q,p| = \int dx|x\rangle\langle x| = \mathbb{1} = A_\zeta[1], \tag{16}$$

using $2\pi\hbar\delta(x-y) = \int e^{ip(x-y)/\hbar}dp$. Useful operators can then be represented, such as powers of q or the momentum p, namely

$$A_\zeta[q^s] = \int_{\mathbb{R}^+\times\mathbb{R}} \frac{dqdp}{2\pi\hbar\rho_0} |q,p\rangle_\zeta q^s\,_\zeta\langle q,p| = \frac{\rho_\zeta(s)}{\rho_0}\hat{Q}^s \tag{17}$$

and
$$A_\xi[p] = \int_{\mathbb{R}^+ \times \mathbb{R}} \frac{dq\,dp}{2\pi\hbar\rho_0} |q,p\rangle_\xi \, p \, _\xi\langle q,p|q,p| = \widehat{P}, \qquad (18)$$

showing that the fiducial state $|\xi\rangle$ should be such that $\rho_\xi(1) = \rho_\xi(0)$ in (14) to ensure the canonical commutation relations $[A_\xi[q], A_\xi[p]] = [\widehat{Q}, \widehat{P}] = i\hbar$. Finally, the compound quantity $q^s p^2$ is quantised to

$$A_\xi[q^s p^2] = \frac{\rho_\xi(s)}{\rho_0} \widehat{P}\widehat{Q}^s\widehat{P} + \hbar^2 \left[\frac{s(1-s)\rho_\xi(s)}{2\rho_0} + \frac{\sigma_\xi(s-2)}{\rho_0} \right] \widehat{Q}^{s-2}, \qquad (19)$$

so that the classical Hamiltonian in (10), namely $H(q,p) = \tfrac{1}{2}p^2$, has an affine quantum counterpart given by

$$A_\xi[H(q,p)] = \widehat{H}(\widehat{Q},\widehat{P}) = \frac{1}{2}\widehat{P}^2 + \hbar^2 \mathcal{K}_\xi \widehat{Q}^{-2} = \frac{1}{2}\widehat{P}^2 + V(\widehat{Q}), \qquad (20)$$

with $\mathcal{K}_\xi = \sigma_\xi(-2)/\rho_0$; given the arbitrariness of the fiducial vector, this coefficient is essentially arbitrary. If instead of the affine quantisation one applies the canonical prescription, it would simply vanish ($\mathcal{K}_{\text{can}} \to 0$). Among the advantages of this quantisation is the fact that it permits us to merely parametrise the well-known operator ordering ambiguity, replacing it by a single unknown number, to be ultimately fixed by experiment.

It should be noted that if $\mathcal{K}_\xi \geq \tfrac{3}{4}$, the Hamiltonian (20) is essentially self-adjoint, so one needs not impose any boundary conditions at $q=0$, the dynamics generated being unique and unitary by construction [5]. In the framework of quantum cosmology that concerns us here, affine quantisation induces a repulsive potential $V(\widehat{Q})$ thanks to which it is natural to expect that the classical GR Big Bang singularity will be resolved by quantum effects, as indeed is found to happen with our choice of clock [3].

4. Clock Transformations

A classically constrained Hamiltonian theory with $H_{\text{full}}(q_{\text{full}}, p_{\text{full}}) \approx 0$ and deparametrised to a reduced phase space (q,p) using an internal degree of freedom t as clock is invariant under the so-called clock transformations. The idea behind the clock transformation is the following: given a clock t and its associated reduced phase space formalism (q,p,t), one seeks, prior to deparametrisation, another choice of clock $\tilde{t}$, say, leading to a similar reduced phase space formalism $(\tilde{q}, \tilde{p}, \tilde{t})$. This involves transformations of both the clock variable $t \mapsto \tilde{t}(q,p,t)$ and the canonical variables $(q,p) \mapsto [\tilde{q}(q,p,t), \tilde{p}(q,p,t)]$ as the change in time generally changes the canonical relations in reduced phase space. These clock transformations can also be understood as canonical transformations in the full phase space $(q_{\text{full}}, p_{\text{full}})$, thereafter restricted to the constraint surface. This restriction is responsible for altering the canonical relations in the reduced phase space. The relation between the full- and reduced-phase-space formulations of the clock transformations was investigated in [6].

Let us start by noticing that the new canonical variables $(\tilde{q}, \tilde{p})$ associated with the new clock $\tilde{t}$ can be chosen conveniently as to satisfy

$$dq \wedge dp - dt \wedge dH(q,p,t) = d\tilde{q} \wedge d\tilde{p} - d\tilde{t} \wedge dH(\tilde{q},\tilde{p},\tilde{t}), \qquad (21)$$

where the form of the reduced Hamiltonian $H(\cdot,\cdot,\cdot)$ is preserved by the clock transformation [7]. The above choice of $\tilde{q}$ and $\tilde{p}$ is convenient because once the solution to the dynamics is known in t as $q = S_q(t), p = S_p(t)$, it is automatically known in all other clocks $\tilde{t}$ as $\tilde{q} = S_q(\tilde{t}), \tilde{p} = S_p(\tilde{t})$. It is obviously the same physical solution but now differently parametrised. It is easy to notice that $\tilde{t} \neq t$ implies $dq \wedge dp \neq d\tilde{q} \wedge d\tilde{p}$, and the clock transformation indeed alters the canonical relations in the reduced phase space. This is a sufficient reason for the existence of unitarily inequivalent quantum dynamics based on different choices of clock (as we shall see shortly). Let us now explain how the clock transformation satisfying the above condition is determined in practice.

One first calculates Dirac observables $C_i(q, p, t)$ ($i = 1, 2$ in the two-dimensional phase space discussed here) by solving their defining equation, namely $\partial_t C_i + \{C_i, H\}_{\text{P.B.}} = 0$. One then demands that for the transformation $(q, p, t) \mapsto (\tilde{q}, \tilde{p}, \tilde{t})$, one has $C_i(q, p, t) = C_i(\tilde{q}, \tilde{p}, \tilde{t})$, thus leading to the required relationship between $(\tilde{q}, \tilde{p})$ and (q, p) for an arbitrary change of clock time $t \mapsto \tilde{t}$. It should be emphasised at this point that the clock transformation provides an actual invariance provided there is an underlying Hamiltonian, even a time-dependent one.

Consider first the Hamiltonian in (10), namely $H_0 = \frac{1}{2}p^2$. The Dirac observable requirement then reads $\partial_t C_i + p \partial C_i / \partial V = 0$. One set of solution is $C_1 = p$ and $C_2 = pt - q$, leading to $\tilde{p} = p$, and $\tilde{p}\tilde{t} - \tilde{q} = pt - q$, which implies $\tilde{q} = q + (\tilde{t} - t)p = q + \Delta p$, thereby defining the function

$$\Delta(t, q, p) := \tilde{t} - t. \qquad (22)$$

The effects of this transformation was studied in Ref. [3] for various arbitrary Δ. In phase space, the solutions for H_0 are $\dot{p} = 0$, and therefore $p = p_0$ constant, with $\dot{q} = p = p_0$ so that $q = p_0 t + q_0$: these are straight lines in the (q, p) space, labelled by t. Changing to $\tilde{t}$ yields the same Hamiltonian, now in the new variables, and therefore the same equations of motion, and thus the same formal solutions, namely $\tilde{p} = \tilde{p}_0$ and $\tilde{q} = \tilde{p}_0 \tilde{t} + \tilde{q}_0$. Applying the transformation implies $\tilde{q}_0 = q_0$ and $\tilde{p}_0 = p_0$. For one particular solution, Equation (22) provides $\tilde{t}(t)$, which must be monotonic and invertible, yielding $t(\tilde{t})$: $q(t)$ now transforms into $q(\tilde{t})$, and because p is constant, one recovers straight lines, now labelled in a different way.

Let us now turn to the more complicated example of the quantum Hamiltonian (20), now considered classical and written as $H = \frac{1}{2}p^2 + \mathcal{K}/q^2$. One now needs to find the solution to

$$\frac{\partial C_i}{\partial t} + \frac{2\mathcal{K}}{q^3} \frac{\partial C_i}{\partial p} + p \frac{\partial C_i}{\partial q} = 0,$$

which is solved by the set

$$C_1 = \frac{1}{2}p^2 + \frac{\mathcal{K}}{q^2} = H(q, p) \quad \text{and} \quad C_2 = qp - 2H(q, p)t. \qquad (23)$$

For the clock transformation, one derives from (23) the relations

$$\tilde{q}^2 = q^2 + Z \quad \text{and} \quad \tilde{p}^2 = p^2 + \frac{2\mathcal{K}Z}{q^2(q^2 + Z)} = 2\left(H - \frac{\mathcal{K}}{q^2 + Z}\right), \qquad (24)$$

where we have set $Z = 2\Delta(pq + H\Delta)$. Two conditions must be imposed for the choice of Δ. First, it must be made such that it satisfies $Z \geq -q^2$ to ensure $\tilde{q}^2 \geq 0$ and hence $\tilde{q} \in \mathbb{R}$. Second, the inequality

$$1 + \frac{\partial \Delta}{\partial t} + \{\Delta, H\}_{\text{P.B.}} = 1 + \frac{\partial \Delta}{\partial t} + p\frac{\partial \Delta}{\partial q} + \frac{2\mathcal{K}}{q^3}\frac{\partial \Delta}{\partial p} \neq 0$$

must hold in order that the time delay function Δ ensures monotony of the new time with respect to the old one, i.e., $\mathrm{d}\tilde{t}/\mathrm{d}t > 0$. Note that the transformation (24) gives back that corresponding to $\mathcal{K} = 0$ in both limits $\mathcal{K} \to 0$ and $q \to \infty$.

Our system originates classically from the simplest option, namely $H \to H_0 = \frac{1}{2}p^2$, but that derived from the quantum one (20) can imply semiclassical (or perhaps semiquantum [4]) trajectories that should be invariant under (24). It is therefore important to derive actual trajectories one way or another to be able to estimate the effects a choice of clock can have.

5. Trajectories

There are various ways to implement physically meaningful trajectories in our quantum description of the dynamics of a Bianchi I universe, as illustrated in Figure 1. The first and most obvious consists merely in evaluating expectation values. If the wavefunction is sufficiently narrow, this can provide an effective semiclassical approximation.

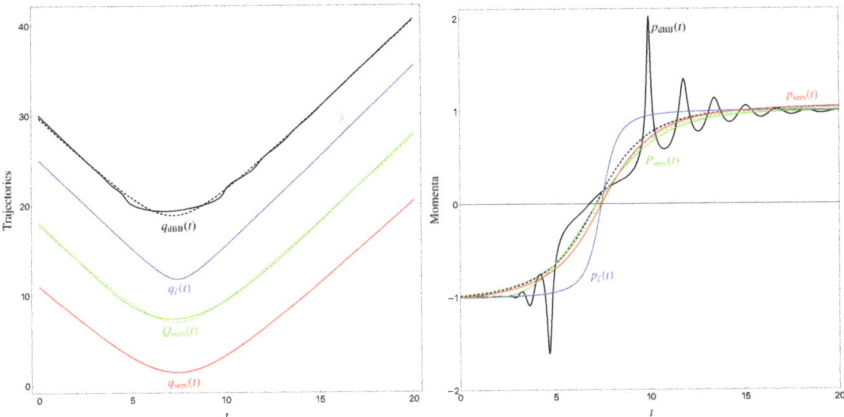

Figure 1. Time developments of the various trajectories proposed in the text (left panel) as obtained from the wavefunctions of Figure 2. Except for the dBB case, all the definitions used for semiclassical trajectories are well fitted (or exactly given) by the solution (29), shown as dashed lines for each curve (these have been arbitrarily displaced up and down for visual purposes; otherwise they are hardly distinguishable). The right panel shows the relevant associated momenta and emphasises the large discrepancy visible only in the dBB case.

With the Hamiltonian (20), it has been shown that an approximate space trajectory can be deduced directly from the quantum version of the algebra [3]: using $[\hat{D}, \hat{H}] = 2i\hat{H}$ and the fact that $\hat{H}$ is a constant operator, one can integrate the Heisenberg equation of motion $d\hat{D}/dt = -i[\hat{D}, \hat{H}] = 2\hat{H}$, leading to $\hat{D}(t) = \hat{D}(0) + 2\hat{H}t$. Even though the operator $\hat{Q}$ itself cannot be integrated directly from the algebra because $[\hat{Q}, \hat{H}] = i\hat{P}$ and $[\hat{P}, \hat{H}] = 2i\mathcal{K}\hat{Q}^{-3}$, its square leads to $[\hat{Q}^2, \hat{H}] = 2i\hat{D}$, so one finds $d\hat{Q}^2/dt = -i[\hat{Q}^2, \hat{H}] = 2\hat{D}(t) = 2\hat{D}(0) + 4\hat{H}t$. This implies $\hat{Q}^2 = \hat{Q}^2(0) + 2[\hat{D}(0)t + \hat{H}t^2]$. A semiclassical trajectory can then be defined in phase space by setting $q_{\text{sem}}(t) = \sqrt{\langle \hat{Q}^2 \rangle}$ and $p_{\text{sem}}(t) = \langle \hat{D} \rangle / q_{\text{sem}}(t)$. Shifting the time to set the minimum of $q_{\text{sem}}(t)$ at $t_B = 0$, one obtains a bouncing behaviour $q_{\text{sem}}(t) = q_B\sqrt{(\omega t)^2 + 1}$.

Another option consists in solving the Schrödinger equation and evaluating the expectation values directly with the relevant wavefunction. This leads to another semiclassical trajectory $Q_{\text{sem}}(t) = \langle \hat{Q} \rangle$ and $P_{\text{sem}}(t) = \langle \hat{P} \rangle$. It turns out that for $t < 0$, one has $Q_{\text{sem}}(t) \simeq q_{\text{sem}}(t)$, although close to the bounce and afterwards, there is a systematic shift between $Q_{\text{sem}}(t)$ and $q_{\text{sem}}(t)$. The phase space trajectories $(q_{\text{sem}}, p_{\text{sem}})$ and $(Q_{\text{sem}}, P_{\text{sem}})$ are in good agreement, with only a difference in their time labelling.

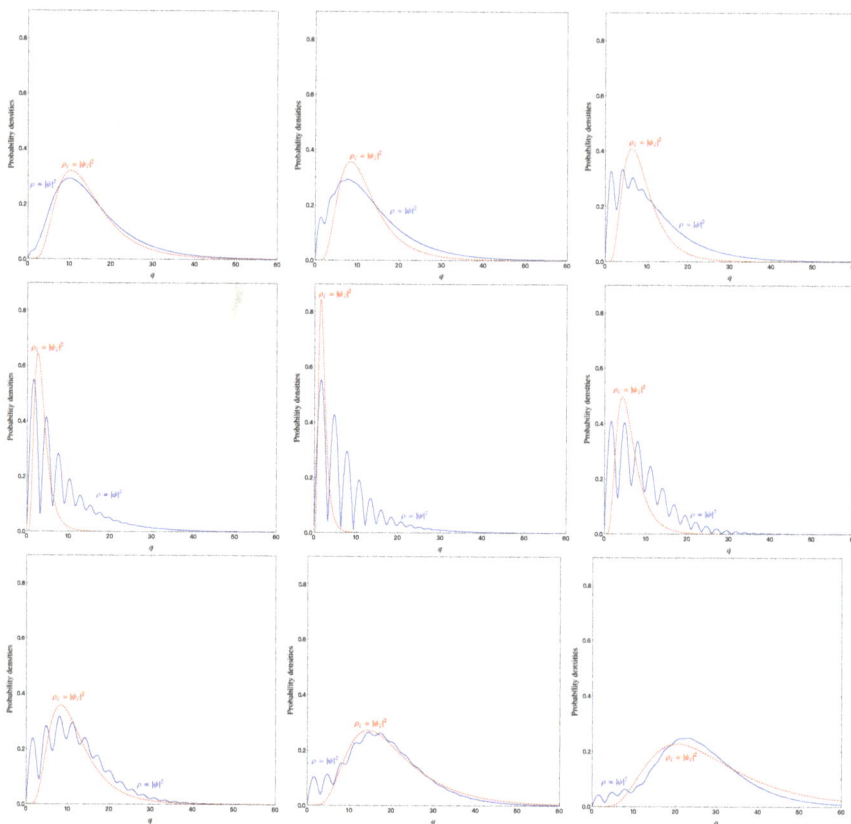

Figure 2. Snapshots of the time evolution of both the full wavefunction $\psi(q,t)$ and its coherent state approximation $\psi_\zeta(q,t)$, with the initial condition shown in Figure 3. As the wavefunctions approach the origin $q \to 0$, the true solution ψ starts oscillating, with the oscillations developing further with time until the wave packet is far enough from the origin and they begin to be damped. On the other hand, the coherent state wavefunction ψ_ζ never oscillates, being merely squeezed at the origin and then bouncing away. In the large time limit, they both evolve in more and more similar ways so that $\lim_{t\to\infty} \psi \sim \psi_\zeta$.

A third way to obtain approximate trajectories consists in considering coherent states, as defined through Equations (12) and (13). Indeed, if one changes the fiducial state $|\zeta\rangle$, satisfying the canonical condition $\rho_\zeta(1) = \rho_\zeta(0)$ below (18), to $|\zeta\rangle$ such that $\langle\zeta|\widehat{Q}|\zeta\rangle = 1$ and $\langle\zeta|\widehat{P}|\zeta\rangle = 0$, the Schrödinger action

$$\mathcal{S}_{\text{sch}}[|\psi\rangle] = \int \langle\psi|\left(i\hbar\frac{\partial}{\partial t} - H\right)|\psi\rangle dt \qquad (25)$$

is transformed into [8]

$$\mathcal{S}_{\text{sch}}[|q(t), p(t)\rangle_\zeta] = \int \left[p_\zeta \dot{q}_\zeta - {}_\zeta\langle q(t), p(t)|H|q(t), p(t)\rangle_\zeta\right] dt$$
$$\to \int \left[p_\zeta(t)\dot{q}(t) - H_{\text{sem}}(q_\zeta(t), q_\zeta(t))\right] dt \qquad (26)$$

once the arbitrary state $|\psi\rangle$ is replaced by the coherent state $|q(t), p(t)\rangle_\zeta$, now defined with a priori unknown functions of time $q_\zeta(t)$ and $p_\zeta(t)$. It is clear from Equation (26) that the initially arbitrary functions $q_\zeta(t)$ and $p_\zeta(t)$ are now, in order to minimise the action, subject to Hamilton equations

$$\dot{q}_\zeta = \frac{\partial H_{\text{sem}}}{\partial p_\zeta} \quad \text{and} \quad \dot{p}_\zeta = -\frac{\partial H_{\text{sem}}}{\partial q_\zeta}. \tag{27}$$

with the original Hamiltonian replaced by the semiclassical one H_{sem}.

Applying the coherent state method to the quantum Hamiltonian (20) yields

$$H_{\text{sem}}(q,p) = \frac{1}{2}p^2 + \frac{\mathcal{K}}{q^2}, \tag{28}$$

in which $\mathcal{K} = \hbar^2 [\mathcal{K}_\zeta \rho_\zeta(1) + \sigma_\zeta(-2)]$. As above, the coefficient $\mathcal{K}$ depends on the choice of fiducial state $|\zeta\rangle$ and is, to a large extent, arbitrary.

Solving Equations (27) with (28) yields

$$q_\zeta(t) = q_B \sqrt{1 + (\omega t)^2} \quad \text{and} \quad p_\zeta(t) = \frac{q_B \omega^2 t}{\sqrt{1 + (\omega t)^2}}, \tag{29}$$

where $q_B = \sqrt{\mathcal{K}/H_{\text{sem}}}$ and $\omega = H_{\text{sem}} \sqrt{2/\mathcal{K}}$. It is interesting to note that the solution (29) is functionally the same as that obtained by using the operator algebra $q_{\text{sem}}(t)$ and $p_{\text{sem}}(t)$, and even though the parameters q_B and ω in both solutions differ in principle, they satisfy $\omega q_B = \sqrt{2H}$ in both cases.

Finally, trajectories can be obtained in the quantum theory of motion [9] formulation of quantum mechanics originally proposed by de Broglie in 1927 [10] and subsequently formalised in more detail by Bohm in 1952 [11,12]; we shall accordingly refer in what follows to this formulation as the de Broglie–Bohm (dBB) approach. Applied to quantum gravity [13], it permits some relevant issues to be reformulated and, in some cases, solved [14].

The basic idea stems from the eikonal approximation in the classical wave theory of radiation for which light rays can be obtained by merely following the gradients of the phase of the wave. Similarly, in quantum mechanics, the wavefunction is understood to represent an actual wave whose phase gradient provides a means to calculate a trajectory. In practice, for a Hamiltonian such as (20), the Schrödinger equation reads $i\hbar \partial_t \psi = -\frac{1}{2}\partial_q^2 \psi + V\psi$, which can be expanded, setting $\psi(q,t) = \sqrt{\rho(q,t)} \exp[iS(q,t)/\hbar]$, into a continuity equation

$$\frac{\partial \rho}{\partial t} + \frac{\partial}{\partial q}\left(\rho \frac{\partial S}{\partial q}\right) = 0, \tag{30}$$

naturally leading to the identification $\dot{q}_{\text{dBB}} = \partial_q S$, and a quantum-modified Hamilton–Jacobi equation

$$\frac{\partial S}{\partial t} + \frac{1}{2}\left(\frac{\partial S}{\partial q}\right)^2 + V(q) + V_Q = \frac{\partial S}{\partial t} + \frac{1}{2}\left(\frac{\partial S}{\partial q}\right)^2 + V(q) \underbrace{- \frac{\hbar^2}{4\rho}\left[\frac{\partial^2 \rho}{\partial q^2} - \frac{1}{2\rho}\left(\frac{\partial \rho}{\partial q}\right)^2\right]}_{V_Q} = 0, \tag{31}$$

which confirms the above identification, while highlighting a new potential adding to the original one. Appropriately called the quantum potential, V_Q, being built out of the wavefunction solving the Schrödinger equation, is in general a time-dependent potential.

With the identification $\dot{q}_{\text{dBB}} = \partial_q S$, one gets $\ddot{q}_{\text{dBB}} = d(\partial_q S)/dt = \partial_t(\partial_q S) + \partial_q^2 S \dot{q}_{\text{dBB}}$, so using the identification again and the Hamilton–Jacobi equation (31), one finds $\ddot{q}_{\text{dBB}} = -\partial_q(V + V_Q)$, i.e., a modified Newton equation that, formally, can be derived from the time-dependent Hamiltonian $H_{\text{dBB}} = \frac{1}{2}p_{\text{dBB}}^2 + V(q_{\text{dBB}}) + V_Q(q_{\text{dBB}}, t)$. These trajectories happen to be very different from those derived above for various reasons. In particular, the coherent state approximation leads to one and only one trajectory $q_\zeta(t)$ once the initial coherent state (including the fiducial state) is given. Similarly, expectation values are unique for a given quantum state, so that $q_{\text{sem}}(t)$ and $Q_{\text{sem}}(t)$ define one semiclassical or semiquantum approximation only, which is entirely fixed by the parameters defining the state, whereas $q_{\text{dBB}}(t)$, stemming from a differential equation, needs an initial value $q_{\text{dBB}}(t_0)$ to be evolved, and therefore there exists, for a given state, an infinite number of acceptable trajectories. One could, however, argue that for the coherent state trajectory, depending on the choice of a particular fiducial state, there remains some amount of ambiguity in this choice, permitting various families of such trajectories to be defined. In that sense, the coherent state approximation and the dBB approach can be compared.

Another crucial difference is that $q_{\text{sem}}(t)$, $Q_{\text{sem}}(t)$ and $q_\zeta(t)$ represent *approximations* supposed to encode the underlying quantum mechanical evolution of the wavefunction. The trajectories $q_{\text{dBB}}(t)$ are, by contrast, an extra degree of freedom in the dBB formulation and thus exact solutions of the equations of motion.

Let us consider beginning with the canonical quantisation case, for which $\mathcal{K} \to 0$. In this case, our Bianchi I vacuum model is formally equivalent, in the minisuperspace limit, to that of a Friedmann universe filled with radiation [15], and one finds that there exists a wavefunction such that the $q_{\text{dBB}}(t)$ has the same functional dependence in time as $q_\zeta(t)$ in Equation (29), except for the fact that the minimum scale factor value is now given not only by the parameters describing the wavefunction, but also depends on an initial condition $q_{\text{dBB}}(t_0)$. In that case, this comes from the fact that the quantum potential happens to be $V_Q \propto q^{-2}$, so one naturally recovers the Hamiltonian (28): one thus finds that all trajectories are similar in shape.

The more relevant model in which $\mathcal{K} \neq 0$ can also be solved analytically under special conditions (see Ref. [3] for details and the solution itself). Our choice in the present work was to assume an initial wavefunction $\psi_{\text{true}}(q, t_{\text{ini}})$ in the far past, with q large, to be in a coherent state $\psi_\zeta(q, t_{\text{ini}})$ (see the right panel of Figure 3) and to evolve it with the Schrödinger equation. Figure 2 shows how $\psi(q, t) = \psi_{\text{true}}(q, t)$ and then very rapidly departs from $\psi_\zeta(q, t)$, although the expectation value trajectories q_{sem} and Q_{sem} remain similar (in shape, if not in actual values) to q_ζ. As it happens, as the wave packets move towards the origin $q \to 0$, ψ starts oscillating, thus producing the oscillations in the dBB trajectory Q_{dBB}, while the coherent state remains smooth at all times, being merely squeezed close to the origin. It is interesting to note that even though the wavefunctions differ drastically at the time of the bounce, the relevant trajectories (except the dBB one) are well described by (29), although with different parameters q_B and ω. We take that as an indication that the coherent state approximation is a valid one in most circumstances as long as one is only interested in expectation values. Given the very significant differences with the true wavefunction, however, it can be assumed that higher order moments are *not* well approximated.

As a result, the trajectories defined through either expectation values or coherent state approximation are invariant under the clock transformation (24), contrary to the dBB ones. However, as can be seen on Figure 4, in which the transformation stemming from the free particle Hamiltonian is applied to the phase space trajectories, they do depend on the choice of clock before quantisation. This is actually even more true for the dBB case, for which these clock transformations can lead to such tremendous modifications of the space space trajectories that the actual predictivity of the underlying theory becomes questionable.

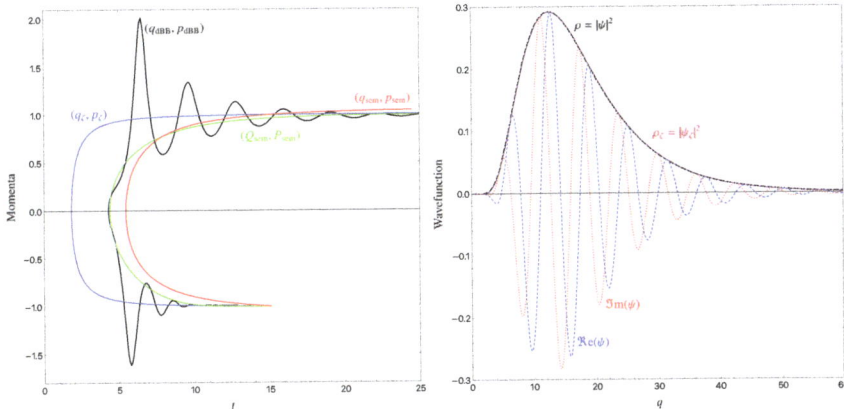

Figure 3. (**Left**): Parametric phase space trajectories built from the data from Figure 1. Except for the dBB case, all are well fitted (if not exactly given) by $p^2 \propto q_B^{-2} - q^{-2}$, as obtained from (29). (**Right**): wavefunction leading to the previous trajectories, at the initial time, at which we assume a a coherent state. Subsequent evolution is shown in Figure 2.

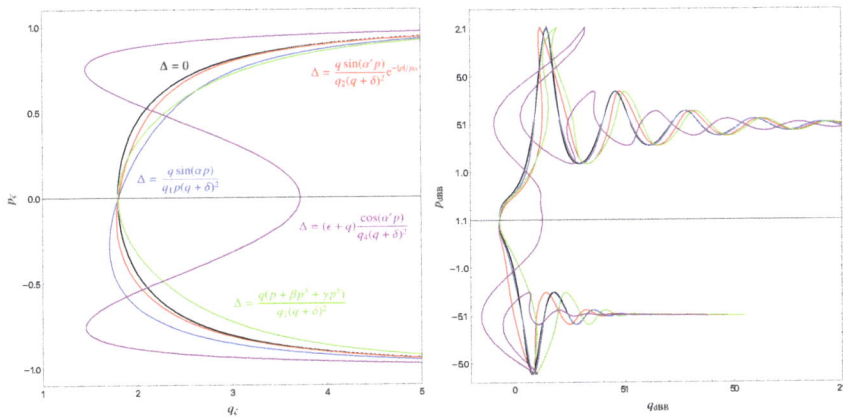

Figure 4. Change of the phase space trajectories when a clock transformation above (22) is applied to Figure 3, with different time functions $\Delta(q, p)$ as indicated in the figure. The numerical parameters are set to $\alpha = 2$, $\alpha' = 3$, $\beta = -0.63$, $\gamma = 0.1$, $\epsilon = 1$, $p_0 = 3/2$, $q_1 = 0.126$, $q_2 = 0.4$, $q_3 = 0.04$, and $q_4 = 0.03$; they have been chosen to yield visually important modifications of the trajectories. Left panel: initial trajectory given by Equation (29). Applying (24) to this trajectory yields the very same trajectory by definition. Right panel: the dBB trajectory, initially having more features, can be modified much more drastically.

6. Conclusions

We have reviewed the question of clock transformation and trajectories in quantum cosmology by means of a simple deparametrised and quantised Bianchi I model. The Wheeler–DeWitt equation in this minisuperspace case reduces to the Schrödinger equation of a free particle or, depending on the quantisation scheme, with a repulsive potential which can be studied using standard techniques. The relevant degree of freedom, from the point of view of cosmology, is the spatial volume $q = V$, i.e., the cube of the scale factor a, while the canonically conjugate momentum is mostly given by the Hubble parameter.

Extending a previous work [3] to include dBB trajectories, we found very substantial differences between those and their counterparts obtained by some averaging processes. In the later case, all trajectories stem from a semiclassical Hamiltonian and are therefore invariant under the corresponding clock transformation (although not for that corresponding to the original classical theory). In the former case, however, unless the wavefunction is restricted to belong to a very special class (for which the coherent state approximation is not valid), we found that the dBB trajectories depend in a much more drastic way on the clock transformations, rendering the ambiguity it stems from extremely serious, to the point that the theory may no longer even be predictive. Calculating the spectrum of primordial perturbations, for instance, involves the second time derivative of the scale factor, and hence of our q, so that the choice of clock and initial conditions can yield tremendously different predictions. For semiclassical trajectories, on the other hand, the choice is mostly irrelevant, and the resulting perturbations might merely depend on a few parameters.

That said, it must be emphasised that the classical limit is, in all cases (hence including dBB), well defined and consistent, so there remains the possibility that whatever dynamical quantity (e.g., perturbations) is evolved through the full quantum phase might be unique. We postpone such a discussion to a forthcoming work [16].

Author Contributions: Writing—review and editing, P.M., P.P. and S.D.P.V. All authors have read and agreed to the published version of the manuscript.

Funding: This research was funded by the Polish National Agency for Academic Exchange and Programme Hubert Curien POLONIUM 2019 grant number 42657QJ.

Institutional Review Board Statement: Not applicable.

Informed Consent Statement: Not applicable.

Data Availability Statement: Not applicable.

Acknowledgments: The authors acknowledge many illuminating discussions with H. Bergeron, J.-P. Gazeau and C. Kiefer.

Conflicts of Interest: The authors declare no conflict of interest.

References

1. Anderson, E. *The Problem of Time*; Springer: Cham, Switzerland, 2017; Volume 190. [CrossRef]
2. Kiefer, C.; Peter, P. Time in quantum cosmology. *Universe* **2022**, *8*, 36. [CrossRef]
3. Małkiewicz, P.; Peter, P.; Vitenti, S.D.P. Quantum empty Bianchi I spacetime with internal time. *Phys. Rev. D* **2020**, *101*, 046012, [CrossRef]
4. de Cabo Martin, J.; Małkiewicz, P.; Peter, P. Unitarily inequivalent quantum cosmological bouncing models. *Phys. Rev. D* **2022**, *105*, 023522, [CrossRef]
5. Vilenkin, A. Quantum Cosmology and the Initial State of the Universe. *Phys. Rev. D* **1988**, *37*, 888. [CrossRef] [PubMed]
6. Małkiewicz, P. Multiple choices of time in quantum cosmology. *Class. Quant. Grav.* **2015**, *32*, 135004, [CrossRef]
7. Małkiewicz, P.; Miroszewski, A. Internal clock formulation of quantum mechanics. *Phys. Rev. D* **2017**, *96*, 046003, [CrossRef]
8. Klauder, J.R. *Enhanced Quantization: Particles, Fields and Gravity*; World Scientific: Hackensack, NJ, USA, 2015. [CrossRef]
9. Holland, P.R. The de Broglie-Bohm theory of motion and quantum field theory. *Phys. Rept.* **1993**, *224*, 95–150. [CrossRef]
10. De Broglie, L. La mécanique ondulatoire et la structure atomique de la matière. *J. Phys. Radium* **1927**, *8*, 225–241. [CrossRef]
11. Bohm, D. A Suggested interpretation of the quantum theory in terms of hidden variables. 1. *Phys. Rev.* **1952**, *85*, 166–179. [CrossRef]
12. Bohm, D. A Suggested interpretation of the quantum theory in terms of hidden variables. 2. *Phys. Rev.* **1952**, *85*, 180–193. [CrossRef]
13. Kiefer, C. *Quantum Gravity*, 3rd ed.; Oxford University Press: Oxford, UK, 2012.
14. Pinto-Neto, N.; Fabris, J.C. Quantum cosmology from the de Broglie-Bohm perspective. *Class. Quant. Grav.* **2013**, *30*, 143001, [CrossRef]
15. Acacio de Barros, J.; Pinto-Neto, N.; Sagioro-Leal, M.A. The Causal interpretation of dust and radiation fluids nonsingular quantum cosmologies. *Phys. Lett.* **1998**, *A241*, 229–239. [CrossRef]
16. Boldrin, A.; Małkiewicz, P.; Peter, P. Problem of time and the generation of primordial structure. 2022, *in preparation*.

Article

Topics in Supersymmetric and Noncommutative Quantum Cosmology

Hugo García-Compeán [1,*], Octavio Obregón [2,*] and Cupatitzio Ramírez [3]

1. Departamento de Física, Centro de Investigación y de Estudios Avanzados del Instituto Politécnico Nacional, P.O. Box 14-740, Ciudad de México 07000, Mexico
2. Departamento de Física, División de Ciencias e Ingeniería, Universidad de Guanajuato, León 37150, Mexico
3. Facultad de Ciencias Físico Matemáticas, Benemérita Universidad Autónoma de Puebla, Puebla 72570, Mexico; cramirez@fcfm.buap.mx
* Correspondence: compean@fis.cinvestav.mx (H.G.-C.); octavio@fisica.ugto.mx (O.O.)

Abstract: In the present article we review the work carried out by us and collaborators on supersymmetric quantum cosmology, noncommutative quantum cosmology and the application of GUPs to quantum cosmology and black holes. The review represents our personal view on these subjects and it is presented in chronological order.

Keywords: quantum cosmology; supersymmetry; noncommutativity; generalized uncertainty principles

---

Citation: García-Compeán, H.; Obregón, O.; Ramírez, C. Topics in Supersymmetric and Noncommutative Quantum Cosmology. *Universe* 2021, 7, 434. https://doi.org/10.3390/universe7110434

Academic Editor: Paulo Vargas Moniz

Received: 16 October 2021
Accepted: 10 November 2021
Published: 12 November 2021

Publisher's Note: MDPI stays neutral with regard to jurisdictional claims in published maps and institutional affiliations.

Copyright: © 2021 by the authors. Licensee MDPI, Basel, Switzerland. This article is an open access article distributed under the terms and conditions of the Creative Commons Attribution (CC BY) license (https://creativecommons.org/licenses/by/4.0/).

1. Introduction

Quantum gravity is a prospect of a physical theory describing the quantum phenomena associated to the gravitational field. At the present time nobody knows with certainty how this theory will look like. There are several proposals in the literature describing the possible nature of the fundamental degrees of freedom and checking the internal consistency by connecting it to the correct limits as the low energy macroscopic general relativity and quantum mechanical laws. These theories have some features which are quite interesting by themselves, as the modification of the spacetime structure near Planck scale (for an exposition of the different approaches, see for instance, [1,2]). String theory and loop quantum gravity are among the most prominent examples. In the present article we review our advances on another approach to quantum gravity as quantum cosmology. This proposal has its origins in the Arnowitt, Deser and Misner (ADM) canonical formalism of quantum gravity [3]. This is an approach to quantum gravity that possess some of the features of the complete theory and it allows to formulate some models easily workable (for some reviews on this topic, see [4,5].)

The study of the universe must be done in the framework of the theories in force at present, depending on the scale to be described, although there is no question that it must rely on general relativity. Cosmology links together, as a theory of the evolution of the universe, small with large scales, hence a natural, theoretical framework for it is quantum cosmology.

Cosmology describes the general laws of the universe, i.e., its evolution and structure formation. Mostly, these laws can be formulated classically, in accordance with the observational bounds. However, assuming the past convergence of matter into a singularity, it emerges the question, still open, of its quantum origin. On the other side, there is large evidence and it is generally accepted that classical physics can be explained from quantum physics. Accordingly, there have been broad efforts to formulate a quantum theory of gravity.

The observation of the universe has lead to the knowledge that it behaves as a classical system, it is no subject to quantum uncertainties. This fact refers to the observable universe, which begins with the time of the Cosmic Microwave Background (CMB). Before this time, the universe is well described after the hot big bang, and there is controversy on what happened before, although it is largely accepted, that an inflationary epoch should have been present, originated during a homogeneous phase. Before this phase not much is known, but it is believed that if the universe has expanded from a previous era above Planck scale, a full quantum gravity description is required. On the other side, one of the successes of inflation is that it produces a growth of quantum fluctuations to the size and density required by gravitational instability, without generating too strong primordial gravitational waves. These fluctuations can be explained by the quantization of small inhomogeneous perturbations of the metric, in a homogeneous background, hence from a semiclassical quantum gravity. Therefore, it is valid to study homogeneous cosmology as a quantum theory. This quantization corresponds to quantum mechanics, with a time independent Klein-Gordon equation, given by the Wheeler-DeWitt equation [5,6].

Among the candidates for a unified theory of quantum gravity, supergravity has an important place, as an effective or possibly fundamental theory. As supergravity includes necessarily fermions, it requires to be a quantum theory. Thus, the study of quantum supergravity cosmology is meaningful and relevant. As a bonus, in the supersymmetric framework, the Wheeler-DeWitt equation turns into a system of first order equations.

Supersymmetric quantum cosmology has been first formulated in an attempt to give a systematic approach for a square root for the Wheeler-DeWitt equation in [7]. In this work, a homogeneous theory following from $N=1$ supergravity theory was considered. In this way, an invariance principle for a square root of the WDW equation has been proposed, at the energy scale of quantum cosmology. As the fields in the action depend only on time, not all the original constraints follow from the variation, and the missing constraints should be implemented by hand for consistency.

If one would like to consider the presence of quantum effects through the Heisenberg uncertainty principle in gravitational systems, it is possible to argue that there is a minimal length, which is precisely the Planck length L_P. If we increase the energy, the length will be start to be increased again. This behavior is typical of noncommutative field theories and noncommutative gravity [8,9]. One of its prominent examples is the description of instantons on noncommutative spaces. It can be appreciated an effective size of the instantons depending on the noncommutative parameter. A novel proposal was carried out by us in the paper [10], where noncommutative deformation was implemented at the level of Wheeler's superspace or more concretely at the level of minisuperspace.

If the uncertainty relations are modified by generalizing them, then the ultraviolet (UV) behavior will change. This is precisely the aim of Generalized Uncertainty Principles (GUP), which asserts that in the UV the usual Heisenberg relations should be modified. These generalizations are consistent with some results of very high energy scattering in string theory [11–18].

However, the observed large scale homogeneity of the universe, indicates a primary description in a minisuperspace, with finitely many degrees of freedom. Such a theory has the indubitable advantage, that it accepts a quantum description, and has served also as a test ground for approaches of quantum gravity. Moreover, this theory has given new insights in the study of the early universe, and of the other realm where it is thought that quantum gravity should manifest, the interior of black holes. Among these approaches, we have made several proposals that we shortly review here: supersymmetric cosmology, effective noncommutativity in cosmology and black holes.

In this work we shortly present some of these developments with collaborators, and concentrate on our own point of view, and main contributions. The article is organized as follows: in Section 2 we review supersymmetric quantum cosmology starting from the original proposal [7] and we describe very briefly the other proposals to supersymmetric quantum cosmology in the literature. Section 3 is devoted to give an overview of the noncommutative quantum cosmology. We give the original idea [10] and briefly discuss its further evolution. In Section 4 we describe a recent proposal of the application of GUPs to quantum cosmology following Ref. [19]. In Section 5 we review the application of the GUPs to black holes, this is based in Ref. [20].

2. Supersymmetric Quantum Cosmology

The starting point is the Freedman-Nieuwenhuizen-Ferrara action [21]. The fields are the tetrad e_μ^a, the spin connection ω_μ^{ab}, and the Rarita-Schwinger Majorana $\frac{3}{2}$-spinor field ψ_μ^a, i.e., it satisfies $\bar{\psi}_\mu = \psi_\mu^T C$, where C is the charge conjugation matrix. The lagrangian is

$$L = \frac{1}{2\kappa^2}[\sqrt{-g}R(\omega) - \varepsilon^{\mu\nu\rho\sigma}\bar{\psi}_\mu \gamma_5 \gamma_\nu D_\rho \psi_\sigma], \tag{1}$$

where the bosonic part is the Palatini lagrangian, which depends on the tetrad and the spin connection. The derivative of the Rarita-Schwinger field is covariant with respect to Lorentz transformations. After its elimination by its equations of motion, the spin connection turns into the Ricci rotation coefficients, modified by the supersymmetric torsion term

$$\omega_{\lambda\mu\nu} = \frac{1}{2}(-\Omega_{\lambda\mu\nu} + \Omega_{\mu\nu\lambda} + \Omega_{\nu\lambda\mu}), \tag{2}$$

where $\Omega_{\lambda\mu\nu} = e_{\lambda a}\left(\partial_\mu e_\nu^a - \partial_\nu e_\mu^a - \frac{i}{2}\bar{\psi}_\mu \gamma^a \psi_\nu\right)$.

Similar to Yang-Mills theories, the components e_0^a, ω_0^{ab}, and ψ_0^a are non-dynamical, and their conjugated momenta p_a, p_{ab} and π_a vanish. The hamiltonian is [22,23]

$$\mathcal{H} = e_0^a H_a + \omega_0^{ab} L_{ab} + \psi_0^\alpha S_\alpha + \lambda^a p_a + \lambda^{ab} p_{ab} + \lambda^\alpha \pi_\alpha, \tag{3}$$

where H_0 is the hamiltonian constraint, H_i, $(i = 1,2,3)$ the momentum constraints, L_{ab} the Lorentz constraints, and S_α the supersymmetry fermionic constraints. All these constraints are first class.

In general the fermionic variables have first order kinetic term, from which follow second class constraints, which require Dirac brackets. Thus, the canonical brackets of ψ_m^α form a Clifford algebra ($m = 1,2,3$). The equal time bracket constraint algebra closes [22], modulo Lorentz transformations,

$$\{S_\alpha(\vec{x}), S_\beta(\vec{x}')\} = -(\gamma^a C)_{\alpha\beta} H_a(\vec{x})\delta(\vec{x} - \vec{x}'). \tag{4}$$

The canonical quantization of the bosonic variables is implemented by derivatives of the conjugated variables, as for canonical quantum gravity [6]. The fermionic variables can be quantized in a similar way [24]. Another way is, as in the case of the spinning particle, representing the fermionic variables by Dirac matrices, leading to the Dirac equation [25]. The constraints are implemented as operator equations $\mathcal{H}\Psi = 0$, on a wave function Ψ that will be a functional of the fields of configuration space, called also superspace. Thus, the solutions do not depend on the non-dynamical fields, are scalar by the Lorentz constraints, and depend on 3-metrics through space diffeomorphisms invariant classes, due to the momentum constraints H_i, see e.g., [5]. On the other side, these solutions satisfy the supersymmetric constraints $S_\alpha \Psi = 0$, then the hamiltonian and momentum constraints will be satisfied, $H_a \Psi = 0$, due to (4).

At the beginning of the nineties, supersymmetric quantum cosmology has attracted the attention of theoretical cosmologists and was developed in several directions, for early developments see [26], and the extensive more recent review [27,28].

As fermionic degrees of freedom of the universe could describe anisotropies, frequently a supergravity description of a homogeneous universes has been done for Bianchi models [7,29], whose metric has the general form, in the ADM formulation, $ds^2 = [N^2(t) - N_i(t)N_i(t)]dt^2 - 2N_m(t)dt\omega^m - h_{mn}(t)\omega^m\omega^n$, where $h_{mn}(t) = e^{2\Omega(t)}(e^{2\beta(t)})_{mn}$, $\beta(t)$ is a 3×3 matrix, and the one-forms ω^m are determined by the Bianchi type. With this metric, a tetrad can be given by $e_0^0 = N$, $e_0^i = e^\Omega(e^{-\beta})_{ij}$, $e_m^0 = 0$, $e_m^i = e^{-\Omega}(e^\beta)_{mi}$, where $i, j, \ldots$ are Lorentz space indices, and $m, n, \ldots$ are world space indices. Usually, the Misner parametrization is taken $\beta = \text{diag}(\beta_+ + \sqrt{3}\beta_-, \beta_+ - \sqrt{3}\beta_-, -2\beta_+)$. These choices amount to a gauge fixing, corresponding to space homogeneity and Lorentz invariance of the tetrad. As a consequence, there are no Lorentz constraints, as noted in [7]. However, as usual in gauge theory QFT, in this case, the Lorentz constraints must be taken into account [29]. In [7,29] the Bianchi I model has been studied, $\omega^m = dx^m$. In [7] a matrix representation has been taken for the fermionic variables, with a spinor wave function, and in [29] a power series in the fermionic variables. In fact, in [30] the Bianchi IX model has been studied, considering the 12 dynamical fermionic degrees of freedom, which require a 64×64 Dirac-matrix representation. In this case, as there is no dependence on space coordinates, the Lorentz constraints act only on the fermionic variables and are algebraic. Their application to the 64-D spinorial wave function is straightforward and, as shown it [30], the wave function reduces to only two non-vanishing components

$$\psi_\pm = C e^{\pm e^{-2\Omega}[2e^{2\beta_+}\cosh(2\sqrt{3}\beta_-)+e^{-4\beta_-}]}. \qquad (5)$$

This result confirms the observation in [29], that the Lorentz constraints restrict strongly the solutions of the WDW equation. In [31], a thorough analysis of this model has been done without considering the Lorentz constraints. In [32] this model has been studied under the inclusion of a cosmological constant, with results indicating that there may be no physical quantum states.

In [33], the Bianchi IX model has been considered by the observation that it has the structure of a sigma model in classical mechanics, on a manifold with three degrees of freedom $q = (\alpha, \beta_+, \beta_-)$, i.e., geometrodynamics with a specific metric for Bianchi IX, $G^{ij} = \text{diag}(-1, 1, 1)$, and whose hamiltonian constraint can be written as

$$H_0 = G^{ij}(q)\left[p_i p_j + \frac{\partial \phi(q)}{\partial q^i}\frac{\partial \phi(q)}{\partial q^j}\right], \qquad (6)$$

where p_i are the conjugate momenta, and $\phi(q) = \frac{1}{6}e^{2\alpha}\text{Tr}e^{2\beta}$. As the hamiltonian (6) is quadratic, a global supersymmetric extension can be given straightforwardly, see e.g., [34–36], with the supersymmetric charges $Q = \psi^i(p_i + i\frac{\partial \phi}{\partial q^i})$ and $\bar{Q} = \bar{\psi}^i(p_i - i\frac{\partial \phi}{\partial q^i})$, and $H = \{Q, \bar{Q}\} = H_0 + \frac{\hbar}{2}\frac{\partial^2 \phi}{\partial q^i \partial q^j}[\bar{\psi}^i \psi^j]$. The fermionic variables satisfy $\{\psi^i, \psi^j\} = \{\bar{\psi}^i, \bar{\psi}^j\} = 0$, and $\{\psi^i, \bar{\psi}^j\} = G^{ij}$. Time reparametrization invariance is restored in [37], by the introduction of the lapse function and its fermionic superpartner in the inverse Legendre transformation. This formulation has been worked out to introduce a cosmological term in [37] and scalar matter in [38].

A superfield formulation has been given in [39,40], by a realization of general time reparametrizations on superspace[1] $(t, \theta, \bar{\theta})$

$$\delta t = \zeta(t) + \frac{i}{2}[\theta \bar{\xi}(t) + \bar{\theta}\xi(t)], \qquad (7)$$

$$\delta \theta = \frac{1}{2}\bar{\xi}(t) + \frac{1}{2}\theta[\dot{\zeta}(t) + ib(t)] + \frac{i}{2}\theta\bar{\theta}\dot{\xi}(t), \qquad (8)$$

$$\delta \bar{\theta} = \frac{1}{2}\xi(t) + \frac{1}{2}\bar{\theta}[\dot{\zeta}(t) - ib(t)] - \frac{i}{2}\theta\bar{\theta}\dot{\bar{\xi}}(t). \qquad (9)$$

where $(\zeta(t), \xi(t), \bar{\xi}(t))$ are the parameters of superspace transformation, and $b(t)$ is the parameter of the $U(1)$ R-symmetry transformation, which acts on the $(\xi, \bar{\xi})$ space. As this is a one dimensional field theory, there are no Lorentz transformations. The fields are replaced by superfields, but their components must be suitably rescaled to have the correct weight under time reparametrizations. There are superfields for the parameters of time and supersymmetry transformations, as in superspace supergravity, see e.g., [41]. The lapse field is also replaced by a superfield, although there is not an invariant volume element for superspace is not considered. However, this formulation allows to write invariant superfield actions, whose bosonic sector agrees with the corresponding non-supersymmetric action. The constraint algebra contains the hamiltonian and supersymmetric constraints, and it closes properly. An interesting feature is that it includes additionally, as a constraint, the fermion number operator, which ensures equal number of bosonic and fermionic degrees of freedom; the corresponding Lagrange multiplier is the highest component of the lapse superfield.

This formalism has been applied in [42,43] to a FRLW cosmology with a scalar field, where the spontaneous symmetry breaking of supersymmetry has been explored. In these works a canonical quantization has been performed, with matrix representations for the fermionic variables, and the wave function has been computed. It has been shown that a similar mechanism of supersymmetry breaking as in supergravity applies here, as the scalar potential is not positive definite, which written in terms of the auxiliary fields is

$$V(\phi) = \frac{1}{2}F^2 - \frac{3}{\kappa^2 a^2}B^2, \qquad (10)$$

where F and B are auxiliary fields, given by their equations of motion $F = 2\frac{\partial g(\varphi)}{\partial \varphi}$, $B = -\kappa^2 a g(\varphi)$, and $g(\varphi)$ is the superfield potential, i.e., the superpotential in supergravity. Thus, supersymmetry is broken if at an extremum φ_0 of the potential, $V(\varphi_0) = 0$, and $F \neq 0$; the superpartner of the lapse, the gravitino, acquires a nonvanishing mass, similar to the Higgs mechanism. The wave function has four components, two of them are not normalizable, and the other two are the 'conjugated' states, one of them normalizable, depending on the sign of the superpotential

$$\psi_\pm(a,\varphi) = Ca^{\frac{3}{4}}e^{\mp 2a^3 g(\varphi) \pm 3\sqrt{k}M_{pl}^2 a^2}. \qquad (11)$$

In order to appreciate phenomenological consequences of the previous model, it has been studied in [44] in a semiclassical setting by the WKB method, with the solutions (11) written as $e^{S_a + S_\varphi}$ see also [45]. For each one of these solutions follow different classical evolution equations for the scale factor and the scalar field. In these equations there are supersymmetric contributions with the behavior of radiation and stiff matter. A detailed numerical analysis has been made for an exponential superpotential and flat space, where, in particular, inflationary stages can be observed. An interesting relation of a $SU(2)$ matrix model with supersymmetric quantum cosmology has been uncovered in [46], arising from the quantization of the 11-dimensional supermembrane, which has a zero energy solution of the same form of a wave function known from supersymmetric quantum cosmology [27].

The superspace approach in supergravity has the drawback, that the component fields of superfields in general are not Lorentz spinors (bosonic or fermionic). This problem has a solution in the formalism of the so called "new" Θ- variables [41,47]. The fermionic power expansion in superfields is redefined as

$$\phi(x,\theta) = \sum \frac{1}{n!}\theta^{\eta_1}\cdots\theta^{\eta_n}\partial_{\eta_1}\cdots\partial_{\eta_n}\phi(x,\theta)|_{\theta=0}$$

$$\to \Phi(x,\Theta) = \sum \frac{1}{n!}\Theta^{\alpha_1}\cdots\Theta^{\alpha_n}\mathcal{D}_{\alpha_1}\cdots\mathcal{D}_{\alpha_n}\phi(x,\theta)|_{\theta=0}, \qquad (12)$$

where $\mathcal{D}_\alpha\phi(x,\theta)$ are Lorentz covariant fermionic derivatives under local superspace coordinate transformations $(x,\theta) \to (x',\theta')$ and local Lorentz transformations. The supergravity transformations are field dependent local superspace plus Lorentz transformations $\delta_\xi \Phi_A(x,\theta) = -\xi^B \mathcal{D}_B \Phi_A(x,\theta)$, where $A = (\mu,\alpha)$ are multiindices formed by a spacetime index, and a Lorentz $\frac{1}{2}$-spin index. For consistency, this formulation requires a covariant formulation of the Wess-Zumino gauge [47]. In [48] it is shown how to apply this formalism to homogeneous spaces, and how to construct invariant supergravity actions. In this framework, in [49,50], the FRLW model with a scalar field has been worked out, with a wave function similar to (11), differing only by the power of the scalar factor in front of it, due to a different operator ordering. Further, in these works the interpretation of the scalar field as time is considered, with an effective time dependent wave function, which corresponds to the conditional probability for a given value of the scalar field

$$\Psi(a,t) = \frac{\psi(a,\phi)}{\sqrt{\int_0^\infty da\, |\psi(a,\phi)|^2}}\bigg|_{\phi=t}.\quad (13)$$

With this wave function, the mean value of the scalar factor gives a classical evolution [49]

$$a(t) = \int_0^\infty a|\Psi(a,t)|^2 da = \Gamma(4/3)\left[\frac{\hbar c}{2|W(t)|}\right]^{1/3}.\quad (14)$$

As a consistency check, a computation of the uncertainty relation gives $\Delta a \Delta \pi_a \approx 0.53\,\hbar$.

Moreover, with an exponential superpotential an inflationary scenario can be obtained, which lasts enough *e*-folds [50]. The exit of inflation requires fine tuning of a constant which must be added to the superpotential. To illustrate it, we consider a Gaussian superpotential

$$W(t) = \frac{c^4 M_p^3}{\hbar^2} t^{-2}\left(\frac{1}{\lambda}e^{-t}+1\right)\quad (15)$$

with $\lambda = 10^{-80}$, and $c = 1$, $\hbar = 1$, $M_p = 1$. With this potential it can be seen that $a(0) = 0$, $\dot{a}(0) = \infty$, and $\ddot{a}(0) = -\infty$. The acceleration increases quickly, and becomes positive at $t \approx 0.45$, and stays positive up to $t \approx 183$. In this time interval, there are $\log(a(183)/a(0.45)) \approx 65$ *e*-folds, see Figures 1 and 2. We show also the potentials corresponding to this behavior in the analog FLRW model with a scalar field ϕ, for $k = 0, 1$, Figure 3.

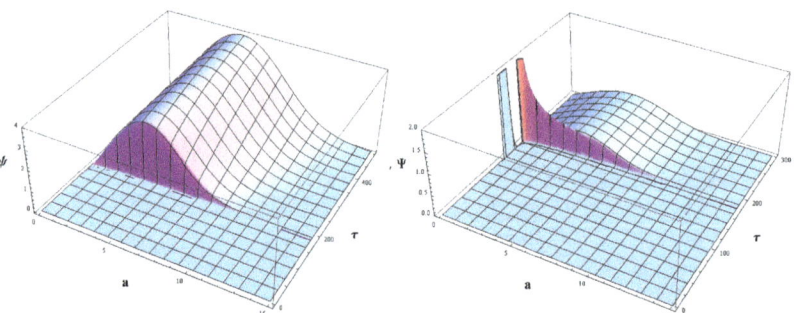

Figure 1. Wave function. Left: wave function. Right: effective wave function (at $a = 0$ there are numerical issues).

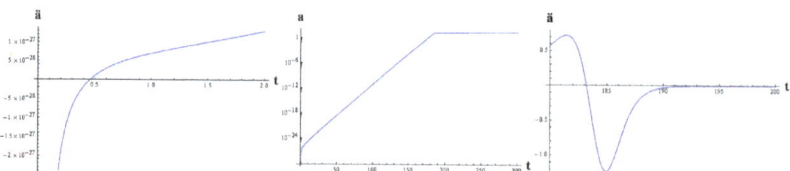

Figure 2. Scale factor. Left: acceleration at inflation begin. Center: scale factor (log scale). Right: acceleration at inflation exit.

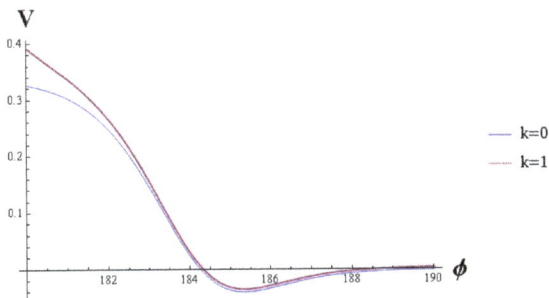

Figure 3. Analog potential for FLRW model.

3. Noncommutative Quantum Cosmology

Noncommutative field theories [8,9], have many remarkable properties as the UV/IR mixing. Some similar features were studied in the case of gravitational theories. There are many articles on the subject, in particular, our group participated with a number of gravity proposals in various versions [51–53]. In the context of some noncommutative gravity theories, one natural question was to carry out a canonical quantization analysis following ADM formalism [3]. However the situations turns out quite complicated. Instead of that in Ref. [10], we proposed to carry out the standard ADM formalism to get the WDW equation and at that level, to propose a non-commutative deformation of the WDW equation and look for solutions to this deformed equation.

The first example considered in this context was an anisotropic universe, the Kantowski-Sach model [10]. Further development (without pretending to be exhaustive) on this subject can be found in Refs. [54–71]. In Misner's parametrization the metric is written as

$$ds^2 = -N^2 dt^2 + e^{2\sqrt{3}\beta} dr^2 + e^{-2\sqrt{3}\beta} e^{-2\sqrt{3}\Omega} (d\theta^2 + \sin^2\theta d\varphi^2) \tag{16}$$

The quantum model of Kantowski-Sach cosmology is implemented through the quantization of its WDW equation, i.e., the hamiltonian constraint

$$e^{\sqrt{3}\beta + 2\sqrt{3}\Omega} \left[P_\Omega^2 - P_\beta^2 + 48 e^{-2\sqrt{3}\Omega} \right] \psi(\Omega, \beta) = 0, \tag{17}$$

where $P_\Omega = -i\frac{\partial}{\partial \Omega}$ and $P_\beta = -i\frac{\partial}{\partial \beta}$. The physical observables in coordinate representation are the position operators q_j and the conjugate momenta p_k, where $q_1 = \Omega$, $q_2 = \beta$ and $p_1 = P_\Omega$, $p_2 = P_\beta$. These operators satisfy the commutation relations

$$[q_j, p_k] = i\delta_{jk}, \quad [q_j, q_k] = 0, \quad [p_j, p_k] = 0. \tag{18}$$

The noncommutative Wheeler-DeWitt equation is

$$e^{\sqrt{3}\beta + 2\sqrt{3}\Omega} \star \left\{ -\frac{\partial^2}{\partial \Omega^2} + \frac{\partial^2}{\partial \beta^2} + 48 e^{-2\sqrt{3}\Omega} \right\} \star \psi(\Omega, \beta) = 0, \tag{19}$$

where $\star$ is the Moyal product (for a complete review, see for instance, [72])

$$f(\Omega, \beta) \star g(\Omega, \beta) = f(\Omega, \beta) \exp\left\{ i\frac{\theta}{2} \left(\overleftarrow{\partial}_\Omega \overrightarrow{\partial}_\beta - \overleftarrow{\partial}_\beta \overrightarrow{\partial}_\Omega \right) \right\} g(\Omega, \beta), \tag{20}$$

for a constant noncommutativity parameter θ.

Thus the Moyal deformed WDW equation is given by

$$\left\{ -\frac{\partial^2}{\partial \Omega^2} + \frac{\partial^2}{\partial \beta^2} + V(\Omega, \beta) \right\} \star \psi(\Omega, \beta) = 0, \tag{21}$$

where $V(\Omega, \beta) = 48e^{-2\sqrt{3}\Omega}$ is the potential. Using the properties of the star product, Equation (20), we have $V(\Omega, \beta) \star \psi(\Omega, \beta) = V(\Omega - \frac{1}{2}\theta P_\beta, \beta - \frac{1}{2}\theta P_\Omega)\psi(\Omega, \beta)$, and it can be rewritten as

$$\left\{ -\frac{\partial^2}{\partial \Omega^2} + \frac{\partial^2}{\partial \beta^2} + 48e^{-2\sqrt{3}\Omega + \sqrt{3}\theta P_\beta} \right\} \psi(\Omega, \beta) = 0. \tag{22}$$

A solution to Equation (22) can be found using the solution to the same equation with $\theta = 0$, which is the usual WDW equation for the Kantowski-Sachs model in GR. This solution in GR is given by

$$\psi_\nu^\pm(\Omega, \beta) = e^{\pm i\sqrt{3}\nu\beta} K_{i\nu}\left\{ 4\exp\left[-\sqrt{3}\Omega \right] \right\}. \tag{23}$$

Now assuming an ansatz for the deformed Equation (22) in the form

$$\psi_\nu^\pm(\Omega, \beta) = e^{\pm i\sqrt{3}\nu\beta} \chi(\Omega). \tag{24}$$

Taking into account the translation $e^{\sqrt{3}\theta P_\beta}\psi(\Omega, \beta) = \psi(\Omega, \beta - i\sqrt{3}\theta)$, it is possible to find that the function $\chi(\Omega)$ satisfies the modified Bessel equation and Equation (22). Thus, it was shown in [10] that Equation (22) has a solution given by

$$\psi_\nu^\pm(\Omega, \beta) = e^{\pm i\sqrt{3}\nu\beta} K_{i\nu}\left\{ 4\exp\left[-\sqrt{3}\left(\Omega \mp \frac{\sqrt{3}}{2}\nu\theta\right) \right] \right\}. \tag{25}$$

Equation (22) represents a noncommutative deformation of the ordinary WDW equation for the Kantowski-Sachs cosmological model. This implies an additional correction to the ordinary WDW equation due to an assumed noncommutative structure of the minisuperspace. This noncommutativity is regarded as an indirect consequence of the noncommutive gravity in spacetime, which seems to be a better approach to study microscopic properties of gravity. In Ref. [10] it was plotted the probability, constructed from the wave function solution (25), depending on coordinates Ω and β for various values of θ, including the case of GR with $\theta = 0$. In GR the probability obtained from the solution (weighted with a Gaussian wave packet) has just one peak near $\Omega = 0$ and $\beta = 0$ which indicates where the universe is more probable to be placed. For nonvanishing theta $\theta \neq 0$ it was found a different behavior than that of GR. In this last case there was a big peak together with other many different smaller peaks, which were interpreted as other (baby) additional universes where the universe may to stay. Thus a bold consequence of the noncommutative minisuperspace is the emergence of many other universes in which the universe can carry out vacuum transitions by tunneling.

4. GUP's in Quantum Cosmology

In this section we review an application of Generalized Uncertainty Principles (GUP) to quantum cosmology [19,61,62,64,66]. In order to be concrete in [19] it was considered the Kantowski-Sachs model, an homogeneous and anisotropic cosmological model in the minisuperspace. The GUP involves a modification of the Heisenberg uncertainty relation at very high energies (near the Planck scale), a behavior expected for the very early universe. This implies a modification of the Heisenberg algebra of commutators, by terms with powers of the momentum. As it is shown in [19], this implies a deformation of the Wheeler-DeWitt equation. It is worth mentioning that the application of the GUP in this context involves a minimal uncertainty of the quantum dynamical variables in the minisuperspace. This UV modification of the Heisenberg algebra is for the phase space minisuperspace variables, and not properly for spacetime variables. However, very interesting consequences of this hypothesis arise, such as black holes without singularity as we will review in the next section.

The WDW Equation (17) has four quantum dynamical variables: the operators Ω and β, and their conjugate momenta P_Ω and P_β, which satisfy the commutation relations (18).

Now, if we consider the commutation relation

$$[q_j, p_k] = i\delta_{jk}[1 + \gamma^2 p_\ell p_\ell], \qquad (26)$$

where $q_1 = \Omega$, $q_2 = \beta$, $p_1 = P_\Omega$ and $p_2 = P_\beta$, and γ is a parameter with units of the inverse of the momentum.

The procedure involves to perform a suitable change of variables

$$q_j = (1 + \gamma^2 p_\ell p_\ell) q'_j, \qquad (27)$$

where $q'_j = i\frac{\partial}{\partial p_j}$ such that $[q'_j, p_k] = i\delta_{jk}$. This change does not respect entirely the Heisenberg algebra, but a noncommutative extension since

$$[q_j, q_k] = 2i\gamma^2 (1 + \gamma^2 p_\ell p_\ell)(p_j q'_k - p_k q'_j). \qquad (28)$$

In particular, for the KS model we have that the potential $V(\Omega, \beta)$ will be modified within the deformed algebra. An approximation to order γ^2 for the potential V can be written as

$$V \approx -48 e^{-2\sqrt{3}(1-4\gamma^2)\Omega'} e^{-2\sqrt{3}\gamma^2 \Omega'(-P_\Omega^2 + P_\beta^2)} e^{12i\gamma^2 \Omega' P_\Omega}. \qquad (29)$$

Assume a representation $P_\Omega = -i\frac{\partial}{\partial \Omega'}$ and $P_\beta = -i\frac{\partial}{\partial \beta'}$, it is easy to see that

$$e^{12i\gamma^2 \Omega' P_\Omega} \psi(\Omega', \beta) = \psi(e^{12\gamma^2} \Omega', \beta). \qquad (30)$$

Thus under the same ansatz as in the noncommutative case [10]

$$\psi(e^{12\gamma^2}\Omega', \beta') = e^{\sqrt{3}\nu\beta'} \chi(\Omega'), \qquad (31)$$

the WDW Equation (17) turns out to be

$$\left(\frac{d^2}{d\Omega'^2} + V_{\gamma,\nu} + \gamma^2 \widetilde{V}\right)\chi(\Omega') = 0, \qquad (32)$$

where

$$V_{\gamma,\nu} = 3\nu^2 + 48 e^{-2\sqrt{3}(1-4\gamma^2)\Omega'} \qquad (33)$$

and

$$\widetilde{V} = -4608\sqrt{3}\Omega' e^{-4\sqrt{3}(1-4\gamma^2)\Omega'}. \qquad (34)$$

The potential $V_{\gamma,\nu}$ is the modified potential of the ordinary Kantowski-Sach model. Moreover, $\widetilde{V}$ is the first correction due to the modified uncertainty relation (26). The relevant contribution of the later potential is concentrated about the value of Ω' given by $\Omega' = \frac{1}{4\sqrt{3}(1-4\gamma^2)}$.

It can be observed that the potential $V_{\gamma,\nu}$ dominates over $\widetilde{V}$ for values of γ satisfying $\gamma \ll \frac{1}{2}\sqrt{\frac{e^{3/2}}{24+e^{3/2}}} \approx 0.198$. For values $\gamma \geq \frac{1}{2}\sqrt{\frac{e^{3/2}}{24+e^{3/2}}}$ the potential $\widetilde{V}$ dominates over $V_{\gamma,\nu}$ and it produces a well. One can see that this well has a local minimum at

$$\Omega'_{min} = \frac{\sqrt{3}\left[1 - 2W\left(-\frac{\sqrt{e}(1-4\gamma^2)}{96\gamma^2}\right)\right]}{12(1-4\gamma^2)}, \qquad (35)$$

where $W(z)$ is the Lambert function satisfying the equation $z = We^W$. This function has real values for $\gamma > \frac{1}{2}\sqrt{\frac{e^{3/2}}{24+e^{3/2}}}$. In Figure 4 we show how the potential of Equation (32), $V_{\gamma,\nu} + \gamma^2 \tilde{V}$, changes under the variation of the Barbero-Immirzi parameter γ.

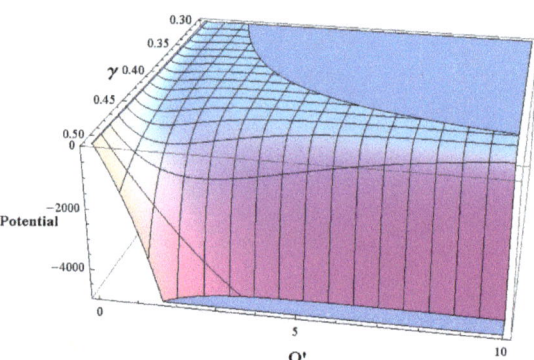

Figure 4. Potential of (32), it is shown how it changes under the variation of the Barbero-Immirzi parameter. For $\gamma \geq 0.5$ it becomes unstable.

Thus, under a variables change $y \equiv \Omega' - \Omega'_{\min}$, Equation (32) turns out into a *quantum mechanical harmonic oscillator* equation

$$\left[-\frac{d^2}{dy^2} + \omega^2 y^2\right]\chi(y) = E\chi(y), \tag{36}$$

where ω is the frequency of the oscillator and E is the energy which can be written as

$$\omega^2 = 3(1-4\gamma^2)^3 \frac{\left[W(-\frac{\sqrt{e}(1-4\gamma^2)}{96\gamma^2}) + 1\right]}{\gamma^2 W^2(-\frac{\sqrt{e}(1-4\gamma^2)}{96\gamma^2})}, \tag{37}$$

$$E = -\frac{3}{2}\left\{\nu - \frac{1-4\gamma^2}{12\gamma^2 W(-\frac{\sqrt{e}(1-4\gamma^2)}{96\gamma^2})} - \frac{1-4\gamma^2}{24\gamma^2 W^2(-\frac{\sqrt{e}(1-4\gamma^2)}{96\gamma^2})}\right\}. \tag{38}$$

The requirement that $E > 0$ to obtain a bounded state from below implies that

$$|\nu| < -\frac{\sqrt{(1-4\gamma^2)[12W(-\frac{\sqrt{e}(1-4\gamma^2)}{96\gamma^2}) + 6]}}{12\gamma W(-\frac{\sqrt{e}(1-4\gamma^2)}{96\gamma^2})}. \tag{39}$$

Thus, the condition of the existence for real values of ν implies that $\gamma \geq \frac{1}{2}\sqrt{\frac{e}{12+e}}$. Then the quantum spectrum of the harmonic oscillator is given by

$$E = \omega\left(n + \frac{1}{2}\right), \tag{40}$$

where n is a natural number. As a consequence of this fact, the parameter ν is quantized in the form

$$\nu = \frac{\sqrt{1-4\gamma^2}}{2\sqrt{6}}\left\{\frac{2W(-\frac{\sqrt{e}(1-4\gamma^2)}{96\gamma^2}) + 1}{\gamma^2 W^2(-\frac{\sqrt{e}(1-4\gamma^2)}{96\gamma^2})} + 8\sqrt{3}(2n+1)\frac{\sqrt{(1-4\gamma^2)[W(-\frac{\sqrt{e}(1-4\gamma^2)}{96\gamma^2}) + 1]}}{\gamma W(-\frac{\sqrt{e}(1-4\gamma^2)}{96\gamma^2})}\right\}^{1/2}. \tag{41}$$

In this derivation it was assumed a quadratic degree of approximation in an expansion of γ. It is possible to obtain some higher order terms, the subsequent terms in the expansion which turn the harmonic oscillator into an anharmonic oscillator. This systems can be solved in perturbation theory, considering the nonlinear terms as a perturbation. Thus it is possible to find perturbatively the first correction to the energy levels and the corresponding Hilbert space.

5. Deformed Dynamics and the Interior of Black Holes

In this section we comment on other application of GUP to gravitational systems. We review the case of the application of GUP to the study of the dynamics in the interior of a Schwarzschild black hole (SBH). SBH is parametrized by the Schwarzschild coordinates (t, r, θ, ϕ). The interior of a SBH is described by a Kantowski-Sach metric (see Equation (16)) defined on a spacetime with topology $\mathbb{R} \times \mathbf{S}^2$, which has an infinite volume since the noncompact nature of it. However, for a *fiducial* finite interval L_0 in $\mathbb{R}$, this spacetime has finite volume. We can define local coordinates in the four dimensional spacetime $\mathbb{R}^2 \times \mathbf{S}^2$ to be (T, x, θ, ϕ).

The idea of [20] consists to apply GUP to the hamiltonian classical dynamics described by the Ashtekar-Barbero hamiltonian H_{AB}, and deform the canonical algebra in terms of the GUP parameter γ. First, we described the classical hamiltonian dynamics. The dynamical variables (b, p_b, c, p_c) are smooth functions depending only on the variable T or t. After gauge fixing the hamiltonian it reads

$$H_{AB} = -\frac{1}{2G\gamma}\left[(b^2 + \gamma^2)\frac{p_b}{b} + 2cp_c\right]. \tag{42}$$

The canonical algebra is given by

$$\{b, p_b\} = G\gamma, \qquad \{c, p_c\} = 2G\gamma. \tag{43}$$

The equations of motion of H_{AB} are

$$\frac{db}{dT} = \{b, H_{AB}\} = -\frac{1}{2}\left(b + \frac{\gamma^2}{b}\right), \tag{44}$$

$$\frac{dp_b}{dT} = \{p_b, H_{AB}\} = \frac{p_b}{2}\left(1 - \frac{\gamma^2}{b^2}\right), \tag{45}$$

$$\frac{dc}{dT} = \{c, H_{AB}\} = -2c, \tag{46}$$

$$\frac{dp_c}{dT} = \{p_c, H_{AB}\} = 2p_c. \tag{47}$$

In the Schwarzschild time t the solutions take the form

$$b(t) = \pm\gamma\sqrt{\frac{2GM}{t} - 1}, \tag{48}$$

$$p_b(t) = \ell L_0 t \sqrt{\frac{2GM}{t} - 1}, \tag{49}$$

$$c(t) = \mp\frac{\gamma GM\ell L_0}{t^2}, \tag{50}$$

$$p_c(t) = t^2, \tag{51}$$

where ℓ satisfies the equation $p_b^2(t) = \ell\left(\frac{2GM}{t} - 1\right)L_0^2 t^2$. Here p_c can be interpreted as the square of the radius of the infalling 2-spheres and consequently is zero at $t = 0$. This interpretation results from the fact that the Kretschmann invariant $K = R_{abcd}R^{abcd}$ is proportional to $\frac{1}{p_c^3}$.

Now, it is possible to deform the classical algebra (43) with the GUP, and find a modified dynamics. This can be achieved imposing minimal uncertainty in p_b and p_c. Thus the modified algebra according to GUP is

$$\{\bar{b}, p_b\} = 1, \quad \{\bar{c}, p_c\} = 1 \tag{52}$$

such that

$$\{b, p_b\}_{\bar{q},p} = 1 + \beta_b b^2, \quad \{c, p_c\}_{\bar{q},p} = 1 + \beta_c c^2, \tag{53}$$

where β_b and β_c are suitable parameters. The modified algebra with a minimal uncertainty in p_b and p_c is

$$[b, p_b] = iG\gamma(1 + \beta_b b^2), \quad [c, p_c] = i2G\gamma(1 + \beta_c c^2). \tag{54}$$

Equivalently this yields

$$\Delta b \Delta p_b \geq \frac{G\gamma}{2}\left[1 + \beta_b(\Delta b)^2\right], \tag{55}$$

$$\Delta c \Delta p_c \geq G\gamma\left[1 + \beta_c(\Delta c)^2\right]. \tag{56}$$

The GUP modified equations of motion are

$$\frac{db}{dT} = \{b, H_{AB}\} = -\frac{1}{2}\left(b + \frac{\gamma^2}{b}\right)(1 + \beta_b b^2), \tag{57}$$

$$\frac{dp_b}{dT} = \{p_b, H_{AB}\} = \frac{p_b}{2}\left(1 - \frac{\gamma^2}{b^2}\right)(1 + \beta_b b^2), \tag{58}$$

$$\frac{dc}{dT} = \{c, H_{AB}\} = -2c(1 + \beta_c c^2), \tag{59}$$

$$\frac{dp_c}{dT} = \{p_c, H_{AB}\} = 2p_c(1 + \beta_c c^2). \tag{60}$$

In the Schwarzschild time t the solutions take the form

$$b(t) = \pm \gamma \frac{\sqrt{2GMt^{\beta_b \gamma^2} - t(2\gamma^2 GM)^{\beta_b \gamma^2}}}{\sqrt{t(2\gamma^2 GM)^{\beta_b \gamma^2} - 2\beta_b \gamma^2 GMt^{\beta_b \gamma^2}}}, \tag{61}$$

$$p_b(t) = \frac{\ell_c}{\sqrt{-\beta_c}} t^{-\beta_b \gamma^2} \sqrt{\left[2GMt^{\beta_b \gamma^2} - t(2\gamma^2 GM)^{\beta_b \gamma^2}\right]\left[t(2\gamma^2 GM)^{\beta_b \gamma^2} - 2\beta_b \gamma^2 GMt^{\beta_b \gamma^2}\right]}, \tag{62}$$

$$c(t) = \mp \frac{\ell_c}{\sqrt{-\beta_c}} \frac{\gamma GM}{\sqrt{t^4 + \ell_c^2 \gamma^2 G^2 M^2}}, \tag{63}$$

$$p_c(t) = \sqrt{t^4 + \ell_c^2 \gamma^2 G^2 M^2}, \tag{64}$$

where it was introduced the fundamental physical length $\ell_c \equiv -\beta_c L_0$. This can be considered as a prescription to cure the dependence on the fiducial length L_0. Moreover, the mentioned interpretation of $p_c(t)$, as the 2-sphere inside the black hole, leads from the last Equation (64) to the existence of a minimum value for p_c and consequently the resolution of the black hole singularity at $t = 0$. See Figure 5.

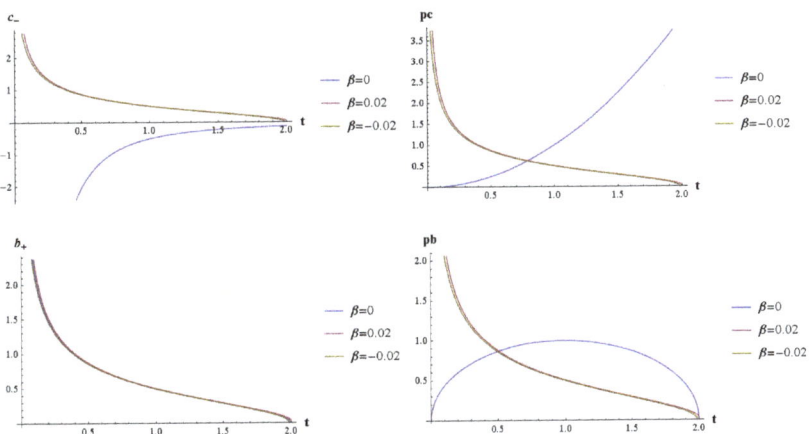

Figure 5. Comparison of the behavior of solutions of the unmodified ($\beta_b = \beta_c = 0$) with the modified cases for the whole interior. $G = M = \ell_c = 1$.

Author Contributions: All authors contributed equally to this paper. All authors have read and agreed to the published version of the manuscript.

Funding: This research was funded by CONACYT 257919, CIIC 188/2019 UGTO, BUAP 11171277-VIEP2021.

Conflicts of Interest: The authors declare no conflict of interest.

Note

1 This is the superspace of supersymmetry, not to be confounded with the superspace of geometrodynamics.

References

1. Oriti, D. (Ed.) *Approaches to Quantum Gravity: Toward a New Understanding of Space, Time and Matter*; Cambridge University Press: Cambridge UK, 2009.
2. Armas, J. (Ed.) *Conversations on Quantum Gravity*; Cambridge University Press: Cambridge UK, 2021. [CrossRef]
3. Arnowitt, R.L.; Deser, S.; Misner, C.W. The Dynamics of general relativity. *Gen. Rel. Grav.* **2008**, *40*, 1997–2027. [CrossRef]
4. Ryan, M.P.; Shepley, L.C. *Homogeneous Relativistic Cosmologies*; Princeton Series in Physics: Princeton, NJ, USA. 1975.
5. Halliwell, J.J. Introductory Lectures on Quantum Cosmology. *arXiv* **1990**, arXiv:0909.2566.
6. DeWitt, B.S. Quantum Theory of Gravity. 1. The Canonical Theory. *Phys. Rev.* **1967**, *160*, 1113–1148. [CrossRef]
7. Macías, A.; Obregón, O.; Ryan, M.P., Jr. Quantum Cosmology: The Supersymmetric Square Root. *Class. Quantum Grav.* **1987**, *4*, 1477. [CrossRef]
8. Douglas, M.R.; Nekrasov, N.A. Noncommutative field theory. *Rev. Mod. Phys.* **2001**, *73*, 977–1029. [CrossRef]
9. Szabo, R.J. Quantum field theory on noncommutative spaces. *Phys. Rept.* **2003**, *378*, 207–299. [CrossRef]
10. Garcia-Compean, H.; Obregon, O.; Ramirez, C. Noncommutative quantum cosmology. *Phys. Rev. Lett.* **2002**, *88*, 161301. [CrossRef]
11. Gross, D.J.; Mende, P.F. The High-Energy Behavior of String Scattering Amplitudes. *Phys. Lett. B* **1987**, *197*, 129–134. [CrossRef]
12. Gross, D.J.; Mende, P.F. String Theory Beyond the Planck Scale. *Nucl. Phys. B* **1988**, *303*, 407–454. [CrossRef]
13. Amati, D.; Ciafaloni, M.; Veneziano, G. Can Space-Time Be Probed Below the String Size? *Phys. Lett. B* **1989**, *216*, 41–47. [CrossRef]
14. Maggiore, M. A Generalized uncertainty principle in quantum gravity. *Phys. Lett. B* **1993**, *304*, 65–69. [CrossRef]
15. Garay, L.J. Quantum gravity and minimum length. *Int. J. Mod. Phys. A* **1995**, *10*, 145–166. [CrossRef]
16. Scardigli, F. Generalized uncertainty principle in quantum gravity from micro-black hole Gedanken experiment. *Phys. Lett. B* **1999**, *452*, 39–44. [CrossRef]
17. Bosso, P.; Das, S. Quantum field theory on noncommutative spaces. *Ann. Phys.* **2017**, *383*, 416–438. [CrossRef]
18. Bosso, P. *Generalized Uncertainty Principle and Quantum Gravity Phenomenology*; University of Lethbridge: Lethbridge, AB, Canada, 2017.
19. Bosso, P.; Obregón, O. Minimal length effects on quantum cosmology and quantum black hole models. *Class. Quantum Grav.* **2020**, *37*, 045003. [CrossRef]

20. Bosso, P.; Obregón, O.; Rastgoo, S.; Yupanqui, W. Deformed algebra and the effective dynamics of the interior of black holes. *Class. Quantum Grav.* **2021**, *38*, 145006. [CrossRef]
21. Freedman, D.Z.; van Nieuwenhuizen, P.; Ferrara, S. Progress Toward A Theory Of Supergravity. *Phys. Rev. D* **1976**, *13*, 3214. [CrossRef]
22. Teitelboim, C. Supergravity and Square Roots of Constraints. *Phys. Rev. Lett.* **1977**, *32*, 1106. [CrossRef]
23. Deser, S.; Kay, J.H.; Stelle, K.S. Hamiltonian Formulation of Supergravity. *Phys. Rev. D* **1977**, *16*, 2448. [CrossRef]
24. D'Eath, P.D. Canonical quantization of supergravity. *Phys. Rev.* **1983**, *29*, 2199.
25. Barducci, A.; Casalbuoni, R.; Lusanna, L. Supersymmetries and the Pseudoclassical Relativistic electron. *Nuovo Cim. A* **1976**, *35*, 377. [CrossRef]
26. D'Eath, P.D. *Supersymmetric Quantum Cosmology*; Cambridge University Press: Cambridge, UK, 1996.
27. Moniz, P.V. *Quantum Cosmology-The Supersymmetric Perspective-Vol. 1 Fundamentals*, Lect. Notes Phys. *803*; Springer: Berlin/Heidelberg, Germany, 2010.
28. Moniz, P.V *Quantum Cosmology-The Supersymmetric Perspective-Vol. 2 Advanced Topics*, Lect. Notes Phys. *804*; Springer: Berlin/Heidelberg, Germany, 2010.
29. D'Eath, P.D.; Hawking, S.W.; Obregón, O. Supersymmetric Bianchi Models and the Square Root of The Wheeler-DeWitt Equation. *Phys. Lett. B* **1993**, *300*, 44. [CrossRef]
30. Obregón, O.; Ramírez, C. Dirac-Like Formulation of Quantum Supersymmetric Cosmology. *Phys. Rev. D.* **1998**, *57*, 1015. [CrossRef]
31. Damour, T.; Spindel, P. Quantum Supersymmetric Bianchi IX Cosmology. *Phys. Rev. D* **2014**, *90*, 103509. [CrossRef]
32. Cheng, A.D.Y.; D'Eath, P.D.; Moniz, P.R.L.V. Quantization of the Bianchi type-IX model in supergravity with a cosmological constant. *Phys. Rev. D* **1994**, *49*, 5246. [CrossRef]
33. Graham, R. Supersymmetric Bianchi Type IX Cosmology. *Phys. Rev. Lett.* **1991**, *67*, 1381. [CrossRef]
34. Witten, E. Dynamical Breaking of Supersymmetry. *Nucl. Phys. B* **1981**, *188*, 513. [CrossRef]
35. Witten, E. Constraints on Supersymmetry Breaking. *Nucl. Phys. B* **2021**, *982*, 253. [CrossRef]
36. Claudson, M.; Halpern, M.B. Supersymmetric Ground State Wave Functions. *Nucl. Phys. B* **1985**, *250*, 689. [CrossRef]
37. Graham, R. Supersymmetric general Bianchi type IX cosmology with a cosmological term. *Phys. Lett. B* **1992**, *277*, 393. [CrossRef]
38. Bene, J.; Graham, R. Supersymmetric Homogeneous Quantum Cosmologies Coupled to a Scalar Field. *Phys. Rev. D* **1994**, *49*, 799. [CrossRef]
39. Obregón, O.; Rosales, J.J.; Tkach, V.I. Superfield Description of FRW Universe. *Phys. Rev. D* **1996**, *53*, R1750. [CrossRef]
40. Tkach, V.I.; Rosales, J.J.; Obregón, O. Supersymmetric Action for Bianchi Type Models. *Class. Quantum Grav.* **1996**, *13*, 2349. [CrossRef]
41. Wess, J.; Bagger, J. *Supersymmetry and Supergravity*; Princeton University Press: Princeton, NJ, USA, 1992.
42. Tkach, V.I.; Obregón, O.; Rosales, J.J. FRW Model and Spontaneous Breaking of Supersymmetry. *Class. Quantum Grav.* **1997**, *14*, 339. [CrossRef]
43. Obregón, O.; Rosales, J.J.; Socorro, J.; Tkach, V.I. Supersymmetry Breaking and a Normalizable Wavefunction for the FRW (k = 0) Cosmological Model. *Class. Quantum Grav.* **1999**, *16*, 2861. [CrossRef]
44. Escamilla-Rivera, C.; Obregon, O.; Urena-Lopez, L.A. Supersymmetric classical cosmology. *JCAP* **2010**, *12*, 11. [CrossRef]
45. Kiefer, C.; Lück, T.; Moniz, P. Semiclassical approximation to supersymmetric quantum gravity. *Phys. Rev. D* **2005**, *72*, 045006. [CrossRef]
46. López, J.L.; Obregón, O. Supersymmetric Quantum Matrix Cosmology. *Class. Quantum Grav.* **2015**, *32*, 235014. [CrossRef]
47. Cupatitzio Ramírez, The Realizations of Local Supersymmetry. *Ann. Phys.* **1988**, *186*, 43. [CrossRef]
48. García-Jiménez, G.; Ramírez, C.; Vázquez-Báez, V. Tachyon potentials from a supersymmetric FRW model. *Phys. Rev. D* **2014**, *89*, 043501. [CrossRef]
49. Ramírez, C.; Vázquez-Báez, V. Quantum supersymmetric FRW cosmology with a scalar field. *Phys. Rev. D* **2016**, *93*, 043505. [CrossRef]
50. Pérez, N.E.M.; Ramírez, C.; Báez, V.V. Inflationary quantum cosmology for FRLW supersymmetric models. *arXiv* **2021**, arXiv:2104.12914.
51. Garcia-Compean, H.; Obregon, O.; Ramirez, C.; Sabido, M. Noncommutative topological theories of gravity. *Phys. Rev. D* **2003**, *68*, 045010. [CrossRef]
52. Garcia-Compean, H.; Obregon, O.; Ramirez, C.; Sabido, M. Noncommutative selfdual gravity. *Phys. Rev. D* **2003**, *68*, 044015. [CrossRef]
53. Estrada-Jimenez, S.; Garcia-Compean, H.; Obregon, O.; Ramirez, C. Twisted Covariant Noncommutative Self-dual Gravity. *Phys. Rev. D* **2008**, *78*, 124008. [CrossRef]
54. Barbosa, G.D.; Pinto-Neto, N. Noncommutative geometry and cosmology. *Phys. Rev. D* **2004**, *70*, 103512. doi:10.1103/PhysRevD.70.103512.
55. Lopez-Dominguez, J.C.; Obregon, O.; Sabido, M.; Ramirez, C. Towards Noncommutative Quantum Black Holes. *Phys. Rev. D* **2006**, *74*, 084024. [CrossRef]
56. Rezaei-Aghdam, A.; Darabi, F.; Rastkar, A.R. Noncommutativity in quantum cosmology and the cosmological constant problem. *Phys. Lett. B* **2005**, *615*, 141–145. [CrossRef]

57. Khosravi, N.; Sepangi, H.R.; Sheikh-Jabbari, M.M. Stabilization of compactification volume in a noncommutative mini-super-phase-space. *Phys. Lett. B* **2007**, *647*, 219–224. [CrossRef]
58. Vakili, B.; Khosravi, N.; Sepangi, H.R. Bianchi spacetimes in noncommutative phase-space. *Class. Quantum Grav.* **2007**, *24*, 931–949. [CrossRef]
59. Ortiz, C.; Mena-Barboza, E.; Sabido, M.; Socorro, J. (Non)commutative isotropization in Bianchi I with barotropic perfect fluid and Lambda cosmological. *Int. J. Theor. Phys.* **2008**, *47*, 1240. [CrossRef]
60. Aguero, M.; Ortiz, J.A.A.S.C.; Sabido, M.; Socorro, J. Non commutative Bianchi type II quantum cosmology. *Int. J. Theor. Phys.* **2007**, *46*, 2928–2934. [CrossRef]
61. Vakili, B.; Sepangi, H.R. Generalized uncertainty principle in Bianchi type I quantum cosmology. *Phys. Lett. B* **2007**, *651*, 79–83. [CrossRef]
62. Bina, A.; Atazadeh, K.; Jalalzadeh, S. Noncommutativity, generalized uncertainty principle and FRW cosmology. *Int. J. Theor. Phys.* **2008**, *47*, 1354–1362. [CrossRef]
63. Bastos, C.; Bertolami, O.; Dias, N.C.; Prata, J.N. Phase-Space Noncommutative Quantum Cosmology. *Phys. Rev. D* **2008**, *78*, 023516. [CrossRef]
64. Vakili, B. Dilaton Cosmology, Noncommutativity and Generalized Uncertainty Principle. *Phys. Rev. D* **2008**, *77*, 044023. [CrossRef]
65. Maceda, M.; Macias, A.; Pimentel, L.O. Homogeneous noncommutative quantum cosmology. *Phys. Rev. D* **2008**, *78*, 064041. [CrossRef]
66. Vakili, B. Cosmology with minimal length uncertainty relations. *Int. J. Mod. Phys. D* **2009**, *18*, 1059–1071. doi:10.1142/S0218271809014935.
67. Guzman, W.; Sabido, M.; Socorro, J. On Noncommutative Minisuperspace and the Friedmann equations. *Phys. Lett. B* **2011**, *697*, 271–274. [CrossRef]
68. Obregon, O.; Quiros, I. Can noncommutative effects account for the present speed up of the cosmic expansion? *Phys. Rev. D* **2011**, *84*, 044005. [CrossRef]
69. Cordero, R.; Garcia-Compean, H.; Turrubiates, F.J. Deformation quantization of cosmological models. *Phys. Rev. D* **2011**, *83*, 125030. [CrossRef]
70. Oliveira-Neto, G.; de Oliveira, M.S.; Monerat, G.A.; Silva, E.V.C. Noncommutativity in the early Universe. *Int. J. Mod. Phys. D* **2016**, *26*, 1750011. [CrossRef]
71. Espinoza-García, A.; Torres-Lomas, E.; Pérez-Payán, S.; Díaz-Barrón, L.R. Noncommutativity in Effective Loop Quantum Cosmology. *Adv. High Energy Phys.* **2019**, *2019*, 9080218. [CrossRef]
72. Curtright, T.L.; Zachos, C.K. Quantum Mechanics in Phase Space. *Asia Pac. Phys. Newslett.* **2012**, *1*, 37–46. doi:10.1142/S2251158X12000069.

Review

Weyl Curvature Hypothesis in Light of Quantum Backreaction at Cosmological Singularities or Bounces

Bei-Lok Hu

Maryland Center for Fundamental Physics and Joint Quantum Institute, University of Maryland, College Park, MD 20742-4111, USA; blhu@umd.edu

Abstract: The Weyl curvature constitutes the radiative sector of the Riemann curvature tensor and gives a measure of the anisotropy and inhomogeneities of spacetime. Penrose's 1979 Weyl curvature hypothesis (WCH) assumes that the universe began at a very low gravitational entropy state, corresponding to zero Weyl curvature, namely, the Friedmann–Lemaître–Robertson–Walker (FLRW) universe. This is a simple assumption with far-reaching implications. In classical general relativity, Belinsky, Khalatnikov and Lifshitz (BKL) showed in the 70s that the most general cosmological solutions of the Einstein equation are that of the inhomogeneous Kasner types, with intermittent alteration of the one direction of contraction (in the cosmological expansion phase), according to the mixmaster dynamics of Misner (M). How could WCH and BKL-M co-exist? An answer was provided in the 80s with the consideration of quantum field processes such as vacuum particle creation, which was copious at the Planck time (10^{-43} s), and their backreaction effects were shown to be so powerful as to rapidly damp away the irregularities in the geometry. It was proposed that the vaccum viscosity due to particle creation can act as an efficient transducer of gravitational entropy (large for BKL-M) to matter entropy, keeping the universe at that very early time in a state commensurate with the WCH. In this essay I expand the scope of that inquiry to a broader range, asking how the WCH would fare with various cosmological theories, from classical to semiclassical to quantum, focusing on their predictions near the cosmological singularities (past and future) or avoidance thereof, allowing the Universe to encounter different scenarios, such as undergoing a phase transition or a bounce. WCH is of special importance to cyclic cosmologies, because any slight irregularity toward the end of one cycle will generate greater anisotropy and inhomogeneities in the next cycle. We point out that regardless of what other processes may be present near the beginning and the end states of the universe, the backreaction effects of quantum field processes probably serve as the best guarantor of WCH because these vacuum processes are ubiquitous, powerful and efficient in dissipating the irregularities to effectively nudge the Universe to a near-zero Weyl curvature condition.

Keywords: weyl curvature hypothesis; early universe cosmology; singularity and bounce; cyclic universe; quantum fields; backreaction effects

Citation: Hu, B.-L. Weyl Curvature Hypothesis in Light of Quantum Backreaction at Cosmological Singularities or Bounces. *Universe* **2021**, *7*, 424. https://doi.org/10.3390/universe7110424

Academic Editor: Paulo Vargas Moniz

Received: 12 October 2021
Accepted: 29 October 2021
Published: 7 November 2021

Publisher's Note: MDPI stays neutral with regard to jurisdictional claims in published maps and institutional affiliations.

Copyright: © 2021 by the author. Licensee MDPI, Basel, Switzerland. This article is an open access article distributed under the terms and conditions of the Creative Commons Attribution (CC BY) license (https://creativecommons.org/licenses/by/4.0/).

Preface

When Professor Moniz invited me last year to write a paper for this special issue with some review or perspective emphasis I could not identify a topic of current interest and importance in quantum cosmology on which I have done enough work to write about. Then in July, Professor Hendrik Ulbricht brought to my attention that this August is Professor Penrose's 90th Birthday (b. 8 August 1931). That got me thinking through some important themes in cosmology from Penrose's viewpoint—his concerns in a broad range of research topics in gravitation, quantum physics, black holes and cosmology, as lucidly presented in many semi-popular books (the last two being [1,2]), are all of fundamental interest. This is how I closed in to the title theme of this paper: The Weyl Curvature Hypothesis (WCH) [3–6].

How does WCH fare when quantum backreaction is added to our considerations? Backreaction in gravitational physics enters at four levels: classical, semiclassical, stochastic and quantum gravity. In classical gravity it refers to the effects of the inhomogeneous modes on the homogeneous or long wavelength (infrared) modes; in quantum cosmology it refers to the effects of the inhomogeneous modes truncated in a mini-superspace approximation. Backreaction at the semiclassical and stochastic levels has a different meaning. It refers to the effects of quantum field processes such as particle creation [7–9] and trace anomaly [10–12] on the structure and dynamics of a classical background spacetime. This is a subject which I explored between 1977–1983–1995 in my work first with Parker [13,14] and Hartle [15,16] from 77–80 (based on [17], continued by Hartle [18,19] and Anderson [20,21] from 1980–1984), then upgraded to a closed-time-path, in-in, or Schwinger-Keldysh treatment with Calzetta in 1987 [22], including noise and fluctuations in 1994 [23], with Matacz in 1994 [24] and with Sinha in 1995 [25]. This was continued by Campos and Verdaguer in 1994–1996 [26,27]. These works explored the backreaction effects of the trace anomaly and particle creation in the early universe near the Planck time, at the threshold of quantum gravity (defined as theories for the microscopic constituents of spacetime and matter). The Weyl curvature hypothesis (WCH) [3] will be examined in the context of semiclassical and stochastic gravity [28] in Section 5.

Taking the time to write this essay has been a good way to update my knowledge since my 1982–1983 papers on vacuum viscosity [29] and gravitational entropy [30] in the areas of gravitational entropy, the Weyl curvature hypothesis, the cosmological singularity or avoidance of singularity, as predicted by different theories, and how vacuum viscosity due to the backreaction of quantum field processes at the Planck time, such as pertaining to the trace anomaly and vacuum particle creation, would bear on these issues. I wanted to see how the WCH fares with other later developed models such as the cyclic cosmology of classical general relativity (GR) (Section 2.1) or particle physics (Section 2.2) origins, singularity avoidance solutions from quantum cosmology (Sections 3.1 and 3.2), in particular, the 'Big Bounce' solutions of loop quantum cosmology (Section 3.3) and quantum phase transitions in asymptotic freedom and causal dynamical triangulation (Section 4.1)—whether there are indications that the WCH holds in these contexts and, in some cases, whether it has proven to be desirable or even necessary. (If the reader only wants to find out how the WCH and gravitational entropy fare in semiclassical gravity, he/she can just selectively read Sections 1.2, 2.3, 3.1 and 5.)

It is not my intention to cover the full range of topics mentioned above, aided with extensive literature, like a review. Far from it. I won't even repeat what Penrose said in his original proposal of the WCH and how he defined gravitational entropy, or how he argued for a conformal cyclic cosmology in classical general relativity. Instead I would urge interested readers to read his original papers for insights which can never be reproduced. For papers by other authors which I find relevant to our central themes here I also prefer to keep their original words rather than add my paraphrase or re-statement. The nature of this article is probably closer to an outline, a selective highlight, a personal guide, as the topics chosen certainly reflect my own interests or what I view as important, which could be very different from other practitioners'.

1. Weyl Curvature Hypothesis and Gravitational Entropy

Penrose pointed out in his 1979 essay [3] that if our Universe had begun close to what the Friedmann–Lemaître–Robertson–Walker (FLRW) models describe, in a state of highest degree of isotropy and homogeneity, corresponding to zero Weyl curvature, then it is something extremely special. He reached this conclusion by comparing the entropy of matter we see today, measured by the Hawking radiation, emitted by black holes, to the entropy of gravity, measured by the Weyl curvature tensor squared integrated over 3-volumes. Gravitational entropy (GEnt) defined as such would increase in time from zero in the beginning (which could be a singularity, the 'Big Bang', or a state of minimal volume, a 'bounce', see below) and provide a cosmological arrow of time. Penrose's proposals for the non-activation of gravitational degrees of freedom registered in the Weyl curvature, at the Big Bang, are referred to as the Weyl Curvature Hypothesis (WCH).

Penrose's position on this theme has not changed in 40 years, as the following statement from his 2018 essay shows [6] "In such a situation (the universe in a collapsing phase), we expect density irregularities to increase as the universe contracts, and the FLRW approximation would get worse and worse. In the late stages of the collapse, these irregularities would result in numerous black holes. Time reversing this picture (and taking into account the invariance of Einstein's equations under time-reversal), we find that the singular origin that we find in the FLRW models is something extremely special." While today we see the Planck distribution of matter suggesting a state of maximum matter entropy, "In contrast with the behaviour of a gas in a box, for example, where maximum entropy would be pictured as something with great spatial uniformity, gravitating bodies, such as systems of stars, would tend to clump more and more in their spatial distribution, as their dynamical time evolution proceeds, representing an increase in the gravitational entropy".

The four themes I listed above are already laden in these descriptions: WCH, GEnt, also the Bang-vs-Bounce issue. The reason is, for people who prefer to see the universe not going to a singularity, they have to come up with some suitable ways to avoid it, such as through processes in particle physics or in quantum gravity. We mention four of them in this essay: pre-Big Bang, loop quantum cosmology (LQC), causal dynamical triangulation (CDT) and asymptotic freedom (AsyF)[1].

For people who want our future universe not just to bounce once or twice, but to continue to bounce forever without suffering much attenuation, or stumbling into a singularity, in the so-called 'cyclic cosmologies', the requirements are even more stringent. What happens to the universe at the end of the contraction phase requires extra careful scrutiny. The highly regular FLRW type of approach to the singularity or bounce would be preferred, because only then would the deviations be kept under control cycle after cycle. This is in turn because gravitational forces tend to clump irregularities, so the next cycle would be more difficult to control than the last.

Most people would agree that quantum effects are important at the beginning and ending stages of the universe, either entering a singularity or with a minimal scale. These developments and considerations add new impetus and interests to Penrose's WCH-GEnt, originally formulated without quantum considerations. To the above mentioned subjects we shall also add semiclassical and quantum backreaction considerations. The former refers to the backreaction of quantum field processes on the dynamics of spacetime. They need be included because these processes were dominant and had exerted significant effects at the Planck scale. The latter pertains to LQC: It concerns the effects of the inhomogeneous quantum modes in the superspace of 3-geometries, so far largely ignored in the minisuperspace formulations of LQC.

1.1. Classical GR

Before we begin our studies of WCH in a quantum setting it is important to remind ourselves of the two very important theoretical achievements in the 60s in classical general relativity (GR), which underlie all investigations in black holes and cosmology built upon the GR theoretical framework: the singularity theorems from 1965–1968 of Penrose [44], Hawking [45] and Geroch [46], and the general solutions of the Einstein equations near the cosmological singularity in the work of Belinsky, Khalatnikov, Lifshitz (BLK) from the early 60s (when the authors preferred a singularity-free scenario, later corrected, after the Penrose-Hawking singularity theorems) to the early 1970s [47–49] with the now famous BKL approach to cosmological singularity, and Misner's mixmaster universe [50,51] and minisuperspace quantum cosmology [52] from 1969 to 1972, and the 'velocity-dominated' approach by Eardley, Liang and Sachs [53].

1.1.1. Singularity Theorems: Penrose-Hawking-Ellis-Geroch

The Penrose 1965 singularity theorem for black holes extended by Hawking to cosmological singularities are known as the Penrose-Hawking singularity theorems in general relativity [54]. They are described in more pedagogical terms in the book by Hawking and

Ellis [55], the premium monograph on global analysis methods in general relativity. These theorems preclude singularity-free "bounce" from occurring in the presence of any type of matter source with stress-energy tensor satisfying a very general condition of local energy non-negativity. For a detailed explanation of its contents with an extensive bibliography on this fundamental topic, see the 2015 review of [56] in a mid-century celebration of Penrose's 1965 paper.

1.1.2. Approach to Cosmological Singularity: BKL-Misner

In addition to the primary sources of the Belinsky-Khalatnikov-Lifshitz -Misner enterprise mentioned above, the 1975 book by Ryan and Shepley [57] has an excellent introduction to spatially-homogeneous (Bianchi) cosmologies. See also the review of Barrow and Tipler [58]. For a discussion of the classical BKL-M mixmaster chaotic dynamics (billiard ball in minisuperspace), see, e.g., [59–63][2]. For the latest developments of this respect in relation to indefinite Kac-Moody algebra, see [66]. The role of chaos in the decoherence of quantum cosmology and the appearance of a classical spacetime has been pointed out by Calzetta [67].

Since many topics we shall discuss are related to the state of the universe near the singularity (or how it negotiates into a bounce), we need some basic knowledge of the BKL-M behavior. The most general solutions to the Einstein equation near and toward the cosmological singularity of the Kasner type (not the Friedmann class where matter does matter) is the inhomogeneous Kasner universe (with two directions contracting and one direction expanding), interspersed with the mixmaster behavior, i.e., switching of the direction of expansion from one axis to another in very short time intervals. This so-called "inhomogeneous mixmaster solution" represents the generic behavior of the universe toward the cosmological singularity. This result has been reworked by many authors employing different methods, e.g., Uggla et al. [68] constructed an invariant set which forms the local past attractor for the evolution equations, a framework for proving rigorous theorems concerning the asymptotic behavior of spatially inhomogeneous cosmological models. A more direct approach relies on methods in dynamical systems [69]. The BKL-M behavior is also supported by Garfinkle's numerical simulations in vacuum spacetimes with no symmetries [70].

Three main features in the approach to cosmological singularity are, therefore: (1) "Matter doesn't matter" in the BKL-Kasner classes of extrinsic curvature ('velocity')- dominated spacetimes[3] while matter does matter in the FLRW classes. (2) Extrinsic curvature (from expansion or contraction) dominates over the intrinsic (3-geometry) curvature, except in brief intervals, which are known as the 'mixmaster bounces' (the universe point bouncing against the 3-fold symmetric potential walls, receding during contraction) in mini-superspace; (3) Inhomogeneous mixmaster behavior means that every point on the 3-geometry undergoes an independent mixmaster dynamics.

We begin with some basics of **Bianchi cosmology**. Bianchi classified all spatially homogeneous spacetimes into nine types [73] (some types are further distinguished by a sub 0 or h, such as Type VII$_h$, etc.) [57]. The class of Bianchi Types I–IX spaces, especially the Type I and Type IX spaces, encompasses many important classical cosmological models for the early universe.

The line elements of the Bianchi universes are given by

$$ds^2 = -dt^2 + \sum_{a,b=1}^{3} \gamma_{ab}(t)\sigma^a(x)\sigma^b(x), \qquad (1)$$

where $\gamma_{ab}(t)$ is the metric tensor and $\sigma^a(x)$ are the invariant basis one-forms on the homogeneous hypersurfaces, satisfying the structure condition $d\sigma^a = \frac{1}{2}C^a_{bc}\sigma^b \wedge \sigma^c$, where C^a_{bc} are the structure constants of the underlying Lie group. For a diagonal type-IX (mixmaster) universe, $C^a_{bc} = \epsilon_{abc}$ the totally antisymmetric tensor, and $\gamma_{ab} = l_a^2 \delta_{ab}$, where l_a are the three principal radii of curvature. The case when two of the three l_a's are equal is called the Taub universe, the case when all three are equal gives the closed Friedmann–Lemaître–Robertson–Walker (FLRW) universe.

The nonrotating, diagonal **mixmaster universe** with three principal radii of curvature $\ell_{a(=1,2,3)}$ form a minisuperspace [50] which can be represented pictorially by the shape θ and deformation β parameters, related to ℓ_a by

$$\ell_{1,2} = \ell_0 \exp[\beta \cos(\theta \mp \pi/3)], \quad \ell_3 = \ell_0 \exp(-\beta \cos \theta)]. \tag{2}$$

where $\ell_0^3 \equiv \ell_1 \ell_2 \ell_3$. With threefold symmetry, the origin of this space gives the FLRW universe, while points along the three axes ($\theta = 0, 120, 240°$) constitute the family of the Taub universes. Along any axis, say $\theta = 0$, increasing β in the positive sense gives an "oblate" configuration while negative β gives a "prolate" configuration. The dynamics of the mixmaster universe is depicted by trajectories in this minisuperspace. The velocity of the universe point is determined by the extrinsic curvature, with free motion depicting the Kasner universe. The influence of spatial anisotropy (intrinsic curvature) is depicted by a set of moving potential walls. The universe point bouncing off these walls signifies a shifting in the contracting Kasner axis, while wandering into one of the three corners signifies an oscillating and "flattening pancake" (oblate) configuration There is also one situation when the universe point bounces off the three walls at a special incidence angle so that it can continue to hover loosely around the origin. These characteristic regimes of the mixmaster universe are called, respectively, (a) the "bounce" (b) the "channel run" and (c) the "quasi-isotropic" solutions. These solutions have the approximate geometric configurations as the three limiting cases of the Taub universe: i.e., (a) $\beta > 0$, (b) $\beta < 0$, and (c) $\beta = 0$, respectively. For any massless field in a spacetime which is a vacuum solution to Einstein's equations R = 0, quantum effects of curvature are important irrespective of the type of coupling.

The Bianchi Type I universe has zero intrinsic curvature but nonzero extrinsic curvature from its expansion and contraction, whose rates differ in three directions. The line element is given by

$$ds^2 = -dt^2 + \sum_{i,j=1}^{3} \ell_{ij}^2(t) dx^i dx^j = a^2(\eta)\left[-d\eta^2 + \sum_{i,j=1}^{3} e^{2\beta(\eta)_{ij}} dx^i dx^j\right] \tag{3}$$

where $\eta = \int dt/a$ is the conformal time and β_{ij} is a symmetric, traceless 3×3 matrix describing the anisotropy of the geometry. For Bianchi Type I one can choose a coordinate such that β_{ij} is diagonal[4] with the three elements β_i. The line element of Bianchi Type I universe is thus

$$ds^2 = -dt^2 + \sum_{i=1}^{3} \ell_i^2(t)(dx^i)^2 \tag{4}$$

where $\ell_i = a(t)e^{\beta_i}$ is the scale factor in the x_i direction. This is a generalization of the spatially flat isotropic FLRW universe to the case where the universe expands anisotropically. A useful quantity for this measure is the anisotropy parameter $Q_\beta \equiv -\frac{1}{2}\sum_{i>j}(\dot{\ell}_i/\ell_i - \dot{\ell}_j/\ell_j)^2$, where an overdot denotes d/dt. When there is no matter present a solution to the vacuum Einstein equation exists, called the **Kasner universe**:

$$\ell_i(t) = t^{p_i}, \quad \text{where} \quad \sum_{i=1}^{3} p_i^2 = \sum_{i=1}^{3} p_i = 1. \tag{5}$$

We see from this relation that the universe expands when two of the p_is are positive and one negative, and contracts when two of the p_is are negative and one is positive. The Kasner solution is important because it is a generic behavior of the universe at every point in space near the singularity where the most general solution of the Einstein equation [47–49] is found to be an inhomogeneous Kasner solution. It is also known as a 'velocity-dominated solution' [53] reflecting the fact that near the cosmological singularity the extrinsic curvature (measuring the time rate of change of the scale factors) dominates over the intrinsic curvature (in the 3-geometry), thus the saying "matter doesn't matter" in the BKL-M generalized Kasner class of solutions near the cosmological singularity, in contradistinction to the FLRW class, which requires the presence of matter. Defining the isotropic expansion rate $\alpha = \ln a$ then we see $p_i = \alpha + \beta_i$. The spatially flat FLRW universe corresponds to the case $\beta = 0$ with scale factor $a = e^{\alpha}$.

1.1.3. Weyl Curvature

For an analysis of the Weyl curvature tensor in spatially-homogeneous cosmological models of classical general relativity, see Barrow and Hervik 2002 [75]. For a review of the entropy of universe with gravitational and thermal contributions, the entropy of black holes and cosmic horizons entropy, see, Grøn 2012 [76].

1.1.4. Gravitational Entropy: Weyl and Other Measures

Assuming the interested reader would have read Penrose's original paper on the WCH, we will not repeat how he defines the gravitationl entropy via the Weyl curvature. Instead, to expand the horizon somewhat we mention a few alternative definitions of gravitational entropy.

Kinematic vs. Weyl singularity:

Lim et al. 2007 [77] and Coley et al. 2009 [78] studied what they called the 'kinematic singularities', namely, the Hubble expansion scalar, the acceleration vector, the shear tensor and the vorticity tensor in the decomposition of the covariant derivative of the 4-velocity of a congruence of worldlines.

Bel–Robinson tensor

Clifton, Ellis and Tavakol [79] proposed a measure of gravitational entropy based on the square-root of the Bel–Robinson tensor of free gravitational fields, which is the unique totally symmetric traceless tensor constructed from the Weyl tensor. (For the square-root to exist as a unique factorization of the Bel–Robinson, the spacetime is required to contain gravitational fields that are only Coulomb-like or only wavelike.) These authors show that their entropy measure (i) is in keeping with Penrose's Weyl curvature hypothesis, (ii) reduces to the usual Bekenstein–Hawking value measure for Schwarzschild black hole and (iii) evolves like the Hubble weighted anisotropy of the free gravitational field for scalar perturbations of a Robertson–Walker geometry thus increases as structure formation occurs, and (iv) for the Lemaître Tolman-Bondi inhomogeneous cosmological models, they found conditions under which the entropy increases.

Kullback-Leibler relative information entropy

While most works try to place the gravitational entropy defined by the Weyl curvature in relation to thermodynamic entropy, Hosoya, Buchert and Morita [80] try to relate it to information entropy. They introduced the Kullback-Leibler relative information entropy to express the distinguishability of the local inhomogeneous mass density field from its spatial average on arbitrary compact domains. Comparing the Kullback–Leibler entropy to the Weyl curvature tensor invariants. Li et al. [81,82] calculated these two measures by perturbing the standard cosmology. Up to the second order in the deviations from a homogeneous and isotropic spacetime they find that they are correlated and can be linked via the kinematical backreaction of a spatially averaged universe model.

1.2. Semiclassical Gravity: Entropy of Gravitons and WCH

The Weyl curvature tensor measures the anisotropy and inhomogeneities of spacetime. Since it is the radiative sector of the Riemann tensor, it measures the curvature associated with gravitational waves, which, of course, are the weakly inhomogeneous perturbations off a background spacetime. Quantizing the linear perturbations from a background gives rise to gravitons. The short wavelength sector under the Brill-Hartle-Isaacson average behaves like radiation fluid with the same equation of state as photons, namely, $p = \rho/3$, with p the pressure and ρ the energy density. Gravitons, like photons, have two polarizations but instead of spin-1 they are spin-2 particles. For these reasons gravitons are the easiest entry point to see the connection between WCH and gravitational entropy.

By semiclassical we mean a quantum field, here, the graviton spin-2 field, propagating in a classical background spacetime. (We also distinguish between quantum field theory in curved spacetime, under the test field situation where the background is prescribed, and semiclassical gravity where the background spacetime and the quantum field are determined in a self-consistent manner). By quantum we mean both the matter field and the spacetime are quantized. We reserve the term 'quantum gravity' to refer to theories about the microscopic structures of spacetime at the Planck energy. We will explain what quantum cosmology means in individual cases later, e.g., quantum cosmology of the minisuperspace as in [52] refers to the situation where the 3-geometry is quantized, regardless of whether the spacetime is an emergent entity or a fundamental one. However, 'loop quantum cosmology' or 'string cosmology' is trickier, because 'quantum' could mean quantizing the metric or the connection of GR, and if GR is an effective low energy theory, the quantized collective excitations are like phonons, a far cry from the fundamental constituents, or the 'atoms', of spacetime. It could also mean the cosmology (after the Planck time) derived from a more fundamental theory of spacetime based on strings, loops, sets or simplices. Explaining how the spacetime manifold emerges from the interaction of these fundamental constituents is a highly nontrivial task, probably the most severe challenge facing the proponents of these theories of quantum gravity [83]. We shall return to this point in the concluding remarks.

1.2.1. Graviton Backreaction in FLRW Universes

We consider this issue at two levels: (1) At the quantum field theory in curved spacetime (QFTCST) [84–87] or test field level, we ask, under what conditions would there be particle creation from the vacuum. It is easy to see that there is no particle production for a massless conformally coupled field in an isotropic and homogeneous universe, because such a spacetime is conformally-flat. But graviton production can happen. This is because each of the two polarizations of the graviton behaves like a massless minimally coupled scalar field [88,89]. Graviton production and its effects have been studied by many authors [13,17,90]. As far as the *matter* entropy budget is concerned, as mentioned above, with gravitons acting like photons (notwithstanding the big difference in their scattering cross sections), their production increases the thermal entropy of the matter content.

(2) To see how this is related to the *gravitational* entropy measured by the Weyl curvature tensor, one needs to find out the backreaction of matter fields on the dynamics of spacetime. This entails finding a self-consistent solution of the equations of motion for both the matter fields and the spacetime they live in, namely, the semiclassical Einstein equation. This backreaction constitutes a channel where the energy and entropy of spacetime can be related to that of the quantum fields, with particle creation acting like a 'transducer' between the gravity and matter sectors. This way of thinking was first proposed by Hu in 1983 [30], having worked out the anisotropy damping scenario due to conformal massless particle creation in Bianchi type I universes earlier with Hartle and Parker. We shall discuss these results in Section 4.

Here, with the FLRW universe, the problem is simpler, since the background spacetime is not only homogenous but also isotropic. In the beginning the Weyl curvature tensor is nonzero because of the perturbations, the gravitational waves or the gravitons, imparting the universe with some gravitational entropy. At the end, after graviton production ceases, the spacetime returns to the FLRW universe with zero Weyl curvature tensor. One can see that the increase of matter entropy from the production process is accompanied by a decrease of gravitational entropy. This is discussed in [91,92].

Now, what causes the graviton production to cease? The answer comes from solving the semiclassical Einstein equations. Doing so, Hu and Parker 77 [13] show that the backreaction of created gravitons leads to a change in the power expansion law $a = t^q$ towards $q = 1/2$, i.e., radiation dominated. And a radiation-dominated RW universe admits no particle creation because the time-dependent frequency of the normal modes of the quantum field contains a term $d^2a/d\eta^2$ where η is the conformal time. A radiation-dominated RW universe behaves like $a = \eta$, rendering $d^2a/d\eta^2 = 0$. Thus, after a certain relaxation time the behavior of the gravitons created corresponds to the evolution of classical radiation in an isotropic universe. With backreaction we see that as the gravitational entropy density goes to zero, graviton production will cease and no further backreaction will take place.

The fact that the backrection of graviton production can alter the equation of state of matter was pointed out by Grishchuk in 1975 [90] and shown with field theory calculations by Hu and Parker [13] and Hartle [17]. We will continue this description of the backreaction of particle creation on background spacetime dynamics in Sec. V and discuss how the damping of anisotropy and inhomogeneities of spacetime bears on gravitational entropy and the WCH.

1.2.2. Entropy of Gravitons in the Gowdy Universe

Husain [93] paraphrased Penrose's ideas of gravitational entropy based on the WCH into three conditions: "If the Weyl curvature is to be identified with entropy, it should satisfy the following criteria that are normally associated with entropy. (1) It should increase monotonically with respect to a time coordinate from the initial singularity where it is finite or zero. (2) Its value should increase with the number of quanta of the gravitational field. (3) Clumped configurations of matter or gravitational quanta should correspond to higher Weyl curvature values than unclumped configurations."

Conditions (1) and (2) are illustrated in the example above, regarding graviton production and its backreaction on the spacetime in a FLRW universe. For conditions (2) and (3), we can see an example for the Gowdy universe [94,95]. The Gowdy three-torus (T^3) universe is an exact solution to the vacuum Einstein equations interpreted to be a single polarization of gravitational waves propagating in an anisotropic, spatially inhomogeneous background.

Appying Hamiltonian quantization via the Arnowitt-Deser-Misner (ADM) method Berger [96,97] studied the production of gravitons from empty space. At large times, graviton number is well defined since the solution is in WKB form. The creation process produces the anisotropic collisionless classical radiation. Near the singularity, the model behaves like an empty Bianchi Type I universe at each point in space (locally Kasner).

Beginning with the square of the curvature as an operator, Hussain calculated its expectation values in states of clumped and unclumped gravitons. These results led him to conclude that the curvature contains information about the entropy of the gravitational field in this class of quantum cosmology models.

Finally, in the context of quantum cosmology based on the Wheeler-DeWitt (WdW) equation we mention the work of Grøn and Hervik [98] These authors introduced a measure of 'gravitational entropy' behaving in accordance with the second law of thermodynamics and investigated its evolution in Bianchi type I and Lemaître–Tolman universe models. They work with the quantity $\Pi \equiv P\sqrt{h} = \sqrt{h}(\frac{C^2}{R^2})^{\frac{1}{2}}$ which is the ratio between the Weyl tensor squared C^2 and the Ricci tensor squared R^2, integrated over $dV = \sqrt{h}d^3x$ in a comoving coordinate x^i, where h is the 3-volume element. They also considered the expectation value of the Weyl curvature squared and the Ricci curvature squared, taken with respect to the wave functions of the universe satisfying the Wheeler-DeWitt equation. They then investigated whether a quantum calculation of initial conditions for the universe based upon the Wheeler–DeWitt equation supports Penrose's Weyl curvature conjecture or not for the Bianchi type I universe models with dust and a cosmological constant and the

Lemaître–Tolman universe models. They investigated two versions of the hypothesis. The local version based on P^2 fails to support the conjecture whereas a non-local entity based on Π^2 showed more promise concerning the conjecture. Grøn and Hervik commented that their findings seem to be corroborated by that of Pelvas and Lake [99] on the one hand, who showed that local entities like P^2 cannot be a measure of gravitational entropy, and on the other hand, by that of Rothman and Anninos [100,101], who showed that a nonlocal quantity they constructed using the volume of the phase space could. The investigations of these authors seem to suggest that while one cannot say that the Weyl tensor is directly linked to the gravitational entropy, a certain non-local entity which is constructed from it has an entropic behavior, and reflects the tendency of the gravitational field to produce inhomogeneities.

We shall say some more about ADM and WdW, and about loop quantum cosmology in Section 3. But we will not discuss quantum geometry and gravitational entropy (see, e.g., [102]) applied to black holes or in the much studied holographic entanglement entropy vein [103] for spacetimes with boundaries.

2. Cyclic Cosmology and Trace-Anomaly Induced Bounces

2.1. Penrose's Conformal Cyclic Cosmology (CCC)

This theory is Penrose's recent favorite in terms of cosmological models–favorite in that after his 2005 proposal he has gathered his thoughts on entropy, black holes, time and cosmology in a 300 page 2010 book [1] which dwells on these foundational issues of Nature in a lucid and inspiring way. In his more recent 2017 book [2] of over 500 pages, "Fashion, Faith, and Fantasy in the New Physics of the Universe" Penrose refers (in Section 4.3) to his "conformal cyclic cosmology" as an idea so fantastic that it could be called "conformal crazy cosmology"[5].

To see what CCC is, the reader may benefit from reading just 4 pages of exposition in [4]. For the background it is useful to know something about singularities in GR, e.g., in [104,105], and in particular, the class of isotropic cosmological singularities, where the Weyl tensor is finite or preferably zero in the WCH. At the isotropic singularity even though the Ricci tensor related to matter may approach infinity, the Weyl curvature remains finite. Goode and Wainwright [106–108] proposed a definition of isotropic cosmological singularities which is adopted by Tod [109,110] whose work Penrose refers to. The essential idea is that the physical metric which is singular at cosmic time $T = 0$, is conformally related to a metric which is regular at $T = 0$, i.e., the singularity arises solely due to the vanishing of the conformal factor. The conformal transformation which is used to define an isotropic singularity is an essential ingredient in Penrose's conformal cyclic cosmology. Namely, the Big Bang can be conformally represented as a smooth spacelike 3-surface, across which the space-time is, in principle, extendable into the past in a conformally smooth way.

According to Penrose [6], the Big Bang (BB) of our 'aeon' is the *conformal continuation* of the remote exponential expansion of a previous aeon. There is a pre-BB aeon, and after our future Big Crunch (BC)—actually Penrose prefers to see a long exponential expansion (EE) to our future—there is a post-BC or post-EE aeon. In this sense a cyclic cosmology scenario naturally emerges, where the aeons are linked by conformal transformations.

2.2. Cyclic Cosmology from Particle/String Models. Big Crunch

No classical matter can exist with a pressure p to energy density ρ ratio $w \equiv p/\rho > 1$. The stiffest Zel'dovich equation of state with $w = 1$ is the upper limit of classical matter because that is when the speed of sound equals the speed of light. In the recent two decades cosmological models with a scalar field providing an effective equation of state $w > 1$ have drawn some interest, motivated by the possibility of cyclic cosmologies based on string theory-related models, notably the Pre-Big Bang model of Gasperini and Veneziano [111–113] and the ekpyrotic cosmologies of Steinhardt and Turok [114,115]. The designers of these theories want to see the universe contract towards a big crunch singularity in a smooth homogeneous and isotropic (FLRW) manner, so it can be reborn in an equally smooth manner to be able to sustain a large number of cycles, ideally ad infinitum.

2.2.1. Cyclic and Ekpyrotic Cosmologies

Ekpyrotic cosmologies solve the flatness problem in the contraction phase in a way similar to how inflation solves it in the expansion phase. Both are looking for the conditions where the time rate of change of the scale factor (Hubble) becomes close to a constant for a period of time (around 65 e-folding time to be compatible with the entropy contents of the observed universe). It is easy to see this from two simple equations, the Einstein G_{00} equation for the dynamics of spacetime and the continuity equation (divergence-free condition) for matter. In a FLRW universe with scale factor a and Hubble expansion rate $H = \dot{a}/a$, where an overdot denotes taking the derivative with respect to cosmic time, the Einstein $G_{00} = 8\pi G\rho$ equation reads: $H^2 = -k/a^2 + \frac{8\pi G}{3}\rho$, where $k = 0, 1, -1$ correspond to the flat, closed and open universes and ρ, the total energy density, contains different kinds of classical matter and quantum fields, represented by equations of state with different values for $w \equiv p/\rho$. The familiar cases of $w = 0, 1/3$ in classical cosmology are for dust and radiation, respectively. We shall first show (i) why $w = -1$ for inflaton field, and then explain (ii) how a special kind of matter with an equation of state $w > 1$ affects the Hubble rate, and what it does for the ekpyrotic cosmologies.

It is well-known how inflation solves the flatness problem, namely, by assuming that the potential $V(\phi)$ of the inflaton field ϕ is sufficiently flat so that it rolls down the slope very slowly. That imparts a constant energy density in the Einstein equation which dominates over the other sources, which all drop as power laws of the scale factor during an expansion phase. The pressure-energy density ratio for a quantum field is $w \equiv p/\rho = \frac{\frac{1}{2}\dot{\phi}^2 - V}{\frac{1}{2}\dot{\phi}^2 + V}$. When $\frac{1}{2}\dot{\phi}^2 \ll V$ as in slow-roll inflation, $w \to -1$.

From the continuity equation we easily get $\rho \propto a^{-3(1+w)}$, namely, for non-relativistic matter $\rho_{nrel} \propto a^{-3}$, for radiation $\rho_{rad} \propto a^{-4}$, for curvature anisotropy $\rho_{anis} \propto a^{-6}$, and for some 'ultra-stiff' nonclassical matter with $w > 1$, $\rho_{ust} \propto a^{-r}$ where $r > 6$. Now it is clear why during contraction, as a gets smaller, contributions from this ultra-stiff matter with $w > 1$ dominate over other forms in governing the expansion rate.

This is why the cyclic and ekpyrotic cosmological models focus on the Big Crunch singularity and why scalar fields with equations of state $w > 1$ are of special interest. This 'ultra-stiff' matter in a homogeneous and isotropic universe can be realized by a scalar field ϕ with a steep negative potential energy $V(\phi) < 0$. The wits and skills in designing the potential of a quantum field to fit one's special needs– these activities which dominate the theoretical stage in inflationary cosmology–will also be active for this newer vein of cosmology. We refer interested readers to the nice review of Lehners [116] from which they can trace back to the original papers.

Of special interest to our concern is whether the mixmaster behavior might be altered by the effect of a scalar field with equation of state $w > 1$. The authors of Ref. [117] show that if $w > 1$, the chaotic mixmaster oscillations due to anisotropy (extrinsic curvature) and (intrinsic) curvature are suppressed, and the contraction is described by a homogeneous and isotropic Friedmann equation. Referring to string theory related models, they show that chaotic oscillations are suppressed in Z_2 orbifold compactification and contracting heterotic M-theory models if $w > 1$ at the crunch.

However, as pointed out by Barrow and Yamamoto [118] this result stands only when the ultrastiff pressures are isotropic. The inclusion of simple anisotropic pressures when pressures can exceed the energy density stops the isotropic Friedmann universe from being a stable attractor as an initial or final singularity is approached. Thus the situation with isotropic pressures, studied earlier in the context of cyclic and ekpyrotic cosmologies, is not generic, and Kasner-like behavior occurs when simple pressure anisotropies are present.

Including the consideration of stochastic effects, as invoked in Linde's 'chaotic' inflation or Starobinsky's stochastic inflation scenarios [37,39], the effective scalar field can climb up the potential in some regions of space, which leads to an increase in the energy density. However, as Brandenberger et al. [119] show, the backreaction of fluctuations that have already exited the Hubble radius will lead to a decrease in the effective energy density, which is strong enough to prevent eternal expansion in the late-time dark-energy phase of ekpyrotic cosmology.

2.2.2. Big Crunch with Dark Energy in Semiclassical Cosmology

As was mentioned above, how the universe evolves in the future– whether it approaches a singularity (Big Crunch), or it can avoid it (a Bounce), and how it approaches the singularity or the minimal radius, are important factors to consider in any cyclic cosmology proposal. Without assuming any exotic matter, Kennedy et al. [120] studied the influence of the vacuum expectation value of the energy-momentum tensor of a conformally-invariant free quantized field without particle creation on the spherically symmetric gravitational collapse of a dust cloud. Using the same method as used by Fischetti et al. [15] and Anderson [20] for the early universe, they found qualitatively similar results, namely, for certain values of an arbitrary parameter the collapse proceeds to singularity, although in a manner which differs from the classical (Oppenheimer-Snyder) behaviour; for other values of the parameter the collapse terminates before singularity and thereafter the dust cloud expands.

Ever since the discovery of a sizable fraction (0.7) of dark energy in the cosmological matter-energy contents, theories about the future of our universe need to take into account the bounds provided by stringent cosmological data. The combined 2018 Planck data together with other major observations give the dark-energy equation of state parameter $w_{de} = -1.03 \pm 0.03$, consistent with a cosmological constant ($w = -1$). One can regard this as weakly favoring 'phantom' dark energy ($w < -1$), which can lead to a future singularity called 'Big Rip' [121]. (A rather complex classification of different possible future singularities is given in, e.g., Ref. [122] Sec. IA.) Assuming that the dark energy which causes these singularities to behave like a perfect fluid, Carlson et al. [122] studied the effects which quantum fields and an $\alpha_0 R^2$ term in the gravitational Lagrangian have on future singularities. Special emphasis was placed on those values of α_0 which are compatible with the universe having undergone Starobinsky inflation [123] in the past, the same values which allow for stable solutions to the semiclassical backreaction equations in the present universe.

2.3. Semiclassical Gravity: Trace Anomaly Induced Bounces

For conformally-invariant fields in conformally-flat spacetimes in the conformal vacuum state, the regularized stress energy tensor is given by [20]

$$<0|\hat{T}_{ab}|0> = \frac{\alpha}{3}(g_{ab}R^c{}_{;c} - R_{;ab} + RR_{ab} - \frac{1}{4}g_{ab}R^2) \qquad (6)$$
$$+ \beta(\frac{2}{3}RR_{ab} - R_a{}^c R_{bc} + \frac{1}{2}g_{ab}R_{cd}R^{cd} - \frac{1}{4}g_{ab}R^2),$$

where R_{ab} is the Ricci tensor, R is the scalar curvature, and α and β are constants which depend on the number and types of fields present. Dimensional regularization gives

$$\alpha = \frac{1}{2880\pi^2}(N_S + 6N_v + 12N_V); \quad \beta = \frac{1}{2880\pi^2}(N_S + 11N_v + 62N_V) \qquad (7)$$

where N_S, N_v, N_V are respectively the number of scalar fields, four-component neutrino fields, and Maxwell fields included in one's theory.

Taking the trace of (6) we see that it is non-vanishing even though for conformal fields the classical stress energy tensor is traceless. This is why it is called the trace anomaly. The semiclassical Einstein equations have been solved for the effects of backreaction in the 70s in various cases by the following authors: Ruzmaikina & Ruzmaikin [124] were the first to work on this problem, they, and later, Gurovich & Starobinsky [125] studied the case $\beta = 0$, Wald [11,12] investigated the case $\alpha = 0$, and Fischetti, Hartle, and Hu [15] the cases $\alpha \leq 0$, all values of β and $\alpha > 0, \beta < 3\alpha$. Each of these was done for spatially-flat FLRW spacetimes containing classical radiation and having no cosmological constant. Starobinsky [123] investigated the case $\alpha < 0, \beta > 0$ for FLRW spacetimes with no classical radiation or matter and no cosmological constant.

The most complete investigation of the back-reaction problem for conformally-invariant free quantum fields in spatially-flat homogeneous and isotropic spacetimes containing classical radiation was carried out by Anderson [20]. He considered only solutions which at late times approach the appropriate solution to the field equations of general relativity, the so-called asymptotically classical solutions, with the following results: For all α, β with $\alpha > 0$ there are many such solutions, while for all α, β with $\alpha < 0$ there is only one such solution. For $\beta > 3\alpha > 0$, there is always one solution which undergoes a "time-symmetric bounce" and which contains no singularities or particle horizons. For $\alpha > 0$ there is always at least one solution which begins with an initial singularity and which has no particle horizons. For all α, β there is always at least one solution which begins with an initial singularity.

In a sequel paper [21], Anderson added the consideration of nonzero spatial curvatures and/or a nonzero cosmological constant. The qualitative behaviors are the same as for spatially-flat spacetimes with zero cosmological constant. Instead, for $\beta > 3\alpha > 0$, there is always one solution which undergoes a "time-symmetric bounce" and which contains no singularities or particle horizons. The differences caused by the spatial curvature and cosmological constant include the initial behavior of the time-symmetric bounce solution and, if the spatial curvature is nonzero, the initial behavior of many solutions for the cases $\beta \approx 3\alpha > 0$ and $\beta \approx 3\alpha < 0$. For further work on semiclassical backreaction of conformal quantum fields in a universe with a cosmological constant, see, e.g., Ref. [126].

3. Singularity Avoidance in Quantum Cosmology

Readers interested in having a broader perspective of this issue might want to first learn of the major milestones in the developments of quantum gravity. A summary is given by C. Rovelli [127].

The first stage of quantum gravity began with a series of papers by Bergmann [128] pursuing a program concerned with the quantization of field theories which are covariant with respect to general coordinate transformations, like the general theory of relativity. All these theories share the property that the existence and form of the equations of motion is a direct consequence of the covariant character of the equations. Dirac's work [129,130] on the quantization of constrained systems is a cornerstone in this pursuit. A more sophisticated and systematic treatment can be found in the book by Henneaux and Teitelboim [131].

ADM Quantization.

As most researchers would agree, the canon of canonical quantization of general relativity is the ADM formalism for quantizing 3-geometries in a 3+1 (space-time) decomposition, developed by Arnowitt, Deser and Misner in a series of papers 1959-1961. See their summary in [132]. Read also Kuchař [133], Ashtekar [134], the monographs of Rovelli [135] and Thiemann [136].

Wheeler-DeWitt equation.

The Wheeler-DeWitt equation [137,138] is a Hamilton-Jacobi equation formulation of the ADM quantization on the superspace, the space of all 3-geometries. (For a short introduction with some historical tidbits, read [139].) Minisuperspace refers to the truncation of the infinite dimensional superspace to a few dimensions, such as the scale factors of the Bianchi universes. Quantum cosmology was studied by Misner [50,52] by applying the ADM quantization to the mini-superspaces of Bianchi Type I and IX universes. Read also Ryan [74].

The second wave of quantum cosmology arrived in the mid-80s with the considerations of the initial conditions for the wave functions of the universe, notably, Hartle and

Hawking's no boundary condition [140] and Vilenkin's tunneling wave function [141–143]. That was also the time when the two major programs of quantum gravity took off, namely, superstring theory [144,145] and loop QG [146,147] (and its modern version of spin foam network [148–150] which has a more obvious microscopic appeal). Supersymmetric quantum cosmology is well represented by the monographs of D'Eath [151] and of Moniz [152]. Other major approaches to quantum gravity are nicely represented in Oriti's edited book [153]. A state-of-the-art overview of multiple approaches from the words of 37 prominent practitioners is recorded in this most recent book [154]. Please refer to the original papers, reviews and monographs on this subject.

3.1. Quantum Cosmology: Bounce in Gowdy T^3 Universe

Here we summarize Berger's work [96,97]. The Gowdy T^3 universe [94,95] can be interpreted as consisting of a single polarization of gravitational waves whose amplitudes satisfy a linear wave equation propagating in a non-linear background spacetime. One can then view these wave amplitudes acting quadratically as a source on the first-order equations for the non-linear background part of the gravitational field as a backreaction problem. Classically the model has an initial "big bang" singularity near which it is velocity-dominated behaving as a different Kasner solution at each value of the spatial coordinate orthogonal to the symmetry plane. Far from the singularity, the method of Isaacson [155,156] may be used to identify the gravitational waves. Their effective stress-energy tensor is that for a $p = \rho$ fluid in two dimensions propagating in a spatially-homogeneous background spacetime. At the quantum level, Berger performed a canonical quantization of the dynamical degrees of freedom. An adiabatic vacuum state was introduced and adiabatic regularization used to obtain non-divergent stress-energy tensor vacuum expectation values. The regularized expectation value was used as a source for the classical background spacetime in the spirit of semi-classical gravity. Berger found that "The effect of the regularization of the stress-energy tensor expectation value is to replace the classical singularity with a symmetric bounce. The semi-classical spatially-averaged Gowdy T^3 cosmological model thus collapses from a state of large volume to a minimum volume near $r = 0$ and then re-expands. The classical singularity is replaced by a symmetric bounce".

3.2. Quantum BKL-Mixmaster Scenarios

Earlier we mentioned applications of dynamical system methods and chaos theories to cosmology, in particular, to the analysis of the BKL-mixmaster behavior [157–159]. For quantum chaos applied to the BKL-M models, we mention the numerical solutions of the Wheeler-DeWitt equation by Furusawa [160,161] and Monte Carlo simulations of the vacuum Bianchi IX universe by Berger [162]. For more recent work we mention two groups of authors on this subject. The first group's authors are Bergeron, Czuchry, Gazeau, Małkiewicz and Piechocki. Of particular interest to our theme here is their results on singularity avoidance [163,164]. How to get the universe to enter an inflationary stage without soliciting the aid of an inflaton field (cf. Starobinsky inflation [123] invokes a scalar field, at least the trace anomaly of it, as was also studied in [15]) is certainly of interest [165]. (See [166,167] for an earlier work on mixmaster inflation via an interacting quantum field). These authors constructed the quantum mixmaster solutions using affine and Weyl–Heisenberg covariant integral quantizations. According to these authors these quantization methods can regularize classical singularities which, they claim, the commonly implemented canonical quantization cannot. From this they show some promising physical features such as the singularity resolution, smooth bouncing, the excitation of anisotropic oscillations and a substantial amount of post-bounce inflation as the backreaction to the latter. For a recent summary of these authors' work on this subject, see [168].

The other group's authors are Góźdź, Kiefer, Kwidzinski, Piechocki, Piontek and G. Plewa. Kiefer et al. [169,170] formulated the criteria of singularity avoidance for general Bianchi class A models in the framework of quantum geometrodynamics based on the Wheeler–DeWitt equation and give explicit and detailed results for Bianchi I models with and without matter. The singularities in these cases are big bang and big rip. They find that the classical singularities can generally be avoided in these models.

Góźdź et al. [171] use affine coherent states quantization in the physical phase space and show that during quantum evolution the expectation values of all considered observables are finite. The classical singularity of the BKL scenario is replaced by the quantum bounce that presents a unitary evolution of the considered gravitational system. In a later paper [172] they asked whether their claims may change because the affine coherent states quantization method depends on the choice of the group parametrization. Using the two simplest parameterizations of the affine group, they show that qualitative features of their quantum system do not depend on the choice. They maintain that the quantum bounce replacing a singular classical scenario is expected to be a generic feature of the considered system.

3.3. Loop Quantum Cosmology (LQC): 'Big Bounce'?

We devote this space to LQC because singularity avoidance is supposed to be a major accomplishment of LQG[6]. This is referred to as the 'Big Bounce', in capital letters. There are many reviews of LQC, e.g., Ashtekar et al. [178–182], Bojowald [183,184], and a recent book [185]. (Note the loop QG community does not sing in unison in regard to many major claims in LQC. Read Ashtekhar and Singh's 2011 status report [182] and the critiques of Bojowald [186,187]).

Without getting into the detailed claims and critiques we ask two key physical questions: (1) What role would genuine quantum degrees of freedom play in the cosmological singularity proven to exist in classical general relativity? (2) Does quantization of gravity alter the Big Bang? If it does, will there be a minimum finite scale in the quantum geometry—the quantum bounce? (Here, the canonical formulation of cosmic 'time' is often provided by a relational variable such as the scale factor in the '3-volume' or a scalar field.)

There seems to be some consensus in the LQC community on the existence of a quantum bounce. On the first issue Bojowald asked the question [188]: How Quantum is the Big Bang? Studying isotropic models with an interacting scalar field he concluded that quantum fluctuations do not affect the bounce much. Quantum correlations, however, do play an important role and could even eliminate the bounce.

In [189] a new model is studied which describes the quantum behavior of transitions through an isotropic quantum cosmological bounce in loop quantum cosmology sourced by a free and massless scalar field. Using dynamical coherent states Bojowald provided a demonstration that in general quantum fluctuations before and after the bounce are unrelated. "Thus, even within this solvable model the condition of classicality at late times does not imply classicality at early times before the bounce without further assumptions. Nevertheless, the quantum state does evolve deterministically through the bounce. These analytical bouncing solutions corroborate results from numerical simulations attesting to robustly smooth bounces under the assumption of semiclassicality".

There seems to be no controversy in this last statement within the LQC community. However, how much of the state of the universe before the bounce is retained, is a subject of debates.

Cosmic memory over bounces

For theorists favoring the bounce scenario, especially the cyclic cosmologies, an important question to ask is, does the universe retain, after the bounce, its memory about the previous phase.

Bojowald is of the opinion of 'cosmic forgetfulness' [190], meaning, not all the fluctuations (and higher moments) of a state before the bounce can be recovered after the bounce, and values depend very sensitively on the late time state. In [191] he showed that quantum fluctuations before the big bang are generically unrelated to those after the big bang. A reliable determination of pre-big bang quantum fluctuations of geometry would thus require exceedingly precise observations.

Countering this claim, Corichi and Singh [192,193], in one exactly solvable model, found that a semiclassical state at late times on one side of the bounce, peaked on a pair of canonically conjugate variables, strongly bounds the fluctuations on the other side, implying semiclassicality. From this result these authors assert that cosmic recall is almost perfect. See Bojowald's Comments [194] and Corichi & Singh's Reply [195].

Bianchi I model as a prototype for a cyclical Universe

For more recent work on this topic, we mention two. Montani, Marchi and Moriconi [196] investigated the classical and quantum behavior of a Bianchi I model with stiff matter, an ultra-relativistic component, and a small negative cosmological constant (at the turnover point). They apply the Vilenkin wave function, using the volume of the universe, a quasi-classical variable, as the dynamical clock for the pure quantum degrees of freedom. They found that in a model where the isotropic variable is considered to be on a lattice, the big-bang singularity is removed at a semiclassical level in favor of a big bounce. The mean value of the Universe anisotropy variables remains finite during the whole evolution across the big bounce, assuming a value that depends on the initial conditions fixed far from the turning point.

An even simpler picture is obtained by Wilson-Ewing [197–199]: "In the loop quantum cosmology effective dynamics for the vacuum Bianchi type I and type IX space-times, a non-singular bounce replaces the classical singularity. The bounce can be approximated as an instantaneous transition between two classical vacuum Bianchi I solutions, with simple transition rules relating the solutions before and after the bounce. These transition rules are especially simple when expressed in terms of the Misner variables: the evolution of the mean logarithmic scale factor Ω is reversed, while the shape parameters $\beta_\pm$ are unaffected. As a result, the loop quantum cosmology effective dynamics for the vacuum Bianchi IX space-time can be approximated by a sequence of classical vacuum Bianchi I solutions, following the usual Mixmaster transition maps in the classical regime, and undergoing a bounce with this new transition rule in the Planck regime." At the level of effective dynamics in loop quantum cosmology, for this type of bounce solutions, Gupt and Singh [200] found the selection rules and underlying conditions for all allowed and forbidden transitions.

Open Issues of LQC

Bojowald raised a few cautionary points about the bounce solutions in LQC [201]: (a) A conceptual gap exists in the form of several unquestioned links between bounded densities, bouncing volume expectation values, and singularity avoidance, made commonly in arguments in favor of a generic bounce in LQC. (b) While bouncing solutions exist and may even be generic within a given quantum representation, they are not generic if quantization ambiguities such as choices of representations are taken into account. Small-volume BKL behavior in a collapsing universe is shown to be crucially different from the large-volume behavior exclusively studied so far in LQC.

4. Quantum Phase Transition and Quantum Backreaction

In addition to the popularly known string and loop theories, there is a handful of proposals of quantum gravity theories describing sub-Planckian (length scale) physics [153]. Our present aim is to see how the WCH fares in these theories, i.e., whether some of these theories predict that the Weyl curvature diminishes by comparison with other components of the Riemann tensor near the singularity or bounce. Note that this question may not be answerable by all of them because many of these theories have not yet presented a full account of how their basic constituents interact, and at the Planck length and above, give rise to the familiar manifold structure of spacetime described by GR. What makes this non-trivial and non-straightforward is the possibility of the emergence of mesoscopic structures between the micro (QG) and the macro (GR), or the occurrence of phase transitions which, in our opinion, is very likely. Quantum gravity not being discussed here since it is a topic covered by many books and reviews, we will limit our attention to identifying those microtheories which predict a robust and smooth (near-isotropic and homogeneous) geometry at the Planck scale rising from sub-Planckian scale by phase transitions or other processes or mechanisms (e.g., polymerization, crystalization).

4.1. Quantum Phase Transition: Continuum to Discrete

Asymptotic Safety

First proposed by Weinberg [202], the asymptotic safety program in quantum gravity [203–211] asserts that the short-distance behavior of gravity is governed by a nontrivial renormalization group fixed point. As describd by Bonano and Saueressig [212] It provides an elegant mechanism for completing the gravitational force at sub-Planckian scales. At high energies, the fixed point controls the scaling of couplings such that unphysical divergences are absent, while the emergence of classical low-energy physics is linked to a crossover between two renormalization group fixed points. These features make asymptotic safety an attractive framework for the building of a cosmological model.

D'Odorico and Saueressig [213] studied quantum corrections to the classical Bianchi I and Bianchi IX universes. The correction terms induce a phase transition in the dynamics of the model, changing the classical, chaotic Kasner oscillations into a uniform approach to a point singularity. This seems to be consistent with the results obtained from the backreaction of particle creation from quantum fields, namely, the isotropization of anisotropy, which we mentioned in Section 2.3 and which will be described in Section 5.1.

Causal Dynamical Triangulation CDT

A systematic and vigorous treatment of discrete quantum geometry is the Causal Dynamical Triangulations (CDT) program pursued by Ambjorn, Jurkiewicz, Loll et al. extensively for decades. CDT is a lattice model of gravity that has been used to study non-perturbative aspects of quantum gravity. Read the review of Loll [214], and for the most recent development, Ambjorn et al. [215]: "CDT has a built-in time foliation but is coordinate-independent in the spatial directions. The *higher-order phase transitions* observed in the model may be used to define a continuum limit of the lattice theory. Some aspects of the transitions are better studied when the topology of space is toroidal rather than spherical. In addition, a toroidal spatial topology allows us to understand more easily the nature of typical quantum fluctuations of the geometry. In particular, this topology makes it possible to use massless scalar fields that are solutions to Laplace's equation with special boundary conditions as coordinates that capture the *fractal structure of the quantum geometry*. When such scalar fields are included as dynamical fields in the path integral, they can have a dramatic effect on the geometry." One can view this as a discrete modeling of the Gowdy T^3 universe we described in Section 3.1. The quantum phase transition is of course the interestingly new and exciting development.

Causal Set: Continuum to crystalline phase transition

Causal set theory, first proposed by Sorkin [216] and developed extensively by Dowker [217,218] and their associates, has seen significant developments both in the number of adherents and in scope of its research topics. How the classical continuum spacetime as we know it evolved from discrete causal sets is one of the key issues. Brightwell et al. [219] noted that non-perturbative theories of quantum gravity inevitably include configurations that fail to resemble physically reasonable spacetimes at large scales. Yet these configurations are entropically dominant and pose an obstacle to obtaining the desired classical limit. These authors examine this 'entropy problem' in a model of causal set quantum gravity corresponding to a discretization of 2D spacetimes. Using results from the theory of partial orders they show that, in the large volume or continuum limit, its partition function is dominated by causal sets which approximate a region of 2D Minkowski space.

Surya [220] continued this investigation and presented evidence for a continuum phase in a theory of 2D causal set quantum gravity, which contains a dimensionless non-locality parameter $\epsilon \in (0,1]$. She also found a phase transition between this continuum phase and a new crystalline phase which is characterized by a set of covariant observables. For more recent developments, see her review [221].

Berezenskii-Kosterlitz-Thouless (BKT) transition

Antoniadis, Mazur, and Mottola [222] presented a simple argument which determines the critical value of the anomaly coefficient in four dimensional conformal factor quantum gravity, at which a phase transition between a smooth and elongated phase should occur. The argument is based on the contribution of singular configurations ("spikes") which dominate the partition function in the infrared. The critical value is the analog of c = 1 in the theory of random surfaces, and the phase transition is similar to the Berezenskii-Kosterlitz-Thouless transition. The critical value they obtain is in agreement with the previous canonical analysis of physical states of the conformal factor and may explain why a smooth phase of quantum gravity has not yet been observed in simplicial simulations.

2D Quantum Gravity and Random Geometry

This was an active topic of research in the second half of the 80s. As a representative work, Gross and Migdal [223] proposed a nonperturbative definition of two-dimensional quantum gravity, based on a double scaling limit of the random matrix model which they solve to all orders in the genus expansion. They derived an exact differential equation for the partition function of two-dimensional gravity as a function of the string coupling constant that governs the genus expansion of two-dimensional surfaces. The works of the principal contributors to this subject are collected in a book [224].

4.2. Quantum Backreaction of Inhomogeneous Superspace Modes

We have mentioned the backreaction problem earlier. It may refer to several different effects. We use different symbols to denote them: (C) the backreaction of inhomogeneous modes of classical gravitational perturbations on the homogeneous modes in classical general relativity, as in the works of Buchert, Clifton, Ellis et al. versus Green & Wald. (Q) The backreaction of the inhomogeneous modes of quantum geometry (in the full superspace of 3-geometries) on the mini-superspace quantum cosmology as studied in [225,226]. In between the classical and quantum levels lie the semiclassical and stochastic. (S) The backreaction in semiclassical and stochastic gravity refers to the effects of quantum matter field processes, such as the trace anomaly or particle creation, back-reacting on a classical background geometry: (SC) Semiclassical gravity when the mean value of the stress energy tensor of matter fields act as the source in the semiclassical Einstein equation; (ST) stochastic gravity when the fluctuations of the quantum fields are included as sources of the Einstein-Langevin equation. The 'quantum backreaction' in the title of this paper refers to the middle two levels between the classical and the quantum.

These 4 levels (C-SC-ST-Q) of backreaction are summarized in a recent review with an extensive bibliography [225]. These authors start by assessing the question of backreaction, i.e., whether cosmological inhomogeneities have an effect on the large scale evolution of the Universe, focusing on the purely quantum mechanical backreaction. They present one recent approach based on mathematical tools inspired by the Born–Oppenheimer approximation to include backreaction in quantum cosmology.

We mention three groups of representative work on this issue in the 80s to 90s, then leave this topic for the readers to explore with the help of this excellent review.

(1) Using path-integral quantization, Hawking and Halliwell [227] assume that the Universe is in the quantum state defined by a path integral over compact four-metrics. This can be regarded as a boundary condition for the wave function of the Universe on superspace, the space of all three-metrics and matter field configurations on a three-surface, the same as was proposed by Hartle and Hawking earlier. They treated the homogeneous and

isotropic degrees of freedom of the Friedmann Universe exactly and the inhomogeneous and anisotropic degrees of freedom in superspace to second order in the perturbations. For the same model Kiefer [228] calculated explicitly the wave functions for all multipoles of the matter field and for the tensor modes of the metric. D'Eath and Halliwell [229] have considered the backreaction of the fermionic perturbations on the homogeneous modes. Origin of structure in supersymmetric quantum cosmology was studied by Moniz [230]. By employing the supersymmetry and Lorentz constraint equations a set of quantum states was obtained and a particular quantum state which has properties typical of the conventional no-boundary Hartle-Hawking solution was identified.

(2) In canonical quantum gravity Kuchař and Ryan [231] questioned the validity of physical predictions based on minisuperspace quantization of Einstein's theory of gravitation. They proposed to investigate a hierarchy of models with higher symmetry embedded in models of lesser symmetry, to spell out the criteria under which minisuperspace quantum results can be expected to make meaningful predictions about full quantum gravity. As a concrete example they studied a homogeneous, anisotropic cosmological model of higher symmetry (the Taub model) embedded in one of lesser symmetry (the mixmaster model) and showed that the respective behavior is widely different.

(3) With quantum field theory placed in an open systems setting, Sinha and Hu [232,233] used the example of an interacting $\lambda\Phi^4$ scalar quantum field in a closed Robertson-Walker universe, where the scale factor and the homogeneous mode of the scalar field model the minisuperspace degrees of freedom. They explicitly computed the back-reaction of the inhomogeneous modes on the minisuperspace sector using a coarse-grained effective action and show that the minisuperspace approximation is valid only when this backreaction is small.

For backreaction due to quantized matter fields, its decoherence effects, and how the semiclassical limits are reached in quantum cosmology, see, e.g., [67,234–237].

Concerning quantum backeaction, specifically the question of how the inhomogeneous quantum modes backreact on the lower dimensional superspace modes, and whether the results from mini- or midi-superspace can give a fair representation of the full picture, a recent paper by Bojowald [238] also expressed such a concern: "Even though the BKL scenario allows us to use the classical dynamics of homogeneous models to understand space-time near a spacelike singularity, it is a poor justification of minisuperspace models in quantum cosmology. Using a minisuperspace model to evolve from a nearly homogeneous geometry at late times to a BKL-like geometry at early times means that we begin with a well-justified, approximate infrared contribution of the full theory, but then push the infrared scale all the way into the ultraviolet."

Meanwhile, we shall wait to see further research results from the practitioners of loop and spin-foam quantum cosmology to see how the singularity avoidance in LQG might be altered by including the backreaction of inhomgeneous quantum modes.

In the next section we shall discuss the backreaction of quantum field processes and how that affects the dynamics of classical background spacetimes. Before ending, we mention two backreaction effects, one in inflation, the other in classical GR. Finelli et al. [239] studied the backreaction of scalar fluctuations on the space-time dynamics in the long wavelength limit using a set of gauge invariant variables. Below, we add a short interlude in the classical backreaction research programs for completeness, as it bears on gravitational entropy, albeit less so on the WCH at the cosmological singularity or bounce.

Backreaction of inhomogeneous mode in classical GR

There are two schools of thought on classical backreaction: This research began around 2000, with the work of Buchert [240] and Ellis with collaborators [241]. The central themes are summarized in this letter [242] where these authors make the observation that "A large-scale smoothed-out model of the universe ignores small-scale inhomogeneities, but the averaged effects of those inhomogeneities may alter both observational and dynamical relations at the larger scale". It also comments briefly on the relation to gravitational entropy.

A broader representation of their work can be found in these two reviews [243,244] See also [245,246].

Green and Wald, on the other hand, showed in a series of papers from 2011–2014 [247] that the backreaction effects of classical inhomogenous modes are insignificant. These issues are related to how one should treat the inhomogeneous modes of shorter wavelengths, such as using the Brill-Hartle-Isaacson average [155,156,248,249], and how much they impact on the behavior of the theory if only the lowest modes are retained. Buchert, Ellis et al. [250] countered the Green-Wald arguments, reasserting their claims. In turn Green and Wald replied to this [251,252] with a simple derivation of their no backreaction results.

5. Semiclassical Backreaction Supports WCH

In this last section we examine to what degree is WCH consistent or contradictory with the results of semiclassical backreaction, namely, the backreaction of quantum field processes involving vacuum polarization or fluctuations on the geometrodynamics near the Planck time, like those due to the trace anomaly or cosmological particle creation. These processes have been studied rather thoroughly in the twenty years 1977–1996 as we mentioned before. Earlier, in Sections 1.2, 2.3 and 3.1, we focused on the avoidance of singularity due to the backreaction effects of the trace anomaly and particle creation. Here we complete our narrative by adding the considerations of the damping of anisotropy and inhomogeneity.

5.1. Damping of Irregularities Ensures a Smooth Transition, Critical for Cyclic Cosmologies

Sparing the reader the details of the derivations of regularized stress energy tensors for particle creation and the calculations of the backreaction effects from the solutions of the semiclassical Einstein equation (which can be found in the papers referred to in earlier sections or in a recent book [28]), we shall just highlight the key features with relevance to WCH in the following. More comments about gravitational entropy in this context can be found in [30].

(1) Vacuum fluctuations of a quantum field are ubiquitous, one need not add them in by hand, nor can one wish them away. Their effects on the background spacetime have to be included in considerations of whether the universe contracts to a singularity or undergoes a bounce.

(2) Many vigorous quantum field- theoretical studies of particle creation in (weakly) anisotropic and inhomogeneous spacetimes show that vacuum particle production is abundant near the Planck time, their backreaction effects are strong and the dissipation of these irregularities happens swiftly. (Note the significant qualitative differences between semiclassical backreaction of quantum field processes and classical backreaction of inhomogeneous modes in GR mentioned earlier).

(3) The damping of irregularities drives down the Weyl curvature with their strong and swift actions. This seems to suggest that quantum field processes act as an effective protector, maybe even a guarantor, of WCH.

(4) These quantum field processes act in such a way as to draw the spacetime into a *fixed point* in superspace, a three-geometry of high isotropy and homogeneity, the special status bestowed by the WCH.

(5) The overall backreaction effects of quantum field processes can be summarized by something resembling a *Lenz law*: the production of particles acts in such a way as to diminish their production. This is because, when the universe is isotropized and homogenized, it becomes conformally-flat, whence no more conformal particles are produced. (There is ground to believe that in the very early universe massless particles are overwhelmingly more abundant.)

(6) For any aficionado of cyclic cosmology, in addition to showing that a bounce is possible, the 'reining-in' of the irregularities near the bounce–in fact, near every bounce– is particularly important [118]: If the isotropic expansion were unstable as the scale of the universe approaches zero, then huge irregularities and anisotropies could accumulate and

the successive cycles would be very different and increasingly anisotropic. Some powerful and ever-ready damping mechanisms need be there to prevent this from happening.

In summary, if we ponder upon the ubiquitous nature of these quantum field processes, their strongly dissipative effects on the spacetime irregularities, and the isotropized and homogenized outcome 'locked-in' by the Lenz law, we can see that WCH is more likely to be validated when these factors are included in our considerations. Add to this the fact that these processes apply both in the expansion and contraction phases of the universe, rendering both the beginning and ending of a cycle to be smooth, we can see that quantum field processes are good facilitators of cyclic cosmology.

5.2. Quantum Gravity: Macro Spacetime Manifold from the Interaction of Micro Constituents

Going from the Planck scale down toward the singularity or bounce, even from our rather skimpy description earlier, it may appear to the reader that some quantum gravity theories, e.g, those based on the WdW equation, Hamiltonian quantum gravity, and loop quantum cosmology are more capable of providing a detailed description of quantum spacetime than others, e.g., strings, spin foams, causal sets. This is not the case. One needs to first scrutinize whether some tacit, commonplace, yet unproven assumption has crept in, that *the macro variables are the same as the micro variables*. One should ask, "Are they?". This is a pivotal question, because by assuming there is no change of variables from micro to macro, one only needs to deal with the quantum to classical transition issue. In my opinion, a more important and challenging question is that of micro to macro transition.

Theoretically, this cuts into the key issue of whether the dynamical variables in GR are 'fundamental' all the way from the scale of our universe now, captured amazingly well by Einstein's theory, to the sub-Planckian scales, and, for the former group of theories, even all the way down toward the singularity. What if the variables are collective variables, that the macroscopic spacetime is emergent, and that general relativity is an effective theory valid only at large scales and low energies? We know Nature has many, possibly an infinite number of, levels of structure, separated by characteristic scales, and each level can be described by a suitably constructed effective theory. If so, the theories we now know of the large scale structures of spacetime will very likely become inapplicable at scales below the Planck length.

Are we to believe that once a macro-variable is quantized, it becomes a micro-variable? This is a big leap of faith. In Nature, a quantized macro-variable can be a very different physical entity from a micro-variable, which can be of a quantum nature. E.g., sound is a collective excitation of atoms. Quantizing them yields phonons. But quantization of sound will never lead to the microvariables we want–atoms. In fact, these collective variables lose their meanings once the atomic scale is reached. If the metric or connection forms are collective variables, quantizing them is the wrong way to discover the sub-Planckian microscopic constituents. In contrast, those theories which propose certain entities as the microscopic constituents of spacetime, such as strings or causets, need to show that their interactions can indeed produce the macroscopic spacetime we know, thus is the difference between quantum and emergent gravity [83].

5.3. Quantum Cosmology: Singularity or Bounce Ruled by the Interaction of Micro-Constituents

Varied as the many schools of thought and practice are, there is agreement that quantum gravity refers to *theories of the microscopic structures of spacetime*. Quantum cosmology is supposed to be the application of quantum gravity to the description of the universe from the Planck scale down to the singularity or the bounce. Thus, some knowledge of the underlying theory of quantum gravity is required. Many theories of quantum cosmology assume the existence of a metric or connection, and many papers, such as in string cosmology, start with a statement like, "Let us consider strings in the FLRW or de Sitter universe". Right there, I must say, an ontological issue already shows up. Where does your spacetime manifold come from? If we believe that strings or loops are the fundamental microscopic constituents of spacetime at scales below the Planck length, then, before one talks about string or loop cosmology in terms of the dynamics of the metric or connection, one needs to explain how the spacetime manifold, this macroscopic structure of our universe, befittingly described by general relativity, comes into being.

It appears that a clear demonstration of how the spacetime manifold emerges from string or loop interactions is still largely lacking. If one takes the dynamical variables of a macroscopic theory of spacetime and quantizes them, thinking that one would get the microscopic degrees of freedom in this way, while disregarding the possibly huge structural and behavioral differences between the micro and the meso- and macro-scopic emergent structures, the claims made pertaining to the nature of a singularity or a bounce in the sub-Planckian length regions may not make much sense or even be outright inconsistent.

Therefore, unless one can provide a proof that the micro-variables describing sub-Planckian physics are the same as the macro variables of the large scale spacetime we are familiar with, in our opinion, all predictions and claims about bounces made in quantum cosmology based on quantizing the macro geometric variables in general relativity need be reassessed.

It is with this broader perspective that we should look at the meaning and applicability of WCH. It may very well be that it is enough to examine the validity and implications of WCH at the Planck scale and not extend it to the sub-Planckian scale, because, after all, the Weyl curvature, like all the geometric objects, is well defined only when the spacetime it lives in has a manifold structure, which could be an emergent entity of the underlying QG theories for the microscopic constituents of spacetime.

Funding: This research received no external funding.

Acknowledgments: I would like to thank Paul Anderson and Paulo Moniz for a careful proofreading of the completed manuscript with helpful suggestions, and for providing the references for the approach to the Big Crunch and for supersymmetic quantum cosmology, respectively. The assistance of Cici Xia at MDPI Wuhan to put the references in the MDPI format is much appreciated.

Conflicts of Interest: The author declares no conflict of interest.

Notes

[1] This list is far from exhaustive. We are not treating inflationary cosmology [31–34] here, since it already has a wide coverage and how the WCH fits in with inflation is rarely discussed. On the relation of inflation and singularity, Borde and Vilenkin [35] made a categorical statement that a physically reasonable spacetime that is eternally inflating to the future must possess an initial singularity. Vilenkin [36] also questioned the necessity of quantum cosmology in the face of eternal inflation. As for the speculations into our universe's future, there is considerable amount of work based on two types of inflationary models, one is the 'stochastic inflation' of Starobinsky [37,38] which allows the noise associated with the fluctuations of a quantum field to drive the universe to inflation. The other is the 'eternal inflation' proposed and developed by Linde, Vilenkin, Guth and others [39,40], where a universe can live forever with the continual births of baby universes in the sequential and parallel generation of multiple branches. For a recent assessment of the likelihood for eternal inflation in a variety of popular models, including the swampland of string theory, see, e.g., [41–43].

[2] Chaotic dynamics of minisuperspace cosmology is not limited to the mixmaster type. As Calzetta and Hasi show [64,65] it can also occur in the dynamics of a spatially-closed FLRW universe conformally coupled to a real, free, massive scalar field for large enough field amplitudes.

[3] An important exception is the case when a massless scalar field is present, which obeys an equation of state $p = \rho$ like a stiff fluid, and, in Friedmann models, has the same dependence of the density on the scale factor as anisotropies ($\rho \approx a^{-6}$). As is rigorously shown in [71], such a scalar field will suppress the BKL oscillations during the evolution towards the singularity. The relevance of stiff matter in the early universe is noted by Barrow [72].

[4] This diagonal form has full generality because there is no spatial curvature. In a Bianchi Type IX universe where the spatial curvature is present, spacetimes represented by the full matrix β_{ij} are more general than that of the diagonal metric which is Misner's mixmaster universe [50]; the off-diagonal components signify rotation [74].

[5] An element of craziness needs to creep in before an idea ascends to the order of the three Fs: Fashion, Faith, and Fantasy. I read this as a warning of an enlightened guru speaking to his present adherents and future believers: Safeguard your independent thinking before you become completely converted.

⁶ We could include group field theory (GFT) [173–176] in the present considerations (only) to the extent that (partial) equivalence of models employed or results reported can be shown to exist between these two theories [177]. Of course we need to hear the views of the GFT community on this.

References

1. Penrose, R. *Cycles of Time: An Extraordinary New View of the Universe*; A. Knopf: New York, NY, USA, 2010.
2. Penrose, R. *Fashion, Faith, and Fantasy in the New Physics of the Universe*; Princeton University Press: Princeton, NJ, USA, 2016.
3. Penrose, R. *Singularities and Time-Asymmetry, in General Relativity: An Einstein Centenary*; Hawking, S.W., Israel, W., Eds.; Cambridge University Press: Cambridge, UK, 1979.
4. Penrose, R. Before the big bang: An outrageous new perspective and its implications for particle physics. In Proceedings of the EPAC2006, Edinburgh, Scotland, 26–30 June 2006; Prior C.R., Ed.; pp. 2759–2762.
5. Penrose, R. Causality, quantum theory and cosmology. In *On Space and Time*; Majid, S., Ed.; Cambridge University Press: Cambridge, UK, 2008; pp. 141–195. ISBN 978-0-521-88926-1.
6. Penrose, R. The Big Bang and its Dark-Matter Content: Whence, Whither, and Wherefore. *Found. Phys.* **2018**, *48*, 1177–1190. [CrossRef]
7. Parker, L. Quantized fields and particle creation in expanding universes. I. *Phys. Rev.* **1969**, *183*, 1057. [CrossRef]
8. Zel'dovich, Y.B. Particle production in cosmology. *JETP Lett.* **1970**, *12*, 307. [Erratum in *Pis'ma Zh. Eksp. Teor. Fiz.* **1970**, *12*, 443.]
9. Zeldovich, Y.B.; Starobinsky, A.A. Particle production and vacuum polarization in an anisotropic gravitational field. *Sov. Phys. JETP* **1971**, *34*, 1159. [Erratum in *Zh. Eksp. Teor. Fiz.* **1971**, *61*, 2161.]
10. Duff, M.J. Observations on conformal anomalies. *Nucl. Phys. B* **1977**, *125*, 334. [CrossRef]
11. Wald, R.M. Trace anomaly of a conformally invariant quantum field in curved spacetime. *Phys. Rev. D* **1978**, *110*, 472.
12. Wald, R.M. Axiomatic renormalization of the stress tensor of a conformally invariant field in conformally flat spacetimes. *Ann. Phys.* **1978**, *110*, 472. [CrossRef]
13. Hu, B.L.; Parker, L. Effect of gravitation creation in isotropically expanding universes. *Phys. Lett. A* **1977**, *63*, 217–220. [CrossRef]
14. Hu, B.L.; Parker, L. Anisotropy damping through quantum effects in the early universe. *Phys. Rev. D* **1978**, *17*, 933. [CrossRef]
15. Fischetti, M.V.; Hartle, J.B.; Hu, B.L. Quantum effects in the early universe. I. Influence of trace anomalies on homogeneous, isotropic, classical geometries. *Phys. Rev. D* **1979**, *20*, 1757. [CrossRef]
16. Hartle, J.B.; Hu, B.L. Quantum effects in the early universe. II. Effective action for scalar fields in homogeneous cosmologies with small anisotropy. *Phys. Rev. D* **1979**, *20*, 1772. [CrossRef]
17. Hartle, J.B. Effective-potential approach to graviton production in the early universe. *Phys. Rev. Lett.* **1977**, *39*, 1373. [CrossRef]
18. Hartle, J.B. Quantum effects in the early Universe. IV. Nonlocal effects in particle production in anisotropic models. *Phys. Rev. D* **1980**, *22*, 2091. [CrossRef]
19. Hartle, J.B. Quantum effects in the early universe. V. Finite particle production without trace anomalies. *Phys. Rev. D* **1981**, *23*, 2121. [CrossRef]
20. Anderson, P.R. Effects of quantum fields on singularities and particle horizons in the early universe. *Phys. Rev. D* **1983**, *28*, 271–285. [CrossRef]
21. Anderson, P.R. Effects of quantum fields on singularities and particle horizons in the early universe. II. *Phys. Rev. D* **1984**, *29*, 615–627. [CrossRef]
22. Calzetta, E.A.; Hu, B.L. Closed time path functional formalism in curved space-time: Application to cosmological backreaction problems. *Phys. Rev. D* **1987**, *35*, 495–509. [CrossRef]
23. Calzetta, E.A.; Hu, B.L. Noise and fluctuations in semiclassical gravity. *Phys. Rev. D* **1994**, *49*, 6636–6655. [CrossRef]
24. Hu, B.L.; Matacz, A. Back reaction in semiclassical cosmology: The Einstein–Langevin equation. *Phys. Rev. D* **1995**, *51*, 1577–1586. [CrossRef]
25. Hu, B.L.; Sinha, S. A fluctuation–dissipation relation for semiclassical cosmology. *Phys. Rev. D* **1995**, *51*, 1587–1606. [CrossRef]
26. Campos, A.; Verdaguer, E. Semiclassical equations for weakly inhomogeneous cosmologies. *Phys. Rev. D* **1994**, *49*, 1861–1880. [CrossRef]
27. Campos, A.; Verdaguer, E. Stochastic semiclassical equations for weakly inhomogeneous cosmologies. *Phys. Rev. D* **1996**, *53*, 1927–1937. [CrossRef] [PubMed]
28. Hu, B.L.; Verdaguer, E. *Semiclassical and Stochastic Gravity: Quantum Field Effects on Curved Spacetime*; Cambridge University Press: Cambridge, UK, 2020.
29. Hu, B.L. Vacuum viscosity description of quantum processes in the early universe. *Phys. Rev. D* **1982**, *90*, 375–380. [CrossRef]
30. Hu, B.L. Quantum dissipative processes and gravitational entropy of the universe. *Phys. Lett. A* **1983**, *97*, 368–374. [CrossRef]
31. Guth, A.H. Inflationary universe: A possible solution to the horizon and flatness problems. *Phys. Rev. D* **1981**, *23*, 347. [CrossRef]
32. Linde, A.D. *Particle Physics and Inflationary Cosmology*; Volume 5 of Contemporary Concepts in Physics; Harwood: Chur, Switzerland; New York, NY, USA, 1990.
33. Linde, A. Inflationary cosmology. *Phys. Scr.* **2000**, *T85*, 168. [CrossRef]
34. Mukhanov, V. *Physical Foundations of Cosmology*; Cambridge University Press: Cambridge, UK, 2005.
35. Borde, A.; Vilenkin, A. Eternal inflation and the initial singularity. *Phys. Rev. Lett.* **1994**, *72*, 3305. [CrossRef] [PubMed]
36. Vilenkin, A. Quantum cosmology and eternal inflation. *arXiv* **2002**, arXiv:gr-qc/0204061.

37. Starobinsky, A.A. Stochastic de Sitter (Inflationary) Stage in the Early Universe. *Lect. Notes Phys.* **1986**, *246*, 107–126.
38. Starobinsky, A.A. Future and origin of our universe: Modern view. *Grav. Cosmol.* **2000**, *6*, 157.
39. Linde, A.D. Eternal chaotic inflation. *Mod. Phys. Lett. A.* **1986**, *1*, 81–85. [CrossRef]
40. Guth, A.H. Inflation and eternal inflation. *Phys. Rep.* **2000**, *333*, 555–574. [CrossRef]
41. Rudelius, T. Conditions for (no) eternal inflation. *arXiv* **2019**, arXiv:1905.05198
42. Wang, Z.; Brandenberger, R.; Heisenberg, L. Eternal inflation, entropy bounds and the swampland. *Eur. Phys. J. C* **2020**, *80*, 864. [CrossRef]
43. Blanco-Pillado, J.J.; Deng, H.; Vilenkin, A. Eternal inflation in swampy landscapes. *J. Cosmol. Astropart. Phys.* **2020**, *2020*, 014. [CrossRef]
44. Penrose, R. Gravitational collapse and space-time singularities. *Phys. Rev. Lett.* **1965**, *14*, 57. [CrossRef]
45. Hawking, S.W. Singularities in the Universe. *Phys. Rev. Lett.* **1966**, *17*, 443. [CrossRef]
46. Geroch, R.P. Singularities in closed universes. *Phys. Rev. Lett.* **1966**, *17*, 445. [CrossRef]
47. Belinskii, V.A.; Khalatnikov, I.M.; Lifshitz, E.M. Oscillatory approach to a singular point in relativistic cosmology. *Adv. Phys.* **1970**, *19*, 523–573. [CrossRef]
48. Belinskii, V.A.; Lifshitz, E.M.; Khalatnikov, I.M. On a general cosmological solution of the Einstein equations with a time singularity. *Adv. Phys.* **1972**, *31*, 639–667.
49. Belinskii, V.A.; Khalatnikov, I.M.; Lifshitz, E.M. A general solution of the Einstein equations with a time singularity. *Adv. Phys.* **1982**, *13*, 639–667. [CrossRef]
50. Misner, C.W. Mixmaster universe. *Phys. Rev. Lett.* **1969**, *22*, 1071–1074. [CrossRef]
51. Misner, C.W. Quantum cosmology I. *Phys. Rev.* **1970**, *186*, 1328. [CrossRef]
52. Misner, C.W. Minisuperspace. In *Magic without Magic: John Archibald Wheeler*; W. H. Freeman: San Francisco, CA, USA, 1972; pp. 441–473.
53. Eardley, D.; Liang, E.; Sachs, R. Velocity-dominated singularities in irrotational dust cosmologies. *J. Math. Phys.* **1972**, *13*, 99–107. [CrossRef]
54. Hawking, S.W.; Penrose, R. The singularities of gravitational collapse and cosmology. *Proc. R. Soc. Lond. A* **1970**, *314*, 529–548.
55. Hawking, S.W.; Ellis, G.F.R. *The Large Scale Structure of Space–Time*; Cambridge University Press: Cambridge, UK, 1973.
56. Senovilla, J.M.M.; Garfinkle, D. The 1965 Penrose singularity theorem. *Class. Quantum Gravity* **2015**, *32*, 124008. [CrossRef]
57. Ryan, M.; Shepley, L. *Homogeneous Relativistic Cosmologies*; Princeton University: Princeton, NJ, USA, 1975.
58. Barrow, J.D.; Tipler, F.J. Analysis of the generic singularity studies by Belinskii, Khalatnikov, and Lifshitz. *Phys. Rep.* **1979**, *56*, 371–402. [CrossRef]
59. Barrow, J.D. Chaos in the Einstein equations. *Phys. Rev. Lett.* **1981**, *46*, 963. [CrossRef]
60. Barrow, J.D. Chaotic behaviour in general relativity. *Phys. Rep.* **1982**, *85*, 1. [CrossRef]
61. Chernoff, D.; Barrow, J.D. Chaos in the mixmaster universe. *Phys. Rev. Lett.* **1983**, *50*, 134. [CrossRef]
62. Cornish, N.J.; Levin, J.J. The Mixmaster Universe is Chaotic. *Phys. Rev. Lett.* **1997**, *78*, 998. [CrossRef]
63. Cornish, N.J.; Levin, J.J. The mixmaster universe: A chaotic farey tale. *Phys. Rev. D* **1997**, *55*, 7489. [CrossRef]
64. Calzetta, E.; El Hasi, C. Chaotic friedmann-robertson-walker cosmology. *Class. Quantum Gravity* **1993**, *10*, 1825. [CrossRef]
65. Calzetta, E. Homoclinic chaos in relativistic cosmology. In *Deterministic Chaos in General Relativity*; Springer: Boston, MA, USA, 1994; pp. 203–235.
66. Belinski, V.; Henneaux, M. *The Cosmological Singularity*; Cambridge University Press: Cambridge, UK, 2017.
67. Calzetta, E. Chaos, decoherence and quantum cosmology. *Class. Quantum Gravity* **2012**, *29*, 143001. [CrossRef]
68. Uggla, C.; van Elst, H.; Wainwright, J.; Ellis, G.F.R. Past attractor in inhomogeneous cosmology. *Phys. Rev. D* **2003**, *68*, 103502. [CrossRef]
69. Heinzle, J.M.; Uggla, C. Mixmaster: Fact and belief. *Class. Quantum Gravity* **2009**, *26*, 075016. [CrossRef]
70. Garfinkle, D. Numerical simulations of generic singularities. *Phys. Rev. Lett.* **2004**, *93*, 161101. [CrossRef]
71. Andersson, L.; Rendall, A.D. Quiescent cosmological singularities. *Commun. Math. Phys.* **2001**, *218*, 479. [CrossRef]
72. Barrow, J.D. Quiescent cosmology. *Nature* **1978**, *272*, 211. [CrossRef]
73. Bianchi, L. *Lezioni Sulla Teoria dei Gruppi Continui Finiti di Transformazioni*; Enrico Spoerri: Pisa, Italy, 1903.
74. Ryan, M.P. *Hamiltonion Cosmology*; Springer: Berlin, Germany, 1972.
75. Barrow, J.D.; Hervik, S. Weyl tensor in spatially homogeneous cosmological models. *Class. Quantum Gravity* **2002**, *19*, 5173. [CrossRef]
76. Grøn, Ø. Entropy and Gravity. *Entropy* **2012**, *14*, 2456–2477. [CrossRef]
77. Lim, W.C.; Coley, A.A.; Hervik, S. Kinematic and Weyl singularities. *Class. Quantum Gravity* **2007**, *24*, 595–604. [CrossRef]
78. Coley, A.A.; Hervik, S.; Lim, W.C.; MacCallum, M.A.H. Properties of kinematic singularities. *Class. Quantum Gravity* **2009**, *26*, 215008. [CrossRef]
79. Clifton, T.; Ellis, G.F.R.; Tavakol, R. A gravitational entropy proposal. *Class. Quantum Gravity* **2013**, *30*, 125009. [CrossRef]
80. Hosoya, A.; Buchert, T.; Morita, M. Information Entropy in Cosmology. *Phys. Rev. Lett.* **2004**, *92*, 141302. [CrossRef] [PubMed]
81. Li, N.; Buchert, T.; Hosoya, A.; Morita, M.; Schwarz, D.J. Relative information entropy and Weyl curvature of the inhomogeneous Universe. *Phys. Rev. D* **2012**, *86*, 083539. [CrossRef]

82. Li, N.; Li, X.-L.; Song, S.-P. Kullback–Leibler entropy and Penrose conjecture in the Lemaître–Tolman–Bondi model. *Eur. Phys. J. C* **2015**, *75*, 114. [CrossRef]
83. Hu, B.L. Emergent/quantum gravity: Macro/micro structures of spacetime. *J. Phys. Conf. Ser.* **2009**, *174*, 012015. [CrossRef]
84. Birrell, N.D.; Davies, P.C.W. *Quantum Fields in Curved Space*; Cambridge University Press: Cambridge, UK, 1982.
85. Fulling, S.A. *Aspects of Quantum Field Theory in Curved Spacetime*; Cambridge University Press: Cambridge, UK, 1989.
86. Wald, R.M. *Quantum Field Theory in Curved Spacetime and Black Hole Thermodynamics*; University of Chicago Press: Chicago, IL, USA, 1994.
87. Parker, L.; Toms, D. *Quantum Field Theory in Curved Spacetime: Quantized Fields and Gravity*; Cambridge University Press: Cambridge, UK, 2009.
88. Grischuk, L.P. Amplification of gravitational waves in an isotropic universe. *J. Exp. Theor. Phys.* **1975**, *40*, 409. [Erratum in *Z. Eksp Teo. Fi.* **1974**, *67*, 825.]
89. Ford, L.H.; Parker, L. Quantized gravitational wave perturbations in Robertson–Walker universes. *Phys. Rev. D* **1977**, *16*, 1601–1608. [CrossRef]
90. Grishchuk, L.P. Graviton creation in the early universe. *Ann. N. Y. Acad. Sci.* **1977**, *302*, 439–444. [CrossRef]
91. Nesteruk, A.V.; Ottewill, A.C. Graviton production as a measure of gravitational entropy in an isotropic universe. *Class. Quantum Gravity* **1995**, *12*, 51. [CrossRef]
92. Nesteruk, A.V. The Weyl curvature hypothesis and a choice of the initial vacuum for quantum fields at the cosmological singularity. *Class. Quantum Gravity* **1994**, *11*, L15. [CrossRef]
93. Husain, V. Weyl tensor and gravitational entropy. *Phys. Rev. D* **1988**, *38*, 3314. [CrossRef]
94. Gowdy, R.H. Gravitational waves in closed universes. *Phys. Rev. Lett.* **1971**, *27*, 826. [CrossRef]
95. Gowdy, R.H. Vacuum spacetimes with two-parameter spacelike isometry groups and compact invariant hypersurfaces: topologies and boundary conditions. *Ann. Phys.* **1974**, *83*, 203. [CrossRef]
96. Berger, B.K. Quantum graviton creation in a model universe. *Ann. Phys.* **1974**, *83*, 458. [CrossRef]
97. Berger, B.K. Quantum effects in the Gowdy T3 cosmology. *Ann. Phys.* **1984**, *156*, 155. [CrossRef]
98. Grøn, Ø.; Hervik, S. Gravitational entropy and quantum cosmology. *Class. Quant. Grav.* **2001**, *18*, 601–618. [CrossRef]
99. Pelavas, N.; Lake, K. Measures of gravitational entropy: Self-similar spacetimes. *Phys. Rev. D* **2000**, *62*, 044009. [CrossRef]
100. Rothman, T.; Anninos, P. Hamitonian dynamics and the entropy of the gravitational field. *Phys. Lett. A* **1997**, *224*, 227. [CrossRef]
101. Rothman, T. A phase space approach to gravitational entropy. *Gen. Rel. Grav.* **2000**, *32*, 1185. [CrossRef]
102. Balasubramanian, V.; Czech, B.; Larjo, K. Quantum geometry and gravitational entropy. *J. High Energy Phys.* **2007**, *12*, 067. [CrossRef]
103. Ryu, S.; Takayanagi, T. Aspects of holographic entanglement entropy. *J. High Energy Phys.* **2006**, *2006*, 045. [CrossRef]
104. Geroch, R.P. Local characterization of singularities in general relativity. *J. Math. Phys.* **1968**, *9*, 450. [CrossRef]
105. Geroch, R.P. What is a singularity in general relativity? *Ann. Phys.* **1968**, *48*, 526. [CrossRef]
106. Goode, S.W.; Wainwright, J. Isotropic singularities in cosmological models. *Class. Quantum Gravity* **1985**, *2*, 99. [CrossRef]
107. Goode, S.W. Isotropic singularities and the Penrose-Weyl tensor hypothesis. *Class. Quantum Gravity* **1991**, *8*, L1. [CrossRef]
108. Goode, S.W.; Coley, A.A.; Wainwright, J. The isotropic singularity in cosmology. *Class. Quantum Gravity* **1992**, *9*, 445–455. [CrossRef]
109. Tod, K.P. Isotropic cosmological singularities: Other matter models. *Class. Quantum Gravity* **2003**, *20*, 521–534. [CrossRef]
110. Tod, K.P. Isotropic cosmological singularities in spatially, homogeneous models with a cosmological constant. *Class. Quantum Gravity* **2007**, *24*, 2415. [CrossRef]
111. Veneziano, G. A Simple/Short Introduction to Pre-Big-Bang Physics/Cosmology. *arXiv* **1998**, arXiv:hep-th/9802057.
112. Gasperini, M.; Veneziano, G. The pre-big bang scenario in string cosmology. *Phys. Rep.* **2003**, *373*, 1–212. [CrossRef]
113. Gasperini, M. Quantum String Cosmology. *Universe* **2021**, *7*, 14. [CrossRef]
114. Steinhardt, P.J.; Turok, N. A cyclic model of the universe. *Science* **2002**, *296*, 1436–1439. [CrossRef] [PubMed]
115. Steinhardt, P.J.; Turok, N. Cosmic evolution in a cyclic universe. *Phys. Rev. D* **2002**, *65*, 126003. [CrossRef]
116. Lehners, J.-C. Ekpyrotic and cyclic cosmology. *Phys. Rep.* **2008**, *465*, 223–263. [CrossRef]
117. Erickson, J.K.; Wesley, D.H.; Steinhardt, P.J.; Turok, N. Kasner and mixmaster behavior in universes with equation of state $w \geq 1$. *Phys. Rev. D* **2004**, *69*, 063514. [CrossRef]
118. Barrow, J.D.; Yamamoto, K. Anisotropic pressures at ultrastiff singularities and the stability of cyclic universes. *Phys. Rev. D* **2010**, *82*, 063516. [CrossRef]
119. Brandenberger, R.; Costa, R.; Franzmann, G. Can backreaction prevent eternal inflation? *Phys. Rev. D* **2015**, *92*, 043517. [CrossRef]
120. Kennedy, G.; Keith, A.F.; Sanders, J.H. A model for the influence of quantised fields on the gravitational collapse of a dust cloud. *Class. Quantum Gravity* **1989**, *6*, 1697. [CrossRef]
121. Caldwell, R.R.; Kamionkowski, M.; Weinberg, N.N. Phantom Energy: Dark Energy with $w < -1$ Causes a Cosmic Doomsday. *Phys. Rev. Lett.* **2003**, *91*, 071301.
122. Carlson, E.D.; Anderson, P.R.; Einhorn, J.R.; Hicks, B.; Lundeen, A.J. Future singularities if the universe underwent Starobinsky inflation in the past. *Phys. Rev. D* **2017**, *95*, 044012. [CrossRef]
123. Starobinsky, A.A. A new type of isotropic cosmological models without singularity. *Phys. Lett. B* **1980**, *91*, 99–102. [CrossRef]

124. Ruzmaikina, T.V.; Ruzmaikin, A.A. Quadratic corrections to the Lagrangian density of the gravitational field and the singularity. *Sov. Phys. JETP* **1970**, *30*, 372. [Erratum in *Zh. Eksp. Teor. Fiz.* **1969**, *57*, 680].
125. Gurovich, V.T.; Starobinsky, A.A. Quantum effects and regular cosmological models. *Sov. Phys. JETP* **1979**, *50*, 844. [Erratum in *Zh. Eksp. Teor. Fiz.* **1979**, *77*, 1683].
126. Azuma, T.; Wada, S. Solutions in the Presence of the Cosmological Constant and Backreaction of Conformally Invariant Quantum Fields. *Prog. Theor. Phys.* **1986**, *75*, 845–861. [CrossRef]
127. Rovelli, C. Notes for a brief history of quantum gravity. *arXiv* **2002**, arXiv:gr-qc/0006061v3.
128. Bergmann, P.G. Non-Linear Field Theories. *Phys. Rev.* **1949**, *75*, 680. [CrossRef]
129. Dirac, P.A.M. Generalized hamiltonian dynamics. *Proc. R. Soc.* **1958**, *A246*, 326–333.
130. Dirac, P.A.M. Fixation of coordinates in the Hamiltonian theory of gravitation. *Phys. Rev.* **1959**, *114*, 924. [CrossRef]
131. Henneaux, M.; Teitelboim, C. *Quantization of Gauge Systems*; Princeton University Press: Princeton, NJ, USA, 1992.
132. Arnowitt, R.; Deser, S.; Misner, C.W. The Dynamics of General Relativity. In *Gravitation: An Introduction to Current Research*; Witten, L., Ed.; Wiley: New York, NY, USA, 1962.
133. Kuchař, K. Canonical Quantization of Gravity. In *Relativity, Astrophysics and Cosmology*; Astrophysics and Space Science Library; Israel, W., Ed.; Springer: Dordrecht, The Netherlands, 1973; Volume 38.
134. Ashtekar, A. *Lectures on Non-Perturbative Canonical Gravity*; World Scientific: Singapore, 1991.
135. Rovelli, C. *Quantum Gravity*; Cambridge University Press: Cambridge, UK, 2007.
136. Thiemann, T. *Modern Canonical Quantum General Relativity*; Cambridge University Press: Cambridge, UK, 2008.
137. DeWitt, B.S. Quantum Theory of Gravity. 1. The Canonical Theory. *Phys. Rev.* **1967**, *160*, 1113–1148. [CrossRef]
138. Wheeler, J.A. Superspace and the nature of quantum geometrodynamics. In *Batelles Rencontres*; DeWitt, B.S., Wheeler, J.A., Eds.; Benjamin: New York, NY, USA, 1968.
139. Rovelli, C. The strange equation of quantum gravity. *Class. Quantum Gravity* **2015**, *32*, 124005. [CrossRef]
140. Hartle, J.B.; Hawking, S.W. Wave function of the Universe. *Phys. Rev. D Part. Fields* **1983**, *28*, 2960–2975. [CrossRef]
141. Vilenkin, A. Quantum creation of universes. *Phys. Rev. D* **1984**, *30*, 509. [CrossRef]
142. Vilenkin, A.; Yamada, M. Tunneling wave function of the universe. *Phys. Rev. D* **2018**, *98*, 066003. [CrossRef]
143. Vilenkin, A.; Yamada, M. Tunneling wave function of the universe. II. The backreaction problem. *Phys. Rev. D* **2019**, *99*, 066010. [CrossRef]
144. Green, M.B.; Schwarz, J.H.; Witten, E. *Superstring Theory*; Cambridge University Press: Cambridge, UK, 1990; Volumes 1–2.
145. Polchinski, J. *String Theory*; Cambridge University Press: Cambridge, UK, 1998; Volumes 1–2.
146. Ashtekar, A.; Pullin, J. *Loop Quantum Gravity*; World Scientific: London, UK, 2017.
147. Rovelli, C. Loop quantum gravity. *Living Rev. Relativ.* **2008**, *11*, 5. [CrossRef]
148. Rovelli, C.; Smolin, L. Spin networks and quantum gravity. *Phys. Rev. D* **1995**, *52*, 5743. [CrossRef]
149. Freidel, L.; Krasnov, K. A new spin foam model for 4D gravity. *Class. Quantum Gravity* **2008**, *25*, 125018. [CrossRef]
150. Perez, A. The spin-foam approach to quantum gravity. *Living Rev. Relativ.* **2013**, *16*, 3. [CrossRef]
151. D'Eath, P.D. *Supersymmetric Quantum Cosmology*; Cambridge University Press: Cambridge, UK, 1996.
152. Moniz, P.V. *Quantum Cosmology-The Supersymmetric Perspective-Vol. 1: Fundamentals & Vol. 2: Advanced Topics*; Springer: Berlin/Heidelberg, Germany, 2010.
153. Oriti, D. (Ed.) *Towards Quantum Gravity*; Cambridge University Press: Cambridge, UK, 2006.
154. Armas, J. *Conversations on Quantum Gravity*; Cambridge University Press: Cambridge, UK, 2021.
155. Isaacson, R.A. Gravitational radiation in the limit of high frequency. I. The linear approximation and geometrical optics. *Phys. Rev.* **1968**, *166*, 1263. [CrossRef]
156. Isaacson, R.A. Gravitational radiation in the limit of high frequency. II. Nonlinear terms and the effective stress tensor. *Phys. Rev.* **1968**, *166*, 1272. [CrossRef]
157. Bogoyavlensky, O.I. *Methods in the Qualitative Theory of Dynamical Systems in Astrophysics and Gas Dynamics*; Springer: Berlin, Germany, 1985.
158. Wainwright, J.; Ellis, G.F.R. (Eds.) *Dynamical Systems in Cosmology*; Cambridge University Press: Cambridge, UK, 2005.
159. Coley, A.A. *Dynamical Systems and Cosmology*; Springer Science & Business Media: Heidelberg, Germany; 31 October 2003.
160. Furusawa, T. Quantum chaos of Mixmaster universe I. *Prog. Theor. Phys.* **1986**, *75*, 59–67. [CrossRef]
161. Furusawa, T. Quantum chaos of Mixmaster universe II. *Prog. Theor. Phys.* **1986**, *76*, 67–74. [CrossRef]
162. Berger, B.K. Quantum chaos in the mixmaster universe. *Phys. Rev. D* **1989**, *39*, 2426. [CrossRef]
163. Bergeron, H.; Czuchry, E.; Gazeau, J.P.; Małkiewicz, P.; Piechocki, W. Singularity avoidance in a quantum model of the Mixmaster universe. *Phys. Rev. D* **2015**, *92*, 124018. [CrossRef]
164. Bergeron, H.; Czuchry, E.; Gazeau, J.P.; Małkiewicz, P.; Piechocki, W. Smooth Quantum Dynamics of Mixmaster Universe. *Phys. Rev. D* **2015**, *92*, 061302. [CrossRef]
165. Bergeron, H.; Czuchry, E.; Gazeau, J.-P.; Malkiewicz, P. Nonadiabatic bounce and an inflationary phase in the quantum mixmaster universe. *Phys. Rev. D* **2016**, *93*, 124053. [CrossRef]
166. Hu, B.L.; O'Connor, D.J. Mixmaster Inflation. *Phys. Rev. D* **1986**, *34*, 2535. [CrossRef]
167. Hu, B.L.; O'Connor, D.J. Infrared behavior and finite size effects in inflationary cosmology. *Phys. Rev. Lett.* **1986**, *56*, 1613. [CrossRef]

168. Bergeron, H.; Czuchry, E.; Gazeau, J.P.; Małkiewicz, P. Quantum Mixmaster as a model of the Primordial Universe. *Universe* **2020**, *6*, 7. [CrossRef]
169. Kiefer, C.; Kwidzinski, N.; Piechocki, W. On the dynamics of the general Bianchi IX spacetime near the singularity. *Eur. Phys. J. C* **2018**, *78*, 691. [CrossRef]
170. Kiefer, C.; Kwidzinski, N.; Piontek, D. Singularity avoidance in Bianchi I quantum cosmology. *Eur. Phys. J. C* **2019**, *79*, 686. [CrossRef]
171. Góźdź, A.; Piechocki, W.; Plewa, G. Quantum Belinski–Khalatnikov–Lifshitz scenario. *Eur. Phys. J. C* **2019**, *79*, 1–16. [CrossRef]
172. Góźdź, A.; Piechocki, W. Robustness of the quantum BKL scenario. *Eur. Phys. J. C* **2020**, *80*, 142. [CrossRef]
173. Calcagni, G.; Gielen, S.; Oriti, D. Group field cosmology: A cosmological field theory of quantum geometry. *Class. Quantum Gravity* **2012**, *29*, 105005. [CrossRef]
174. Gielen, S.; Oriti, D.; Sindoni, L. Cosmology from Group Field Theory Formalism for Quantum Gravity. *Phys. Rev. Lett.* **2013**, *111*, 031301. [CrossRef]
175. Gielen, S.; Oriti, D.; Sindoni, L. Homogeneous cosmologies as group field theory condensates. *J. High Energy Phys.* **2014**, *1406*, 013. [CrossRef]
176. Gielen, S.; Oriti, D. Quantum cosmology from quantum gravity condensates: Cosmological variables and lattice-refined dynamics. *N. J. Phys.* **2014**, *16*, 123004. [CrossRef]
177. Bayta, B.; Bojowald, M.; Crowe, S. Equivalence of Models in Loop Quantum Cosmology and Group Field Theory. *Universe* **2019**, *5*, 41. [CrossRef]
178. Ashtekar, A.; Bojowald, M.; Lewandowski, J. Mathematical structure of loop quantum cosmology. *Adv. Theor. Math. Phys.* **2003**, *7*, 233268. [CrossRef]
179. Ashtekar, A.; Pawlowski, T.; Singh, P. Quantum nature of the big bang. *Phys. Rev. Lett.* **2006**, *96*, 141301. [CrossRef] [PubMed]
180. Ashtekar, A.; Pawlowski, T.; Singh, P. Quantum nature of the big bang: Improved dynamics. *Phys. Rev. D* **2006**, *74*, 084003. [CrossRef]
181. Ashtekar, A.; Corichi, A.; Singh, P. Robustness of key features of loop quantum cosmology. *Phys. Rev. D* **2008**, *77*, 024046. [CrossRef]
182. Ashtekar, A.; Singh, P. Loop quantum cosmology: A status report. *Class. Quantum Gravity* **2011**, *28*, 213001. [CrossRef]
183. Bojowald, M. Loop quantum cosmology. *Living Rev. Relativ.* **2005**, *8*, 11. [CrossRef] [PubMed]
184. Bojowald, M. Quantum cosmology: A review. *Rep. Prog. Phys.* **2015**, *78*, 023901. [CrossRef]
185. Bojowald, M. *Foundation of Quantum Cosmology*; AAS/IOP Publishing: Bristol, UK, 2020. [CrossRef]
186. Bojowald, M. Critical Evaluation of Common Claims in Loop Quantum Cosmology. *Universe* **2020**, *6*, 36. [CrossRef]
187. Bojowald, M. Cosmic Tangle: Loop Quantum Cosmology and CMB Anomalies. *Universe* **2021**, *7*, 186. [CrossRef]
188. Bojowald, M. How Quantum is the Big Bang? *Phys. Rev. Lett.* **2008**, *100*, 221301. [CrossRef]
189. Bojowald, M. Dynamical coherent states and physical solutions of quantum cosmological bounces. *Phys. Rev. D* **2007**, *75*, 123512. [CrossRef]
190. Bojowald, M. What happened before the big bang? *Nat. Phys.* **2007**, *3*, 523. [CrossRef]
191. Bojowald, M. Harmonic cosmology: How much can we know about a universe before the big bang? *Proc. R. Soc. A* **2008**, *464*, 2135. [CrossRef]
192. Corichi, A.; Singh, P. Quantum Bounce and Cosmic Recall. *Phys. Rev. Lett.* **2008**, *100*, 161302. [CrossRef] [PubMed]
193. Corichi, A.; Singh, P. Is loop quantization in cosmology unique? *Phys. Rev. D* **2008**, *78*, 024034. [CrossRef]
194. Bojowald, M. Comment on "Quantum Bounce and Cosmic Recall". *Phys. Rev. Lett.* **2008**, *101*, 209001. [CrossRef]
195. Corichi, A.; Singh, P. Reply to Bojowald's comment *Phys. Rev. Lett.* **2008**, *101*, 209002. [CrossRef]
196. Montani, G.; Marchia, A.; Moriconi, R. Bianchi I model as a prototype for a cyclical Universe. *Phys. Lett. B* **2018**, *777*, 191–200. [CrossRef]
197. Wilson-Ewing, E. Loop quantum cosmology of Bianchi type IX models. *Phys. Rev. D* **2010**, *82*, 043508. [CrossRef]
198. Wilson-Ewing, E. The loop quantum cosmology bounce as a Kasner transition. *Class. Quantum Gravity* **2018**, *35*, 065005. [CrossRef]
199. Wilson-Ewing, E. A quantum gravity extension to the Mixmaster dynamics. *Class. Quantum Gravity* **2019**, *36*, 195002. [CrossRef]
200. Gupt, B.; Singh, P. Quantum gravitational Kasner transitions in Bianchi-I spacetime. *Phys. Rev. D* **2012**, *86*, 024034. [CrossRef]
201. Bojowald, M. Non-bouncing solutions in loop quantum cosmology. *arXiv* **2019**, arXiv:1906.02231.
202. Weinberg, S. Ultraviolet divergences in quantum theories of gravitation. In *General Relativity. An Einstein Centenary Survey*; Hawking, S.W., Israel, W., Eds.; Cambridge University Press: Cambridge, UK, 1979; pp. 790–831.
203. Reuter, M. Nonperturbative evolution equation for quantum gravity. *Phys. Rev. D* **1998**, *57*, 971. [CrossRef]
204. Bonanno, A.; Reuter, M. Cosmology of the Planck era from a renormalization group for quantum gravity. *Phys. Rev. D* **2002**, *65*, 043508. [CrossRef]
205. Bonanno, A.; Reuter, M. Cosmology with self-adjusting vacuum energy density from a renormalization group fixed point. *Phys. Lett. B* **2002**, *527*, 9. [CrossRef]
206. Niedermaier, M.; Reuter, M. The asymptotic safety scenario in quantum gravity. *Living Rev. Relativ.* **2006**, *9*, 5. [CrossRef] [PubMed]
207. Codello, A.; Percacci, R.; Rahmede, C. Investigating the ultraviolet properties of gravity with a Wilsonian renormalization group equation. *Ann. Phys.* **2009**, *324*, 414. [CrossRef]

208. Litim, D.F. Renormalization group and the Planck scale. *Philos. Trans. R. Soc. Lond. A* **2011**, *369*, 2759. [CrossRef]
209. Reuter, M.; Saueressig, F. Quantum einstein gravity. *New J. Phys.* **2012**, *14*, 055022. [CrossRef]
210. Reuter, M.; Saueressig, F. *Quantum Gravity and the Functional Renormalization Group: The Road Towards Asymptotic Safety*; Cambridge University Press: Cambridge, UK,2019.
211. Bonanno, A.; Eichhorn, A.; Gies, H.; Pawlowski, J.M.; Percacci, R.; Reuter, M.; Saueressig, F.; Vacca, G.P. Critical Reflections on Asymptotically Safe Gravity. *Front. Phys.* **2020**, *8*, 269. [CrossRef]
212. Bonanno, A.; Saueressig, F. Asymptotically safe cosmology—A status report. *C. R. Phys.* **2017**, *18*, 254–264. [CrossRef]
213. D'Odorico, G.; Saueressig, F. Quantum phase transitions in the Belinsky-Khalatnikov-Lifshitz universe. *Phys. Rev. D* **2015**, *92*, 124068. [CrossRef]
214. Loll, R. Discrete Approaches to Quantum Gravity in Four Dimensions. *Living Rev. Relativ.* **1998**, *1*, 13. [CrossRef] [PubMed]
215. Ambjorn, J.; Drogosz, Z.; Gizbert-Studnicki, J.; Görlich, A.; Jurkiewicz, J.; Németh, D. CDT Quantum Toroidal Spacetimes: An Overview. *Universe* **2021**, *7*, 79. [CrossRef]
216. Sorkin, R.D. Causal sets: Discrete gravity. In *Lectures on Quantum Gravity*; Springer: Boston, MA, USA, 2005; pp. 305–327.
217. Dowker, F. Causal sets and the deep structure of spacetime. *arXiv* 2005 arXiv:gr-qc/0508109.
218. Dowker, F. Introduction to causal sets and their phenomenology. *Gen. Relativ. Gravitation.* **2013**, *45*, 1651–67. [CrossRef]
219. Brightwell, G.; Henson, J.; Surya, S. A 2D model of causal set quantum gravity: The emergence of the continuum. *Class. Quantum Gravity* **2008**, *25*, 105025. [CrossRef]
220. Surya, S. Evidence for the continuum in 2D causal set quantum gravity. *Class. Quantum Gravity* **2012**, *29*, 132001. [CrossRef]
221. Surya, S. The causal set approach to quantum gravity. *Living Rev. Relativ.* **2019**, *22*, 5. [CrossRef]
222. Antoniadis, I.; Mazur, P.O.; Mottola, E. *Phys. Lett. B* **1997**, *394*, 49–56. [CrossRef]
223. Gross, J.D.; Migdal, A.A. A nonperturbative treatment of two-dimensional quantum gravity. *Nucl. Phys. B* **1990**, *340*, 333–365. [CrossRef]
224. Gross, D.J.; Piran, T.; Weinberg, S. (Eds.) Two Dimensional Quantum Gravity And Random Surfaces. In *Proceedings of the 8th Jerusalem Winter School For Theoretical Physics*; World Scientific: London, UK, 1991.
225. Schander, S.; Thiemann, T. Backreaction in Cosmology. *Front. Astron. Space Sci.* **2021**, *8*, 113. [CrossRef]
226. Brunnemann, J.; Thiemann, T. On (cosmological) singularity avoidance in loop quantum gravity. *Class. Quantum Gravity* **2006**, *23*, 1395. [CrossRef]
227. Halliwell, J.J.; Hawking, S.W. Origin of structure in the Universe. *Phys. Rev. D* **1985**, *31*, 1777–1791. [CrossRef]
228. Kiefer, C. Continuous measurement of mini-superspace variables by higher multipoles. *Class. Quantum Gravity* **1987**, *4*, 1369. [CrossRef]
229. D'Eath, P.D.; Halliwell, J.J. Fermions in quantum cosmology. *Phys. Rev. D* **1987**, *35*, 1100–1123. [CrossRef] [PubMed]
230. Moniz, P.V. Origin of structure in supersymmetric quantum cosmology. *Phys. Rev. D* **1998**, *57*, R7071. [CrossRef]
231. Kuchař, K.V.; Ryan, M.P. Is minisuperspace quantization valid? Taub in mixmaster. *Phys. Rev. D* **1989**, *40*, 3982–3996. [CrossRef]
232. Sinha, S.; Hu, B.L. Validity of the minisuperspace approximation: An example from interacting quantum field theory. *Phys. Rev. D* **1991**, *44*, 1028–1037. [CrossRef]
233. Hu, B.L.; Paz, J.P.; Sinha, S. Minisuperspace as a quantum open system. In *Directions in General Relativity*; Hu, B.L., Ryan, M.P., Vishveswara, C.V., Eds; Cambridge University Press: Cambridge, UK; New York, NY, USA, 1993; Volume 1, pp. 145–165.
234. Wada, S. Quantum cosmological perturbations in pure gravity. *Nucl. Phys. B.* **1986**, *276*, 729–743. [CrossRef]
235. Paz, J.P.; Sinha, S. Decoherence and back reaction: The origin of the semiclassical Einstein equations. *Phys. Rev. D* **1991**, *44*, 1038. [CrossRef] [PubMed]
236. Paz, J.P.; Sinha, S. Decoherence and back reaction in quantum cosmology: Multidimensional minisuperspace examples. *Phys. Rev. D.* **1992**, *45*, 2823–2842. [CrossRef] [PubMed]
237. Padmanabhan, T.; Singh, T.P. On the semiclassical limit of the Wheeler–DeWitt equation. *Class. Quantum Gravity* **1990**, *7*, 411. [CrossRef]
238. Bojowald, M. The BKL scenario, infrared renormalization, and quantum cosmology. *arXiv* 2019, arXiv:1810.00238.
239. Finelli, F.; Marozzi, G.; Vacca, G.P.; Venturi, G. Backreaction during Inflation: A Physical Gauge Invariant Formulation. *Phys. Rev. Lett.* **2011**, *106*, 121304. [CrossRef] [PubMed]
240. Buchert, T. On Average Properties of Inhomogeneous Fluids in General Relativity: Dust Cosmologies. *Gen. Relativ. Gravit.* **2000**, *32*, 105. [CrossRef]
241. Ellis, G.F.R. Inhomogeneity Effects in Cosmology. *Class. Quan. Grav.* **2011**, *28*, 164001. [CrossRef]
242. Ellis, G.F.R.; Buchert, T. The universe seen at different scales. *Phys. Lett. A* **2005**, *347*, 38. [CrossRef]
243. Clarkson, C.; Ellis, G.; Larena, J.; Umeh, O. Does the growth of structure affect our dynamical models of the Universe? The averaging, backreaction, and fitting problems in cosmology. *Rep. Prog. Phys.* **2011**, *74*, 112901. [CrossRef]
244. Buchert, T.; Räsänen, S. Backreaction in Late-Time Cosmology. *Annu. Rev. Nucl. Part. Sci.* **2012**, *62*, 57–79. [CrossRef]
245. Sussman, R.A. Weighed scalar averaging in LTB dust models I & II. *Class. Quantum Gravity* **2013**, *30*, 065015–065016.
246. Clifton, T.; Sussman, R.A. Cosmological backreaction in spherical and plane symmetric dust-filled space-times. *Class. Quantum Gravity* **2019**, *36*, 205004. [CrossRef]
247. Green, S.R.; Wald, R.M. New framework for analyzing the effects of small scale inhomogeneities in cosmology. *Phys. Rev. D* **2011**, *83*, 084020. [CrossRef]

248. Brill, D.R.; Hartle, J.B. Method of the self-consistent field in general relativity and its application to the gravitational geon. *Phys. Rev.* **1964**, *135*, B271. [CrossRef]
249. Burnett, G.A. The high-frequency limit in general relativity. *J. Math. Phys.* **1989**, *30*, 90. [CrossRef]
250. Buchert, T.; Carfora, M.; Ellis, G.F.R.; Kolb, E.W.; MacCallum, M.A.H.; Ostrowski1, J.J.; Räsänen, S.; Roukema, B.F.; Andersson, L.; Coley, A.A.; et al. Is there proof that backreaction of inhomogeneities is irrelevant in cosmology? *Class. Quantum Gravity* **2015**, *32*, 215021. [CrossRef]
251. Green, S.R.; Wald, R.M. Comments on Backreaction. *arXiv* **2015**, arXiv:1506.06452v2.
252. Green, S.R.; Wald, R.M. A simple, heuristic derivation of our 'no backreaction' results. *Class. Quantum Gravity* **2016**, *33*, 125027. [CrossRef]

Review

Quantum Cosmology with Third Quantisation

Salvador J. Robles-Pérez

Departamento de Matemáticas, Universidad Carlos III de Madrid, Avda. de la Universidad 30, 28911 Leganés, Spain; sarobles@math.uc3m.es

Abstract: We reviewed the canonical quantisation of the geometry of the spacetime in the cases of a simply and a non-simply connected manifold. In the former, we analysed the information contained in the solutions of the Wheeler–DeWitt equation and showed their interpretation in terms of the customary boundary conditions that are typically imposed on the semiclassical wave functions. In particular, we reviewed three different paradigms for the quantum creation of a homogeneous and isotropic universe. For the quantisation of a non-simply connected manifold, the best framework is the third quantisation formalism, in which the wave function of the universe is seen as a field that propagates in the space of Riemannian 3-geometries, which turns out to be isomorphic to a (part of a) $1 + 5$ Minkowski spacetime. Thus, the quantisation of the wave function follows the customary formalism of a quantum field theory. A general review of the formalism is given, and the creation of the universes is analysed, including their initial expansion and the appearance of matter after inflation. These features are presented in more detail in the case of a homogeneous and isotropic universe. The main conclusion in both cases is that the most natural way in which the universes should be created is in entangled universe–antiuniverse pairs.

Keywords: quantum cosmology; multiverse; superspace; third quantisation; universe–antiuniverse pair

---

Citation: Robles-Pérez, S.J. Quantum Cosmology with Third Quantisation. *Universe* **2021**, *7*, 404. https://doi.org/10.3390/universe7110404

Academic Editor: Jaime Haro Cases

Received: 25 June 2021
Accepted: 25 October 2021
Published: 27 October 2021

Publisher's Note: MDPI stays neutral with regard to jurisdictional claims in published maps and institutional affiliations.

Copyright: © 2021 by the author. Licensee MDPI, Basel, Switzerland. This article is an open access article distributed under the terms and conditions of the Creative Commons Attribution (CC BY) license (https://creativecommons.org/licenses/by/4.0/).

Contents

1 Introduction 2
2 Quantisation of a Simply Connected Spacetime Manifold 3
   2.1 Quantisation of the Spacetime Geometry 3
   2.2 Boundary Conditions 7
   2.3 Semiclassical Quantum Gravity 9
   2.4 Minisuperspace Model 11
      2.4.1 Inflationary Universe 13
      2.4.2 Small Perturbations and Backreaction 15
   2.5 Paradigms for the Creation of the Universe in Quantum Cosmology . 20
      2.5.1 Creation of the Universe from *Nothing* 22
      2.5.2 Creation of the Universe from *Something* 26
      2.5.3 Creation of Universes in Pairs 28
3 Third Quantisation Formalism 29
   3.1 Historical Review 29
   3.2 Quantum Field Theory in $M \equiv \text{Riem}(\Sigma)$ 33
      3.2.1 Geometrical Structure of M 33
      3.2.2 Classical Evolution of the Universe 35
      3.2.3 Quantum Field Theory in M 36
      3.2.4 Boundary Conditions and the Creation of the Universes in Pairs . . . 39
      3.2.5 Semiclassical Regime 41
   3.3 Minisuperspace Model 43
      3.3.1 Geometrical Structure of the Minisuperspace 43
      3.3.2 Field Quantisation of a FRW Spacetime 47

		3.3.3 Reheating and the Matter–Antimatter Content of the Entangled Universe 49
4	Observable Effects of Quantum Cosmology 54
5	Conclusions 56
	References 60

1. Introduction

Quantum cosmology is the application of the quantum theory to the universe as a whole. However, it was clear from the beginning that the customary formalism of the Copenhagen interpretation cannot be applied to the quantisation of the universe because the Schrödinger equation and the measurement process on which the Copenhagen formalism is based cannot be fundamental elements of a quantum theory of the spacetime. Let us notice that if the quantum theory must represent the quantum state of the spacetime, then, as Wheeler showed [1], its quantum state must be at the Planck length with a superposition of spacetime geometries that is impossible to visualise or represent and where one cannot even *use the word "observation" at all* (cf. [1]). However, as we approach the macroscopic scale, the quantum state of the universe must represent the approximately stable spacetime where we live and perform measurements of particles and other matter fields. It means that the description of the universe that we observe must be an emergent feature of the quantum representation of the universe.

In this article, we review the canonical formulation of quantum cosmology. We start from the foliation of the spacetime into space and time that allows us to express the Einstein–Hilbert action as a functional action of the components of the 3-metric of the spatial sections of the spacetime. The evolution of the universe turns out to then be a trajectory in the space of 3-Riemannian metrics, M, and its quantum state is represented by a wave function that is the solution of the Wheeler–DeWitt equation, which, in principle, contains all the information about the spacetime and the matter fields that propagate therein. However, as we have already said, the full quantum state is a superposition of solutions that correspond to different paths, i.e., different evolutions, in the space M. It is only in the semiclassical regime where a particular kind of solution emerges by a decoherence process[1]. This kind of solution includes the semiclassical solutions that represent a fixed classical spacetime background with matter fields propagating therein and in which the Schrödinger equation appears as an approximated equation at order $\hbar^1$. In particular, we analyse the semiclassical wave function of a homogeneous and isotropic universe with small inhomogeneities that can be treated as perturbations. In that case, explicit solutions can be given for which it is easier to analyse the different boundary conditions that can be imposed on the state of the universe. They give rise to different scenarios for the creation of the universe, which are analysed in detail.

On the other hand, the space of 3-dimensional space-like metrics, M, defined at any point in the space, turns out to be isomorphic to a $1+5$ dimensional Minkowski spacetime[2]. This analogy between the space M and the spacetime allows us to consider the wave function of the universe as a field that propagates in M, and the Wheeler–DeWitt equation as the field equation. In that case, a procedure of quantisation called *third quantisation* can formally be performed in a similar way as it is done in a quantum field theory. For instance, we can define quantum operators representing the creation and annihilation of particular modes of the spacetime, i.e., different universes, and the corresponding Fock space will then allow us to represent the quantum state of a multiply connected spacetime manifold. It turns out to then be the appropriate framework to describe the quantum state of the multiverse. Moreover, as it happens in a quantum field theory, the isotropy of the background space implies that the creation of universes must be in pairs with opposite values of the components of the momentum conjugated to the configuration variables. We shall analyse the charge and parity relation between the matter fields that propagate in one of these pairs to see that the matter content of one of the universes must be CP inversely related with the content of the other universe. They thus form a universe–antiuniverse pair. We analyse this pair creation in detail in the case of a homogeneous and isotropic universe where the period of reheating after inflation is investigated. The decay of the inflaton field into the particles of the Standard Model is produced in a CP-conjugated way in the two universes, so any excess of matter over antimatter in one of the universes of the entangled pair is balanced with the excess of the antimatter over matter in the partner universe, these two concepts (matter and antimatter) always having a relative meaning, i.e., an internal observer in any of the universe always interprets the content of his/her universe as matter.

This proposal is however still far from being directly testable. The effects of quantum cosmology and, in particular, the effects of the existence of an entangled universe are mainly restricted to the very early stage of the universe. Perhaps with future advances in the detection of gravitational waves or of a cosmic neutrino background we will be able to test the pre-inflationary stage of the universe where the effects of quantum gravity may be significant. Moreover, the third quantisation formalism can also be a proposal for the quantisation of the spacetime, and thus a better understanding of the formalism and its application to other gravitational scenarios can provide us with a new line of research for the search of a quantum theory of gravity.

2. Quantisation of a Simply Connected Spacetime Manifold

2.1. Quantisation of the Spacetime Geometry

Following the customary approach[3], the spacetime can be foliated in space and time by assuming a global time function t such that each surface $t = constant$ is a spacelike Cauchy hypersurface, Σ_t. The proper distance between the point x_0 in Σ_t and the point $x_0 + dx$ of Σ_{t+dt} is given by [7,8] (see, Figure 1)

$$ds^2 = g_{\mu\nu}dx^\mu dx^\nu = \left(N_a N^a - N^2\right)dt^2 + 2N_a dx^a dt + h_{ab} dx^a dx^b, \tag{1}$$

where $N_a N^a = h_{ab} N^a N^b$ and h_{ab} is the three-dimensional metric induced on each hypersurface Σ_t, with unit normal n_μ, satisfying, $n^\mu n_\mu = -1$. The functions N and N^a are called the lapse and the shift functions, respectively. They are the normal and tangential components of the vector field t^μ, which is the vector field that transport the point x_0 from Σ_t to Σ_{t+dt}.

With the split of the spacetime in space and time, the spacetime can be seen as a spacelike hypersurface evolving in time. The geometry of the hypersurface Σ_t at a given time t_0 is determined by the metric tensor $h_{ab}(t_0)$, so eventually the evolution of the universe is encoded in the evolution of the metric $h_{ab}(t)$. It is then remarkable that in the end, as Wheeler says [9], *Eintein's geometrodynamics deals with the dynamics of 3-geometry, not 4-geometry!* (emphasis his). From this $3 + 1$ viewpoint of the spacetime, we can cast the Einstein-Hilbert action[4] [7]

$$S_{EH} \int_{\mathcal{M}} d^4x \sqrt{-g}\left({}^4R - 2\Lambda\right) - 2\int_{\partial\mathcal{M}} d^3x \sqrt{h}K, \tag{2}$$

and the action of the matter fields[5]

$$S_{\text{matter}} = \int_{\mathcal{M}} d^4x \sqrt{-g}\left(\frac{1}{2}g^{\mu\nu}\partial_\mu\varphi\partial_\nu\varphi - V(\varphi)\right), \tag{3}$$

into the standard Lagrangian form,

$$S = \int dt\, L(q^i, \dot{q}^i, t), \tag{4}$$

where q_i and $\dot{q}_i$ will be here the spatial metric, $h_{ab}(x)$, the matter field(s), encoded in the variable $\varphi(x)$, and their corresponding velocities. In (2), ${}^4R$ is the Ricci scalar associated to the 4-dimensional metric $g_{\mu\nu}$, Λ is the cosmological constant, $K = h_{ab}K^{ab}$, is the trace of the extrinsic curvature, which can be written as,

$$K_{ab} = \frac{1}{2N}\left(\dot{h}_{ab} - D_a N_b - D_b N_a\right), \tag{5}$$

where, D_a is the covariant derivative on the spatial section Σ_t, and $\partial\mathcal{M}$ is the boundary of the manifold $\mathcal{M}$. After some manipulation (see Ref. [7]), one finds that the Einstein–Hilbert action (2) can be written as

$$S_{EH} = \int_\mathcal{M} dt d^3x N\left(G^{abcd}K_{ab}K_{cd} + \sqrt{h}\left(^3R - 2\Lambda\right)\right), \tag{6}$$

where h is the determinant of the spatial metric h_{ab}, and

$$G^{abcd} = \frac{\sqrt{h}}{2}\left(h^{ac}h^{bd} + h^{ad}h^{bc} - 2h^{ab}h^{cd}\right), \tag{7}$$

is the DeWitt's metric [10]. The structure of the action (6) is very interesting. First, it is of the standard form (4),

$$S_{EH} = \int dt\, L_{EH} = \int dt d^3x\, \mathcal{L}_{EH}, \tag{8}$$

with L_{EH} the Lagrangian associated to the Einstein–Hilbert action (6) and $\mathcal{L}_{EH}$ the Lagrangian density. Second, from (5) one can see that the extrinsic curvature K_{ab} contains the time derivative of the metric tensor h_{ab}, and ^3R only depends on h_{ab}. Thus, the action (6) presents the customary structure of a kinetic term that is quadratic in the *velocities* plus a potential term. Furthermore, the supermetric (7) defines a metric structure on the space of spacelike metrics, called the superspace[6]. Thus, the action (6) looks like the action of a particle that moves in a curved space; the coordinates of the "particle" are the time-dependent values of the components of the metric tensor, $h_{ab}(t)$ and the curved space where this particle moves in the space of symmetric Riemannian 3-metrics. That is, the evolution of the universe can be seen as the *trajectory* in the superspace[7] (see Figure 2).

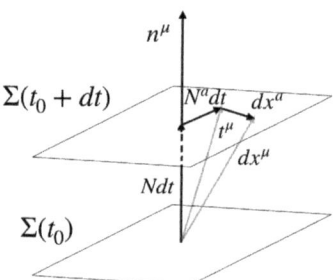

Figure 1. Splitting the spacetime into space and time.

The momenta conjugated to the metric tensor components, h_{ab}, are given by [7]

$$p^{ab} \equiv \frac{\partial \mathcal{L}_{EH}}{\partial \dot{h}_{ab}} = G^{abcd}K_{cd} = \sqrt{h}\left(K^{ab} - Kh^{ab}\right). \tag{9}$$

Thus, in terms of the momenta, the total action (the gravitational action plus the action of the matter field) can be written as

$$S = S_{EH} + S_{\text{matter}} = \int dt d^3x \left(p^{ab}\dot{h}_{ab} + p_\varphi \dot{\varphi} - N\mathcal{H} - N^a\mathcal{H}_a\right), \tag{10}$$

where the lapse and the shift functions act as Lagrange multipliers, with [7]

$$\mathcal{H} = G_{abcd}p^{ab}p^{cd} - \sqrt{h}\left(^3R - 2\Lambda\right) + \mathcal{H}_{matter}, \tag{11}$$

$$\mathcal{H}_a = -2D_b p_a^b + \sqrt{h}J_a, \tag{12}$$

where [7], $J_a \equiv h_a^\mu T_{\mu\nu} n^\nu$, and G_{abcd} is the inverse of the DeWitt metric (7),

$$G_{abcd} = \frac{1}{2\sqrt{h}}(h_{ac}h_{bd} + h_{ad}h_{bc} - h_{ab}h_{cd}), \tag{13}$$

with [10],

$$G^{abcd}G_{cdef} = \frac{1}{2}\left(\delta_e^a \delta_f^b + \delta_f^a \delta_e^b\right). \tag{14}$$

Therefore, variation of the action (10) with respect to the lapse and the shift functions yields the classical Hamiltonian and momentum constraints, respectively, i.e.,

$$\mathcal{H} = 0 \ , \ \mathcal{H}_a = 0. \tag{15}$$

Let us now focus for a moment on the gravitational part of these constraints. The Hamiltonian constraint, $H = 0$, can be seen as the analogue of the momentum constraint of a particle that propagates in the spacetime,

$$g^{\mu\nu}p_\mu p_\nu + m^2 = 0, \tag{16}$$

in which the mass is substituted by a non-constant potential (this analogy will be further exploited in Section 3). On the other hand, the function $\mathcal{H}_a$ generates infinitesimal diffeomorphisms (change of coordinates) in the spatial hypersurfaces, Σ. Thus, the second constraint in (15), $\mathcal{H}_a = 0$, means that the Einstein–Hilbert action (6) is invariant under such diffeomorphisms, which turns out to be like a gauge freedom [7]. Thus, the real configuration space is the quotient space of all Riemannian 3-metrics, $M \equiv \text{Riem}(\Sigma)$, in which all three metrics related by diffeomorphisms correspond to the same class, i.e.

$$\mathcal{S}(\Sigma) = \frac{M}{\text{Diff}(\Sigma)}, \tag{17}$$

which is called the *superspace* [7–10].

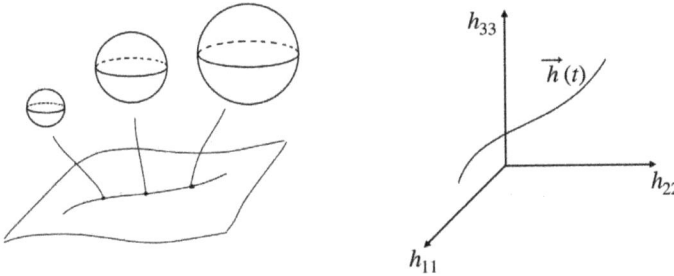

Figure 2. Left: the evolution of the universe can be seen as a path in the abstract space spanned by the component of the spatial metric tensor, h_{ab}. Right: an example of evolution is depicted for the case that the spatial metric is diagonal, i.e., given by, $h_{ab}(t) = \text{diag}(h_{11}(t), h_{22}(t), h_{33}(t))$.

Following Dirac [11], the canonical procedure of quantisation consists in assuming the quantum version of the classical constraints (15) by promoting the classical variables and their conjugate momenta into quantum operators and applying them to a wave function, ϕ, that is defined in the configuration space,

$$\hat{\mathcal{H}}\phi = 0 \ , \ \hat{\mathcal{H}}_a \phi = 0. \tag{18}$$

The quantum version of the momentum constraint in (15), $\hat{\mathcal{H}}_a \phi = 0$, assures that the wave function ϕ is invariant under spatial diffeomorphisms in the 3-dimensional slices Σ_t [7]. For our purposes, much more interesting is the constraint, $\hat{\mathcal{H}}\phi = 0$, which under canonical quantisation becomes the Wheeler–DeWitt equation [7,8,10],

$$\left(-\hbar^2 \nabla \cdot \vec{\nabla} + \sqrt{h}\left(-{}^{(3)}R + 2\Lambda + \hat{T}^{00} \right) \right) \phi(h_{ab}, \varphi) = 0, \tag{19}$$

where, $\nabla \cdot$ and $\vec{\nabla}$, are the divergence and the gradient, respectively, defined in the space of 3-dimensional Riemannian metrics, M, and for a scalar field $\hat{T}^{00}$ reads

$$\hat{T}^{00} = \frac{-\hbar^2}{2h} \frac{\delta^2}{\delta \varphi^2} + \frac{1}{2} h^{ij} \varphi_{,i} \varphi_{,j} + V(\varphi). \tag{20}$$

The Wheeler–DeWitt Equation (19) is the keystone of the canonical formalism of quantum cosmology. The solution, $\phi(h_{ab}, \varphi)$, is usually called the *wave function of the universe* [12] because it represents the quantum state of the spacetime and the matter fields that propagate therein. This is usually applied to the case of a single universe. However, as we shall see in Section 3, the whole spacetime manifold can present a more complicated structure and represent something more than what is typically called a universe. In that sense, the name can be misleading. However, for historical reasons we shall retain sometimes the name *wave function of the universe* even in the cases where it may represent the state of many different universes.

The Wheeler–DeWitt Equation (19) can be seen as a Schrödinger-like equation with no time variable, which is a consequence of the invariance of the quantum state of the universe with respect to the time variable. In that sense, there is no preferred time in the quantum description of the universe (additionally, there are no preferred spatial coordinates because the constraint, $\mathcal{H}_a = 0$). It is then sometimes stated that there is no time evolution of the quantum state of the universe. However, this is not true, or at least it is not accurate. As we have already pointed out, the evolution of the universe can be seen as the trajectory in the superspace. The spacetime coordinates are the parameters that parametrise the trajectory, which is therefore invariant under reparametrisations, but that does not mean that there is no evolution. It is similar to the description of the path followed by a particle in the spacetime, which is independent of the parametrisation of the path, but that does not mean that the particle does not move.

There is also a path integral approach to quantum gravity and quantum cosmology [7,8,12]. It is a generalisation of Feynman's idea that the amplitude for a particle to go from one to another point point is given by a functional integral that weighs all the paths that start from the point x_0 at time t_0 and end in the point x_1 at t_1 (see Figure 3). Following a parallel reasoning, Hartle and Hawking propose [12] that the amplitude for the universe to change from the hypersurface Σ, in which the spatial geometry and the field configuration are given by h_{ab} and φ, respectively, to the hypersurface Σ', where they are given by the values h'_{ab} and φ', is given by [12]

$$\langle h'_{ab}, \varphi' | h_{ab}, \varphi \rangle = \int \delta g \, \delta \varphi \, e^{iS[g,\varphi]}, \tag{21}$$

where the integral must be performed over all 4-geometries and field configurations that match the given values on the two spacelike hypersurfaces [12], Σ and Σ'. Following that approach, the wave function of the universe is then given by,

$$\phi(h_{ab}, \varphi) = \sum \int_C \delta g \delta \varphi \, e^{iS(g,\varphi)}, \qquad (22)$$

where C denotes the class of spacetimes and matter configurations that fulfil the boundary requirements on the hypersurfaces Σ, and the sum is performed over all kind of topologies. In order to make well defined the path integral in (22) one has to rotate to Euclidean time. However, that does not remove all the technical problems, which we are not going to deal with here. In practice, as it happens with the Wheeler–DeWitt equation, the path integral can only be performed for spacetimes and matter field configurations with a high degree of symmetry. Moreover, both formulations become equivalent because the requirement of invariance of the wave function $\phi(h_{ab}, \varphi)$ under reparametrizations of the time variable implies the constraint [12], $\frac{\delta S}{\delta N} = 0$, whose quantised version is the Wheeler–DeWitt equation.

Regardless, the path integral formulation has two interesting points. First, the analogy with the Feynman's path integral formulation of the trajectory of a particle in spacetime makes it very intuitive. As we have seen, the classical evolution of the universe can be seen as a path in the superspace. Applying Feynman's idea to gravity means that the quantum state of the universe is given by a quantum superposition of all the paths that go from one to another configuration of the spatial hypersurfaces. As it happens in the spacetime, the *classical path* (i.e., the classical evolution) emerges in some specific limit because of the constructive interference between the paths of the quantum superposition, and in the same limit the non-classical paths suffer from a destructive interference or *decoherence* [3,4,13,14]. The second interesting point of the path integral approach is the following: let us assume that we already know how to construct the quantum amplitude for the universe to go from one to another given configuration of the spacelike section. Then, what is the amplitude for the birth of the universe? What are the boundary conditions that one must impose on the state of the universe to obtain the appropriate probability amplitude for the universe to be created?

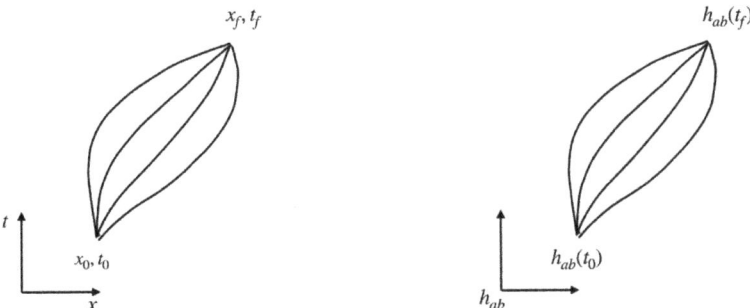

Figure 3. (**Left**): path integral in spacetime. (**Right**): path integral in quantum gravity.

2.2. Boundary Conditions

In classical mechanics, as well as in quantum mechanics, we usually work with some given conditions that we know or assume for certain at some initial time. Then, knowing the law of evolution, the initial conditions determine the state of the system at any later time. In the universe the thing is a bit different. What we only know for certain is the state of the observable universe, say from the inflationary period[8] to the current stage of accelerated expansion, and we have to make some guess about the initial conditions that give rise to a universe like that. However, this is all classical cosmology. The question in quantum cosmology is: what are the conditions at the quantum level that give rise to a specific initial boundary, Σ_0, that is propitious to inflate? The probability for the creation of such universe would be given by the modulus squared of the wave function of the universe, when this is evaluated at the initial hypersurface[9]. From that point of view, one can say that the initial state of the (classical) universe is the final state of the amplitude for the universe to be created ... *from what?* That is actually the issue behind the question of the boundary condition of the universe.

Using the path integral approach, Hartle and Hawking proposed [12] that the class C over which the integral (22) has to be performed is the class of compact geometries (in principle of all topologies) that have Σ_0 as their only boundary [7,12] (see Figure 4) and matter fields that are regular on those geometries. It means that the *boundary of the universe is that the universe has no boundary*, or equivalently, that the boundary Σ_0 is created from *nothing*. We shall see later on that for the case of an inflationary spacetime, the quantum state that results from the no-boundary condition is [7,12,15]

$$\phi_{NB} \propto \exp\left(\frac{1}{3V(\varphi)}\right)\cos\left(\frac{(a^2V(\varphi)-1)^{\frac{3}{2}}}{3V(\varphi)} - \frac{\pi}{4}\right), \qquad (23)$$

where $V(\varphi)$ is the potential of the inflaton field and a is the scale factor, which goes from the initial value, $a_0 = V(\varphi)^{-1/2}$, to infinity. It is important to notice that the wave function (23) can be written as

$$\phi_{NB} \propto e^{\frac{1}{3V(\varphi)}}\left(e^{iS} + e^{-iS}\right), \qquad (24)$$

where,

$$S = \frac{(a^2V(\varphi)-1)^{\frac{3}{2}}}{3V(\varphi)} - \frac{\pi}{4}. \qquad (25)$$

We shall see in the next section that in terms of the same time variable, one of the two terms in (24), say the *branch* with e^{-iS}, describes an expanding universe, and the branch with e^{iS} describes a contracting universe. It means that the result of imposing the no-boundary condition on the quantum state of the universe is that it is given by the linear combination of two states: one representing an expanding universe and one representing a contracting universe. Typically, one considers the expanding branch of the universe as representing our universe and disregards the contracting branch for being unphysical. However, we shall see that there is another interpretation. In terms of the *physical* time variable of each universe, the two universes can both be seen as expanding universes but with their matter fields being CP-conjugated. They can be interpreted then as an expanding universe–antiuniverse pair (see Sections 2.5.3 and 3). From that point of view, the no-boundary proposal would yield the creation of universes in entangled universe–antiuniverse pairs.

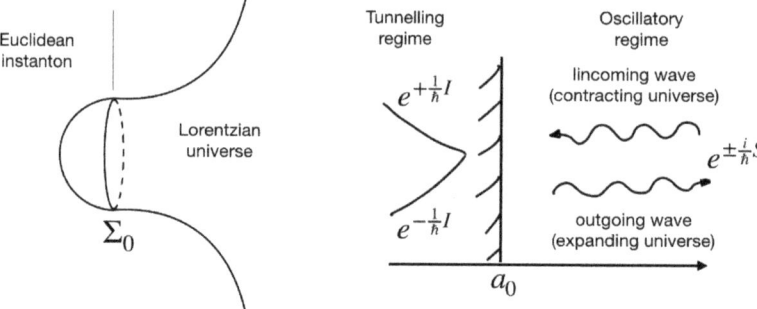

Figure 4. Left: the class C over which the path integral has to be performed is the class of compact geometries that have Σ_0 as their only boundary [12]. Right: the tunnelling proposal states that the only modes that survive the quantum barrier are the "outgoing" modes that represent expanding universes.

Vilenkin's tunneling proposal is quite different[10]. Perhaps more based on practical grounds, he proposes that the only mode that survives the tunnelling from the Euclidean region (the region located at, $a < V(\varphi)^{-1/2}$) is the one that represents an expanding universe. Imposing the tunnelling boundary condition, the resulting wave function of the universe is

$$\phi_T \propto \exp\left(-\frac{1}{3V(\varphi)}\right) \exp\left(\frac{-i}{3V(\varphi)}(a^2 V(\varphi) - 1)^{\frac{3}{2}}\right). \tag{26}$$

The main difference with respect to the Hartle–Hawking wave function (23) is the negative sign in the exponent of the exponential pre-factor, which may have important consequences. Let us notice that the probability for the universe to be created from nothing is $P \propto |\phi|^2$; so, in the case of the no-boundary condition, we have

$$P_{HH} \propto e^{\frac{2}{3V(\varphi)}}, \tag{27}$$

while in the case of the Vilenkin's tunnelling condition, we have

$$P_T \propto e^{-\frac{2}{3V(\varphi)}}. \tag{28}$$

One immediate consequence is that the Vilenkin's condition seems to favour the creation of a universe with a large value of the potential, which is a necessary condition for the initial hypersurface Σ_0 to inflate. That would in principle reject the Hartle–Hawking proposal because, on the contrary, the no-boundary proposal seems to favour the creation of a universe with a small (or zero) value of the potential ($P_{HH} \to \infty$, as $V \to \infty$). However, this result changes when high-order corrections are taken into account, so the result is not conclusive [7].

From a purely theoretical point of view, it seems that the Hartle–Hawking proposal is more fundamental in the sense that these authors put the focus on the *natural* condition that one should impose on the Euclidean region of the spacetime. As a consequence, they obtain that the universe is represented by two branches, one corresponding to an expanding universe and the other describing a contracting branches. These two branches can be considered independently once they suffer a process of decoherence, so in practice they may represent two different universes. Vilenkin's proposal seems to be more practical (although it is also based on a parallelism with some processes of quantum mechanics).

2.3. Semiclassical Quantum Gravity

The first thing that the quantum state of the spacetime must provide us with is a consistent explanation of how the classical background can emerge from the full quantum state of the spacetime. Fortunately, not only can this be done in a beautiful manner but it is in fact one of the greatest achievements of quantum cosmology.

Again, the path integral formulation of the spacetime supplies a clear picture of how it can be obtained. At some appropriate limit, which is generally a large length or mass scale compared with their corresponding Planck values, the contribution of most of the paths in the integral vanishes because of their destructive interference. The only paths that survive the interference are those that are *in phase*, i.e., those for which $\delta S \approx 0$. These are actually the trajectories of the superspace given by the classical constraints. Therefore, much in a similar way as the classical trajectory of a particle emerges from the constructive interference in the path integral approach of the quantum mechanics of a particle, the classical background spacetime emerges as the constructive interference among the paths in the superspace. In the quantum mechanics of a particle that we can then compute the quantum corrections to the trajectory of the particle in terms of *quantum uncertainties*. We shall see in this section that the quantum corrections to the classical background of the universe are caused by the matter fields.

Following the customary approach [5,7,8,15], let us consider the following semiclassical wave function [5],

$$\phi(h_{ab}, \varphi) = \Delta(h_{ab}) e^{\pm \frac{i}{\hbar} S_0(h_{ab})} \psi(h_{ab}, \varphi), \tag{29}$$

where Δ and ψ are slowly varying functions of the metric tensor h_{ab}, and $S_0(h_{ab})$ is the classical action for gravity alone, given by (2). Essentially, the wave function (29) contains two parts: one that only depends on the geometric variables of the spacetime and another part that contains all the matter degrees of freedom in the wave function ψ. The basic idea is that we expect that the quantum fluctuations of the spacetime will weaken more rapidly than the quantum fluctuations in the state of the matter fields. In that case, we shall find a regime where the spacetime behaves nearly classically with quantum matter fields propagating therein. That is exactly what we need to describe the universe we observe. Also notice the presence of the Planck constant in (29). It means that the classical behaviour of the spacetime will be present whenever the gravitational action is large with respect to the Planck constant.

Now, insert the semiclassical wave function (29) into the Wheeler–DeWitt Equation (19) and solve it order by order in $\hbar$ in the geometrical degrees of freedom. At order $\hbar^0$, one finds

$$G_{abcd} \frac{\delta S_0}{\delta h_{ab}} \frac{\delta S_0}{\delta h_{cd}} - \sqrt{h}\left(^3R - 2\Lambda\right) = 0, \tag{30}$$

which is the gravitational part of the Hamiltonian constraint (11) if we make the identification,

$$p^{ab} = \frac{\delta S_0}{\delta h_{ab}}. \tag{31}$$

At first order in $\hbar$, neglecting second derivatives of the slowly varying terms with respect to the 3-metric, two equations are obtained. The first equation [5]

$$G_{abcd} \frac{\delta}{\delta h_{ab}} \left(\Delta^2 p^{cd}\right) = 0, \tag{32}$$

is actually the condition for the function $\Delta(h_{ab})$ to be a *slowly varying* function. In other words, whenever (32) is satisfied (or to the extent it is satisfied) the wave function (29) can be a good candidate to describe the observable universe. Equation (32) is also the equation of the conservation of the probability current, $\Delta^2 p^{cd}$. The other equation that is obtained at order $\hbar$ is

$$\mp 2i G_{abcd} \frac{\delta S_0}{\delta h_{ab}} \frac{\delta \psi}{\delta h_{cd}} + \sqrt{h} \hat{T}^{00} = 0, \tag{33}$$

where the $\mp$ signs correspond to the $\pm$ signs of the exponent of (29). It suggests the identification of a time variable t, given by

$$\frac{\partial}{\partial t} = \pm 2 G_{abcd} \frac{\delta S_0}{\delta h_{ab}} \frac{\delta}{\delta h_{cd}}. \tag{34}$$

In that case, (33) becomes the Schrödinger-like equation of the matter fields

$$i \frac{\partial \psi}{\partial t} = \sqrt{h}\, \hat{T}^{00}(\varphi, -i\partial/\partial\varphi)\psi. \tag{35}$$

Therefore, in the semiclassical regime we obtain at order $\hbar^0$ the classical behaviour of the spacetime and, at first order in $\hbar$, the quantum evolution of the matter fields. Thus, the semiclassical wave function (29) contains all the physical information of the observable universe. One could say that recovering the classical equations for the spacetime degrees of freedom and the Schrödinger equation for the matter fields does not add anything to what we already knew before the quantisation of the universe. That is true; these two features are nothing more than a test of consistency for quantum cosmology. After all, the recovering of the classical spacetime must not be surprising. We started the process of quantisation from the classical action of the spacetime and the matter fields, and the quantisation procedure consists basically in promoting the classical variables to non-commuting quantum operators, $[\hat{h}_{ab}, \hat{p}^{ab}] \propto \hbar$. Then, it should not be surprising that in the limit $\hbar \to 0$ we recover the classical behaviour (this is the essence of the correspondence principle) and something similar for the quantum behaviour of the matter fields.

The canonical quantisation of the whole universe that we have seen in this section is interesting for several reasons. First, it suggests that the quantisation processes that leads to the wave function of the universe is consistent and, in that case, one can assume that the wave function of the universe, $\phi(h_{ab}, \varphi)$, would contain in principle all the physical (classical and quantum) information of all the degrees of freedom of the universe. Any physical process should be describable within the formalism of quantum cosmology. Of course, this reductionist point of view is not practical at all, but from the conceptual point of view it results appealing. A more interesting feature is that it allows us to analyse higher-order corrections to the semiclassical universe, and this should give novel features that cannot be foreseen in the classical scenario. It might help us to go beyond the quantum description of matter fields in a classical spacetime background. In particular, it can help us to find some exclusive features of the quantum regime of the spacetime, i.e., small deviations from the known behaviour caused by the high-order corrections of quantum gravity [17–20].

Another interesting feature of quantum cosmology is the appearance of time. From the point of view of the superspace, time is just the parameter that parametrises the curve that describes a particular trajectory. In the picture given by the path integral, the quantum state of the universe is given by the set of all paths that join together the initial and final states. If the quantum wave-packet is spread, no definite time variable can be chosen mainly because there is no definite curve that describes the evolution of the universe. It is only when the wave-packet is peaked around a particular solution (ideally becoming a delta function) when we have a definite curve, the classical evolution of the universe, that it can therefore be parametrised in terms of a parameter that we can call time[11]. It therefore appears as the result of a decoherence process between the different *histories* of the universe [3–5,21].

Furthermore, quantum cosmology relates the two concepts of time of contemporary physics: the one of the theory of relativity and the one of quantum mechanics. This is a very subtle point. Both the theory of relativity and the quantum theory work with a time variable, say t_r and t_q, respectively. We usually assume that both time variables are the same, $t_r = t_q = t$, but this is an assumption that is not guaranteed from the beginning. Of course, they (must) coincide in the Newtonian limit of both theories, but in general, they only coincide if the time variable of theory of relativity would be measured with an actual clock, which is made of matter fields. However, the theory of relativity deals with "ideal clocks," and the consideration of an actual clock may entail some problems[12].

2.4. Minisuperspace Model

Despite its conceptual importance, it is not hard to see that the Wheeler–DeWitt equation found in the previous chapter is very difficult if not impossible to solve for a general configuration of the spacetime and the matter field(s). In order to make computations, one generally has to assume some symmetries in the underlying spacetime of the universe. This process, called *symmetry reduction* [7], reduces the number of variables of the superspace and the so-reduced superspace is called *minisuperspace*[13].

Furthermore, a minisuperspace model is not necessarily an unrealistic model. On the contrary, the observational data indicates that for most of the history of the universe the spacetime presents a high degree of symmetry. Even more, if the initial hypersurface Σ_0 from which the universe starts evolving is large enough compared with the Planck length, a minisuperspace model could describe the whole history of the universe[14]. In other cases, it can be taken as a *toy model* from which we can obtain relevant information about some (classical and quantum) aspects of the universe like, for instance, the creation of the initial hypersurface Σ_0 or the *quantum-to-classical* transition and the appearance of time, among others.

They can also allow us to study the effect of small deviations from the symmetric picture. For instance, we shall consider later on small departures from the homogeneous spacetime and the matter field in the form of gravitons and matter particles, respectively. These are local perturbations of the otherwise homogeneous and isotropic background. The picture then becomes quite realistic and allows us to analyse potentially observable effects like the kind of correlations between the modes of the matter fields in different regions of the spacetime or the quantum gravitational corrections to the Schrödinger equation of the matter fields. In all those cases, the study of the minisuperspace model turns out to be justified.

Therefore, let us consider the minisuperspace that is obtained from the foliation of a 4-dimensional spacetime with closed homogeneous and isotropic spatial sections. The geometry of the spacetime is then characterised by a Freedman–Robertson–Walker (FRW) metric

$$ds^2 = -N^2(t)dt^2 + a^2(t)d\Omega_3^2, \qquad (36)$$

where $d\Omega_3^2$ is the line element on the unit three sphere. The foliation of the spacetime into space and time is in that case parametrised by just two functions, the scale factor $a(t)$ and the lapse function $N(t)$. The geometry of the spatial sections, h_{ab}, is then fully characterised by the scale factor $a(t)$ that parametrises the variation in the distance between two fixed points of the space along the evolution of the universe. It parametrises therefore the expansion or the contraction of the universe. On the other hand, the lapse function $N(t)$ determines the time parametrisation of the foliation. Different values of $N(t)$ entail different time variables (i.e., different time parametrisations), with some special cases. For instance, if $N = 1$, the time variable t is called cosmic time, and if $N = a(t)$, t is customarily renamed with the Greek letter η and is called conformal time because in terms of η the metric becomes conformal to the metric of a closed static spacetime. We know that the evolution of the universe is invariant under the choice of time reparametrisation, so, at the end of the process, we can take the preferable time variable[15].

The line element of the spacetime is then fully determined by these two functions, $a(t)$ and $N(t)$. The total action, i.e., the Einstein–Hilbert action (2) plus the action of the scalar field (3), can be written as [7]

$$S = S_{EH} + S_m = \frac{1}{2}\int dt N \left(-\frac{a\dot{a}^2}{N^2} + a - \frac{\Lambda a^3}{3} + \frac{a^3\dot{\varphi}^2}{N^2} - 2a^3 V(\varphi) \right), \qquad (37)$$

where an integration over the spatial variables has been performed and absorbed with a definition of units in which, $2G/3\pi = 1$, and the rescalings, $\varphi \to \varphi/\sqrt{2\pi}$ and $V \to V/2\pi^2$. The total action has been simplified considerably. The only dynamical degrees of freedom are the scale factor, $a(t)$, and the scalar field, $\varphi(t)$, which according to the homogeneity condition must only depend on the time variable. The superspace has then been reduced to a two dimensional space. Now, we can proceed as described in the preceding sections. The momenta conjugated to the configuration variables are,

$$p_a \equiv \frac{\delta L}{\delta \dot{a}} = -\frac{a\dot{a}}{N}, \quad p_\varphi \equiv \frac{\delta L}{\delta \dot{\varphi}} = \frac{a^3 \dot{\varphi}}{N}, \qquad (38)$$

and the Hamiltonian then reads

$$H = N\mathcal{H} = \frac{N}{2}\left(-\frac{1}{a}p_a^2 + \frac{1}{a^3}p_\varphi^2 - a + \frac{\Lambda a^3}{3} + 2a^3 V(\varphi) \right). \qquad (39)$$

The momentum constraint is automatically satisfied by the symmetries of the spacetime, and the Hamiltonian constraint, $\frac{\delta H}{\delta N} = 0$, then becomes, $\mathcal{H} = 0$. Promoting the momenta into quantum operators and applying the quantum version of the Hamiltonian constraint, $\mathcal{H} = 0$, to the wave function of the universe, $\phi(a, \varphi)$, which depends now on the two variables of the minisuperspace, a and φ, we obtain the Wheeler–DeWitt equation, $\hat{\mathcal{H}}\phi(a, \varphi) = 0$.

2.4.1. Inflationary Universe

Let us first analyse the initial stage of the universe where the scalar field is assumed to be approximately constant on the small time scale in which the universe rapidly undergoes an inflationary expansion [25], $\dot{\varphi} \approx 0$ and $\varphi_0 \gg 1$. In that case, the kinetic term of the scalar field in (37) can be neglected and the potential turns out to be approximately constant, $V(\varphi_0)$. From (37), it can be seen that a constant value of the potential is equivalent to a cosmological constant term, so the universe effectively behaves like a DeSitter spacetime. Then, we can write (37) as

$$S = \frac{1}{2}\int dt N \left(-\frac{a\dot{a}^2}{N^2} + a - H_0^2 a^3 \right), \tag{40}$$

where, $H_0^2 = 2V(\varphi_0)$, in the case of the inflationary universe or, $H_0^2 = \Lambda/3$, in the case of a "pure" DeSitter spacetime, or the sum of both in a general case. The corresponding Hamiltonian constraint turns out to be

$$\mathcal{H} = -\frac{1}{a}p_a^2 - a + H_0^2 a^3 = 0, \tag{41}$$

which is nothing more than the Friedmann equation expressed in terms of the momentum conjugated to the scale factor, p_a. In terms of the time derivative of the scale factor, using $p_a = a\dot{a}$ (in cosmic time[16], with $N = 1$ in (38)), the Hamiltonian constraint (41) can be written as

$$\dot{a} = \sqrt{H_0^2 a^2 - 1}, \tag{42}$$

whose solutions can easily be obtained,

$$a(t) = \frac{1}{H_0}\cosh H_0 t. \tag{43}$$

If $t \in (-\infty, \infty)$, the scale factor (43) describes a universe that starts shrinking from an infinite volume, bounces at the minimum value, $a_0 = H_0^{-1}$, and ends up in an eternal expansion (see, Figure 5 Left). However, it does not seem quite plausible that the universe is created with an infinite volume, so the most reasonable possibility consist in restricting ourselves to the domain, $t \in (0, \infty)$, which describes a "bubble" of spacetime that is created with radius $a_0 = H_0^{-1}$ at $t = 0$ (the origin of time), and starts expanding exponentially in an inflationary like expansion. If the length scale of the initial "bubble" is some orders of magnitude greater than the Planck scale, i.e., $H_0^{-1} \gg l_P$, then, the quantum fluctuations of the spacetime would be small, and the homogeneous and isotropic picture described here could reasonably represent the creation of a universe like ours[17].

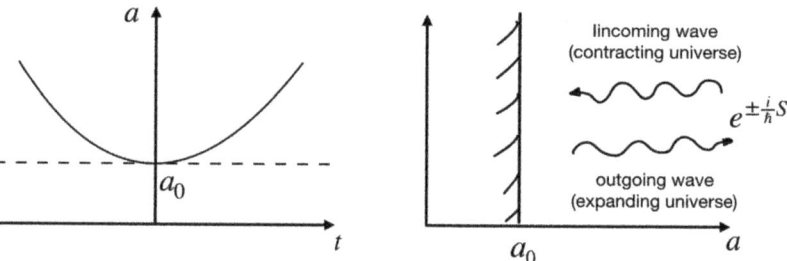

Figure 5. *Left*: the closed DeSitter spacetime describes a universe that starts shrinking from an infinite volume, bounces at the minimum value, $a_0 = H_0^{-1}$, and ends up in an eternal expansion. *Right*: the value a_0 can be seen as a barrier where the incoming waves are reflected and converted into outgoing waves.

Let us now analyse the solutions of the corresponding Wheeler–DeWitt equation. Making the substitution, $p_a \to -i\hbar \partial_a$, and leaving aside the ambiguity of the factor ordering, the Hamiltonian constraint (41) can conveniently be written as

$$\hbar^2 \frac{d^2\phi(a)}{da^2} + \omega^2(a)\phi(a) = 0, \tag{44}$$

where,

$$\omega(a) = \sqrt{H_0^2 a^4 - a^2}. \tag{45}$$

Written in this way, the Wheeler–DeWitt Equation (44) resembles the equation of a harmonic oscillator with time-dependent frequency (such parallelism will be further exploited in Section 3). Equation (44) is not exactly solvable. However, far from the turning point, $a_0 = H_0^{-1}$, we can approximate the solutions by the WKB wave functions

$$\phi^\pm(a) = \frac{N^\pm}{\sqrt{\omega(a)}} e^{\pm \frac{i}{\hbar} S(a)}, \tag{46}$$

where $N^\pm$ is a normalisation constant and $S(a)$ is given by

$$S(a) = \int da\, \omega(a) = \frac{(H_0^2 a^2 - 1)^{\frac{3}{2}}}{3H_0^2}. \tag{47}$$

The two wave functions $\phi^\pm$ in (46) represent incoming and outgoing wave functions. Let us see it by inserting them into the Wheeler–DeWitt Equation (44). In that case, at order $\hbar^0$ the Hamilton–Jacobi equation is obtained

$$\left(\frac{\partial S}{\partial a}\right)^2 + a^2 - H_0^2 a^4 = 0, \tag{48}$$

which corresponds to the Hamiltonian constraint (41) if one makes the identification

$$p_a = \pm \frac{\partial S}{\partial a}. \tag{49}$$

In fact, let us note that at leading order in $\hbar$ it is satisfied

$$-a\dot{a} \equiv p_a \approx \langle \phi^\pm | \hat{p}_a | \phi^\pm \rangle = \pm \frac{\partial S}{\partial a} + \mathcal{O}(\hbar^1). \tag{50}$$

Therefore, it is obtained

$$\dot{a} = \mp \frac{\omega(a)}{a} = \mp \sqrt{H_0^2 a^2 - 1}, \tag{51}$$

which is the Friedmann Equation (42), with two signs: the $-$ sign that corresponds to ϕ^+ and the $+$ sign that corresponds to ϕ^-. Thus, ϕ^- describes an outgoing wave, i.e., a wave that travels towards greater values of the scale factor—it thus represents an expanding universe—and ϕ^+ an incoming wave, i.e., a wave that travels towards smaller values of the scale factor, which corresponds therefore to a contracting universe. Let us notice however that the interpretation of $\phi^{\pm}$ in terms of incoming and outgoing waves must be taken carefully [26]. The Friedmann Equation (41) is invariant under the time reversal change, $t \to -t$, and in terms of $-t$ ($\dot{a} \to -\dot{a}$), ϕ^+ would represent the expanding universe (i.e., an outgoing wave) and ϕ^- the contracting universe (i.e., an incoming wave). Clearly, it depends on the (time) parametrisation. The important thing is that the general quantum state of the universe (far from the turning point) can be written as

$$\phi(a) = A_0 \phi^+(a) + B_0 \phi^-(a), \tag{52}$$

which represents a quantum superposition of incoming and outgoing waves, irrespective of the particular wave function that represents each one. Another important feature that is worth noticing is that the minimum value a_0 constitutes a classical barrier below which the universe cannot go through. Let us notice that for a value $a < a_0$, there is no real solution of the Friedmann Equation (51), so an incoming wave function representing a contracting universe is classically reflected (bounced) into an outgoing wave (see Figure 5 Right).

2.4.2. Small Perturbations and Backreaction

Let us now consider a more realistic scenario by introducing two important changes. First, we shall not neglect the kinetic term of the scalar field in the action (3), and we shall consider a general form (i.e., not necessarily a constant) for the potential $V(\varphi)$. Contrary to what may be thought, that will not introduce qualitative changes. In return, it will allow us to represent other stages of the evolution of our universe as well as many other types of universes. The second change that we are going to make is the introduction of small perturbations around the homogeneous and isotropic background spacetime. This will help us to analyse several phenomena like the behaviour of the matter fields in the semiclassical universe or the appearance of a physical time variable.

Regarding the last question, we can assume that except for the very beginning, the inhomogeneities of the universe are relatively small[18]. Therefore, to a good order of approximation, the universe can be represented by a homogeneous and isotropic background with relatively small inhomogeneities propagating in the background. In that case, it seems reasonable to expand the variables of the spacetime and the matter fields around the homogeneous and isotropic values and study the inhomogeneities as perturbations of the homogeneous and isotropic background. We can still consider the 3 + 1 splitting of the spacetime given in (1). However, the idea is now to expand the configuration variables (h_{ab}, N, N_a and φ) around their homogeneous and isotropic values and retain just the first-order terms. Then, let us consider the following expansions [6,27]

$$h_{ab}(t,\mathbf{x}) = a^2(t)\Omega_{ab} + a^2(t)\sum_{\mathbf{n}} 2d_{\mathbf{n}}(t) G_{ab}^{\mathbf{n}}(\mathbf{x}) + \ldots, \tag{53}$$

$$\varphi(t,\mathbf{x}) = \frac{1}{\sqrt{2\pi}}\varphi(t) + \sum_{\mathbf{n}} f_{\mathbf{n}}(t) Q^{\mathbf{n}}(\mathbf{x}), \tag{54}$$

where Ω_{ab} is the metric on the unit three-sphere, $\varphi(t)$ is the homogeneous mode of the scalar field, $Q^{\mathbf{n}}(\mathbf{x})$ are the scalar harmonics on the three-sphere, and $G_{ij}^{\mathbf{n}}(\mathbf{x})$ the transverse traceless tensor harmonics [27], with, $\mathbf{n} \equiv (n,l,m)$. More harmonics can be present in (53). However, we shall only focus on the tensor modes, $d_{\mathbf{n}}$, as the representative of the perturbation of the spacetime. Eventually, these modes will represent gravitons propagating in the background spacetime, and, analogously, the perturbation modes f_n will represent the particles of the scalar field(s). The lapse and shift functions must also be expanded in terms of the spherical harmonics. Then, all these perturbed functions are inserted in the action (10). The configuration variables are now the scale factor $a(t)$, the homogeneous mode of the scalar field $\varphi(t)$, and the infinite number of modes $d_\mathbf{n}(t)$ and $f_\mathbf{n}(t)$, denoted generically by $x_\mathbf{n}(t)$. The minisuperspace is then the infinite dimensional space spanned by the variables $(a, \varphi, x_\mathbf{n})$, and the time evolution of the universe is represented by a parametrised trajectory in that space, $(a(t), \varphi(t), x_\mathbf{n}(t))$. We should have included the modes of the perturbed lapse and shift functions. However, as it happens with their homogeneous counterparts, they are not dynamical variables but Lagrange multipliers that generate a set of constraints that can be used to simplify the equations [27]. After a cumbersome computation, one arrives at the Hamiltonian constraint

$$\mathcal{H} = \mathcal{H}_0 + \mathcal{H}_m = 0, \tag{55}$$

where the Hamiltonian $\mathcal{H}_0$ contains only degrees of freedom of the homogeneous and isotropic background, and $\mathcal{H}_m$ is the Hamiltonian density that contains also the degrees of freedom of the perturbation modes.

Let us first consider the Hamiltonian of the background spacetime, which is given by the Hamiltonian density of (39),

$$\mathcal{H}_0 = \frac{1}{a}\left(-p_a^2 - a^2 + \frac{1}{a^2}p_\varphi^2 + 2a^4 V(\varphi)\right) = 0. \tag{56}$$

After canonical quantisation, the Wheeler–DeWitt equation reads

$$\left(\hbar^2 \frac{\partial^2}{\partial a^2} - \frac{\hbar^2}{a^2}\frac{\partial^2}{\partial \varphi^2} + a^4 H^2(\varphi) - a^2\right)\phi_0(a, \varphi) = 0, \tag{57}$$

where, $H^2(\varphi) \equiv 2V(\varphi)$, is not necessarily constant now. Let us consider the semiclassical solutions,

$$\phi_0^\pm(a, \varphi) = \Delta(a, \varphi) e^{\pm \frac{i}{\hbar} S(a, \varphi)}, \tag{58}$$

where $\Delta(a, \varphi)$ is a slow varying field of the scale factor. Inserting the wave function (58) into the Wheeler–DeWitt Equation (57), as we did in the previous section, and disregarding second-order derivatives with respect to the background variables, one obtains at order $\hbar^0$ the Hamilton–Jacobi equation [6]

$$-\left(\frac{\partial S}{\partial a}\right)^2 + \frac{1}{a^2}\left(\frac{\partial S}{\partial \varphi}\right)^2 + a^4 H^2(\varphi) - a^2 = 0. \tag{59}$$

Following the semiclassical development of the previous sections, let us define a WKB-time parameter t_w as [6]

$$\frac{\partial}{\partial t_w^\pm} \equiv \pm\left(-\frac{1}{a}\frac{\partial S}{\partial a}\frac{\partial}{\partial a} + \frac{1}{a^3}\frac{\partial S}{\partial \varphi}\frac{\partial}{\partial \varphi}\right), \tag{60}$$

in terms of which,

$$\dot{a}^2 = \frac{1}{a^2}\left(\frac{\partial S}{\partial a}\right)^2, \quad \dot{\varphi}^2 = \frac{1}{a^6}\left(\frac{\partial S}{\partial \varphi}\right)^2, \tag{61}$$

and the Hamilton–Jacobi Equation (59) turns out to be

$$\dot{a}^2 + 1 - a^2\left(\dot{\varphi}^2 + 2V(\varphi)\right) = 0, \tag{62}$$

which is the Friedmann equation of the background spacetime. The WKB wave functions $\phi_0^\pm$ describe universes with a background spacetime that evolves according to the Friedmann Equation (62). The wave function ϕ_0 may thus represent the quantum state of a large variety of classical models of the universe. The two signs in (60) are irrelevant at the classical level because the Friedman Equation (62) is invariant under the reversal change of the time variable. However, they will play an important role at the semiclassical level.

Let us now consider the complete Hamiltonian constraint (55). The corresponding Wheeler–DeWitt equation is

$$(\hat{\mathcal{H}}_0 + \hat{\mathcal{H}}_m)\phi(a,\varphi;x_\mathbf{n}) = 0, \quad (63)$$

where $\hat{\mathcal{H}}_0$ is the Hamiltonian of the background that we have already seen (56) and (57), and $\hat{\mathcal{H}}_m$ is the Hamiltonian of the perturbation modes, which for the moment we do not need to specify. The wave function of the universe can be separated in two factors, the wave function $\phi_0(a,\varphi)$ of the background (58) and another wave function that contains the matter degrees of freedom[19]. Therefore, the semiclassical wave function can be written now as

$$\phi^\pm(a,\varphi;x_\mathbf{n}) = \Delta(a,\varphi) e^{\pm \frac{i}{\hbar} S(a,\varphi)} \psi_\pm(a,\varphi;x_\mathbf{n}), \quad (64)$$

If we insert the wave function (64) into the Wheeler–DeWitt Equation (63) and we solve it order by order in $\hbar$, it at order $\hbar^0$ the Hamilton–Jacobi Equation (59) is obtained. Therefore, the wave function (64) still describes the background spacetime that evolves according to (62). On the other hand, at order $\hbar^1$ in $\mathcal{H}_0$, one obtains

$$\pm i\hbar \left(-\frac{1}{a}\frac{\partial S}{\partial a}\frac{\partial}{\partial a} + \frac{1}{a^3}\frac{\partial S}{\partial \varphi}\frac{\partial}{\partial \varphi} \right)\psi_\pm = \hat{\mathcal{H}}_m \psi_\pm. \quad (65)$$

Here comes a subtle but crucial point of the semiclassical regime. In terms of the initial proper time, t, the two branches of the universe represent an expanding and a contracting universe because, from (38),

$$-a\frac{da}{dt} = p_a \approx \langle \phi_0^\pm | \hat{p}_a | \phi_0^\pm \rangle \sim \pm \frac{\partial S}{\partial a}, \quad (66)$$

so ϕ_0^- describes a universe whose spacetime background is expanding and ϕ_0^+ a universe whose background spacetime is contracting. In that case, in order for the WKB-time (60) to represent the proper time variable t in the two branches, we have to choose for the branch ϕ^- the WKB-time variable t_- defined by

$$\frac{\partial}{\partial t_-} \equiv -\left(-\frac{1}{a}\frac{\partial S}{\partial a}\frac{\partial}{\partial a} + \frac{1}{a^3}\frac{\partial S}{\partial \varphi}\frac{\partial}{\partial \varphi} \right), \quad (67)$$

in which case,

$$\frac{\partial a}{\partial t_-} = \frac{1}{a}\frac{\partial S}{\partial a}, \quad (68)$$

which represents an expanding universe. For the WKB-time variable in the ϕ^+ branch, we must choose

$$\frac{\partial}{\partial t_+} \equiv -\frac{1}{a}\frac{\partial S}{\partial a}\frac{\partial}{\partial a} + \frac{1}{a^3}\frac{\partial S}{\partial \varphi}\frac{\partial}{\partial \varphi}, \quad (69)$$

in terms of which,

$$\frac{\partial a}{\partial t_+} = -\frac{1}{a}\frac{\partial S}{\partial a}, \quad (70)$$

describes a contracting universe. It is worth noticing that this assignation is somehow arbitrary [26,28]. If we would have started with a time variable, $t \to -t$, then, ϕ_0^+ would represent the contracting universe and ϕ_0^- the expanding one, and the assignations of the WKB-time variables would have been the other way around. However, in terms of the definitions (69) and (67), the Equation (65) reads,

$$i\hbar \frac{\partial}{\partial t_\pm}\psi_\pm(t_\pm,\varphi) = \hat{\mathcal{H}}_m \psi_\pm(t_\pm,\varphi), \quad (71)$$

where $\psi_\pm(t_\pm, \varphi) \equiv \psi_\pm[a(t_\pm), \varphi]$, evaluated in the background solutions of (70) and (68). Therefore, we have ended up with two universes, one contracting and another expanding, both filled with matter.

There is however a different interpretation. One may assume that the physical time variable, i.e., the time variable measured by actual clocks that are made of matter, is the time variable that appears in the Schrödinger equation. In that case, it is worth noticing that the physical time variable of the two universes is reversely related, $t_+ = -t_-$. Let us assume that we fix the time variable by fixing the time that a particular observer measures and consider thus t_- as the physical time. Then, in terms of the time variable t_- the evolution of the scale factor is given by (70) so the two wave functions, ϕ^+ and ϕ^-, represent both with a universe with an expanding background spacetime, i.e from the point of view of this hypothetical observer both universes are expanding. The Schrödinger Equation (65) in the observer's universe becomes,

$$i\hbar \frac{\partial}{\partial t_-}\psi_-(t_-, x_n) = \hat{\mathcal{H}}_m \psi_-(t_-, x_n). \qquad (72)$$

However, the Schrödinger equation for the fields of the partner universe is

$$-i\hbar \frac{\partial}{\partial t_-}\psi_+(t_-, x_n) = \hat{\mathcal{H}}_m \psi_+(t_-, x_n), \qquad (73)$$

for the wave function ψ_+. The "wrong sign" in (73) is not problematic. It is only indicating that (73) is the Schrödinger equation of the complex conjugated wave function ψ_+^* with a CP-transformed Hamiltonian [26]. Let us notice that (73) can be written as

$$i\hbar \frac{\partial}{\partial t_-}\psi_-(t_-, \bar{x}_n) = \hat{\mathcal{H}}_m(\bar{x}_n)\psi_-(t_-, \bar{x}_n). \qquad (74)$$

It is therefore the Schrödinger equation for the antimatter fields of the observer's universe. In this case, we have ended up in the description of two universes, both expanding but one filled with matter and the other filled with antimatter[20]. The two interpretations raised here for the linear combination of incoming and outgoing wave look very similar to the two interpretations made in QED of an electron–positron pair (see Figure 6). In Section 3.2.4, we shall interpret this combination as the creation of a universe–antiuniverse pair.

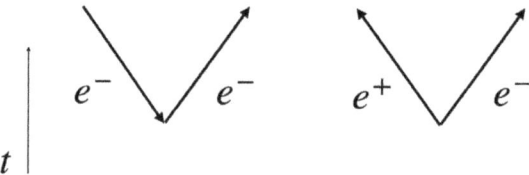

Figure 6. An electron–positron pair can equivalently be seen as either an electron propagating backwards in time, bouncing, and then propagating forward in time or as a particle–antiparticle pair propagating forward in time.

Let us now specify the Hamiltonian of the perturbation modes. If we restrict to small linear perturbations, the different modes do not interact, and $\mathcal{H}_m$ turns out to be the sum of the Hamiltonian of a set of harmonic oscillators [4,29]

$$\mathcal{H}_m = \sum_n \mathcal{H}_n, \qquad (75)$$

with

$$\mathcal{H}_n = \frac{1}{2M}p_{x_n}^2 + \frac{M\omega_n^2}{2}x_n^2, \qquad (76)$$

where, $M(t) = a^3(t)$, and [6]

$$\omega_n^2 = \frac{n^2 - 1}{a^2}, \qquad (77)$$

for the tensorial modes of the spacetime ($x_\mathbf{n} \equiv d_\mathbf{n}$), and

$$\omega_n^2 = \frac{n^2 - 1}{a^2} + m^2, \qquad (78)$$

for the perturbation modes of the scalar field ($x_\mathbf{n} \equiv f_\mathbf{n}$). Thus, for small perturbations, the Hamiltonian of the modes turns out to be the Hamiltonian of a set of uncoupled harmonic oscillators with time-dependent mass, $M(t) = M[a(t)]$, and frequency given by, $\omega_n(t) = \omega_n[a(t)]$, where $a(t)$ is the solution of the Friedmann equation of the background spacetime (possibly including the backreaction).

The Schrödinger Equation (71) of the perturbation modes is then the Schrödinger equation of a set of uncoupled harmonic oscillators, whose general solution can be written as [4,6,30]

$$\psi_\pm = \prod_\mathbf{n} \psi_\mathbf{n}(t_\pm, x_\mathbf{n}), \qquad (79)$$

where the function $\psi_\mathbf{n}(t, x_\mathbf{n})$ is the wave function of a harmonic oscillator with time-dependent mass and frequency, which can be written in terms of the wave function of a harmonic oscillator with constant mass and frequency [31–34]. The general solution of $\psi_\mathbf{n}(t, x_\mathbf{n})$ can then be expanded in the basis of number eigenstates of the invariant representation, $\psi_{N,\mathbf{n}}$, as

$$\psi_\mathbf{n} = \sum_N c_N \, \psi_{N,\mathbf{n}}, \qquad (80)$$

where c_N are constants coefficients, and the wave function of the invariant number state, $\psi_{N,\mathbf{n}}$, is given by [32,35]

$$\psi_{N,\mathbf{n}}(a, \phi; x_\mathbf{n}) \equiv \langle a, \phi; x_\mathbf{n} | N_\mathbf{n} \rangle = \frac{1}{\sigma^{\frac{1}{2}}} \exp\left\{ \frac{i M}{2} \frac{\dot\sigma}{\sigma} x_\mathbf{n}^2 \right\} \bar\psi_N\left(\frac{x_\mathbf{n}}{\sigma}, \tau \right) \qquad (81)$$

where $\bar\psi_N(q, \tau)$ is the customary wave function of the harmonic oscillator,

$$\bar\psi_N(q, \tau) = \left(\frac{1}{2^N N! \pi^{\frac{1}{2}}} \right)^{\frac{1}{2}} e^{-i(N+\frac{1}{2})\tau} e^{-\frac{q^2}{2}} H_N(q) \qquad (82)$$

with $H_N(q)$ the Hermite polynomial of order N, $q \equiv \frac{x_\mathbf{n}}{\sigma}$,

$$\tau(t) = \int^t \frac{1}{M(t')\sigma^2(t')} dt', \qquad (83)$$

and $\sigma(t)$ is an auxiliary function that satisfies the non-linear equation [31,32]

$$\ddot\sigma + \frac{\dot M}{M} \dot\sigma + \omega_n^2 \sigma = \frac{1}{M^2 \sigma^3}, \qquad (84)$$

plus some boundary condition [36]. The interesting property of the invariant representation is that once the *field*[21] is in a number state of the invariant representation, it remains in the same state along the entire evolution of the field. For instance, let us assume that the perturbation modes are in the vacuum state of the invariant representation, $|0\rangle = \prod_\mathbf{n} |0_\mathbf{n}\rangle$. In that case, the mean value of the energy of the perturbations reads[22]

$$\langle \mathcal{H}_m \rangle = \sum_\mathbf{n} \hbar \omega_n \left(\langle \hat N_\mathbf{n} \rangle + \frac{1}{2} \right) = \sum_n \frac{\hbar \omega_n}{2} \to \frac{\hbar}{2} \int^{n_{max}} dn \, n^2 \, \omega_n, \qquad (85)$$

where ω_n is given by (77) and (78) for matter particles or spacetime gravitons, respectively. The sum in (85) diverges and some cut-off $n_{\max}$ must be taken. The energy $\langle \mathcal{H}_m \rangle$ is the backreaction of the perturbations on the homogeneous and isotropic background that would induce a modification of the Friedmann Equation (62),

$$\left(\frac{\dot{a}}{a}\right)^2 = 2\rho_\varphi - \frac{1}{a^2} + \frac{\langle \mathcal{H}_m \rangle}{a^3}, \qquad (86)$$

where, $\rho_\varphi = \frac{1}{2}\dot{\varphi}^2 + V(\varphi)$ is the energy of the homogeneous mode of the matter field and $\langle \mathcal{H}_m \rangle \propto a^{-1}$ for the massless tensor modes or $\langle \mathcal{H}_m \rangle \propto m$ for the perturbations of the scalar field.

2.5. Paradigms for the Creation of the Universe in Quantum Cosmology

In the preceding section we ended up with a Friedman equation that was corrected by the backreaction of the perturbation modes of the spacetime. We shall see in this section that such a term may induce important consequences in the way in which the universes can be created. However, for historical reasons, instead of considering the backreaction of the perturbation modes, we shall consider the model of a massless scalar field conformally coupled to gravity and a cosmological constant, which eventually raises the same term in the Friedmann equation, so the two models effectively entail similar effects. Later on, we shall briefly comment on the nature of this term. The conformally coupled massless scalar field is the field used by Hartle and Hawking in Ref. [12] to describe the quantum state of the universe and the process of quantum creation. Besides, it will also help us to introduce different paradigms for the creation of the universe.

Therefore, let us consider the following action for the spacetime and the massless, i.e., $V(\varphi) = 0$, scalar field,

$$S = S_{EH} + S_m = \frac{1}{2} \int dt N \left(-\frac{a\dot{a}^2}{N^2} + a - \frac{\Lambda a^3}{3} + \frac{a^3 \dot{\varphi}^2}{N^2} - \frac{1}{6} a^3 \, {}^4R\varphi^2 \right), \qquad (87)$$

where the last term represents the conformal coupling of the scalar field, and ${}^4R$ is the Ricci scalar. Then, with the change, $\chi = a\varphi$, and after an integration by parts, the total action (87) can be written as [12]

$$S = S_{EH} + S_m = \frac{1}{2} \int dt N \left(-\frac{a\dot{a}^2}{N^2} + a - H_0^2 a^3 + \frac{a\dot{\chi}^2}{N^2} - \frac{\chi^2}{a} \right), \qquad (88)$$

where H_0^2 can be a pure cosmological constant, $H_0^2 = \Lambda/3$, or the constant value of the potential of an auxiliary inflaton field (different from the scalar field φ that we are considering in (88)). In any case, it will be assumed that it is a constant. Now, the momenta conjugated to the scale factor and the conformally coupled massless field χ can be easily obtained, and the Hamiltonian constraint associated to the action (88) reads

$$H = N\mathcal{H} = \frac{N}{2a}\left(-p_a^2 - a^2 + H_0^2 a^4 + p_\chi^2 + \chi^2\right) = 0, \qquad (89)$$

which, taking into account that $p_\chi^2 + \chi^2$ is nothing more than (twice) the energy of the scalar field [36], $2E$, can also be written as

$$\frac{1}{2}p_a^2 + \mathcal{U}(a) = E, \qquad (90)$$

with

$$\mathcal{U}(a) = \frac{1}{2}\left(a^2 - H_0^2 a^4\right). \qquad (91)$$

which is formally similar to the energy equation of a particle that propagates under the action of the potential (91). The first term of the l.h.s. of (90) would be the kinetic energy, the second term would be the potential energy, and E would be the total energy[23]

Quantum-mechanically, the wave function of the universe, $\phi(a,\chi)$, is the solution of the Wheeler–DeWitt equation associated to the Hamiltonian constraint (89),

$$\left(\frac{\partial^2}{\partial a^2} + H_0^2 a^4 - a^2 - \frac{\partial^2}{\partial \chi^2} + \chi^2\right)\phi(a,\chi) = 0, \tag{92}$$

which can be solved by the method of the separation of variables. Making, $\phi(a,\chi) = \xi(\chi)\psi(a)$, the Wheeler–DeWitt Equation (92) can be split into the two following equations,

$$\left(-\frac{d^2}{d\chi^2} + \chi^2\right)\xi(\chi) = 2E\xi(\chi), \tag{93}$$

$$\frac{d^2\psi(a)}{da^2} + \left(H_0^2 a^4 - a^2\right)\psi(a) = -2E\psi(a). \tag{94}$$

The first of these equations is the equation of a quantum harmonic oscillator with unit mass and frequency. It can be solved in terms of Hermite polynomials, $H_n(x)$,

$$\xi(\chi) \equiv \xi_n(\chi) = \frac{1}{\sqrt{2^n n!}}\left(\frac{1}{\pi}\right)^{\frac{1}{4}} e^{-\frac{\chi^2}{2}} H_n(\chi), \tag{95}$$

with a quantised energy given by

$$E \equiv E_n = \left(n + \frac{1}{2}\right). \tag{96}$$

On the other hand, Equation (94) can be written as

$$\frac{1}{2}\frac{d^2\psi(a)}{da^2} + (E - \mathcal{U}(a))\psi(a) = 0, \tag{97}$$

with $\mathcal{U}(a)$ given by (91). Equation (97) is formally similar to the Schrödinger equation of a particle of energy E moving under the action of the potential $\mathcal{U}(a)$ (see Figure 7). For the value $E \in (0, \mathcal{U}_{max})$, where $\mathcal{U}_{max}$ is the maximum value of the potential, we can distinguish three regions: two classically allowed regions (regions I and III in Figure 7) separated by a classically forbidden region (region II). We have already analysed in the preceding section the behaviour of the wave function in region I. There are incoming and outgoing waves that represent a contracting universe that shrinks to the minimum value of the scale factor a_+ (see, Figure 7) and bounces (or it is reflected), becoming an expanding universe. In the other classically allowed region, region III, we shall see that there are also incident and reflected waves that represent a universe that is confined to oscillate between $a = 0$ and the maximum value a_-.

In the case of a particle moving under the action of a similar potential, the particle that is placed in the region III is classically confined to move in that region like the universes in our example. However, we know that quantum-mechanically there is a non-zero probability for the particle to tunnel though the quantum barrier and appear in the region I. Something similar happens with the universe. If $E > 0$, the small oscillating universe of region III can tunnel out through the Euclidean barrier appearing in region I as a new-born universe. The universe is then said to be created from *something*. In the limiting case, $E = 0$, which was the case analysed by Hartle and Hawking [12,37–39] and by Vilenkin [40–42] for the creation of the universe, there is no region III from which to tunnel out to region I. The universe appears in region I from a pure tunnelling phenomena (like the creation of particles from the quantum vacuum). In that case, the universe is said to be created from *nothing*[24]. Let us analyse the two cases separately.

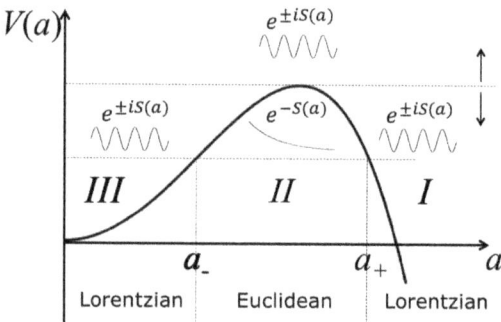

Figure 7. Potential $\mathcal{U}(a)$. For $E > 0$, there are three regions: two Lorentzian regions separated by an Euclidean one. Compare it with the case $E = 0$ analysed in Figure 5.

2.5.1. Creation of the Universe from *Nothing*

Let us first analyse the creation of the universe from nothing, i.e., $E = 0$ in (90). In that case, in terms of the time derivative of the scale factor, $p_a = -a\dot{a}$, the Hamiltonian constraint (90) reduces to the Friedmann Equation (42) already studied in Section 2.4.1,

$$\dot{a} = \sqrt{H_0^2 a^2 - 1}, \tag{98}$$

which yields the well-known solution,

$$a(t) = a_0 \cosh H_0 t, \tag{99}$$

with, $a_0 = H_0^{-1}$. If we restrict ourselves to the "expanding branch," $t \geq 0$, the scale factor (99) represents a universe that starts expanding from the initial boundary $\Sigma(a_0)$ until infinity. For values $a < a_0$, there is no real solution, and the value $a = a_0$ constitutes a classical barrier for the universe (see Figure 5). However, one can perform a Wick rotation into Euclidean time, $t = -i\tau$, in terms of which the Friedmann Equation (98) becomes

$$\frac{da_E}{d\tau} = \sqrt{1 - H_0^2 a_E^2}. \tag{100}$$

Now, the Euclidean Equation (100) has the solution

$$a_E(\tau) = a_0 \cos H_0 \tau, \tag{101}$$

where $\tau \in (-\frac{\pi}{2H_0}, 0)$. Let us notice that, in Euclidean time, $-dt^2 \to +d\tau^2$, the line element turns to be

$$ds^2 = d\tau^2 + a_E^2(\tau) d\Omega_3^2, \tag{102}$$

which is the line element of a 4-sphere of radius $1/H_0$ embedded in a 5-dimensional flat Euclidean spacetime. The "spatial section" starts expanding from a single point of the sphere (at $\tau = -\pi/2H_0$) until it reaches the value H_0^{-1} at the Euclidean time, $\tau = 0$ (see, Figure 8). It is called a DeSitter *instanton*, and it gives the maximum contribution to the tunnelling wave function (it is the extremal solution of the Euclidean action [12]). At the boundary hypersurface $\Sigma(a_0)$, it appears in the Lorentzian region as a new born DeSitter universe that starts expanding exponentially.

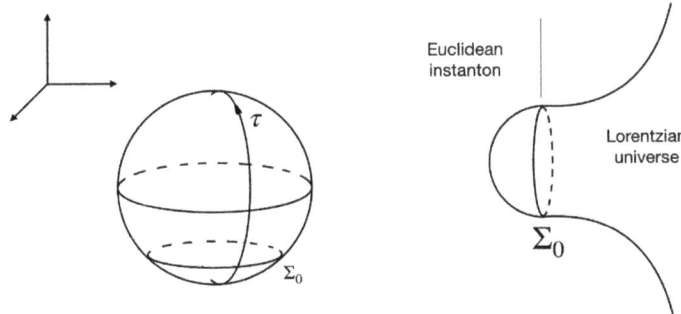

Figure 8. Left: a DeSitter instanton can be seen as a 4-dimensional sphere in the 5-dimensional Euclidean flat spacetime. **Right**: the creation of a DeSitter spacetime from a DeSitter instanton of the Euclidean sector of the spacetime.

Quantum-mechanically, the situation is the following. The initial boundary $\Sigma(a_0)$ separates the Lorentzian region of the spacetime (see, Figure 8), where classical solutions are allowed, from the Euclidean region, where only tunnelling solutions can exist. The wave function that represents a classical spacetime in the Lorentzian region is given in terms of the oscillatory wave functions e^{+iS} and e^{-iS}, where S is given by (47)

$$S(a,\varphi) = \frac{(H^2 a^2 - 1)^{\frac{3}{2}}}{3H^2}. \tag{103}$$

In the Euclidean sector, $a < a_0$, the wave function of the universe, can be written in terms of the tunnelling wave functions, e^I and e^{-I}, where I is given by the integral (47) with $\omega(a)$ being replaced by $|\omega(a)|$, i.e.,

$$I(a,\varphi) = \frac{(1 - H^2 a^2)^{\frac{3}{2}}}{3H^2}. \tag{104}$$

The picture is then similar to the problem of a wave-particle tunnelling through a quantum barrier. The oscillatory wave functions $e^{\pm iS}$ can be seen as the incoming and the reflected waves that may represent a photon or another quantum mechanical particle. Classically, the boundary $\Sigma(a_0)$ acts as a barrier that cannot be crossed (see Figure 9). Quantum-mechanically, however, there is a non-null probability of penetrating into the barrier, although the amplitude is exponentially suppressed in the Euclidean region. Analogously, one can say in cosmology that there is a non-zero probability for the universe to appear *from nothing*, i.e., from the Euclidean barrier of the spacetime. An essential difference is that here the tunnelling is not from another classically allowed region of the spacetime. It is therefore more similar to the creation of virtual particles from the quantum vacuum in a quantum field theory. In a quantum field theory, the pair of virtual particles can only exist at a small amount of time compatible with Heisenberg's uncertainty relations, otherwise the principle of energy conservation would be violated. In the universe, however, the energy is zero (the negative gravitational energy balances the positive energy of the matter fields) so the creation of the universe from nothing does not violate the conservation of energy.

From the above reasoning it is clear that the name "creation from nothing" does not refer to the absolute meaning of nothing, i.e., to something to which we can ascribe no properties. As we have seen, the Euclidean region of the spacetime has geometrical properties. In the standard literature (see, for instance, Refs. [7,8,16,37,38,40]), it usually refers to two meanings. One, perhaps the most consistent, is that the universe is created from a region of the spacetime where nothing real exists; in particular, there is no actual time (i.e., time measured by clocks). In that sense, there is nothing. Another sense with which the term "nothing" is used in the creation of the universe is that the universe, in the paradigmatic case of the creation of a DeSitter spacetime from the Euclidean instanton (101), begins from a single non-singular point of the Euclidean 4-sphere (which is geometrically equivalent to any other point in the sphere). In that case, the single point is meant to be "nothing".

Vilenkin's vs. Hartle–Hawking's Versions

The general quantum state of the universe is therefore given in the region I by the linear combination

$$\phi_I(a, \varphi) = A_I \frac{1}{\sqrt{\omega(a)}} e^{+\frac{i}{\hbar}S(a)} + B_I \frac{1}{\sqrt{\omega(a)}} e^{-\frac{i}{\hbar}S(a)}, \qquad (105)$$

and in the tunnelling region II by

$$\phi_{II}(a, \varphi) = A_{II} \frac{1}{\sqrt{|\omega(a)|}} e^{+\frac{1}{\hbar}I(a)} + B_{II} \frac{1}{\sqrt{|\omega(a)|}} e^{-\frac{1}{\hbar}I(a)}. \qquad (106)$$

The particular combination of WKB wave functions in the Lorentzian and in the Euclidean regions of the spacetime, i.e., the particular values of the constants $A_{I,II}$ and $B_{I,II}$, depend on the boundary condition that we impose on the state of the universe.

Hartle and Hawking [12,37,39] propose as the boundary condition that the path integral must be performed over compact Euclidean geometries that fit with the "final" values[25] a_0 and φ in the boundary $\Sigma(a_0)$ (see Figure 8). In the homogeneous and isotropic minisuperspace, it is equivalent to the conditions [15]

$$a(\tau_0) = 0, \frac{da}{d\tau}(\tau_0) = 1, \frac{d\varphi}{d\tau}(\tau_0) = 1. \qquad (107)$$

The wave function obtained for the Euclidean sections is e^{-I}, with

$$I = \int_0^a da\, a\, (1 - H^2 a^2)^{\frac{1}{2}} = \frac{-1}{3H^2}\left(1 - (1 - H^2 a^2)^{\frac{3}{2}}\right). \qquad (108)$$

It yields [7,12,15]

$$\phi_{HH}^{II}(a, \varphi) = \frac{1}{(1 - H^2 a^2)^{\frac{1}{4}}} \exp\left(\frac{1}{3H^2}\left(1 - (1 - H^2 a^2)^{\frac{3}{2}}\right)\right). \qquad (109)$$

From the matching conditions of the WKB method, the Hartle–Hawking wave function (109) implies a linear combination of oscillatory wave functions in the Lorentzian region ($a > a_0$) [7,12,15],

$$\phi_{HH}^{I}(a, \varphi) = \frac{1}{(H^2 a^2 - 1)^{\frac{1}{4}}} \exp\left(\frac{1}{3H^2}\right) \cos\left(\frac{(H^2 a^2 - 1)^{\frac{3}{2}}}{3H^2} - \frac{\pi}{4}\right). \qquad (110)$$

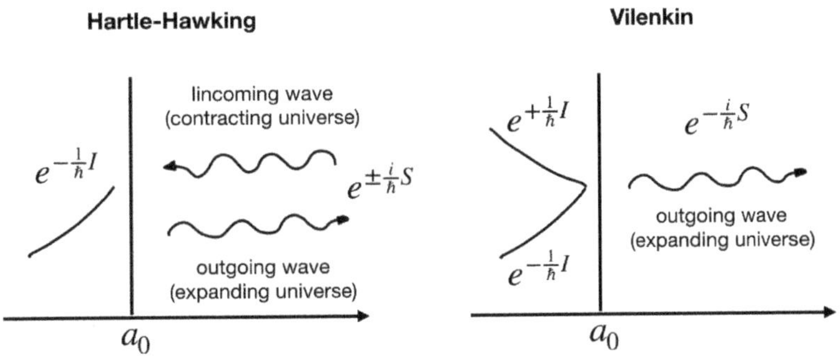

Figure 9. Hartle–Hawking "no-boundary" boundary proposal vs. Vilenkin's tunnelling boundary proposal. The H–H state implies a linear combination of expanding and contracting wave functions in the Lorentzian region, where the tunnelling wave function corresponds only to expanding universes.

On the other hand, Vilenkin [40–42] proposes as the boundary condition, in an analogy with the tunnelling process of quantum mechanics, that the only modes that survive the quantum barrier are the "outgoing" modes, i.e., those that represent an expanding universe[26]. It means that in the region I, the wave function of the universe is given by [7,15,41,43]

$$\phi_T^I(a,\varphi) = \frac{A_I(\varphi)}{(H^2 a^2 - 1)^{\frac{1}{4}}} \exp\left(-i \frac{(H^2 a^2 - 1)^{\frac{3}{2}}}{3H^2}\right), \tag{111}$$

where $A_I(\varphi)$ is a normalisation "constant" that can be found by imposing the regularity conditions [15], $\partial\phi/\partial\varphi \to 0$ as $a \to 0$. Then, $A_I(\varphi) = \exp(-1/3H^2)$, for which $\phi_T \sim e^{-\frac{1}{2}a^2}$, that is regular at $a \to 0$ for any value of the scale factor. By following the same WKB procedure of matching conditions, we found in the Euclidean sector [15]

$$\phi_T^{II}(a,\varphi) = \frac{1}{(1-H^2a^2)^{\frac{1}{4}}}\left(e^I - \frac{i}{2}e^{-I}\right) \approx \frac{1}{(1-H^2a^2)^{\frac{1}{4}}} e^I, \tag{112}$$

where I is given by (108). Except for the values of the scale factor close to a_0, the second term in (112) is exponentially smaller than the first, so it is usually neglected.

Besides their conceptual meaning, the main difference between the wave function of the two proposals is the different sign in the exponent of the pre-factor, $\exp(-1/3H^2)$, in the case of the tunnelling wave function, and $\exp(1/3H^2)$, in the case of the no-boundary wave function. The probability measures are different in both cases. Because the similarity of the Wheeler–DeWitt equation with the Klein–Gordon equation, Vilenkin proposes to use the probability current[27] [43,44]

$$J = \frac{i}{2}(\phi^* \nabla \phi - \phi \nabla \phi^*), \tag{113}$$

which is conserved because in virtue of the Wheeler–DeWitt equation it satisfies $\nabla \cdot J = 0$. The Hartle–Hawking wave function is real, so the probability measure (113) would yield zero. These authors propose instead to use the customary probability measure of quantum mechanics,

$$J = |\phi|^2. \tag{114}$$

With these two choices, the probability for the creation of the universe reads [15]

$$P = J \cdot d\Sigma \approx \exp\left(\pm \frac{2}{3H^2(\varphi)}\right) d\varphi, \tag{115}$$

where the $+$ sign is for the no-boundary wave function and the $-$ sign for the tunnelling wave function. Thus, as we have already noticed in Section 2.2, the no-boundary proposal seems to favour small values of the potential ($P_{HH} \to \infty$ for $H(\varphi) \to 0$), and the tunnelling proposal seems to favour the creation of universe with a large value of the potential ($P_T \to 0$ for $H(\varphi) \to 0$). Therefore, it is usually stated that Vilenkin's tunnelling condition fits better with the inflationary scenario [25,43].

2.5.2. Creation of the Universe from *Something*

Let us now analyse the case $E \in (0, \mathcal{U}_{max})$ in (90). The corresponding Friedman equation is obtained by substituting the value, $p_a = -a\dot{a}$, in (90). It yields

$$\left(\frac{\dot{a}}{a}\right)^2 = H_0^2 - \frac{1}{a^2} + \frac{2E}{a^4}. \tag{116}$$

We can see that the conformally massless scalar field can reproduce the effect of a radiation content of the universe ($\rho \sim a^{-4}$, $\langle \mathcal{H}_m \rangle \propto a^{-1}$ in (86)). The Friedman Equation (116) can also be written as

$$\dot{a} = \frac{H_0}{a}\sqrt{(a^2 - a_+^2)(a^2 - a_-^2)}, \tag{117}$$

with [26,45]

$$a_\pm^2 = \frac{1}{2H_0^2}\left(1 \pm \left(1 - 8EH_0^2\right)^{\frac{1}{2}}\right). \tag{118}$$

For the value, $\frac{1}{8H_0^2} = \mathcal{U}_{max} > E > 0$, the two Lorentzian regions are located at $a > a_+$ and $a < a_-$, respectively. These two sectors represent two separated regions where the universe may exist. In between, there is the Euclidean sector, which is a classically forbidden region. We have therefore two different types of universes separated by a quantum barrier (see Figure 7). In region III, the solution of (117) can be written as [36]

$$a(t) = \left(a_-^2 \cosh^2 H_0 \Delta t - a_+^2 \sinh^2 H_0 \Delta t\right)^{\frac{1}{2}}, \tag{119}$$

where, $\Delta t = t - t_0$, with

$$\tilde{t}_0 = \frac{1}{H_0} \operatorname{arctanh} \frac{a_-}{a_+}. \tag{120}$$

It represents a small universe that starts in a big-bang-like singularity at $t = 0$, expands to the maximum value a_-, at $t = \tilde{t}_0$, and then re-collapses to a big-crunch-like singularity, at $t = 2\tilde{t}_0$. For a value $H_0 \ll 1$, the evolution of this type of universe is like the evolution of a radiation-dominated universe (see Figure 10). This kind of universes are called *baby universes* [46], which are typically associated with quantum fluctuations of the spacetime.

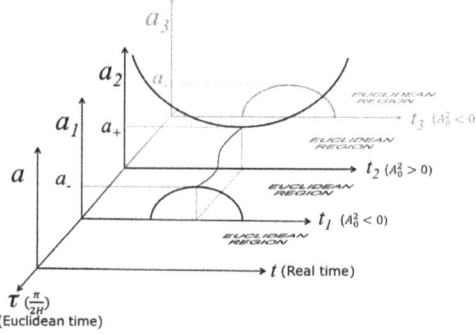

Figure 10. The universes of region III are cyclic universes that start from a big-bang singularity and end in a big-crunch one. In region I, the universe effectively behaves like a closed DeSitter spacetime.

On the other hand, the solution of (116) in region I can be written as [36]

$$a(t) = \left(a_+^2 \cosh^2 H_0 \Delta t - a_-^2 \sinh^2 H_0 \Delta t\right)^{\frac{1}{2}}, \qquad (121)$$

with $\Delta t \in (-\infty, \infty)$. It represents a universe that contracts from infinity to the minimum value a_+, reached at $\Delta t = 0$, and then, it expands again to infinity (see, Figure 10). This solution is essentially very similar to the closed DeSitter universe. In fact, it can continuously be transformed into the customary solution of the closed DeSitter spacetime in the limit $E \to 0$, for which $a_- \to 0$ and $a_+ \to 1/H_0$.

In between, there is a tunnelling region, where the solution of the Euclidean version of the Friedman equation is the Euclidean instanton (102) with the scale factor given by

$$a_E(\tau) = \left(a_+^2 \sin^2 H_0 \Delta \tau + a_-^2 \cos^2 H \Delta \tau\right)^{\frac{1}{2}}, \qquad (122)$$

where $a_\pm$ is given by (118), $a \in (a_-, a_+)$, and

$$\Delta \tau = \tau - \tau_0 \in \left(0, \frac{\pi}{2H_0}\right). \qquad (123)$$

The Euclidean instanton with scale factor (122) connects the maximum expansion hypersurface $\Sigma(a_-)$ of the baby universes of region I with the initial hypersurface $\Sigma(a_+)$ of the large parent universe in region I (see Figure 11). It therefore connects the two regions, and it also provides the first-order contribution to the probability of crossing the quantum barrier and appearing in region I as a new-born universe (see, Ref. [36]). The universe is then said to be created *from something*.

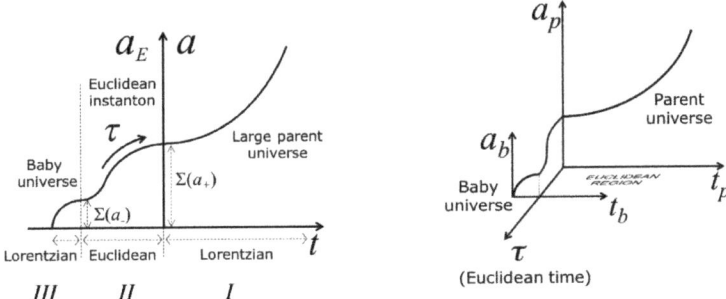

Figure 11. The creation of a large parent universe from a baby universe.

Gott and Li: The Universe Is Its Own Mother

The question of whether the universe is created from *nothing* or from *something* seems to be rooted in the value of the energy E in the Friedman Equation (90). If $E = 0$, the universe must be created from nothing, and if $E > 0$, the universe must be created from something. However, the value $E = 0$ is controversial because from (96), $E = n + 1/2$, so the value $E = 0$ would violate the uncertainty principle of quantum mechanics [45]. In their original study [12], Hartle and Hawking suggested that this term might be cancelled by some renormalisation procedure. However, Gott and Li [45] argue that there is no expectation for such an exact cancellation, and in fact, Barvinsky and Kamenshchik [47–49] computed the renormalisation corrections and not only is the energy term not cancelled but new similar terms appear. Even more, we have seen that the backreaction of the perturbation modes of the spacetime produces a similar term in the Friedmann equation (their energy density is given by $\langle \mathcal{H}_m \rangle / V \propto a^{-4}$, where $\langle \mathcal{H}_m \rangle \propto a^{-1}$ in (85) for the perturbations of the spacetime). Therefore, a term with $E > 0$ seems to be unavoidable. One may argue that it can be effectively small, and thus it could be neglected. However, regardless of how small it can be, we have seen that the consequences for the creation of the universe are very important.

A conceptual problem arises however if the universe is created from "something" because then we need the existence of that "something" prior to the creation of the universe, i.e., if the universe is created from the tunnelling of a quantum fluctuation of a pre-existing spacetime, then, one should explain how the *first* spacetime has been created from which the rest of the universes have subsequently been generated. Gott and Li give an apparently exotic although quite interesting explanation. They argue and show [45] that in region *III* there can exist closed temporal curves (CTC's). In that case, the spacetime fluctuations of a large parent spacetime can travel through a CTC and become the baby universe that "tunnelled out" through the Euclidean barrier to give rise to the parent universe in an atemporal process in which terms like "after" or "before" become meaningless. They are only meaningful within the large parent regions of the spacetime where CTCs do not exist. Thus, according to these authors, the universe could be its own mother.

2.5.3. Creation of Universes in Pairs

There is a way out of this paradoxic explanation. Even with the value $E > 0$ in (116), there is still room for the universes to be created from *nothing*, i.e., from the Euclidean region without the need of a pre-existing spacetime. However, as we have seen, the process cannot be the one studied in Section 2.5.1 for the creation of a single universe from the single Euclidean instanton (101). Instead, one has to consider more elaborated instantons. For instance, one can consider the double Euclidean instanton that is formed by joining together two single Euclidean instantons through their minimal hypersurfaces $\Sigma_i(a_-)$ (see, Refs. [47,48] and Figure 12). The result is the creation *from nothing* of a pair of entangled universes in the region *I* [50–52] (see Figure 12).

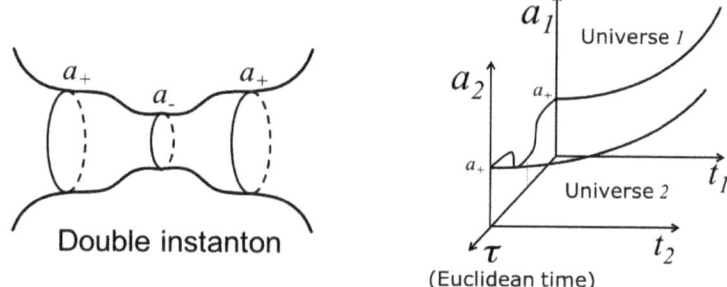

Figure 12. (**Left**) a double Euclidean instanton can be formed by matching two single Euclidean instantons. (**Right**) The creation of a pair of entangled universes from *nothing*, i.e., from a double Euclidean instanton.

Let us notice that, in terms of the same time variable, one of the universes of the entangled pair is a contracting universe, and the other is an expanding universe so the situation is very similar to the case of coexisting incoming and outgoing waves. The wave function $\phi(a,\chi)$ can therefore be written as

$$\phi(a,\chi) = \phi^+(a,\chi) + \phi^-(a,\chi), \qquad (124)$$

with $\phi^\pm(a,\chi)$ given by [36]

$$\phi^\pm(a,\chi) = \frac{N}{\sqrt{\omega_{DS}}} e^{\pm \frac{i}{\hbar} \int \omega_{DS}(a) da} \xi_\pm(\eta_\pm,\chi), \qquad (125)$$

where ω_{DS} is the potential of the Wheeler–DeWitt equation for the DeSitter spacetime,

$$\omega_{DS} = \sqrt{H^2 a^4 - a^2}, \tag{126}$$

and

$$\xi_\pm(\eta_\pm, \chi) \equiv \xi(a = a(\eta), \chi) = \sum_n c_n e^{-\frac{i}{\hbar}(n+\frac{1}{2})\eta_\pm} \xi_n(\chi), \tag{127}$$

where $\xi_n(\chi)$ are the eigenfunctions of the harmonic oscillator[28]. As we have already seen in Section 2.4.2, the two newborn universes can also be interpreted as two expanding universes filled with matter and antimatter, respectively. Let us notice that if one inserts the wave functions (125) into the Wheeler–DeWitt Equation (92), it is obtained at order $\hbar^1$

$$\pm 2i\hbar\omega_{DS}\frac{\partial \xi}{\partial a} - \hbar^2 \frac{\partial^2 \xi}{\partial \chi^2} + \chi^2 \xi = 0, \tag{128}$$

which is the time-dependent Schrödinger equation

$$i\hbar \frac{\partial}{\partial \eta_\pm} \xi(\eta_\pm, \chi) = \frac{1}{2}\left(-\hbar^2 \frac{\partial^2}{\partial \chi^2} + \chi^2\right)\xi(\eta_\pm, \chi), \tag{129}$$

provided that one identifies the (conformal) time variable of the background spacetime of the two universes, $\phi^\pm$, by

$$\frac{\partial}{\partial \eta_\pm} = \mp \omega_{DS}\frac{\partial}{\partial a} \quad \Rightarrow \quad \eta_\pm = \mp \int \frac{da}{\omega_{DS}} = \mp \int \frac{dt}{a}. \tag{130}$$

If we assume that the physical time variable is the variable measured by real clocks, which are made up of matter, and thus that it is the time that appears in the Schrödinger equation, then, in terms of the physical time of an observer in one of the universes, the Schrödinger equation for the fields in the partner universe turns out to be the Schrödinger equation of a field $\bar{\varphi}$ that is CP-conjugated with respect the field in the observer's universe (see Section 3). The two universes form then a universe–antiuniverse pair. The process can be compared with the creation of an electron–positron pair (see Figure 1 of Ref. [40], and Figure 6), which can be seen as the creation of an electron moving backward in time and an electron moving forward in time. Here, the "time" variable is the scale factor, so "moving forward in time" means an expanding universe, and "moving backward in time" means a contracting universe. Therefore, the creation of a contracting–expanding pair of universes can be paralleled with the creation of an electron–positron pair.

3. Third Quantisation Formalism
3.1. Histoirial Review

There was a great excitation in the 1980s with the formulation of the third quantisation formalism and the associated description of topology change in quantum gravity [28,46,53–58]. At that time, the cosmological paradigm was a universe in a decelerating expansion with a zero value of the cosmological constant. The third quantisation formalism, which was initially proposed [53] as an analogue to the second quantisation formalism of a quantum field theory, fit well with the description of the quantum fluctuations of the spacetime. As a consequence of the interaction with the fluctuations of the spacetime, it turns out that the coupling constants become dynamical functions, and this was seen as a possible explanation for the expected vanishing value of the cosmological constant.

On the other hand, general relativity is a (local) geometrical theory, and therefore it does not account for the global topology of the spacetime manifold. However, it may perfectly happen that the foliation of the spacetime gives rise, at some given value t, to a collection of simply connected spatial sections (see Figure 13 (Left)). From the point of view of the evolution of the universe, the topology change represented in Figure 13 (Middle) can be seen as the creation of a universe (universe B) from the pre-existing one (universe

A). In fact, the process is formally similar to the QED process depicted in Figure 13 (Right), which represents the creation of a photon from the propagation of an electron. As we know from QED, this kind of process is better explained in the formalism of quantum field theory where we can define the creation and annihilation operators of particles. Therefore, it seems reasonable to develop a field theoretical approach too to describe the creation (and annihilation) of the universe(s).

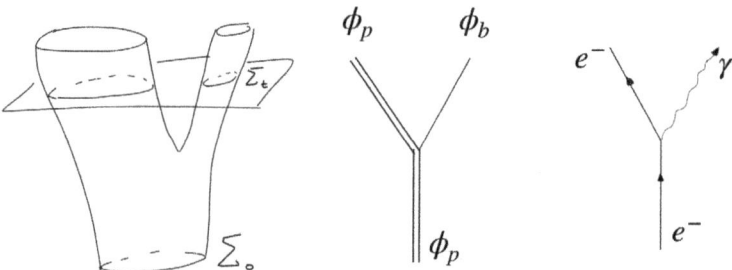

Figure 13. (**Left**) The foliation of a given spacetime can give rise, at some value t_0, to two disconnected spatial sections. (**Middle**) Schematic representation. (**Right**) Creation of a photon from the scattering of an electron.

As an introductory example, let us consider the Wheeler–DeWitt equation of a closed DeSitter spacetime (see (44)), which can be written as

$$\ddot{\phi} + \omega^2(a)\phi = 0, \tag{131}$$

where, $\phi = \phi(a)$, $\dot{\phi} \equiv \frac{d\phi}{da}$, and

$$\omega^2(a) = \frac{H_0^2 a^4 - a^2}{\hbar^2}, \tag{132}$$

with $H^2 = \Lambda/3$. Clearly, (131) is the equation of a harmonic oscillator with the scale factor a playing the role of the time variable. One can assume then that (131) is the result of the variational principle of the action of a harmonic oscillator with time-dependent frequency,

$$S_3 = \int da \left(\dot{\phi}^2 - \omega^2 \phi^2 \right), \tag{133}$$

from which one can obtain the conjugate momentum, $P_\phi = \dot{\phi}$, and construct a the corresponding Hamiltonian,

$$H_3 = \frac{1}{2} P_\phi^2 + \frac{\omega^2}{2} \phi^2. \tag{134}$$

The third quantisation procedure consists in promoting the variables ϕ and P_ϕ to quantum operators in the usual way, $\hat{\phi} \to \phi$ and $\hat{P}_\phi \to -i\hbar \partial_\phi$, and describing the quantum state of the whole spacetime manifold by the use of a new wave function, $\Psi(\phi)$, constructed as

$$\Psi(\phi) = \int \delta\phi e^{\frac{i}{\hbar} S_3}, \tag{135}$$

in the path integral approach, or via the Schrödinger equation,

$$H_3 |\Psi\rangle = i\hbar \frac{\partial |\Psi\rangle}{\partial a}. \tag{136}$$

One can also define the *ladder* operators, $\hat{b}$ and $\hat{b}^\dagger$, of this particular harmonic oscillator in terms of the operators $\hat{\phi}$ and $\hat{P}_\phi$, as usual, and construct the state of the whole spacetime manifold, whatever the topology it has[29], in terms of the eigenstates of the number operator, $|N, a\rangle$, so

$$|\Psi\rangle = \sum_N C_N |N, a\rangle. \tag{137}$$

The number states, $|N, a\rangle$, would represent N universes with a scale factor a, and the state of the whole spacetime manifold, $|\Psi\rangle$, would be a quantum superposition of different number states (see Section 5). Classically, these N universes are disconnected, and therefore one should just consider one of them as representing our universe and disregard the rest as being physically irrelevant. However, from the quantum mechanical standpoint, new phenomena may appear as quantum correlations and other collective behaviour, so it seems interesting, at least in principle, to analyse the quantum description of the whole many-universe state.

Parent and Baby Universes: The Hybrid Action

The third quantisation formalism was mainly applied in the 1980s–1990s to the description of the quantum fluctuations of the spacetime. Let us notice that the quantum gravity of multiply connected spacetime manifolds can be applied to two well-distinguished cases: one that accounts for the properties of large regions of the spacetime, called *parent universes*[30], and another that focus on a local region of the spacetime where small pieces of length of the order of the Planck length can branch off and disconnect from the parent spacetime and become small *baby universes* [46]. In both cases, the spacetime manifold under study turns out to be non-simply connected. However, in the 1980s–1990s, the idea of a multiverse was not seriously considered, and the main problem at that time was to explain the supposedly zero value of the cosmological constant[31].

Then, from the point of view of the third quantisation formalism, our universe can then be seen as a large parent universe propagating in a plasma of baby universes. The effects of the baby universes could be measured by their influence on the observable properties of the parent universe, and the most representative picture of the baby–parent universe interaction becomes the *hybrid action* [46], in which the parent universe is described by a second quantised wave function, $\phi_p(q^a)$, and the baby universes are described by the third quantised wave functions, $\hat{\phi}_b$, i.e., the behaviour of the spacetime is assumed to be classical, and its fluctuations are seen as small particles propagating in the parent spacetime. The total action is then given by

$$S_T = S_0(a, \varphi) + S_b(\hat{\phi}_b) + S_I(a, \varphi; \hat{\phi}_b), \tag{138}$$

where $S_0(a, \varphi)$ is the Einstein–Hilbert action of the homogenous and isotropic parent spacetime with scale factor a, and matter field φ, $S_b(\hat{\phi}_b)$ is the third quantised action of the baby universes, while S_I is the action of interaction,

$$S_I(a, \varphi; \hat{\phi}_b) = \int dt \mathcal{N} \sum_i \mathcal{L}_i(t, \vec{x}) \hat{\phi}_b^i = \int dt \mathcal{N} \sum_i \mathcal{L}_i(t, \vec{x}) (\hat{b}_i^+ + \hat{b}_i^-), \tag{139}$$

where the index i labels the different modes of the baby universe field (i.e., it labels different species of baby universes), and $\mathcal{L}_i(t, \vec{x})$ is called the insertion operator at the nucleation event [46]. It defines the space–time points of the parent universe in which the baby universes effectively nucleate.

Two main problems were addressed in the 1980s using the third quantisation formalism: the dynamical value of the coupling constants and the loss of quantum coherence (decoherence) produced by the plasma of baby universes.

As an example of the former, let us consider a universe with a matter field $\varphi(t)$ that is coupled to the spacetime through the interaction with the baby universes. The parent universe two point function becomes [46]

$$G(\varphi_f, \varphi_i) \propto \langle \hat{\phi}_b, 0 | \phi_p(\varphi_f) \phi_p(\varphi_i) | \hat{\phi}_b, 0 \rangle$$
$$= \langle \hat{\phi}_b, 0 | \int_{\varphi_i}^{\varphi_f} d\varphi(t) \int_0^\infty dN\, e^{iS_p + iS_I} | \hat{\phi}_b, 0 \rangle, \quad (140)$$

where S_I is given by (139) (for simplicity, let us assume just one specie of baby universes). Now, suppose the baby universes are in a wave function eigenstate $|\alpha\rangle$, with $\hat{\phi}_b |\alpha\rangle = \alpha |\alpha\rangle$, where α satisfies

$$\left(\nabla_q^2 + m_b^2 \right) \alpha = 0. \quad (141)$$

In that case, the function (140) becomes

$$G(\varphi_f, \varphi_i) \propto \int_{\varphi_i}^{\varphi_f} d\varphi(t) \int_0^\infty dN\, e^{i\tilde{S}_p}, \quad (142)$$

where

$$\tilde{S}_p = \int dt N a^3 \left(\frac{1}{2N^2} \dot{\varphi}^2 - V(\varphi) - \kappa \alpha \right). \quad (143)$$

The effects of the baby universes are thus encoded in an addition of an ordinary potential to the second quantised action. The new term is dynamical in the sense that it must satisfy the dynamical Equation (141). That was used as an argument for a possible mechanism for the vanishing value of the cosmological constant, which was the expected value at that time.

The other question addressed with the third quantisation formalism was the loss of quantum coherence of the matter fields caused by their propagation in the plasma of baby universes. Basically, the argument was the following [54]. Let us suppose the composite state between a matter field φ and the baby universes. Let $|\varphi, n\rangle$ be the state in which the matter field is in the state $|\varphi\rangle$ and there are n baby universes. Then, the initial state is

$$|\text{in}\rangle = |\varphi^{\text{in}}, 0\rangle, \quad (144)$$

where $|\varphi^{\text{in}}\rangle$ is the initial state of the matter field. If we do not measure the state of the baby universes, and we therefore integrate out their quantum state from the composite state (144), then, the initial state is described by the density matrix [54]

$$\rho^{\text{in}} = |\varphi^{\text{in}}\rangle \langle \varphi^{\text{in}}|. \quad (145)$$

After the interaction with the baby universes, the final state becomes a linear combination of the states of the fluctuations and the corresponding states of the matter field that come out from the interaction with the $|n\rangle$ states. For simplicity, let us consider just two $|n\rangle$ states, $|0\rangle$ and $|1\rangle$. The composite state after the interaction would be

$$|\text{out}\rangle = |\varphi_0, 0\rangle + |\varphi_1, 1\rangle, \quad (146)$$

where φ_0 and φ_1 are in general different, and the linear combination in (146) can be weighted accordingly. Then, the reduced density matrix that describes the state of the matter field alone becomes after the interaction

$$\rho^{\text{out}} = |\varphi_0\rangle \langle \varphi_0| + |\varphi_1\rangle \langle \varphi_1|. \quad (147)$$

The field turns out to be in a statistical mixture of two states. The initial state was a pure state, i.e., a state of total information with zero entropy, $S = 0$. The final state, instead, becomes a mixed state with entropy, $S > 0$, so information (quantum coherence) has been lost. Coleman's argument was that the operators of the baby universes must be independent of the coordinates of the parent spacetime, and thus, the coupling with the matter fields is independent of their evolution. In that case, the state of the field does not change along the time evolution because the states of the baby universes do not change *in time*[32]. However, counterarguments were also given for the loss of quantum coherence [60,61].

3.2. Quantum Field Theory in $M \equiv \text{Riem}(\Sigma)$

3.2.1. Geometrical Structure of M

We have seen in Section 2 that the evolution of the universe can be seen as the time evolution of the 3-dimensional metric that is induced on the spatial hypersurfaces by the 4-dimensional metric that is the solution of the Einstein's equations. Therefore, the evolution of the universe is a trajectory in the space of Riemannian symmetric 3-metrics with components, h_{ab}. Let us call it M. With the DeWitt metric (7), M becomes a metric space, where we can define the line element as

$$ds^2 = G^{abcd} dh_{ab} dh_{cd}. \tag{148}$$

However, not all of the h_{ab} components are independent. A symmetric 3-metric has only 6 independent components, so it turns out that M is isomorphic to $\mathbb{R}^6$. Thus, we can make the following choice for the coordinates[33] in M,

$$q^A = \{h_{11}, h_{22}, h_{33}, \sqrt{2}h_{23}, \sqrt{2}h_{13}, \sqrt{2}h_{12}\}, \tag{149}$$

in terms of which the line element (148) can be written

$$ds^2 = G_{AB} dq^A dq^B, \tag{150}$$

where G_{AB} is a 6-dimensional metric tensor that is related to the components of DeWitt's metric, G^{abcd}. The signature of M is $(-,+,+,+,+,+)$, which is easy to check for the case of the flat metric, $h_{ab} = \delta_{ab}$, and because the signature remains invariant under a change of coordinates, it holds then for the general case too. Thus, DeWitt showed [10] that the 6-dimensional space M is indeed a $5+1$ dimensional space with a 1 *time-like* dimension and an orthogonal 5-dimensional *space-like* subspace. As the coordinate of the time-like subspace, it is appropriate to take the coordinate τ defined by [7,10,62]

$$\tau = \left(\frac{32}{3}\right)^{\frac{1}{2}} h^{1/4}, \tag{151}$$

where, $h = \det h_{ij}$, which essentially represents the volume of an infinitesimal volume element of the spatial sections of the spacetime ($V \propto \int dx^3 \sqrt{h}$). The hypersurfaces of constant τ are the space-like sections of M, labelled by $\bar{M}$ [10]. Then, in terms of the variables, $q^\mu = \{\tau, \bar{q}^A\}$, where $\bar{q}^A$, with $A = 1, \ldots, 5$, are the five coordinates in $\bar{M}$, the line element (150) becomes

$$ds^2 = -d\tau^2 + h_0^2 \tau^2 d\bar{s}^2 = -d\tau^2 + h_0^2 \tau^2 \bar{G}_{AB} d\bar{q}^A d\bar{q}^B, \tag{152}$$

where, $h_0^2 = 3/32$, and $d\bar{s}$ is the line element in $\bar{M}$, with [10]

$$\bar{G}_{AB} = \text{tr}\left(h^{-1} h_{,A} h^{-1} h_{,B}\right) \equiv h^{ij} \frac{\partial h_{jk}}{\partial \bar{q}^A} h^{kl} \frac{\partial h_{li}}{\partial \bar{q}^B}. \tag{153}$$

The metric (153) is invariant under a conformal transformation of the metric, and in particular, it is invariant under the change, $h_{ab} \to \xi(h) h_{ab}$, so it is convenient labelling the points of $\bar{M}$ with the five independent components of the transformed metric,

$$\bar{h}_{ab} = h^{-\frac{1}{3}} h_{ab}, \tag{154}$$

which has a unit determinant. Furthermore, $\bar{M}$ is noncompact and diffeomorphic to Euclidean 5-space [10], with the Ricci tensor $\bar{R}_{AB}$ given by[34]

$$\bar{R}_{AB} = -\frac{3}{4}\bar{G}_{AB}, \qquad (155)$$

and scalar curvature

$$\bar{R} = -\frac{15}{4} \equiv \frac{k}{a^2}, \qquad (156)$$

with $k = -1$. $\bar{M}$ is then an "Einstein space" of constant negative curvature. In a homogeneous universe, the time-like variable τ represents the volume of the spatial sections of the universe, and the five coordinates, $\bar{h}_{ab}$, represent the *shape* of a unit volume. A fixed point in $\bar{M}$ represents therefore the evolution of a universe that scales the volume of the spatial sections without changing their shape, and lines of constant τ represent different shapes of the spatial sections of the universe with a fixed given volume. Thus, the line element (152) reveals the space M as *a set of "nested" 5-dimensional submanifolds, all having the same intrinsic shape* [10]. From that point of view, the 5-dimensional submanifold $\bar{M}$ can be seen as a proper realisation of what is called the "shape space" [63].

On the other hand, the space M with the metric (152) has the same formal structure of a Friedmann–Robertson–Walker spacetime with the hyperbolic 5-space H^5 as the "spatial" section. In particular, it has the same formal structure as the Milne spacetime[35]. Therefore, one can find a set of coordinates $(\chi, \theta, \phi, \psi, \zeta)$ in $\bar{M}$ in terms of which the metric (152) can be written as[36]

$$ds^2 = -d\tau^2 + \tau^2\left(d\chi^2 + \sinh^2\chi\, d\Omega_4^2\right), \qquad (157)$$

where, $\chi \in [0, \infty)$, and $d\Omega_4^2$ is the line element on the 4-sphere of unit radius

$$d\Omega_4^2 = d\theta^2 + \sin^2\theta\left(d\phi^2 + \sin^2\phi(d\psi^2 + \sin^2\psi d\zeta^2)\right). \qquad (158)$$

The Milne spacetime is a particular coordination of part of the Minkowski spacetime. It does not cover the whole Minkowski spacetime but only the interior of the upper (or the lower, with a time reversal change) light cone of the Minkowski spacetime. Something similar occurs in M. Let us introduce the variables

$$T = \tau \cosh \chi, \quad R = \tau \sinh \chi, \qquad (159)$$

in terms of which the line element (157) becomes

$$ds^2 = -dT^2 + dR^2 + R^2 d\Omega_4^2, \qquad (160)$$

with $0 < T < \infty$ and $0 < R < \infty$. The metric (160) is nothing more than the metric of a 6-dimensional Minkowski space, and the Milne space only covers the upper light cone (see Figure 14). The interior of the lower light cone is covered by a time reversal change of coordinates, $\tau \to -\tau$ (let us notice that the metric (152) is invariant under this change). However, although the manifold $\bar{M}$ is geodesically complete, the manifold M is not. The scalar curvature of M,

$$R = -\frac{20}{\tau^2} \qquad (161)$$

presents a singular frontier of infinite curvature, located at $\tau = 0$, where all geodesics in M eventually hit [10]. It means that the upper and the lower light cones of the 6-dimensional Minkowski space must be considered independently. They represent two time-reversed copies of the universe, i.e., two universes related by a time reversal change of the time (volume) variable. We shall see that, quantum-mechanically, this can be seen as a universe–antiuniverse pair.

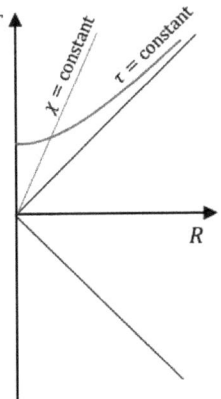

Figure 14. The space of three metrics, M, turns out to be a particular coordination of upper (lower) light cone of a 6-dimensional Minkowki space. Every point in the T, R plane is a four-sphere of unit radius. Lines of constant τ are lines of constant volume of the spatial sections of the spacetime (with different shapes). Lines of constant χ correspond to different volumes of the same shape (a scaling universe). Something similar occurs in the lower light cone, which would represent a time-reversed copy of the universe.

3.2.2. Classical Evolution of the Universe

Let us now analyse the evolution of the universe in the space, $M \equiv \mathrm{Riem}(\Sigma)$. From a geometrical point of view, the evolution of the universe is the trajectory that extremizes the Einstein–Hilbert action (6), which can conveniently be written as[37][7]

$$S_{EH} = \frac{1}{2} \int_M dt d^3x N \left(\frac{1}{N^2} G^{abcd} \dot{h}_{ab} \dot{h}_{cd} - m^2(h_{ab}) \right), \tag{162}$$

where the rescale, $G^{abcd} \to \frac{1}{32\pi G} G^{abcd}$, was made with G the Newton's constant, and the potential terms of the Einstein–Hilbert action have been gathered in a *mass term*

$$m^2(h_{ab}) = \frac{\sqrt{h}}{8\pi G} \left(2\Lambda - {}^{(3)}R \right). \tag{163}$$

In term of the variables, $q^A = (\tau, \bar{q}^A)$, the action (162) can be written

$$S_{EH} = \frac{1}{2} \int_M dt d^3x N \left(\frac{1}{N^2} G_{AB} \dot{q}^A \dot{q}^B - m^2(q^A) \right), \tag{164}$$

where G_{AB}, is given by (152) or (157). The Einstein–Hilbert action has been written in the form of (164) to make clear the formal resemblance with respect to the action of a particle that moves in the spacetime,

$$S[x^\mu(\lambda)] = \frac{1}{2} \int \left(\frac{1}{N^2} g_{\mu\nu} \dot{x}^\mu \dot{x}^\nu - m^2 \right) N d\lambda, \tag{165}$$

with $\dot{x}^\mu = \frac{dx^\mu}{d\lambda}$, for which the trajectory is given by the geodesic equation,

$$\ddot{x}^\mu + \Gamma^\mu_{\alpha\beta} \dot{x}^\alpha \dot{x}^\beta = 0, \tag{166}$$

with

$$\Gamma^\mu_{\alpha\beta} = \frac{1}{2} g^{\mu\nu} \left(\frac{\partial g_{\alpha\nu}}{\partial x^\beta} + \frac{\partial g_{\beta\nu}}{\partial x^\alpha} - \frac{\partial g_{\alpha\beta}}{\partial x^\nu} \right). \tag{167}$$

The case of the universe is formally similar. The evolution of the universe is a trajectory in M. The only difference is the trajectory is not a geodesic[38] because of the non-constant potential, $m^2(h_{ab})$. Instead, it is given by

$$\ddot{q}^A + \Gamma^A_{BC}\dot{q}^B\dot{q}^C = -G^{AB}\frac{\partial V(q)}{\partial q^B}, \tag{168}$$

where $\dot{q}^A = \frac{dq^A}{dt}$ and $2V = m^2(h_{ab})$, and the Christoffel's symbols, are defined analogously in terms of the metric components as

$$\Gamma^A_{BC} = \frac{1}{2}G^{AD}\left(\frac{\partial G_{DC}}{\partial B} + \frac{\partial G_{BD}}{\partial C} - \frac{\partial G_{BC}}{\partial D}\right). \tag{169}$$

In terms of the variables (τ, q^A), the Equation (168) turn out to be

$$\ddot{\tau} + h_0^2 \tau \bar{G}_{AB}\mathring{q}^A \mathring{q}^B = \frac{\partial V}{\partial \tau}, \tag{170}$$

$$\mathring{\mathring{q}}^A + \frac{2\dot{\tau}}{\tau}\mathring{q}^A + \Gamma^A_{BC}\mathring{q}^A\mathring{q}^B = -\frac{1}{h_0^2 \tau^2}\bar{G}^{AB}\frac{\partial V}{\partial \bar{q}^B}, \tag{171}$$

As we have seen in the preceding section, a geodesic in M eventually hits the singular frontier located at $\tau = 0$ (the zero volume hypersurface). However, because of the potential term, which may also include the Lagrangian of the matter fields, the universe does not follow a geodesic in M and may thus avoid the singular frontier. The paradigmatic case is the closed DeSitter spacetime. In addition to (170) and (171), one can use the Hamiltonian constraint (11), which in terms of the $(\tau, \bar{q}^A)$ coordinates reads,

$$\dot{\tau}^2 = h_0^2 \tau^2 \bar{G}_{AB}\mathring{q}^A\mathring{q}^B + m^2 = h_0^2 \tau^2 \mathring{s}^2 + m^2. \tag{172}$$

In the case that the right hand side of (171) is zero, it can be shown that [10], $\mathring{s} = \alpha/\tau^2$, with α a constant of integration[39], and then

$$\dot{\tau}^2 = \frac{h_0^2 \alpha^2}{\tau^2} + m^2. \tag{173}$$

In the case for which, $\Lambda \gg {}^3R$, the potential is proportional to τ, $m^2 = h_0^2 H_0^2 \tau^2$, and

$$\dot{\tau}^2 = h_0^2\left(\frac{\alpha^2}{\tau^2} + H_0^2 \tau^2\right), \tag{174}$$

whose solutions are given by

$$\tau^2(t) = \alpha \sinh(2H_0 \Delta \tilde{t} - \ln H_0\alpha), \tag{175}$$

for $\alpha \neq 0$ and

$$\tau(t) \propto e^{H_0 \Delta \tilde{t}}, \tag{176}$$

for $\alpha = 0$ (FRW spacetime) with $\Delta \tilde{t} = h_0 \Delta t$. From (175) and (176), one can see that, as expected, for a universe stage in which $\Lambda \gg {}^3R$ the expansion of the volume element is exponential.

3.2.3. Quantum Field Theory in M

As we have seen above, the third quantisation[40] procedure consists in promoting the field $\phi(h_{ab})$ and its conjugate momentum to quantum operators. One can then pose another wave function, Ψ, as in (135)–(137), and work with the corresponding Schrödinger equation (see, (136)). However, it turns out to be much more interesting to develop and study the quantum field theory (QFT) of the field $\phi(h_{ab})$ propagating in the 6-dimensional space M.

Let us first notice that, in terms of the coordinates $q^A = \{\tau, \tilde{q}^A\}$ (see Equation (152)), the Hamiltonian constraint (11) can be written as

$$\mathcal{H} = G^{AB} p_A p_B + m^2(q, \varphi) = 0, \tag{177}$$

where G^{AB} is the inverse of (152), and in $m^2(q, \varphi)$ we have also included the Hamiltonian of the matter fields (which for simplicity have not been considered so far) that can generically be encapsulated in a variable φ,

$$m^2(q, \varphi) = m_g^2(q) + 2\mathcal{H}_m(q, \varphi), \tag{178}$$

with, $m_g^2(q)$, given by (see, Equation (163))

$$m_g^2(q) = \frac{h_0^2 \tau^2}{8\pi G}\left(2\Lambda - {}^3R(q)\right), \tag{179}$$

and

$$\mathcal{H}_m = \frac{1}{2 h_0^2 \tau^2} p_\varphi^2 + \ldots + h_0^2 \tau^2 V(\varphi), \tag{180}$$

where the dots indicate terms that contain spatial derivatives of the matter fields, which for simplicity we shall consider negligible.

Under canonical quantisation of the momenta in the Hamiltonian constraint (206), one obtains the Wheeler–DeWitt equation, which, with an appropriate choice of factor ordering, can also be written as,

$$\left(-\hbar^2 \Box_q + m^2(q, \varphi)\right)\phi(q, \varphi) = 0, \tag{181}$$

where $\phi(q, \varphi)$, is the wave function of the universe [12], and

$$\Box_q = \nabla \vec{\nabla} = \frac{1}{\sqrt{-G}} \frac{\partial}{\partial q^A}\left(\sqrt{-G} G^{AB} \frac{\partial}{\partial q^B}\right), \tag{182}$$

where $G = \det G_{AB}$; we have used the customary definitions of the gradient and the divergence in a curved space,

$$\vec{\nabla}\phi = G^{AB} \frac{\partial \phi}{\partial q^B}, \quad \nabla \cdot \vec{F} = \frac{1}{\sqrt{-G}} \frac{\partial}{\partial q^A}\left(\sqrt{-G} F^A\right). \tag{183}$$

With these definitions, the Wheeler–DeWitt Equation (181) can be obtained from the variational principle of the third quantised action

$$S_{(3)} = \frac{1}{2} \int dq \sqrt{-G}\left(-\hbar^2 \vec{\nabla}\phi \cdot \vec{\nabla}\phi - m^2(q)\phi^2\right), \tag{184}$$

Variation of (184) with respect to the wave function ϕ gives rise to the wave Equation (181). Now, following the analogy with a QFT, we can define a conserved current in the superspace,

$$\vec{J} = i\hbar\left(\phi^* \vec{\nabla}\phi - (\vec{\nabla}\phi^*)\phi\right), \tag{185}$$

from which it can easily be checked that $\nabla \cdot \vec{J} = 0$ and the following inner product

$$(\phi_m, \phi_n) = i\hbar \int d\vec{\Sigma}\left(\phi_m^* \vec{\nabla}\phi_n - (\vec{\nabla}\phi_m^*)\phi_n\right), \tag{186}$$

where $d\vec{\Sigma}$ is the future-oriented surface element of the 5-dimensional spacelike subspace.

The procedure of field quantisation consists in promoting the wave function $\phi(q)$ to an operator and expanding it into modes that are orthonormal with respect to the inner product (186). Then [56]

$$\hat{\phi}(q) = \sum_n \phi_n(q)\hat{A}_n + \phi_n^*(q)\hat{A}_n^\dagger, \tag{187}$$

where $\phi_n(q)$ is a complete set of orthonormal solutions of the Wheeler–DeWitt Equation (181), the index n symbolises the particular set of quantum number associated to that state [56], and $\hat{A}_n^\dagger$ and $\hat{A}_n$ are the creation and annihilation operators of modes ϕ_n, respectively, satisfying the customary commutation relations,

$$[\hat{A}_n, \hat{A}_m^\dagger] = \delta_{nm}, \ [\hat{A}_n, \hat{A}_m] = 0 = [\hat{A}_n^\dagger, \hat{A}_m^\dagger]. \tag{188}$$

In terms of the variables $(\tau, \bar{q})$, the label n of the modes ϕ_n in (187) can be associated to the 5-dimensional spacelike momentum of the *particles* that propagate in the space M. In particular, we have seen that M has the geometrical structure of a $5+1$-dimensional Friedmann–Robertson–Walker universe, so we can use this information to develop the quantisation of the field $\phi(q)$. In terms of the coordinates $q^A = (\tau, \bar{q}^A)$, the Laplace–Beltrami operator (182) can be written as,

$$\Box_q = -\frac{1}{\tau^5}\frac{\partial}{\partial \tau}\left(\tau^5 \frac{\partial}{\partial \tau}\right) + \frac{1}{\tau^2}\Box_{\bar{q}}, \tag{189}$$

where $\Box_{\bar{q}}$, is the corresponding 5-dimensional Laplacian (that with $\Box_q$ is given by (182) with the 5-dimensional metric $\bar{G}_{AB}$ instead of G_{AB} and without the minus sign in the square roots). In *conformal time*, $\lambda = \ln \tau$, and with the rescale, $\phi(q) = e^{-2\lambda}\tilde{\phi}(\lambda, \bar{q})$, the wave Equation (181) (i.e., the Wheeler–DeWitt equation) becomes [75]

$$\left\{\frac{\partial^2}{\partial \lambda^2} - \Box_{\bar{q}} + \left(\frac{m^2}{\hbar^2}e^{2\lambda} - 4\right)\right\}\tilde{\phi}(\lambda, \bar{q}) = 0. \tag{190}$$

However, the "mass" of the field (260) is not a constant. Even considering just the geometrical degrees of freedom, it continues being a non-constant function of the components of the metric tensor h_{ij} (or, equivalently, of the variables q^A) through the dependence on the ^3R curvature (see (179)). In that case, the space M turns out to be a dispersive medium for the wave function of the universe. It does not invalidate the formalism, but it becomes more complicated from a technical point of view. For that reason, let us focus on the case for which $^3R \ll 2\Lambda$, which, on the other hand, is a very plausible condition for the initial state of the universe[41], and consider only the geometrical degrees of freedom plus the constant Λ. We can then assume the value[42]

$$m_g^2(q) \approx \frac{h_0^2 \Lambda}{4\pi G}\tau^2 \equiv \hbar^2 m_0^2 e^{2\lambda}, \tag{191}$$

in the Wheeler–DeWitt Equation (181). In that case, the mass (260) only depends on the *time* variable τ, and we can perform the quantisation of the field ϕ in the customary way (see, for instance, Refs. [76,77]). Then, we can decompose the wave function of the universe $\phi(q)$ in normal modes as

$$\phi(q) = \int_0^\infty dk \sum_{\vec{j}} [a_{\mathbf{k}} u_{\mathbf{k}}(q) + a_{\mathbf{k}}^\dagger u_{\mathbf{k}}^*(q)], \tag{192}$$

where, $\mathbf{k} = (k, \vec{j})$,

$$u_{\mathbf{k}}(q) = e^{-2\lambda}\chi_{k,J}(\lambda)\mathcal{Y}_{J,\vec{M}}(\bar{q}), \tag{193}$$

and $\mathcal{Y}_{k,\vec{j}}(\vec{q})$ are the eigenfunctions of the Laplacian defined on the 5-dimensional hyperboloid, which satisfy [78]

$$\Box_{\vec{q}} \mathcal{Y}_{k,\vec{j}}(\vec{q}) = -(k^2 + 4)\mathcal{Y}_{k,\vec{j}}(\vec{q}), \tag{194}$$

with $0 < k < \infty$, and $\vec{j}$ denotes the four indices that distinguish the four components of the generalisation of the angular momentum on the four spheres[43]. Thus, the wave Equation (181) (i.e., the Wheeler–DeWitt equation) reduces to

$$\chi_k'' + \left(m_0^2 e^{2\lambda} + k^2\right)\chi_k = 0, \tag{195}$$

where $\chi' \equiv \frac{d\chi}{d\lambda}$. One interesting thing is that the frequency squared of the oscillator (195), $\omega_k^2(\lambda) \equiv m_0^2 e^{2\lambda} + k^2$, is never negative. The other interesting thing is that (195) is readily solvable in terms of Bessel functions. With the customary normalisation condition

$$\chi_k \partial_\lambda \chi_k^* - \chi_k^* \partial_\lambda \chi_k = i. \tag{196}$$

we easily find two set of orthonormal modes given by [75]

$$\tilde{\chi}_k(\tau) = \left(\frac{2}{\pi}\sinh(\pi k)\right)^{-\frac{1}{2}} \mathcal{J}_{-ik}(m_0\tau), \tag{197}$$

$$\chi_k(\tau) = \frac{\sqrt{\pi}}{2} e^{\frac{k\pi}{2}} \mathcal{H}_{ik}^{(2)}(m_0\tau). \tag{198}$$

3.2.4. Boundary Conditions and the Creation of the Universes in Pairs

In order to choose the particular set of modes, we have to impose some boundary condition. For this, we shall consider the multiverse as a really closed system, so no external influence is expected to modify its state. Therefore, it seems appropriate to describe the state of the multiverse in a representation that is invariant under the evolution of the third quantised Hamiltonian, which is an extension of the invariant representation used in quantum mechanics [31–33,79–86].

An invariant representation can be given in terms of creation and annihilation operators, $\hat{b}_\mathbf{k}$ and $\hat{b}_\mathbf{k}^\dagger$, defined as [83]

$$\hat{b}_\mathbf{k} = \frac{i}{\sqrt{\hbar}}(\xi_k^* \hat{p}_\phi - (\xi_k^*)'\hat{\phi}) \ , \ \hat{b}_\mathbf{k}^\dagger = -\frac{i}{\sqrt{\hbar}}(\xi_k \hat{p}_\phi - (\xi_k)'\hat{\phi}), \tag{199}$$

where, $\hat{\phi}$ and $\hat{p}_\phi$, are the operator version of the wave function and the conjugate momentum, respectively, and ξ_k is a solution of the wave Equation (190), or equivalently with (195), the orthonormality condition (196), which ensures the usual commutation relations,

$$[\hat{b}_\mathbf{k}, \hat{b}_\mathbf{k}^\dagger] = 1. \tag{200}$$

The operators $\hat{b}_\mathbf{k}$ and $\hat{b}_\mathbf{k}^\dagger$ in (252) are *time*-dependent operators, but the dependence is such that the the eigenstates of the corresponding number operator, $\hat{N}_\mathbf{k} = \hat{b}_\mathbf{k}^\dagger \hat{b}_\mathbf{k}$, remain invariant under the action of the third quantised Hamiltonian. It means, for example, that once the multiverse is in the vacuum state of an invariant representation it remains in the same vacuum state irrespective of the internal histories of the connected pieces of the whole spacetime manifold. From this point of view, the multiverse does not evolve in a proper sense, although the time dependence of the vacuum state makes that the vacuum state at (conformal) time, λ_0, is functionally different than the vacuum state at λ_1, being both however the same vacuum state of the same invariant representation.

The conditions (196)–(252) do not fix the vacuum state. There is in fact an infinite number of solutions that fit with (196)–(252), each of which define a particular representation and the associated vacuum state. For instance, the modes (259)–(264) define two vacuum states, $|\bar{0}_\mathbf{k}\bar{0}_{-\mathbf{k}}\rangle$ and $|0_\mathbf{k} 0_{-\mathbf{k}}\rangle$, respectively. The modes (259) can be identified with the Hartle–Hawking no-boundary condition, first, because they are regular at the Euclidean origin[44], $\tau \to i\tau \to 0$, and, second, because that holds for large values of the variable τ, $\bar{\chi}_k(\tau) \propto \cos m_0 \tau$, which essentially matches with the result (23). On the other hand, Vilenkin's tunnelling wave function can be identified with the mode (264), because at large values of τ it reads, $\chi_k(\tau) \propto e^{-im_0\tau}$, which represent, in terms of the time variable t, an expanding universe (see, Sections 2.2 and 2.3). One can also follow the analysis made in Ref. [76] to conclude that the state $|\bar{0}_\mathbf{k}\bar{0}_{-\mathbf{k}}\rangle$ is the conformal vacuum state and that the state $|0_\mathbf{k}0_{-\mathbf{k}}\rangle$ is the vacuum state of the 6-dimensional Minkowski space, M.

Regardless, these two set of modes are related by a Bogolyubov transformation,

$$\bar{\chi}_\mathbf{k} = \alpha_k \chi_\mathbf{k} + \beta_k \chi_\mathbf{k}^*, \tag{201}$$

where (see, for instance, Ref. [76])

$$\alpha_k = \left[\frac{e^{\pi k}}{2 \sinh(\pi k)} \right]^{\frac{1}{2}}, \beta_k = \left[\frac{e^{-\pi k}}{2 \sinh(\pi k)} \right]^{\frac{1}{2}}, \tag{202}$$

with, $|\alpha_k|^2 - |\beta_k|^2 = 1$. It means that the vacuum state of the $\bar{\chi}_\mathbf{k}$ modes, $|\bar{0}_\mathbf{k}\bar{0}_{-\mathbf{k}}\rangle$ can be written as [75,77]

$$|\bar{0}_\mathbf{k}\bar{0}_{-\mathbf{k}}\rangle = \prod_\mathbf{k} \frac{1}{|\alpha_k|^{1/2}} \left(\sum_{n=0}^{\infty} \left(\frac{\beta_k}{\alpha_k} \right)^n |n_\mathbf{k} n_{-\mathbf{k}}\rangle \right), \tag{203}$$

with a number of universes in the no bar representation given by

$$N_k = |\beta_k|^2 = \frac{1}{e^{2\pi k} - 1}, \tag{204}$$

which corresponds to a thermal distribution with generalised temperature

$$T = \frac{1}{2\pi}. \tag{205}$$

Then, one can state that, in this case, the Hartle–Hawking no-boundary version of the vacuum state is full of (Vilenkin's) universes (and antiuniverses) [75]. This result is very interesting because it implies that the consideration of universe–antiuniverse pairs seems to be quite unavoidable. It is formally similar to what happens in the quantum field theory of a matter field in an isotropic background spacetime, where the isotropy of the space makes that the particles are created in pairs with opposite values of the field modes, $\mathbf{k}$ and $-\mathbf{k}$ (see (203)), and in the case of a complex field in particle–antiparticle pairs with opposite momenta. In the third quantisation formalism, the space-like subspace $\bar{M}$ is homogeneous and isotropic. Therefore, if the potential of the Wheeler–DeWitt equation is also isotropic in $\bar{M}$, i.e., invariant under rotations in $\bar{M}$, the universes should be created in pairs with opposite values of the 5-dimensional $\mathbf{k}$ ($\equiv \bar{\mathbf{k}}_{ab}$) modes. This is not the most general case, but it is a quite plausible one provided that we assume a high value of the potential of the inflaton field, which can be identified at the initial stage of the universe with Λ, or equivalently, a small value of the spatial curvature ^3R of the newborn universe. In both cases, the potential term of the Wheeler–DeWitt equation can be approximated by (191), and small deviations can be treated as perturbations, which should not significantly violate the isotropy of the space $\bar{M}$.

This could be confirmed as well from a more geometrical point of view. Let us notice that the Milne spacetime can separately cover the interior of the upper and the lower light cones of the Minkowski spacetime. These two sections of the full light cone can be seen as the regions of the spacetime where future-oriented particles and past-oriented particles, or antiparticles, are propagated, which turn out to be entangled [87]. One would expect something similar in the case of the space M, which also covers the upper- and lower-half light cones of the 6-dimensional Minkowski space. These two regions would describe expanding and contracting universes (created in pairs as we have seen above), which are equivalent to the future- and past-oriented particles in Minkowski spacetime. Thus, much in a similar way as particles propagating backwards in time can be interpreted as antiparticles propagating forward in time [88], we have seen in Section 2, and we will see it again in the next section, contracting universes can be seen as expanding antiuniverses with a time variable that is reversely related with respect to the time variable of the partner universe. It means that the fields that propagate in one of the two entangled universes appear, from the point of view of an hypothetical observer in the partner universe, as moving backward in time. This is an illusory effect created by the relative definition of the time variables in the two universes, i.e., internal observers always define the fields that propagate in their universes as matter and the fields that propagate in the partner universe as antimatter. Furthermore, the value of each mode of the Fourier decomposition in (192) is proportional to the momentum conjugated to the components of the scaled metric tensor. It means that any change that is produced by the momentum associated to $+\bar{\mathbf{k}}_{ab}$ in the shape of the universe with metric $\bar{h}_{ab}$ is being also produced in the shape of the partner universe with opposite sign, $-\bar{\mathbf{k}}_{ab}$, so it is the parity of the two spatial sections tat is reversely related too, and so it is the relative parity of the fields that propagate in the two universes. In the next section, we shall see that the fields that propagate in the two universes are also charge-conjugated as a consequence of the reversely relation of their time variables. It turns out therefore that the field of the two universes is CP-conjugated. One can then conclude that, quite generally, the universes of the multiverse are created in symmetric universe–antiuniverse pairs whose composite quantum state is also expected to be entangled [50,89,90].

3.2.5. Semiclassical Regime

If one takes into account the matter fields, the total Hamiltonian constraint (206) can be written as,

$$\hat{H}_T \phi = (\hat{H}_G + \hat{H}_{SM})\phi = 0, \tag{206}$$

where $\hat{H}_G$ is the Hamiltonian operator that yields the WDW equation of the spacetime geometry alone (181) with Λ related to the constant part of the potential of the field that drives the inflationary period, $2\Lambda = 2V_0 \equiv H_0^2$, and $\hat{H}_{SM}$ ($\mathcal{H}_m$ in (260)) is the Hamiltonian operator of the matter fields, which essentially are the fields of the Standard Model (SM) with their corresponding potentials and interactions. Following the procedure described in Section 2.3, the wave function of the universe can be written as the product of two components, a wave function ϕ_0 that depends only on the gravitational degrees of freedom and the value of the constant Λ and a wave function that contains all the dependence on the fields of the SM, collectively denoted by the variable, φ, i.e

$$\phi^{\pm}(h_{ij}, \Lambda; \varphi) = \phi_0^{\pm}(h_{ij}, \Lambda) \psi_{\pm}(h_{ij}, \Lambda; \varphi), \tag{207}$$

where the two signs have been introduced for later convenience and $\phi^+ = (\phi^-)^*$. The wave function ϕ_0 is the solution of the WDWE of the geometrical degrees of freedom, computed in the preceding section. In general, it can be written in the semiclassical approach as

$$\phi_0^{\pm}(h_{ij}, \Lambda) \propto e^{\pm \frac{i}{\hbar} S(h_{ij}, \Lambda)}. \tag{208}$$

If one introduces the wave function (207) into the complete WDW equation and use the classical constraint (30) one obtains, at order $\hbar^1$, the following equation (see (33))

$$\mp 2i\hbar \vec{\nabla} S \cdot \vec{\nabla} \psi_{\pm} = H_{SM} \psi_{\pm}, \tag{209}$$

where $\vec{\nabla}$ is the gradient in M, and the negative and the positive signs correspond, respectively, to ϕ^+ and ϕ^- in (207). The Schrödinger equation for the matter fields is then obtained if one defines the (WKB) time parameter t through the condition,

$$\frac{\partial}{\partial t} = \mp 2\vec{\nabla} S \cdot \vec{\nabla} \equiv \mp 2 G^{\alpha\beta} \frac{\partial S}{\partial q^\alpha} \frac{\partial}{\partial q^\beta}, \qquad (210)$$

where $q^\alpha = (\tau, \bar{q}^A)$, and $\bar{q}^A$ are the coordinates of $\bar{M}$ given in (152). We have now two choices. Typically, the positive sign in (210) is chosen for the spacetime represented by the wave function ϕ_0^- and the negative sign for the spacetime represented by the wave function ϕ_0^+. With these choices, the Schrödinger equation in the two branches turns out to be

$$i\hbar \frac{\partial \psi_\pm}{\partial t_\pm} = H_{HSM}(\varphi) \psi_\pm, \qquad (211)$$

where it can now be written $\psi_\pm = \psi_\pm(t_\pm; \varphi)$. From (210), one easily gets

$$\frac{\partial \tau}{\partial t_\pm} = \pm 2 \frac{\partial S}{\partial \tau}, \qquad (212)$$

so the wave functions $\psi_\pm$ represent two universes, one expanding and one contracting (recall that the variable τ is proportional to the volume of the space), which from (211) are both filled with matter. An alternative although equivalent interpretation is to choose the positive sign in (210) for both universes, i.e., $t \equiv t_+$. In that case, both wave functions represent expanding universes, but then the corresponding Schrödinger equations for the internal fields are given by

$$i\hbar \frac{\partial \psi_+}{\partial t} = H_{HSM}(\varphi) \psi_+, \qquad (213)$$

$$-i\hbar \frac{\partial \psi_-}{\partial t} = H_{HSM}(\varphi) \psi_-, \qquad (214)$$

respectively. The last of which can be written as,

$$i\hbar \frac{\partial \psi_+}{\partial t} = H_{HSM}(\bar{\varphi}) \psi_+, \qquad (215)$$

where we have used that $\psi_-^*(\varphi) = \psi_+(\bar{\varphi})$. It is therefore the Schrödinger equation of a field that is charge-conjugated with respect to the field given in (213). The wave functions ϕ^+ and ϕ^- represent then two expanding universes, but, from the point of view of the same time variable, one is filled with matter and the other with antimatter, these two concepts having always a relative meaning.

Let us focus on the wave function of the matter fields in one of the universes, say, ψ_+. If we consider that the modes of the field are decoupled, then, the Schrödinger equation for the scalar field φ_+, which generically denotes any of the polarisations of the $W^\pm$ and Z bosons, can be written as the product of the wave functions of the modes, i.e.

$$\psi_+(t, \varphi) = \prod_k \psi_+^{(k)}(t, \varphi_k), \qquad (216)$$

where $\psi_+^{(k)}(t_+, \varphi_k)$ is the solution of the Schrödinger Equation (213) for each mode, whose general solution can be expressed in the basis of number eigenfunctions of the time-dependent harmonic oscillator [35].

The wave function in the time reversely symmetric universe, $\psi_-(t, \bar{\varphi})$, can be obtained from the relation $\psi_-(\bar{\varphi}) = \psi_+^*(\varphi)$, so the eigenfunctions of the basis for the state of the boson fields in the symmetric universe turn out to be given by (283) with the replacements, $t \to -t$ and $\varphi_k \to \bar{\varphi}_k$. Thus, the field φ that represents the matter content of one of the universes is the charge conjugated of the field $\bar{\varphi}$ that represents the matter content of the partner universe. We have seen in the preceding section that their parity is also reversely related, so $\bar{\varphi}$ turns out to be a CP-conjugated field of the field φ. Thus, the matter content of one of the universes is the CP conjugated of the matter in the partner universe, and they form thus a universe–antiuniverse pair. It does not necessarily mean that one of the universes is completely made up of matter and the other is made up of antimatter. In fact, the two universes can contain matter as well as antimatter but exactly in the opposite ratio; so, from the global point of view, the total amount of matter in the two universes is balanced with the total amount of antimatter.

3.3. Minisuperspace Model

3.3.1. Geometrical Structure of the Minisuperspace

Let us now apply the third quantisation formalism to the case of the minisuperspace of homogeneous and isotropic metrics with small perturbations that represent the matter content of the universe. The formalism greatly simplifies, and one can still obtain a clear picture of the scenario described by the third quantisation formalism. On the other hand, we have seen that the minisuperspace description of the universe, although not complete, is a good approximation for most of the evolution of the universe provided that the universe is created with a length scale of some orders of magnitude above from the Planck length. In that case, the small deviations from the homogeneity and the isotropy of the universe can be treated as perturbations described as particles propagating in the homogeneous and isotropic background.

Let us therefore consider the homogeneous and isotropic Friedmann–Robertson–Walker (FRW) metric as the background spacetime (36)

$$ds^2 = -N^2(t)dt^2 + a^2(t)d\Omega_3^2,$$

where $a(t)$ is the scale factor, and $d\Omega_3^2$ is the line element on the three sphere[45]. We saw in Section 2 that the lapse function is not a dynamical variable, so the only dynamical variable turns out to be the scale factor, $a(t)$. In this case, all the components of the spatial metric are fixed except for the value of the scale factor. $\bar{M}$ turns out to then be a 0-dimensional space, where the spatial sections of the universes are represented by single points and their evolution by (curved) lines in the $1 + 0$-dimensional space M.

This picture can easily be extended by considering as well the homogeneous mode of some matter fields, represented by a set of scalar fields, $\vec{\varphi}(t) = (\varphi_1(t), \ldots, \varphi_n(t))$, minimally coupled to gravity. We will see that these fields enter as space-like variables in the configuration space. For simplicity, we shall consider only one single scalar field representing the matter of the universe, so the configuration space, M, will be a $1 + 1$-dimensional space. In addition, one can also consider the inhomogeneous modes of these fields, so the total configuration space would be the $1 + n \cdot \infty$-dimensional space spanned by the variables, $(a(t), \varphi_{1,k}(t.x), \ldots, \varphi_{n,k}(t,x))$. For simplicity, we shall only consider[46] the homogeneous mode of a single scalar field, φ, and its inhomogeneities will be treated as a perturbation described by particles propagating in the spacetime. Therefore, by now let us consider the $1 + 1$-dimensional configuration space M of coordinates, $q^A \equiv (a, \varphi)$.

The total action, i.e., the Einstein–Hilbert action of gravity plus the action of the scalar field, given by (37), can be written as

$$S = S_g + S_m = \frac{1}{2} \int dt N \left(G_{AB} \frac{\dot{q}^A \dot{q}^B}{N^2} - \mathcal{V}(q) \right), \tag{217}$$

where, $q^A = (a, \varphi)$, with the supermetric G_{abcd} in (13) given now by [7]

$$G_{AB} = \text{diag}(-a, a^3), \tag{218}$$

from which one can clearly see that the scale factor (i.e., the first component) is a time-like variable and the scalar field (the second component) is a space-like variable. The potential term, $\mathcal{V}(q)$ in (221), reads

$$\mathcal{V}(q) \equiv \mathcal{V}(a, \varphi) = -a + 2a^3 V(\varphi). \tag{219}$$

The first term in (219) comes from the closed geometry of the three space, and $V(\varphi)$ is the potential of the scalar field. The case of a spacetime with a cosmological constant, Λ, is implicitly included if we consider a constant value of the potential of the scalar field, $V(\varphi) = \Lambda/6$. As we showed in Section 3.2.2, the evolution of the universe can be seen as a parametrised trajectory of the superspace with the variable $\tau \propto \sqrt{h}$ formally playing the role of a time variable. In the case of the minisuperspace, $\sqrt{h} = a^3$; but it is interesting to change to a *conformal* scale factor, $\alpha = \ln a$, in terms of which the metric G_{AB} turns out to be conformal to the 2-dimensional Minkowski space,

$$G_{AB} = e^{3\alpha} \eta_{AB}, \tag{220}$$

and the action (221) can be written as,

$$S = S_g + S_m = \frac{1}{2} \int dt N e^{3\alpha} \left(\eta_{AB} \frac{\dot{q}^A \dot{q}^B}{N^2} - \left(H^2(\varphi) - e^{-2\alpha} \right) \right), \tag{221}$$

where $e^{3\alpha} = a^3$ is essentially the volume of the spatial sections, and $H^2(\varphi) = 2V(\varphi)$ is the Hubble function. From the signature of (220), it can be seen that the scale factor formally plays the role of the time variable and the matter field(s) the role of the space-like component(s), and the minisupermetric (220) provides the minisuperspace with a complete metric structure with a line element given by

$$ds^2 = G_{AB} dq^A dq^B = -a\, da^2 + a^3 d\varphi^2 = e^{3\alpha} \left(-d\alpha^2 + d\varphi \right). \tag{222}$$

It also allows us to define the usual machinery of a geometric manifold. For instance, we can define the Christoffel symbols associated to the minisupermetric G_{AB}, defined as usual by

$$\Gamma^A_{BC} = \frac{G^{AD}}{2} \left\{ \frac{\partial G_{BD}}{\partial q^C} + \frac{\partial G_{CD}}{\partial q^B} - \frac{\partial G_{BC}}{\partial q^D} \right\}, \tag{223}$$

which, in terms of the variables (a, φ), the non-zero values are

$$\Gamma^a_{aa} = \frac{1}{2a}, \; \Gamma^a_{\varphi\varphi} = \frac{3a}{2}, \; \Gamma^\varphi_{\varphi a} = \Gamma^\varphi_{a\varphi} = \frac{3}{2a}, \tag{224}$$

or in terms of the variables (α, φ),

$$\Gamma^\alpha_{\alpha\alpha} = \Gamma^\alpha_{\varphi\varphi} = \Gamma^\varphi_{\varphi\alpha} = \Gamma^\varphi_{\alpha\varphi} = \frac{3}{2}. \tag{225}$$

In any case, we could compute other geometrical properties of the minisuperspace like the corresponding Riemann tensor, the curvature scalar, etc. (see Ref. [10]).

From the geometrical point of view, the evolution of the universe can be seen as a trajectory in the minisuperspace (see Figure 15), with $a(t)$ and $\varphi(t)$ being the parametric coordinates of the universe along the *worldline* of the universe, and the time variable t is the parameter that parametrises the trajectory. From that point of view, it is easy to see that the evolution of the universe, i.e., the trajectory of the universe in the minisuperspace, cannot depend on the particular choice of time variable, i.e., the trajectory must be independent of the parametrisation used to describe it.

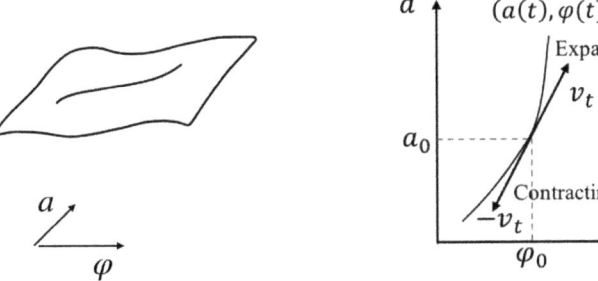

Figure 15. (**Left**): the evolution of the universe can be seen as a trajectory in the minisuperspace. (**Right**): a trajectory in the minisuperspace that is positively oriented with respect to the scale factor component describes an expanding universe. Similarly, a negatively oriented trajectory describes a contracting universe.

However, because of the presence of the potential $\mathcal{V}(a,\varphi)$ in the action (221), the trajectory of the universe along the minisuperspace manifold is not a geodesic. It is instead given by the equation

$$\ddot{q}^A + \Gamma^A_{BC}\dot{q}^B\dot{q}^C = -G^{AB}\frac{\partial \mathcal{V}}{\partial q^B}, \quad (226)$$

which, with the help of (224), yields the customary field equations (see, for instance, Refs. [7,25])

$$\ddot{a} + \frac{\dot{a}^2}{2a} + \frac{3a}{2}\dot{\varphi}^2 = -\frac{1}{2a} + 3aV(\varphi)\ , \ \ddot{\varphi} + 3\frac{\dot{a}}{a}\dot{\varphi} = -\frac{\partial V(\varphi)}{\partial \varphi}. \quad (227)$$

The fact that the curve $(a(t),\varphi(t))$ is not a geodesic is not a big deal. As we have said, the trajectory of the universe is invariant under reparametrisations of time, so we can make the following change of time variable

$$d\tilde{t} = \mathrm{m}^{-2}\mathcal{V}(q)dt, \quad (228)$$

where m is some constant. Now, if we also perform the following conformal transformation of the minisupermetric

$$\tilde{G}_{AB} = \mathrm{m}^{-2}\mathcal{V}(q)G_{AB}, \quad (229)$$

the action (221) becomes

$$S = \frac{1}{2}\int d\tilde{t} N \left(\frac{1}{N^2}\tilde{G}_{AB}\frac{dq^A}{d\tilde{t}}\frac{dq^B}{d\tilde{t}} - \mathrm{m}^2\right), \quad (230)$$

which is a similar action but with a constant potential. The new time variable, $\tilde{t}$, turns out to be the affine parameter of the minisuperspace geometrically described by the metric tensor $\tilde{G}_{AB}$, and the trajectory of the universe in this minisuperspace is given by the geodesic equation

$$\frac{d^2 q^A}{d\tilde{t}^2} + \tilde{\Gamma}^A_{BC}\frac{dq^B}{d\tilde{t}}\frac{dq^C}{d\tilde{t}} = 0. \quad (231)$$

Thus, the classical trajectory of the universe can equivalently be seen as either a geodesic of the minisuperspace geometrically determined by the minisupermetric $\tilde{G}_{AB}$ or a non-geodesic of the minisuperspace geometrically determined by G_{AB}.

We can also define the momenta conjugated to the minisuperspace variables

$$\tilde{p}_A \equiv \frac{\delta L}{\delta \frac{dq^A}{d\tilde{t}}}, \quad (232)$$

and the Hamiltonian constraint associated to the action (230) turns out to be

$$\tilde{G}^{AB}\tilde{p}_A\tilde{p}_B + \mathrm{m}^2 = 0, \tag{233}$$

or in terms of the metric G_{AB} and the time variable t,

$$G^{AB}p_A p_B + \mathrm{m}_{\mathrm{ef}}^2(q) = 0, \tag{234}$$

where for convenience we have written $\mathrm{m}_{\mathrm{ef}}^2(q) = \mathcal{V}(q)$, with $\mathcal{V}(q)$ given by (219). It is worth noticing that the phase space does not change in the transformation $\{G_{AB}, t\} \to \{\tilde{G}_{AB}, \tilde{t}\}$, because

$$\tilde{p}_A = \tilde{G}_{AB}\frac{dq^B}{d\tilde{t}} = G_{AB}\frac{dq^B}{dt} = p_A, \tag{235}$$

where $p_A = \{p_a, p_\varphi\}$ and $q^A \equiv \{a, \varphi\}$, and the Hamiltonian constraints (233) and (234) are related by the inverse of the conformal transformation (229),

$$\tilde{G}^{AB} = \frac{\mathrm{m}^2}{\mathcal{V}(q)}G^{AB}. \tag{236}$$

The field equations given either by (226) or by (231) are invariant under the reversal change in the time variable, $t \to -t$. From the geometrical point of view, it only changes the direction along which the curved has travelled, i.e., the direction of the tangent vector $\frac{\partial}{\partial t}$. It means that for any given solution $a(t)$ and $\varphi(t)$, one may also consider the symmetric solution, $a(-t)$ and $\varphi(-t)$.

In our case the momenta conjugated to the variables of the minisuperspace, given in (235), turn out to be

$$p_a = -\frac{a\dot{a}}{N}, \quad p_\varphi = \frac{a^3\dot{\varphi}}{N}, \tag{237}$$

in terms of which the Hamiltonian constraint (234) reads

$$-\frac{1}{a}p_a^2 + \frac{1}{a^3}p_\varphi^2 + m_{\mathrm{eff}}^2(a,\varphi) = 0, \tag{238}$$

which is the Friedmann equation expressed in terms of the momenta instead of in terms of the time derivatives of the minisuperspace variables. As pointed out before, the geodesic equation and the momentum constraint (238) are invariant under a reversal change of the time variable. Let us notice however that the momenta (237) are not invariant, but they turn out to be reversely changed, $p_a \to -p_a$ and $p_\varphi \to -p_\varphi$. Nevertheless, they appear squared in the Hamiltonian constraint (238), so it is not affected by the change.

However, by conservation of the momenta, one would expect that the cosmological solutions should come in symmetric pairs with opposite values of the associated momenta. From (237) and (238), it is easy to see that in terms of the cosmological time ($N = 1$), the two symmetric solutions are given by

$$a\frac{da}{dt} = -p_a = \pm\sqrt{\frac{1}{a^2}p_\varphi^2 + am_{\mathrm{eff}}^2(a,\varphi)}. \tag{239}$$

It clearly reminds to the solutions of the trajectory of a test particle moving in the spacetime [66]. For instance, in Minkowski spacetime[47], the time component of the geodesics satisfies

$$\frac{dt}{d\mu} = -p_t = \pm\sqrt{\vec{p}^2 + m^2}, \tag{240}$$

where μ is an affine parameter and $p_t = \pm E$, with E the energy of the test particle. The two solutions are eventually associated to particles and antiparticles in a quantum field theory.

In the case of the universe, the two solutions given in (239) also represent two universes: one universe moving forward in the scale factor component and the other moving backward in the scale factor component (see Figure 16). In the minisuperspace, however, moving forward in the scale factor component means evolving with an increasing value of the scale factor, so the associated solution represents an expanding universe, and moving backward in the scale factor component means evolving with a decreasing value of the scale factor, so the symmetric solution represents a contracting universe. Therefore, the two symmetric solutions form an expanding–contracting pair of universes (see Figure 16). However, we have already showed in previous sections that an expanding–contracting pair filled with matter can also be interpreted as two expanding universes, one of them filled with matter and the other filled with antimatter [26,91], i.e., it can be interpreted as a universe–antiuniverse pair [90,91].

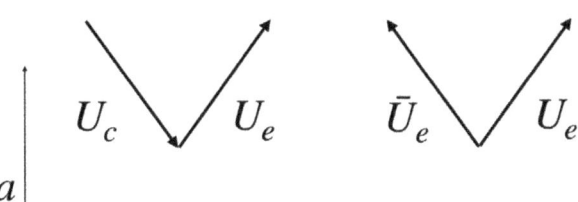

Figure 16. A contracting and an expanding universe, both made of matter, can also be seen as a pair of expanding universes, one of them made up of matter and the other made up of antimatter, i.e., they can be seen as a universe–antiuniverse pair.

3.3.2. Field Quantisation of a FRW Spacetime

As we have already seen, the procedure of third quantisation parallels that of a second quantisation in a curved spacetime (see Section 3.2.3). Now, the field is the wave function $\phi(a, \varphi)$ that satisfies the corresponding Wheeler–DeWitt equation, which is now seen as a wave equation. With the minisupermetric (220) in (181), the Wheeler–DeWitt equation turns out to be

$$a\frac{\partial}{\partial a}\left(a\frac{\partial \phi}{\partial a}\right) - \frac{\partial^2 \phi}{\partial \varphi^2} + a^2\omega^2(a)\phi = 0, \qquad (241)$$

where,

$$\omega^2(a, \varphi) = H^2 a^4 - a^2, \qquad (242)$$

with $H^2 = V(\varphi_0)$ evaluated at the moment of the creation of the (inflationary) universe, where it can be approximated by a constant. In that case, following the procedure shown in Section 3.2.4, we can decompose the wave function $\phi(a, \varphi)$ in Fourier modes,

$$\phi(a, \varphi) = \int \frac{dK}{2\pi} e^{iK\varphi} \phi_K(a), \qquad (243)$$

where $\phi_K(a)$ must satisfy

$$\ddot{\phi}_K + \frac{1}{a}\dot{\phi}_K + \omega_K^2(\alpha)\phi_K = 0, \qquad (244)$$

with [50,89]

$$\omega_K^2(a) = H^2 a^4 - a^2 + \frac{K^2}{a^2}. \qquad (245)$$

Let us notice that the inner product turns out to be here by (186) with [92], $d\Sigma^A = n^A d\Sigma$, where $n^A = (a^{-\frac{1}{2}}, 0)$ is a timelike unit vector, and $d\Sigma = d\varphi$, which defines the orthogonal hypersurfaces (one dimensional curves) of constant a. It then becomes [50,92]

$$(u_1, u_2) = -i \int_{-\infty}^{+\infty} d\varphi\, a\, \left(u_1(a, \varphi) \overset{\leftrightarrow}{\partial}_a u_2^*(a, \varphi)\right). \qquad (246)$$

We can now define the operator version of the field, $\hat{\phi}$, and write it as

$$\hat{\phi}(a,\varphi) = \frac{1}{\sqrt{2}} \int \frac{dK}{2\pi} \left(e^{iK\varphi} v_K^*(a) \hat{A}_K^- + e^{-iK\varphi} v_K(a) \hat{A}_K^+ \right), \qquad (247)$$

where $\hat{A}_K^+$ and $\hat{A}_K^-$ are the creation and annihilation operators, respectively, of universes with momentum K conjugated to the scalar field; and the modes are normalised according to the condition

$$\frac{dv_K}{da} v_K^* - v_K \frac{dv_K^*}{da} = \frac{2i}{a}. \qquad (248)$$

We can now define the ground state of the invariant representation, $\hat{A}_K^+$ and $\hat{A}_K^-$, by

$$|0\rangle_I = \prod_K |0_K, 0_{-K}\rangle_I, \qquad (249)$$

where $|0_K\rangle_I$ ($|0_{-K}\rangle_I$) is the state annihilated by the operator $\hat{A}_K^-$ ($\hat{A}_{-K}^-$). An excited state, i.e., a state representing different number of universes with momenta $K_1, K_2, \ldots$, is then given by [92]

$$|m_{K_1}, n_{K_2}, \ldots\rangle = \frac{1}{\sqrt{m!n!\ldots}} \left[\left(\hat{A}_{K_1}^+\right)^m \left(\hat{A}_{K_2}^+\right)^n \ldots \right] |0\rangle_I, \qquad (250)$$

which represents m universes in the mode K_1, n universes in the mode K_2, etc. In the case of a field that propagates in a homogeneous and isotropic spacetime, the value of the mode k represents the value of the spatial momentum of the particle. In a homogeneous and isotropic minisuperspace, the value of the mode K labels the eigenvalues of the momentum conjugated to the scalar field φ, which formally plays the role of a spacelike variable in the minisuperspace. In that case, the values $K_1, K_2, \ldots$, in (250), label the different initial values of the time derivatives of the scalar field in the universes. Thus, the state (250) represents m universes with a scalar field with $\dot\varphi \propto K_1$, n universes with a scalar field with $\dot\varphi \sim K_2$, etc. They represent different energies of the matter fields, which would correspond to different numbers of particles in the universes. The general quantum state of the field ϕ, which represents the quantum state of the spacetime and the matter fields, all together, is then given by

$$|\phi\rangle = \sum_{m,n,\ldots} C_{mn\ldots} |m_{K_1} n_{K_2} \ldots\rangle_I, \qquad (251)$$

which represents therefore the *quantum state of the multiverse* [93] in the model of the minisuperspace that we are considering.

Here, it follows the subtle subject of the boundary conditions in quantum cosmology. From a QFT we know that the vacuum state of a given representation may contain a certain number of particles of another representation, so the question then asks which representation is the appropriate one. We have already imposed that the representation of the field that represents the state of the multiverse should be an invariant representation because, in that case, once that field is in a given state, it will then remain in the same state along the entire evolution of the universe. Furthermore, one would expect that the field would be in the ground state of such an invariant representation provided that we assume that no external *force* is exciting the state of the multiverse. However, that condition does not completely fix the state of the field ϕ because there are many different invariant representations. In general, an invariant representation, $\hat{A}_K^+$ and $\hat{A}_K^-$, can be defined as [83]

$$\hat{A}_K^- = \frac{i}{\sqrt{\hbar}} (v_K^* \hat{p}_\phi - \dot{v}_K^* \hat{\phi}), \qquad (252)$$

$$\hat{A}_K^+ = -\frac{i}{\sqrt{\hbar}} (v_K \hat{p}_\phi - \dot{v}_K \hat{\phi}), \qquad (253)$$

where $\hat{\phi}$ and $\hat{p}_\phi$ are the operator version of the wave function and the conjugate momentum in the Schrödinger picture, respectively, and v_K is a solution of the wave Equation (244) satisfying the orthonormality condition (248), which ensures the usual commutation relations, $[\hat{A}_K^-, \hat{A}_K^+] = 1$. However, there are many different solutions of the wave Equation (244) satisfying the orthonormality condition (248), so any of them provides an invariant representation. In fact, we have seen in Section 3.2.4 that the Hartle–Hawking's boundary condition and the Vilenkin's boundary condition provide two sets of solutions.

One might say that the Hartle–Hawking boundary condition has a more fundamental character because it is rooted on a more ontological reasoning. It essentially rests on the idea that *the boundary conditions of the universe are that it has no boundary* [39], i.e., that the universe, and therefore the multiverse as well, comes from no prior configuration of the space. In that case, it seems consistent to impose or to assume that the multiverse is *always* in the ground state of the Hartle–Hawking invariant representation. However, single universes are better represented by the representation obtained by imposing the Vilenkin's tunnelling condition. In fact, this boundary condition is specifically imposed to assure that it describes single universes created in the Lorentzian region of the (mini)superspace (see Section 2.5.1). In that case, as we have seen in Section 3.2.4, it turns out that the multiverse is full of Vilenkin's universes [50], which due to the isotropy of the superspace should come in universe-antiuniverse pairs [75] (see, Section 3.2.4, and the next section). Thus, it seems quite unavoidable to assume that our universe has been created in an entangled pair.

3.3.3. Reheating and the Matter–Antimatter Content of the Entangled Universe

In Section 3.2.5, we saw that the matter fields of the two universes of an entangled pair are CP reversely related and that the two universes thus form a universe–antiuniverse pair. Let us now apply the same semiclassical formulation to the period after inflation called (p)reheating (see, for instance, Ref. [94]), where the inflation field[48], χ, eventually decays into the particles of the Standard Model (SM). In that period, the spacetime can largely be considered homogeneous and isotropic, and the inhomogeneities of both the matter fields and the spacetime can be analysed as small perturbations propagating in a homogeneous and isotropic background.

We are not going to repeat the development of Section 3.2.5 but only to present a particular and detailed example that will show the consequences of the complex conjugated relation between these two wave functions. It is worth noticing that the CP conjugated relation between the matter fields of the universe–antiuniverse pair is based on the fundamental considerations described in Section 3.2.5, and it is therefore independent of the model chosen for the reheating scenario after inflation, so similar steps can be followed in any other reheating scenario. For concreteness, we shall describe this period in the appealing model of the Higgs-inflaton [22,23], in which the field that drives the inflationary expansion of the space decays after inflation into the particles of the Standard Model (SM). The idea rest on the form of the potential of the Higgs inflaton field. At high energy scales, during the first stages of the inflationary period, the functional form of the potential can be approximated by an exponential, and it can thus drive inflation. When the field has rolled down the exponential slope of the potential, it finds a minimum around which it starts oscillating. Then, inflation ends and the Higgs-inflaton field behaves like the rest of fields of the SM, with interactions that allow the decays of the Higgs-inflaton into the particles of the SM (see below). Finally, in the low-energy regime, the functional form of the potential can be approximated by the customary double-well potential of the Higgs that gives the expected masses to the particles of the SM [22,23].

Therefore, after the inflationary period, the potential of the inflaton field, $V(\chi)$, can no longer be considered a constant. However, we saw in Section 2.4.2 that this does not introduce a large qualitative change. The solutions of the Wheeler–DeWitt equation are still expected to come in pairs, $\phi = \phi^+ + \phi^-$, with conjugate complex phases that are the solutions of the Hamilton–Jacobi Equation (59). In the semiclassical regime, they are given by (64), i.e.,

$$\phi^\pm(a;\chi,\varphi) = \Delta(a)e^{\pm\frac{i}{\hbar}S(a)}\psi_\pm(a;\chi,\varphi), \tag{254}$$

where χ is the inflaton field, and φ collectively denotes all the fields of the SM. Following the development of Section 2.3, the complex phase in (254) determines the dynamics of the homogeneous and isotropic background spacetime, and the wave functions of the matter fields in the two universes, $\psi_\pm(\chi,\varphi)$, satisfy the Schrödinger equation of two sets of CP-conjugated fields. They are related by the condition $\psi^*_-(\chi,\varphi) = \psi_+(\chi,\bar\varphi)$.

As we have said, at the end of the inflationary period, the Higgs-inflaton field χ has slow-rolled down the potential, and it then approaches the minimum of the potential located at χ_m, for which $V'(\chi_m) = 0$. The expansion rate of the spacetime also slows down, and the field starts oscillating around the minimum like a weakly damped harmonic oscillator with mass $m^2 = V''(\chi_m)$. The total Hamiltonian constraint can be written during this period as (206), with a gravitational part given by

$$H_G = -\frac{1}{2M_P^2}p_a^2 - \frac{M_P^2 a^2}{2} + B(a), \tag{255}$$

where $B(a)$ contains the backreaction of the Higgs-field and eventually the backreaction of the rest of fields of the SM that will contribute to the dynamics of the background spacetime. It may also contain some residual constant term, which is expected to be subdominant at least until the advent of the dark-energy period. The Hamiltonian of the Higgs-SM sector, H_{SM} in (206), can now be written as

$$H_{HSM} = H_\chi + H_{SM}, \tag{256}$$

with

$$H_\chi = \frac{1}{2a^3}p_\chi^2 + \frac{1}{2}a^3 M^2 \chi^2 + \Delta V(\chi), \tag{257}$$

where [23] $M^2 = \lambda M_P^2 / 3\xi^2$ with ξ a coupling constant of the theory, p_χ is the the momentum conjugated to the Higgs-inflaton field, $p_\chi = a^3 \dot\chi$, and $\Delta V(\chi)$ contains high-order correction terms that can be neglected in a first approach [23]. The interactions between the Higgs and the matter and gauge fields of the SM have been included in the Hamiltonian H_{SM}. The Klein–Gordon equation of the Higgs field can then be written,

$$\ddot\chi + 3\frac{\dot a}{a}\dot\chi + M^2 \chi = 0, \tag{258}$$

where we assumed that the Higgs is essentially in the zero mode. For instance, for a power-law evolution of the background spacetime, $a(t) \propto t^p$, Equation (258) is a Bessel equation that can be solved analytically. With the appropriate boundary conditions, and assuming $Mt \gg 1$ and $p \approx 2/3$, it can be written as [23]

$$\chi(t) = \frac{\chi_{end}}{Mt}\sin(Mt), \tag{259}$$

where $\chi(t=0) = \chi_{end}$ is the value of the Higgs field at the end of the inflationary period, which coincides with the beginning of the appearance of the (p)reheating mechanisms ($t=0$).

Different channels can now be considered for the decaying of the Higgs field into the particles of the SM (see Refs. [23,95] for the details). It turns out that the perturbative decay of the Higgs field is only effective when the amplitude of the Higgs is below a critical value that depends on the mass of the final particles. This, together with the time dependence of the decay rate of the Higgs into the particles of the SM, makes it that the Higgs needs to oscillate a large number of times before decaying into the massive gauge bosons and fermions and many more times to decay into the less massive fermions, so the perturbative decay becomes ineffective during the first oscillations of the Higgs. In that period, the most effective channel turns out to be the parametric resonance [23,95]. However, this channel is enhanced by the effect of Bose stimulation, so the production of fermions through this channel is highly restricted. These will be mainly produced later on through the perturbative channel or through the subsequent decay of the intermediate bosons into fermions.

Therefore, for the purpose of the present analysis, it is enough to focus on the production of the intermediate gauge bosons, $W^\pm$ and Z. In the customary SSB mechanism, the fields of the SM acquire a constant value of their masses. However, during the reheating period, the potential still depends on the value of the Higgs field, χ, and thus the mass acquired from the interaction with the Higgs depends on its value. In that period, it can be approximated by [23]

$$m_W^2 \simeq \frac{g_2^2 |\chi|}{4\sqrt{6\xi}}, \quad m_Z^2 \simeq \frac{m_W^2}{\cos^2\theta_W}, \quad m_f \simeq \frac{y_f^2 |\chi|}{2\sqrt{6\xi}}, \qquad (260)$$

where g_2 is the coupling of the intermediate gauge bosons, θ_W is the weak mixing angle, and y_f are the Yukawa couplings of the fermion sector [23]. Eventually, after the period of reheating, in the low-energy regime, the potential takes the customary form of a double-well potential, and the masses of the particles of the SM become the customary ones [23]. Thus, in the low-energy limit, the Higgs-inflationary scenario is indistinguishable from the Higgs scenario of particle physics, as expected. However, it is in this mid-energy regime, during the reheating period, when the basic components of matter are created in the two universes and the one in which we are mainly interested now.

On the other hand, the quantisation of the intermediate gauge bosons $W^\pm$ and Z followed as usual by decomposing them into normal modes as

$$\hat{\varphi}(t, \mathbf{x}) = \int \frac{d^3k}{(2\pi)^{3/2}} \left(e^{-i\mathbf{k}\mathbf{x}} \varphi_k(t) \hat{b}_k + e^{i\mathbf{k}\mathbf{x}} \varphi_k^*(t) \hat{b}_k^\dagger \right), \qquad (261)$$

where $\varphi \equiv W^\pm, Z$, and $\hat{b}_k, \hat{b}_k^\dagger$ are the annihilation and creation operators. As we saw in Section 2.4.2, the inhomogeneous modes of the matter fields can be treated as particles propagating in the background spacetime. The mode amplitude $\varphi_k(t)$ satisfies

$$\ddot{\varphi}_k + 3\frac{\dot{a}}{a}\dot{\varphi}_k + \omega_k^2(t) \varphi_k = 0, \qquad (262)$$

with (see (78))

$$\omega_k^2 = \frac{k^2}{a^2} + m_\varphi^2(t), \qquad (263)$$

where m_φ is now given by (260) with the value of the Higgs given in (259). In terms of the number of times that it crosses zero, $j = \frac{Mt}{\pi}$, the field (259) can be written as

$$\chi(j) = \frac{\chi_{\text{end}}}{j\pi} \sin(\pi j), \qquad (264)$$

so that the frequency (263) turns out to be

$$\omega_k^2(j) = \frac{k^2}{a^2} + \frac{\tilde{m}_0^2 \sin(\pi j)}{\pi j}, \qquad (265)$$

where $\tilde{m}_0$ is the effective mass of the gauge bosons at the beginning of the first oscillation ($j = 0$). The time-dependence of the frequency entails the production of particles from two different sources. The first one is the expansion of the background spacetime, which can be neglected during the first oscillations of the Higgs. The other one is the time-dependence of the Higgs field. The Bogolyubov transformation that relates the creation and annihilation operators of the j-crossing, $\hat{b}_j$ and $\hat{b}_j^\dagger$, with those of the initial oscillation $\hat{b}_0$ and $\hat{b}_0^\dagger$ is

$$\hat{b}_j = \alpha \hat{b}_0 + \beta \hat{b}_0^\dagger, \qquad (266)$$

$$\hat{b}_j^\dagger = \alpha \hat{b}_0^\dagger + \beta \hat{b}_0, \qquad (267)$$

where

$$\alpha = \frac{1}{2}\left(a_j^{3/2}\sqrt{\frac{\omega_j}{\omega_0}} + a_j^{-3/2}\sqrt{\frac{\omega_0}{\omega_j}}\right), \quad (268)$$

$$\beta = \frac{1}{2}\left(a_j^{3/2}\sqrt{\frac{\omega_j}{\omega_0}} - a_j^{-3/2}\sqrt{\frac{\omega_0}{\omega_j}}\right), \quad (269)$$

with $|\alpha|^2 - |\beta|^2 = 1$, and $a_j = \frac{a(j)}{a(0)}$ is the ratio between the value of the scale factor at the initial oscillation and the value of the scale factor at the oscillation j. The number of particles is then

$$n_j = |\beta|^2 = \frac{1}{4}\left(\frac{\omega_j}{\omega_0} + \frac{\omega_0}{\omega_j} - 2\right), \quad (270)$$

where we have neglected the expansion of the universe during the first part of the reheating period. From Figure 17, it can be seen that the production of particles is resonant near the points where the Higgs crosses zero. This is why this channel is called narrow resonance [95].

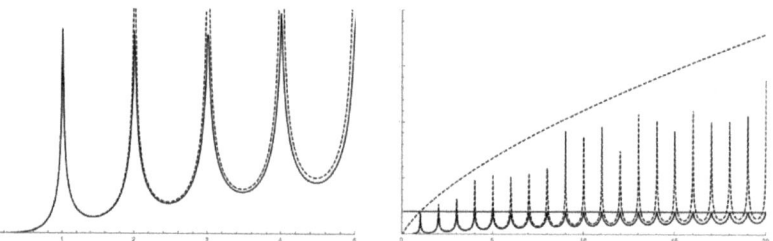

Figure 17. Particle production (270) in terms of the number of crossings of zero, $j = \frac{Mt}{\pi}$. Dashed line accounts for the expansion of the universe, with $a(t) = t^{\frac{2}{3}}$, which can be neglected in the first few oscillations. The production of particles is resonant in the points where the Higgs crosses zero (**left**). If the expansion of the background spacetime is neglected, the production of particles in the peaks rapidly tends to a constant value. However, when the expansion is taken into account, the number of particles in the peaks scale with the scale factor (**right**).

During the first few oscillations, and mainly in the adiabatic regime, the influence of the expansion of the background spacetime can be neglected, and the number of particles at the peaks, which coincide with the value for which the Higgs crosses zero, rapidly tends to the constant value $n_p(a=1)$, given by

$$n_p(a=1) = \frac{1}{4}\left(\sqrt{\frac{k^2}{k^2+\tilde{m}_0^2}} + \sqrt{\frac{k^2+\tilde{m}_0^2}{k^2}} - 2\right). \quad (271)$$

However, when the expansion of the background spacetime is taken into account, the number of particles created at the peaks scales with the scale factor, as (see Figure 17, Right.)

$$n_p(t) \approx \frac{\tilde{m}_0}{4k}a(t) - \frac{1}{2}. \quad (272)$$

In that case, the contribution to the production of intermediate bosons $W^\pm$ and Z is much more enhanced, and the expansion of the spacetime cannot be neglected.

From the very beginning, the intermediate gauge bosons start decaying into the fermions of the SM through their mutual interaction given by the Hamiltonian [23]

$$H_I = -\frac{g_2}{\sqrt{2}}\left(W_\mu^+ J_\mu^- + W_\mu^- J_\mu^+\right) - \frac{g_2}{\cos\theta_W} Z_\mu J_Z^\mu, \tag{273}$$

where $J_\mu^- \equiv \bar{d}_L \gamma^\mu u_L$ and $J_\mu^+ \equiv \bar{u}_L \gamma^\mu d_L$ are the charged currents that couple to the boson W^+ and to the boson W^-, respectively, and the neutral current

$$J_Z^\mu \equiv \kappa_1 \bar{u}_L \gamma^\mu u_L + \kappa_2 \bar{d}_L \gamma^\mu d_L, \tag{274}$$

with,

$$\kappa_1 = \frac{1}{2} - \frac{2\sin^2\theta_W}{3}, \kappa_2 = \frac{1}{2} - \frac{\sin^2\theta_W}{3}. \tag{275}$$

These interactions lead to the charged decays

$$W^+ \to u + \bar{d}, \ W^- \to \bar{u} + d, \tag{276}$$

and the neutral decays

$$Z \to u + \bar{u}, \ Z \to d + \bar{d}, \tag{277}$$

where d and u stands= for the down- and up-type quarks, respectively, and similar decays can also be considered for the rest of quarks. Analogously, we can consider the following decays in the lepton sector

$$W^+ \to e^+ + \nu_e, \ W^- \to e^- + \bar{\nu}_e, \tag{278}$$

as well as the neutral decays

$$Z \to e^- + e^+, \ Z \to \nu_e + \bar{\nu}_e, \tag{279}$$

all of them with their respective decay widths, $\Gamma_{W^\pm, Z \to i}$. Let us then notice that an asymmetry in the decay of the Higgs into the intermediate gauge bosons would entail the asymmetry in the production of quarks and leptons and therefore an asymmetry in the creation of primordial matter during the reheating period without the need of any other mechanism[49].

In the scenario of universes created in correlated pairs (see Section 3), the two universes can be seen as expanding universes with the wave functions of their matter fields CP-conjugated. Let us focus on one of the two wave functions, say, $\psi_+(\varphi)$. If we consider that the modes of the expansion (261) are decoupled, then, the Schrödinger wave function for the scalar field, φ, which generically denote any of the polarisations of the $W^\pm$ and Z bosons, can be written as the product of the wave functions of the modes, i.e.,

$$\psi_+(t,\varphi) = \prod_k \psi_+^{(k)}(t,\varphi_k), \tag{280}$$

where $\psi_+^{(k)}(t,\varphi_k)$ is the solution of the Schrödinger equation

$$i\hbar \frac{\partial \psi_+^{(k)}}{\partial t_+} = H_{\varphi_k} \psi_+^{(k)}, \tag{281}$$

with

$$H_{\varphi_k} = \frac{1}{2a^3} p_{\varphi_k}^2 + \frac{a^3 \omega_k^2}{2} \varphi_k^2, \tag{282}$$

where $a = a(t)$ is the scale factor of the background spacetime, and $\omega_k = \omega_k(t)$ is given by (263). The general solution of the Schrödinger Equation (281) can then be expressed in the basis of the number eigenfunctions given by [35]

$$\psi_{+,N_k}(t, \varphi_k) = \frac{e^{-i(N+1/2)\tau}}{\sqrt{2^N N!} \pi^{1/4} \sqrt{\sigma}} e^{-\frac{\Omega}{2}\varphi_k^2} H_N(\varphi_k/\sigma), \qquad (283)$$

where $H_N(x)$ is the Hermite polynomial of degree, $N \equiv N_k$, which is the number occupation of the mode k, and $\tau = \tau(t)$ is given by

$$\tau_+(t) = \int^t \frac{1}{a^3 \sigma^2} dt, \qquad (284)$$

the function Ω is given by

$$\Omega = \frac{1}{\sigma^2} - i\frac{a^3 \dot{\sigma}}{\sigma}, \qquad (285)$$

and σ is a real function that satisfies the auxiliary equation [35]

$$\ddot{\sigma} + 3\frac{\dot{a}}{a}\dot{\sigma} + \omega_k^2 \sigma = \frac{k^2}{m^2 \sigma^3}. \qquad (286)$$

The wave function in the time reversely symmetric universe is given by (see Section 3.2.5) $\psi_+(t, \bar{\varphi})$ so the eigenfunctions of the basis for the state of the boson fields in the symmetric universe turns out to be given by (283) with the replacement $\varphi_k \to \bar{\varphi}_k$. Therefore, if the scalar field φ represents the boson field W^- in one of the universes, then $\bar{\varphi}$ represents the boson field, $\bar{W}^- = W^+$, in the symmetric universe[50]. The decay of the Higgs into the boson W^+ and W^- can then be produced separately in the two symmetric universes.

Then, one can make the hypothesis that the intermediate gauge boson W^+ and W^- are created in different universes, or at least at different rates in the two universes, without violating the global matter–antimatter asymmetry, an appealing scenario that is also suggested in [71,96]. It is not mandatory that the asymmetry is complete, but a small asymmetry in the decay of the Higgs into the W^+ and W^- bosons in the two universes would eventually derive into an asymmetry in the production of fermions in the two universes due to the different decays of the $W^\pm$ bosons into fermions (see (276) and (278)). In the universe in which the boson W^+ predominates, there would be an excess of the up-quark with respect to the up-antiquark, and accordingly, there would be an excess of protons over antiprotons, and matter would therefore dominate over antimatter. From the global picture of the two correlated universes, the total amount of matter is always balanced with the total amount of antimatter, so there is no global matter–antimatter asymmetry. It is worth noticing that the creation of a universe–antiuniverse pair does not assure that the content of one of the universes is completely matter and that the content of the partner universe is completely antimatter. It is not therefore a mechanism for creating the matter–antimatter asymmetry but a mechanism to restore or explain the apparent asymmetry [91]. In a multiverse scenario, one may expect a whole range of matter–antimatter distributions along the pairs of universes in the multiverse. In some of them, there would be the needed asymmetry to form matter and therefore galaxies and planets like in our universe, without violating any physical law.

4. Observable Effects of Quantum Cosmology

Testing the predictions of a theory with the observational data is a fundamental keystone of any physical proposal. However, it is not the unique consideration, theoretical consistency must also be taken into account, and, in fact, it may help us to break through new paradigms. Well-known examples in contemporary physics are the study of the unobservable black holes from the theoretical consistency of the perturbed motion of an observable companion or the prediction of the charm quark from symmetry consistencies of the Standard Model of particles physics, not mention the unobserved "dark matter," which is basically supported by consistency arguments. Furthermore, observability and falsifiability are not the same thing, as has been clearly argued[51] in Ref. [97] (see also Ref. [98] for a recent review).

Nevertheless, any theory must eventually be tested. In principle, the effects of quantum gravity are expected to be relevant at a very small length, or equivalent to very large scale of energy, and that makes them to be hardly measurable. For that reason, it is quite difficult to propose practical tests in quantum cosmology. However, we may expect some quantum corrections or deviations from the classical behaviour due to quantum cosmological effects that, at least from a theoretical point of view, could be detected. Among these effects, let us here briefly mention two: the pre-inflationary stage induced by the backreaction of the perturbation modes and the corrections due to the high-order terms in the WKB approximation [17–19].

In Section 2.4.2, we saw that the vacuum state of the perturbation modes possesses an energy that permeates the whole universe. In principle, that *backreaction* energy is expected to be a small correction to the energy of the unperturbed background spacetime. However, despite being small, it might produce some observable effects. Let us notice that the effective value of the Hamiltonian constraint, which is obtained by tracing out from (55) the degrees of freedom of the perturbation modes, is,

$$\mathcal{H} = \mathcal{H}_0 + \langle \mathcal{H}_m \rangle = 0, \tag{287}$$

which in terms of the time derivative of the scale factor yields the modified Friedmann equation,

$$\frac{\dot{a}^2}{a^2} = H^2(\varphi) - \frac{1}{a^2} + \frac{\langle \mathcal{H}_m \rangle}{a^3}, \tag{288}$$

where $H^2(\varphi)$ is here the energy of the homogeneous mode of the scalar field, which for simplicity we shall consider constant, and

$$\langle \mathcal{H}_m \rangle \propto m\, n_{\max}^3, \tag{289}$$

for the particles of the matter fields, and

$$\langle \mathcal{H}_m \rangle \propto \frac{n_{\max}^4}{a}, \tag{290}$$

in the massless case. In the former case, the last term in (288) represents a matter-like content in the universe ($\sim a^{-3}$), and in the latter it mimics a radiation energy content ($\sim a^{-4}$). In both cases, it can be shown [99–101] that a matter- or radiation-predominated pre-inflationary period might, under some conditions, leave some observable imprints in the power spectrum of the CMB. However, it is sometimes considered [3,4,102] that $n_{\max} \approx Ha$, which means accounting only for the backreaction of the superhorizon modes. In that case, the back reaction becomes equivalent to a cosmological constant that effectively shifts the value of the potential [89] (see also Refs. [102–104]),

$$\varepsilon = \frac{H^4}{8}\left\{1 - \frac{m^2}{H^2}\log\frac{b^2}{H^2} + \left(1 + \frac{m^2}{H^2}\right)\left(1 - \frac{b^2}{H^2}\right)\right\}, \tag{291}$$

where terms of higher order have been disregarded. The energy shift (291) can be seen as a correction to the effective value of the potential of the scalar field, an effect that is expected to produce observable imprints in the properties of the CMB [105–107].

A different effect from quantum cosmology can be obtained by considering higher-order terms in the WKB wave functions (64). Following [17–20], let us assume a WKB wave function

$$\phi(q; x_\mathbf{n}) = C(q) e^{\pm \frac{i}{\hbar} S(q)} \psi_\pm^{(1)}(q; x_\mathbf{n}), \tag{292}$$

with,

$$S(q) = S^{(0)}(q) + \hbar^2 S^{(2)}(q) + \ldots, \qquad (293)$$

$$\psi_{\pm}^{(1)}(q; x_\mathbf{n}) = \psi_{\pm}^{(0)}(q; x_\mathbf{n}) e^{i\hbar S_\mathbf{n}^{(2)} + \cdots}, \qquad (294)$$

where q is the variable of the background, and $x_\mathbf{n}$ are the perturbation modes. The first-order terms give rise to the corresponding Hamilton–Jacobi of the background spacetime and a modified Schrödinger equation for the corrected wave function of the perturbation modes, $\psi_{\pm}^{(1)}$, which can be written as the Schrödinger equation for a set of uncoupled harmonic oscillator with a perturbed frequency with respect to the unperturbed frequencies (77) and (78) that can be written as

$$\omega_n^2 \to \tilde{\omega}_n^2 = \omega_n^2 + \mathcal{F}_n, \qquad (295)$$

where $\mathcal{F}_n = \mathcal{F}_n(t)$ is a time-dependent function. A term like that is expected to produce a variation in the power spectrum of the perturbation modes that would be in principle measurable [18–20,35]. However, the effect is too small to be distinguishable from the statistical uncertainty implied by cosmic variance [19,20].

Perhaps the application to different inflationary models or the expected advances in the field of astronomical detectors and associated space missions, with the detection and analysis of gravitational waves or the :cosmic neutrino background" (CNB), might make directly testable in the future the deviations from classicality predicted from quantum cosmology. Nevertheless, even though they may be difficult to be observed these effects may have important conceptual consequences. For instance, we saw in Section 2.5 that an energy term like the one produced by the backreaction in the Friedman Equation (288) might drastically change the way in which the universes can be created.

On the other hand, the creation of the universe in entangled pairs[52] may also add new features to be tested in the future. For instance, in an entangled universe, the fields of the matter content in the two universe stop being a vacuum state. If one computes the state of the matter field in one single universe of the entangled pair by tracing out from the composite state the degrees of freedom of the matter in the partner universe, then the resulting state turns out to be a quasi-thermal state with a temperature that depends on the degree of entanglement (which eventually depends on the size of the universe). In Ref. [89], the ratio between the fluctuations of the perturbation modes of a field that is initially in a thermal state and those corresponding to a initial vacuum state was computed, yielding [89]

$$\frac{\delta\phi_\mathbf{n}^{th}}{\delta\phi_\mathbf{n}^{I}} = \sqrt{\frac{1}{2}\left(1 + \frac{x^2}{(1+x^2)(1+\frac{m^2}{H^2 x^2})}\right)}, \qquad (296)$$

with,

$$x \equiv \frac{n}{Ha} = \frac{n_{ph}}{H} \sim \frac{H^{-1}}{L_{ph}}, \qquad (297)$$

where L_{ph} is the physical wave length, and H^{-1} is the distance to the Hubble horizon. The large modes ($x \gg 1$) are in the vacuum state, and then $\delta\phi_\mathbf{n}^{th} \approx \delta\phi_\mathbf{n}^{I}$. However, the departure is significant for the horizon modes, $x \sim 1$. This would be a distinctive effect of the creation of the universes in entangled pairs, and it should leave an observable imprint in the properties of the CMB.

5. Conclusions

Quantum geometrodynamics provides us with a consistent framework for the quantum description of the universe in terms of a wave function that contains, at least in principle, all the information about both the spacetime and the matter fields that propagate therein. In the semiclassical regime, the complex phase of the wave function contains the information about the dynamics of the background spacetime, and the wave function of the matter fields satisfy a Schrödinger equation that depends on the geometry of the subjacent spacetime. The dynamics of the spacetime turns out to be invariant under the complex conjugation of the semiclassical wave function. That gives rise to two different solutions that have been typically interpreted as representing the expanding and the contracting branches of the universe, both filled with matter. We have seen that a different, more consistent interpretation is that the two solutions represent expanding universes with their matter contents being CP reversely related, so from the point of view of an internal observer of any of the universes, the partner universe is always made up of antimatter, these two terms having therefore a relative meaning.

On the other hand, the Wheeler–DeWitt equation can be seen as the wave equation of a field that propagates in the space of Riemannian 3-dimensional geometries, M, where we can describe the evolution of the universe as a trajectory parametrised by a parameter that we can call *time*. The quantum mechanical counterpart is a quantum field that, following the customary approach of a quantum field theory, can be expressed in terms of creation and annihilation operators that satisfy the usual commutation relations. These operators represent the creation and the annihilation of modes for the spatial sections of the universe. Thus, the third quantisation formalism allows us to describe the quantum state of a whole spacetime manifold that can, in general, be a disconnected collection of simply connected manifolds. It is thus an appropriate framework to describe a multiverse scenario, in which the most natural boundary condition turns out to be that the field that represents the whole spacetime manifold remains in the ground state of an invariant representation along the entire history of the universes. This boundary condition implements the idea that the multiverse is the true isolated system, and therefore no external interaction may excite its quantum state. However, as we have seen, this invariant boundary condition does not fix completely the quantum representation of the universe and, in fact, the ground state of one invariant representation is, in general, full of pairs of universes in another invariant representation. For instance, we have seen that the Hartle–Hawking no-boundary condition can be seen as more fundamental because it is based on a more ontological argument. In that case, one may assume that the field that represents the multiverse is in the ground state of the invariant Hartle–Hawking representation. Because the invariance of the invariant representation, the field remains then in this ground state irrespective of the evolution of the universes. However, in terms of the invariant representation associated to the Vilenkin's tunnelling boundary condition, which represents the state of single universes, it turns out that the ground state of the Hartle–Hawking no-boundary state is full of Vilenkin's pairs of universes. It means that the ground state of the multiverse is full of pairs of universes whose matter contents turn out to be CP-conjugated. The charge conjugation comes from the complex conjugation relation between the Schrödinger wave functions of the matter fields in the two universes, and the reversely parity relation comes from the opposite signs of the momentum conjugated to the geometrical variables of the spatial sections of the spacetime. The two universes form thus a universe–antiuniverse pair.

We have seen all these features in a general model but particularly in the model of a homogeneous and isotropic spacetime with particles and matter fields propagating therein, where explicit examples can be analysed. In particular, we have seen that there are three main paradigms for the creation of the universe in quantum cosmology. The universes can be created from *nothing*, i.e., from no preexisting spacetime. However, it requires a precise fine-tuning that seems to be quite unnatural. The other possibility is then that the universe is created from the quantum fluctuation of the spacetime of a preexisting spacetime. However, to then be consistent, one should also give an explanation for the creation of the *first* spacetime. We have seen that Gott and Li's explanation is that a vacuum fluctuation of our spacetime can travel back along one of the closed temporal curves that are allowed to exist in the vacuum state of the gravitational field to become the seminal spacetime from which our spacetime has been created. From that point of view, the universe would be the mother of itself. There is yet another possibility that avoids this paradoxical conclusion. The universes can be created from nothing, i.e., with no need of any prior spacetime, but they must then be created in pairs, from double Euclidean instantons. That turns out to be the most natural and self-consistent way in which the universes can be quantum mechanically created.

We have also seen in a very specific model that the creation of matter after the period of inflation would be correlated in the two universes of an entangled pair. The decay of the inflaton field into the particles of the Standard Model after inflation in the reheating period is produced in such a way that the matter and antimatter of the two universes is perfectly balanced. The matter–antimatter asymmetry observed in our universe would only be therefore an apparent asymmetry, and, in fact, it might be considered as *evidence* of the existence of an entangled companion of our universe that would contain the amount of antimatter that is left in our universe. However, the observational test of this and other quantum cosmological hypothesis seems to be still far from the current state of observation as the effects of an entangled partner would be in the domain of quantum gravity or at least in a pre-inflationary stage of the universe. Perhaps in the future the advances in the detection of primordial gravitational waves or a possible cosmic neutrino background may shed some light onto these intriguing questions.

Funding: This work was supported by Comunidad de Madrid (Spain) under the Multiannual Agreement with UC3M in the line of Excellence of University Professors (EPUC3M23), in the context of the 5th. Regional Programme of Research and Technological Innovation (PRICIT).

Acknowledgments: This review is based on a series of lectures given at the Institute of Physics of the University of Szczecin, in September 2019. I thank M. Dabrowski for his kind invitation to give these lectures and to the members of the Szczecin Cosmology Group for their hospitality at the University of Szczecin.

Conflicts of Interest: The author declares no conflict of interest.

Notes

[1] We shall not deal here with such processes of decoherence, which can be seen in thebibliography [2–6].
[2] More exactly, it is isomorphic to a Milne spacetime, which is a particular coordination of the light cones of the Minkowski spacetime.
[3] I shall closely follow Refs. [7,8].
[4] Following Ref. [10], we are going to use throughout units in which, $\hbar = c = 16\pi G = 1$, although we will leave the constant $\hbar$ in some expressions to remark their quantum character.
[5] For the matter fields we shall generally consider a scalar field.
[6] We shall be more precise later on.
[7] Spacetime becomes a 'trajectory of spaces', cfr. p. 107, Ref. [7].
[8] Let us assume that the period of inflation is fully supported by the current observational data.
[9] Let us notice however that the creation of this initial boundary hypersurface Σ_0 is not a process occurring in time but it corresponds to the creation of the spacetime itself [7], actually.
[10] There is also the so-called DeWitt's boundary condition that states that the wave function of the spacetime must be zero as the curvature approaches the initial singularity [7,10], which might have some interesting properties on the cosmic entanglement [16].
[11] A classical path of the superspace is invariant under reparametrisations so by *time* we mean any time variable.
[12] For instance, the march of a material clock would be given by the matter fields that are solutions of Einstein's equations, in a circular argument.
[13] Sometimes it is distinguished between *minisuperspace* and *midisuperspace* models depending on the number of variables of the reduced superspace.
[14] In some cosmological scenarios [22–24], the length scale of the initial hypersurface can be some orders of magnitude greater than the Planck length, enough for the fluctuations of the spacetime to be subdominant.
[15] Unless otherwise indicated we shall always use cosmic time, t, for which $N = 1$.
[16] Recall that after doing the variation with respect to N we can fix any particular value.

17. However, one would generally expect that the creation of the universe comes from a quantum fluctuation of the spacetime and therefore be of order of the Planck scale, $H^{-1} \sim l_P$. In that case, new elements should be incorporated, although the picture described here and in the next section would still be instructive.
18. This is especially clear in the case of the cosmic microwave background radiation (CMB), in which the relative scale of the energy fluctuations in the last scattering surface are of order 10^{-5}.
19. By 'matter degrees of freedom' we mean the perturbation modes that can represent matter, radiation or even fluctuations of the gravitational field (gravitons).
20. In terms of t_+ we would have ended up with two contracting universes, one made of matter and the other made up of antimatter. However, that case is not interesting because the two newborn contracting universes would rapidly delve into the spacetime foam from which they came up
21. Here, the field is the Schrödinger wave function $\psi_\mathbf{n}$.
22. In Ref. [29] it is shown that the distribution of a large number harmonic oscillators becomes highly peaked around its average value.
23. Let us note however that this is only a formal analogy. In fact, the Hamiltonian constraint (89) indicates is that the total energy of the universe is zero, i.e., the (negative) energy of the spacetime exactly balances the (positive) energy of the matter fields.
24. Let us note however that this process does not violate the conservation of the energy because the total energy, i.e., the gravitational energy plus the energy of the matter fields is, as we have already said, balanced.
25. These are the final values of the Euclidean regime. From the point of view of the Lorentzian sections, these are the "initial" values.
26. As we have seen, it is somehow arbitrary determining which solution describes an expanding universe and accordingly there is an ambiguity in determining which modes are the 'outgoing' modes.
27. In Section 3.3 we shall define more concretely the operator ∇ in the minisuperspace.
28. In the superposition (127) it should appear $\zeta_n^\pm$, with $(\zeta_n^+)^* = \zeta_n^-$. However, the eigenfunctions of the harmonic oscillator are real functions so, $\zeta_n^+ = \zeta_n^- \equiv \zeta_n$.
29. In general, a non simply connected manifold can be divided into N simply connected parts [59] and this N parts can be seen as N classically independent universes.
30. Typically, large regions of order of the Hubble length of our universe.
31. In the 1980s–1990s, the paradigm was a universe in a non-accelerated expansion.
32. It means that the state (146) would actually be,

$$|\text{out}\rangle = |\varphi_0\rangle(|0\rangle + |1\rangle).$$

In that case, when we trace out the state of the baby universes, the state of the field remains unaffected in the initial state, $\rho^{\text{out}} = |\varphi_0\rangle\langle\varphi_0|$

33. We have followed the normalisation applied in Ref. [10].
34. This result is corrected from the one given in Ref. [10] by a factor $\frac{1}{2}$, which is already noted in Ref. [62].
35. For the Milne spacetime, see Ref. [64]
36. A rescale, $\chi \to a\chi$, $\theta \to a\theta$, ..., has been made to absorb the constant a.
37. Assuming the value, $N^i = 0$.
38. The fact that the trajectory is not a geodesic is not really determinant. In fact, using a generalisation of the Maupertuis principle [65,66], one can compute the metric where the trajectory of the universe is a geodesic. Let us consider the reparametrisation given by, $d\bar{t} = m^2(h_{ab})dt$ and $G^{abcd} \to \bar{G}^{abcd} = m^2(h_{ab})G^{abcd}$. In that case, the action (162) turns out to be

$$S_{EH} = \int_\mathcal{M} d\bar{t}d^3x N\left(\frac{1}{2N^2}\bar{G}^{abcd}h'_{ab}h'_{cd} - 1\right), \tag{298}$$

where, $h'_{ab} = \frac{dh_{ab}}{d\bar{t}}$. In the superspace determined by the supermetric $\bar{G}^{abcd}$ the evolution of the universe turns out to be a geodesic.

39. For a FRW spacetime, $\alpha = 0$ (because, $\dot{s} = 0$).
40. For recent works on the third quantisation, see Refs. [67–74].
41. Let us notice that the condition, $^3R \ll 2\Lambda$, does not assume that the universe is homogeneous.
42. The factor $\hbar^2$ has been introduced for later convenience.
43. In the 2 sphere, $\vec{j} = \{l, m\}$.
44. There is here no Euclidean region because we have assumed, $2\Lambda \gg ^3R$, in (191). Otherwise, the condition $^3R > 2\Lambda$ defines an Euclidean region where, in the no-boundary proposal, the universe would be created *from nothing* (see Section 2).
45. We are considering geometrically closed spatial sections.

46 The inhomogeneities of the spacetime can also be considered as fields propagating in the spacetime (see Section 2).
47 A similar procedure can be followed in a curved spacetime.
48 We shall use now the variable χ to represent the inflaton field and leave the variable φ to represent collectively the rest of fields of the SM.
49 Although other mechanisms of baryon asymmetry can simultaneously be present.
50 Typically, φ would represent a linear combination of the W^+ and W^+ fields. In that case, $\bar{\varphi}$ would represent the corresponding conjugated combination.
51 Tegmark poses the following example: *a theory stating that there are 666 parallel universes, all of which are devoid of oxygen, makes the testable prediction that we should observe no oxygen here, and is therefore ruled out by observation*, cfr. Ref. [97], p. 105.
52 And, in general, the creation of universes in N-entangled states, see [36,108].

References

1. Wheeler, J.A. On the nature of quantum geometrodynamics. *Ann. Phys.* **1957**, *2*, 604–614. [CrossRef]
2. Gell-Mann, M.; Hartle, J.B. Quantum mechanics in the light of quantum cosmology. In *Complexity, Entropy and the Physics of Information*; Zurek, W.H., Ed.; Addison-Wesley: Reading, PA, USA, 1990.
3. Halliwell, J.J. Decoherence in quantum cosmology. *Phys. Rev. D* **1989**, *39*, 2912–2923. [CrossRef]
4. Kiefer, C. Decoherence in quantum electrodynamics and quantum gravity. *Phys. Rev. D* **1992**, *46*, 1658–1670. [CrossRef]
5. Hartle, J.B. The quantum mechanics of cosmology. In *Quantum Cosmology and Baby Universes*; Coleman, S., Hartle, J.B., Piran, T., Weinberg, S., Eds.; World Scientific: London, UK, 1990; Volume 7.
6. Kiefer, C. Continuous measurement of mini-superspace variables by higher multipoles. *Class. Quant. Grav.* **1987**, *4*, 1369–1382. [CrossRef]
7. Kiefer, C. *Quantum Gravity*; Oxford University Press: Oxford, UK, 2007.
8. Wiltshire, D.L. An Introduction to Quantum Cosmology. In *Cosmology: The Physics of the Universe*; Robson, B., Visvanathan, N., Woolcock, W., Eds.; World Scientific: Singapore, 1996; pp. 473–531.
9. Wheeler, J.A. Superspace and the nature of quantum geometrodynamics. In *Battelle Rencontres*; DeWitt, C.M., Wheeler, J.A., Eds.; W. A. Benjamin, Inc.: New York, NY, USA, 1968; Chapter 9.
10. De Witt, B.S. Quantum Theory of Gravity. I. The Canonical Theory. *Phys. Rev.* **1967**, *160*, 1113–1148. [CrossRef]
11. Dirac, P.A.M. Lectures on Quantum Mechanics. In *Belfer Graduate School of Science Monographs Series*; Number 2; Belfer Graduate School of Science: New York, NY, USA, 1964.
12. Hartle, J.B.; Hawking, S.W. Wave function of the Universe. *Phys. Rev. D* **1983**, *28*, 2960. [CrossRef]
13. Joos, E.; Zeh, H.D.; Kiefer, C.; Giulini, D.J.; Kupsch, J.; Stamatescu, I.O. *Decoherence and the Appearance of a Classical World in Quantum Theory*; Springer: Berlin, Germany, 2003.
14. Schlosshauer, M. *Decoherence and the Quantum-to-Classical Transition*; Springer: Berlin, Germany, 2007.
15. Halliwell, J.J. Introductory lectures on quantum cosmology. In *Quantum Cosmology and Baby Universes*; Coleman, S., Hartle, J.B., Piran, T., Weinberg, S., Eds.; World Scientific: London, UK, 1990; Volume 7.
16. Davidson, A.; Ygael, T. From DeWitt initial condition to cosmological quantum entanglement. *Class. Quant. Grav.* **2015**, *32*, 152001. [CrossRef]
17. Kiefer, C.; Singh, T.P. Quantum gravitational corrections to the functional Schrödinger equation. *Phys. Rev. D* **1991**, *44*, 1067. [CrossRef]
18. Kiefer, C.; Krämer, M. Quantum gravitational contributions to the CMB anisotropy spectrum. *Phys. Rev. Lett.* **2012**, *108*, 021301. [CrossRef]
19. Brizuela, D.; Kiefer, C.; Krämer, M. Quantum-gravitational effects on gauge-invariant scalar and tensor perturbations during inflation: The de Sitter case. *Phys. Rev. D* **2016**, *93*, 104035. [CrossRef]
20. Brizuela, D.; Kiefer, C.; Krämer, M. Quantum-gravitational effects on gauge-invariant scalar and tensor perturbations during inflation: The slow-roll approximation. *Phys. Rev. D* **2016**, *94*, 123527. [CrossRef]
21. Hartle, J.B. Spacetime quantum mechanics and the quantum mechanics of spacetime. *arXiv* **1995**, arXiv:gr-qc/9304006.
22. Bezrukov, F.L.; Shaposhnikov, M.E. The Standard Model Higgs boson as the inflaton. *Phys. Lett. B* **2008**, *659*, 703. [CrossRef]
23. Garcia-Bellido, J.; Figueroa, D.G.; Rubio, J. Preheating in the Standard Model with the Higgs-Inflaton coupled to gravity. *Phys. Rev. D* **2009**, *79*, 063531. [CrossRef]
24. Garay, I.; Robles-Pérez, S. Effects of a scalar field on the thermodynamics of interuniversal entanglement. *Int. J. Mod. Phys. D* **2014**, *23*, 1450043. [CrossRef]
25. Linde, A. *Particle Physics and Inflationary Cosmology*; Contemporary Concepts in Physics; Harwood Academic Publishers: Chur, Switzerland, 1993; Volume 5.
26. Rubakov, V.A. Quantum Cosmology. In *Lecture at NATO ASI 'Structure Formation in the Universe'*; Springer: Cambridge, MA, USA, 1999.
27. Halliwell, J.J.; Hawking, S.W. Origin of structure in the Universe. *Phys. Rev. D* **1985**, *31*, 1777–1791. [CrossRef]
28. Rubakov, V.A. On third quantization and the cosmological constant. *Phys. Lett. B* **1988**, *214*, 503–507. [CrossRef]
29. Halliwell, J.J. Correlations in the wave function of the Universe. *Phys. Rev. D* **1987**, *36*, 3626–3640. [CrossRef] [PubMed]

30. Grishchuk, L.P.; Sidorov, Y.V. Squeezed quantum states of relic gravitons and primordial density fluctuations. *Phys. Rev. D* **1990**, *42*, 3413–3421. [CrossRef] [PubMed]
31. Lewis, H.R.; Riesenfeld, W.B. An Exact Quantum THeory of the Time-Dependent Harmonic Oscillator and of a Charged Particle in a Time-Dependent Electromagnetic Field. *J. Math. Phys.* **1969**, *10*, 1458–1473. [CrossRef]
32. Leach, P.G.L. Harmonic oscillator with variable mass. *J. Phys. A* **1983**, *16*, 3261–3269. [CrossRef]
33. Kanasugui, H.; Okada, H. Systematic treatments of general time-dependent harmonica oscillator in classical and quantum mechanics. *Prog. Theor. Phys.* **1995**, *93*, 949–960. [CrossRef]
34. Sheng, D.; Khan, R.D.; Jialun, Z.; Wenda, S. Quantum Harmonic Oscillator with Time-Dependent Mass and Frequency. *Int. J. Theor. Phys.* **1995**, *34*, 355–368. [CrossRef]
35. Brizuela, D.; Kiefer, C.; Krämer, M.; Robles-Pérez, S. Quantum-gravity effects for excited states of inflationary perturbations. *Phys. Rev. D* **2019**, *99*, 104007. [CrossRef]
36. Robles-Pérez, S. Quantum cosmology of a conformal multiverse. *Phys. Rev. D* **2017**, *96*, 063511. [CrossRef]
37. Hawking, S.W. The boundary conditions of the universe. In *Astrophysical Cosmology*; Pontificia Academiae Scientarium: Vatican City, Vatican, 1982; pp. 563–572.
38. Hawking, S.W. Quantum cosmology. In *Relativity, groups and topology II, Les Houches, Session XL, 1983*; De Witt, B.S., Stora, R., Eds.; Elsevier Science Publishers B.V.: Amsterdam, The Netherlands, 1984.
39. Hawking, S.W. The quantum state of the universe. *Nucl. Phys. B* **1984**, *239*, 257–276. [CrossRef]
40. Vilenkin, A. Creation of universes from nothing. *Phys. Lett. B* **1982**, *117*, 25–28. [CrossRef]
41. Vilenkin, A. Quantum creation of universes. *Phys. Rev. D* **1984**, *30*, 509–511. [CrossRef]
42. Vilenkin, A. Boundary conditions in quantum cosmology. *Phys. Rev. D* **1986**, *33*, 3560–3569. [CrossRef]
43. Vilenkin, A. Predictions from Quantum Cosmology. *Phys. Rev. Lett.* **1995**, *74*, 846–849. [CrossRef] [PubMed]
44. Vilenkin, A. Interpretation of the wave function of the universe. *Phys. Rev. D* **1989**, *D*, 1116. [CrossRef]
45. Gott, J.R.I.; Li, L.X. Can the universe create itself? *Phys. Rev. D* **1998**, *58*, 023501. [CrossRef]
46. Strominger, A. Baby Universes. In *Quantum Cosmology and Baby Universes*; Coleman, S., Hartle, J.B., Piran, T., Weinberg, S., Eds.; World Scientific: London, UK, 1990; Volume 7.
47. Barvinsky, A.O.; Kamenshchik, A.Y. Cosmological Landscape From Nothing: Some Like It Hot. *JCAP* **2006**, *0609*, 014. [CrossRef]
48. Barvinsky, A.O.; Kamenshchik, A.Y. Cosmological landscape and Euclidean quantum gravity. *J. Phys. A* **2007**, *40*, 7043–7048. [CrossRef]
49. Barvinsky, A.O. Why there is something rather than nothing (out of everything)? *Phys. Rev. Lett.* **2007**, *99*, 071301. [CrossRef] [PubMed]
50. Robles-Pérez, S.; González-Díaz, P.F. Quantum entanglement in the multiverse. *JETP* **2014**, *118*, 34. [CrossRef]
51. Robles-Pérez, S.J. Creation of entangled universes avoids the Big Bang singularity. *J. Gravity* **2014**, *2014*, 382675. [CrossRef]
52. Chen, P.; Hu, Y.C.; Yeom, D.H. Fuzzy Euclidean wormholes in de Sitter space. *JCAP* **2017**, *07*, 001. [CrossRef]
53. Caderni, N.; Martellini, M. Third quantization formalism for Hamiltonian cosmologies. *Int. J. Theor. Phys.* **1984**, *23*, 233. [CrossRef]
54. Coleman, S. Black holes as red herrings: Topological fluctuations and the loss of quantum coherence. *Nucl. Phys. B* **1988**, *307*, 867. [CrossRef]
55. Coleman, S. Why there is nothing rather than something? A theory of the cosmological constant. *Nucl. Phys. B* **1988**, *310*, 643–668. [CrossRef]
56. McGuigan, M. Third quantization and the Wheeler-DeWitt equation. *Phys. Rev. D* **1988**, *38*, 3031. [CrossRef]
57. McGuigan, M. Universe creation from the third quantized vacuum. *Phys. Rev. D* **1989**, *39*, 2229. [CrossRef] [PubMed]
58. McGuigan, M. Universe decay and changing the cosmological constant. *Phys. Rev. D* **1990**, *41*, 418. [CrossRef] [PubMed]
59. Hawking, S.W. Wormholes and Non-simply Connected Manifolds. In *Quantum Cosmology and Baby Universes*; Coleman, S., Hartle, J.B., Piran, T., Weinberg, S., Eds.; World Scientific: London, UK, 1990; Volume 7.
60. González-Díaz, P.F. Nonclassical states in quantum gravity. *Phys. Lett. B* **1992**, *293*, 294. [CrossRef]
61. González-Díaz, P.F. Regaining quantum incoherence for matter fields. *Phys. Rev. D* **1992**, *45*, 499. [CrossRef]
62. Higuchi, A.; Wald, R.M. Applications of a new proposal for solving the problem of time to some simple quantum cosmological models. *Phys. Rev. D* **1995**, *51*, 544–561. [CrossRef]
63. Barbour, J. Shape Dynamics. An introduction. In *Quantum Field Theory and Gravity*; Finster, F., Müller, O., Nardmann, M., Tolksdorf, J., Zeidler, E., Eds.; Springer: Basel, Switerland, 2012; pp. 257–297.
64. Griffiths, J.B.; Podolsky, J. *Exact Space-Times in Einstein's General Relativity*; Cambridge Monographs on Mathematical Physics; Cambridge University Press: Cambridge, UK, 2009.
65. Biesiada, M.; Rugh, S. Maupertuis principle, Wheeler's superspace and an invariant criterion for local instability in general relativity. *arXiv* **1994**, arXiv:gr-qc/9408030.
66. Garay, I.; Robles-Pérez, S. Classical geodesics from the canonical quantisation of spacetime coordinates. *arXiv* **2019**, arXiv:1901.05171.
67. Pimentel, L.O.; Mora, C. Third quantization of Brans-Dicke Cosmology. *Phys. Lett. A* **2001**, *280*, 191–196. [CrossRef]
68. Kim, S.P. Third quantization and quantum universes. *arXiv* **2012**, arXiv:1212.535.

69. Ohkuwa, Y.; Ezawa, Y. Third quantization of f(R)-type gravity II—General f(R) case. *Class. Quantum Gravity* **2013**, *20*, 235015. [CrossRef]
70. Calgani, G.; Gielen, S.; Oriti, D. Group field theory cosmology: A cosmological field theory of quantum geometry. *Class. Quantum Gravity* **2012**, *29*, 105005. [CrossRef]
71. Faizal, M. Multiverse in the third quantized formalism. *Commun. Theor. Phys.* **2014**, *62*, 697. [CrossRef]
72. Balcerzak, A.; Marosek, K. Emergence of multiverse in third quantized varying constants cosmologies. *Eur. Phys. J. C* **2019**, *79*, 563. [CrossRef]
73. Balcerzak, A.; Marosek, K. Doubleverse entanglement in third quantized non-minimally coupled varying constants cosmologies. *Eur. Phys. J. C* **2020**, *80*, 709. [CrossRef]
74. Campanelli, L. Creation of universes from the third-quantized vacuum. *Phys. Rev. D* **2020**, *102*, 043514. [CrossRef]
75. Robles-Pérez, S.J. Hartle-Hawking vacuum is full of Vilenkin's universe-antiuniverse pairs. *arXiv* **2021**, arXiv:2110.06521.
76. Birrell, N.D.; Davies, P.C.W. *Quantum Fields in Curved Space*; Cambridge University Press: Cambridge, UK, 1982.
77. Mukhanov, V.F.; Winitzki, S. *Quantum Effects in Gravity*; Cambridge University Press: Cambridge, UK, 2007.
78. Bander, M.; Itzykson, C. Group theory and the hydrogen atom (II). *Rev. Mod. Phys.* **1966**, *38*, 346. [CrossRef]
79. Lewis, H.R. Class of exact invariants for classical and quantum time dependent harmonic oscillators. *J. Math. Phys.* **1968**, *9*, 1976. [CrossRef]
80. Pedrosa, I.A. Comment on "Coherent states for the time-dependent harmonic oscillator". *Phys. Rev. D* **1987**, *36*, 1279. [CrossRef]
81. Dantas, C.M.A.; Pedrosa, I.A.; Baseia, B. Harmonic oscillator with time-dependent mass and frequency and a perturbative potential. *Phys. Rev. A* **1992**, *45*, 1320. [CrossRef] [PubMed]
82. Song, D.Y. Unitary relation between a harmonic oscillator of time-dependent frequency and a simple harmonic oscillator with and withuot an inverse-square potential. *Phys. Rev. A* **2000**, *62*, 014103. [CrossRef]
83. Kim, S.P.; Page, D.N. Classical and quantum action-phase variables for time-dependent oscillators. *Phys. Rev. A* **2001**, *64*, 012104. [CrossRef]
84. Park, T.J. Canonical Transformations for Time-Dependent Harmonic Oscillators. *Bull. Korean Chem. Soc.* **2004**, *25*.
85. Robles-Pérez, S. Invariant vacuum. *Phys. Lett. B* **2017**, *774*, 608–615. [CrossRef]
86. Rajeev, K.; Chakraborty, S.; Padmanabhan, T. Inverting a normal harmonic oscillator: Physical interpretation and applications. *Gen. Rel. Grav.* **2018**, *50*, 116. [CrossRef]
87. Olson, S.J.; Ralph, T.C. Entanglement between the future and past in the quantum vacuum. *Phys. Rev. Lett.* **2011**, *106*, 110404. [CrossRef]
88. Feynman, R.P. The theory of positrons. *Phys. Rev.* **1949**, *76*, 749. [CrossRef]
89. Robles-Pérez, S.J. Cosmological perturbations in the entangled inflationary universe. *Phys. Rev. D* **2018**, *97*, 066018. [CrossRef]
90. Robles-Pérez, S.J. Time reversal symmetry in cosmology and the creation of a universe-antiuniverse pair. *Universe* **2019**, *5*, 150. [CrossRef]
91. Robles-Pérez, S.J. Restoration of matter-antimatter symmetry in the multiverse. *arXiv* **2017**, arXiv:1706.06304.
92. Robles-Pérez, S.J. Quantum cosmology in the light of quantum mechanics. *Galaxies* **2019**, *7*, 50. [CrossRef]
93. Robles-Pérez, S.; González-Díaz, P.F. Quantum state of the multiverse. *Phys. Rev. D* **2010**, *81*, 083529. [CrossRef]
94. Mukhanov, V.F. *Physical Foundations of Cosmology*; Cambridge University Press: Cambridge, UK, 2008.
95. Kofman, L.; Linde, A.; Starobinsky, A.A. Towards the theory of reheating after inflation. *Phys. Rev. D* **1997**, *56*, 3258. [CrossRef]
96. Boyle, L.; Finn, K.; Turok, N. CPT-Symmetric universe. *Phys. Rev. Lett.* **2018**, *121*, 251301. [CrossRef]
97. Tegmark, M. The multiverse hierarchy. In *Universe or Multiverse*; Carr, B., Ed.; Cambridge University Press: Cambridge, UK, 2007; Chapter 7.
98. Alonso-Serrano, A.; Jannes, G. Conceptual challenges on the road to the multiverse. *Universe* **2019**, *5*, 212. [CrossRef]
99. Scardigli, F.; Gruber, C.; Chen, P. Black hole remnants in the early universe. *Phys. Rev. D* **2011**, *83*, 063507. [CrossRef]
100. Bouhmadi-Lopez, M.; Chen, P.; Liu, Y. Cosmological imprints of a generalized Chaplygin gas model for the early universe. *Phys. Rev. D* **2011**, *84*, 023505. [CrossRef]
101. Morais, J.; Bouhmadi-Lopez, M.; Krämer, M.; Robles-Pérez, S. Pre-inflation from th emultiverse: Can it solve the quadrupole problem in the cosmic microwave background? *Eur. Phys. J. C* **2018**, *78*, 240. [CrossRef]
102. Holman, R.; Mersini-Houghton, L.; Takahashi, T. Cosmological avatars of the Landscape II. *Phys. Rev. D* **2008**, *77*, 063511. [CrossRef]
103. Mersini-Houghton, L. Thoughts on defining the multiverse. *arXiv* 2008.
104. Holman, R.; Mersini-Houghton, L.; Takahashi, T. Cosmological avatars of the Landscape I. *Phys. Rev. D* **2008**, *77*, 063510. [CrossRef]
105. Mersini-Houghton, L. Predictions of the quantum landscape multiverse. *Class. Quantum Gravity* **2017**, *34*, 047001. [CrossRef]
106. Di Valentino, E.; Mersini-Houghton, L. Testing predictions of the quantum landscape multiverse 1: The Starobinsky inflationary potential. *JCAP* **2017**, *03*, 002. [CrossRef]
107. Di Valentino, E.; Mersini-Houghton, L. Testing predictions of the quantum landscape multiverse 2: The exponential inflationary potential. *JCAP* **2017**, *03*, 020. [CrossRef]
108. Alonso, J.; Carmona, J. Before spacetime: A proposal of a framework for multiverse quantum cosmology based on three cosmological conjectures. *Class. Quantum Gravity* **2018**, *36*, 185001. [CrossRef]

Review

Fuzzy Instantons in Landscape and Swampland: Review of the Hartle–Hawking Wave Function and Several Applications

Dong-han Yeom [1,2]

1 Department of Physics Education, Pusan National University, Busan 46241, Korea; innocent.yeom@gmail.com
2 Research Center for Dielectric and Advanced Matter Physics, Pusan National University, Busan 46241, Korea

Abstract: The Euclidean path integral is well approximated by instantons. If instantons are dynamical, they will necessarily be complexified. Fuzzy instantons can have multiple physical applications. In slow-roll inflation models, fuzzy instantons can explain the probability distribution of the initial conditions of the universe. Although the potential shape does not satisfy the slow-roll conditions due to the swampland criteria, the fuzzy instantons can still explain the origin of the universe. If we extend the Euclidean path integral beyond the Hartle–Hawking no-boundary proposal, it becomes possible to examine fuzzy Euclidean wormholes that have multiple physical applications in cosmology and black hole physics.

Keywords: quantum cosmology; no-boundary proposal; instantons

1. Introduction: Preliminaries

In modern physics, understanding the nature of the origin of the universe is one of the most fundamental problems. Due to the singularity theorem [1], if we move backward in time and assume reasonable physical conditions, it appears that there must exist an initial singularity. At this singularity, all the laws of general relativity break down; hence, a quantum gravitational prescription is required.

To understand the initial singularity, the quantum gravitational description must be non-perturbative. The most conservative approach is to quantize the gravitational degrees of freedom as per the canonical quantization method [2]. Using this approach, one can obtain the quantized Hamiltonian constraint; or the so-called Wheeler–DeWitt equation. If we solve the equation, we can in principle obtain the probability for a given hypersurface and the corresponding field configurations.

One of the limitations of canonical quantization is that the probability depends on the selection of boundary conditions [3]. By selecting a certain boundary condition, one may or may not provide a reasonable probability distribution for the early universe. There is no fundamental principle that can be used to select the boundary condition; in principle, the boundary condition must be confirmed by the possible observational consequences [4].

1.1. Hartle–Hawking Wave Function

Now, we can ask what the most natural assumption regarding the boundary conditions of the universe is. It might be considered that the ground state wave function corresponds to the most natural choice of boundary conditions, although one potential problem with this is that the ground state is not defined in the context of quantum gravity. However, one may reasonably argue that the Euclidean path integral might be the ground state wave function of the Wheeler–DeWitt Equation [5]. The mathematical form, the so-called Hartle–Hawking wave function, is listed below (we use the convention $c = G = \hbar = 1$):

$$\Psi[h_{\mu\nu}, \psi] = \int \mathcal{D}g_{\mu\nu} \mathcal{D}\phi \; e^{-S_E[g_{\mu\nu}, \phi]}, \tag{1}$$

where $g_{\mu\nu}$ is the metric, ϕ is a matter field, and S_E is the Euclidean action; we sum over all regular and compact Euclidean geometries and field configurations satisfying conditions $\partial g_{\mu\nu} = h_{\mu\nu}$ and $\partial \phi = \psi$. One interesting feature of this wave function is that there is only one boundary (the final boundary) of the path integral; however, the path integral usually must have two boundaries (the initial and final boundaries). As the wave function has no initial boundary, it is known as the *no-boundary* wave function. Although there is no guarantee regarding the convergence of this path integral (this might diverge for Minkowski or anti-de Sitter background), it will still be useful in understanding the physics of de Sitter background.

1.2. Steepest-Descent Approximation and Fuzzy Instantons

In cosmology, it is reasonable to assume $O(4)$-symmetry as follows:

$$ds_E^2 = d\tau^2 + a^2(\tau)d\Omega_3^2, \tag{2}$$

where τ is the Euclidean time, $d\Omega_3^2$ is the 3-sphere, and $a(\tau)$ is the scale factor. In addition to this symmetry, if we impose the on-shell condition to the metric and matter field; or we restrict to *instantons*, we can approximate the wave function based on the steepest-descent approximation:

$$\Psi[b,\psi] \simeq \sum_{\text{on-shell}} e^{-S_E^{\text{on-shell}}}, \tag{3}$$

where b and ψ are the boundary values of $a(\tau)$ and $\phi(\tau)$, respectively. Finally, the probability for each instanton is approximately:

$$P[b,\psi] = |\Psi[b,\psi]|^2 \simeq e^{-2\,\text{Re}\,S_E^{\text{on-shell}}}, \tag{4}$$

where

$$S_E^{\text{on-shell}} = \text{Re}\,S_E^{\text{on-shell}} + i\,\text{Im}\,S_E^{\text{on-shell}}. \tag{5}$$

Due to the analyticity, at the point of the Wick-rotation $\tau = \tau_0 + it$, we must impose the continuity of fields

$$a(t=0) = a(\tau = \tau_0), \tag{6}$$
$$\phi(t=0) = \phi(\tau = \tau_0), \tag{7}$$

as well as the *Cauchy-Riemann conditions*

$$\dot{a}(t=0) = i\dot{a}(\tau = \tau_0), \tag{8}$$
$$\dot{\phi}(t=0) = i\dot{\phi}(\tau = \tau_0). \tag{9}$$

therefore, in general, if the fields are dynamical, the on-shell solutions will be complex-valued; the instantons will be *fuzzy*. However, the boundary values b and ψ must be real-valued [6]. In some sense, this is a type of boundary condition of the instantons. The reality at the boundary of the wave function is related to the *classicality* of the solution. Once the solution becomes classical, the probability must slowly vary along the steepest-descent path. If the solution is real-valued, or if the real component of each function is at least dominant over that of the imaginary part, the probability must slowly vary compared with the phase part, and hence the history will be sufficiently classical. Furthermore, the history should satisfy the classical equations of motion (e.g., the Hamilton–Jacobi equation). This condition can be summarized as follows:

$$|\nabla_\alpha \text{Re}\,S_E| \ll |\nabla_\alpha \text{Im}\,S_E|, \tag{10}$$

where $\alpha = a, \phi$ is the canonical direction [7]. In many practical cases, this classicality condition can be easily demonstrated by verifying whether the real parts of the functions dominate the imaginary parts after the Wick-rotation and a sufficient Lorentzian time.

1.3. Scope of This Paper

The question that naturally arises is this: for which physical situations can the classicality condition be satisfied? The answer is that *inflation is required* to satisfy the classicality condition. This is very important: if our universe was created from the Hartle–Hawking wave function, a small amount of inflation is required [8]. However, there still remain several questions:

1. Does the Hartle–Hawking wave function prefer sufficient inflation?
2. Which type of inflation allows classicalization: slow-roll or fast-roll?
3. Is the Hartle–Hawking wave function a unique choice for quantum cosmology; or can there be additional generalization from the Euclidean path integral approach?
4. Is the Hartle–Hawking wave function compatible with the recent progress of quantum gravity?

In this study, we review several interesting developments about the Hartle–Hawking wave function and its potential applications. Furthermore, we answer a number of previous questions and provide certain possible future applications and research directions.

2. Fuzzy Instantons with Slow-Roll Inflation

The first issue is to obtain classicalized fuzzy instantons based on slow-roll inflation.

2.1. Simplest Model

To discuss the generic properties of slow-roll inflation and fuzzy instantons, we consider the following model [7]:

$$S_E = -\int d^4x \sqrt{+g} \left[\frac{1}{16\pi}(R - 2\Lambda) - \frac{1}{2}(\nabla \Phi)^2 - V(\Phi) \right], \quad (11)$$

where R is the Ricci scalar, Λ is the cosmological constant, Φ is a scalar field, and

$$V(\phi) = \frac{1}{2}m^2 \Phi^2 \quad (12)$$

is the potential. For simplicity, one can define the metric and several other variables as follows:

$$ds_E^2 = \frac{3}{\Lambda}\left(d\tau^2 + a^2(\tau) d\Omega_3^2\right), \quad (13)$$

$$\phi \equiv \sqrt{\frac{4\pi}{3}} \Phi, \quad (14)$$

$$\mu \equiv \sqrt{\frac{3}{\Lambda}} m. \quad (15)$$

The equations of motion are as follows:

$$\ddot{a} + a + a\left(2\dot{\phi}^2 + \mu^2 \phi^2\right) = 0, \quad (16)$$

$$\ddot{\phi} + 3\frac{\dot{a}}{a}\dot{\phi} - \mu^2 \phi = 0. \quad (17)$$

These are two second-order differential equations but we will consider complexified instantons. Hence, each equation has two parts with one being the real part and the other being the imaginary part. Therefore, there are basically eight initial conditions (at $\tau = 0$) that determine the solution; however, because of the Hamiltonian constraint, two of them are restricted. If we assume the no-boundary condition $a(\tau = 0) = 0$ with the Hamiltonian constraint, we must require that

$$a(\tau = 0) = 0, \tag{18}$$
$$\dot{a}(\tau = 0) = 1, \tag{19}$$
$$\dot{\phi}(\tau = 0) = 0. \tag{20}$$

The above equations already fix six of the initial conditions. There are thus two free parameters Re $\phi(\tau = 0)$ and Im $\phi(\tau = 0)$, or we present

$$\phi(\tau = 0) = \phi_0 e^{i\theta}, \tag{21}$$

where both ϕ_0 and θ are real values (see a recent analytic review in [9]).

Physically, ϕ_0 corresponds to the initial condition of the inflaton field, i.e., the initial condition of an inflationary universe. Therefore, we will eventually examine the probability distribution as a function of ϕ_0. On the other hand, θ is merely a free parameter. This must be used to satisfy the boundary condition after the Wick-rotation (i.e., to achieve classicality). If we select an appropriate θ, it may be possible that after the Wick-rotation $\tau = \tau_0 + it$, the imaginary parts of both a and ϕ will approach zero, and the real parts will dominate (e.g., see Figure 1 [10]). Therefore, in other words, θ is a tuning parameter for the classicality at a future infinity.

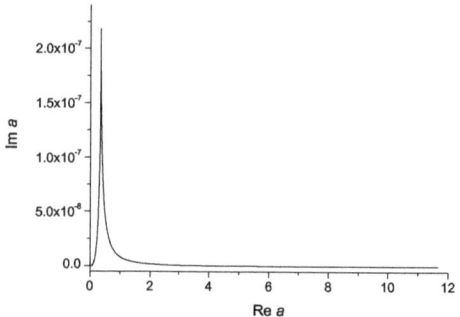 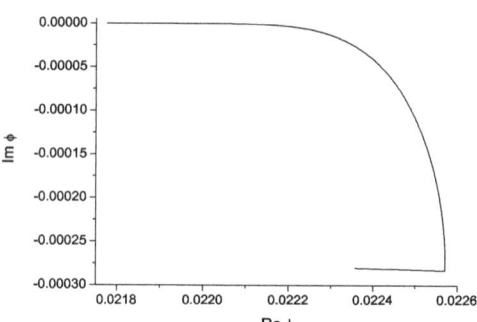

Figure 1. Example of a fuzzy instanton solution with $m^2/V_0 = 0.2$ and $m\phi_0/\sqrt{V_0} = 0.02$, where the left side is the metric a, and the right side is the scalar field ϕ. Here, the cusp is the turning point from a Euclidean to a Lorentizan signature [10].

Then, one may ask what the role of θ is in detail [11]. To determine this, let us first assume slow-roll inflation given a classical background metric, i.e., $\dot{\phi}^2 \ll 1$, $\mu^2 \phi^2 \ll 1$, and Im $a \ll$ Re a. In this case, it can be stated that

$$\text{Re } a = C_a e^t + D_a e^{-t} \simeq C_a e^t, \tag{22}$$

where C_a and D_a are integration constants. The equations of motion for the scalar fields are approximately

$$\text{Re } \ddot{\phi} + 3\text{Re } \dot{\phi} + \mu^2 \text{Re } \phi \simeq 0, \tag{23}$$
$$\text{Im } \ddot{\phi} + 3\text{Im } \dot{\phi} + \mu^2 \text{Im } \phi \simeq 0. \tag{24}$$

Hence,

$$\text{Re } \phi \simeq C_\phi^{\text{Re}} e^{-\frac{3}{2}t + \omega t} + D_\phi^{\text{Re}} e^{-\frac{3}{2}t - \omega t}, \tag{25}$$
$$\text{Im } \phi \simeq C_\phi^{\text{Im}} e^{-\frac{3}{2}t + \omega t} + D_\phi^{\text{Im}} e^{-\frac{3}{2}t - \omega t}, \tag{26}$$

where C_ϕ^{Re}, D_ϕ^{Re}, C_ϕ^{Im}, and D_ϕ^{Im} are constants, and

$$\omega^2 \equiv \left(\frac{3}{2}\right)^2 - \mu^2. \tag{27}$$

From these equations, we can easily obtain the following conclusion. If $\mu < 3/2$, the solutions satisfy the *over-damped motion*. In other words, there are two linearly independent solutions with different exponents. Therefore, it is possible to select θ such as to make $C_\phi^{\text{Im}} \ll 1$ (and hence finely tune the integration constants). Then,

$$\frac{\text{Im }\phi}{\text{Re }\phi} \simeq e^{-2\omega t} \to 0, \tag{28}$$

and hence after the Wick-rotation, the solution will eventually satisfy the classicality conditions. On the other hand, if $\mu > 3/2$, the solutions satisfy *under-damped motion*. For any choice of initial conditions, Im ϕ/Re ϕ will be proportional to a trigonometric function, and hence,

$$\frac{\text{Im }\phi}{\text{Re }\phi} \simeq \mathcal{O}(1). \tag{29}$$

Therefore, the imaginary part and real part of the scalar field oscillate in a similar order.

One might query the consequences of this if there is no approach to classicalize a scalar field. Note that the imaginary part of the scalar field provides the negative kinetic term, indicating it is ghost-like. The energy of the scalar field will contribute to the matter content of the universe. If the imaginary part of the scalar field is of the same order as the real part even after the Wick-rotation, we cannot avoid the instability of the ghost-like imaginary part of the field. This is a catastrophic consequence, and we cannot physically allow this possibility. (However, for the possibility of observing restricted contributions from the ghost-like term, please refer to [12]).

For a generic scalar field potential V, the criterion for a classical universe is $\mu^2 < 9/4$, or

$$\left|\frac{V''}{V}\right| < 6\pi. \tag{30}$$

If $m^2/V < 6\pi$, $\phi = 0$ (local minimum) can allow for a classical universe. If $m^2/V > 6\pi$, there exists a cutoff $\phi_{\text{cutoff}} > 0$ such that $\phi > \phi_{\text{cutoff}}$ only allows for classical universes [7].

2.2. Probabilities and Preferences of Large e-Foldings

If one is able to construct a classical universe, it is possible to obtain the probability of that universe. For the slow-roll potential, the probability of a classical universe with the initial condition ϕ_0 is approximately

$$\log P \simeq \frac{3}{8V(\phi_0)}. \tag{31}$$

Note that this is positive definite. Hence, the probability is exponentially enhanced. The most highly favored initial condition is a field value with the smallest possible potential. Considering the no-boundary proposal [8], this indicates that $\phi \simeq \phi_{\text{cutoff}}$ corresponds to the most probable initial condition.

However, the limitation is that the *e*-foldings of the most probable initial condition are, in general, insufficient. For example, if we have a quadratic potential with $\Lambda = 0$, the most preferred *e*-folding number will be ~ 0.62, while more than ~ 50 *e*-foldings will be required [13].

There have been several proposals to resolve or understand this problem. First, the simplest suggestion is that the no-boundary proposal is simply wrong. For example, if we do not trust the steepest-descent approximation for theoretical reasons, we may obtain an alternative probability distribution [14]. Alternatively, if we begin from a new fundamental wave function, it is possible to obtain a different probability distribution that may prefer a large vacuum energy [3]. However, it is fair to say that there are still several theoretical arguments to support the consistency of the original approaches put forward by Hartle and Hawking [15]. Thus, we can consider several viable possibilities to rescue the Hartle–Hawking wave function [16]:

1. *We require a number of ad hoc terms to measure the probability.* For example, Hartle, Hawking, and Hertog introduced the volume-weighting factor to the probability measure [8]. Consequently, there is competition between the volume-weighting component and the Euclidean probability component. If the vacuum energy is sufficiently large, the volume-weighting component is dominated, and large e-foldings are eventually preferred. However, this assumption cannot be justified from first principles. Furthermore, this leads to eternal inflation, while this eternal inflation goes beyond the scope of our understanding because of the infinite volume and subsequent quantum tunneling.

2. *Our universe began from $V \sim 1$ (Planck scale) vacuum energy.* If this is the case, there are no significant probability differences between the cutoff and other field values. However, based on observational constraints, this Planck-scale inflation cannot be the primordial inflation of our universe.

3. *Unknown physical degrees of freedom are required.* For example, if there exists a very long field space or a large number of fields that contribute to inflation [13], such degeneracy of the field space can compete with the Euclidean probability component. However, in multiple cases, this requires too many degrees of freedom. Hence, in terms of quantum field theory, these possibilities may be unnatural.

4. *Certain modifications of the theory of gravity can explain large e-foldings.* For example, massive gravity models [17,18] could provide certain interesting possibilities; however, this approach goes beyond the regime of Einstein gravity.

2.3. Rescue from the Secondary Scalar Field

However, the most reasonable rescue is to probably introduce one additional massive field [11]. The primary idea is that the classicality condition requires the classicality of *all* fields not only the inflaton field but also the other matter fields. If this is not the case (i.e., if the only inflaton field is classicalized while the second field is not classicalized), the ghost-like modes of the secondary field cannot be controlled after the Wick-rotation. Hence, this possibility must be avoided.

For simplicity, let us consider the following model

$$S_E = -\int d^4x \sqrt{+g}\left[\frac{1}{16\pi}R - \frac{1}{2}(\nabla\Phi_1)^2 - \frac{1}{2}(\nabla\Phi_2)^2 - \frac{1}{2}m_1^2\Phi_1^2 - \frac{1}{2}m_2^2\Phi_2^2\right], \quad (32)$$

where m_1 and m_2 are mass parameters of Φ_1 and Φ_2, respectively. In particular, we assume that $m_1 \ll m_2$, and hence Φ_1 is the inflaton field and Φ_2 is only an assisting field. Similar to the previous section, one can select the metric ansatz as follows:

$$ds_E^2 = \frac{1}{m_2^2}\left(d\tau^2 + a^2(\tau)d\Omega_3^2\right). \quad (33)$$

Due to the slow-roll condition, the variation of Φ_1 is negligible along the field direction of Φ_2. Hence, it is possible to approximate that $(1/2)m_1\Phi_1^2 \simeq V_0$ is a constant; therefore, the classicality condition of the potential (Equation (30)) is

$$\frac{m_2^2}{V_0} \simeq \frac{m_2^2}{\frac{1}{2}m_1^2\Phi_1^2} \leq 6\pi. \quad (34)$$

The results of numerical investigation are consistent with this expectation (Figure 2 [11]). The shadowed box region becomes increasingly narrow as m_1/m_2 decreases. Hence, in the $m_1/m_2 \ll 1$ limit, if $\Phi_2 \simeq 0$, the genuine cutoff of the Φ_1 direction will satisfy $\Phi_1 \gg 1$.

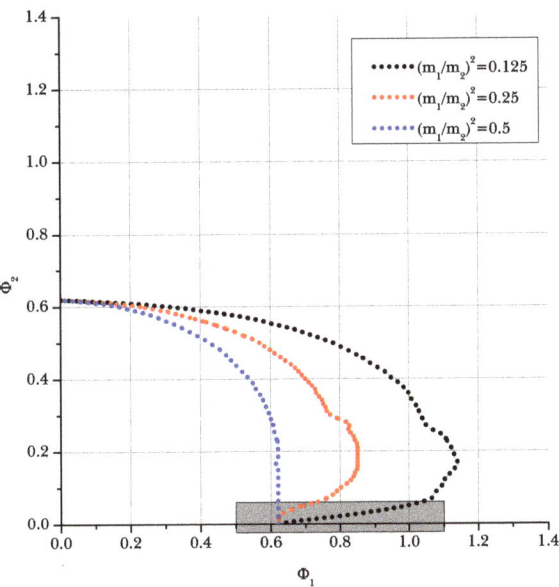

Figure 2. Numerical calculations of the cutoffs for $(m_1/m_2)^2 = 0.125, 0.25,$ and 0.5 [11].

Now, we are required to ask what the most probable initial condition over the field space (Φ_1, Φ_2) is. The smallest potential energy is the most preferred initial condition. As we assumed $m_1 \ll m_2$, the potential varies very sensitively along the Φ_2 direction. Hence, the initial conditions with $\Phi_2 \simeq 0$ must be the most preferred. If we assume $\Phi_2 \simeq 0$, the most probable initial condition of the Φ_1 direction is $\Phi_{1,\text{cutoff}}$. However, to classicalize the Φ_2 field, the following condition must be satisfied:

$$\frac{m_2^2}{3\pi m_1^2} \leq \Phi_{1,\text{cutoff}}^2, \tag{35}$$

where the details regarding the constants on the left-hand-side are not extremely important. If there is a mass hierarchy $m_1 \ll m_2$, the cutoff of Φ_1 will be sufficiently large while the initial condition of the inflaton field must have large e-foldings $\mathcal{N}$:

$$\mathcal{N} \simeq \mathcal{O}(1) \times \frac{m_2^2}{m_1^2}. \tag{36}$$

In this case, it is easy to make $\mathcal{N}$ greater than 50.

It might be asked why massive particles play an important role in the no-boundary wave functions, as per our physical intuitions, massive particles can be integrated out based on low energy effective theory. This is an interesting feature of *Euclidean* quantum gravity. Massive particles will have *greater* stability in Lorentzian signatures and *lower* stability in Euclidean signatures. If the Hartle–Hawking wave function is a fundamental prescription of quantum gravity, it must classicalize all fundamental fields, including the most massive (probably Planck-scale) particles. If the massive fields are not classicalized, there remains an imaginary degree of freedom, which is detrimental for providing a consistent description.

This idea is not limited to quadratic potential models. It might be interesting to apply it to realistic inflation models, in addition to models with various interactions among fields.

3. Fuzzy Instantons with Fast-Roll Potential

The second issue is to obtain classicalized fuzzy instantons even if the slow-roll condition is not guaranteed. Indeed, this issue has been highlighted in recent discussions in string theory.

3.1. Landscape vs. Swampland

To understand the cosmological constant problem and multiple fine-tuning issues regarding the universe, the *cosmic landscape* was a highly sophisticated hypothesis [19]. String theory allows for a wide variety (almost all possible) constants of nature, including the cosmological constant, as well as detailed shapes of the inflaton potential with these being referred to as the cosmic landscape. These possible parameter spaces are physically realized via eternal inflation and the quantum tunneling of bubble universes. Eventually, any fine-tuned parameters can be realized at a certain location in the *multiverse*.

Although there have been several criticisms of this approach, the most significant criticism was suggested by the string theory community [20]. As per the authors, the landscape where string theory is allowed is indeed a very restricted region among possible parameter spaces, e.g., it was conjectured that the inflaton potential must be restricted by

$$\left|\frac{V''}{V}\right| > \mathcal{O}(1), \tag{37}$$

$$\left|\frac{V'}{V}\right| > \mathcal{O}(1), \tag{38}$$

where these conditions are known as the *swapland criteria*. Of course, there are several subtle issues here. First, there is no fundamental proof of the criteria. Hence, the order-one constant is tricky to define. Perhaps slow-roll inflation can be marginally allowed [21]; however, there can be no fundamental cosmological constant if we seriously accept the swampland criteria.

In this study, we do not agree or disagree on the details of the swampland criteria. However, it has been established that they are harmful to the Hartle–Hawking proposal. In particular, there is a tension with Equation (30). On seeing more details, we may identify a run-away quintessence model, which is typical for string-inspired models [22]:

$$V(\phi) = Ae^{-C\phi}. \tag{39}$$

For each point near ϕ_0, we can approximate the potential as

$$V(\phi) \simeq \frac{1}{2}AC^2 e^{-C\phi_0}\left(\phi - \phi_0 - \frac{1}{C}\right)^2 + \frac{Ae^{-C\phi_0}}{2}. \tag{40}$$

Therefore, it is not surprising that $\phi = \phi_0$ has a classical history only if

$$C^2 \lesssim 3\pi. \tag{41}$$

From numerical computations, we can confirm that $C \lesssim 4$ is the condition for the existence of a classical solution (Figure 3, [22]). On the other hand, if $C > 4$ happens, which is extremely natural for string-inspired models, there will be no classicalized instantons along the runaway direction. Hence, such a quintessence model is not compatible with the Hartle–Hawking wave function.

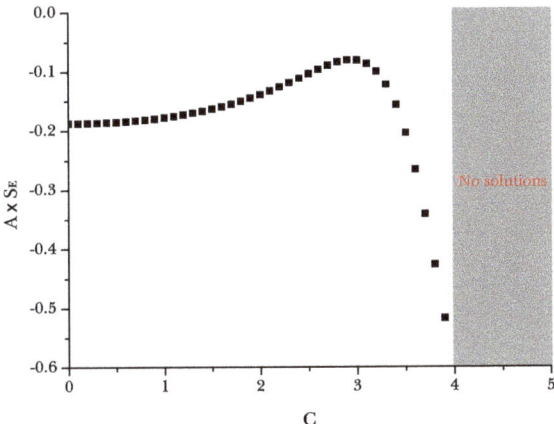

Figure 3. Euclidean action for exponential potential. If C > 4, classicalized solutions are not allowed [22].

The question then arises as to whether, even in the context of the swampland criteria, there is any way to rescue the Hartle–Hawking wave function.

3.2. Rescue Using Hwang-Sahlmann-Yeom Instantons

Although the swampland criteria do not favor the local minimum of the potential, they do not exclude the unstable local maximum of the potential. We describe the hilltop potential near the hilltop ($\phi = 0$) as follows:

$$V(\phi) = V_0 \left(1 - \frac{1}{2}\mu^2 \phi^2\right). \tag{42}$$

If $\mu \ll 1$, the slow-roll condition is satisfied; moreover, the usual fuzzy instanton can exist. On the other hand, if $\mu \gg 1$, which is more natural for string-inspired models, the slow-roll condition is no longer satisfied. Thus, close to the hilltop, the initial field values rapidly rise to the local maximum of the potential during the Euclidean time. By selecting proper initial conditions, one can obtain the fuzzy instantons close to the fast-rolling hilltop potentials [23] (for example, see Figure 4). We name these solutions Hwang-Shalmann-Yeom (HSY) instantons to contrast them with the slow-roll fuzzy instantons proposed by Hartle–Hawking–Hertog (HHH).

The physical difference is attributed to the probability (Figure 5 [22]). If $\mu \ll 1$, the probability is approximately

$$\log P_{\text{HHH}} \simeq \frac{3}{8V_0 \left(1 - \frac{1}{2}\mu^2 \phi_0^2\right)}. \tag{43}$$

On the other hand, if $\mu \gg 1$, the field quickly approaches the local maximum of the potential, and hence the dependence on the initial condition is negligible:

$$\log P_{\text{HSY}} \simeq \frac{3}{8V_0}. \tag{44}$$

Therefore, once there is a hilltop with $\mu \gg 1$, the probability of left-rolling and right-rolling are almost the same.

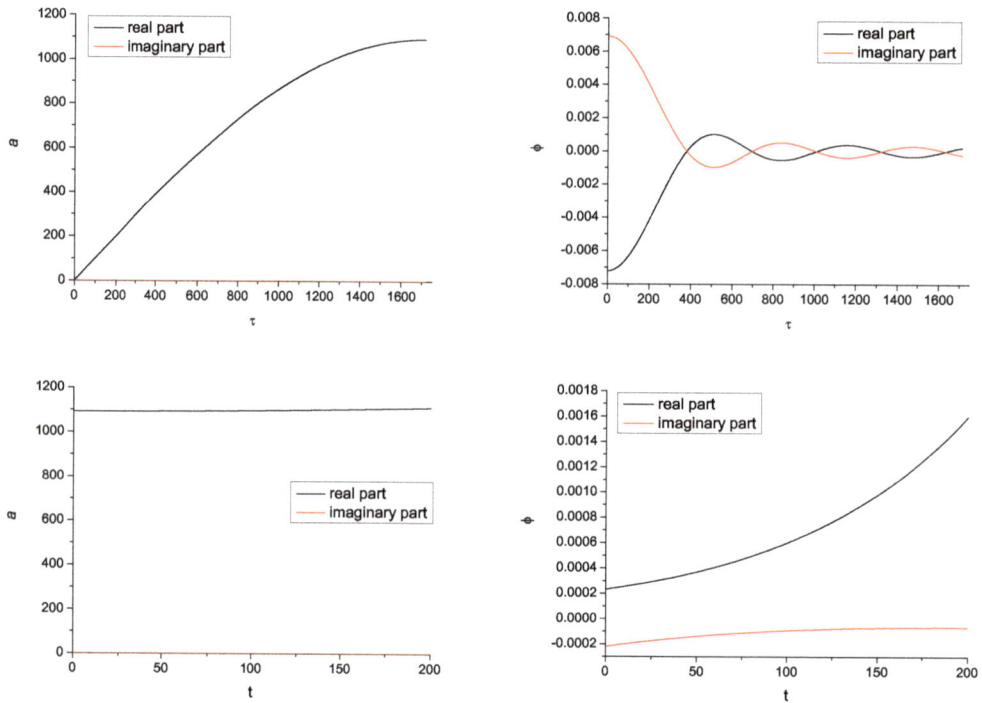

Figure 4. A fuzzy instanton solution ($\phi_0 = 0.01$) of a toy potential $V = V_0 - (1/2)m^2\phi^2 + (1/24)\lambda\phi^4$, where $V_0 = 10^{-7}$, $m^2 = 10^{-4}$, and $\lambda = 2 \times 10^{-2}$. The top figures depict Euclidean time, while the bottom figures depict Lorentzian time. During Euclidean time, the field rapidly oscillates along the hilltop ($\phi = 0$ in this potential, **top right**). After the Wick-rotation, the field rolls in the left or right direction (**bottom right**).

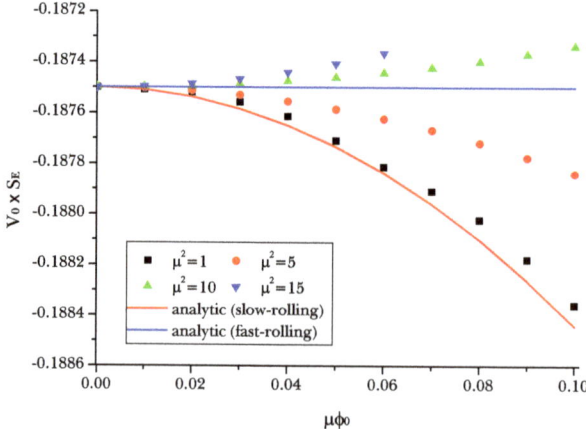

Figure 5. Euclidean action $V_0 S_E$ as a function of $\mu\phi_0$. The red curve is $\log P_{\text{HHH}}$; however, the blue line is $\log P_{\text{HSY}}$ [22].

This HSY instanton can rescue the Hartle–Hawking wave function even in the context of the swampland because the classicalized universes can be created close to the narrow and unstable hilltop of the field space.

3.3. Cosmological Applications

If fuzzy instantons and the swampland criteria both exist in the context of the cosmic landscape, there can be multiple cosmological applications of HSY instantons [22].

1. As per the moduli or dilaton stabilization issue, there is a probability competition between the stable and unstable directions. It may be that the only possible starting point from the no-boundary wave function is the hilltop of the potential. As per HSY instantons, there is no preference between left-rolling and right-rolling. Hence, given a reasonable probability, the moduli or dilaton stabilization can be explained using quantum cosmology [23].
2. The universe starts from the local maximum rather than the local minimum. The cosmological constant depends on the local *minimum*; however, the probability of the HSY instanton depends on the local *maximum*. Therefore, although the cosmological constant varies from anti-de Sitter to de Sitter space, there may be no singular changes in the a priori probability because there is no singular change in the local maximum [22].
3. HSY instantons can rescue the Hartle–Hawking wave function even in the context of the swampland criteria because classicalized universes can be created close to the narrow and unstable hilltop of the field space.

In terms of embedding a consistent inflation model with the swampland criteria and the trans-Planckian censorship conjecture, if we consider only a single-field inflation model, the no-boundary wave function will not be compatible with the criteria [21]. On the other hand, if we include one additional field, there may be the possibility of rescuing the no-boundary proposal. Alternatively, if the universe began from a hilltop of a very sharp potential, it can be explained from the no-boundary wave function. However, its smooth connection to a successful inflation model must be explained.

4. Extensions

In the previous sections, we examined the no-boundary proposal in a single scalar field model with Einstein gravity. However, in principle, there are possible additional extensions such as the following:

1. The Euclidean path integral does *not* necessarily indicate the no-boundary proposal, which is a specific choice of the Euclidean path integral. In more generic cases, there can be two boundaries (initial and final boundaries). However, because of the ambiguity of time in quantum gravity, one may make the interpretation that two universes are created from nothing. These solutions are known as Euclidean wormholes (Figure 6 [24]).
2. The theory can be extended by or embedded with *quantum gravitational models*. For example, string-inspired models can be used to introduce a number of additional terms, e.g., the Gauss–Bonnet term with dilaton coupling [25]. Furthermore, loop quantum cosmological models suggest the big bounce near the putative singularity [26]. These corrections suggest a new type of solution.

4.1. Fuzzy Euclidean Wormholes

As a simple extension, we consider fuzzy Euclidean wormholes in Einstein gravity [24]. Indeed, in terms of instantons, Euclidean wormholes are more natural than compact instantons. The intuitive reason for this is listed below [27].

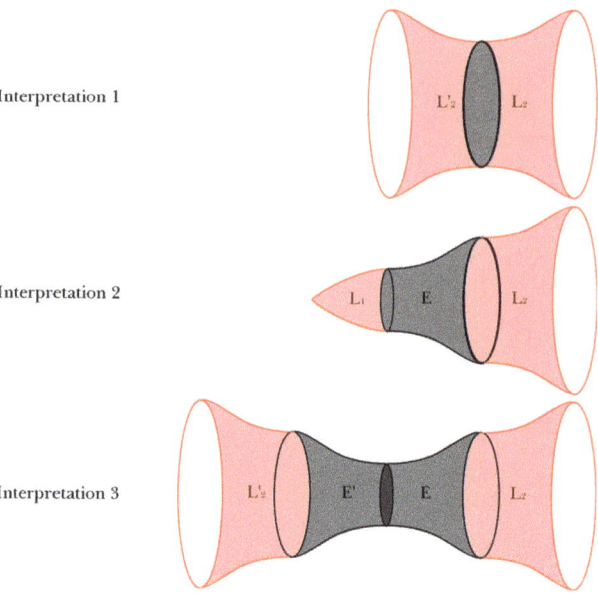

Figure 6. Possible interpretations of Euclidean wormholes [24]. Top: A collapsing universe is bounced (Interpretation 1). Middle: A small universe tunnels to a large universe or one contracting and one expanding universes are created (Interpretation 2). Bottom: Two entangled universes are created or a contracting universe is bounced to an expanding universe (Interpretation 3).

Let us first consider a free scalar field model with an $O(4)$-symmetric metric ansatz. The following will then be the generic solution of the scalar field in a Lorentzian signature:

$$\frac{d\phi}{dt} = \frac{\mathcal{A}}{a^3}. \tag{45}$$

Of course, due to classicality, $\mathcal{A}$ is a real-valued number. If we Wick-rotate this solution to Euclidean time, then we obtain

$$\frac{d\phi}{d\tau} = -i\frac{\mathcal{A}}{a^3}. \tag{46}$$

Therefore, if the velocity of the scalar field is non-vanishing in Lorentzian signatures and if the solution is classical in Lorentzian signatures, it is necessary that the velocity of the scalar field must be purely imaginary in Euclidean signatures. However, purely imaginary scalar fields in Euclidean signatures are perfectly acceptable in terms of the formalism of the Euclidean path integral. Then, the corresponding Euclidean metric satisfies

$$\dot{a}^2 = 1 - \frac{a^2}{\ell^2} + \frac{a_0^4}{a^4}, \tag{47}$$

where $\ell \equiv \sqrt{3/\Lambda}$ and $a_0^4 = 4\pi\mathcal{A}^2/3$.

If $a_0 \ll \ell$, $\dot{a}$ has two zeros. Hence, the Euclidean solution has two turning points say $a_{\max}$ and $a_{\min}$. If we consider a solution that covers $a_{\max}$ to $a_{\min}$ to $a_{\max}$, this becomes the Euclidean wormhole solution where there are two boundaries arising from the solution, and the Wick-rotation can be applied for these two boundaries. The compact instantons are available only if $\mathcal{A} = 0$, or when the velocity of the scalar field is zero. On the other hand, the non-compact instantons occur in more general situations when the Lorentzian solutions have non-trivial velocities.

What we have considered is the case in which the scalar field is free. The natural question that arises is what happens if we generalize to a specific inflation model. For this purpose, we can introduce the ansatz of the initial condition of the Euclidean wormholes as follows [27]:

$$\text{Re } a(0) = a_{\min} \cosh \eta, \tag{48}$$

$$\text{Im } a(0) = a_{\min} \sinh \eta, \tag{49}$$

$$\text{Re } \dot{a}(0) = \sqrt{\frac{4\pi}{3} \frac{\mathcal{B}}{a_{\min}^2}} \sqrt{\sinh \zeta \cosh \zeta}, \tag{50}$$

$$\text{Im } \dot{a}(0) = \sqrt{\frac{4\pi}{3} \frac{\mathcal{B}}{a_{\min}^2}} \sqrt{\sinh \zeta \cosh \zeta}, \tag{51}$$

$$\text{Re } \phi(0) = \phi_0 \cos \theta, \tag{52}$$

$$\text{Im } \phi(0) = \phi_0 \sin \theta, \tag{53}$$

$$\text{Re } \dot{\phi}(0) = \frac{\mathcal{B}}{a_{\min}^3} \sinh \zeta, \tag{54}$$

$$\text{Im } \dot{\phi}(0) = \frac{\mathcal{B}}{a_{\min}^3} \cosh \zeta, \tag{55}$$

where $a_{\min}$, $\mathcal{B}$, ϕ_0, η, ζ, and θ are free parameters. However, these free parameters are not entirely free, but should satisfy the real-part and imaginary-part equations of the Hamiltonian constraint:

$$0 = 1 + \frac{8\pi}{3} a_{\min}^2 (-V_r + \sinh 2\eta \, V_i) - \frac{4\pi \mathcal{B}^2}{3 a_{\min}^4} (1 + \sinh 2\zeta \sinh 2\eta), \tag{56}$$

$$0 = a_{\min}^6 + \frac{\mathcal{B}^2 \sinh 2\eta}{2(V_r \sinh 2\eta + V_i)}, \tag{57}$$

where V_r and V_i are the real and the imaginary part of $V(\phi)$ at $\tau = 0$. Based on these two equations, two parameters among the five free parameters are determined, say, $a_{\min}$ and η.

Now, the remaining free parameters are $\mathcal{B}$, ϕ_0, ζ, and θ. However, $\mathcal{B}$ determines the amplitude of the imaginary part of the scalar field, and hence the size of the wormhole throat. ζ is the parameter that determines the symmetry between the left and right sides of the wormhole. These two parameters do nothing but determine the shape of the wormhole. ϕ_0 corresponds to the initial field value of the solution. Therefore, the only tuning parameter that can be used to fulfill the classicality condition is θ.

This situation is the same as the compact instanton case; however, there is a serious problem. In the compact instanton case, there is only one boundary (future boundary); hence, the classicality must be imposed given only one boundary. Using θ, we can make one boundary classical. However, in the Euclidean wormhole case, there are two boundaries that we need to classicalize. In general, it is impossible to classicalize two boundaries at the same time.

However, there are a number of exceptional cases, e.g., let us consider the following potential:

$$V(\phi) = \frac{3}{8\pi \ell^2} \left(1 + A \tanh^2 \frac{\phi}{\alpha}\right), \tag{58}$$

where ℓ, A, and α are free parameters (see Figure 7 [27]). This model provides a flat hilltop [28] that is consistent with the Starobinsky model [29], which is preferred by the recent Planck data analysis [30]. In this model, the scalar field at the end of the wormhole rolls down to the local minimum, and hence the primordial inflation is naturally terminated. By tuning θ, we classicalize this end. On the other hand, it is not possible to tune the other end; however, if the scalar field rolls up to the hilltop, it is possible it will come to an automatic stop as long as the hilltop is sufficiently flat. If the field stops at the hilltop, the field will not be classicalized though the metric will be classicalized because the kinetic terms of the scalar field provide no contribution. Furthermore, because of the flat potential, there will be a local shift symmetry. By shifting the field along the complex direction, one can classicalize the field value at the hilltop. In any case, the point is that we can definitely classicalize one end; the other end is a little bit subtle, but the Euclidean action does not vary after the Wick-rotation at the hilltop. As we are observing only one universe, we do not require to worry about the details at the other end of the wormhole as long as the real part of the Euclidean action is bounded well.

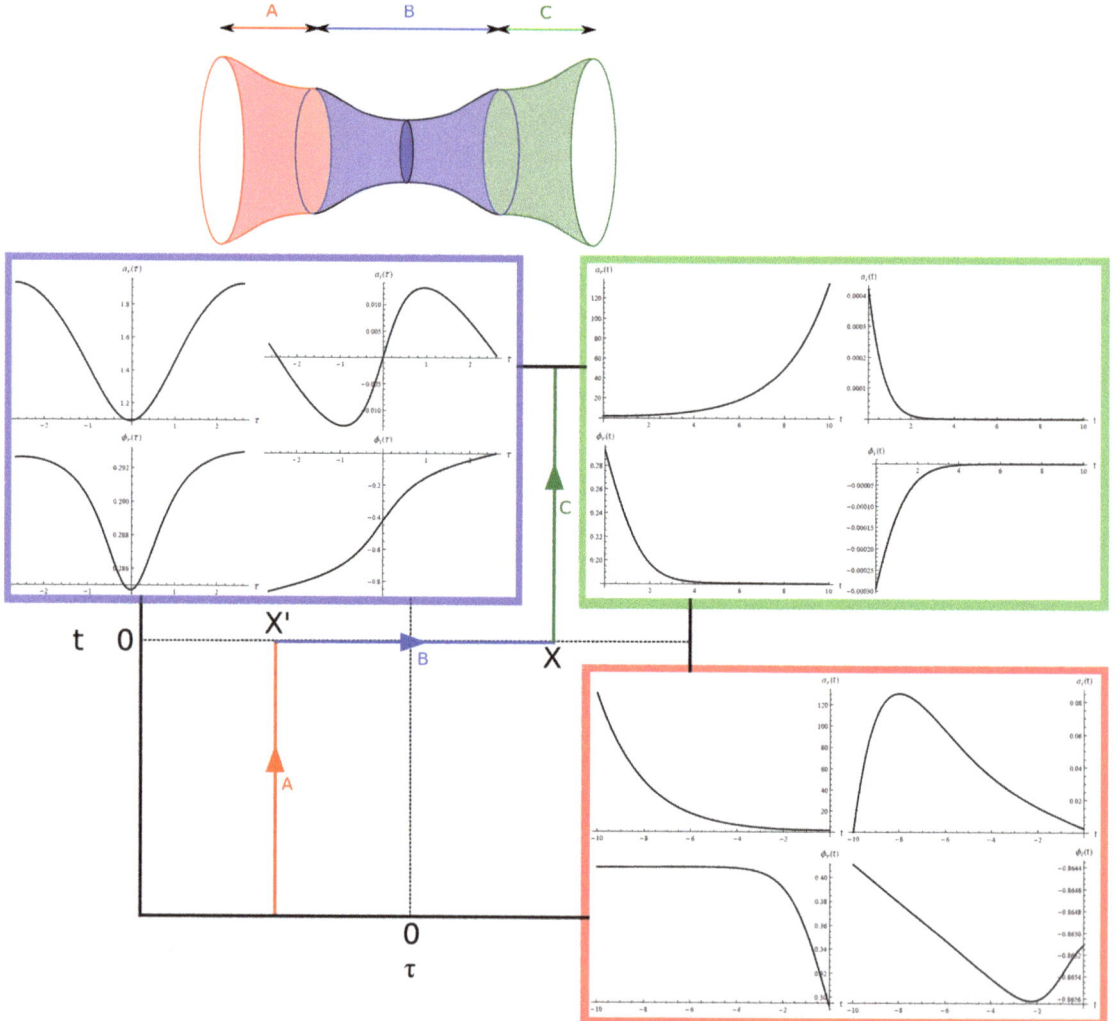

Figure 7. Complex time contours and numerical solution of Re a, Im a, Re ϕ, and Im ϕ for Equation (58). The top figure is the physical interpretation of the wormhole. Part A (red) and C (green) are Lorentzian, and Part B (blue) is Euclidean [27].

In terms of this mechanism, we have three important comments:

1. This mechanism cannot be applied to convex inflaton potential (e.g., quadratic potential). Therefore, the Euclidean wormhole selects the concave inflaton potential [27].

2. For a given vacuum energy scale V_0 or ℓ, the probability of the Euclidean wormhole is larger than that of the compact instanton [24] because the maximum probability of the Euclidean wormhole is

$$\log P \simeq \pi \ell^2 \left(1 + 0.16 \left(\frac{a_0}{\ell} \right)^{5/2} \right), \tag{59}$$

where that of the compact instanton is $\log P = \pi \ell^2$.

3. For a given concave potential, there may be competition between the compact instantons and Euclidean wormholes. The largest probability of compact instantons occurs close to the cutoff, where it is generally larger than that of the Euclidean wormholes that appear only near the hilltop. On the other hand, if we assume a mechanism that enhances the large *e*-foldings (e.g., introducing a massive field direction), Euclidean wormholes will be more highly favored than compact instantons [28]. Therefore, as long as we assume that our universe experienced more than 50 *e*-foldings, a *Euclidean wormhole* with a *concave* potential will be preferred over compact instantons with convex or concave potentials.

In conclusion, Euclidean wormholes can interestingly answer the question of *why our universe started from the concave part of a potential rather than the convex part*? However, there is a point of warning that is worth remarking upon. For all computations of Euclidean wormholes, we implicitly assumed the ultraviolet(UV)-completion of the inflaton potentials. In general, the Euclidean wormhole requires the potential to have a flat direction. However, in the context of the swampland criteria, this might be an unjustifiable assumption. Identifying a sufficiently flat field direction within a UV-completed model would make an interesting future research topic.

4.2. Euclidean Wormholes in Gauss–Bonnet-Dilaton Gravity

If we consider the string theory as the UV-completion of quantum gravity, it is reasonable to include higher-order corrections of string-inspired models and observe their physical applications. The most famous model in this regard is known as Gauss–Bonnet-dilaton gravity:

$$S = \int d^4 x \sqrt{-g} \left(\frac{R}{16\pi} - \frac{1}{2} (\nabla \phi)^2 - V(\phi) + \frac{1}{2} \xi(\phi) R_{\text{GB}}^2 \right), \tag{60}$$

where

$$R_{\text{GB}}^2 = R_{\mu\nu\rho\sigma} R^{\mu\nu\rho\sigma} - 4 R_{\mu\nu} R^{\mu\nu} + R^2 \tag{61}$$

is the Gauss–Bonnet term, ϕ is the dilaton field, and

$$\xi(\phi) = \lambda e^{-c\phi} \tag{62}$$

is the coupling function of the dilaton field. Note that λ and c are model-dependent parameters.

Due to the corrections of the Gauss–Bonnet-dilaton term, although the null energy condition is satisfied, the null curvature condition is effectively violated. Equivalently, if we consider the Gauss–Bonnet-dilaton term as an effective contributor to matter, the null energy condition will be effectively violated. Accordingly, it is not surprising that a Lorentzian or Euclidean wormhole solution might exist [31].

To obtain a Euclidean wormhole solution, we consider the following initial condition at $\tau = 0$ [32]:

$$a(0) = \sqrt{\frac{3}{4\pi(2V_0 - \dot{\phi}^2(0))}}, \quad (63)$$

$$\dot{a}(0) = 0, \quad (64)$$

$$\phi(0) = \phi_0, \quad (65)$$

$$\dot{\phi}(0) = 0, \quad (66)$$

where $a(0)$ is obtained from the Hamiltonian constraint equation. By tuning ϕ_0, we must satisfy the boundary condition at $\tau = \tau_{\text{end}}$:

$$a(\tau_{\text{end}}) = 0, \quad (67)$$

$$\dot{a}(\tau_{\text{end}}) = -1, \quad (68)$$

to achieve a regular end.

In general, this solution penetrates over a sharp potential barrier. Furthermore, because the volume is greater than that of the usual compact instanton, the probability is higher than that of the pure de Sitter instanton. Therefore, once there exists a string-inspired term, although there exists a potential barrier, it can be used to create a universe with a higher probability.

If we Wick-rotate at $\tau = 0$, we can apply this solution to quantum cosmology [25]. However, if we extend the Euclidean time to $\tau < 0$ and Wick-rotate the solution along the anisotropic direction, this can explain a (expanding) Lorentzian wormhole based on quantum tunneling [32]. The examination of the quantum tunneling of the Lorentzian wormhole is another interesting issue and requires further investigation.

4.3. Hartle–Hawking Wave Function with Loop Quantum Cosmology

For loop quantum gravity, we consider the generic quantum state that satisfies the (quantum) Hamiltonian constraint equation, in addition to the (quantum) momentum constraint equations. The generic states that satisfy the momentum constraint equations should follow the loop representations. Due to the loop representation, there must be a correction to the Hamiltonian constraint at the classical level [33]. By including these corrections, we can examine the effects of quantum gravity.

In general, it is believed that, in a cosmological context, the beginning of the universe can be explained by the big bounce. The Lorentzian dynamics of the scale factor satisfy the equation $\dot{a}^2 + \mathcal{V}(a) = 0$, where $\mathcal{V}(a)$ has a zero at a minimum value $a_{\min}$. This corresponds to the bouncing point of the universe [26]. However, there remains a conceptual question: As a goes to $a_{\min}$, the universe approaches the deep quantum regime, and it must be asked how we can select the arrow of time. It would seem strange if we were able to determine a definite direction for time even in this quantum gravitational regime.

Perhaps this conceptual tension might be explained if we introduce the Hartle–Hawking wave function [26]. To compute the Euclidean Lagrangian L_E from the loop quantum gravity modified Euclidean Hamiltonian H_E, we follow the relation:

$$L_E = p_a \dot{a} - H_E, \quad (69)$$

where p_a is the canonical momentum of a. However, because of the Hamiltonian constraint, $H_E = 0$ in the on-shell level description. Therefore, the Euclidean action is simply

$$S_E = -\frac{3\pi}{2}\int a\dot{a}^2 d\tau = -\frac{3\pi}{2}\int_0^{a_{\min}} a\sqrt{|\mathcal{V}(a)|}da. \quad (70)$$

Interestingly, in Euclidean signatures, as a approaches to zero, $\mathcal{V}(a)$ approaches to zero. This indicates that the instanton explains the infinitely stretched solution as a function of τ, although the probability is well-defined [34]. Except for this feature, the interpretation is the same as that of the usual Hartle–Hawking wave function. Therefore, close to the quantum bouncing point, the bouncing interpretation is not the only possible explanation, as a universe can be created from nothing (Figure 8 [26]). Furthermore, in certain parameter regimes, a Euclidean wormhole solution is possible. Accordingly, there is an ambiguity in defining the arrow of time around the quantum bouncing point; either a contracting phase bounces to an expanding phase or two expanding universes are created via a Euclidean wormhole solution.

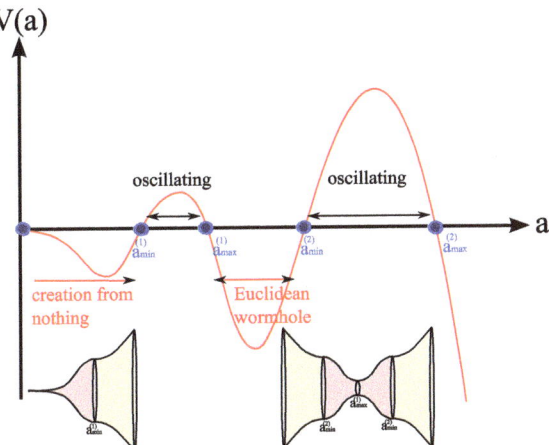

Figure 8. A conceptual interpretation of $\mathcal{V}(a)$ [26].

4.4. Fuzzy Instantons in Anti-De Sitter Space

Finally, we report a number of discussions of the anti-de Sitter space. Let us consider the following potential:

$$V(\Phi) = V_0\left(-1 - \frac{1}{2}\mu^2\Phi^2 + \lambda\Phi^4\right). \tag{71}$$

In the Euclidean domain, it is not surprising to have a complex-valued solution. Thus, we consider the following pure imaginary field: $\Phi \to i\phi$. Then, the potential is effectively

$$U(\phi) = V_0\left(-1 + \frac{1}{2}\mu^2\phi^2 + \lambda\phi^4\right), \tag{72}$$

while the kinetic term has an opposite sign. Therefore, it is possible to determine a solution according to which the scalar field asymptotically approaches zero while there may exist a throat at the center [35].

One potential issue is whether the Euclidean action is well-defined or not. As the volume of Euclidean anti-de Sitter space is infinite, the Euclidean action itself is infinite. However, by subtracting to the pure anti-de Sitter background, one may obtain a finite action difference. The sufficient condition to obtain a finite action difference is that the field should approach zero sufficiently quickly near the infinity [36]. This can be achieved if we tune the shape of the potential [35].

If this finite action difference is allowed, this instanton can explain a case of tunneling from two separate anti-de Sitter spaces to a connected anti-de Sitter wormhole after the Wick-rotation (Figure 9 [35]). This solution satisfies the classicality at the time-like infinity. One potential question is whether any effects from the fuzzy core of the solution can reach a future infinity or not. However, in principle, this solution is embedded in the Euclidean path integral formalism. Therefore, we can very easily extend this technique to anti-de

Sitter fuzzy Euclidean wormholes in a black hole background. After the Wick-rotation, this explains a Lorentzian (probably unstable) wormhole in anti-de Sitter space. The existence of this structure can then cause conceptual trouble with the ER = EPR conjecture [37]. We leave these interesting connections to the information loss paradox for future research [38,39].

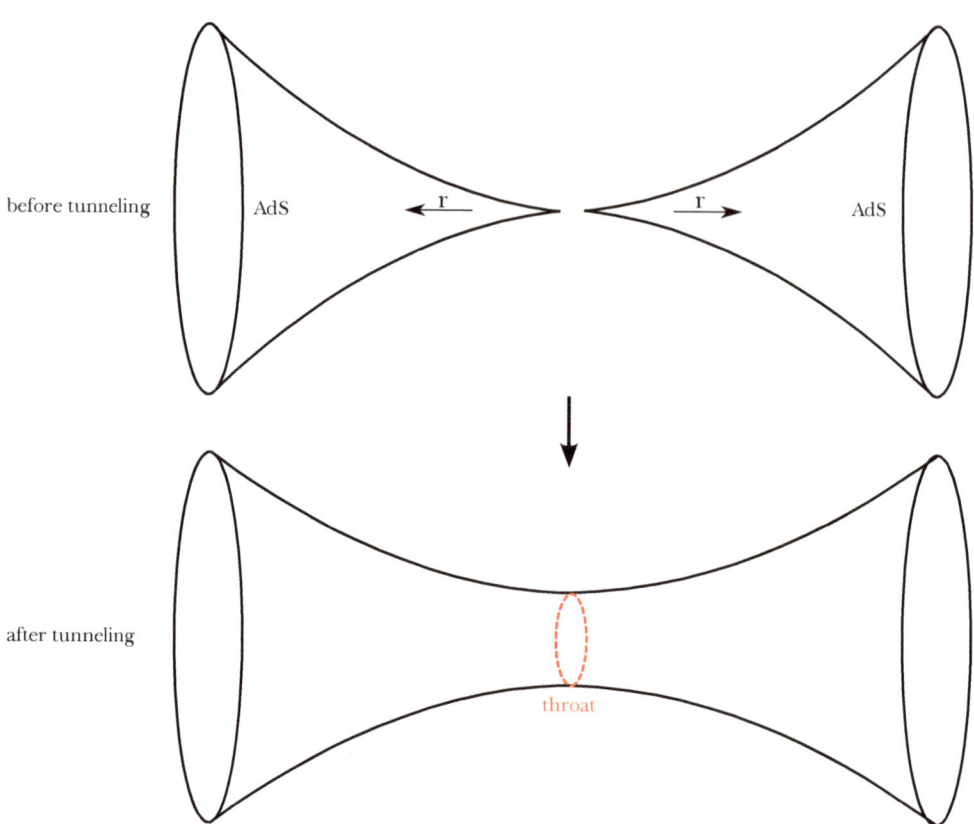

Figure 9. Conceptual picture of tunneling in anti-de Sitter space [35].

5. Future Perspectives

In this study, we discussed various aspects of the fuzzy instantons of the Hartle–Hawking wave function.

First, we examined the slow-roll inflation models. Due to the fuzzy instanton analysis and classicality condition, the universe should experience a small amount of inflation. However, the amount is not sufficient, and we require several routes to rescue the Hartle–Hawking wave function to fit the observations. Perhaps the most natural approach is to introduce a massive field and impose the classicalization of all matter fields.

Second, we investigated the cases in which the slow-roll conditions break down. This is very natural from the point of view of a UV-completed theory such as string theory. However, a classical universe can be created even in these cases. We named these new types of solutions HSY instantons.

Third, we extended this to divergent situations, e.g., the Euclidean path integral formalism does not necessarily indicate that there is only one boundary. In principle, there can be two boundaries. This case is related to fuzzy Euclidean wormholes, which in de Sitter space can explain the preference of a concave inflaton potential. Fuzzy Euclidean wormholes in anti-de Sitter space may be related to the information loss paradox, in addition to a possible criticism of the ER=EPR conjecture. Furthermore, the no-boundary wave function can be applied to string-inspired models or loop quantum cosmology models. These models allow for Euclidean wormhole solutions.

There are several interesting possible directions for future research:

1. Traditionally, we assumed cosmological landscapes and considered slow-roll inflation models. However, in recent discussions, models that are consistent with the swampland criteria might be more interesting. The Hartle–Hawking wave function is definitely useful for both problems, though it might be more interesting to provide possible observational consequences [40,41] that reveal issues related to the swampland criteria.
2. Fuzzy Euclidean wormholes can be realized in various systems, but the application of the associated techniques might be complicated beyond Einstein gravity. This might include the Gauss–Bonnet-dilaton gravity model or the loop quantum cosmological model. Some fuzzy extensions of oscillating instantons are interesting [42]. In any case, this will be a challenging topic.
3. Fuzzy instantons in anti-de Sitter backgrounds or black hole backgrounds are another interesting topic. This issue may cover a number of topics regarding Hawking radiation [43] as well as the information loss problem [44]. However, it is fair to say that it is not easy to impose the classicality condition at a future infinity if the symmetry is less than the $O(4)$-symmetry. The generalization of dynamical instantons in spherical symmetry will be an important topic.

In conclusion, the Euclidean path integral is approximated well by instantons. If the instantons are dynamical, they must be fuzzy or complexified. An investigation of dynamical wormholes is a challenging and fruitful future research topic. This is necessarily related to the study of fuzzy instantons not only in the context of cosmology but also in black hole physics. We leave these fascinating topics for future research.

Funding: This research received no external funding.

Acknowledgments: This work is supported by the National Research Foundation of Korea (Grant no.:2021R1C1C1008622, 2021R1A4A5031460).

Conflicts of Interest: The author declares no conflict of interest.

References

1. Hawking, S.W.; Penrose, R. The Singularities of gravitational collapse and cosmology. *Proc. R. Soc. Lond. A* **1970**, *314*, 529–548.
2. DeWitt, B.S. Quantum Theory of Gravity. 1. The Canonical Theory. *Phys. Rev.* **1967**, *160*, 1113–1148. [CrossRef]
3. Vilenkin, A. Quantum Cosmology and the Initial State of the Universe. *Phys. Rev. D* **1988**, *37*, 888. [CrossRef] [PubMed]
4. Halliwell, J.J.; Hawking, S.W. The Origin of Structure in the Universe. *Phys. Rev. D* **1985**, *31*, 1777. [CrossRef]
5. Hartle, J.B.; Hawking, S.W. Wave Function of the Universe. *Phys. Rev. D* **1983**, *28*, 2960–2975. [CrossRef]
6. Halliwell, J.J.; Hartle, J.B. Integration Contours for the No Boundary Wave Function of the Universe. *Phys. Rev. D* **1990**, *41*, 1815. [CrossRef]
7. Hartle, J.B.; Hawking, S.W.; Hertog, T. The Classical Universes of the No-Boundary Quantum State. *Phys. Rev. D* **2008**, *77*, 123537. [CrossRef]
8. Hartle, J.B.; Hawking, S.W.; Hertog, T. No-Boundary Measure of the Universe. *Phys. Rev. Lett.* **2008**, *100*, 201301. [CrossRef]
9. Janssen, O. Slow-roll approximation in quantum cosmology. *Class. Quant. Gravity* **2021**, *38*, 095003. [CrossRef]
10. Hwang, D.; Lee, B.H.; Stewart, E.D.; Yeom, D.; Zoe, H. Euclidean quantum gravity and stochastic inflation. *Phys. Rev. D* **2013**, *87*, 063502. [CrossRef]
11. Hwang, D.; Kim, S.A.; Yeom, D. No-boundary wave function for two-field inflation. *Class. Quant. Gravity* **2015**, *32*, 115006. [CrossRef]
12. Chen, P.; Qiu, T.; Yeom, D. Phantom of the Hartle–Hawking instanton: Connecting inflation with dark energy. *Eur. Phys. J. C* **2016**, *76*, 91. [CrossRef]
13. Hwang, D.; Kim, S.A.; Lee, B.H.; Sahlmann, H.; Yeom, D. No-boundary measure and preference for large e-foldings in multi-field inflation. *Class. Quant. Grav.* **2013**, *30*, 165016. [CrossRef]
14. Feldbrugge, J.; Lehners, J.L.; Turok, N. Lorentzian Quantum Cosmology. *Phys. Rev. D* **2017**, *95*, 103508. [CrossRef]
15. Dorronsoro, J.D.; Halliwell, J.J.; Hartle, J.B.; Hertog, T.; Janssen, O. Real no-boundary wave function in Lorentzian quantum cosmology. *Phys. Rev. D* **2017**, *96*, 043505. [CrossRef]

16. Hwang, D.; Yeom, D. Toward inflation models compatible with the no-boundary proposal. *J. Cosmol. Astropart. Phys.* **2014**, *2014*, 7. [CrossRef]
17. Sasaki, M.; Yeom, D.; Zhang, Y.L. Hartle-Hawking no-boundary proposal in dRGT massive gravity: Making inflation exponentially more probable. *Class. Quant. Gravity* **2013**, *30*, 232001. [CrossRef]
18. Zhang, Y.L.; Sasaki, M.; Yeom, D. Homogeneous Instantons in Bigravity. *J. High Energy Phys.* **2015**, *4*, 16. [CrossRef]
19. Susskind, L. The Anthropic landscape of string theory. *arXiv* **2003**, arXiv:hep-th/0302219.
20. Obied, G.; Ooguri, H.; Spodyneiko, L.; Vafa, C. De Sitter Space and the Swampland. *arXiv* **2018**, arXiv:1806.08362.
21. Brahma, S.; Brandenberger, R.; Yeom, D. Swampland, Trans-Planckian Censorship and Fine-Tuning Problem for Inflation: Tunnelling Wavefunction to the Rescue. *JCAP* **2020**, *2020*, 37. [CrossRef]
22. Hwang, D.; Lee, B.H.; Sahlmann, H.; Yeom, D. The no-boundary measure in string theory: Applications to moduli stabilization, flux compactification, and cosmic landscape. *Class. Quant. Gravity* **2012**, *29*, 175001. [CrossRef]
23. Hwang, D.; Sahlmann, H.; Yeom, D. The No-boundary measure in scalar-tensor gravity. *Class. Quant. Gravity* **2012**, *29*, 095005. [CrossRef]
24. Chen, P.; Hu, Y.C.; Yeom, D. Fuzzy Euclidean wormholes in de Sitter space. *JCAP* **2017**, *2017*, 1. [CrossRef]
25. Chew, X.Y.; Tumurtushaa, G.; Yeom, D. Euclidean wormholes in Gauss–Bonnet-dilaton gravity. *Phys. Dark Univ.* **2021**, *32*, 100811. [CrossRef]
26. Brahma, S.; Yeom, D. No-boundary wave function for loop quantum cosmology. *Phys. Rev. D* **2018**, *98*, 083537. [CrossRef]
27. Chen, P.; Yeom, D. Why concave rather than convex inflaton potential? *Eur. Phys. J. C* **2018**, *78*, 863. [CrossRef]
28. Chen, P.; Ro, D.; Yeom, D. Fuzzy Euclidean wormholes in the inflationary universe. *Phys. Dark Univ.* **2020**, *28*, 100492. [CrossRef]
29. Starobinsky, A.A. A New Type of Isotropic Cosmological Models Without Singularity. *Phys. Lett. B* **1980**, *91*, 99–102. [CrossRef]
30. Akrami, Y.; Arroja, F.; Ashdown, M.; Aumont, J.; Baccigalupi, C.; Ballardini, M.; Banday, A.J.; Barreiro, R.B.; Bartolo, N.; Basak, S. et al. Planck 2018 results. X. Constraints on inflation. *Astron. Astrophys.* **2020**, *641*, A10.
31. Kanti, P.; Kleihaus, B.; Kunz, J. Wormholes in Dilatonic Einstein–Gauss–Bonnet Theory. *Phys. Rev. Lett.* **2011**, *107*, 271101. [CrossRef] [PubMed]
32. Tumurtushaa, G.; Yeom, D. Quantum creation of traversable wormholes ex nihilo in Gauss–Bonnet-dilaton gravity. *Eur. Phys. J. C* **2019**, *79*, 488. [CrossRef]
33. Bojowald, M.; Brahma, S.; Yeom, D. Effective line elements and black-hole models in canonical loop quantum gravity. *Phys. Rev. D* **2018**, *98*, 046015. [CrossRef]
34. Brahma, S.; Yeom, D. On the geometry of no-boundary instantons in loop quantum cosmology. *Universe* **2019**, *5*, 22. [CrossRef]
35. Kang, S.; Yeom, D. Fuzzy Euclidean wormholes in anti–de Sitter space. *Phys. Rev. D* **2018**, *97*, 124031. [CrossRef]
36. Kanno, S.; Sasaki, M.; Soda, J. Destabilizing Tachyonic Vacua at or above the BF Bound. *Prog. Theor. Phys.* **2012**, *128*, 213–226. [CrossRef]
37. Maldacena, J.; Susskind, L. Cool horizons for entangled black holes. *Fortsch. Phys.* **2013**, *61*, 781–811. [CrossRef]
38. Chen, P.; Wu, C.H.; Yeom, D. Broken bridges: A counter-example of the ER=EPR conjecture. *JCAP* **2017**, *6*, 40. [CrossRef]
39. Chen, P.; Sasaki, M.; Yeom, D. A path(-integral) toward non-perturbative effects in Hawking radiation. *Int. J. Mod. Phys. D* **2020**, *29*, 2050086. [CrossRef]
40. Chen, P.; Lin, Y.H.; Yeom, D. Suppression of long-wavelength CMB spectrum from the no-boundary initial condition. *Eur. Phys. J. C* **2018**, *78*, 930. [CrossRef]
41. Chen, P.; Yeh, H.H.; Yeom, D. Suppression of the long-wavelength CMB spectrum from the Hartle–Hawking wave function in the Starobinsky-type inflation model. *Phys. Dark Univ.* **2020**, *27*, 100435. [CrossRef]
42. Lee, B.H.; Lee, W.; Yeom, D. Oscillating instantons as homogeneous tunneling channels. *Int. J. Mod. Phys. A* **2013**, *28*, 1350082. [CrossRef]
43. Chen, P.; Sasaki, M.; Yeom, D. Hawking radiation as instantons. *Eur. Phys. J. C* **2019**, *79*, 627. [CrossRef]
44. Sasaki, M.; Yeom, D. Thin-shell bubbles and information loss problem in anti de Sitter background. *J. High Energy Phys.* **2014**, *12*, 155. [CrossRef]

Review

An Overview on the Nature of the Bounce in LQC and PQM

Gabriele Barca [1,*], Eleonora Giovannetti [1] and Giovanni Montani [1,2]

1. Department of Physics, "Sapienza" University of Rome, P.le Aldo Moro, 00185 Roma, Italy; giovannetti.1612404@studenti.uniroma1.it (E.G.); giovanni.montani@enea.it (G.M.)
2. Fusion and Nuclear Safety Department, European Nuclear Energy Agency (ENEA), C.R. Frascati, Via E. Fermi, 00044 Frascati, Italy
* Correspondence: gabriele.barca@uniroma1.it

Abstract: We present a review on some of the basic aspects concerning quantum cosmology in the presence of cut-off physics as it has emerged in the literature during the last fifteen years. We first analyze how the Wheeler–DeWitt equation describes the quantum Universe dynamics, when a pure metric approach is concerned, showing how, in general, the primordial singularity is not removed by the quantum effects. We then analyze the main implications of applying the loop quantum gravity prescriptions to the minisuperspace model, i.e., we discuss the basic features of the so-called loop quantum cosmology. For the isotropic Universe dynamics, we compare the original approach, dubbed the μ_0 scheme, and the most commonly accepted formulation for which the area gap is taken as physically scaled, i.e., the so-called $\bar{\mu}$ scheme. Furthermore, some fundamental results concerning the Bianchi Universes are discussed, especially with respect to the morphology of the Bianchi IX model. Finally, we consider some relevant criticisms developed over the last ten years about the real link existing between the full theory of loop quantum gravity and its minisuperspace implementation, especially with respect to the preservation of the internal $SU(2)$ symmetry. In the second part of the review, we consider the dynamics of the isotropic Universe and of the Bianchi models in the framework of polymer quantum mechanics. Throughout the paper, we focus on the effective semiclassical dynamics and study the full quantum theory only in some cases, such as the FLRW model and the Bianchi I model in the Ashtekar variables. We first address the polymerization in terms of the Ashtekar–Barbero–Immirzi connection and show how the resulting dynamics is isomorphic to the μ_0 scheme of loop quantum cosmology with a critical energy density of the Universe that depends on the initial conditions of the dynamics. The following step is to analyze the polymerization of volume-like variables, both for the isotropic and Bianchi I models, and we see that if the Universe volume (the cubed scale factor) is one of the configurational variables, then the resulting dynamics is isomorphic to that one emerging in loop quantum cosmology for the $\bar{\mu}$ scheme, with the critical energy density value being fixed only by fundamental constants and the Immirzi parameter. Finally, we consider the polymer quantum dynamics of the homogeneous and inhomogeneous Mixmaster model by means of a metric approach. In particular, we compare the results obtained by using the volume variable, which leads to the emergence of a singularity- and chaos-free cosmology, to the use of the standard Misner variable. In the latter case, we deal with the surprising result of a cosmology that is still singular, and its chaotic properties depend on the ratio between the lattice steps for the isotropic and anisotropic variables. We conclude the review with some considerations of the problem of changing variables in the polymer representation of the minisuperspace dynamics. In particular, on a semiclassical level, we consider how the dynamics can be properly mapped in two different sets of variables (at the price of having to deal with a coordinate dependent lattice step), and we infer some possible implications on the equivalence of the μ_0 and $\bar{\mu}$ scheme of loop quantum cosmology.

Keywords: quantum cosmology; loop quantum cosmology; polymer quantum mechanics; bounce

Citation: Barca, G.; Giovannetti, E.; Montani, G. An Overview on the Nature of the Bounce in LQC and PQM. *Universe* **2021**, *7*, 327. https://doi.org/10.3390/universe7090327

Academic Editor: Paulo Vargas Moniz

Received: 2 August 2021
Accepted: 23 August 2021
Published: 1 September 2021

Publisher's Note: MDPI stays neutral with regard to jurisdictional claims in published maps and institutional affiliations.

Copyright: © 2021 by the authors. Licensee MDPI, Basel, Switzerland. This article is an open access article distributed under the terms and conditions of the Creative Commons Attribution (CC BY) license (https://creativecommons.org/licenses/by/4.0/).

1. Introduction

Despite the fact that no self-consistent theory has been developed in quantum gravity (for the most interesting approaches, see [1–13]), along the years, the arena of primordial cosmology has constituted a valuable test to estimate the predictivity of the proposed theories on the birth of the Universe and quantum evolution [14].

The most significant change in the point of view on how to approach the quantization of the gravitational degrees of freedom took place with the formulation of the so-called loop quantum gravity (LQG) [11], especially because this formulation was able to construct a kinematical Hilbert space and to justify spontaneously the emergence of discrete area and volume spectra. LQG relies on the possibility to reduce the gravitational phase space to that of a $SU(2)$ non-Abelian theory [6–9], and then the quantization scheme is performed by using "smeared" (non-local) variables, such as the holonomy and flux variables, as suggested by the original Wilson loop formulation and by non-Abelian gauge theories on a lattice. Indeed, when adopting Astekar–Barbero–Immirzi (first order) variables, the invariance of the gravitational action under the local rotation of the triad adapted to the spacetime foliation is expressed in the form of a Gauss constraint.

The implementation of this new approach to the cosmological setting leads to define, in a rather rigorous, mathematical way, the concept of a primordial Big Bounce, already hypothesized in the seventies. However, the cosmological implementation of LQG, commonly dubbed loop quantum cosmology (LQC), has the intrinsic limitation that the basic $SU(2)$-symmetry underlying the LQG formulation is unavoidably lost [15,16] when the minisuperspace dynamics is addressed. This is due to the fact that the homogeneity constraint reduces the cosmological problem to a finite number of degrees of freedom; in particular, it becomes impossible to perform the local rotation and preserve the structure constants of the Lie algebra associated to the specific isometry group. In this respect, we could say that LQC requires a sort of gauge fixing of the full $SU(2)$ invariance; see the analysis in [17], where this question is explicitly addressed. In addition, the problem of translating the quantum constraints from the full to the reduced level remains still open [11].

Despite these limitations, LQC remains an interesting attempt to regularize the cosmological singularity, opening a new perspective on the origin and evolution of the Universe. Furthermore, the so-called "effective formulation" of LQC is isomorphic to the implementation of polymer quantum mechanics (PQM) [18–21] to the minisuperspace variables, typically the Universe scale factors. This correspondence allows to investigate some features of the LQC formulation by applying simplified formalisms to more complicated models, thus making them viable [22–45].

Here, we provide a review of basic well-established results of LQC and more recent analyses in polymer quantum cosmology, with the purpose of better outlining the reliable achievements and the open questions in this sector of the quantum cosmological problem formulation. In order to better compare it with LQC, when presenting the cosmological implementation of PQM, we mainly focus on its effective dynamics, except for the less involved models, where a full quantum analysis is possible. In particular, two recent studies [44,45] applied the polymer framework to the formulation of the flat Friedmann–Lemaître–Robertson–Walker (FLRW) and Bianchi I models, respectively. The peculiarities of these two analyses lie in the comparison between the polymer quantum dynamics in terms of the Ashtekar–Barbero–Immirzi connections and in terms of volume-like configurational variables.

In this review, we highlight how the evolution of the quantum Universe is sensitive to the considered set of configurational variables: when real connections are polymerized, the resulting picture resembles that which is commonly dubbed the μ_0 scheme of LQC [46–48]; on the other hand, the use of volume coordinates can be associated to the so-called $\bar{\mu}$ scenario [49–51]. In LQC, the difference in these two schemes is due to the cosmological implementation of the area element as a kinematical or a dynamical quantity (in the $\bar{\mu}$

scenario the area gap is rescaled for the momentum variable, i.e., the squared cosmic scale factor).

Actually, the μ_0 and $\bar{\mu}$ schemes lead to very different pictures of the primordial quantum Universe: both correspond to a Big Bounce, but, while in the $\bar{\mu}$ scheme the critical energy density is fixed by fundamental constants and the Immirzi parameter only, in the μ_0 dynamics it depends on the initial conditions for the wave packets. The fact that these two very different representations of the early Universe are associated with the polymerization of the two different sets of variables cited above offers an intriguing perspective to better interpret the real physics of the two scenarios and shows how PQM could shed light on the possible shortcomings of the LQC formalism.

We will also provide an interesting comparison of the cosmological implementation of the PQM in the metric representation. In particular, we will compare the analyses in [34,38,42], where the homogeneous and inhomogeneous Mixmaster dynamics is studied through the polymerization of the standard Misner isotropic variable α and of the Universe volume, respectively: the difference is only in using or not using a logarithm in the definition of the isotropic variable, but the implication is very deep since only the volume representation ensures a bouncing cosmology. Furthermore, questions concerning the chaotic or non-chaotic nature of the semiclassical Bianchi IX dynamics are addressed in some detail.

Then, we will further present a coherent and detailed discussion of the relations existing between LQC and polymer quantum cosmology, also discussing some of the most relevant open questions, especially concerning the equivalence or non-equivalence of the resulting dynamics in different sets of configurational variables.

We would like to stress that in this work, we will not consider the basic problem of the implementation of the Copenhagen School interpretation to the Universe quantum dynamics, common to all quantum cosmological formulations. We will briefly present this issue because we think that it is important to keep in mind such difficulties since they could perhaps drive the investigation toward a fully consistent theory of quantum gravity and, hence, quantum cosmology. However, our presentation will escape this puzzling basic interpretative question, as is often implicitly done in the literature. We must remark that there are many other approaches of a different nature that are able to replace the singularity with a Bounce, such as, for example, the Ekpyrotic scenario, massive gravity, and other modified gravity theories; for general reviews of these models, see [52,53].

The paper is organized as follows. In Section 2, we introduce general features of the cosmological dynamics. We present the difficulties in implementing the Copenhagen interpretation to cosmology, then we describe the classical dynamics of homogeneous models (both isotropic and anisotropic) with some attention to the problem of time and to the definition of a cosmological clock. In Section 3, we present LQC: first, we briefly summarize the features of LQG that are relevant for its cosmological implementation; then, we discuss in detail the two formulations of isotropic LQC that are the μ_0 and $\bar{\mu}$ schemes, and also the implementation of the latter to the anisotropic sector; and finally, we conclude with a summary of critics and shortcomings that show the need for a different quantum mechanical approach to cosmology. In Section 4, we present polymer cosmology: we first introduce the formalism of polymer quantum mechanics, and then we focus on its implementation to both the isotropic Universe and the anisotropic Bianchi models in different sets of variables (namely, the Ashtekar variables, the volume-like variables and the Misner-like ones). We conclude this section with a discussion on the results obtained with different sets and also present a possible way to recover an equivalence between them. Finally, in Section 5 we summarize the review and provide some final remarks. Throughout all the paper, we use the natural units $8\pi G = c = \hbar = 1$.

2. Cosmological Quantum Dynamics

The first attempt to implement the canonical quantum gravity approach developed in [1–3] to the cosmological setting was due to the analysis proposed in [54], where the

Bianchi IX Universe was studied within certain approximations, and the most relevant properties of its quantum dynamics were elucidated (for extensions of this approach to generic inhomogeneous models, see [55–62], and for a refined numerical study of the Bianchi IX quantization, see [63]).

Before entering some technical aspects of canonical quantum cosmology in the metric formulation, it is mandatory to fix our attention to some intrinsic conceptual difficulties that we meet on the interpretative level, even if we could assume the construction of a Hilbert space and the determination of a suitable time variable to describe the quantum dynamics as solved.

The standard interpretation of canonical quantum mechanics is due to the so-called Copenhagen school, which postulated some general prescriptions, validated by the analysis of atomic and molecular spectra and was never contradicted by experiments in modern relativistic particle physics. We briefly summarize here the Copenhagen school interpretation via its main statements (very difficult to implement in quantum cosmology).

- The concept of probability to find a physical system into a given state is the "large numbers" limit of the frequency by which that state is registered in repeated experiments.
- The "measure" operation on a given quantum system must be performed by a classical (or better, quasi-classical) observer, who induces a "collapse" of the wave function into a specific eigenstate by physically interacting with the quantum environment.

When referred to the cosmological setting, both the statements above have a very critical implementation. In fact, on one hand, we observe only one realization of the Universe, and no frequency approximating the probability can be determined; on the other hand, in a quantum Universe, it appears impossible (or at least ambiguous) to speak of a quasi-classical observer. The possibility to recover both the concepts postulated above would require that at least a portion of the Universe be in a quasi-classical state, so that the interaction of these degrees of freedom with the fully quantum ones offers an arena to recover the basic notion of the Copenhagen school interpretation (see, in this respect, the Universe wave function interpretation provided in [64], where such a picture is investigated).

We conclude by observing that the classical portion of the Universe mentioned above cannot be identified with the present classical Universe thought of as an "observer" of the primordial quantum phases. This claim is supported by the following two considerations.

- The information we receive from the quantum Universe in the Planck regime (mediated by the physics of the cosmic microwave background radiation (CMB)) is already a single classical determination of the quantum system, among all the possible ones.
- That information cannot be induced by a direct measurement on the primordial Universe, simply because it lives in our past light cone, and no physical interaction between our classical apparatus and the Planckian Universe can take place (even if we were able to detect photons directly emitted in the quantum phase).

Thus, in what follows, we will think of the Universe wave function as if it were associated to physical notions in principle, according to the Copenhagen school interpretation, without entering further into how it can be really demonstrated, or which alternative interpretation could be addressed.

2.1. The Isotropic Universe

The Robertson–Walker (RW) geometry describing the isotropic Universe is a very simple model, which has only one dynamical degree of freedom due to the high level of symmetry. If on a classical level its employment is well justified by a large number of phenomenological evidences (above all, the isotropy of the CMB temperature), on a quantum level, it appears very close to be just a "toy model" deprived of many basic features that more general cosmological models outline. We elucidate the reliability of this apparently strong claim in this subsection and in the following one.

The main failure of the canonical quantum cosmology as depicted by the Wheeler–DeWitt (WDW) equation in the metric approach is that no removal of the initial singularity emerges in general when the nature of the Universe wave function is elucidated. Let us now develop some simple technical considerations for the isotropic Universe (for a pioneering analysis, see [65]), limiting our attention to the spatially flat model and adopting as a configurational variable the cubed cosmic scale factor $v(t) = a^3(t)$ (for example, the Universe volume).

In the Arnowitt–Deser–Misner (ADM) formulation [66], the RW line element reads as follows:

$$ds^2 = N^2 dt^2 - v^{\frac{2}{3}}\left(dx^2 + dy^2 + dz^2\right), \qquad (1)$$

where we set the speed of light equal to one, and $N = N(t)$ denotes the so-called lapse function. The action describing the Hamiltonian dynamics of the isotropic FLRW model takes the following form:

$$S_{\text{FLRW}} = \int dt (P_v \dot{v} - N\mathcal{C}_{\text{FLRW}}), \qquad (2)$$

where we have set the space integration on a fiducial volume to unity, P_v is the conjugate momentum to v, and the super-Hamiltonian $\mathcal{C}_{\text{FLRW}}$ reads as follows:

$$\mathcal{C}_{\text{FLRW}} \equiv -\frac{9}{4} v P_v^2 + \frac{1}{3} v \rho(v). \qquad (3)$$

When the equation of state for the cosmological fluid takes the form $P = w\rho$ (with P being the pressure and w a constant parameter), the matter energy density ρ reads as follows:

$$\rho(v) = \frac{\rho_0^{(w)}}{v^{1+w}}, \qquad (4)$$

with $\rho_0^{(w)} > 0$. Clearly, varying the action with respect to N, we obtain the constraint $\mathcal{C}_{\text{FLRW}} = 0$, which reduces to the Friedmann equation for the isotropic Universe, using the Hamilton equation $\dot{v} = -9v P_v/2$ to express the momentum P_v. The existence of a Hamiltonian constraint reflects the possibility to freely choose the time variable (i.e., the form of the lapse function) to describe the system dynamics, according to the general relativity principle.

The Dirac prescription for the canonical quantization of a constrained theory consists of implementing the phase space variables to canonical operators [67], leading to the following WDW equation for the isotropic Universe (3):

$$\partial_v^2 \psi(v) + \frac{4}{27} \rho(v) \psi(v) = 0, \qquad (5)$$

in which we adopt the natural operator ordering, where momenta are always at the right of coordinate variables. This equation clearly resembles a time-independent Schrödinger equation in the space-like coordinate v, and no evolution emerges for the Universe wave function $\psi(v)$.

If we instead adopt the symmetric operator ordering of [68], i.e., $v P_v^2 \to \hat{P}_v v \hat{P}_v$, and introduce the variable $\xi \equiv \ln v$, we arrive to an equation of the following form:

$$\partial_\xi^2 \psi = -\frac{4}{27} \rho_0^{(w)} e^{-(w-1)\xi} \psi. \qquad (6)$$

For the relevant cosmological case of a "stiff matter", corresponding to $w = 1$ and de facto mimicking a massless free scalar field (the kinetic component of an inflaton field), we obtain the simple solutions as follows:

$$\psi(v) \propto \exp\left(\pm i \sqrt{\frac{4\rho_0^{(1)}}{27}} \ln v \right). \tag{7}$$

Indeed, the potential term of many inflationary models can be neglected at the high temperatures of the Planckian regime [69,70] (we recall that the transition phase responsible for inflation takes place in a classical Universe). The stiff matter is the most rapidly increasing contribution allowed by a causal fluid when the zero volume limit is approached, and it is therefore expected to dominate during the Planckian era. However, we stress that the singularity is also present for all natural values of the parameter w [14,70].

The classical Universe has a singular behavior for $v \to 0$ (the Big Bang singularity), and it regularly expands indefinitely for $v \to \infty$; here, we see that the Universe wave function singles out a qualitatively similar behavior in these two different regimes. Thus, no indications emerge from the WDW equation about the singularity removal, and this turns out to be a general feature of the canonical metric approach.

2.2. Internal Clock

It is clear that it is not possible to construct a Hilbert space for the isotropic Universe discussed above, and therefore, any precise notion of probability density is forbidden in the absence of a well-defined time variable.

In general, any component of a gravitational system could be identified as a time variable for the classical dynamics, as soon as a specific time gauge is assigned. Such a concept can be retained also at the quantum level in a fully covariant form. Indeed, in the WDW equation, it is possible to identify a given internal degree of freedom as a "relational time", by promoting it to the role of a physical clock for the quantum evolution of the remaining gravitational or matter degrees of freedom. The most natural relational time for the isotropic Universe dynamics, and in general for quantum cosmology, is a free massless scalar field $\phi = \phi(v)$, which is expected to be present in the primordial phases of the Universe because of the inflationary paradigm; its energy density increases as $\sim v^{-2}$ toward the singularity, which is the fastest growth allowed before the fluid acquires a superluminal sound speed.

If we replace the stiff matter in Equation (6) with the energy density of ϕ, i.e.,

$$\rho_\phi = \frac{P_\phi^2}{2v^2} \tag{8}$$

(P_ϕ being the conjugate momentum to ϕ), then we arrive to the following $1 + 1$ Klein–Gordon-like equation:

$$\left(\partial_\zeta^2 - \frac{2}{27}\partial_\phi^2\right)\psi(\zeta,\phi) = 0. \tag{9}$$

The general solution of this equation reads as follows:

$$\psi = \int dk_\zeta A(k_\zeta) \exp\left\{ik_\zeta\left(\zeta \pm \sqrt{\frac{27}{2}}\phi\right)\right\}. \tag{10}$$

Now, it is possible to adopt ϕ as a physical clock for the Universe dynamics, so we can construct localized wave packets of the form (10), for instance with a Gaussian weight function $A(k_\zeta)$. If we compare the peak of the Klein–Gordon probability density $i(\psi^*\partial_\phi\psi - \psi\partial_\phi\psi^*)$ with the classical trajectory $v = v(\phi)$, it is easy to check that there is a very good correspondence, leading to the fact that the singularity is not removed by the canonical quantization of the model.

Thus, the introduction of the concept of a relational internal matter clock provides a good solution to the problem of time since the zero-eigenvalue Schrödinger equation can be interpreted as a Klein–Gordon-like operator in the configurational space. The similarities of the relativistic case and the present WDW equation allow to define a conserved probability density, which retains its positive nature when it is possible to perform the frequency separation (violated when a non-zero potential for the scalar field ϕ is present).

We also observe that there exists a clear correspondence between the quantity ρ_0^1 in Equation (6) and the quantum number k_ξ, namely, the following:

$$\rho_0^1 \equiv \frac{27}{4} k_\xi^2. \tag{11}$$

We conclude by observing that the considerations above regarding the absence of a singularity removal when comparing the classical and quantum evolution are particularly reliable in the present case, in view of the linearity of the dispersion relation for the $1+1$ Klein–Gordon-like equation. Such a property allows to construct localized non-spreading wave packets up to the initial singularity. It is immediate to realize (see below) that a linear dispersion relation is a feature that clearly does not survive when a higher dimensional problem is faced.

2.3. The Bianchi Universes

A better understanding of the minisuperspace formulation of canonical quantum gravity in the metric approach is provided by the investigation of the Bianchi Universes. These models generalize the isotropic Universe by preserving the homogeneity constraint and allowing for three different independent scale factors along the three spatial directions.

The ADM line element of the Bianchi Universes reads as follows:

$$ds^2 = N^2 dt^2 - e^{2\alpha} \left(e^{2\beta}\right)_{ab} \sigma^a \sigma^b, \tag{12}$$

where the variable α is related to the Universe volume v by the relation $\alpha = (1/3) \ln v$, the matrix β parametrizes the anisotropies and has the diagonal form $\beta = \mathrm{diag}(\beta_+ + \sqrt{3} \beta_-, \beta_+ - \sqrt{3} \beta_-, -2\beta_-)$, while σ_a and σ_b are the 1-forms describing the specific isometry group under which that Bianchi model is invariant. The variables $(\alpha, \beta_\pm)$ are known as Misner variables, and their usefulness lies in the fact that they make the kinetic term in the Hamiltonian diagonal.

The homogeneity of the space allows to deal with the functions N, α, β_+ and β_- as depending on time only. The isotropic limit is recovered for $\beta_+ \equiv \beta_- \equiv 0$, and it is possible only for the three Bianchi models of type I, type V and type IX, corresponding to the flat, negatively and positively curved FLRW model respectively.

The action of the Bianchi Universes in vacuum reads as follows:

$$S_B = \int dt \left(P_\alpha \dot{\alpha} + P_+ \dot{\beta}_+ + P_- \dot{\beta}_- - N\mathcal{C}_B\right), \tag{13}$$

with

$$\mathcal{C}_B \equiv \frac{e^{-3\alpha}}{3(8\pi)^2} \left[-P_\alpha^2 + P_+^2 + P_-^2 + 3(4\pi)^4 e^{4\alpha} U_B(\beta_+, \beta_-)\right]. \tag{14}$$

Above, we set to unity the space integral on the fiducial volume and denoted the conjugate momenta to the corresponding variables α, β_+ and β_- with P_α, P_+ and P_- respectively. The potential U_B is provided by the spatial curvature of the specific model, and it is identically zero for the Bianchi I model only. It is immediate to recognize that the WDW equation for the Bianchi Universes takes the following form:

$$\left[\partial_\alpha^2 - \partial_{\beta_+}^2 - \partial_{\beta_-}^2\right]\psi + 3(4\pi)^4 e^{4\alpha} U_B(\beta_+, \beta_-)\psi = 0, \tag{15}$$

with $\psi = \psi(\alpha, \beta_+, \beta_-)$. In this case, there is no need to add a free massless scalar field to identify an internal time variable since the variable α (related to the three-metric determinant) has a different signature with respect to the two anisotropy degrees of freedom β_+ and β_-. This is a very general feature of the WDW equation, first investigated in [1–3].

It is worth noting that the same signature of α and the same wave equation could be found by using the Universe volume $v = e^{3\alpha}$ and adopting the symmetric operator ordering, as done above in Equation (6). Hence, we can realize how misleading it was to use the volume of the Universe as a space-like coordinate in the isotropic Universe. This interpretation is clearly possible, but as far as we introduce β_+ and β_- (the real physical degrees of freedom of the cosmological gravitational field), we are naturally led to consider v or α as the most natural internal time variables to describe the quantum Universe evolution.

The first classical Hamilton equation $\dot{\alpha} = -2NP_\alpha e^{-3\alpha}$ shows that we must require that $P_\alpha < 0$ in order to deal with the expanding Universe ($\dot{\alpha} > 0$). On the contrary, for the collapsing Universe, i.e., $\dot{\alpha} < 0$, we need $P_\alpha > 0$. However, we note that P_α is a constant of motion only when the potential term is negligible or when it is exactly zero, as in the Bianchi I model.

If we set $U_B \equiv 0$ in the WDW Equation (15), we can easily perform the frequency separation, and we obtain the following general solutions:

$$\psi^\pm(\alpha, \beta_\pm) = \int dk_+ dk_- \, A(k_+, k_-) \, e^{i(k_+\beta_+ + k_-\beta_- \mp \sqrt{k_+^2 + k_-^2}\,\alpha)}, \tag{16}$$

where the suffixes $(+)$ and $(-)$ refer to positive and negative frequency wave functions, respectively.

Since the mean value of the operator $\hat{P}_\alpha$ is negative for the positive frequency solution and positive for the negative one (see the sign of its eigenvalues), we are led to identify the expanding Universe with the positive frequency wave packet and vice-versa for the collapsing one.

Actually, if we consider Gaussian weight $A(k_+, k_-)$, it is possible to construct localized wave packets, both representing the expanding (ψ^+) or collapsing (ψ^-) dynamics of the Universe. The localized states follow the classical trajectories $\beta_+(\alpha)$ and $\beta_-(\alpha)$, so we are naturally led to claim that the initial singularity is not removed by the canonical quantization of the system also for a Bianchi I model. However, now the dispersion relation contains a square root; therefore, it is no longer linear as it was for the isotropic Universe. As a result, the wave packet spreads toward the singularity (for $\alpha \to -\infty$), and the localized state cannot be extrapolated asymptotically. This fact prevents a definitive word on what the initial singularity resembles in such a non-localized picture of the Universe. However, we can surely claim that in the Planckian era, the Universe unavoidably becomes a fully quantum system.

The Bianchi IX Model

The peculiarity of the Bianchi IX model lies in the chaotic dynamics near the singularity; this has earned it the nickname of the *Mixmaster* model.

The explicit form of the potential $U_{BIX}(\beta_\pm)$ is the following:

$$U_{BIX}(\beta_\pm) = 2e^{4\beta_+}\left(\cosh\left(4\sqrt{3}\beta_-\right) - 1\right) - 4e^{-2\beta_+}\cosh\left(2\sqrt{3}\beta_-\right) + e^{-8\beta_+}. \tag{17}$$

The potential walls are steeply exponential and define a closed domain with the symmetry of an equilateral triangle [71]. These walls move outwards while approaching the cosmological singularity due to the term $e^{4\alpha}$ in front of the potential in (14) that increases for $\alpha \to -\infty$.

The implementation of the ADM reduction allows to describe the Mixmaster dynamics by means of the motion of a pinpoint particle, named the point-Universe, moving in the triangular potential well. So, we solve the constraint (14) with respect to the momentum conjugate to the chosen time coordinate, here α, and then we obtain the reduced ADM Hamiltonian:

$$\mathcal{C}_{\text{BIX}}^{\text{ADM}} := -P_\alpha = \sqrt{P_+^2 + P_-^2 + e^{4\alpha} 3 (4\pi)^4 U_{\text{BIX}}(\beta_\pm)}. \tag{18}$$

Because of the steepness of the walls, we can consider the point-Universe as a free particle for most of its motion, except when a rebound against one of the three walls occurs. So, by using the free particle approximation $U_{\text{BIX}}(\beta_\pm) \sim 0$, we can derive the velocity of the point-Universe as follows:

$$\beta' \equiv \sqrt{\beta'^2_+ + \beta'^2_-} = 1. \tag{19}$$

where in this picture, the anisotropies have the role of the coordinates of the point-Universe. On the other hand, it can be shown that the potential walls move outwards with velocity $|\beta'_{\text{wall}}| = \frac{1}{2}$, so a rebound is always possible. In particular, every rebound occurs according to the following reflection law:

$$\frac{1}{2} \sin\left(\theta^i + \theta^f\right) = \sin \theta^i - \sin \theta^f, \tag{20}$$

where θ^i and θ^f are the incidence angle and the reflection one to the potential wall normal, respectively. The maximum incidence angle results to be the following:

$$\theta_{\max} \equiv \arccos\left(\frac{1}{2}\right) = \frac{\pi}{3}, \tag{21}$$

so the point-Universe always experiences a rebound against one of the three potential walls, thanks to the triangular symmetry of the system.

In conclusion, the ADM reduction procedure in the Misner parametrization maps the dynamics of the Mixmaster Universe into the motion of a pinpoint particle inside a closed two-dimensional domain. The particle undergoes an infinite series of rebounds against the potential walls while approaching the singularity, and the motion between two subsequent rebounds is a uniform rectilinear one. Once the particle is reflected off one of the walls, the values assumed by the constants of motion change, as well as, thus, the direction of the particle. This way, the trajectory of the point-Universe assumes all possible directions regardless of the initial conditions, giving rise to the chaotic behavior of the Bianchi IX dynamics near the singularity.

3. Loop Quantum Cosmology

The name *loop quantum cosmology* refers to a specific quantum cosmological model, i.e., the quantization of the FLRW spacetime, according to the methods of LQG [46–51,72–74]. More in general, it is often also used to indicate all cosmological models that are quantized through LQG procedures [75–86]. Note, however, that this implies that LQC is *not* the cosmological sector of LQG: the internal symmetries of the formalism used to derive the Loop quantization of general relativity do not allow the usual reduction of the Wheeler Superspace to the cosmological minisuperspaces. However, it is possible to implement the quantization procedure of LQG to a spacetime that is already reduced to a minisuperspace model; this is, indeed, the scope of LQC.

In this section, we briefly introduce the formalism of kinematical LQG and show in detail its implementation to the isotropic Universe in both the old ("standard") and new ("improved") prescriptions of LQC; we also present the work that was done on the Loop quantization of anisotropic models and then conclude with a short description of critiques and shortcomings.

3.1. Loop Quantum Gravity

LQG was developed in the 1990s [6–8,87–91] and remains today the best attempt at a background-independent quantization of general relativity (GR) (for recent, more comprehensive reviews, see [92,93]). The requirement of background independence calls for a reformulation of GR through new formalisms that allow this quantization process: the formalisms of geometrodynamics and of Gauge theories.

GR was reformulated as a $SU(2)$ Gauge theory by Ashtekar [5,94] by performing a $3+1$ splitting of spacetime and using as fundamental conjugate variables a connection A_a^i and an electric field E_i^a, which take values in the Lie algebra $su(2)$ of $SU(2)$. The symmetry group is generated by the local $SU(2)$ gauge transformations that leave points of the manifold invariant, and the theory is covariant with respect to diffeomorphisms. The constraints represent the simplest covariant functions that contain (A_a^i, E_i^a) at most, quadratically, and that do not reference any background quantity:

$$\mathcal{G}_i = \mathcal{D}_a E_i^a = 0, \quad \mathcal{D}_a E_i^a = \partial_a E_i^a + \epsilon_{ij}^k A_a^j E_k^a, \tag{22a}$$

$$\mathcal{C}_a^{\text{LQC}} = E_i^b F_{ab}^i = 0, \quad F_{ab}^i = 2\partial_{[a} A_{b]}^i + \epsilon_{jk}^i A_a^j A_b^k, \tag{22b}$$

$$\mathcal{C}^{\text{LQC}} = \epsilon_k^{ij} E_i^a E_j^b F_{ab}^k = 0, \tag{22c}$$

which are respectively the Gauss constraint (generator of the $SU(2)$ rotations), the diffeomorphism constraint (generator of spatial diffeomorphisms) and the scalar Hamiltonian constraint (generator of time evolution).

Before moving on to quantization, the canonical fields (A_a^i, E_i^a) must be appropriately smeared, also because it is not possible to construct an operator corresponding to the connection [87]. This smearing is achieved, defining holonomies of the connections along an edge ℓ and fluxes of the electric field across a bidimensional surface S:

$$h_\ell[A] = \mathbb{P} \exp\left(\int_\ell A_a^i \tau_i d\ell^a\right), \quad \Phi_S[E] = \int_S E_i^a \tau^i dS_a, \tag{23}$$

where τ^i are the $SU(2)$ generators. Note that the holonomies have a one-dimensional support; their trace for a closed edge results in the so-called Wilson loop that gives the theory its name.

Now, the quantum kinematics is obtained by promoting these objects to operators and defining their commutator; a very important consequence of the requirement of background independence, i.e., of diffeomorphism invariance, is that the holonomy-flux algebra results in having a unique representation and, therefore, a unique Hilbert space $\mathcal{H}^{\text{kin}}$. This is called a *spin network* space, defined as a graph Γ, made of a finite number L of edges (each with a half-integer spin-quantum number j_L) and a finite number n of nodes (each with an intertwiner i_n). The basis vectors of this Hilbert space are, therefore, *spin network states* denoted as $|S\rangle = |\Gamma, j_L, i_n\rangle$; wave functions on the spin network are cylindrical functionals $\Psi_\Gamma[A] = \psi(h_{\ell_1}[A], h_{\ell_2}[A], \ldots, h_{\ell_L}[A])$, which depend on the connections only through holonomies and are square-integrable with respect to the Haar measure.

A key result of the kinematical framework of LQG is the quantization of the geometrical operators of area and volume. For example, the area operator and its action on a functional can be defined through the flux operator (23), and the eigenvalues result in being dependent on how many edges of Γ intersect the considered surface. In particular, the smallest non-zero eigenvalue of the area operator is a constant quantity depending on fundamental constants and on the Immirzi parameter only; it is called the *area gap* Δ, and is a key parameter of the theory. Note that this result is purely kinematical [7,8,90].

The dynamics is derived through the implementation of the operators corresponding to the constraints (22); in order to do this, they must first be expressed in terms of the fundamental variables, i.e., holonomies and fluxes, and then quantized, usually through the Dirac procedure [67]. We will not implement the dynamics here, but will show the procedure directly in the cosmological sector of the following sections.

3.2. Standard Loop Quantum Cosmology

We now introduce the "old" procedure to implement the quantization methods of LQG on the homogeneous and isotropic FLRW model [46–48]. Note that the Gauss constraint (22a) and the diffeomorphism one (22b) are automatically satisfied by the symmetries of the model; therefore, we have to deal only with the scalar constraint (22c) which will be given the suffix "grav" to distinguish it from the matter Hamiltonian C_ϕ.

3.2.1. Classical Phase Space

The standard classical procedure in a flat, isotropic, open model is to introduce an elementary cell $\mathcal{V}$ and restrict all integrations to its volume V_0 calculated with respect to a fiducial metric $^0q_{ab}$. Given the symmetries of the model, the gravitational phase space variables (A^i_a, E^a_i) can be expressed as follows:

$$A^i_a = c\, V_0^{-\frac{1}{3}}\, {}^0\omega^i_a, \quad E^a_i = p\, V_0^{-\frac{2}{3}} \sqrt{\det({}^0q_{ab})}\, {}^0e^a_i, \tag{24}$$

where $({}^0\omega^i_a, {}^0e^a_i)$ are a set of orthonormal co-triads and triads adapted to $\mathcal{V}$ and compatible with $^0q_{ab}$. Therefore, the gravitational phase space becomes two-dimensional with fundamental variables (c, p), defined to be insensitive to (positive) rescaling transformations of the fiducial metric and whose physical meaning is obtained through their relation with the cosmic scale factor $a(t)$: $c \propto \dot{a}$, $|p| \propto a^2$. The fundamental Poisson brackets are independent on the fiducial volume V_0 and are given by the following:

$$\{c, p\} = \frac{\gamma}{3}. \tag{25}$$

This is the classical cosmological phase space that constitutes the starting point of LQC.

3.2.2. Kinematics

The quantum theory is constructed, following Dirac, by firstly giving a kinematical description through the identification of elementary observables that have unambiguous operator analogs. LQC can be constructed following the procedure of the full theory: the elementary variables of LQG are holonomies of the connections and fluxes of the fields, and their natural equivalent in this setting are holonomies h^λ along straight edges ($\lambda\, {}^0e^a_k$) and the momentum p itself. Since the holonomy along the ith edge is given by

$$h^\lambda_i(c) = \cos\frac{\lambda c}{2} \mathbb{I} + 2\sin\frac{\lambda c}{2} \tau_i, \tag{26}$$

where $\mathbb{I}$ is the identity matrix, the elementary configurational variables can be taken to be the almost periodic functions $N_\lambda(c) = e^{i\frac{\lambda c}{2}}$ and the momentum p.

The Hilbert space $\mathcal{H}^{\text{kin}}_{\text{grav}}$ is the space $L^2(\mathbb{R}_B, d\mu_H)$ of square integrable functions on the Bohr compactification of the real line endowed with the Haar measure. It is convenient to work in the p-representation, in which eigenstates of $\hat{p}$ are kets $|\mu\rangle$ labeled by a real number and are orthonormal; the fundamental variables are promoted to operators acting as follows:

$$\widehat{N_\lambda(c)}|\mu\rangle = \widehat{e^{i\frac{\lambda c}{2}}}|\mu\rangle = |\mu + \lambda\rangle, \tag{27a}$$

$$\hat{p}|\mu\rangle = \frac{\gamma}{6}\mu|\mu\rangle. \tag{27b}$$

3.2.3. Dynamics

The dynamics is defined by the introduction of an operator on $\mathcal{H}^{\text{kin}}_{\text{grav}}$ corresponding to the Hamiltonian constraint $C^{\text{LQC}}_{\text{grav}}$ shown in (22c). Given the absence of the operator $\hat{c}$, this must be done by returning to the integral expression of the constraint and expressing it as function of our fundamental variables before quantization. The gravitational Hamiltonian constraint of GR in the flat case becomes the following:

$$\mathcal{C}_{\text{grav}}^{\text{LQC}} = -\frac{1}{\gamma^2} \int_{\mathcal{V}} d^3x \, N \, \epsilon_i^{jk} \, F_{ab}^i \, e^{-1} \, E_j^a \, E_k^b, \tag{28}$$

where $e = \sqrt{|\det E|}$, $N_i = 0$ due to isotropy, and N does not depend on spatial coordinates so it can be set to 1 without loss of generality. Using the Thiemann strategy [95], the term $\epsilon_i^{jk} e^{-1} E_j^a E_k^b$ can be written as follows:

$$\sum_k \frac{4\,\text{sgn}(p)}{\gamma \lambda V_0^{\frac{1}{3}}} \, {}^0\epsilon^{abc} \, {}^0\omega_c^k \, \text{Tr}\left(h_k^\lambda \left\{ (h_k^\lambda)^{-1}, V \right\} \tau_i \right), \tag{29}$$

where $V = |p|^{\frac{3}{2}}$ is the volume function on the phase space; for the field strength F_{ab}^i we follow the standard strategy used in gauge theory of considering a square of side $\lambda V_0^{\frac{1}{3}}$ in the ij plane spanned by two of the triad vectors and defining the curvature component as follows:

$$F_{ab}^k = -2 \lim_{\lambda \to 0} \text{Tr}\left(\frac{h_{ij}^\lambda - 1}{\lambda^2 V_0^{\frac{2}{3}}} \right) \tau^k \, {}^0\omega_a^i \, {}^0\omega_b^j, \tag{30}$$

where the holonomy around the square is simply the product of the holonomies along its sides: $h_{ij}^\lambda = h_i^\lambda h_j^\lambda (h_i^\lambda)^{-1} (h_j^\lambda)^{-1}$.

Given these expressions, the gravitational constraint can be written as the limit of a λ-dependent constraint that is now expressed entirely in terms of holonomies and p, and can therefore be now promoted to the operator as follows:

$$\mathcal{C}_{\text{grav}}^{\text{LQC}} = \lim_{\lambda \to 0} \mathcal{C}_{\text{grav}}^\lambda, \tag{31}$$

$$\mathcal{C}_{\text{grav}}^\lambda = -\frac{4\,\text{sgn}(p)}{\gamma^3 \lambda^3} \sum_{ijk} \epsilon^{ijk} \, \text{Tr}\left(h_{ij}^\lambda \left\{ (h_k^\lambda)^{-1}, V \right\} \right), \tag{32}$$

$$\hat{\mathcal{C}}_{\text{grav}}^\lambda = \frac{24 i\,\text{sgn}(p)}{\gamma^3 \lambda^3} \sin^2(\lambda c) \, \hat{O}(\lambda), \tag{33}$$

$$\hat{O}(\lambda) = \sin\frac{\lambda c}{2} \hat{V} \cos\frac{\lambda c}{2} - \cos\frac{\lambda c}{2} \hat{V} \sin\frac{\lambda c}{2}, \tag{34}$$

where the action of the volume operator (acting simply as $\widehat{|p|^{\frac{3}{2}}}$), of the holonomy operators and of sine and cosine functions can be easily derived from (27). Note that in the promotion of the Hamiltonian constraint to a quantum operator, a specific discretization choice is made among many possibilities. This is a delicate point for the derivation of LQC, and as explained in later sections, it is addressed in [16,96].

Now, in LQC, the limit $\lambda \to 0$ does not exist by construction. This can be interpreted as a reminder of the underlying quantum geometry, where the area operator has a discrete spectrum with a smallest non-zero eigenvalue corresponding to the area gap Δ. It is, therefore, incorrect to let λ go to zero because in full LQG, the area of the ij square cannot be zero; as a consequence, λ must be set to a fixed positive value μ_0 that can be appropriately related to the area gap by considering that the holonomies are eigenstates of the area operator $\hat{A} = \widehat{|p|}$ and demanding that the eigenvalue be exactly equal to Δ:

$$\hat{A} \, h_{ij}^{\mu_0}(c) = \frac{\gamma \mu_0}{6} h_{ij}^{\mu_0}(c) = \Delta \, h_{ij}^{\mu_0}(c). \tag{35}$$

The operator corresponding to the Hamiltonian constraint can be now defined as the λ-dependent operator (33) with $\lambda = \mu_0$:

$$\hat{\mathcal{C}}_{\text{grav}}^{\text{LQC}} = \hat{\mathcal{C}}_{\text{grav}}^{\mu_0}. \tag{36}$$

The final step is to make this operator self-adjointed by either taking its self-adjoint part or by symmetrically redistributing the sine operator as follows:

$$\hat{\mathcal{C}}_{\text{grav}}^{\text{LQC}(1)} = \frac{1}{2}\left(\hat{\mathcal{C}}_{\text{grav}}^{\text{LQC}} + (\hat{\mathcal{C}}_{\text{grav}}^{\text{LQC}})^\dagger\right), \tag{37a}$$

$$\hat{\mathcal{C}}_{\text{grav}}^{\text{LQC}(2)} = \frac{24i\,\text{sgn}(p)}{\gamma^3 \mu_0^3} \sin(\mu_0 c)\, \hat{O}(\mu_0)\, \sin(\mu_0 c). \tag{37b}$$

The Ashtekar school uses the second one, but both are equivalent and yield similar results.

Now we can introduce matter in the form of a massless scalar field ϕ obeying an Hamiltonian constraint of the following form:

$$\hat{\mathcal{C}}_\phi = \widehat{|p|^{-\frac{3}{2}}}\, \hat{P}_\phi^2, \tag{38}$$

where P_ϕ is the momentum conjugate to ϕ. Physical states $\Psi(\mu, \phi)$ are the solutions of the total constraint as follows:

$$(\hat{\mathcal{C}}_{\text{grav}}^{\text{LQC}} + \hat{\mathcal{C}}_\phi)\, \Psi(\mu, \phi) = 0. \tag{39}$$

In the classical theory, the field does not appear in the matter part of the Hamiltonian; this leads to its conjugate momentum P_ϕ being a constant of motion and to ϕ being able to play the role of emergent internal time. In quantum cosmology in general, this choice of relational time is the most natural one because near the classical singularity, a monotonic behavior of ϕ as a function of the isotropic scale factor $a(t)$ always appears. The constraint (39) can then be considered an evolution equation with respect to this internal time ϕ and can be recast in a Klein–Gordon-like form, thus allowing for the usual separation into positive and negative frequency subspaces. Once that is done, the procedure to extract physics from the model is: to introduce an inner product on the space of solutions of the constraint to obtain the physical Hilbert space $\mathcal{H}^{\text{phy}}$; to isolate classical Dirac observables to be promoted to a self-adjoint operator on $\mathcal{H}^{\text{phy}}$; to use them to construct wave packets that are semiclassical at late times; and to evolve them backwards in time using the constraint itself.

After the internal time procedure, the constraint (39) takes the following form:

$$\frac{\partial^2 \Psi}{\partial \phi^2} = \frac{1}{B}\left(C^+(\mu)\Psi(\mu + 4\mu_0, \phi) + C^0(\mu)\Psi(\mu, \phi) + C^-(\mu)\Psi(\mu - 4\mu_0, \phi)\right) = -\Theta(\mu)\Psi(\mu, \phi), \tag{40}$$

$$C^+(\mu) = \frac{1}{72|\mu_0|^3}\left||\mu + 3\mu_0|^{\frac{3}{2}} - |\mu + \mu_0|^{\frac{3}{2}}\right|, \tag{41a}$$

$$C^-(\mu) = C^+(\mu - 4\mu_0), \tag{41b}$$

$$C^0(\mu) = -C^+(\mu) - C^-(\mu), \tag{41c}$$

where $B = B(\mu)$ is the eigeinvalue of the inverse volume operator appearing in the matter constraint (38):

$$\widehat{|p|^{-\frac{3}{2}}}\Psi(\mu, \phi) = \left(\frac{6}{\gamma}\right)^{\frac{3}{2}} B(\mu), \tag{42a}$$

$$B(\mu) = \left(\frac{2}{3\mu_0}\right)^6 \left(|\mu + \mu_0|^{\frac{3}{4}} - |\mu - \mu_0|^{\frac{3}{4}}\right)^6. \tag{42b}$$

The operator $\Theta(\mu)$ on the right-hand side of (40) is a difference operator, as opposed to the differential character of the operator that appears in the equivalent equation of the WDW theory [1–3]. This allows for the space of physical states to be naturally superselected into different sectors that can be analyzed separately.

In the choice of observables, classical considerations are helpful: it is possible to choose the conjugate momentum to the field since it is a constant of motion, and the value of p at a fixed instant ϕ_0. The set $(P_\phi, p|_{\phi_0})$ uniquely determines a classical trajectory; therefore, it constitutes a complete set of Dirac observables in the quantum theory. The operators act as follows:

$$\widehat{|p|_{\phi_0}} \Psi(\mu,\phi) = e^{i\sqrt{\Theta(\mu)}\,(\phi-\phi_0)} |\mu| \,\Psi(\mu,\phi_0), \tag{43a}$$

$$\hat{P}_\phi \Psi(\mu,\phi) = -i\frac{\partial \Psi(\mu,\phi)}{\partial \phi}, \tag{43b}$$

where $\Psi(\mu,\phi_0)$ is the initial configuration, i.e., the wave function calculated at a fixed initial time ϕ_0 and the absolute value $|\mu|$ is due to the fact that states are symmetric under the action of the parity operator $\hat{\Pi}$.

The evolution of wave packets is then carried out numerically. In the following, we briefly summarize the results that are relevant for the resolution of the singularity. For a more detailed analysis of all resulting properties, see [47,48].

- Singularity resolution: an initially semiclassical state remains sharply peaked around the classical trajectories and the expectation values of the Dirac observables are in good agreement with their classical counterparts for most of the evolution when coherent states are considered. However, when the matter density approaches a critical value, the state bounces from the expanding branch to a contracting one with the same value of $\langle \hat{P}_\phi \rangle$, as shown in Figure 1. This occurs in every sector and for any choice of $P_\phi \gg 1$, universally solving the singularity by replacing the Big Bang with a Big Bounce.
- Critical density: the critical value of the matter density results in being inversely proportional to the expectation value $\langle \hat{P}_\phi \rangle$ and can, therefore, be made arbitrarily small by choosing a sufficiently large value for P_ϕ. This fact, besides being physically unreasonable because it could imply departures from the classical trajectories well away from the Planck regime, becomes even more problematic in the case of a closed model: the point of maximum expansion depends on P_ϕ as well. In order to have a bounce density comparable with that of Planck, a very small value is needed, but in that case, the Universe would never become big enough to be considered classical; on the other hand, a closed Universe that grows to become classical needs a large value of P_ϕ but would have a bounce density comparable with, for example, that of water.

This framework, although it successfully solves the singularity, has, therefore, a very important drawback and needs to be substantially improved.

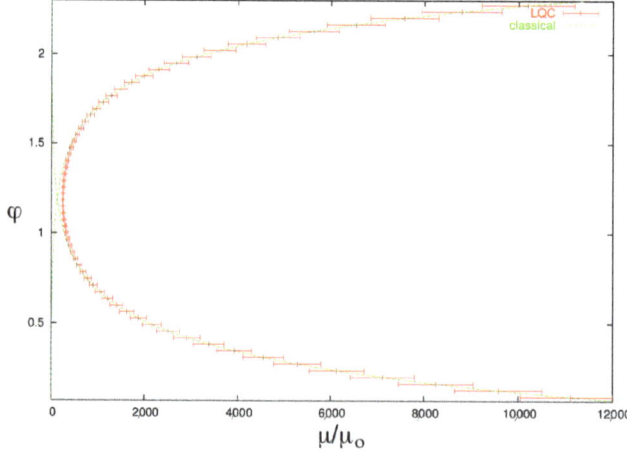

Figure 1. Expectation values and dispersion of $|\mu|$ (red horizontal bars) in terms of μ_0 as function of time φ for a coherent state, compared with classical trajectories (green dashed lines). Image from [48].

3.3. Improved Loop Quantum Cosmology

In this section, we present the new scheme introduced by the Ashtekar school in [49,50] that improves on the standard LQG procedure.

The idea is that the quantization of the area operator must refer to *physical* geometries. Therefore, when performing the limit (31) needed to construct the gravitational constraint, we should shrink the ij square until its area reaches Δ as measured with respect to the physical metric instead of the fiducial one. The area of the faces of the elementary cell is simply $|p|$, and each side of the square is λ times the edge of the cell; with this consideration, the parameter λ now becomes a function $\bar{\mu}(p)$ given by the following:

$$\bar{\mu}^2 |p| = \Delta. \tag{44}$$

This means that the curvature operator now depends both on the connection and the geometry, whereas with the previous μ_0 scheme, it depended on the connection only. As a consequence, more care is needed in the definition of the exponential operator because now, $e^{i\frac{\bar{\mu}c}{2}}$ depends also on p.

By using geometric considerations, we can make a comparison with the Schrödinger representation and set the following:

$$\widehat{e^{i\frac{\bar{\mu}c}{2}}} \Psi(\mu) = e^{\bar{\mu}\frac{d}{d\mu}} \Psi(\mu), \tag{45}$$

i.e., the exponential operator translates the state by a unit affine parameter distance along the integral curve of the vector field $\bar{\mu}\frac{d}{d\mu}$. The affine parameter along this vector field is given by the following:

$$\nu = K \operatorname{sgn}(\mu) |\mu|^{\frac{3}{2}}, \tag{46}$$

with $K = \frac{2\sqrt{2}}{3\sqrt{3\sqrt{3}}}$. Since $\nu(\mu)$ is an invertible and smooth function of μ, the action of the exponential operator is well-defined; however, its expression in the μ-representation is very complicated because the variable μ is not well-adapted to the vector field $\bar{\mu}\frac{d}{d\mu}$. It is therefore useful to change the basis from $|\mu\rangle$ to $|\nu\rangle$; in this representation, the action of the exponential operator takes the following extremely simple form:

$$\widehat{e^{i\frac{\bar{\mu}c}{2}}} \Psi(\nu) = \Psi(\nu + 1). \tag{47}$$

The kets $|\nu\rangle$ still constitute an orthonormal basis on $\mathcal{H}^{\text{kin}}_{\text{grav}}$ and are eigenvectors of the volume operator:

$$\hat{V}|\nu\rangle = \left(\frac{\gamma}{6}\right)^{\frac{3}{2}} \frac{|\nu|}{K} |\nu\rangle. \tag{48}$$

The gravitational constraint can now be constructed in the same way as before.

The matter constraint has the same form (38) of the standard case; therefore, it is sufficient to express the inverse volume eigenvalues (42b) in terms of ν:

$$B(\nu) = \left(\frac{3}{2}\right)^3 K |\nu| \left||\nu+1|^{\frac{1}{3}} - |\nu-1|^{\frac{1}{3}}\right|^3. \tag{49}$$

Repeating the same steps of the standard case, the total constraint can again be expressed as a difference operator but this time in terms of ν:

$$\frac{\partial^2 \Psi}{\partial \phi^2} = \frac{1}{B}\left(\mathcal{C}^+(\nu)\Psi(\nu+4,\phi) + \mathcal{C}^0(\nu)\Psi(\nu,\phi) + \mathcal{C}^-(\nu)\Psi(\nu-4,\phi)\right) = -\Theta(\nu)\Psi(\nu,\phi), \tag{50}$$

$$\mathcal{C}^+(\nu) = \frac{3K}{64} |\nu+2| \left||\nu+1| - |\nu+3|\right|, \tag{51a}$$

$$\mathcal{C}^-(\nu) = \mathcal{C}^+(\nu - 4), \tag{51b}$$

$$\mathcal{C}^0(\nu) = -\mathcal{C}^+(\nu) - \mathcal{C}^-(\nu). \tag{51c}$$

The old operator $\Theta(\mu)$ in (40) involves steps that are constant in the eigenvalues of $\hat{p}$, while the new one $\Theta(\nu)$, called *improved constraint*, involves steps that are constant in eigenvalues of the volume operator $\hat{V}$. In the $|\mu\rangle$ basis, these steps vary, becoming larger for smaller μ and diverging for $\nu = 0$; however the constraint is well-defined since the operators acting on the state $|\nu = 0\rangle$ are well-defined as well.

Regarding the Dirac observables, it is sufficient to substitute $p|_{\phi_0}$ with the volume $\nu|_{\phi_0}$, and the set $(P_\phi, \nu|_{\phi_0})$ is again complete. Therefore, the action of the correspondent operators is

$$\widehat{|\nu|_{\phi_0}} \Psi(\nu,\phi) = e^{i\sqrt{\Theta(\nu)}\,(\phi - \phi_0)} \, |\nu| \, \Psi(\nu,\phi_0), \tag{52a}$$

$$\hat{P}_\phi \Psi(\nu,\phi) = -i \frac{\partial \Psi(\nu,\phi)}{\partial \phi}. \tag{52b}$$

After numerical calculations, the improved framework yields the following results.

- Singularity resolution: also in this case, the states remain sharply peaked throughout all the evolution, and the expectation values of the Dirac observables calculated on coherent states follow the classical trajectory up to a critical value of the energy density; when that value is approached, the states jump to a contracting branch and undergo a quantum bounce instead of following the classical trajectory into the singularity, as shown in Figure 2.

- Critical density: the real improvement of the new scheme is that the numerical value of the bounce density is independent of $\langle \hat{P}_\phi \rangle$ and is the same in all simulations, given by $\rho_{\text{crit}} \approx 0.82 \rho_P$. The behavior of the energy density was also studied independently from the evolution of wave packets by analyzing the evolution of the density operator defined as follows:

$$\hat{\rho}_\phi = \widehat{\left(\frac{P_\phi^2}{2V^2}\right)}, \tag{53}$$

and it was found that in all quantum solutions, the expectation value $\langle \hat{\rho}_\phi \rangle$ is bounded from above by the same value ρ_{crit}.

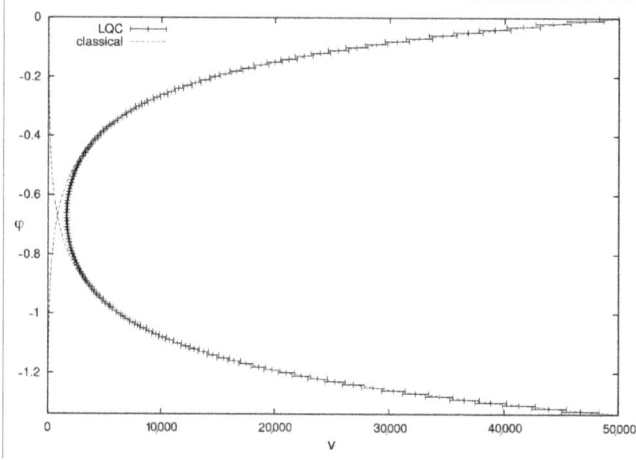

Figure 2. Expectation values and dispersion of v as function of time φ (horizontal bars) for a coherent state, compared with the classical trajectories (dashed lines). Image from [49].

It is shown that the absolute value of the critical density is not modified even when a non-zero cosmological constant is included in the model [49]. The physical understanding of this phenomenon is given by an effective description obtained through a semiclassical limit.

The improved scheme is able to overcome the main weakness of old standard LQC through a physically motivated modification in the construction procedure of the quantum gravitational constraint. This is the model currently referred to when talking about LQC and on which all subsequent literature is based. Indeed, this model has allowed for a series of phenomenological predictions; they are not part of the aim of this paper, but we present some examples below.

The $\bar{\mu}$ scheme made it possible to perform thermodynamical analyses where the Loop-quantized FLRW model is considered a thermodynamical system, and the energy density and pressure are given a precise thermodynamical meaning [97,98]; the computations for the duration of the inflationary de Sitter phase give results consistent with the minimum amount of e-folds necessary to solve the paradoxes (although slightly higher, depending on some parameters, such as the shear at the Bounce in anisotropic models), predict a phase of deflation in the contracting branch and also allow the extension of the standard inflationary paradigm from the Planck scale up to the onset of slow roll inflation, yielding novel effects, such as non-Gaussianities [99–104]. Most importantly, the improved scheme allows for a computation of the primordial power spectrum through two main methods: in the first, the implementation of holonomy corrections as a deformed algebra yields a slightly blue-tilted scale invariant spectrum followed by oscillations and an exponential behavior in the ultraviolet [105,106]; in the other, a test field approximation is used such that the evolution of tensor modes on any background quantum geometry is completely equivalent to that of the same modes propagating on a smooth but quantum-corrected metric called "dressed metric", and it is able to recover the red-titled ultraviolet power spectrum of classical cosmology [99,107]. For a more detailed comparison between the two methods, see [108–110].

3.4. Effective Dynamics

The semiclassical limit of LQC, i.e., the inclusion of quantum corrections in the classical dynamics, can be obtained through a geometric formulation of quantum mechanics where the Hilbert space is treated as an infinite-dimensional phase space [21]. In simpler cases with coherent states that are preserved by the full quantum dynamics, the resulting Hamiltonian coincides with the classical one; however, in more general systems it is possible to choose suitable semiclassical states that are preserved up to a desired accuracy (e.g., in a $\hbar$ expansion), and the corresponding effective Hamiltonian preserving this evolution is generally different from the classical one [111].

3.4.1. Effective μ_0 Scheme

In our model with a massless scalar field, the leading order quantum corrections yield an effective Hamiltonian constraint for the μ_0 scheme in the following form:

$$\frac{C_{\text{eff}}^{\mu_0}}{2} = -\frac{3}{\gamma^2 \mu_0^2}|p|^{\frac{1}{2}}\sin^2(\mu_0 c) + \frac{1}{2}B(\mu)P_\phi^2, \tag{54}$$

where $B(\mu)$ is given by (42b) and for $\mu \gg \mu_0$ can be approximated as follows:

$$B(\mu) = \left(\frac{6}{\gamma}\right)^{\frac{3}{2}}|\mu|^{-\frac{3}{2}}\left(1 + \frac{5}{96}\frac{\mu_0^2}{\mu^2} + O\left(\frac{\mu_0^4}{\mu^4}\right)\right). \tag{55}$$

Since quantum corrections are significant only in the quantum region near $\mu = 0$, we can ignore them and, through Hamilton equations, obtain a modified Friedmann equation:

$$H^2 = \left(\frac{\dot{p}}{2p}\right)^2 = \frac{1}{3}\rho\left(1 - \frac{\rho}{\rho_{\text{crit}}}\right), \tag{56a}$$

$$\rho_{\text{crit}}^{\mu_0} = \left(\frac{3}{\gamma^2 \mu_0^2}\right)^{\frac{3}{2}} \frac{\sqrt{2}}{P_\phi}. \tag{56b}$$

As in the full quantum dynamics, the critical density at the bounce is inversely proportional to the value of the constant of motion P_ϕ.

3.4.2. Effective $\bar{\mu}$ Scheme

Applying the same procedure to the $\bar{\mu}$ scheme, the improved effective Hamiltonian reads as follows:

$$\frac{\mathcal{C}_{\text{eff}}^{\bar{\mu}}}{2} = -\frac{3}{\gamma^2 \bar{\mu}^2}|p|^{\frac{1}{2}}\sin^2(\bar{\mu}c) + \frac{1}{2}B(\nu)P_\phi^2, \tag{57}$$

where $B(\nu)$ is the eigenvalue of the inverse volume operator expressed in terms of ν as given by (49). Again, for $|\nu| \gg 1$, $B(\nu)$ quickly approaches its classical value:

$$B(\nu) = \left(\frac{6}{\gamma}\right)^{3/2}\frac{K}{|\nu|}\left(1 + \frac{5}{9}\frac{1}{|\nu|^2} + O\left(\frac{1}{|\nu|^4}\right)\right). \tag{58}$$

Neglecting the higher order quantum corrections as before and given that the Poisson bracket between ν and c is easily derived from (25), the modified Friedmann equation in this case is as follows:

$$H^2 = \left(\frac{\dot{\nu}}{3\nu}\right)^2 = \frac{\rho}{3}\left(1 - \frac{\rho}{\rho_{\text{crit}}}\right), \tag{59a}$$

$$\rho_{\text{crit}}^{\bar{\mu}} = \frac{4\sqrt{3}}{\gamma^3}. \tag{59b}$$

The critical density does not depend on P_ϕ anymore, and that is the main reason for which the improved model is much more appealing than the standard one.

3.5. Loop Quantization of the Anisotropic Sector

Let us now show the work that was done on the implementation of the LQG quantization procedures to the anisotropic sector of cosmology, i.e., to the Bianchi models. The classical phase space in this case is six-dimensional since there are three spatial directions that evolve independently (i.e., three different scale factors $a_1(t), a_2(t), a_3(t)$); therefore, in the Hamiltonian formulation, the fundamental variables will be (c_i, p_i) with $i = 1, 2, 3$ where the momenta c_i are dependent on the velocities $\dot{a}_i$ while the variables p_i are proportional to the (comoving) areas perpendicular to the direction i: $|p_i| \propto a_j a_k$ with $i \neq j \neq k$. We will briefly present the results obtained by Ashtekar and Wilson-Ewing on the improved method of Loop quantization of the Bianchi type I, II and IX models [79–81]; their work simplifies and improves the previous analyses on the Loop quantum homogeneous models by Bojowald [75–78].

3.5.1. Bianchi Type I

The Bianchi type I model corresponds to the simplest anisotropic model; its classical Hamiltonian constraint in the Ashtekar variables reads as follows:

$$\mathcal{C}_{\text{BI}} = -\frac{1}{\gamma^2 V}\sum_{i \neq j} c_i p_i c_j p_j, \tag{60}$$

where $V = a_1 a_2 a_3 = \sqrt{|p_1 p_2 p_3|}$ is the Universe volume. The Hamiltonian equations yield a system of six coupled differential equations that can be easily solved by recognizing that the quantities $\mathcal{K}_i = c_i p_i$ are constants of motion. The solution in the void case is the famous Kasner solution where $a_i(t) \propto t^{k_i}$, where k_i are the constant Kasner indices that obey $\sum_i k_i = \sum_i k_i^2 = 1$ [112]; usually, these indices are parametrized through a variable $u \in (1, +\infty)$. Repeating the procedure of the isotropic sector, we introduce matter in the form of a scalar field ϕ obeying a Hamiltonian similar to (38) and playing the role of relational time; then, we quantize the system according to the Dirac procedure, following [79].

Now, when implementing the $\bar{\mu}$ scheme, we are naturally induced to use three different parameters $\bar{\mu}_i$ relating to the three different directions. An analogous reasoning to the previous section yields the following:

$$\bar{\mu}_i \bar{\mu}_j = \frac{\Delta}{|p_k|} \quad \Longrightarrow \quad \bar{\mu}_i = \sqrt{\Delta \frac{|p_i|}{|p_j p_k|}} \tag{61}$$

with $i \neq j \neq k$. This implies that, when constructing the holonomies and extracting the operators corresponding to the quasi-periodic functions $\widehat{e^{i\frac{\bar{\mu}_i c_i}{2}}}$, their action would depend on all the p_j and be unmanageable. A solution can be obtained using a generalization of (44), i.e., $\bar{\mu}_i = \sqrt{\frac{\Delta}{|p_i|}}$ so that it could be possible to define volume-like variables $\nu_i \propto |p_i|^{\frac{3}{2}}$ and implement the operators as in (47): $\widehat{e^{i\frac{\bar{\mu}_i c_i}{2}}} \Psi(\nu_i, \nu_j, \nu_k) = \Psi(\nu_i + 1, \nu_j, \nu_k)$. This process allows to find the quantum dynamics of the three spatial directions separately, and each results to be a copy of the isotropic model; however, the description of the evolution of the whole model is not viable in this framework.

In order to keep the correct expression (61) of the parameters $\bar{\mu}_i$, a new representation was developed through the introduction of dimensionless variables $q_i \propto \text{sgn}(p_i)\sqrt{|p_i|}$; this allows for the definition of a new basis in $\mathcal{H}^{\text{kin}}_{\text{grav}}$ comprised of vectors that are still eigenvectors of the operators p_i, and the action of the exponential operators depend on these variables:

$$\widehat{e^{i\frac{\bar{\mu}_i c_i}{2}}} \Psi(q_i, q_j, q_k) = \Psi(q_i + \frac{\text{sgn}(q_i)}{q_j q_k}, q_j, q_k). \tag{62}$$

Note how the shift along the q_i direction depends on q_j and q_k. In order to make this more manageable, there is a further possible substitution: to define a volume variable $\nu \propto q_1 q_2 q_3$ and use a basis $|q_1, q_2, \nu\rangle$ (note that it is possible substitute any of the q_i with ν and obtain the same results). This way, after some calculations, it is possible to write the action of the Hamiltonian gravitational constraint as dependent only on the volume: it will give a combination of shifted wave functionals of the following form:

$$\Psi(f_1(\nu) q_1, f_2(\nu) q_2, \nu \pm 4) \tag{63}$$

where $f_1(\nu), f_2(\nu)$ are simple rescaling functions of ν. This is more easily comparable with the dynamics of the isotropic model; indeed, it is possible to construct a projection that maps the anisotropic wave functional $\Psi(q_1, q_2, \nu)$ into the isotropic state $\Psi(\nu)$ of previous sections, as well as making the gravitational constraints become the same.

Finally, let us address the issue of singularity resolution. It is possible to decompose the Hilbert space in the two subspaces $\mathcal{H}^{\text{grav}}_{\text{singular}}$ and $\mathcal{H}^{\text{grav}}_{\text{regular}}$, where the first contains all states with support on points with $\nu = 0$, and the second contains the states without; since all terms in the gravitational constraint contain a factor proportional to a power of ν (depending on the chosen factor ordering), the two subspaces are invariant under time evolution and remain decoupled. Therefore, a state that starts as regular will remain regular throughout all its evolution; in this sense, the singularity is avoided. This behavior is again captured by the effective dynamics that predicts the classical Kasner solution away from the singularity and a Bounce in the Planck regime that jumps to the contracting branch in a similar way to the isotropic model. However, in this case, we need to keep track also of the three Hubble functions H_i that undergo different Bounces separately; still, for conclusive evidence that this effective evolution correctly reproduces the exact quantum dynamics, one would need numerical simulations of the exact quantum model.

3.5.2. Bianchi Type II

The Bianchi type II model augments the Bianchi I with a curvature term (coming from the full expression of the connection) and the introduction of a potential along only one direction (here the direction 1). Its classical Hamiltonian constraint is as follows:

$$\mathcal{C}_{\text{BII}} = \mathcal{C}_{\text{BI}} + \mathcal{C}_{\text{BII}}^{\text{curv}} + \mathcal{U}_{\text{BII}}, \tag{64a}$$

$$\mathcal{C}_{\text{BII}}^{\text{curv}} = -\frac{1}{\gamma^2 V} \epsilon\, p_2 p_3 c_1, \tag{64b}$$

$$\mathcal{U}_{\text{BII}} = \frac{1 + \gamma^2 V}{\gamma^2} \frac{p_2^2 p_3^2}{4 p_1^2}, \tag{64c}$$

where $\epsilon = \pm 1$ is an internal pseudoscalar parametrizing the orientation of the triad. The classical dynamics is comprised of two different Kasner epochs bridged by a transition that changes the value of the Kasner indices [112]; in terms of the variable u introduced before, the second Kasner epoch is parametrized by $u_2 = u_1 - 1$, where u_1 is the value of the first epoch.

When implementing the quantization procedure, it is possible to follow the same steps of Bianchi I, but more care is needed for the new terms that appear in the constraint [80]. In particular, after the implementation of the $\bar{\mu}$ scheme through (61) and the definition of the variables q_i, the term $\mathcal{C}_{\text{BII}}^{\text{curv}}$ contains a power $|p_1|^{-\frac{1}{2}}$, that usually becomes $|p_1|^{-\frac{1}{4}}$ after a symmetric factor ordering; this is handled through a variation of the Thiemann inverse triad identities [95], that in the q_i representation yield the following:

$$\widehat{|p_1|^{-\frac{1}{4}}}|q_1, q_2, q_3\rangle \propto \text{sgn}(q_1)\sqrt{|q_2 q_3|}\, g(\nu, q_2, q_3)\, |q_1, q_2, q_3\rangle, \tag{65a}$$

$$g(\nu, q_2, q_3) = \sqrt{\nu + \text{sgn}(q_2 q_3)} - \sqrt{\nu - \text{sgn}(q_2 q_3)}. \tag{65b}$$

On the other hand, the term $\mathcal{U}_{\text{BII}}$ contains only a power $\widehat{|p_1|^{-2}}$ whose action can be simply defined as the eighth power of the operator (65). This form suggests that, in this case, the simplest representation is to substitute ν to q_1 and use the basis $|\nu, q_2, q_3\rangle$.

The quantum dynamics of the Bianchi II model is analogous to that of Bianchi I (and therefore of FLRW) because the regular and singular Hilbert spaces decouple in this model as well, and a state that starts away from $\nu = 0$ will never reach it. Far from the singularity, the classical dynamics of the two bridged Kasner-like solutions is recovered, while near the Planck regime, there is a Bounce that joins with the contracting branch.

3.5.3. Bianchi Type IX

The Bianchi type IX is the most complex homogeneous model (together with type VIII). The Hamiltonian is as follows:

$$\mathcal{C}_{\text{BIX}} = \mathcal{C}_{\text{BI}} + \mathcal{C}_{\text{BIX}}^{\text{curv}} + \mathcal{U}_{\text{BIX}}, \tag{66a}$$

$$\mathcal{C}_{\text{BIX}}^{\text{curv}} = -\frac{\epsilon}{\gamma^2 V}(c_1 p_2 p_3 + c_2 p_3 p_1 + c_3 p_1 p_2), \tag{66b}$$

$$\mathcal{U}_{\text{BIX}} = \frac{1}{\gamma^2 V}\left(\frac{p_2^2 p_3^2}{4 p_1^2} + \frac{p_1^2 p_3^2}{4 p_2^2} + \frac{p_1^2 p_2^2}{4 p_3^2} - \frac{p_1^2 + p_2^2 + p_3^2}{2}\right); \tag{66c}$$

the classical dynamics is a chaotic evolution from one Kasner solution to the next, each changing the Kasner indices and each closer in time to the next, until the singularity is reached. In terms of u, each transition from one Kasner epoch (with a specific value of u) to another one lowers the previous value by 1, until a value $u_{1_{\text{final}}} < 1$ is reached that puts an end to the first Kasner era; at that point, the following epoch starts a second era with $u_{2_{\text{init}}} = u_{1_{\text{final}}}^{-1}$ and again each transition lowers the previous value of u by 1. The cycle continues indefinitely toward the singularity, giving rise to the same chaotic evolution that appears in the point-Universe description outlined in Section 2.3. However, chaos is tamed when introducing matter in the form of a massless scalar field ϕ (which is exactly what is done in order to implement the quantization procedure): the Friedmann equation becomes asymptotically velocity-term dominated (AVTD), so that a standard Kasner-like dynamics is recovered, and the singularity is approached through a single, stable Kasner epoch.

The quantization procedure is the same as the other models. The scalar field plays the role of relational time. The $\bar{\mu}_i$ are defined as in (61), and we introduce the variables (q_1, q_2, q_3) and then change to (q_1, q_2, ν) so that the action of the gravitational constraint involves constant shifts in ν and rescalings of q_i dependent only on ν (note that in this model the symmetries are in place again, so any q_i could be substituted with ν); the singular and regular Hilbert spaces decouple and the singularity is still solved.

We must now address the issue of chaoticity. It can be shown that already for the classical dynamics, the presence of the scalar field tames the chaos [113]; this is true also in the quantum model, as long as the AVTD regime is reached before the quantum gravitational effects become relevant, but if the value of the momentum conjugate to the scalar field is too small, this will not happen. However one could argue that the quantum gravity effects giving rise to the Bounce make it so that the model evolves away from the high curvature region toward the classical contracting branch, and there is not a sufficient number of Kasner epochs for chaos to appear before the AVTD regime is unavoidably reached, even if this happens after the Bounce. Further arguments for the removal of chaos in the Loop quantum Bianchi type IX model are given in [76–78]. Still, numerical simulations of the complete quantum system are needed in order to give a definitive word on the chaoticity of these models.

3.6. Criticisms and Shortcomings of LQC

Over the years, many criticisms have been made on the LQC framework, mainly about the following points: whether the Bounce can be regarded as a semiclassical phenomenon or must be considered a purely quantum effect; the fact that the quantum dynamics is not derived by a symmetry reduction of the full LQG theory, but by quantizing cosmological models that are reduced before quantization; the use of the area gap as a parameter to construct the dynamics of the reduced theory and its effective description. In this section, we briefly summarize these issues.

As already mentioned, LQC is not the cosmological sector of LQG, but rather the implementation of the latter's quantization procedures on a cosmological spacetime; a good symmetry reduction of full LQG would require that some degrees of freedom be frozen out, but this procedure conflicts with the quantum character of the $SU(2)$ variables. The spatial geometry of a cosmological spacetime is fixed, and during the quantization on invariant variables, any kind of spatial structure, such as the possibility to perform local $SU(2)$ transformations, is lost. Furthermore, as it is well known, in LQG the implementation of the scalar constraint is not yet a viable task [11], and it is worth noting how this problem is somehow bypassed in LQC, where the dynamics for the cosmological models is constructed; however, this procedure is far from being completely clear. Another way to see the problem of the $SU(2)$ symmetry is that the resulting algebra on the reduced model is different from the holonomy-flux algebra of the full theory; therefore, LQC is not equivalent to LQG [15]. Taking this a step further, in [16], it is claimed that the resulting algebra of LQC has several different representations, among which the Ashtekar school implicitly chooses the one that favors bouncing solutions, while in [96], it is argued that the mechanism for the resolution of the singularity lies in the regularization of the constraint rather than in the quantization procedure itself (and indeed the singularity in LQC is avoided already at a semiclassical level). Alternatively, in [17] an $SU(2)$-invariant gauge fixing is considered, which yields a modified holonomy-flux algebra that reproduces the original one of LQG only when holonomies are evaluated along the triad vectors; the quantization procedure is then performed according to the full theory, and the resulting model is a quantum cosmology that manages to better preserve the $SU(2)$ structure. A different approach to derive a consistent description of the Loop quantum cosmological sector is provided in [114], where through the introduction of local patches, it is possible to define local cosmological variables that properly take into account the presence and the properties in full LQG of both holonomy corrections and inverse-volume operators. Another interesting and more developed approach that tries to solve this problem is that of quantum reduced loop gravity (QRLG). In this approach, inspired by the criticisms in [15,17], some gauge-fixing conditions are implemented on the kinematical Hilbert space of LQG that restrict the full gravitational model to a diagonal metric tensor and to diagonal triads; then, the cosmological reduction is performed by considering only that part of the scalar constraint that generates the evolution of the homogeneous part of the metric. Finally, the dynamics is obtained by performing a cubation (instead of a triangulation) on the reduced spin-foam graphs. This way, QRLG gives a quantum description of the Universe in terms of a cuboidal graph and it provides a framework for deriving the cosmological setting from full LQG. For a more detailed presentation, see [115–120]. In this context, also the formalism of group field theory (GFT) for quantum gravity contributes to clarify the link between the effective cosmological equations and LQC when applied to the cosmological sector. In this approach, the effective cosmological dynamics emerges as an hydrodynamic-like approximation of the multi-condensate quantum states, i.e., the fundamental quantum gravitational degrees of freedom. In particular, a second order quantization and the basis to the idea of lattice refinement are provided, showing the dependence of the effective cosmological connection on the number of spin network vertices (a quantity of a purely quantum origin) and thus, on the scale factor. For more details regarding the GFT cosmology and the emergent bouncing dynamics, see [121–125].

Another problem that is often raised, linked to the previous one, is that an external parameter fixing the discretization scale must be introduced from the full theory by hand because LQC is derived independently from LQG; this leads to some issues. For example, in [16] it is stated that an effective description will have a scale of validity (given by the area gap itself), while the Ashtekar school uses the same effective equations across very different regimes (namely, it follows the evolution of a wave packet from the classical regime up to the Planck region near the singularity; this is connected also to the issue about the nature of the Bounce.

LQG and LQC attempt to provide a promising framework for a quantum mechanical description of general relativity and of cosmological models, but as outlined in this section, both—the latter in particular—need to be substantially improved.

A good achievement toward this goal is the formulation of polymer quantum mechanics (PQM), a new quantum mechanical framework that is able to reproduce LQC effects but can be derived independently from LQG and is much more versatile and easily applicable to any Hamiltonian system. Its implementation on the cosmological minisuperspaces is the focus of the next sections.

4. Polymer Cosmology

The power of the polymer formulation stands in its capability of introducing regularization effects typical of the LQG quantum gravity approach by means of a simpler mathematical framework with respect to the LQC theory. Therefore, its employment in the cosmological sector has great relevance in trying to overcome the singularity issue of GR and making also a comparison with the LQC main results regarding the presence of an initial Big Bounce and its properties. In this sense, the principal feature of the polymer approach is making clear that the properties of the cosmological dynamics are strongly dependent on the set of variables on which the polymer quantization is implemented [27,34,38,42,44,45].

In this section, we focus on the main applications of the polymer formulation to the FLRW [44], Bianchi I [45] and Bianchi IX models [38,42] that represent the main cosmological scenarios on which the polymer-modified dynamics of the primordial Universe are tested. We will start with a discussion on the main results obtained by treating the polymer quantization of these models in the Ashtekar variables, that constitute the setting more connected to the original LQC formulation from which the polymer formulation is derived. Indeed, as originally affirmed by Ashtekar in [126], at the Planck scale, the implementation of LQG shows that quantum geometry has a close similarity with polymers and that the continuum picture arises only upon a coarse-graining procedure by means of suitable semiclassical states. Then, we will proceed by applying the polymer quantization to the volume-like variables, obtained after doing a canonical transformation from the Ashtekar connections to new generalized coordinates. In particular, we will implement both a semiclassical and a quantum treatment for the FLRW and the Bianchi I models, i.e., the homogeneous and isotropic model and its simplest anisotropic generalization. Finally, we will apply the semiclassical polymer framework in the Misner-like variables (the isotropic variable α or the Universe volume V plus the anisotropies) to the Bianchi IX model that represent the most general candidate for the primordial Universe from which even the properties of the general cosmological solution can be extrapolated.

4.1. Polymer Quantum Mechanics

In this section, we introduce PQM as described in the main paper written by Corichi in 2007 [18], where a complete mathematical framework is developed. PQM is an alternative representation of the canonical commutation relations non-unitarily connected to the ordinary Schrödinger one. In fact, it can be introduced as a limit of the Fock representation, where the continuity hypothesis of the Stone–Von Neumann theorem is violated. However, PQM can also be derived without recurring to its connection with the Schrödinger representation as follows.

4.1.1. Polymer Kinematics

In order to introduce the polymer representation without any reference to the Schrödinger one, let us consider the abstract kets $|b\rangle$ labeled by the real parameter $b \in \mathbb{R}$ and taken from the Hilbert space $\mathcal{H}_{\text{poly}}$.

A generic cylindrical state can be defined through the finite linear combination as follows:

$$|\psi\rangle = \sum_{i=1}^{N} n_i |b_i\rangle, \tag{67}$$

where $b_i \in \mathbb{R}$, $i = 1, \ldots, N \in \mathbb{N}$. We choose the inner product so that the fundamental kets are orthonormal as follows:

$$\langle b_i | b_j \rangle = \delta_{ij}. \tag{68}$$

From this choice, it follows that the inner product between two cylindrical states $|\psi\rangle = \sum_i n_i |b_i\rangle$ and $|\phi\rangle = \sum_j m_j |b_j\rangle$ is as follows:

$$\langle \phi | \psi \rangle = \sum_{i,j} m_i^* n_j \delta_{ij} = \sum_i m_i^* n_i. \tag{69}$$

It can be demonstrated that the Hilbert space $\mathcal{H}_{\text{poly}}$ is the Cauchy completion of the finite linear combination of the form (67) with respect to the inner product (68) and that it results to be non-separable.

Two fundamental operators can be defined on this Hilbert space: the symmetric label operator $\hat{e}$ and the shift operator $\hat{s}(\zeta)$ with $\zeta \in \mathbb{R}$. They act on the kets $|b\rangle$ as follows:

$$\hat{e}|b\rangle := b\,|b\rangle, \quad \hat{s}(\zeta)|b\rangle := |b+\zeta\rangle. \tag{70}$$

The shift operator defines a one-parameter family of unitary operators on $\mathcal{H}_{\text{poly}}$. However, since the kets $|b\rangle$ and $|b+\zeta\rangle$ are orthogonal for any $\zeta \neq 0$, the shift operator $\hat{s}(\zeta)$ is discontinuous in ζ, and there is no Hermitian operator that can generate it by exponentiation.

Now, the abstract structure of the Hilbert space is described, so we can proceed by defining the physical states and operators. In the following, we will consider a one-dimensional system identified by the phase-space coordinates (Q, P), and we will separate the discussion into two cases referred to the two possible polarizations for the wave function. We suppose also that the configurational coordinate Q has a discrete character, due to the relation that it often possess with geometrical quantities. This is a way to investigate the physical effects of discreteness at a certain scale, for example, when introducing quantum gravity effects on the cosmological dynamics.

P-Polarization

In the momentum polarization, the wave function is written as follows:

$$\psi(P) := \langle P|\psi\rangle \tag{71}$$

where

$$\psi_b(P) := \langle P|b\rangle = e^{ibP}. \tag{72}$$

The shift operator $\hat{s}(\zeta)$ is identified with the multiplicative exponential operator $\hat{T}(\zeta)$:

$$\hat{T}(\zeta)\psi_b(P) = e^{i\zeta P}e^{ibP} = e^{i(b+\zeta)P} = \psi_{b+\zeta}(P); \tag{73}$$

$\hat{T}(\zeta)$ is discontinuous by definition and as a result, the momentum P cannot be promoted to a well-defined operator. On the other hand, $\hat{Q}$ corresponds to the label operator $\hat{e}$ and in this polarization acts differentially:

$$\hat{Q}\psi_b(P) = -i\frac{\partial}{\partial P}\psi_b(P) = b\psi_b(P). \tag{74}$$

Additionally, it has to be considered as a discrete operator since $\psi_b(P)$ are orthonormal for all b, even though b belongs to a continuous set.

By means of the C*-algebra it can be seen that $\mathcal{H}_{\text{poly}}$ is isomorphic to the following:

$$\mathcal{H}_{\text{poly},P} := L^2(\mathbb{R}_B, d\mu_H) \tag{75}$$

where $\mathbb{R}_B$ is the Bohr compactification of the real line, i.e., the dual group of the real line equipped by the discrete topology, and $d\mu_H$ the Haar measure. Moreover, the wave functions are quasi-periodic with the inner product as follows:

$$\left\langle \psi_{b_i}\middle|\psi_{b_j}\right\rangle := \int_{\mathbb{R}_B} d\mu_H\, \psi_{b_i}^\dagger(P)\psi_{b_j}(P) = \lim_{L\to\infty} \frac{1}{2L}\int_{-L}^{L} dP\, \psi_{b_i}^\dagger(P)\psi_{b_j}(P) = \delta_{ij}. \tag{76}$$

Q-Polarization

In the position polarization, the wave functions depend on the configurational variable Q and a generic state, written as follows:

$$\psi(Q) := \langle Q|\psi\rangle \tag{77}$$

where the basis functions can be derived using a Fourier-like transform as follows:

$$\tilde{\psi}_b(Q) := \langle Q|b\rangle = \langle Q|\int_{\mathbb{R}_B} d\mu_H\, |P\rangle\langle P|b\rangle = \langle Q|\int_{\mathbb{R}_B} d\mu_H\, |P\rangle\psi_b(P) = \int_{\mathbb{R}_B} d\mu_H e^{-iQP}e^{ibP} = \delta_{Qb} \tag{78}$$

through which we can easily see that the $\hat{P}$ operator does not exist since the derivative of the Kronecker delta is not well defined. However, for the operator $\hat{T}(\zeta)$ we have the following:
$$\hat{T}(\zeta)\psi(Q) = \psi(Q+\zeta). \tag{79}$$

As in the previous case, the $\hat{Q}$ operator corresponds to $\hat{e}$, but in this polarization, it acts in a multiplicative way, i.e., the following:
$$\hat{Q}\tilde{\psi}_b(Q) := b\tilde{\psi}_b(Q). \tag{80}$$

The Hilbert space has analogous features as before:
$$\mathcal{H}_{\text{poly},Q} := L^2(\mathbb{R}_d, d\mu_c) \tag{81}$$

where $\mathbb{R}_d$ is the real line equipped with the discrete topology and $d\mu_c$ is the counting measure. The inner product is as follows:
$$\langle \tilde{\psi}_{b_i}(Q), \tilde{\psi}_{b_j}(Q) \rangle = \delta_{ij} \tag{82}$$

so, it is clear how the $\hat{Q}$ operator is discrete also in this polarization.

4.1.2. Polymer Dynamics

In the previous section, the polymer kinematic Hilbert space was introduced. In particular, the discussion above has highlighted that it is not possible to well define the $\hat{Q}$ and $\hat{P}$ operators simultaneously in the polymer framework. So, it is necessary to understand how to implement the dynamics in order to apply the polymer framework to a physical system.

Let us consider a one-dimensional system described by the Hamiltonian as follows:
$$\mathcal{C} = \frac{P^2}{2m} + U(Q) \tag{83}$$

in the P-polarization. If we assume that $\hat{Q}$ is a discrete operator, we have to find an approximate form for $\hat{P}$. For this reason, the required regularization procedure consists of introducing a lattice with a constant spacing b_0:
$$\gamma_{b_0} = \{Q \in \mathbb{R} : Q = Nb_0, \forall N \in \mathbb{Z}\}. \tag{84}$$

In order to remain in the lattice, the only states permitted are as follows:
$$|\psi\rangle = \sum_N n_N |b_N\rangle \in \mathcal{H}_{\gamma_{b_0}} \tag{85}$$

where $b_N = Nb_0$ and $\mathcal{H}_{\gamma_{b_0}}$ is a subspace of $\mathcal{H}_{\text{poly}}$ that contains all the functions ψ so that $\sum_N |b_N|^2 < \infty$.

Now, we have to find an approximate form for $\hat{P}$ in order to have a well-defined Hamiltonian operator through which implement the dynamics in both the polarizations. We notice that the operator $\widehat{e^{i\zeta P}}$ is well defined and acts as the shift operator on the kets $|b\rangle$. In particular, its action is restricted to the lattice only if ζ is a multiple of b_0 and the simplest choice corresponds to $\zeta = b_0$, so that its actions reads as follows:
$$\hat{T}(b_0)|b_N\rangle := |b_N + b_0\rangle = |b_{N+1}\rangle. \tag{86}$$

Therefore, it is possible to use the shift operator to introduce the following approximation:
$$P \sim \frac{1}{b_0}\sin(b_0 P) = \frac{1}{2ib_0}(e^{ib_0 P} - e^{-ib_0 P}), \tag{87}$$

valid in the limit $b_0 P \ll 1$, so that the regularized $\hat{P}$ operator acts as follows:

$$\hat{P}_{b_0}|b_N\rangle = \frac{1}{2ib_0}[\hat{T}(b_0) - \hat{T}(-b_0)]|b_N\rangle = \frac{1}{2ib_0}(|b_{N+1}\rangle - |b_{N-1}\rangle). \tag{88}$$

It is possible to introduce also an approximate version of $\hat{P}^2$ as follows:

$$\hat{P}^2_{b_0}|b_N\rangle \equiv \hat{P}_{b_0} \cdot \hat{P}_{b_0}|b_N\rangle = \frac{1}{4b_0^2}[-|b_{N-2}\rangle + 2|b_N\rangle - |b_{N+2}\rangle] = \frac{1}{b_0^2}\sin^2(b_0 p)|b_N\rangle. \tag{89}$$

We remind that $\hat{Q}$ is a well-defined operator, so the regularized version of the Hamiltonian is written as follows:

$$\hat{\mathcal{C}}^{\text{poly}} := \frac{1}{2m}\hat{P}^2_{b_0} + \hat{U}(Q) \tag{90}$$

that represents a symmetric and well-defined operator on $\mathcal{H}_{\gamma_{b_0}}$.

We notice that this Hamiltonian provides an effective description at the given scale b_0. More specifically, the question about the consistency between the effective theories at different scales and the existence of the continuum limit is deeply investigated in [127]. In particular, it is demonstrated that the continuum Hamiltonian can be represented in a Hilbert space unitarily equivalent to the ordinary L^2 space of the Schrödinger theory by means of a renormalization procedure that involves coarse graining as well as rescaling, following Wilson's renormalization group ideas.

When implementing PQM on the cosmological minisuperspaces, the geometrical variables (namely, the areas p, p_i in the Ashtekar variables, the scale factor α and the anisotropies β_i in the Misner variables, and both the volumes v, v_i and the lengths q_i in the volume-like variables) will be discretized and therefore will play the same role as the position Q; therefore, all the conjugate variables, i.e., $c, \tilde{c}, c_i, \eta_i, P_\alpha, P_v, P_\pm$, will play the same role as the momentum P and will be subjected to the polymer substitution (87).

4.2. Polymer Cosmology in the Ashtekar Variables

In this section, we present the main results about the polymer semiclassical quantization of the FLRW and Bianchi I models in the Ashtekar variables, following [44,45]. In particular, the emergence of a Big Bounce regularizing the initial singularity is a solid prediction in the polymer framework, but the physical properties of the dynamics result in being dependent on the initial conditions on the motion.

4.2.1. The FLRW Universe in the Ashtekar Variables

In this subsection, we treat the semiclassical and quantum polymer dynamics of the FLRW model [44]. Additionally, some analogies with the original LQC scheme are highlighted.

The classical Hamiltonian constraint for this configuration is as follows:

$$\mathcal{C}_{\text{FLRW}} = -\frac{3}{\gamma^2}\sqrt{p}\, c^2 + \frac{P_\phi^2}{2|p|^{\frac{3}{2}}} = 0, \tag{91}$$

where a massless scalar field is included so that ϕ can be chosen as the internal clock for the dynamics.

On a semiclassical level, the polymer paradigm is implemented by considering the variable p as discrete in view of its geometrical character (i.e., it has the dimension of an area) and so a regularized version for the momentum c in the form $c \to \frac{\sin(b_0 c)}{b_0}$ is introduced, obtaining the following:

$$\mathcal{C}_{\text{FLRW}}^{\text{poly}} = -\frac{3}{\gamma^2 b_0^2}\sqrt{p}\, \sin^2(b_0 c) + \frac{P_\phi^2}{2|p|^{\frac{3}{2}}} = 0, \tag{92}$$

in which for the square of the momentum c, we have used the semiclassical version of (89).

Given the Poisson brackets $\{c, p\} = \frac{\gamma}{3}$, we can obtain the equations of motion for p and c as follows:

$$\dot{p} = -\frac{2N}{\gamma b_0}\sqrt{|p|}\sin(b_0 c)\cos(b_0 c), \tag{93a}$$

$$\dot{c} = \frac{N}{3}\left(\frac{3}{\gamma b_0^2}\frac{1}{2\sqrt{p}}\sin^2(b_0 c) + \frac{3\gamma}{4}\frac{P_\phi^2}{|p|^{5/2}}\right). \tag{93b}$$

The analytical expression of the Friedmann equation can be derived as follows:

$$H^2 = \left(\frac{\dot{a}}{a}\right)^2 = \left(\frac{\dot{p}}{2p}\right)^2 = \frac{1}{\gamma^2 b_0^2}\frac{1}{|p|}\sin^2(b_0 c)\left(1 - \sin^2(b_0 c)\right); \tag{94}$$

then, by using (92), we obtain the following:

$$H^2 = \left(\frac{\dot{p}}{2p}\right)^2 = \frac{\rho}{3}\left(1 - \frac{\rho}{\rho_{\text{crit}}}\right), \tag{95}$$

where

$$\rho_{\text{crit}} = \frac{3}{\gamma^2 b_0^2 |p|}. \tag{96}$$

Let us now consider the scalar field ϕ as the internal time for the dynamics, by requiring the lapse function to be as follows:

$$1 = \dot{\phi} = N\frac{\partial \mathcal{C}_{\text{FLRW}}^{\text{poly}}}{\partial P_\phi} = N\frac{P_\phi}{p^{\frac{3}{2}}}, \quad N = \frac{|p|^{\frac{3}{2}}}{P_\phi} = \frac{1}{\sqrt{2\rho}}; \tag{97}$$

therefore, the effective Friedmann equation in the (p, ϕ) plane reads as follows:

$$\left(\frac{1}{|p|}\frac{dp}{d\phi}\right)^2 = \frac{2}{3}\left(1 - \frac{\gamma^2 b_0^2}{6}\frac{P_\phi^2}{|p|^2}\right), \tag{98}$$

and it is analytically solvable after rewriting it in a dimensionless form. The expression of $p(\phi)$ can be written as follows:

$$p(\phi) = \sqrt{\frac{\gamma^2 b_0^2}{6}}P_\phi \cosh\left(\sqrt{\frac{2}{3}}\phi\right). \tag{99}$$

As shown in Figure 3 the polymer trajectory follows the classical one until it reaches a purely quantum era where the effects of quantum geometry become dominant and the resulting dynamics is that of a bouncing Universe replacing the classical Big Bang.

However, the critical energy density depends on the initial conditions on the momentum conjugate to the scalar field, as we can see in the following expression:

$$\rho_{\text{crit}} = \left(\frac{3}{\gamma^2 b_0^2}\right)^{\frac{3}{2}}\frac{\sqrt{2}}{P_\phi}, \tag{100}$$

obtained by putting together the Equations (96) and (99). The non-universal character of the bouncing dynamics in this set of variables has the following consequence:

$$\rho_{\text{crit}} \to \infty \quad \text{for} \quad P_\phi \to 0, \tag{101}$$

Thus, the initial singularity can be asymptotically approached, and the quantum corrections become irrelevant (see Figure 4). Thus, in the Ashtekar variables, the non-diverging behavior of the energy density at the Bounce ensures the regularization of the singularity, due to its scalar nature, but it can assume arbitrarily large values and is not a fixed feature of the dynamics. This result is very similar to the μ_0 scheme of LQC presented in Section 3.2.

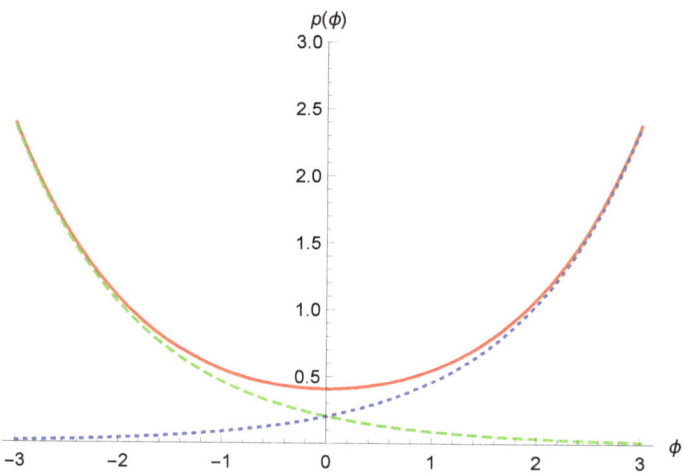

Figure 3. The polymer trajectory (red continuous line) is compared to the classical ones for the flat FLRW model. The Big Bang solution is the blue dotted line and the Big Crunch solution is the green dashed line.

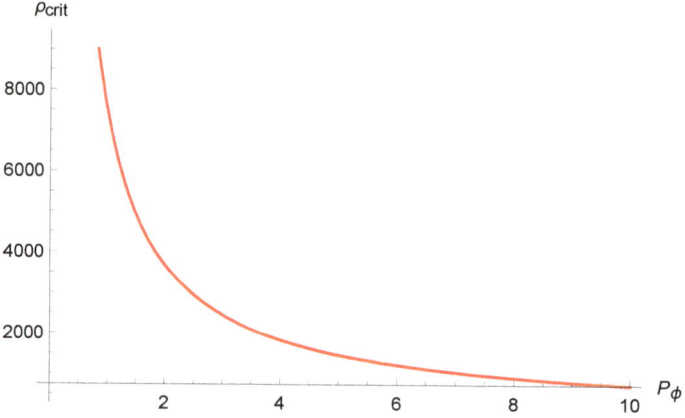

Figure 4. Dependence of the critical energy density on the momentum of the scalar field. For $P_\phi \to 0$ the Bounce approaches the singularity.

Now we want to extend the semiclassical results obtained above to a full quantum level. In order to implement the Dirac quantization method [67], we have to promote variables to quantum operators, i.e., the following:

$$\hat{p} = -\frac{i\gamma}{3}\frac{d}{dc}, \quad \hat{c} = \frac{1}{b_0}\sin(b_0 c), \quad \hat{P}_\phi = -i\frac{d}{d\phi}; \qquad (102)$$

Thus, the Hamiltonian constraint operator in the momentum representation is as follows:

$$\mathcal{C}_{FLRW}^{poly} = \left[-\frac{2}{3b_0^2}\frac{d^2}{dc^2}\sin^2(b_0 c) + \frac{d^2}{d\phi^2}\right] = 0. \qquad (103)$$

In particular, the Hamiltonian constraint selects the physical states by annihilation, giving rise to the following WDW equation:

$$\left[-\frac{2}{3b_0^2}\left(\sin(b_0 c)\frac{d}{dc}\right)^2 + \frac{d^2}{d\phi^2}\right]\Psi(c,\phi) = 0, \quad (104)$$

where we have used a mixed factor ordering that will lead us to a solvable differential equation through the following substitution:

$$x = \sqrt{\frac{3}{2}}\ln\left[\tan\left(\frac{b_0 c}{2}\right)\right] + \bar{x}. \quad (105)$$

Thus, (104) assumes the form of a massless Klein–Gordon-like equation:

$$\frac{d^2}{dx^2}\Psi(x,\phi) = \frac{d^2}{d\phi^2}\Psi(x,\phi), \quad (106)$$

where Ψ is the wave function of the Universe. The solution can be written as follows:

$$\Psi(x,\phi) = \int_0^\infty dk_\phi \, \frac{e^{-\frac{|k_\phi - \bar{k}_\phi|^2}{2\sigma^2}}}{\sqrt{4\pi\sigma^2}} k_\phi \, e^{ik_\phi x} e^{-ik_\phi \phi}, \quad (107)$$

where we have considered only positive energy-like eigenvalues k_ϕ and have used a Gaussian-like weighing function peaked on the initial value $\bar{k}_\phi$.

Now, in order to investigate the non-singular behaviour of the model, we can evaluate the expectation value of the energy density operator as follows:

$$\hat{\rho} = \frac{\hat{p}_\phi^2}{2|\hat{p}|^3} \quad (108)$$

using the basic operators (102) and the substitution (105) to compute the Klein–Gordon scalar product:

$$\langle \Psi | \hat{O} | \Psi \rangle = \int_{-\infty}^\infty dx \, i\left(\Psi^* \partial_\phi(\hat{O}\Psi) - (\hat{O}\Psi)\partial_\phi \Psi^*\right), \quad (109)$$

where we consider Ψ to be normalized.

In Figure 5, we show the time dependence of $\langle \hat{\rho}(\phi) \rangle$ for a fixed value of $\bar{k}_\phi$, while the maximum $\langle \hat{\rho}(\phi_B) \rangle$, i.e., the expectation value of the density at the Bounce is presented in Figure 6 from which we can appreciate the inversely proportional relation with $\bar{k}_\phi$ in accordance with the semiclassical critical density given in (100). The points representing the quantum expectation values are obtained through numerical integration and are fitted with the continuous lines; they are in good accordance with the semiclassical trajectories when taking into account numerical effects and quantum fluctuations.

The quantum analysis performed here has highlighted the non-diverging nature (although strongly dependent on the initial conditions) of the energy density expectation value at the Bounce. This result ensures the existence of a minimum non-zero volume in view of the scalar and physical nature of the energy density, thus confirming the replacement of the singularity with a Big Bounce also at a quantum level. Clearly, a more precise assessment of the nature of the Bounce would require a non-trivial calculation of the expectation value of the Universe volume operator and also a careful analysis of the variance of the energy density operator on the Universe wave function (see for instance [128]).

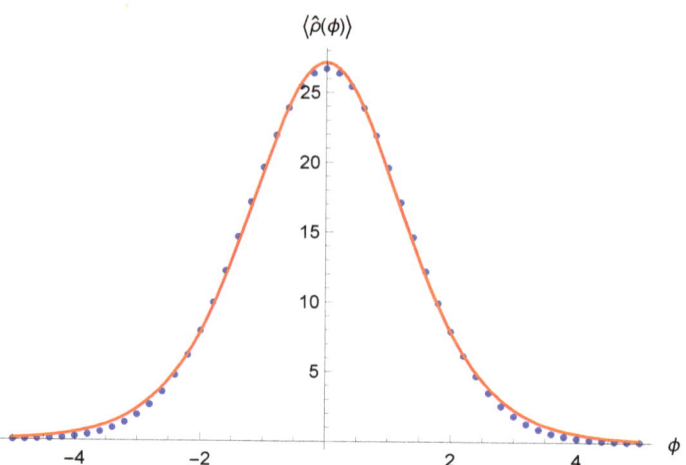

Figure 5. The expectation value of the energy density as function of time (blue dots), fitted with a function in accordance with the semiclassical evolution (continuous red line).

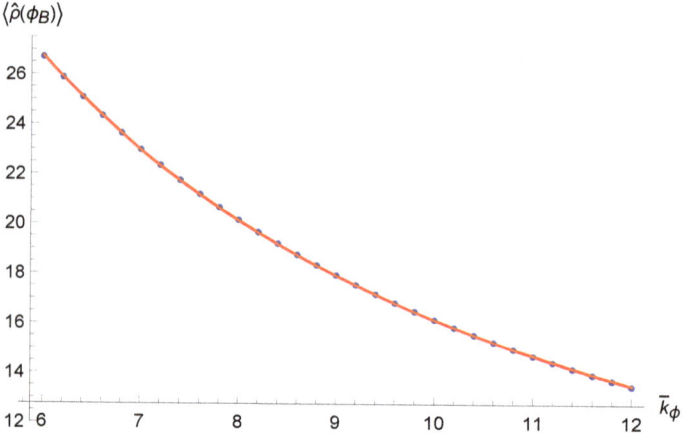

Figure 6. The expectation value of the energy density at the time ϕ_B of the Bounce as function of $\bar{k}_\phi$ (blue dots), fitted with a function in accordance with the semiclassical evolution (continuous red line).

4.2.2. The Bianchi I Universe in the Ashtekar Variables

In this section, we extend all the results obtained for the FLRW model by considering its simplest anisotropic generalization, i.e., the Bianchi I model, as done in [45]. Firstly, we develop the polymer semiclassical analysis of the model in the Ashtekar variables, and then we study the dynamics of the Universe wave packet at a quantum level.

We proceed by discretizing the areas p_i and imposing in (60) the polymer substitution for the connections c_i:

$$c_i \to \frac{1}{b_i} \sin(b_i c_i), \tag{110}$$

where b_i ($i = 1, 2, 3$) refer to the three independent polymer lattices and $\{c_i, p_j\} = \gamma \delta_{ij}$. Then, the polymer Hamiltonian takes the following form:

$$\mathcal{C}_{BI}^{poly} = -\frac{1}{\gamma^2 V} \sum_{i \neq j} \frac{\sin(b_i c_i) p_i \sin(b_j c_j) p_j}{b_i b_j} + \frac{P_\phi^2}{2V} = 0, \tag{111}$$

where $i, j = 1, 2, 3$ and $V = \sqrt{p_1 p_2 p_3}$.

We derive the dynamics after choosing ϕ as the internal time, i.e., imposing the gauge $N = \frac{\sqrt{p_1 p_2 p_3}}{P_\phi}$, so the equations of motion take the following form:

$$\frac{dp_i}{d\phi} = -\frac{p_i \cos(b_i c_i)}{\gamma P_\phi} \left[\frac{p_j}{b_j} \sin(b_j c_j) + \frac{p_k}{b_k} \sin(b_k c_k) \right], \quad (112a)$$

$$\frac{dc_i}{d\phi} = \frac{\sin(b_i c_i)}{\gamma b_i P_\phi} \left[\frac{p_j}{b_j} \sin(b_j c_j) + \frac{p_k}{b_k} \sin(b_k c_k) \right], \quad (112b)$$

for $i, j, k = 1, 2, 3$, $i \neq j \neq k$. It is possible to solve this system by establishing the initial conditions on the variables (c_i, p_i), which also satisfy the Hamiltonian constraint (111). In this respect, we make the following choice:

$$c_1(0) = c_2(0) = c_3(0) = \frac{\pi}{2 b_0},$$
$$p_1(0) = \bar{p}_1, \quad p_2(0) = \bar{p}_2, \quad (113)$$
$$p_3(0) = \bar{p}_3 = \frac{P_\phi^2 \gamma^2 b_0 - 2 \bar{p}_1 \bar{p}_2}{2(\bar{p}_1 + \bar{p}_2)},$$

where $b_1 = b_2 = b_3 = b_0$ without loss of generality. Moreover, it can be easily seen that the momentum conjugate to the scalar field is a first integral since the variable ϕ is cyclic in (111). Other constants of motion can be obtained by combining the Hamilton Equation (112), so we obtain the following:

$$\frac{p_i \sin(b_0 c_i)}{b_0} = \mathcal{K}_i, \quad P_\phi = \mathcal{K}_\phi, \quad (114)$$

where the considered values of $\mathcal{K}_\phi$ and $\mathcal{K}_i$ depend on the initial conditions. Identifying these first integrals allows to transform the six-equations system shown in (112) in three closed systems along the three spatial directions as follows:

$$\frac{dp_i}{d\phi} = -\frac{p_i \cos(b_0 c_i)}{\gamma P_\phi} \left[\mathcal{K}_j + \mathcal{K}_k \right], \quad (115a)$$

$$\frac{dc_i}{d\phi} = \frac{\sin(b_0 c_i)}{\gamma b_0 P_\phi} \left[\mathcal{K}_j + \mathcal{K}_k \right], \quad (115b)$$

where $i \neq j \neq k$.

Thanks to this procedure, the equations of motion can be solved analytically, leading to the following solutions:

$$p_1(\phi) = \bar{p}_1 \cosh\left[\frac{(\gamma^2 P_\phi^2 b_0^2 + 2 \bar{p}_2^2)\phi}{2 \gamma P_\phi b_0 (\bar{p}_1 + \bar{p}_2)} \right],$$
$$p_2(\phi) = \bar{p}_2 \cosh\left[\frac{(\gamma^2 P_\phi^2 b_0^2 + 2 \bar{p}_1^2)\phi}{2 \gamma P_\phi b_0 (\bar{p}_1 + \bar{p}_2)} \right], \quad (116)$$
$$p_3(\phi) = \bar{p}_3 \cosh\left[\frac{(\bar{p}_1 + \bar{p}_2)\phi}{\gamma P_\phi b_0} \right].$$

We have reported only the explicit expression of the variables p_i as functions of ϕ since we are interested in the Universe volume behavior $V(\phi) = \sqrt{p_1(\phi) p_2(\phi) p_3(\phi)}$ that is shown in Figure 7. The resulting trajectory highlights that a quantum Big Bounce replaces the classical Big Bang thanks to the polymer effects, which are expected to become dominant near the Planckian region.

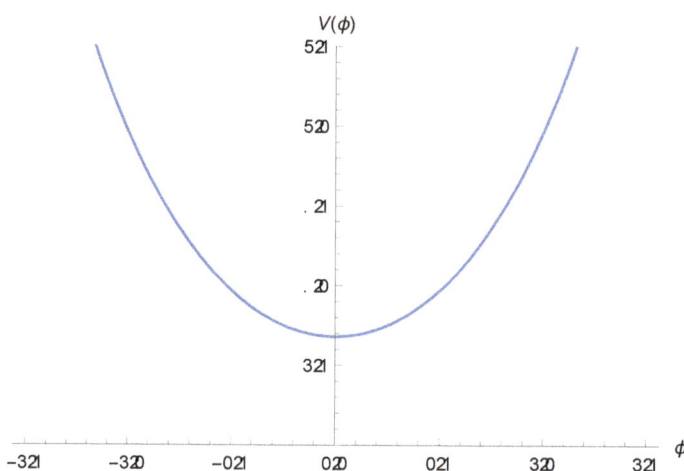

Figure 7. Polymer trajectory of the Universe volume $V = \sqrt{p_1 p_2 p_3}$ as function of ϕ: the Big Bounce replaces the classical singularity of the Bianchi I model.

Now, we focus our attention on the analysis of the critical energy density of matter at the Bounce:

$$\rho_{\text{crit}}^\phi = \frac{P_\phi^2}{2V(\phi)^2}\bigg|_{\phi_B} = \frac{P_\phi^2}{2\bar{p}_1 \bar{p}_2 \bar{p}_3} = \frac{P_\phi^2(\bar{p}_1 + \bar{p}_2)}{\bar{p}_1 \bar{p}_2 (\gamma^2 P_\phi^2 b_0^2 - 2\bar{p}_1 \bar{p}_2)}, \quad (117)$$

i.e., the energy density of the matter scalar field in correspondence of the minimum Universe volume at the time of the Bounce $\phi_B = 0$. As we can see from (117), ρ_{crit}^ϕ clearly depends on the initial conditions on the motion. Moreover, in the simplest case $\mathcal{K}_1 = \mathcal{K}_2 = \mathcal{K}_3$ it reduces to the following:

$$\rho_{\text{crit}}^\phi = \left(\frac{3}{\gamma^2 b_0^2}\right)^{\frac{3}{2}} \frac{\sqrt{2}}{P_\phi}, \quad (118)$$

reproducing a consistent behavior with that one obtained in the FLRW model in the same variables; see (100). Indeed, it is possible to see that for $P_\phi \ll 1$ the critical matter energy density increases until it diverges, while on the other hand, it approaches zero when $P_\phi \gg 1$, highlighting that in the Ashtekar formulation of the Bianchi I model, the Big Bounce has no universal features.

Let us now study this model on a quantum level by implementing the Dirac quantization. In the momentum representation of the polymer formulation, the fundamental operators act as follows:

$$\hat{p}_i := -i\gamma \frac{d}{dc_i}, \quad \hat{c}_i := \frac{\sin(b_i c_i)}{b_i}, \quad \hat{P}_\phi := -i\frac{d}{d\phi}. \quad (119)$$

Before quantization, we rewrite the WDW equation in the form of a Schrödinger one in the attempt of defining a positive and conserved probability density. Therefore, we recall the semiclassical scalar constraint and perform an ADM reduction of the variational principle:

$$P_\phi^2 - \Theta = 0, \quad (120)$$

where

$$\Theta = \frac{2}{\gamma^2}\left[\frac{\sin(b_1 c_1)p_1 \sin(b_2 c_2)p_2}{b_1 b_2} + \frac{\sin(b_1 c_1)p_1 \sin(b_3 c_3)p_3}{b_1 b_3} + \frac{\sin(b_2 c_2)p_2 \sin(b_3 c_3)p_3}{b_2 b_3}\right]. \quad (121)$$

After choosing the scalar field ϕ as the temporal parameter, we derive the ADM Hamiltonian by solving the scalar constraint (120) with respect to the momentum associated to the scalar field:

$$P_\phi \equiv \mathcal{C}_{BI}^{ADM\text{-}poly} = \sqrt{\Theta}. \tag{122}$$

where we choose the positive root in order to guarantee the positive character of the lapse function. Thanks to this procedure, the WDW equation can be rewritten in the form of a Schrödinger one by promoting the ADM Hamiltonian to a quantum operator:

$$-i\partial_\phi \Psi = \sqrt{\hat{\Theta}}\, \Psi, \tag{123}$$

where the operator $\sqrt{\hat{\Theta}}$, that we assume well-defined, can be written as follows:

$$\sqrt{\hat{\Theta}} = \left[\frac{2}{\gamma^2}\left(\partial_{x_1}\partial_{x_2} + \partial_{x_1}\partial_{x_3} + \partial_{x_2}\partial_{x_3}\right)\right]^{1/2}, \tag{124}$$

where the new variables x_i are defined from the connections c_i as follows:

$$x_i = \ln\left[\tan\left(\frac{\mu c_i}{2}\right)\right] + \bar{x}_i. \tag{125}$$

Now, we are allowed to introduce the probability density as follows:

$$\mathcal{P}(\vec{x}, \phi) = \Psi^*(\vec{x}, \phi)\Psi(\vec{x}, \phi), \tag{126}$$

where

$$\Psi(\vec{x}, \phi) = \int_{-\infty}^{\infty} dk_1\, dk_2\, dk_3\, A(k_1, k_2, k_3)\, e^{i(k_1 x_1 + k_2 x_2 + k_3 x_3 + \sqrt{2|k_1 k_2 + k_1 k_3 + k_2 k_3|}\, \phi)}, \tag{127}$$

$$A(k_1, k_2, k_3) = \exp\left(\sum_{i=1}^{3} -\frac{(k_i - \bar{k}_i)^2}{2\sigma_{k_i}^2}\right). \tag{128}$$

This quantity is positive everywhere by definition, and its spatial integral remains constant through time.

Hence, we have discarded the covariant formulation of the WDW equation in favor of a Schrödinger-like one in order to define a probability density through which we analyze the quantum dynamics of the Bianchi I wave packet presented in (127). This way, we have avoided all the issues regarding the sign of the probability density that would be encountered in the interpretation of the WDW formulation as a Klein–Gordon-like theory.

In Figure 8, some different sections of the probability density $\mathcal{P}$ are shown at different times in order to verify how its shape and its maximum evolve. The present sections were obtained by fixing two of the three coordinates through the values that they assume in the semiclassical trajectories[1]. As we can see from Figure 8, the normalized quantum distributions of x_1, x_2, x_3 are shown in sequence, and their spreading behavior over time is evident. Additionally, in Figure 9, the position of the peaks of $\mathcal{P}$ is represented by the red dots that are fitted by means of a linear interpolation and compared with the semiclassical trajectories. We can affirm that there is a good correspondence between the quantum behavior of the wave packet and the solutions of the semiclassical dynamics since all the three slopes of the functions resulting from the fit of the red dots are consistent with the semiclassical ones with a confidence level of 3 standard deviations (we have supposed a relative error of 10%, accounting for quantum fluctuations and numerical integration errors). Moreover, we want to highlight that this quantum analysis based on the sections of $\mathcal{P}$ is justified by the semiclassical decoupling of the equations of motion. In conclusion, we remark that this feature of our analysis in the Ashtekar variables suggests the presence of a bouncing dynamics with non-universal properties also at a quantum level.

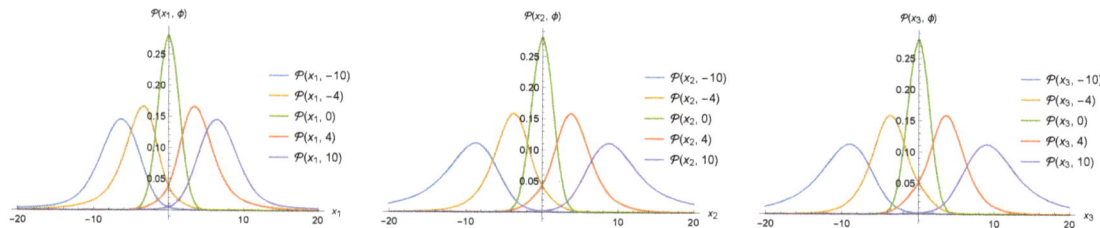

Figure 8. The normalized sections $\mathcal{P}(x_i, \phi)$ are shown in sequence for $i = 1, 2, 3$ respectively at different times. Their spreading behavior over time is evident together with the Gaussian-like shape.

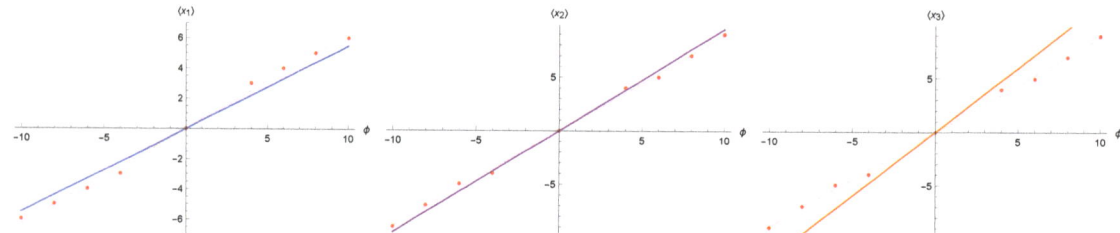

Figure 9. The three pictures show the position of the peaks of $\mathcal{P}(x_i, \phi)$ for $i = 1, 2, 3$ respectively in function of time ϕ (red dots). The resulting fitting functions (red dashed straight lines) overlap the semiclassical trajectories (continuous lines) with a confidence level of the order of 3 standard deviations.

4.3. Polymer Cosmology in the Volume-like Variables

In this section, we describe the FLRW dynamics in the polymer framework when the polymer lattice is implemented on the Universe volume v. Then, we generalize the analysis by taking under consideration the Bianchi I model, which admits two different sets of volume-like variables thanks to its anisotropic structure. In particular, when the Universe volume itself is considered to be one of the configurational variables, it is possible to derive a polymer modified Friedmann equation for the Bianchi I model. We will see that the total critical energy density has universal features, as demonstrated in [37] for the isotropic case.

4.3.1. The FLRW Universe in the Volume Variables

Following [44], we outline the semiclassical dynamics of the model and then we perform a full quantum analysis by analyzing the properties of the Universe wave packet. In order to apply the polymer semiclassical framework to the flat FLRW model in the volume variables, we introduce the Hamiltonian constraint as follows:

$$\mathcal{C}_{\text{FLRW}} = -\frac{27}{4\gamma^2} v \, \tilde{c}^2 + \frac{p_\phi^2}{2v}, \tag{129}$$

where the phase space variables $\tilde{c}, v$ are linked to the Ashtekar variables c, p through the following canonical transformation:

$$v = |p|^{\frac{3}{2}} = a^3, \quad \tilde{c} = \frac{2}{3} \frac{c}{\sqrt{|p|}} \propto \frac{\dot{a}}{a} \tag{130}$$

that preserves the Poisson brackets $\{\tilde{c}, v\} = \frac{\gamma}{3}$. Additionally, a massless scalar field has been added to the dynamics with the role of a relational time. Thus, the modified polymer Hamiltonian is as follows:

$$\mathcal{C}_{\text{FLRW}}^{\text{poly}} = -\frac{27}{4\gamma^2 b_0^2} v \sin^2(b_0 \tilde{c}) + \frac{p_\phi^2}{2v} = 0, \tag{131}$$

where the variable v is defined as discrete and so the semiclassical polymer substitution (87) $\tilde{c} \to \frac{\sin(b_0 \tilde{c})}{b_0}$ is used for the generalized coordinate $\tilde{c}$.

The equations of motion for these variables are as follows:

$$\dot{v} = -\frac{9N}{2\gamma b_0} v \sin(b_0 \tilde{c}) \cos(b_0 \tilde{c}), \tag{132a}$$

$$\dot{\tilde{c}} = \frac{N}{3}\left(\frac{27}{4\gamma b_0^2} \sin^2(b_0 \tilde{c}) + \gamma \frac{P_\phi^2}{2v^2}\right). \tag{132b}$$

As expected in the polymer framework, a modified Friedmann equation appears; in particular, we can obtain its analytical expression by using (132a) and the vanishing Hamiltonian constraint (131):

$$H^2 = \left(\frac{\dot{v}}{3v}\right)^2 = \frac{\rho}{3}\left(1 - \frac{\rho}{\rho_{\text{crit}}}\right), \quad \rho_{\text{crit}} = \frac{27}{4\gamma^2 b_0^2}. \tag{133}$$

The presence of a bouncing point for the Hubble parameter represents the mechanism through which the initial singularity is regularized. In addition, the critical energy density at which the Big Bounce occurs is fixed by the spacing b_0 associated to the polymer lattice, giving the dynamics a universal character. This result is in strong contact with the analogous considerations made in the context of the $\bar{\mu}$ approach in the LQC formulation in Section 3.3.

Considering now the scalar field ϕ as the internal time of the dynamics, we fix the time gauge, i.e.,

$$1 = \dot{\phi} = N \frac{\partial \mathcal{C}_{\text{FLRW}}^{\text{poly}}}{\partial P_\phi} = N \frac{P_\phi}{v}, \quad N = \frac{v}{P_\phi} = \frac{1}{\sqrt{2\rho}}; \tag{134}$$

thus the effective Friedmann equation in the (v, ϕ) plane reads as follows:

$$\left(\frac{1}{v}\frac{dv}{d\phi}\right)^2 = \frac{3}{2}\left(1 - \frac{4\gamma^2 b_0^2}{54}\frac{P_\phi^2}{v^2}\right), \tag{135}$$

whose analytical solution is as follows:

$$v(\phi) = \sqrt{\frac{4\gamma^2 b_0^2}{54}} P_\phi \cosh\left(\sqrt{\frac{3}{2}}\phi\right). \tag{136}$$

As shown in Figure 10 the Bounce is clearly visible since the Universe possesses a minimum volume.

Now, we promote the system to a full quantum level by applying the Dirac procedure [67]. In particular, the Hamiltonian operator annihilates and selects the physical wave functions when applied to the generic quantum states, yielding the WDW equation:

$$\hat{\mathcal{C}}_{\text{FLRW}}^{\text{poly}} |\Psi\rangle = 0. \tag{137}$$

If we promote the classical variables to a quantum level in the momentum representation, we obtain the following:

$$\hat{v} = -\frac{i\gamma}{3}\frac{d}{d\tilde{c}}, \quad \hat{\tilde{c}} = \frac{1}{b_0}\sin(b_0 \tilde{c}), \quad \hat{P}_\phi = -i\frac{d}{d\phi}, \tag{138}$$

recalling that v is discrete and therefore $\tilde{c}$ must be regularized. So, the quantum version of the Hamiltonian constraint (131) is as follows:

$$\left[-\frac{3}{2b_0^2}\left(\sin\left(b_0\tilde{c}\right)\frac{d}{d\tilde{c}}\right)^2 + \frac{d^2}{d\phi^2}\right]\Psi(\tilde{c},\phi) = 0; \tag{139}$$

Using again a mixed factor ordering and the substitution

$$\tilde{x} = \sqrt{\frac{2}{3}}\ln\left[\tan\left(\frac{b_0\tilde{c}}{2}\right)\right] + \tilde{x}, \tag{140}$$

we can recast recast the expression (139) in the form of a massless Klein–Gordon-like equation:

$$\frac{d^2}{d\tilde{x}^2}\Psi(\tilde{x},\phi) = \frac{d^2}{d\phi^2}\Psi(\tilde{x},\phi). \tag{141}$$

This way, the wave packet representing the wave function of the Universe can be written as follows:

$$\Psi(\tilde{x},\phi) = \int_0^\infty dk_\phi \frac{e^{-\frac{|k_\phi - \bar{k}_\phi|^2}{2\sigma^2}}}{\sqrt{4\pi\sigma^2}} k_\phi e^{ik_\phi\tilde{x}}e^{-ik_\phi\phi} \tag{142}$$

in the $\tilde{x}$-representation. We have constructed a superposition of plane waves in ϕ by means of Gaussian-like coefficients and we have restricted the analysis only to the positive energy-like eigenvalues k_ϕ.

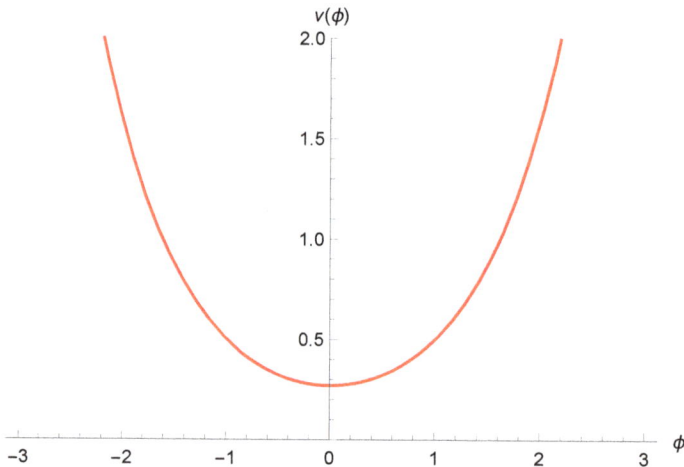

Figure 10. The polymer trajectory of the volume in $(v,\tilde{c})$ for the flat FLRW Universe. The volume presents a minimum, showing that a Big Bounce takes place.

In this case, to study the bouncing behavior of the model, we can evaluate the expectation values of both the volume and the energy density operators:

$$\hat{V} = \hat{v}, \quad \hat{\rho} = \frac{\hat{p}_\phi^2}{2\hat{v}^2}. \tag{143}$$

In what follows, we express all the quantities as functions of the new variable $\tilde{x}$ using (140), so that we can calculate the expectation values through the Klein–Gordon scalar product as follows:

$$\langle \Psi | \hat{O} | \Psi \rangle = \int_{-\infty}^{\infty} d\tilde{x}\, i\big(\Psi^* \partial_\phi (\hat{O}\Psi) - (\hat{O}\Psi)\partial_\phi \Psi^*\big), \tag{144}$$

where we assume normalized wave functions. The action of the operators (143) is derived from the basic operators shown in (138). In Figures 11 and 12, we show, respectively, the expectation values $\langle \hat{V}(\phi) \rangle$ and $\langle \hat{\rho}(\phi) \rangle$ as functions of time. In Figure 13, the value $\langle \hat{V}(\phi_B) \rangle$ of the volume at the Bounce is shown as a function of the initial value $\bar{k}_\phi$; it is clear how the minimum volume scales linearly with the energy-like eigenvalue, in accordance with (136). Then, in Figure 14, we show the Bounce density $\langle \hat{\rho}(\phi_B) \rangle$ for different values of $\bar{k}_\phi$. In accordance with the semiclassical critical density (133), choosing the volume itself as the configurational variable for the quantization of the system makes the density at the Bounce independent from the initial conditions on the quantum solution (142). This quantum analysis allows a direct comparison with the $\bar{\mu}$ scheme of LQC, and it shows how polymer quantum cosmology is unable to reproduce the inverse triad corrections. We notice that in Figures 11–14, the numerically integrated points are fitted with the continuous lines to check for consistency with the semiclassical solutions.

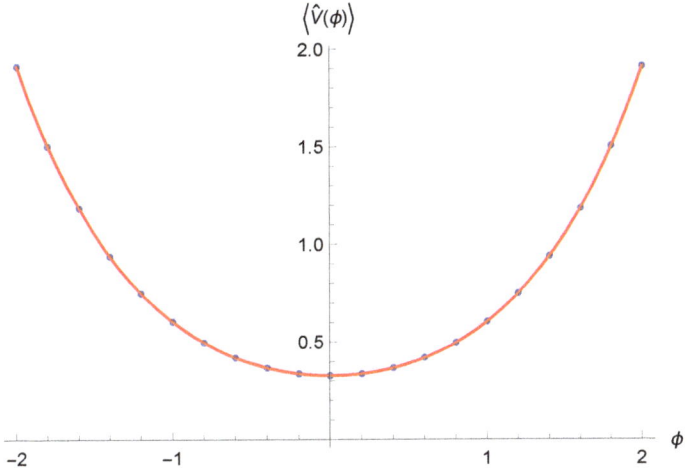

Figure 11. The expectation value of the Universe volume as a function of time (blue dots), fitted with a function in accordance with the semiclassical evolution (continuous red line).

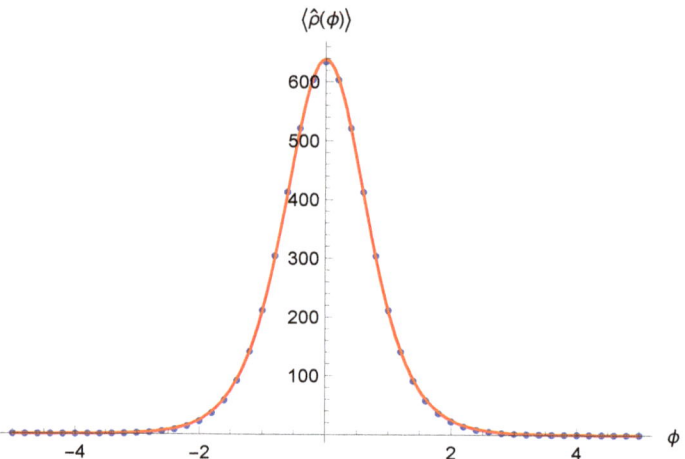

Figure 12. The expectation value of the energy density as a function of time (blue dots), fitted with a function in accordance with the semiclassical evolution (continuous red line).

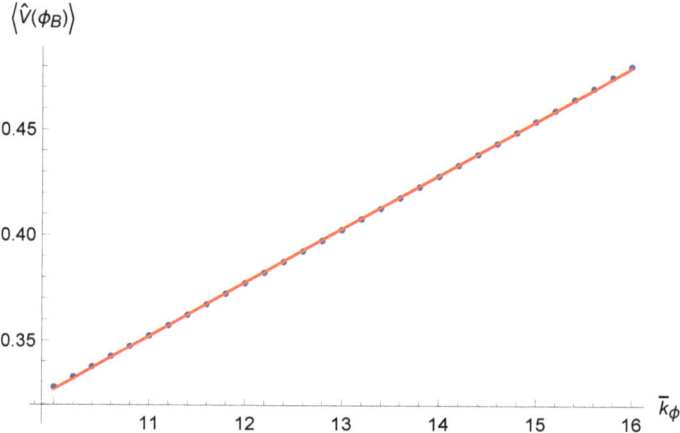

Figure 13. The expectation value of the Universe volume at the time ϕ_B of the Bounce as a function of $\bar{k}_\phi$ (blue dots), fitted with a function in accordance with the semiclassical evolution (continuous red line).

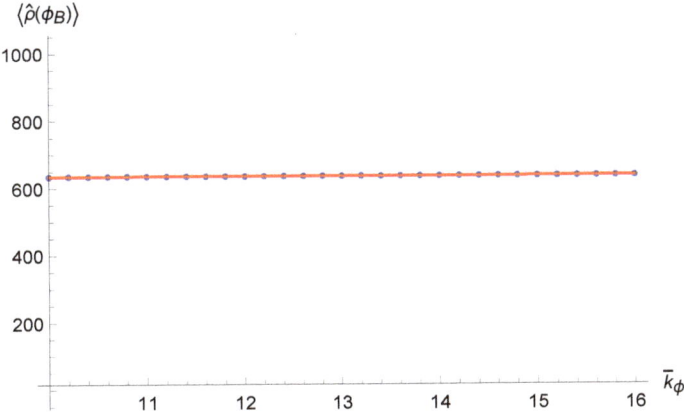

Figure 14. The expectation value of the energy density at the time ϕ_B of the Bounce as a function of $\bar{k}_\phi$ (blue dots), fitted with a function in accordance with the semiclassical evolution (continuous red line).

4.3.2. The Bianchi I Model in the Anisotropic Volume Variables: (v_1, v_2, v_3)

In this subsection, we study the polymer semiclassical dynamics of the Bianchi I model in complete analogy with the analysis performed for the FLRW model in the volume variables, as presented in [45]. More specifically, the anisotropic character of the Bianchi I model leads to the possibility of taking into account two different sets of volume-like variables that coincide in the case of the isotropic model. The first that we take under consideration consists of three equivalent generalized coordinates (see [129]):

$$v_i = \text{sgn}(p_i)|p_i|^{\frac{3}{2}}, \quad \eta_i = \frac{c_i}{\sqrt{|p_i|}} \tag{145}$$

for $i = 1, 2, 3$, where η_i are the conjugate momenta and the new symplectic structure for the system is characterized by the following Poisson brackets:

$$\{\eta_i, v_j\} = \frac{3}{2}\gamma \delta_{ij}. \tag{146}$$

In this case, we are not promoting one of the configurational variables to represent the Universe volume. On the contrary, we are imposing that the three independent coordinates are isomorphic to the isotropic volume for each direction so that $V = (v_1 v_2 v_3)^{\frac{1}{3}}$. In particular, each coordinate v_i reduces to the Universe volume in the isotropic limit.

The Hamiltonian constraint for this framework in the polymer representation reads as follows:

$$\mathcal{C}_{BI}^{\text{poly}} = -\frac{1}{4\gamma^2 V} \sum_{i \neq j} \frac{v_i \sin(b_i \eta_i) v_j \sin(b_j \eta_j)}{b_i b_j} + \frac{P_\phi^2}{2V} = 0, \tag{147}$$

where $i, j = 1, 2, 3$ and the substitution $\eta_i \to \frac{\sin(b_i \eta_i)}{b_i}$ is performed.

If we fix the gauge through the choice $N = \frac{V}{P_\phi}$, the Hamilton equations will describe the dynamics of the model with respect to ϕ playing the role of the time variable:

$$\frac{dv_i}{d\phi} = -\frac{v_i \cos(b_i \eta_i)}{4\gamma P_\phi}\left[\frac{v_j}{b_j}\sin(b_j \eta_j) + \frac{v_k}{b_k}\sin(b_k \eta_k)\right] \tag{148a}$$

$$\frac{d\eta_i}{d\phi} = \frac{\sin(b_i \eta_i)}{4\gamma b_i P_\phi}\left[\frac{v_j}{b_j}\sin(b_j \eta_j) + \frac{v_k}{b_k}\sin(b_k \eta_k)\right] \tag{148b}$$

for $i, j, k = 1, 2, 3$ and $i \neq j \neq k$. In analogy with the previous treatment, we can identity the following constants of motion:

$$\frac{v_i \sin(b_0 \eta_i)}{b_0} = \mathcal{K}_i, \quad P_\phi = \mathcal{K}_\phi \tag{149}$$

where $b_1 = b_2 = b_3 = b_0$. Taking general initial conditions that satisfy the constraint (147), the analytical solutions for the anisotropic volume coordinates read as follows:

$$\begin{aligned}
v_1(\phi) &= \bar{v}_1 \cosh\left[\frac{(2\gamma^2 P_\phi^2 b_0^2 + \bar{v}_2^2)\phi}{4\gamma P_\phi b_0 (\bar{v}_1 + \bar{v}_2)}\right], \\
v_2(\phi) &= \bar{v}_2 \cosh\left[\frac{(2\gamma^2 P_\phi^2 b_0^2 + \bar{v}_1^2)\phi}{4\gamma P_\phi b_0 (\bar{v}_1 + \bar{v}_2)}\right], \\
v_3(\phi) &= \bar{v}_3 \cosh\left[\frac{(\bar{v}_1 + \bar{v}_2)\phi}{4\gamma P_\phi b_0}\right],
\end{aligned} \tag{150}$$

where

$$\begin{aligned}
v_1(0) &= \bar{v}_1, \quad v_2(0) = \bar{v}_2, \\
v_3(0) &= \bar{v}_3 = \frac{(2\gamma^2 b_0^2 P_\phi^2 - \bar{v}_1 \bar{v}_2)}{(\bar{v}_1 + \bar{v}_2)}.
\end{aligned} \tag{151}$$

By combining the solutions (150), we can find the Universe volume behavior in function of ϕ as $V(\phi) = (v_1(\phi) v_2(\phi) v_3(\phi))^{1/3}$. As shown in Figure 15, the Big Bounce appears as a polymer regularization effect in place of the classical Big Bang. Analogously, it is possible to investigate the properties of the critical energy density of matter by computing the energy density of the scalar field in correspondence of the minimum volume at the time $\phi_B = 0$ of the Bounce, i.e., the following:

$$\rho_{crit}^\phi = \frac{P_\phi^2}{2V(\phi)^2}\bigg|_{\phi_B} = \frac{P_\phi^2}{2(\bar{v}_1 \bar{v}_2 \bar{v}_3)^{2/3}} = \frac{P_\phi^2(\bar{v}_1 + \bar{v}_2)^{2/3}}{2[\bar{v}_1 \bar{v}_2 (2\gamma^2 P_\phi^2 b_0^2 - \bar{v}_1 \bar{v}_2)]^{2/3}}; \tag{152}$$

this results to be dependent on the initial conditions also in this set of anisotropic volume variables. In the next section, we implement directly the Universe volume as a generalized coordinate, trying to overcome this issue.

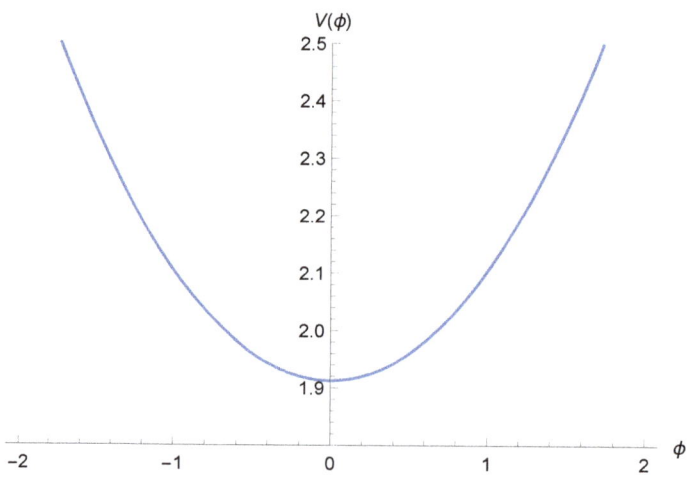

Figure 15. Bouncing trajectory of the Universe volume $V = (v_1 v_2 v_3)^{1/3}$ as function of ϕ in the semiclassical polymer framework.

4.3.3. The Bianchi I Model in the Volume Variables: (v, q_1, q_2)

Now, we consider the set of volume variables introduced in [45]:

$$q_i = \text{sgn}(p_i)\sqrt{|p_i|}, \quad v = q_1 q_2 q_3 \tag{153}$$

where η_i (with $i = 1, 2, 3$) are the conjugate momenta and v represents the Universe volume. The Poisson brackets are as follows:

$$\{\eta_i, q_j\} = \gamma \delta_{ij}. \tag{154}$$

and the Hamiltonian takes the following form:

$$\mathcal{C}_{BI} = -\frac{1}{4\gamma^2 V}(q_1 \eta_1 q_2 \eta_2 + q_1 \eta_1 v \eta_3 + q_2 \eta_2 v \eta_3) + \frac{P_\phi^2}{2V} = 0. \tag{155}$$

When we rewrite this expression using the polymer substitution $\eta_i \to \frac{\sin(b_i \eta_i)}{b_i}$, we obtain the following:

$$\mathcal{C}_{BI}^{poly} = -\frac{1}{4\gamma^2 V}\left(\sum_{i=1,2} \frac{q_i \sin(b_i \eta_i) v \sin(b_3 \eta_3)}{b_i b_3} + \frac{q_1 \sin(b_1 \eta_1) q_2 \sin(b_2 \eta_2)}{b_1 b_2}\right) + \frac{P_\phi^2}{2V} = 0. \tag{156}$$

In order to derive the dynamics of the model in function of the relational time ϕ, we impose the following:

$$\dot{\phi} := N\frac{\partial \mathcal{C}_{BI}^{poly}}{\partial P_\phi} = 1, \quad N = \frac{V}{P_\phi}, \tag{157}$$

so the Hamilton equations for the couple (v, η_3) take the following expressions:

$$\frac{dv}{d\phi} = -\frac{v \cos(b_3 \eta_3)}{4\gamma P_\phi}\left[\frac{q_1}{b_1}\sin(b_1 \eta_1) + \frac{q_2}{b_2}\sin(b_2 \eta_2)\right], \tag{158a}$$

$$\frac{d\eta_3}{d\phi} = \frac{\sin(b_3 \eta_3)}{4\gamma b_3 P_\phi}\left[\frac{q_1}{b_1}\sin(b_1 \eta_1) + \frac{q_2}{b_2}\sin(b_2 \eta_2)\right], \tag{158b}$$

while for the conjugate variables (q_1, η_1), (q_2, η_2) we have the following:

$$\frac{dq_i}{d\phi} = -\frac{q_i \cos(b_i \eta_i)}{4\gamma P_\phi}\left[\frac{v}{b_3}\sin(b_3 \eta_3) + \frac{q_j}{b_j}\sin(b_j \eta_j)\right], \tag{159a}$$

$$\frac{d\eta_i}{d\phi} = \frac{\sin(b_i \eta_i)}{4\gamma b_i P_\phi}\left[\frac{v}{b_3}\sin(b_3 \eta_3) + \frac{q_j}{b_j}\sin(b_j \eta_j)\right], \tag{159b}$$

for $i, j = 1, 2$ and $i \neq j$.

Once fixed the initial conditions on the variables (q_1, η_1), (q_2, η_2), (v, η_3) according to (156), we can solve this system analytically since the 3D motion is decoupled in three one-dimensional trajectories, thanks to the use of the constants of motion as follows:

$$\frac{q_i \sin(b_0 \eta_i)}{b_0} = \mathcal{K}_i, \quad P_\phi = \mathcal{K}_\phi \tag{160}$$

where $i = 1, 2, 3$ and $b_1 = b_2 = b_3 = b_0$. Differently from the previous analyses, we fix the constants of motion as follows:

$$\mathcal{K}_1 = \sqrt{\frac{2\gamma^2 P_\phi^2 + \mathcal{K}^2}{3}} + \mathcal{K},$$

$$\mathcal{K}_2 = \sqrt{\frac{2\gamma^2 P_\phi^2 + \mathcal{K}^2}{3}} - \mathcal{K}, \quad (161)$$

$$\mathcal{K}_3 = \sqrt{\frac{2\gamma^2 P_\phi^2 + \mathcal{K}^2}{3}},$$

where $\mathcal{K}$ is an arbitrary constant. As we will see below, this choice allows to write a convenient form for the Friedmann equation in order to develop a more rigorous analysis of the bouncing behavior of the model. Indeed, the existence of a non-trivial solution to the equation $H^2 = 0$ identifies the expression of the critical energy density for which the scale factor velocity becomes null. More precisely, this information allows to identify also the anisotropy contribution to the total critical energy density added to the standard one associated to the matter fields. This way, it is possible to rigorously analyze the physical properties of the critical point.

In this set of volume variables, the Hubble parameter can be written as follows:

$$H^2 = \left(\frac{1}{3v}\frac{dv}{dt}\right)^2 = \frac{(\mathcal{K}_1 + \mathcal{K}_2)^2}{144\gamma^2 v^2} \cos^2(b_0 \eta_3) =$$
$$= \frac{(\mathcal{K}_1 + \mathcal{K}_2)^2}{144\gamma^2 v^2}[1 - \sin^2(b_0 \eta_3)] = \frac{(\mathcal{K}_1 + \mathcal{K}_2)^2}{144\gamma^2 v^2}\left(1 - \frac{b_0^2}{v^2}\mathcal{K}_3^2\right) \quad (162)$$

where we restored the synchronous time-gauge $N = 1$ in the equation for the volume written in (158a). Now, if we substitute the conditions expressed above in (161), we obtain the following:

$$H^2 = \frac{P_\phi^2 + \tilde{\mathcal{K}}^2}{54v^2}\left[1 - \frac{4\gamma^2 b_0^2}{3}\left(\frac{P_\phi^2 + \tilde{\mathcal{K}}^2}{2v^2}\right)\right], \quad \tilde{\mathcal{K}} = \frac{\mathcal{K}}{\sqrt{2\gamma^2}}. \quad (163)$$

This expression represents the polymer modified Friedmann equation for the Bianchi I model in the volume variables that, written in this convenient form, allows to derive the total critical energy density of the model. In particular, the additional term $\tilde{\mathcal{K}}^2/2v^2$ reasonably mimics the anisotropic contribution ρ^{aniso}, so that we can compute ρ^{tot}_{crit} as follows:

$$\rho^{tot}_{crit} = \rho^\phi_{crit} + \rho^{aniso}_{crit}, \quad (164)$$

where the term regarding the matter scalar field takes the usual expression $P_\phi^2/(2v^2)$. Moreover, from (163) the total critical energy density results to be as follows:

$$\rho^{tot}_{crit} = \frac{3}{4\gamma^2 b_0^2}. \quad (165)$$

The solution for the Universe volume $v(\phi)$ when the initial conditions on the motion satisfy (161) is as follows:

$$v(\phi) = \sqrt{\frac{2\gamma^2 P_\phi^2 + \mathcal{K}^2}{3}}\, b_0 \cosh\left(\frac{\sqrt{2\gamma^2 P_\phi^2 + \mathcal{K}^2}\,\phi}{2\sqrt{3}\,\gamma P_\phi}\right), \quad (166)$$

clearly resembling a bouncing behavior as shown in Figure 16.

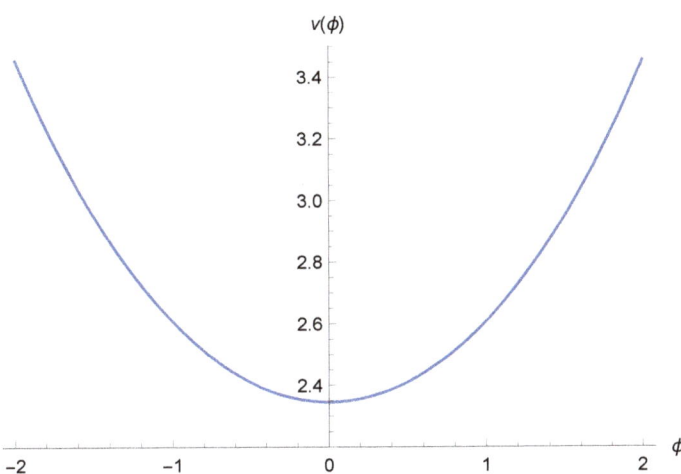

Figure 16. Semiclassical polymer trajectory of the Universe volume $v(\phi)$.

Now, we can verify whether the total critical energy density computed as (164) acquires the expression (165). In particular, we have the following:

$$\rho_{\text{crit}}^{\text{tot}} = \left.\frac{P_\phi^2}{2v(\phi)^2}\right|_{\phi_B} + \left.\frac{\bar{\mathcal{K}}^2}{2v(\phi)^2}\right|_{\phi_B} = \frac{3P_\phi^2}{2b_0^2(2\gamma^2 P_\phi^2 + \mathcal{K}^2)} + \frac{3\mathcal{K}^2}{4\gamma^2 b_0^2(2\gamma^2 P_\phi^2 + \mathcal{K}^2)}, \quad (167)$$

which clearly reduces to (165); therefore, the total critical density computed from the laws of motion for the Universe volume is consistent with the expression derived from the Friedmann Equation (163). Imposing a cut-off on the volume variables makes the critical energy density independent from the initial conditions on the scalar field, and therefore, produces a Big Bounce with universal properties. This result shows that taking the Universe volume itself as a configurational variable makes the Big Bounce acquire universal physical properties, in agreement with the behavior obtained for the set $(v, \bar{c})$ in the FLRW model.

We notice that, even if the polymer-modified Friedmann equation in this convenient form is derived by considering the particular choice (161) for the constants of motion, the physical properties of the bouncing behavior of our model as derived from (163) have a general meaning since they are independent from the value assigned to the constant $\mathcal{K}$.

4.4. Polymer Cosmology in the Misner-like Variables

In this section, we treat the semiclassical polymer dynamics of the most general anisotropic Universe, i.e., the Bianchi IX model. First, we implement the polymer paradigm in the Misner variables [42], and then we generalize the analysis by considering a inhomogeneous extension of the model in Misner-like variables, i.e., the anisotropies together with the Universe volume $v \propto e^{3\alpha}$ [38].

4.4.1. The Bianchi IX Model in the Misner Variables

The aim of this section is to discuss the main features of the Mixmaster semiclassical dynamics in the polymer representation [42]. We choose to define the Misner variables $(\alpha, \beta_\pm)$ as discrete, so we perform the following formal substitutions:

$$P_\pm^2 \to \frac{1}{b^2}\sin^2(bP_\pm), \quad (168a)$$

$$P_\alpha^2 \to \frac{1}{b_\alpha^2}\sin^2(b_\alpha P_\alpha), \quad (168b)$$

where b is the polymer parameter for the anisotropies, while b_α is that one related to the isotropic variable α.

The super Hamiltonian constraint (14) becomes the following:

$$\mathcal{C}_{\text{BIX}}^{\text{poly}} = \frac{N}{3(8\pi)^2} e^{-3\alpha} \left[-\frac{1}{b_\alpha^2} \sin^2(b_\alpha P_\alpha) + \frac{1}{b^2} \sin^2(bP_+) + \frac{1}{b^2} \sin^2(bP_-) + \frac{\mathcal{U}_{\text{BIX}}(\beta_\pm)}{b_\alpha^2} \right] = 0, \tag{169}$$

while the ADM Hamiltonian has the following expression:

$$\mathcal{C}_{\text{BIX}}^{\text{ADM-poly}} = \frac{\arcsin\sqrt{\frac{b_\alpha^2}{b^2}[\sin^2(bP_+) + \sin^2(bP_-)] + \mathcal{U}_{\text{BIX}}}}{b_\alpha}, \tag{170}$$

with the condition $0 \leq \frac{b_\alpha^2}{b^2}[\sin^2(bP_+) + \sin^2(bP_-)] + \mathcal{U}_{\text{BIX}}(\beta_\pm) \leq 1$, due to the presence of the inverse sine function. In both (169) and (170), we have performed the substitution $\mathcal{U}_{\text{BIX}}(\beta_\pm) = b_\alpha^2 3(4\pi)^4 e^{4\alpha} U_{\text{BIX}}(\beta_\pm)$.

The dynamics of the model is described by the following Hamilton equations:

$$\beta'_\pm = \frac{\partial \mathcal{C}_{\text{BIX}}^{\text{ADM-poly}}}{\partial P_\pm} = \frac{b_\alpha}{b} \frac{\sin(2bP_\pm)}{\sin\left(2b_\alpha \mathcal{C}_{\text{BIX}}^{\text{ADM-poly}}\right)}, \tag{171a}$$

$$P'_\pm = -\frac{\partial \mathcal{C}_{\text{BIX}}^{\text{ADM-poly}}}{\partial \beta_\pm} = -\frac{3b_\alpha(4\pi)^4 e^{4\alpha}}{\sin\left(2b_\alpha \mathcal{C}_{\text{BIX}}^{\text{ADM-poly}}\right)} \frac{\partial U(\beta_\pm)}{\partial \beta_\pm}, \tag{171b}$$

$$\left(\mathcal{C}_{\text{BIX}}^{\text{ADM-poly}}\right)' = 4e^{4\alpha - 8\beta_+} \frac{3b_\alpha(4\pi)^4}{\sin\left(2b_\alpha \mathcal{C}_{\text{BIX}}^{\text{ADM-poly}}\right)}, \tag{171c}$$

where the symbol $'$ denotes the derivative with respect to α.

Due to the steepness of the potential walls, we initially study the dynamics in the regime $U \sim 0$. Under this condition, it can be demonstrated that there is a logarithmic relation between the time variable t and the isotropic one α, so we have that $\alpha \sim \ln(t) \to -\infty$ for $t \to 0$, even if α is described in the polymer formulation. This result points out that the discrete nature of the isotropic variable α does not prevent the Universe volume from vanishing, so the cosmological singularity is still present. We remark that the hypothesis $U \sim 0$ becomes more reliable near the singularity because for $\alpha \to -\infty$, the walls move outwards, and the region where the approximation is valid asymptotically covers the whole (β_+, β_-) plane.

Regarding the study of the chaotic features in the polymer modified picture, we have to analyze the relative motion between the particle-Universe and the potential walls, whose velocity is still $|\beta'_{\text{wall}}| = \frac{1}{2}$ since the polymer representation leaves the potential unchanged. On the other hand, if we study the anisotropy velocity of the particle while varying the values of the polymer parameters, we can see that the following holds:

$$\beta'(b_\alpha, b, P_\pm) \geq 1 \ \forall \ P_\pm \in \mathbb{R} \Leftrightarrow \frac{b_\alpha}{b} \geq 1, \tag{172}$$

where

$$\beta' \equiv \sqrt{\beta'^2_+ + \beta'^2_-} = \sqrt{\frac{\sin^2(bP_+)\cos^2(bP_+) + \sin^2(bP_-)\cos^2(bP_-)}{r^2(bP_+, bP_-)\left[1 - \frac{b_\alpha^2}{b^2}r^2(bP_+, bP_-)\right]}}, \tag{173}$$

and $r(x,y) = \sqrt{\sin^2(x) + \sin^2(y)}$. We notice that (P_+, P_-) (and consequently $\mathcal{C}_{\text{BIX}}^{\text{ADM-poly}}$) are constants of motion in the free particle regime. In addition, it is easy to verify that the relation (19) is recovered for $b, b_\alpha \to 0$.

The figures represented in Figure 17 show that the anisotropy velocity represented on the vertical axis is always greater than 1 only if we choose the ratio b_α/b to be greater than or equal to one. Therefore, the dynamics of the Mixmaster model is expected to be still chaotic because of the existence of the singularity and the presence of a never-ending series of rebounds against the potential walls. Instead, if we choose the polymer parameters such that $b_\alpha/b < 1$ (Figure 18), the series of rebounds occurs until the particle velocity becomes smaller than the velocity of the potential walls; then, the point-Universe reaches the singularity with no other rebound.

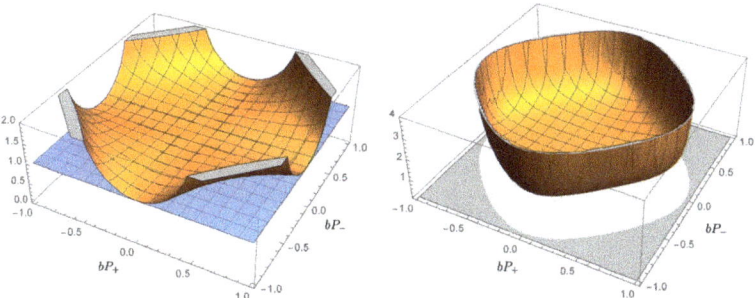

Figure 17. The 3D-profiles of the anisotropy velocity (173) with $b_\alpha/b = 1$ (left panel) and $b_\alpha/b = 1.5$ (right panel).

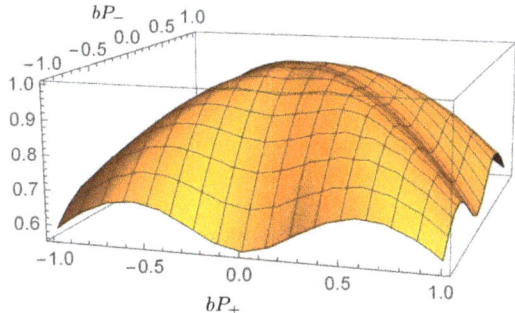

Figure 18. The 3D-profile of the anisotropy velocity (173) with $b_\alpha/b = 0.1$.

Moreover, it can be demonstrated that for $b_\alpha/b < 1$, we have $\theta_{\text{poly}}^{\max} < \frac{\pi}{3}$, and this implies the absence of rebounds, even if the particle moves towards a specific wall. On the other hand, if $b_\alpha/b \geq 1$, we have $\frac{\pi}{3} \leq \theta_{\text{poly}}^{\max} < \frac{\pi}{2}$; this means that a rebound is always possible, given the triangular symmetry of the system.

Regarding the polymer modified reflection law, by using analogous constants of motion with respect to the standard Misner case, we are able to derive the following law:

$$\frac{1}{4b}\left[\arcsin\left(\cos\theta^f \frac{\sin\theta^i}{\sin\theta^f} r(2bP_+^i, 2bP_-^i)\right) + \arcsin\left(\cos\theta^i r(2bP_+^i, 2bP_-^i)\right)\right] = $$
$$= \frac{1}{b_\alpha}\left[\arcsin\left(\frac{b_\alpha}{b} r(bP_+^f, bP_-^f)\right) - \arcsin\left(\frac{b_\alpha}{b} r(bP_+^i, bP_-^i)\right)\right], \quad (174)$$

where $r(x,y) = \sqrt{\sin^2(x) + \sin^2(y)}$.

In order to make a comparison with the standard case (20), an expansion up to the second order for b and b_α is required:

$$\frac{\sin\left(\theta^i + \theta^f\right)}{2} = \sin\theta^i(1 + \Pi_f^2)(1 + R_f) - \sin\theta^f(1 + \Pi_i^2)(1 + R_i) \quad (175)$$

where

$$\Pi^2 = \frac{1}{6}b_\alpha^2(P_+^2 + P_-^2),\tag{176a}$$

$$R = \frac{2}{3}b^2\frac{(P_+)^4 + (P_-)^4}{(P_+)^2 + (P_-)^2},\tag{176b}$$

and it is easy to show that in the limit $b, b_\alpha \to 0$, we find the standard reflection law (20) obtained by Misner.

Recovering this limit leads us to infer that the map above still admits stochastic properties; therefore, when $b_\alpha < b$, sooner or later the parameter region where no rebound takes place is reached, as in [27].

Given the results of this analysis and that of previous sections, we can infer that not only the physical nature of the Big Bounce, but also its very existence depends on the geometrical dimension of the variable chosen as discrete: we have seen here that using a logarithmic variable does not indeed avoid the singularity. Therefore, in the last model that we present in the next subsection, we choose to use a mixed representation involving the anisotropies as they are defined by Misner, together with an isotropic volume variable instead of $\alpha \propto \ln v$.

4.4.2. The Inhomogeneous Mixmaster Model in the Volume Variable

In [38], the semiclassical polymer dynamics of the inhomogeneous Mixmaster model is analyzed by choosing the cubed scale factor (i.e., the Universe volume) alone as the discretized configurational variable. In addition, a massless scalar field and a cosmological constant are included in the dynamics, accounting respectively for a quasi-isotropization and inflationary-like mechanisms. The resulting dynamics is a singularity-free Kasner-like phase that is linked with a homogeneous and isotropic de Sitter evolution. Moreover, the chaotic character of the Mixmaster dynamics is absent in this framework. Thus, the study presented in [38] demonstrates that the generic cosmological solution is singularity-free and non-chaotic once the polymer discretization of the Universe volume variable is performed at a semiclassical level. This result is alike to that achieved also in LQC, as presented in Section 3.5.

In order to summarize more in detail all the main results presented in [38], let us start by characterizing the generic cosmological problem and its relationship with the homogeneous Mixmaster model. The most general line element can be written as follows:

$$ds^2 = N^2 dt^2 - h_{ab}(dx^a + N^a dt)(dx^b + N^b dt),\tag{177}$$

where $N^{a,b}(t,\vec{x})$ is the shift vector and the following holds:

$$h_{ab}(t,\vec{x}) = e^{m_1(t,\vec{x})} l_a^1 l_b^1 + e^{m_2(t,\vec{x})} l_a^2 l_b^2 + e^{m_3(t,\vec{x})} l_a^3 l_b^3,\tag{178}$$

with $a, b = 1, 2, 3$. Rigorously, this is not the most general picture since we are not considering the rotation of the Kasner vectors $l_{a,b}^{1,2,3}(t,\vec{x})$. Differently from the homogeneous Bianchi line element (12), the lapse function $N(t,\vec{x})$ depends explicitly on the spatial coordinates, as well as the Misner-like variables, that in their generalized version are defined as follows:

$$m_a(t,\vec{x}) = \frac{2}{3}\ln[v(t,\vec{x})] + 2\beta_a(t,\vec{x}),\tag{179}$$

where $a = 1, 2, 3$ and $\beta_a = (\beta_+ + \sqrt{3}\beta_-, \beta_+ - \sqrt{3}\beta_-, -2\beta_+)$. Here, we are considering v as the isotropic variable proportional to the spatial volume of the Universe (instead of α as in the proper Misner variables), while β_+, β_- are the anisotropies. Let us include a massless, self-interacting scalar field ϕ in the dynamics, taking into account the slow-roll condition $\dot\phi^2 \ll U(\phi)$ as well as the following hypothesis:

$$|\nabla\phi|^2 \ll U[\phi(t,\vec{x})] \sim \Lambda(\vec{x}),\tag{180}$$

both typical of the inflationary paradigm.

The action in the inhomogeneous case reads as follows:

$$S_B^{inhom} = \int d^3x\, dt \left(P_v \dot{v} + \sum_j P_j \dot{\beta}_j - N\mathcal{C}_B^{inhom} - N^i \mathcal{C}_i \right), \qquad (181)$$

where $j = +, -, \phi$ and $\beta_\phi \equiv \phi$. The super Hamiltonian constraint is as follows:

$$\mathcal{C}_B^{inhom} = \frac{3}{4}\left[-vP_v^2 + \sum_j \frac{P_j^2}{9v^2} + \frac{v^{1/3}}{3} U_B^{inhom} + v\Lambda(\vec{x}) \right], \qquad (182)$$

while $\mathcal{C}_i$ represents the supermomentum one. We remark how the self-interacting scalar field, under the slow-roll condition, is equivalent to a free scalar field plus a cosmological constant $\Lambda(\vec{x})$.

The potential term is due to the three-dimensional scalar curvature and can be split as $U^{inhom} = W + U_B^{inhom}$, where the contribution due to the spatial gradients of the configurational variables is encoded in W, and the term U_B^{inhom} represents the inhomogeneous generalization of the Bianchi model's potential (the Bianchi VIII and Bianchi IX models are the most general choices). By using the Belinskii–Khalatnikov–Lifshitz (BKL) conjecture [130], the term W is considered to be negligible and so the super Hamiltonian (182) reduces to that one of the Bianchi IX model, where the spatial coordinates appear only as parameters. The reliability of this assumption can be verified by explicitly evaluating the term W by means of the supermomentum constraint. Moreover, by means of the gauge choice $N^i = 0$, each point of space evolves independently and can be described, using a homogeneous Bianchi IX model. This way, point by point, the inhomogeneous cosmological evolution is approximated by a Mixmaster-like one. We notice that the corresponding solutions of the dynamics must satisfy also the supermomentum constraint, which identically vanishes for a homogeneous spacetime. In addition, with the hypothesis that the physical scale of the inhomogeneities is much bigger than the average Hubble horizon, each causal connected region verifies a Mixmaster-like evolution, instead of each point of space [131].

Let us now analyze the semiclassical polymer dynamics of the model. Using the substitution (89) for the momentum conjugate to the variable v, i.e., $P_v \to \frac{\sin(b_0 P_v)}{b_0}$, the Hamiltonian constraint becomes the following:

$$\mathcal{C}_{BIX}^{poly} = \frac{3}{4}\left[-v\frac{\sin^2(b_0 P_v)}{b_0^2} + \sum_j \frac{P_j^2}{9v^2} + \frac{v^{1/3}}{3} U_{BIX} + v\Lambda \right], \qquad (183)$$

where the spatial gradients are neglected and the potential term is that one of the Bianchi IX model. Firstly, we can notice that the following condition is imposed by the form of the Hamiltonian when the potential term is negligible:

$$\sin^2(b_0 P_v) = b_0^2 \left(\frac{\sum_j P_j^2}{9v^2} + \Lambda \right). \qquad (184)$$

As a consequence, the right-hand side of (184) must be smaller than 1, and therefore, the Universe volume acquires a lower bound:

$$v > v_B \equiv \frac{b_0}{3}\sqrt{\frac{\sum_j P_j^2}{b_0^2 \Lambda}}, \qquad (185)$$

with the condition $b_0^2 \Lambda < 1$, which is physically reasonable in the Planck units. So, in the polymer semiclassical dynamics, the initial singularity is replaced by a Big Bounce.

We notice that the minimum value of the Universe volume depends not only on the polymer scale b_0, but also on the cosmological constant Λ and on the conjugate momenta P_j, which are constants of motion when the potential term is negligible. This feature is in accordance with the results obtained in LQC [49]. The absence of singularity thanks to the semiclassical polymer formalism can be shown more in general by solving the Hamilton equation for the Universe volume, from which we obtain the following:

$$v(t) = \sqrt{\frac{\sum_j P_j^2 \left[\cosh\left(3\sqrt{\Lambda}\sqrt{1 - b_0^2 \Lambda} t\right) - 1 + 2b_0^2 \Lambda \right]}{18\Lambda(1 - b_0^2 \Lambda)}}. \qquad (186)$$

In particular, as shown in Figure 19 the Bounce occurs when the volume reaches its minimum, which is given by Equation (185).

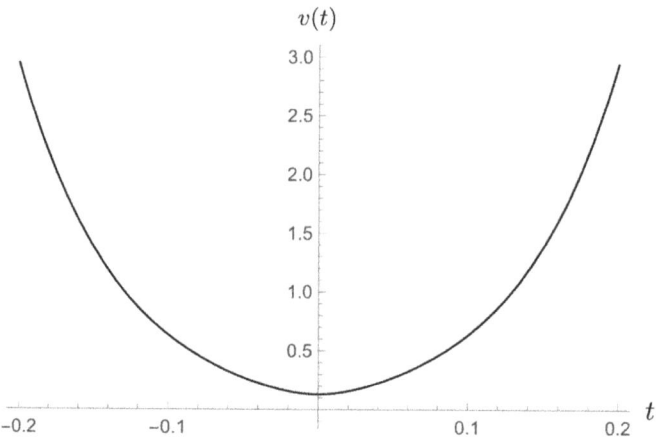

Figure 19. Semiclassical polymer trajectory of the Universe volume v in the synchronous time t. Image from [38].

Moreover, by performing the ADM reduction and choosing v as the relational time, the Hamilton equations for the anisotropies and the scalar field can be solved, showing that the former do not diverge approaching the Big Bounce, as it happens in the standard case. Additionally, far from the Bounce, the anisotropies tend to a constant value that can be absorbed by an appropriate and local rescaling of the coordinate system. In this sense, the quasi-isotropization mechanism described in [62,131] results to be preserved under the polymer modification.

In conclusion, the semiclassical polymer dynamics is well described by a bridge solution connecting an initial non-singular, inhomogeneous and anisotropic Kasner-like regime with a later homogeneous and isotropic de Sitter phase. We notice that this result is local, i.e., valid in a nearly homogeneous region. Nonetheless, with the hypothesis of having an inhomogeneity scale much larger than the average Hubble horizon, each causal connected region undergoes such an isotropization phase and can be expanded by the inflationary mechanism on a scale much larger than the Hubble radius. Thus, if the inflationary phase is long enough, it is possible to obtain a homogeneous and isotropic region larger than the Hubble horizon today.

Finally, in [38], it is also demonstrated that the chaotic properties of the dynamics are suppressed when the polymer representation is implemented at a semiclassical level. This property is a consequence of the following considerations. As we have seen in Section 2.3, the chaoticity of the Bianchi IX model is derived by analyzing the relative velocity between the point-Universe and the potential walls. It is possible to verify that, similar to the analysis in Section 4.4.1, in this model, the velocity of the walls is finite at the Bounce, while that of the point-Universe diverges in the polymer framework. So, rebounds always occur, even in the presence of a scalar field, even though in the classical framework, its presence is able to remove the chaos [113]. However, due to the polymer modifications, the following condition holds:

$$\sum_j P_j^2 + 9\Lambda v^2 > \max\left(3v^{4/3} U_{\text{BIX}}\right). \tag{187}$$

Therefore, there are two possible scenarios for the removal of chaos, similar to the case of the Loop quantized homogeneous Bianchi IX model described in Section 3.5: either the condition (187) is satisfied before the Bounce, resulting in a final stable Kasner-like epoch (similarly to the AVTD regime reached classically when a scalar field is present), or chaos is removed by the fact that the presence of a Big Bounce makes the number of rebounds on the potential walls finite.

4.5. The Link between Polymer Quantum Mechanics and Loop Quantum Cosmology: Canonical Equivalence

In Sections 4.2–4.4, we have presented the main results gained thanks to the application of the polymer representation at the level of the semiclassical dynamics, with some extensions to the full quantum regime in Sections 4.2 and 4.3. Along the presentation, we have proceeded by considering more and more general cosmological models, starting from the FLRW one and then extending the analysis to the Bianchi I and IX models afterwards. For each model, we have compared the obtained results in different sets of variables, always performing the canonical transformation between the phase space variables before implementing the polymer approximation for the variables conjugate to those represented as discrete on the polymer lattices. If, on one hand, the polymer formulation permits to successfully overcome the GR shortcoming of the initial singularity, on the other hand, it reproduces different physical pictures for the same cosmological model when represented in different sets of variables canonically linked in the classical regime. In this respect, the polymer behavior of the Bianchi IX cosmological model in the standard Misner variables is of particular relevance since it shows a singular dynamics in contrast with the bouncing cosmology and the chaos removal ensured by the use of a volume-like variable (note that in the Misner variables the presence of chaos depends on the ratio of the different discretization parameters associated to the isotropic variable and the anisotropy coordinates $\beta_\pm$ [42]).

This fact raises an issue about what it really means to do a canonical transformation in the polymer framework. For a proposal about recovering the equivalence between different sets of variables after the polymer is implemented on the Hamiltonian constraint, see [44]. In this work, it is demonstrated how, by fixing the preferred system of generalized coordinates and assigning a constant lattice step (actually in the considered case, there is only the scale factor, but a generalization of the analysis can be easily inferred), the dynamical features of the model remain the same in any other set of coordinates, at the price of considering the discretization step as a function of the adopted variables. For example, starting from the volume coordinates and then changing to the Ashtekar connections maintains the canonical equivalence between the dynamics thanks to the dependence of the polymer parameter on p. Therefore, the dynamics is ruled by the set of variables in which the polymer spacing results in being constant. We note that this study was performed only at a semiclassical level since the implementation of a translational operator depending on the coordinates is not trivial, so the equivalence in the full quantum theory is still an open question.

Nevertheless, this result achieved in [44] about the recovery of an equivalence class of configurational variables could have relevant implications for the $\bar{\mu}$ scheme of LQC. In particular, in the $\bar{\mu}$ scheme of LQC the translational operator acting on the physical states acquires a parameter depending on the momentum variable (due to the physical rescaling of the area element) that, in the polymer framework, plays exactly the role of p.

This analogy suggests that a change of coordinates is at the basis of the improved LQC approach [49–51] in a way that a translational operator of constant step is restored and the theory is made technically viable.

However, such a change of variables opens a possible criticism about the quantization procedure, which is no longer referred to the natural $SU(2)$ connection so that LQC would be departing from full LQG. While on a quantum level, this question is somewhat an open issue, on the level of an effective LQC theory, the result mentioned above can somewhat support the validity of the change in variables in view of the link between PQM and LQC. Indeed, since the physical picture of the model is dictated by the scheme with a constant polymerization parameter, we can conclude that the semiclassical dynamics of the $\bar{\mu}$ formulation is the same of that obtained in the Ashtekar variables if the lattice parameter is canonically transformed so that no ambiguity arises about the universal character of the Big Bounce, i.e., about its independence on the initial conditions.

5. Concluding Remarks

In this review, we analyzed basic aspects of LQC and presented polymer quantum cosmology in depth in order to trace some solid implications on the early Universe evolution precisely regarding the regularization of the singularity.

In this respect, we can firmly say that a bouncing cosmology always emerges in LQC and in polymer quantum cosmology, when the Universe volume is considered one of the configurational variables. Furthermore, the Big Bounce has also the features of a universal cut-off since the critical energy density of the Universe takes a maximum value depending only on fundamental constants and the Immirzi parameter. This result is obtained in the improved scheme of LQC as shown in Section 3.3 and also demonstrated in the context of PQM for the isotropic model in [37]. Moreover, similar results to those obtained in the $\bar{\mu}$ scheme are outlined in [44,45] when PQM is implemented on volume-like variables.

The situation is slightly different when we polymerize different sets of generalized variables in PQM. In particular, when we consider the polymerization of the Ashtekar–Barbero–Immirzi connections, the Big Bounce still emerges in the semiclassical dynamics, but the critical energy density now depends on the initial conditions on the motion, or equivalently on the wave packets when the full quantum approach is implemented (see [44,45]). In these analyses, the polymerization of the natural connection seems to reproduce a physical picture very similar to the μ_0 scheme of LQC presented in Section 3.2.

The importance of the choice of the adopted variable for the polymer quantization of the Universe is outlined by observing that the polymerization of the Bianchi IX model (both in the homogeneous and inhomogeneous cases) expressed in the standard Misner variables yields a model that is still singular [34,42], while discretizing the Universe volume in the metric formulation does avoid the singularity [38]. This result highlights the link between the dimensionality of the discretized variable and the regularization of the singularity in PQM; also, it opens the non-trivial question about which equivalence class among all possible choices of canonically related coordinates leads to the same physical picture.

An interesting result toward the solution of this puzzling question is elucidated from the analysis in [44]. We have discussed how the equivalence is ensured for the isotropic model on a semiclassical level by the possibility to restate the problem in terms of a new polymerization step dependent on the configurational coordinates. This result can have positive implications in stating the equivalence of the effective dynamics in the μ_0 and $\bar{\mu}$ schemes of LQC [20,50], as discussed in Section 4.5, at least on a semiclassical level.

On the base of the results illustrated here, we can conclude that, if on one hand, some basic features of cut-off quantum cosmology are well traced, many other detailed points must be addressed. For instance, it is necessary to better characterize the physical properties of the Big Bounce, both by providing a pure quantum description instead of just a semiclassical one, and by further investigating the thermodynamical nature of the quantum cosmological fluid. Moreover, it is crucial to understand how a collapsing Universe before our expanding branch is generated and how it can influence the morphology of the present

Universe (see some pioneering works [37,39,132–138]). A typical example of this issue affecting bouncing cosmologies is provided in [37], where it is argued that the flatness paradox is still present, even though the horizon paradox is naturally solved, thanks to the pre-existing collapsing Universe. These kinds of physical questions become even more meaningful if we postulate that the Universe dynamics is associated to a cyclical evolution between a Big Bounce and a later classical turning point [139,140]: in this context, the behavior of the Universe entropy becomes a delicate feature to be addressed.

Author Contributions: All authors contributed to all parts of the research and manuscript. G.M. has been slightly more active than the others in the conceptualization while G.B. and E.G. have contributed slightly more on the software and visualization. All authors have read and agreed to the published version of the manuscript.

Funding: This research received no external funding.

Institutional Review Board Statement: Not applicable.

Informed Consent Statement: Not applicable.

Data Availability Statement: Not applicable.

Conflicts of Interest: The authors declare no conflict of interest.

Note

1. In [45] it is shown that by combining the semiclassical solutions for $c_i(\phi)$ with (125) we obtain that $x_i(\phi) \propto \phi$ with slopes depending on the initial conditions on the motion.

References

1. DeWitt, B.S. Quantum theory of gravity. 1. The canonical theory. *Phys. Rev.* **1967**, *160*, 1113–1148. [CrossRef]
2. DeWitt, B.S. Quantum theory of gravity. 2. The manifestly covariant theory. *Phys. Rev.* **1967**, *162*, 1195–1239. [CrossRef]
3. DeWitt, B.S. Quantum theory of gravity. 3. Applications of the covariant theory. *Phys. Rev.* **1967**, *162*, 1239–1256. [CrossRef]
4. Kuchar, K.V. Canonical Quantum Gravity. *arXiv* **1993**, arXiv:gr-qc/9304012. [CrossRef]
5. Ashtekar, A. New variables for classical and quantum gravity. *Phys. Rev. Lett.* **1986**, *57*, 2244–2247. [CrossRef]
6. Rovelli, C.; Smolin, L. Loop space representation of quantum General Relativity. *Nucl. Phys. B* **1990**, *331*, 80–152. [CrossRef]
7. Rovelli, C.; Smolin, L. Discreteness of area and volume in quantum gravity. *Nucl. Phys. B* **1995**, *442*, 593–619. [CrossRef]
8. Rovelli, C.; Smolin, L. Spin networks and quantum gravity. *Phys. Rev. D* **1995**, *52*, 5743–5759. [CrossRef] [PubMed]
9. Cianfrani, F.; Lecian, O.M.; Lulli, M.; Montani, G. *Canonical Quantum Gravity: Fundamentals and Recent Developments*; World Scientific: Singapore, 2014. [CrossRef]
10. Hartle, J.B.; Hawking, S.W. Wave function of the Universe. *Phys. Rev. D* **1983**, *28*, 2960–2975. [CrossRef]
11. Thiemann, T. *Modern Canonical Quantum General Relativity*; Cambridge Monographs on Mathematical Physics; Cambridge University Press: Cambridge, UK, 2007. [CrossRef]
12. Rovelli, C.; Vidotto, F. *Covariant Loop Quantum Gravity: An Elementary Introduction to Quantum Gravity and Spinfoam Theory*; Cambridge University Press: Cambridge, UK, 2014. [CrossRef]
13. Rovelli, C. *Quantum Gravity*; Cambridge Monographs on Mathematical Physics; Cambridge University Press: Cambridge, UK, 2004. [CrossRef]
14. Montani, G.; Battisti, M.V.; Benini, R.; Imponente, G. *Primordial Cosmology*; World Scientific: Singapore, 2009.
15. Cianfrani, F.; Montani, G. A critical analysis of the cosmological implementation of Loop Quantum Gravity. *Mod. Phys. Lett. A* **2012**, *27*, 1250032. [CrossRef]
16. Bojowald, M. Critical evaluation of common claims in Loop Quantum Cosmology. *Universe* **2020**, *6*, 36. [CrossRef]
17. Cianfrani, F.; Montani, G. Implications of the gauge-fixing in Loop Quantum Cosmology. *Phys. Rev. D* **2012**, *85*, 024027. [CrossRef]
18. Corichi, A.; Vukašinac, T.; Zapata, J.A. Polymer Quantum Mechanics and its continuum limit. *Phys. Rev. D* **2007**, *76*, 044016. [CrossRef]
19. Corichi, A.; Vukašinac, T.; Zapata, J.A.; Macias, A.; Lämmerzahl, C.; Camacho, A. On a Continuum Limit for Loop Quantum Cosmology. In *AIP Conference Proceedings*; American Institute of Physics: College Park, MD, USA, 2008. [CrossRef]
20. Ashtekar, A.; Gupt, B. Generalized effective description of Loop Quantum Cosmology. *Phys. Rev. D* **2015**, *92*, 084060. [CrossRef]
21. Singh, P.; Vandersloot, K. Semiclassical states, effective dynamics, and classical emergence in Loop Quantum Cosmology. *Phys. Rev. D* **2005**, *72*, 084004. [CrossRef]
22. Battisti, M.V.; Lecian, O.M.; Montani, G. Polymer quantum dynamics of the Taub Universe. *Phys. Rev. D* **2008**, *78*, 103514. [CrossRef]
23. Battisti, M.V.; Lecian, O.M.; Montani, G. GUP vs polymer quantum cosmology: The Taub model. *arXiv* **2009**, arXiv:0903.3836.

24. Hossain, G.M.; Husain, V.; Seahra, S.S. Nonsingular inflationary Universe from polymer matter. *Phys. Rev. D* **2010**, *81*, 024005. [CrossRef]
25. Lawrie, I.D. Time evolution in quantum cosmology. *Phys. Rev. D* **2011**, *83*, 043503. [CrossRef]
26. Kreienbuehl, A.; Pawłowski, T. Singularity resolution from polymer quantum matter. *Phys. Rev. D* **2013**, *88*, 043504. [CrossRef]
27. Lecian, O.M.; Montani, G.; Moriconi, R. Semiclassical and quantum behavior of the Mixmaster model in the polymer approach. *Phys. Rev. D* **2013**, *88*, 103511. [CrossRef]
28. Cianfrani, F.; Montani, G.; Pittorino, F. Nonsingular cosmology from evolutionary quantum gravity. *Phys. Rev. D* **2014**, *90*, 103503. [CrossRef]
29. Hassan, S.M.; Husain, V.; Seahra, S.S. Polymer inflation. *Phys. Rev. D* **2015**, *91*, 065006. [CrossRef]
30. Moriconi, R.; Montani, G.; Capozziello, S. Big Bounce cosmology from quantum gravity: The case of a cyclical Bianchi I Universe. *Phys. Rev. D* **2016**, *94*, 023519. [CrossRef]
31. Hassan, S.M.; Husain, V. Semiclassical cosmology with polymer matter. *Class. Quantum Gravity* **2017**, *34*, 084003. [CrossRef]
32. Moriconi, R.; Montani, G. Behavior of the Universe anisotropy in a Big Bounce cosmology. *Phys. Rev. D* **2017**, *95*, 123533. [CrossRef]
33. Ali, M.; Seahra, S.S. Natural inflation from polymer quantization. *Phys. Rev. D* **2017**, *96*, 103524. [CrossRef]
34. Crinò, C.; Montani, G.; Pintaudi, G. Semiclassical and quantum behavior of the Mixmaster model in the polymer approach for the isotropic Misner variable. *Eur. Phys. J. C* **2018**, *78*, 886. [CrossRef]
35. Montani, G.; Marchi, A.; Moriconi, R. Bianchi I model as a prototype for a cyclical Universe. *Phys. Lett. B* **2018**, *777*, 191–200. [CrossRef]
36. Ben Achour, J.; Livine, E.R. Polymer quantum cosmology: Lifting quantization ambiguities using a SL(2,R) conformal symmetry. *Phys. Rev. D* **2019**, *99*, 126013. [CrossRef]
37. Montani, G.; Mantero, C.; Bombacigno, F.; Cianfrani, F.; Barca, G. Semiclassical and quantum analysis of the isotropic Universe in the polymer paradigm. *Phys. Rev. D* **2019**, *99*, 063534. [CrossRef]
38. Antonini, S.; Montani, G. Singularity-free and non-chaotic inhomogeneous Mixmaster in polymer representation for the volume of the Universe. *Phys. Lett. B* **2019**, *790*, 475–483. [CrossRef]
39. Barca, G.; Di Antonio, P.; Montani, G.; Patti, A. Semiclassical and quantum polymer effects in a flat isotropic Universe. *Phys. Rev. D* **2019**, *99*, 123509. [CrossRef]
40. Cascioli, V.; Montani, G.; Moriconi, R. WKB approximation for the polymer quantization of the Taub model. *arXiv* **2020**, arXiv:1903.09417.
41. Achour, J.B.; Livine, E. Protected SL(2,R) symmetry in quantum cosmology. *J. Cosmol. Astropart. Phys.* **2019**, *2019*, 12. [CrossRef]
42. Giovannetti, E.; Montani, G. Polymer representation of the Bianchi IX cosmology in the Misner variables. *Phys. Rev. D* **2019**, *100*, 104058. [CrossRef]
43. Campolongo, A.; Montani, G. Specific entropy as a clock for the evolutionary quantization of the isotropic Universe. *Eur. Phys. J. C* **2020**, *80*, 983. [CrossRef]
44. Mandini, F.; Barca, G.; Giovannetti, E.; Montani, G. Polymer quantum dynamics of the isotropic Universe in the Ashtekar-Barbero-Immirzi and in the volume variables. *arXiv* **2021**, arXiv:2006.10614.
45. Giovannetti, E.; Montani, G.; Schiattarella, S. Semiclassical and quantum features of the Bianchi I cosmology in the polymer representation. *arXiv* **2021**, arXiv:2105.00360.
46. Bojowald, M. Loop Quantum Cosmology. *Living Rev. Relativ.* **2005**, *8*, 11. [CrossRef]
47. Ashtekar, A.; Pawlowski, T.; Singh, P. Quantum nature of the Big Bang. *Phys. Rev. Lett.* **2006**, *96*, 141301. [CrossRef]
48. Ashtekar, A.; Pawlowski, T.; Singh, P. Quantum nature of the Big Bang: An analytical and numerical investigation. *Phys. Rev. D* **2006**, *73*, 124038. [CrossRef]
49. Ashtekar, A.; Pawlowski, T.; Singh, P. Quantum nature of the Big Bang: Improved dynamics. *Phys. Rev. D* **2006**, *74*, 084003. [CrossRef]
50. Ashtekar, A.; Corichi, A.; Singh, P. Robustness of key features of Loop Quantum Cosmology. *Phys. Rev. D* **2008**, *77*, 024046. [CrossRef]
51. Ashtekar, A.; Singh, P. Loop Quantum Cosmology: A status report. *Class. Quantum Gravity* **2011**, *28*, 213001. [CrossRef]
52. Battefeld, D.; Peter, P. A critical review of classical bouncing cosmologies. *Phys. Rep.* **2015**, *571*, 1–66. [CrossRef]
53. Brandenberger, R.; Peter, P. Bouncing cosmologies: Progress and problems. *Found. Phys.* **2017**, *47*, 797–850. [CrossRef]
54. Misner, C.W. Quantum Cosmology. I. *Phys. Rev.* **1969**, *186*, 1319–1327. [CrossRef]
55. Imponente, G.; Montani, G. Inhomogeneous de Sitter solution with scalar field and perturbations spectrum. *Mod. Phys. Lett. A* **2004**, *19*, 1281–1290. [CrossRef]
56. Imponente, G.; Montani, G. Classical and quantum behavior of the generic cosmological solution. In *AIP Conference Proceedings*; American Institute of Physics: College Park, MD, USA, 2006. [CrossRef]
57. Benini, R.; Montani, G. Frame independence of the inhomogeneous Mixmaster chaos via Misner-Chitré-like variables. *Phys. Rev. D* **2004**, *70*, 103527. [CrossRef]
58. Benini, R.; Montani, G. Inhomogeneous quantum Mixmaster: From classical towards quantum mechanics. *Class. Quantum Gravity* **2006**, *24*, 387–404. [CrossRef]

59. Benini, R.; Montani, G. Covariant description of the inhomogeneous Mixmaster chaos. In Proceedings of the XI Marcel Grossmann meeting on Relativistic Astrophysics, Berlin, Germany, 23–29 July 2006. [CrossRef]
60. Benini, R.; Montani, G. Classical and quantum aspects of the inhomogeneous Mixmaster chaoticity. In Proceedings of the MG11 Meeting on General Relativity, Berlin, Germany, 23–29 July 2006. [CrossRef]
61. Kirillov, A.A. Quantum creation of quasihomogeneous inflationary Universe. *Grav. Cosmol.* **1996**, *2*, 35–37.
62. Kirillov, A.A.; Montani, G. Quasi-isotropization of the inhomogeneous Mixmaster Universe induced by an inflationary process. *Phys. Rev. D* **2002**, *66*, 064010. [CrossRef]
63. Graham, R. Chaos and quantum chaos in cosmological models. *Chaos Solitons Fractals* **1995**, *5*, 1103–1122. [CrossRef]
64. Vilenkin, A. The interpretation of the wave function of the Universe. *Phys. Rev. D* **1989**, *39*, 1116. [CrossRef] [PubMed]
65. Blyth, W.F.; Isham, C.J. Quantization of a Friedmann Universe filled with a scalar field. *Phys. Rev. D* **1975**, *11*, 768–778. [CrossRef]
66. Arnowitt, R.; Deser, S.; Misner, C.W. Canonical variables for General Relativity. *Phys. Rev.* **1959**, *117*, 6. [CrossRef]
67. Matschull, H.J. Dirac's Canonical Quantization Program. *arXiv* **1996**, arXiv:quant-ph/9606031.
68. Cianfrani, F.; Lulli, M.; Montani, G. Solution of the noncanonicity puzzle in General Relativity: A new Hamiltonian formulation. *Phys. Lett. B* **2012**, *710*, 703–709. [CrossRef]
69. Weinberg, S. *Cosmology*; Oxford University Press: Oxford, UK, 2008.
70. Kolb, E.W.; Turner, M.S. *The Early Universe*; CRC Press: Boca Raton, FL, USA, 1990; Volume 69.
71. Misner, C.W. Mixmaster Universe. *Phys. Rev. Lett.* **1969**, *22*, 1071–1074. [CrossRef]
72. Szulc, L.; Kaminski, W.; Lewandowski, J. Closed Friedmann–Robertson–Walker model in Loop Quantum Cosmology. *Class. Quantum Gravity* **2007**, *24*, 2621–2635. [CrossRef]
73. Ashtekar, A.; Pawlowski, T.; Singh, P.; Vandersloot, K. Loop Quantum Cosmology of K=1 FRW models. *Phys. Rev. D* **2007**, *75*, 024035. [CrossRef]
74. Pawłowski, T.; Pierini, R.; Wilson-Ewing, E. Loop Quantum Cosmology of a radiation-dominated flat FLRW Universe. *Phys. Rev. D* **2014**, *90*, 123538. [CrossRef]
75. Bojowald, M. Homogeneous Loop Quantum Cosmology. *Class. Quantum Gravity* **2003**, *20*, 2595–2615. [CrossRef]
76. Bojowald, M.; Date, G.; Vandersloot, K. Homogeneous Loop Quantum Cosmology: The role of the spin connection. *Class. Quantum Gravity* **2004**, *21*, 1253–1278. [CrossRef]
77. Bojowald, M.; Date, G. Quantum suppression of the generic chaotic behavior close to cosmological singularities. *Phys. Rev. Lett.* **2004**, *92*, 071302. [CrossRef] [PubMed]
78. Bojowald, M.; Date, G.; Hossain, G.M. The Bianchi IX model in Loop Quantum Cosmology. *Class. Quantum Gravity* **2004**, *21*, 3541–3569. [CrossRef]
79. Ashtekar, A.; Wilson-Ewing, E. Loop Quantum Cosmology of Bianchi type I models. *Phys. Rev. D* **2009**, *79*, 083535. [CrossRef]
80. Ashtekar, A.; Wilson-Ewing, E. Loop Quantum Cosmology of Bianchi type II models. *Phys. Rev. D* **2009**, *80*, 123532. [CrossRef]
81. Wilson-Ewing, E. Loop Quantum Cosmology of Bianchi type IX models. *Phys. Rev. D* **2010**, *82*, 043508. [CrossRef]
82. Martín-Benito, M.; Garay, L.J.; Mena Marugán, G.A.; Wilson-Ewing, E. Loop Quantum Cosmology of the Bianchi I model: Complete quantization. *J. Phys. Conf. Ser.* **2012**, *360*, 012031. [CrossRef]
83. Garay, L.J.; Martín-Benito, M.; Mena Marugán, G. Inhomogeneous Loop Quantum Cosmology: Hybrid quantization of the Gowdy model. *Phys. Rev. D* **2010**, *82*, 044048. [CrossRef]
84. Wilson-Ewing, E. Anisotropic Loop Quantum Cosmology with self-dual variables. *Phys. Rev. D* **2016**, *93*, 083502. [CrossRef]
85. Corichi, A.; Karami, A. Loop Quantum Cosmology of Bianchi IX: Inclusion of inverse triad corrections. *Int. J. Mod. Phys. D* **2016**, *25*, 1642011. [CrossRef]
86. Wilson-Ewing, E. A quantum gravity extension to the Mixmaster dynamics. *Class. Quantum Gravity* **2019**, *36*, 195002. [CrossRef]
87. Ashtekar, A.; Lewandowski, J.; Marolf, D.; Mourão, J.; Thiemann, T. Quantization of diffeomorphism invariant theories of connections with local degrees of freedom. *J. Math. Phys.* **1995**, *36*, 6456–6493. [CrossRef]
88. Thiemann, T. Quantum Spin Dynamics (QSD). *Class. Quantum Gravity* **1998**, *15*, 839–873. [CrossRef]
89. Thiemann, T. Quantum Spin Dynamics (QSD): II. The kernel of the Wheeler - DeWitt constraint operator. *Class. Quantum Gravity* **1998**, *15*, 875–905. [CrossRef]
90. Ashtekar, A.; Lewandowski, J. Quantum theory of geometry. I. Area operators. *Class. Quantum Gravity* **1997**, *14*, A55–A81. [CrossRef]
91. Ashtekar, A.; Lewandowski, J. Quantum theory of geometry. II. Volume operators. *Adv. Theor. Math. Phys.* **1998**, *1*, 388–429. [CrossRef]
92. Ashtekar, A.; Pullin, J. *Loop Quantum Gravity: The First 30 Years*; World Scientific Publishing Co. Pte Ltd.: Singapore, 2017.
93. Ashtekar, A.; Bianchi, E. A short review of Loop Quantum Gravity. *Rep. Prog. Phys.* **2021**, *84*, 042001. [CrossRef] [PubMed]
94. Ashtekar, A. New Hamiltonian formulation of General Relativity. *Phys. Rev. D* **1987**, *36*, 1587–1602. [CrossRef] [PubMed]
95. Thiemann, T. Anomaly-free formulation of non-perturbative, four-dimensional Lorentzian quantum gravity. *Phys. Lett. B* **1996**, *380*, 257–264. [CrossRef]
96. Haro, J.; Elizalde, E. Loop Cosmology: Regularization vs. quantization. *Europhys. Lett.* **2010**, *89*, 69001. [CrossRef]
97. Zhang, X. Thermodynamics in new model of Loop Quantum Cosmology. *Eur. Phys. J. C* **2021**, *81*, 117. [CrossRef]
98. Li, L.F.; Zhu, J.Y. Thermodynamics in Loop Quantum Cosmology. *Adv. High Energy Phys.* **2009**, *2009*, 905705. [CrossRef]

99. Agullo, I.; Ashtekar, A.; Nelson, W. Quantum gravity extension of the inflationary scenario. *Phys. Rev. Lett.* **2012**, *109*, 251301. [CrossRef]
100. Linsefors, L.; Barrau, A. Duration of inflation and conditions at the Bounce as a prediction of effective isotropic Loop Quantum Cosmology. *Phys. Rev. D* **2013**, *87*, 123509. [CrossRef]
101. Linsefors, L.; Barrau, A. Exhaustive investigation of the duration of inflation in effective anisotropic Loop Quantum Cosmology. *Class. Quantum Gravity* **2015**, *32*, 035010. [CrossRef]
102. Bolliet, B.; Barrau, A.; Martineau, K.; Moulin, F. Some clarifications on the duration of inflation in Loop Quantum Cosmology. *Class. Quantum Gravity* **2017**, *34*, 145003. [CrossRef]
103. Martineau, K.; Barrau, A.; Schander, S. Detailed investigation of the duration of inflation in Loop Quantum Cosmology for a Bianchi I Universe with different inflaton potentials and initial conditions. *Phys. Rev. D* **2017**, *95*, 083507. [CrossRef]
104. Assanioussi, M.; Dapor, A.; Liegener, K.; Pawłowski, T. Emergent de Sitter epoch of the loop quantum cosmos: A detailed analysis. *Phys. Rev. D* **2019**, *100*, 084003. [CrossRef]
105. Linsefors, L.; Cailleteau, T.; Barrau, A.; Grain, J. Primordial tensor power spectrum in holonomy corrected Omega-LQC. *Phys. Rev. D* **2013**, *87*, 107503. [CrossRef]
106. Barrau, A.; Bojowald, M.; Calcagni, G.; Grain, J.; Kagan, M. Anomaly-free cosmological perturbations in effective Canonical Quantum Gravity. *J. Cosmol. Astropart. Phys.* **2015**, *2015*, 51. [CrossRef]
107. Agullo, I.; Ashtekar, A.; Nelson, W. The pre-inflationary dynamics of Loop Quantum Cosmology: confronting quantum gravity with observations. *Class. Quantum Gravity* **2013**, *30*, 085014. [CrossRef]
108. Schander, S.; Barrau, A.; Bolliet, B.; Linsefors, L.; Mielczarek, J.; Grain, J. Primordial scalar power spectrum from the Euclidean Big Bounce. *Phys. Rev. D* **2016**, *93*, 023531. [CrossRef]
109. Martineau, K.; Barrau, A.; Grain, J. A first step towards the inflationary trans-Planckian problem treatment in Loop Quantum Cosmology. *Int. J. Mod. Phys. D* **2018**, *27*, 1850067. [CrossRef]
110. Barrau, A.; Jamet, P.; Martineau, K.; Moulin, F. Scalar spectra of primordial perturbations in Loop Quantum Cosmology. *Phys. Rev. D* **2018**, *98*, 086003. [CrossRef]
111. Taveras, V. Corrections to the Friedmann equations from Loop Quantum Gravity for a Universe with a free scalar field. *Phys. Rev. D* **2008**, *78*, 064072. [CrossRef]
112. Kasner, E. Geometrical theorems on Einstein's cosmological equations. *Am. J. Math.* **1921**, *43*, 217–221. [CrossRef]
113. Belinski, V.A.; Khalatnikov, I.M. Effects of scalar and vector fields on the nature of the cosmological singularity. *Sov. Phys. JETP* **1973**, *36*, 591.
114. Bojowald, M. Consistent Loop Quantum Cosmology. *Class. Quantum Gravity* **2009**, *26*, 075020. [CrossRef]
115. Alesci, E.; Cianfrani, F. A new perspective on cosmology in Loop Quantum Gravity. *EPL Europhys. Lett.* **2013**, *104*, 10001. [CrossRef]
116. Alesci, E.; Cianfrani, F. Loop Quantum Cosmology from Loop Quantum Gravity. *arXiv* **2014**, arXiv:1410.4788.
117. Alesci, E.; Cianfrani, F. Quantum Reduced Loop Gravity and the foundation of Loop Quantum Cosmology. *Int. J. Mod. Phys. D* **2016**, *25*, 1642005. [CrossRef]
118. Alesci, E.; Botta, G.; Cianfrani, F.; Liberati, S. Cosmological singularity resolution from quantum gravity: The emergent-bouncing Universe. *Phys. Rev. D* **2017**, *96*, 046008. [CrossRef]
119. Alesci, E.; Barrau, A.; Botta, G.; Martineau, K.; Stagno, G. Phenomenology of Quantum Reduced Loop Gravity in the isotropic cosmological sector. *Phys. Rev. D* **2018**, *98*, 106022. [CrossRef]
120. Alesci, E.; Botta, G.; Luzi, G.; Stagno, G.V. Bianchi I effective dynamics in Quantum Reduced Loop Gravity. *Phys. Rev. D* **2019**, *99*, 106009. [CrossRef]
121. Gielen, S.; Oriti, D. Quantum cosmology from quantum gravity condensates: Cosmological variables and lattice-refined dynamics. *New J. Phys.* **2014**, *16*, 123004. [CrossRef]
122. Oriti, D.; Sindoni, L.; Wilson-Ewing, E. Emergent Friedmann dynamics with a quantum bounce from quantum gravity condensates. *Class. Quantum Gravity* **2016**, *33*, 224001. [CrossRef]
123. Oriti, D.; Sindoni, L.; Wilson-Ewing, E. Bouncing cosmologies from quantum gravity condensates. *Class. Quantum Gravity* **2017**, *34*, 04LT01. [CrossRef]
124. Oriti, D. The Universe as a quantum gravity condensate. *C. R. Phys.* **2017**, *18*, 235–245. [CrossRef]
125. Gerhardt, F.; Oriti, D.; Wilson-Ewing, E. Separate Universe framework in Group Field Theory condensate cosmology. *Phys. Rev. D* **2018**, *98*, 066011. [CrossRef]
126. Ashtekar, A. Polymer geometry at Planck scale and quantum Einstein equations. *Int. J. Mod. Phys. D* **1996**, *5*, 629–648. [CrossRef]
127. Corichi, A.; Vukašinac, T.; Zapata, J.A. Hamiltonian and physical Hilbert space in Polymer Quantum Mechanics. *Class. Quantum Gravity* **2007**, *24*, 1495–1511. [CrossRef]
128. Bojowald, M.; Brizuela, D.; Hernández, H.H.; Koop, M.J.; Morales-Técotl, H.A. High-order quantum back-reaction and quantum cosmology with a positive cosmological constant. *Phys. Rev. D* **2011**, *84*, 043514. [CrossRef]
129. Szulc, L. Loop Quantum Cosmology of diagonal Bianchi type I model: simplifications and scaling problems. *Phys. Rev. D* **2008**, *78*, 064035. [CrossRef]
130. Belinskii, V.; Khalatnikov, I.; Lifshitz, E. A general solution of the Einstein equations with a time singularity. *Adv. Phys.* **1982**, *31*, 639–667. [CrossRef]

131. Cianfrani, F.; Montani, G.; Muccino, M. Semiclassical isotropization of the Universe during a de Sitter phase. *Phys. Rev. D* **2010**, *82*, 103524. [CrossRef]
132. Grain, J.; Vennin, V. Unavoidable shear from quantum fluctuations in contracting cosmologies. *Eur. Phys. J. C* **2021**, *81*, 132. [CrossRef]
133. Lyth, D.H. The primordial curvature perturbation in the ekpyrotic Universe. *Phys. Lett. B* **2002**, *524*, 1–4. [CrossRef]
134. Khoury, J.; Ovrut, B.A.; Steinhardt, P.J.; Turok, N. Density perturbations in the ekpyrotic scenario. *Phys. Rev. D* **2002**, *66*, 046005. [CrossRef]
135. Khoury, J.; Ovrut, B.A.; Seiberg, N.; Steinhardt, P.J.; Turok, N. From Big Crunch to Big Bang. *Phys. Rev. D* **2002**, *65*, 086007. [CrossRef]
136. Steinhardt, P.J.; Turok, N. The cyclic Universe: An informal introduction. *Nucl. Phys. B Proc. Suppl.* **2003**, *124*, 38–49. [CrossRef]
137. Brandenberger, R.H. Unconventional Cosmology. In *Quantum Gravity and Quantum Cosmology*; Calcagni, G., Papantonopoulos, L., Siopsis, G., Tsamis, N., Eds.; Springer: Berlin/Heidelberg, Germany, 2013; pp. 333–374. [CrossRef]
138. Alexander, S.; Cormack, S.; Gleiser, M. A cyclic Universe approach to fine tuning. *Phys. Lett. B* **2016**, *757*, 247–250. [CrossRef]
139. Clifton, T.; Barrow, J.D. Ups and downs of cyclic Universes. *Phys. Rev. D* **2007**, *75*, 043515. [CrossRef]
140. Barrow, J.D.; Ganguly, C. Cyclic Mixmaster Universes. *Phys. Rev. D* **2017**, *95*, 083515. [CrossRef]

Review

A Brief Overview of Results about Uniqueness of the Quantization in Cosmology

Jerónimo Cortez [1], Guillermo A. Mena Marugán [2] and José M. Velhinho [3,*]

[1] Departamento de Física, Facultad de Ciencias, Universidad Nacional Autónoma de México, Ciudad de México 04510, Mexico; jacq@ciencias.unam.mx
[2] Instituto de Estructura de la Materia, IEM-CSIC, Serrano 121, 28006 Madrid, Spain; mena@iem.cfmac.csic.es
[3] Faculdade de Ciências and FibEnTech-UBI, Universidade da Beira Interior, R. Marquês D'Ávila e Bolama, 6201-001 Covilhã, Portugal
* Correspondence: jvelhi@ubi.pt

Abstract: The purpose of this review is to provide a brief overview of recent conceptual developments regarding possible criteria to guarantee the uniqueness of the quantization in a variety of situations that are found in cosmological systems. These criteria impose certain conditions on the representation of a group of physically relevant linear transformations. Generally, this group contains any existing symmetry of the spatial sections. These symmetries may or may not be sufficient for the purpose of uniqueness and may have to be complemented with other remaining symmetries that affect the time direction or with dynamical transformations that are, in fact, not symmetries. We discuss the extent to which a unitary implementation of the resulting group suffices to fix the quantization—a demand that can be seen as a weaker version of the requirement of invariance. In particular, a strict invariance under certain transformations may eliminate some physically interesting possibilities in the passage to the quantum theory. This is the first review in which this unified perspective is adopted to discuss otherwise different uniqueness criteria proposed either in homogeneous loop quantum cosmology or in the Fock quantization of inhomogeneous cosmologies.

Keywords: quantum cosmology; uniqueness of the quantization

Citation: Cortez, J.; Mena Marugán, G.A.; Velhinho, J.M. A Brief Overview of Results about Uniqueness of the Quantization in Cosmology. *Universe* **2021**, *7*, 299. https://doi.org/10.3390/universe7080299

Academic Editor: Paulo Vargas Moniz

Received: 19 July 2021
Accepted: 8 August 2021
Published: 13 August 2021

Publisher's Note: MDPI stays neutral with regard to jurisdictional claims in published maps and institutional affiliations.

Copyright: © 2021 by the authors. Licensee MDPI, Basel, Switzerland. This article is an open access article distributed under the terms and conditions of the Creative Commons Attribution (CC BY) license (https://creativecommons.org/licenses/by/4.0/).

1. Introduction

Quantization is the process of constructing a description that incorporates the principles of Quantum Mechanics starting with a given classical system. In the present work, we consider exclusively the so-called canonical quantization process. This means that the classical system can be described in canonical form (e.g., in terms of variables that form canonical pairs) and that one aims at promoting classical variables to operators in a Hilbert space, preserving the canonical structure as much as possible. The prototypical system is, of course, the phase space $\mathbb{R}^{2n}$, with coordinates $\{(q_i, p_i), i = 1, \cdots, n\}$, equipped with the Poisson bracket $\{q_i, p_j\} = \delta_{ij}$. The standard quantization is realized in the Hilbert space $L^2(R^n)$ of square integrable functions, with configuration variables promoted to multiplicative operators

$$\hat{q}_i \psi = q_i \psi, \qquad (1)$$

and momentum variables acting as derivative operators

$$\hat{p}_i \psi = -i \frac{\partial}{\partial q_i} \psi. \qquad (2)$$

Here, and in the following, we set the reduced Planck constant $\hbar$ equal to one. These operators satisfy the canonical commutation relations (CCRs), thus, implementing Dirac's quantization rule that *Poisson brackets go to commutators* [1]. This quantization is irreducible, in the sense that any operator that commutes with all the $\hat{q}_i$'s and $\hat{p}_i$'s is necessarily proportional to the identity operator. In addition, this quantization satisfies a technical continuity condition, since it provides a continuous representation of the Weyl relations associated with the CCRs (see Section 2 for details).

These two conditions uniquely determine the quantization. This is the celebrated Stone–von Neumann uniqueness theorem (see, e.g., [2]): every irreducible representation of the CCRs coming from a continuous representation of the Weyl relations is unitarily equivalent to the aforementioned quantization, meaning that, given operators $\hat{q}'_i$ and $\hat{p}'_i$ with the above properties in a Hilbert space $\mathcal{H}$, there exists a unitary operator $U : \mathcal{H} \to L^2(\mathbb{R}^n)$ relating the two sets of operators $\hat{q}_i$, $\hat{p}_i$ and $\hat{q}'_i$, $\hat{p}'_i$ (see Equation (6) below).

On the other hand, it has been known since Dirac's work (and it was rigorously proved by Groenewold and van Hove [3,4]) that imposing Dirac's quantization rule on a large set of observables is not viable in the sense that it is impossible to satisfy the relations

$$[\hat{f}, \hat{g}] = i\widehat{\{f, g\}} \tag{3}$$

for a large set of classical observables, and certainly not for the whole algebra of classical observables (see [5] for a thorough discussion).

Nevertheless, in any given physical system, there are certainly observables of interest, other than the coordinate variables $\hat{q}_i$ and $\hat{p}_i$, which need to be quantized as well. Certain canonical transformations typically stand out in a given classical system, e.g., dynamics or symmetries, and these require, as well, a proper quantum treatment (these two aspects are, in fact, related since observables of physical interest often emerge as generators of some special groups of canonical transformations). This poses no problem as far as one considers the standard quantization of linear canonical transformations and the corresponding generators in $\mathbb{R}^{2n}$, precisely because of the above uniqueness theorem. One can easily illustrate this issue by considering a canonical transformation

$$\begin{pmatrix} q \\ p \end{pmatrix} \to \begin{pmatrix} q' \\ p' \end{pmatrix} = S \begin{pmatrix} q \\ p \end{pmatrix} \tag{4}$$

in $\mathbb{R}^{2n}$, where S is a symplectic matrix. Then, the operators $\hat{q}'_i$ and $\hat{p}'_i$, defined as

$$\begin{pmatrix} \hat{q}' \\ \hat{p}' \end{pmatrix} = S \begin{pmatrix} \hat{q} \\ \hat{p} \end{pmatrix}, \tag{5}$$

provide a new representation of the CCRs (in this case, in the same Hilbert space) with the same properties of irreducibility and continuity. Thus, it is guaranteed that there exists a unitary operator U such that

$$\hat{q}'_i = U^{-1} \hat{q}_i U, \quad \hat{p}'_i = U^{-1} \hat{p}_i U. \tag{6}$$

The unitary operator U is naturally interpreted as the quantization of the symplectic transformation S. We will also refer to it as the unitary implementation (at the quantum level) of the canonical transformation S. If, instead of a single linear transformation S, one has a 1-parameter group, generated, e.g., by a quadratic Hamiltonian function H, one obtains a corresponding 1-parameter group of unitary operators $U(t)$, which is typically continuous, so that a self-adjoint generator $\hat{H}$ such that $U(t) = e^{-i\hat{H}t}$ can be extracted. Non-quadratic classical Hamiltonians of the type $H = (1/2) \sum p_i^2 + V(q_i)$ do not fall in this last category; however, there is nevertheless a standard and well-defined procedure to obtain their quantum version (see [6] for details), which is simply to define $\hat{H}$ as $\hat{H} = (1/2) \sum \hat{p}_i^2 + V(\hat{q}_i)$.

Some ambiguities may occur in the quantization of more general functions, involving e.g., products of q's and p's; however, these ambiguities are typically not overly severe. These are precisely the type of ambiguities that may happen in standard homogeneous quantum cosmology (QC). In fact, in a homogeneous cosmological model, the number of both gravitational and matter degrees of freedom (DoF) is hugely reduced, ending up with just a finite number of global DoF, precisely due to homogeneity. In this so-called minisuperspace setup, the configuration variables on the gravitational side are typically given by the different scale factors, the number of which depend on the degree of anisotropy.

The object of interest here is the Hamiltonian constraint, the quantization of which leads to what is often called the Wheeler–de Witt equation $\hat{H}\Psi = 0$. This quantization may not be entirely trivial, owing to a possibly complicated dependence of the constraint with respect to the basic configuration and momentum variables. Nevertheless, the ambiguity that may emerge from the quantization of the Hamiltonian constraint typically involves a choice of factor ordering in $\hat{H}$ with respect to the basic quantum operators. Although different choices may lead to different versions of the Hamiltonian constraint, it is often the case that this does not affect the physical predictions substantially, in the sense that the predictions remain qualitatively the same.

Thus, the formalism of standard homogeneous QC, based on the standard quantization for a finite number of DoF, is to a large extent free of major ambiguities, as follows from the Stone–von Neumann theorem.

This last paradigm can be broken in two different ways, for very distinct reasons. First, the quantization—even of the set of basic configuration and momentum variables—of systems with an infinite number of DoF escapes the conclusions of the Stone–von Neumann theorem. On the contrary, in that case, there are representations of the CCRs leading to physically inequivalent descriptions of the same system. This occurs when local DoF are considered, of which one can mention two distinct situations of interest in cosmology: (i) the quantization of gravitational DoF (and possibly also of matter fields) in inhomogeneous cosmologies (such as in the example of Gowdy models [7]), and (ii) the quantum treatment of fields propagating in a non-stationary curved spacetime (e.g., of the FRW or de Sitter type), which is considered as a classical background.

While the first case embodies genuine applications to QC, i.e., a full quantum treatment of the gravitational DoF (in cases with considerable symmetry, like the Gowdy model, in the so-called midisuperspace setup), the second situation finds applications in the treatment of quantum perturbations in cosmology (both of gravitational and matter DoF), with the homogeneous background kept as a classical entity, such as, e.g., in inflationary scenarios. There is, thus, the need of selecting physically relevant quantizations corresponding to a given system containing an infinite number of DoF. Note that available selection criteria (leading to uniqueness) typically rely on stationarity and are, therefore, not applicable in the above described situations.

Section 4 is devoted to review selection criteria that were recently introduced and proved viable, leading to unique and well defined quantizations in the aforementioned cases. Such criteria are based on remaining symmetries, present in the cosmological system, and crucially on the unitary implementation of the dynamics, which can be seen as a weaker version of the requirement of invariance under time-translations, which can be applied only in stationary settings.

The other avenue for departure from the conclusions of the Stone–von Neumann theorem, which is relevant in the cosmological context, is exemplified by the quantization approach for homogeneous cosmologies known as loop quantum cosmology (LQC)[1]. Although the same models with a finite number of DoF are considered as in standard homogeneous cosmology, the obtained quantizations are not physically equivalent.

The type of quantization used in LQC is not unitarily equivalent to that of standard QC, and the reason why this is possible is that, in LQC, one of the conditions of the Stone–von Neumann theorem is broken in a hard way. The LQC-type of quantization starts from the Weyl relations, which are the exponentiated version of the CCRs and considers representations of the Weyl relations that are not continuous, thus, violating one of the conditions of the Stone–von Neumann theorem. In particular, it is the configuration part of the representation that is not continuous.

As a result, and although the correspondent of the unitary group $e^{it\hat{q}}$ is well defined in the LQC-type of representation as a unitary group $U(t)$, the would-be generator $\hat{q}$ cannot be defined, owing to the lack of continuity. This, in turn, is at the heart of the emergence of the discretization (in the canonically conjugate variable) that is so characteristic of LQC. Together with quantization methods adapted from those of loop quantum gravity (LQG) [8,9], this discretization is responsible, at the end of the day, for the results about singularity avoidance for which LQC is known.

One question that naturally arises is the following: are there other representations of the Weyl relations with different physical properties? Or is this particular LQC-type of representation naturally selected in some way? In Section 3, we discuss and comment on a uniqueness result for isotropic LQC recently put forward by Engle, Hanusch, and Thiemann [10], following a previous discussion concerning the Bianchi I case [11].

We would like to stress that, to the best of our knowledge, this is the first time that the results reviewed in Section 3 and those mentioned in Section 4 have been considered and discussed together. In particular, this joint and integrated review brings about a discussion on the two possible ways to use relevant transformations in order to select a unique quantization. In fact, results like those described in Section 3 are rooted on a requirement of strict invariance, while the results of Section 4 relax that condition (regarding what dynamical transformations are concerned), requiring only the weaker condition of unitary implementation of the transformations in question. These two approaches are discussed, providing a better understanding on the results achieved thus far in cosmology.

For completeness, we will start with a very brief review of the formalism for the study of Weyl algebras and their representations.

Also, we include an Appendix A sketching the proof of the uniqueness of the representation results mentioned in Section 4, in the simplest case of a scalar field in S^1 with time-dependent mass, with the purpose of providing the main steps and typical arguments of the proofs to the interested reader, without overloading the main text.

2. Weyl Algebra and Standard Representations

Let $\mathcal{U}$ and $\mathcal{V}$ be a pair of unitary representations of the commutative group $\mathbb{R}$, in the same Hilbert space $\mathcal{H}$, i.e., $\mathcal{U}(a)$ and $\mathcal{V}(b)$ are unitary operators for all real values of a and b and such that $\mathcal{U}(a+a') = \mathcal{U}(a)\mathcal{U}(a')$ and $\mathcal{V}(b+b') = \mathcal{V}(b)\mathcal{V}(b')$. The pair $\mathcal{U}$, $\mathcal{V}$ is said to satisfy the Weyl relations if

$$\mathcal{V}(b)\mathcal{U}(a) = e^{iab}\mathcal{U}(a)\mathcal{V}(b). \tag{7}$$

The standard representation of the Weyl relations, corresponding to the usual Schrödinger representation of the CCRs, is obtained as follows. Consider the Hilbert space $L^2(\mathbb{R})$ of square integrable functions $\psi(q)$ with respect to the usual Lebesgue measure dq. The expressions

$$(\mathcal{U}(a)\psi)(q) = e^{iaq}\psi(q) \tag{8}$$

and

$$(\mathcal{V}(b)\psi)(q) = \psi(q+b) \tag{9}$$

define unitary representations of $\mathbb{R}$, which clearly satisfy the Weyl relations (7). These representations are, moreover, jointly irreducible and continuous, i.e., $a \mapsto \mathcal{U}(a)$ and $b \mapsto \mathcal{V}(b)$ are continuous functions. The appropriate notion of continuity of these operator valued functions is that of strong continuity, and irreducibility means that no proper subspace of $\mathcal{H}$ supports the action of both $\mathcal{U}(a)$ and $\mathcal{V}(b)$ $\forall a, b$.

It is precisely due to continuity that Stone's theorem guarantees that it is possible to define infinitesimal generators $\hat{q}$ and $\hat{p}$ such that $\mathcal{U}(a) = e^{ia\hat{q}}$ and $\mathcal{V}(b) = e^{ib\hat{p}}$. In this case, it turns out that $\hat{q}$ is the multiplication operator q and $\hat{p} = -i\frac{d}{dq}$.

The celebrated Stone–von Neumann Theorem ensures that any other representation of the Weyl relations on a separable Hilbert space, with the same properties of irreducibility and continuity, is unitarily equivalent to the one above.

In order to make contact with the language of $\star$-algebras, we now introduce the so-called Weyl algebra. This is the algebra of formal products of objects $\mathcal{U}(a)$ and $\mathcal{V}(b)$, subjected to the Weyl relations (7). Note that, thanks to the Weyl relations, a generic element of the Weyl algebra can always be written as a finite linear combination of elements of the form $\mathcal{U}(a)\mathcal{V}(b)$, $a, b \in \mathbb{R}$, or equivalently of elements

$$\mathcal{W}(a,b) = e^{iab/2}\mathcal{U}(a)\mathcal{V}(b), \tag{10}$$

known as Weyl operators.

A representation of the Weyl relations is, thus, tantamount to a representation of the Weyl algebra, and, since this is a $\star$-algebra with identity, its representations can be discussed in terms of states of the algebra[2]. In particular, given a state, one can construct a cyclic representation of the algebra, by means of the so-called GNS construction [12]. Taking into account the above remarks, it follows that states of the Weyl algebra, and therefore the corresponding representations, are uniquely determined by the values assigned to the Weyl operators.

Concerning the unitary implementation of automorphisms of $\star$-algebras, it is a well known fact that, if a state ω is invariant under a given transformation, then a unitary implementation of that transformation is ensured to exist in the GNS representation defined by ω. The above constructions related to the Weyl algebra are straightforwardly generalized to any finite number of DoF, and also without difficulties to field theories. In this respect, let us consider, for instance, a scalar field in $\mathbb{R}^3$.

The starting point in the canonical quantization process is the choice of properly defined variables in phase space, that are going to play, e.g., the role of the q's and the p's. The integration of the field ϕ and of the canonically conjugate momentum π against smooth and fast decaying *test* functions provides just those amenable variables. Thus, given test functions f and g with the above properties, hence, belonging to the so-called Schwartz space $\mathcal{S}$, one defines linear functions in phase space by

$$(\phi, \pi) \mapsto \int \phi f \, d^3x + \int \pi g \, d^3x =: \phi(f) + \pi(g). \tag{11}$$

In particular, the variables $\phi(f)$ and $\pi(g)$, with Poisson bracket

$$\{\phi(f), \pi(g)\} = \int fg \, d^3x, \tag{12}$$

replace in this context the familiar q's and p's. The corresponding Weyl relations are

$$\mathcal{V}(g)\mathcal{U}(f) = e^{i \int fg d^3x}\mathcal{U}(f)\mathcal{V}(g), \qquad f, g \in \mathcal{S}, \tag{13}$$

with seemingly defined Weyl operators

$$\mathcal{W}(f,g) = e^{\frac{i}{2}\int fg d^3x}\mathcal{U}(f)\mathcal{V}(g). \tag{14}$$

Let us focus on a particular type of representations of the Weyl relations (or equivalently of the associated Weyl algebra), namely representations of the Fock type. These are representations defined by complex structures on the phase space or equivalently on the space $\mathcal{S} \oplus \mathcal{S}$ of pairs of test functions (f, g). In this respect, the pairs (f, g) define linear functionals in the phase space, and thus $\mathcal{S} \oplus \mathcal{S}$ is naturally dual to the phase space inheriting, therefore, a symplectic structure that is induced from that originally considered on phase space. We also recall that a complex structure J on a linear space with symplectic form Ω is a linear symplectic transformation such that $J^2 = -\mathbf{1}$ and is compatible with Ω in the sense that the bilinear form defined by $\Omega(J\cdot, \cdot)$ is positive definite.

Let J be a complex structure on the symplectic space $\mathcal{S} \oplus \mathcal{S}$. As mentioned above, a state of the Weyl algebra is defined by the values on the Weyl operators, and, therefore, the assignment

$$\mathcal{W}(f,g) \mapsto e^{-\frac{1}{4}\Omega(J(f,g),(f,g))} \qquad (15)$$

defines a state and an associated cyclic representation of the Weyl algebra. The cyclic vector is, here, physically interpreted as the vacuum of the Fock representation, and therefore the above expression coincides precisely with the expectation values of the Weyl operators on the vacuum of the Fock representation defined by J.

Let us now discuss the question of unitary implementation of symplectic transformations in the specific context of Fock representations. We consider unitary operators $\mathcal{W}(f,g)$ providing a representation of the Weyl algebra and a linear canonical transformation A. There is, then, a new representation $\mathcal{W}_A$, defined (in the same Hilbert space) by $\mathcal{W}_A(f,g) := \mathcal{W}(A^{-1}(f,g))$. If the representation $\mathcal{W}$ is defined by a complex structure J, it follows that $\mathcal{W}_A$ corresponds to a new complex structure $J_A := AJA^{-1}$. In general, A is not unitarily implementable, i.e., there is no unitary operator U_A such that

$$U_A^{-1}\mathcal{W}(f,g)U_A = \mathcal{W}(A^{-1}(f,g)). \qquad (16)$$

In fact, the Fock representations defined by J and J_A are unitarily equivalent if and only if the difference $J_A - J$ is an operator of a special type, namely a Hilbert–Schmidt operator [13]. Two notorious cases where that condition is automatically satisfied are the following. First, every operator in a finite dimensional linear space is of the Hilbert–Schmidt type, and therefore the unitary implementation of symplectic transformations comes for free in finite dimensional phase spaces, as expected from the Stone–von Neumann theorem. On the other hand, the null operator is always of the Hilbert–Schmidt type, regardless of the dimensionality, and therefore a transformation A that leaves J invariant is always unitarily implementable *in the Fock representation defined by J*.

With respect to previous remarks in our exposition, we note that invariance of J immediately translates into invariance of the associated Fock state defined by (15). Of course, the two situations that we have described correspond only to sufficient conditions for unitary implementation, which are by no means necessary. In particular, unitary implementation of a canonical transformation A can be achieved via a non-invariant complex structure J, provided that $J_A - J$ is Hilbert–Schmidt.

In any case, and in clear contrast with the situation found for a finite number of DoF, in field theory, no Fock representation supports the unitary implementation of the full group of linear canonical transformations. Fock representations are, therefore, distinguished by the class of transformations that are unitarily implementable. Now, in a particular theory, specified, e.g., by a given Hamiltonian, a particular set of canonical transformations stands out, namely transformations generated by the Hamiltonian and by possible symmetries.

The requirement of unitary implementation of relevant canonical transformations, therefore, provides a criterion guiding the selection of one representation over another when we are trying to quantize a field theory. In this respect, we notice that the case of the set of transformations corresponding to classical time evolution is particularly relevant, given the role that unitarity plays in the probabilistic interpretation of the quantum theory.

The simplest situation is that of a free field of mass m in Minkowski spacetime. In this case, the representation is completely fixed by the requirement of invariance under spatial symmetries and time evolution (or the full Poincaré group), in the sense that a unique complex structure is selected by that requirement, namely J_m defined by

$$J_m(f,g) = \left((m^2 - \Delta)^{\frac{1}{2}}g, -(m^2 - \Delta)^{-\frac{1}{2}}f\right), \qquad (17)$$

where Δ is the Laplacian.

In addition to the above free field in Minkowski spacetime, there are other known situations of linear dynamics where uniqueness results apply. In fact, provided that the

Hamiltonian is time independent, the criterion of positivity of the energy, together with invariance under the 1-parameter group of canonical transformations generated by the Hamiltonian, is sufficient to select a unique complex structure.

Here, positivity means that the unitary group implementing the dynamics possesses a positive generator, i.e., the quantum Hamiltonian is a positive operator [14,15]. This result finds remarkable applications in the quantization of free fields in stationary curved spacetimes (i.e., with a timelike Killing vector) [16,17]. On the other hand, no general uniqueness results are available for the non-stationary situations typical in cosmology.

3. Loop Quantum Cosmology

Let us now consider the representation of the Weyl relations used in LQC, sometimes referred to as the polymer representation. We will restrict our attention to its simplest version, namely the one associated with the homogeneous and isotropic flat FLRW model[3]. For convenience, we set the speed of light and Newton constant multiplied by 4π equal to the unit, and we make the Immirzi parameter [18] equal to $3/2$ in order to simplify our equations, without loss of generality.

At the classical level, the system is described by a pair of canonically conjugate variables, typically denoted by c and p, with Poisson bracket $\{c, p\} = 1$. The variables c and p parametrize, respectively, the (homogeneous) Ashtekar connection and the densitized triad (see [10] for details in the context of the current uniqueness discussion, and [19,20] for more general introductions to LQC). In particular, p is proportional to the square scale factor of the FLRW spacetime.

Let us consider the Hilbert space $\mathcal{H}_\mathcal{P}$ defined by the discrete measure in $\mathbb{R}$, i.e., the space of complex functions $\psi(p)$ such that

$$\sum_{p \in \mathbb{R}} |\psi(p)|^2 < \infty, \tag{18}$$

with the inner product given by

$$\langle \psi, \psi' \rangle = \sum_{p \in \mathbb{R}} \bar{\psi}(p) \psi'(p), \tag{19}$$

where the overbar denotes complex conjugation. We note that this Hilbert space, also referred to as the polymer Hilbert space, is very different from the standard one, $L^2(\mathbb{R})$. In particular, $\mathcal{H}_\mathcal{P}$ is non-separable[4]. An orthonormal basis of $\mathcal{H}_\mathcal{P}$ is formed, e.g., by the uncountable set of functions Ψ_{p_0}, for all $p_0 \in \mathbb{R}$, where

$$\Psi_{p_0}(p) = \delta_{pp_0}, \tag{20}$$

with δ_{pp_0} as the Kronecker delta.

We, then, define the operators $\mathcal{U}_\mathcal{P}(a)$ and $\mathcal{V}_\mathcal{P}(b)$, for $a, b \in \mathbb{R}$, acting on $\mathcal{H}_\mathcal{P}$ by

$$(\mathcal{U}_\mathcal{P}(a)\psi)(p) = \psi(p - a) \tag{21}$$

and

$$(\mathcal{V}_\mathcal{P}(b)\psi)(p) = e^{ibp}\psi(p). \tag{22}$$

It is clear that these operators satisfy the Weyl relations (7) and that the representation is irreducible. The map $b \mapsto \mathcal{V}_\mathcal{P}(b)$ is continuous, and thus one can define the infinitesimal generator. We will denote it by $\pi(p)$ and not $\hat{p}$, to distinguish it from the standard Schrödinger representation in $L^2(\mathbb{R})$. It follows that

$$(\pi(p)\psi)(p) = p\psi(p). \tag{23}$$

On the other hand, $\mathcal{U}_\mathcal{P}(a)$ is not continuous. To see this, it suffices to note that, for arbitrarily small a, the vector Ψ_{p_0} is mapped by $\mathcal{U}_\mathcal{P}(a)$ to an orthogonal one Ψ_{p_0+a}. The would-be generator of the unitary group $\mathcal{U}_\mathcal{P}(a)$ cannot, therefore, be defined. Taking into account the commutator $[\pi(p), \mathcal{U}_\mathcal{P}(a)]$, one can see that the operators $\mathcal{U}_\mathcal{P}(a)$ can be regarded as providing a quantization of the classical variables e^{iac}. We note that the reality conditions are properly satisfied, since $\mathcal{U}^\dagger(a) = \mathcal{U}(-a)$, with the dagger denoting the adjoint. Thus, the set of operators $\mathcal{U}_\mathcal{P}(a)$, together with $\pi(p)$, provide a quantization, in the usual Dirac's sense, of the Poisson algebra of phase space functions made of finite linear combinations of the functions p and e^{iac}, with $a \in \mathbb{R}$ (see [23] for details).

This Poison algebra is, of course, different from the kinematical algebra usually associated with the linear phase space $\mathbb{R}^2$ with coordinates c and p, which is simply the Heisenberg algebra of linear (non-homogeneous) functions in c and p. At the foundation of LQC, there is, thus, a situation similar to that of LQG: a non-standard choice of basic variables and a representation of the associated algebra that is non-continuous (at least in part of the algebra), thus, obstructing the quantization of the connection itself [8]. However, we note that, contrary to the situation in LQG, the standard Schrödinger quantization gives us a different representation of the very same Poisson algebra, since the variables e^{iac} are trivially quantized in $L^2(\mathbb{R})$ by multiplication operators, e.g., by operators $\widehat{e^{iac}}$ such that

$$\left(\widehat{e^{iac}}\psi\right)(q) = e^{iaq}\psi(q). \qquad (24)$$

In the case of this Schrödinger quantization, the continuity of the representation $\mathcal{U}(a)$ allows us to define the operator $\hat{c}$ itself, and therefore to extend the quantization of configuration variables to a much larger set of functions $f(c)$, simply by defining $\widehat{f(c)} = f(\hat{c})$; whereas, in the polymer quantization, one is restricted to quantize configuration variables of the type e^{iac} (and linear combinations thereof).

In LQG, there is a celebrated result about the uniqueness of the quantization that gives robustness to the loop representation [24,25]. In [10], the authors proved a similar uniqueness result for LQC, that we now discuss. In order to make contact with that work, which uses the language of $\star$-algebras and corresponding states, we first introduce the LQC analogue of the LQG holonomy-flux algebra, which is again denoted in [10] as the quantum holonomy-flux $\star$-algebra $\mathfrak{U}$.

The LQC $\star$-algebra $\mathfrak{U}$ is constructed in [10] as the algebra of formal products of operators corresponding to the variables p and e^{iac} subjected to the conditions coming from the commutator $[\pi(p), \mathcal{U}_\mathcal{P}(a)]$. However, since we have already introduced the Weyl algebra, it is more natural here to follow an alternative procedure, and identify instead the $\star$-algebra $\mathfrak{U}$ with the Weyl algebra, i.e., the algebra of formal products of objects $\mathcal{U}(a)$ and $\mathcal{V}(b)$, subjected to the Weyl relations (7).

Recall now that the group of spatial diffeomorphisms is a "gauge" symmetry in General Relativity (GR). Physical states in quantum gravity should, therefore, be invariant under the quantum operators representing these diffeomorphisms (or annihilated by the quantum diffeomorphism constraint). Thus, a unitary implementation of the group of spatial diffeomorphisms in the quantum Hilbert space is required. Since the LQG holonomy-flux algebra is again a $\star$-algebra with identity, it follows from previous comments that, in order to achieve the required unitary implementation of the group of diffeomorphisms, it is sufficient that the quantization be defined by a diffeomorphism invariant state.

The LQG result of uniqueness of the quantization [24,25] guarantees, precisely, that there exists a unique diffeomorphism invariant state of the LQG holonomy-flux algebra. Moreover, the GNS representation defined by such a unique invariant state is unitarily equivalent to the LQG representation that was previously known. The analogous result for LQC starts from the observation, explained in detail in [10], that a residual gauge group still remains, when descending from full GR to homogeneous and isotropic flat models. In fact, although almost all of the diffeomorphism gauge symmetry is automatically fixed, in homogeneous and isotropic flat models, one is left with a small gauge group, namely the group of isotropic dilations, acting on phase space as

$$(p,c) \mapsto (\lambda p, c/\lambda), \quad \lambda \in \mathbb{R}. \tag{25}$$

Thus, there is the possibility of exploring this residual gauge symmetry in order to select a state in LQC, very much like in the above mentioned LQG uniqueness result. The authors of [10] succeeded in proving that there is a unique dilation invariant state of the LQC holonomy-flux algebra, for which the associated GNS representation is unitarily equivalent to the polymer representation described above.

Although the result of [10] is interesting and rigorous, we argue here that, in a certain sense, its status is not as strong as that of the LQG uniqueness result. From the physical viewpoint, what is really required is a unitary implementation of the group of interest (which may correspond to dynamics, to symmetries, or to gauge transformations), and not necessarily an invariant state. There are even situations (see, e.g., Section 4) where invariant states are not available, and nevertheless a unitary implementation of the relevant group exists.

It is true that constructing the quantization by means of an invariant state is sufficient to achieve unitary implementation—and it is perhaps the "Kings way" of doing it—however, it is by no means necessary. In the present case, the Schrödinger representation, although not (unitarily equivalent to a representation) defined by an invariant state, carries a unitary implementation of the group of dilations (25), which is physically just as good as the one provided by the polymer representation. The existence of this unitary implementation actually follows from the Stone–von Neumann theorem, but let us show it explicitly. We consider the transformations $U_\lambda : L^2(\mathbb{R}) \to L^2(\mathbb{R})$ defined by

$$\psi(q) \xmapsto{U_\lambda} \lambda^{1/2} \psi(\lambda q), \quad \lambda \in \mathbb{R}, \tag{26}$$

which are clearly unitary $\forall \lambda$, with respect to the standard inner product defined by the measure dq. We consider also the standard operator $\hat{p} = -i\frac{d}{dq}$ and the Schrödinger quantization of configuration variables

$$\left(\widehat{f(c)}\psi\right)(q) = f(q)\psi(q), \quad \psi \in L^2(\mathbb{R}), \tag{27}$$

which clearly includes the LQC variables e^{iac}. A straightforward computation shows that

$$\left(U_\lambda^{-1} \hat{p} U_\lambda \psi\right)(q) = -i\lambda \frac{d}{dq}\psi(q), \tag{28}$$

$$\left(U_\lambda^{-1} \widehat{f(c)} U_\lambda \psi\right)(q) = f(q/\lambda)\psi(q), \tag{29}$$

or

$$U_\lambda^{-1} \hat{p} U_\lambda = \widehat{\lambda p}, \quad U_\lambda^{-1} \widehat{f(c)} U_\lambda = \widehat{f(c/\lambda)}, \tag{30}$$

which is the announced unitary implementation of the group of dilations (25) in the Schrödinger representation.

From this perspective, we conclude that the physical criterion of a unitary implementation of the residual group of dilations in homogeneous and isotropic flat cosmology does not fully succeed in selecting a unique quantization, since both the polymer and the Schrödinger representations are viable from this viewpoint. Only the more mathematically stringent requirement of strict invariance selects a unique state. This is perhaps a reminder about the fact that strict invariance is not an unavoidable requirement and that the quantization of groups of interest via unitary implementations that are not *necessarily* based on invariant states is worth exploring.

Let us end with a brief comment regarding the analogous uniqueness result in LQG. In that case, the uniqueness is also proved by requiring the strict invariance of a state of the holonomy-flux algebra under the action of spatial diffeomorphisms. There is, however, a key difference with respect to the above described situation in LQC: no other (irreducible) representation is known, of any kind, admitting a unitary implementation of the diffeomorphism group[5], and thus there is no alternative route that may cast any shadow on the uniqueness.

The situation remains, however, somewhat open, until a stronger uniqueness result is demonstrated that is based exclusively on a unitary implementation of the spatial diffeomorphisms and not just in strict invariance or otherwise until a new representation of the LQG algebra admitting a unitary non-invariant implementation of the diffeomorphism group is constructed.

4. Fock Quantization in Non-Stationary Cosmological Settings

As we have already mentioned, the Stone–von Neumann uniqueness result fails for an infinite number of DoF, and no general result on the uniqueness of the quantization is available for non-stationary situations. This includes, of course, cases of interest in QC. In some of those cases, the underlying theory can be recast in the form of a linear scalar field with a time-dependent mass, propagating in an auxiliary background spacetime, which is both static and spatially compact.

One such situation is the linearly polarized Gowdy model with the spatial topology of a three-torus, where the gravitational degrees of freedom are encoded by a scalar field on S^1, evolving in time precisely as a linear field with a time-dependent mass of the type $m(t) = 1/(4t^2)$ [31,32]. Other situations of interest include free scalar fields in cosmological scenarios, e.g., propagating in (compact) FLRW or the Sitter spacetimes. In those cases, the non-stationarity is transferred from the background to the field effective mass, by means of a simple transformation.

4.1. Gowdy Models

Midisuperspace models are symmetry reductions of full GR that retain an infinite number of degrees of freedom. Typically, these are local degrees of freedom, and thus they often describe inhomogeneous scenarios. Therefore, these midisuperspace models must face the inherent ambiguity that affects the quantization of fields.

One of the simplest inhomogeneous cosmologies obtained with a symmetry reduction is the linearly polarized Gowdy model on the three-torus, T^3 [7]. This model describes vacuum spacetimes with spatial sections of T^3-topology containing linearly polarized gravitational waves, with a symmetry group generated by two commuting, spacelike, and hypersurface orthogonal Killing vector fields. As a consequence, the local physical degrees of freedom can be parametrized by a scalar field corresponding to those waves and effectively living in S^1.

After a partial gauge fixing, the line element of the linearly polarized Gowdy T^3 model can be written as [33]

$$ds^2 = e^{\gamma - \phi/\sqrt{p}}\left(-dt^2 + d\theta^2\right) + e^{-\phi/\sqrt{p}}t^2 p^2 d\sigma^2 + e^{\phi/\sqrt{p}}d\delta^2, \tag{31}$$

where $(\partial/\partial\sigma)^a$ and $(\partial/\partial\delta)^a$ are the two Killing vector fields. The true dynamical field DoF are encoded in $\phi(\theta,t)$, where $t > 0$ and $\theta \in S^1$. On the other hand, p is a homogeneous non-dynamical variable, and the field γ is completely determined by p and ϕ as a result of the gauge-fixing process [33]. There remains a global spatial constraint on the system, giving rise to the symmetry group of (constant) translations in S^1. The time evolution is dictated by the field equation

$$\ddot{\phi} + \frac{1}{t}\dot{\phi} - \phi'' = 0. \tag{32}$$

Here, the prime denotes the derivative with respect to θ, and the dot denotes the time derivative.

A complete quantization of the system was obtained in [34]. Nevertheless, it was soon realized [33,35] that the classical dynamics could not be implemented as a unitary transformation in such quantization. With the purpose of achieving unitarity, and restoring, in particular, the standard probabilistic interpretation of quantum physics within the quantum Gowdy model, an alternative quantization was introduced by Corichi, Cortez, and Mena Marugán [31,32]. A crucial step toward a quantization with unitary dynamics is the following time-dependent transformation performed at the level of the classical phase space. Instead of working with the original field ϕ and its corresponding conjugate momentum P_ϕ, the authors introduced a new canonical pair χ and P_χ related to the first one by means of the canonical transformation

$$\chi = \sqrt{t}\phi, \quad P_\chi = \frac{1}{\sqrt{t}}\left(P_\phi + \frac{\phi}{2}\right), \tag{33}$$

taking advantage in this way of the freedom in the scalar field parametrization of the metric of the Gowdy model.

The evolution of the new canonical pair was found to be governed by a time-dependent Hamiltonian that, in our system of units, adopts the expression

$$H_\chi = \frac{1}{2}\oint d\theta \left[P_\chi^2 + \chi'^2 + \frac{\chi^2}{4t^2}\right]. \tag{34}$$

The corresponding Hamiltonian equations are

$$\dot{\chi} = P_\chi, \quad \dot{P}_\chi = \chi'' - \frac{\chi}{4t^2}, \tag{35}$$

that, combined, give the second-order field equation

$$\ddot{\chi} - \chi'' + \frac{\chi}{4t^2} = 0. \tag{36}$$

We can, therefore, view the system as a linear scalar field with a time-dependent mass of the form $m(t) = 1/(4t^2)$, evolving in an effective static spacetime with one-dimensional spatial sections with the topology of the circle[6]. Another relevant aspect of the model is the invariance of the dynamics under (constant) S^1-translations:

$$T_\alpha : \theta \mapsto \theta + \alpha, \quad \alpha \in S^1. \tag{37}$$

These translations are, moreover, symmetries generated by a remaining constraint, as we already mentioned.

The quantization of the system put forward by Corichi, Cortez, and Mena Marugán in [31,32] starts from the CCRs satisfied by the canonical pair χ and P_χ, or rather by the corresponding Fourier modes. The advantage of using Fourier components, say $\chi_n = (1/\sqrt{2\pi})\oint d\theta e^{-in\theta}\chi(\theta)$, instead of the field $\chi(\theta)$ itself is clear: since the spatial manifold is compact, the set of Fourier modes is discrete, and one, therefore, avoids the issues of dealing with operator valued distributions, such as, e.g., $\hat{\chi}(\theta)$. The aforementioned quantization is of the standard Fock type, with the following remarkable properties. To begin with, the complex structure on phase space that effectively defines the Fock representation is invariant under S^1-translations.

Thus, the corresponding state of the Weyl algebra is S^1-invariant, leading to a natural unitary implementation of these gauge transformations. Secondly and most importantly, the classical dynamics in phase space defined by Equation (35) (i.e., generated by the Hamiltonian (34)) are unitarily implemented at the quantum level. In other words, let t_0 be an arbitrary but fixed initial time, and $S(t, t_0)$ be the linear symplectic transformation corresponding to the classical evolution in phase space from the time t_0 to the arbitrary time t. Then, to each transformation $S(t, t_0)$, there corresponds a unitary operator $U(t, t_0)$, that intertwines between the quantum operators $\hat{\chi}$ and $\hat{P}_\chi$ defined at the initial time t_0 and those obtained from $\hat{\chi}$ and $\hat{P}_\chi$ by application of $S(t,t_0)$, as exemplified in Equations (5) and (6).

This last transformation corresponds to the usual evolution in the Heisenberg picture, which can always be formally defined, once canonical operators are given at some initial time. The key difference with respect to systems with a finite number of DoF is that, whereas, in those cases, the relation between the "initial" and the evolved operators in principle is always unitary, the existence of such unitary operators is far from being guaranteed in an arbitrary representation of the CCRs for field theory (or generally with an infinite number of DoF).

We note that, although a unitary implementation of all the transformations $S(t,t_0)$ is achieved in the Corichi, Cortez, and Mena Marugán representation, the corresponding state is not invariant under these transformations. In fact, no state exists such that it remains invariant under all the transformations $S(t,t_0)$, $\forall t$.

All in all, the quantization of the linearly polarized Gowdy model proposed in [31,32] is one of the few available examples of a rigorous and fully consistent quantization of an inhomogeneous cosmological model. Nonetheless, the eventual robustness of its physical predictions might be affected by the possible existence of major ambiguities in the quantization process. Fortunately, a quantization with the aforementioned properties is, indeed, unique as shown in [36,37], where the uniqueness result that we now discuss was derived.

A source of ambiguity in the process leading to the Corichi, Cortez, and Mena Marugán quantization is the choice of representation for the canonical pair χ and P_χ. However, it was shown in [36] that any other Fock representation of the CCRs that (i) is defined by a S^1-invariant complex structure (or equivalently by a S^1-invariant state of the Weyl algebra) and (ii) allows a unitary implementation of the dynamics defined by Equation (35), is unitarily equivalent to the considered representation (and therefore physically indistinguishable). We note that there actually is an infinite number of S^1-invariant states, leading to many inequivalent representations. It is only after the requirement of unitary dynamics that a unique unitary equivalence class of representations is selected.

Another possible source of ambiguity concerns the choice of the "preferred" canonical pair (χ, P_χ). In this respect, note that a time-dependent transformation different from Equation (33) would lead to classical dynamics that would not reproduce Equation (35), and it is, in principle, conceivable that the new dynamics could be unitarily implemented in a different representation, thus, leading to a distinct quantization. This is, however, not the case.

It was shown in [37] that any other canonical transformation of the type (33) modifies the equations of motion in such a way as to render impossible the unitary implementation of the dynamics, with respect to any Fock representation defined by a S^1-invariant complex structure. The kind of transformations considered in [37] is restricted by the natural requirements of locality, linearity, and preservation of S^1-invariance. In configuration space, these are contact transformations that produce a time-dependent scaling of the field ϕ (see [37] for a detailed discussion). Such scalings can always be completed into a canonical transformation in phase space, of the general form

$$(\phi, P_\phi) \mapsto \left(f(t)\phi, \frac{P_\phi}{f(t)} + g(t)\phi \right), \tag{38}$$

which includes a contribution to the new momentum, which is linear in the field ϕ.

Finally, let us mention that completely analogous results were obtained for the remaining linearly polarized Gowdy models, namely those with spatial sections with topologies $S^1 \times S^2$ and S^3. This analysis was performed in the following independent steps. First, the classical models were addressed in [38], showing that, in these cases, the local gravitational DoF can also be parametrized by a single scalar field, namely an axisymmetric field in S^2. Then, following a procedure similar to the one introduced by Corichi, Cortez, and Mena Marugán, a Fock quantization with unitary dynamics was obtained [39]. In particular, a time-dependent scaling of the original field is again involved, now of the form $\phi \mapsto \sqrt{\sin t}\phi$. Finally, the uniqueness of the quantization obtained in this way was proved in [40].

4.2. Quantum Field Theory in Cosmological Settings

A common feature of the Gowdy models mentioned in the previous section is that the local DoF are parametrized by a scalar field effectively living in a compact spatial manifold. Moreover, after the crucial scaling of the field, the dynamics are those of a linear field with time-dependent mass, i.e., they obeys a second-order equation of the type

$$\ddot{\chi} - \Delta \chi + s(t)\chi = 0, \tag{39}$$

where Δ is the Laplace–Beltrami (LB) operator for the spatial sections in question, e.g., S^1 for the Gowdy model on T^3 and S^2 for the remaining two models.

Remarkably, a whole different type of situations in cosmology can also be described by an equation of the form (39). Let us consider, e.g., a free scalar field in a homogeneous and isotropic FLRW spacetime, with the line element

$$ds^2 = a^2(t)\left[-dt^2 + h_{ab}dx^a dx^b\right], \tag{40}$$

where t is the conformal time, $a(t)$ is the scale factor, and h_{ab} ($a,b = 1,2,3$) is the Riemannian metric, for either flat Euclidean space or the 3-sphere. A minimally coupled scalar field of mass m obeys, in this cosmological spacetime, the equation

$$\ddot{\phi} + 2\frac{\dot{a}}{a}\dot{\phi} - \Delta\phi + m^2 a^2 \phi = 0, \tag{41}$$

where Δ is the LB operator defined by the metric h_{ab} of the spatial sections.

The most obvious situation described by this setup is the propagation of an actual (test) scalar matter field (disregarding the backreaction) in an FLRW background. Nonetheless, the treatment of quantum perturbations, both of matter and of gravitational DoF, also fits the above description. In fact, in the context of cosmological perturbations, the leading-order approximation in the action, together with a neglected backreaction, amounts to keep the homogeneous classical cosmology as the background and treats both matter and gravitational perturbations as fields propagating on that background [41–45].

The quantum treatment of the situations described above, therefore, faces the ambiguity of the choice of quantum representation. In particular, the criterion based on stationarity, mentioned in Section 2, is not available, since all these situations are inherently non-stationary.

There is, however, a natural avenue to address this issue, arriving hopefully at a unique quantum theory, with unitary dynamics. The way forward is actually suggested by a standard procedure, commonly found precisely in QFT in curved spacetimes (see, e.g., [46,47]) and in the treatment of cosmological perturbations (see, for example, [41–44]). This consists in the scaling of the field variable ϕ by means of the scale factor, thus, introducing the rescaled field $\chi = a(t)\phi$, which now obeys an equation of the type (39), with $s(t) = m^2 a^2 - (\ddot{a}/a)$.

Thus, the combined effect of the use of conformal time and the scaling $\phi \mapsto a(t)\phi$ is to recast the field equation in the form (39), which is effectively the field equation of a linear field with a time-dependent quadratic potential $V(\chi) = s(t)\chi^2/2$, propagating in a static background with metric

$$ds^2 = -dt^2 + h_{ab}dx^a dx^b. \tag{42}$$

In this manner, we see that a generalization of the uniqueness result obtained for the Gowdy models would provide a useful criterion to select a unique quantization for fields in non-stationary backgrounds, such as FLRW universes, typically with compact spatial sections with the topology of S^3 or T^3, also allowing applications to the usual flat universe case.

Note that the restriction to spatial compactness is most convenient from the viewpoint of mathematical rigor, as, otherwise, infrared issues would plague the analysis. Nonetheless, the physical effects of the artificially imposed compactness, e.g., in the spatially flat case, should be irrelevant when the physical problem at hand does not involve arbitrarily larges scales, e.g., going beyond the Hubble radius. The case of continuous scales, which corresponds to non-compactness, can be reached in a suitable limit for flat universes, after completing all the demonstrations of uniqueness in the framework of compact spatial sections [48].

The desired generalization of the result about the uniqueness of the representation in the aforementioned context of test fields and perturbations on cosmological backgrounds was obtained in [49–53]. In the rest of this Section, we very briefly explain the implications of these results.

Let $\mathbb{I} \times \Sigma$ be a globally hyperbolic spacetime, where $\mathbb{I} \in \mathbb{R}$ is an interval and Σ is a compact Riemannian manifold of dimension $d \leq 3$ (for cosmological applications, one can think of Σ as being either S^1, S^2, S^3, or T^3). The spacetime is assumed to be static, with a metric given by Equation (42), where h_{ab} is the time-independent, Riemannian metric on Σ.

Consider a linear scalar field in $\mathbb{I} \times \Sigma$ obeying a field equation of the type (39), where $s(t)$ is essentially[7] an arbitrary function, and Δ is the LB operator associated with the metric h_{ab}. Note that any symmetry of this metric is transmitted to the LB operator and, therefore, to the equations of motion. Let us consider the Weyl algebra associated with the field χ (and of course its canonical conjugate momentum) and their Fock representations. Then,

1. there exists a Fock representation defined by a state that is invariant under the symmetries of the metric h_{ab} (or equivalently, a Fock representation with an invariant vacuum) and such that the classical dynamics can be unitarily implemented;
2. that representation is unique, in the sense that any other Fock representation defined by an invariant state and allowing a unitary implementation of the dynamics is unitarily equivalent to the previous one.

Remarkably, it again follows that the rescaling of the field $\phi \mapsto a(t)\phi$ is quite rigid and uniquely determined: no other time-dependent scaling can lead to a(n invariant Fock) quantization with unitary dynamics, and thus no ambiguity remains in the choice of the preferred field configuration variable. Furthermore, the unitarity requirement essentially selects, as well, the canonical momentum field. Physically, this can interpreted as a unique splitting between the time-dependence assigned to the background and that corresponding to the evolution of the scalar field DoF.

Finally, extensions of these uniqueness results were obtained in several directions. First, analogous results were obtained for scalar fields in homogeneous backgrounds of the Bianchi I type [54]. These spacetimes are not isotropic, and therefore the conformal symmetry, which was a common characteristic of the previous cases (at least asymptotically for large frequencies) is no longer present in an obvious way. Even more remarkable are the extensions of the uniqueness results attained for fermions, since they mark the transition to a largely unexplored territory [55–57].

A recent account of these results[8] can be found in [59].

5. Conclusions

We reviewed and discussed several results concerning the quantization of systems with relevant applications in cosmology. In particular, we have focused our attention on results ascertaining the uniqueness of the quantization process. This uniqueness is crucial in order to provide physical robustness to the eventual cosmological predictions of the quantum models in question, which would otherwise be fundamentally affected by ambiguities.

Starting with homogeneous models in this cosmological context, we discussed a uniqueness result from Engle, Hanusch, and Thiemann, concerning the representation of the Weyl relations commonly used in LQC. Turning to recent investigations carried out by our group and collaborators, the quantization of the family of inhomogeneous cosmologies, known as the linearly polarized Gowdy models, was reviewed next. Crucial in this discussion is the requirement of unitary implementation (at the quantum level) of the dynamics. Together with invariance under spatial symmetries, the criterion of unitary dynamics proved very effective in the selection of a unique and physically meaningful quantization in a variety of situations.

A common general mathematical model embodying all these cases is that of a scalar field with a time-dependent mass (with arbitrary time dependence except for some very mild conditions), propagating in a static spacetime with compact spatial sections. The existence and uniqueness of a Fock quantization with unitary dynamics has been proven for such a general model, thus, providing unique quantizations in cosmological systems, ranging from quantum fields in FLRW backgrounds to the quantization of (perturbative) gravitational DoF.

In particular, it follows from our analysis and the aforementioned discussions that the quantization of linear transformations of physical interest via a unitary implementation does not necessarily require the existence of an invariant state, and that it is worthwhile pursuing alternatives that are not based on such invariance.

In fact, what is really required in physical terms is a unitary implementation and not necessarily an invariant state, i.e., unitary implementations via non-invariant states are still physically acceptable and cannot be discarded. Clearly, a proof on the uniqueness of the quantization based solely on the unitary implementation of those transformations, rather than on the existence of an invariant state, is a stronger result inasmuch as the requirement of unitarity is weaker than invariance. Whereas, in some circumstances, there are good reasons to restrict attention to invariant states (for instance, when there is a time-independent Hamiltonian giving rise to a group of transformations), in the general case of arbitrary transformations, the requirement of an invariant state is not so compelling and non-invariant states cannot be simply disregarded.

In this sense, invariance can actually be an overly rigid demand in certain situations, especially if the considered transformations provide a notion of evolution that is crucial to describe the dynamics. For instance, the unitary implementation of the evolution via (dynamical-)invariant states in typically non-stationary scenarios is of doubtful physical use and, in fact, seems hopeless, whereas physically viable states, still leading to unitary dynamics and ensuring uniqueness, are available and have direct cosmological applications. In this respect, we conclude with the following remark.

The uniqueness of the Fock quantization of the scalar field with time-dependent mass was obtained by restricting the attention to states that remain invariant under spatial symmetries. Although the removal of this restriction seems unlikely to lead to physically new representations with unitary dynamics, it is, nevertheless, an open possibility. More precisely, it remains to be disproved the existence of representations with unitary dynamics and a unitary implementation of the spatial symmetries, which are, nevertheless, not unitarily equivalent to the representation defined by an invariant vacuum.

Thus, a conceivable line of future research in this area is to consider also Fock representations that, while not possessing an invariant vacuum, still allow a unitary implementation of the spatial symmetries. The uniqueness result would be strengthened if those representations were shown to be equivalent to the previous one; otherwise, new and potentially interesting representations could emerge. Another possibility, with a clearly higher potential to produce physically inequivalent results, is to use polymer-inspired representations for the scalar field, instead of Fock representations.

Author Contributions: Conceptualization: J.C., G.A.M.M. and J.M.V. Original Draft Preparation: J.C., G.A.M.M. and J.M.V. All authors have read and agreed to the published version of the manuscript.

Funding: This work was supported by the Spanish MINECO grant number FIS2017-86497-C2-2-P, the Spanish MICINN grant number PD2020-118159GB-C41, and the European COST (European Cooperation in Science and Technology) Action number CA16104 GWverse. J.M.V. is grateful for the support given by research unit Fiber Materials and Environmental Technologies (FibEnTech-UBI), on the extent of the project reference UIDB/00195/2020, funded by the Fundação para a Ciência e a Tecnologia (FCT).

Institutional Review Board Statement: Not applicable.

Informed Consent Statement: Not applicable.

Data Availability Statement: Not applicable.

Acknowledgments: The authors are grateful to B. Elizaga Navascués for discussions.

Conflicts of Interest: The authors declare no conflict of interest.

Appendix A. Sketch of the Proof of Uniqueness of the Representation for the Scalar Field with Time-Dependent Mass in S^1

In this appendix, we consider the simplest example of a linear scalar field in a static spacetime with compact spatial sections, namely, the case where the spatial sections are 1-dimensional with the topology of the circle. The field equation reads

$$\ddot{\chi} - \Delta\chi + s(t)\chi = 0, \tag{A1}$$

where the mass term $s(t)$ can be an (essentially) arbitrary function of time.

Since the spatial manifold is compact and the field equation is linear, a Fourier decomposition gives us a discrete set of independent modes:

$$\chi(\theta, t) = \frac{1}{\sqrt{2\pi}} \sum_{n=-\infty}^{\infty} \chi_n(t) e^{in\theta} = \frac{1}{\sqrt{\pi}} \sum_{n=1}^{\infty} \left(q_n \cos(n\theta) + x_n \sin(n\theta)\right) + \frac{q_0}{\sqrt{2\pi}}. \tag{A2}$$

The configuration space for the scalar field is then described by the set of real variables q_n, $n \geq 0$, and x_n, $n > 0$, which are completely decoupled. For simplicity, we drop all the modes x_n (which can be treated like their cosine counterparts q_n for $n > 0$) and q_0, and continue with the infinite set $\{q_n, n > 0\}$. The variable q_0 is dropped to avoid introducing a special treatment in the case $s(t) = 0$, since $n = 0$ corresponds to a zero frequency oscillator or a free particle instead of a regular harmonic oscillator. In any case, it describes a single degree of freedom, which cannot affect the considered matters of unitary implementation.

The equations of motion for the modes are

$$\ddot{q}_n + [n^2 + s(t)]q_n = 0. \tag{A3}$$

The corresponding Hamiltonian equations are

$$\dot{q}_n = p_n, \quad \dot{p}_n = -[n^2 + s(t)]q_n, \tag{A4}$$

where p_n is the momentum canonically conjugate to q_n, i.e., $\{q_n, p_{n'}\} = \delta_{nn'}$.

There are many representations of the CCRs satisfied by the infinite set of pairs $\{(q_n, p_n), n > 0\}$, or of the associated Weyl algebra. For instance, every sequence $\{\mu_n, n > 0\}$ of (quasi-invariant[9]) *probability* measures in $\mathbb{R}$ gives a representation, since it defines a regular product measure in the set of all sequences $(q_1, q_2, \ldots)$, thus, providing a Schrödinger type of quantization in the Hilbert space of square integrable functions in the configuration space.

What is not available, however, is the straightforward generalization of the usual representation in finite dimensions obtained from the Lebesgue measure dq, since no mathematical sense can be made of the formal infinite product $\prod_{n=1}^{\infty} dq_n$. In such a context, Fock representations of the Weyl relations are given by (normalized) Gaussian measures, which still make perfect sense in infinite dimensions. Even after restricting our attention to Gaussian measures, there are endless possibilities, and many of them lead to inequivalent representations.

In the present case, a particularly important measure in our infinite dimensional configuration space is

$$d\mu = \prod_{n=1}^{\infty} e^{-nq_n^2} \frac{\sqrt{n}dq_n}{\sqrt{\pi}}, \tag{A5}$$

which is associated with the quantum operators

$$\hat{q}_n \Psi = q_n \Psi, \qquad \hat{p}_n \Psi = -i\frac{\partial}{\partial q_n}\Psi + inq_n\Psi. \tag{A6}$$

This is, in fact, a particular realization of the Fock representation given by the complex structure J_0, defined as in Equation (17)[10] with m equal to zero. Thus, it should be no surprise that this particular representation allows a unitary implementation of the free massless field dynamics (i.e., with $s(t) = 0$). The quantization is the most natural one for the massless field, since the wave functional $\Psi = 1$ is invariant under the unitary group $U(t)$ implementing the dynamics. Putting it differently, there is a well-defined quantum Hamiltonian, and $\Psi = 1$ is the zero-energy state of the free massless field.

The same representation also allows a unitary implementation of the time-dependent mass case. To see this, let us introduce the annihilation and creation-like variables a_n and $\bar{a}_n$, defined by

$$a_n = \frac{nq_n + ip_n}{\sqrt{2n}}, \tag{A7}$$

which are precisely the ones associated with the complex structure J_0. In particular, this means that J_0 takes the diagonal matrix form $\mathrm{diag}(i, -i)$ when written in terms of the basis in phase space made of the pairs $(a_n, \bar{a}_n)$. With respect to this basis of (complex) variables, the classical evolution (from time t_0 to time t) determined by Equation (A4) is given by non-vanishing 2×2 matrix blocks of the form

$$\mathcal{U}_n(t, t_0) = \begin{pmatrix} \alpha_n(t, t_0) & \beta_n(t, t_0) \\ \bar{\beta}_n(t, t_0) & \bar{\alpha}_n(t, t_0) \end{pmatrix}, \tag{A8}$$

with

$$|\alpha_n(t, t_0)|^2 - |\beta_n(t, t_0)|^2 = 1, \qquad \forall n > 0, \forall t, t_0. \tag{A9}$$

Note that, in the case $s(t) = 0$, precisely because of the way in which the variables a_n are defined, the above parameters β_n are all vanishing (and the parameters α_n are only phases).

For quite general functions $s(t)$ in Equation (A4), the following condition is satisfied [49]:

$$\sum_{n}^{\infty} |\beta_n(t, t_0)|^2 < \infty, \qquad \forall t, t_0. \tag{A10}$$

This is precisely the necessary and sufficient conditions for unitary implementability of the dynamics, in the J_0 representation. Thus, there is a Fock representation, defined by a complex structure that remains invariant under the spatial symmetries, such that a unitary quantum dynamics can be achieved.

The proof that a quantization with the above characteristics is unique is as follows. To begin with, any other invariant (under spatial isometries) complex structure J is related to J_0 by $J = KJ_0K^{-1}$, where K is a symplectic transformation given by a block diagonal matrix with 2×2 blocks of the form

$$\begin{pmatrix} \kappa_n & \lambda_n \\ \bar{\lambda}_n & \bar{\kappa}_n \end{pmatrix}, \qquad |\kappa_n|^2 - |\lambda_n|^2 = 1, \forall n > 0. \tag{A11}$$

Suppose now that the dynamics are unitary in the Fock representation defined by J. This is equivalent to the unitary implementability in the J_0-representation of a modified dynamics, obtained precisely by applying the transformation K to the canonical transformations corresponding to time evolution. A simple computation shows that, for this modified dynamics, the coefficients β_n in Equation (A8) are replaced with

$$\beta_n^J(t,t_0) = 2i\bar{\kappa}_n\lambda_n Im[\alpha_n(t,t_0)] + (\bar{\kappa}_n)^2\beta_n(t,t_0) - \lambda_n^2\bar{\beta}_n(t,t_0), \tag{A12}$$

where the notation Im denotes the imaginary part. Then, it follows from the hypothesis of unitary dynamics that the following condition holds:

$$\sum_n^\infty |\beta_n^J(t,t_0)|^2 < \infty, \quad \forall t, t_0. \tag{A13}$$

Now, a detailed asymptotic analysis [49,50] shows that condition (A13) implies that

$$\sum_n^\infty |\lambda_n|^2 < \infty. \tag{A14}$$

Given the relation between J and J_0, this last condition guarantees precisely that the operator $J - J_0$ is of the Hilbert–Schmidt type, i.e., that the Fock representations defined by J and J_0 are unitarily equivalent.

Notes

1. There are currently also LQC-inspired applications to inhomogeneous cosmologies. We will not consider them in the present work.
2. A state ω of a $\star$-algebra $\mathcal{A}$ is a linear functional such that $\omega(aa^*) \geq 0$, $\forall a \in \mathcal{A}$, and $\omega(\mathbf{1}) = 1$, where $\mathbf{1}$ is the identity of the algebra and the symbol $*$ denotes the involution operation, e.g., complex conjugation in algebras of functions and adjointness in algebras of operators.
3. We also restrict attention to the more usual formulation of LQC, leaving aside the so-called Fleischhack approach, which is also considered in [10].
4. Nevertheless, applications of LQC are effectively performed on a separable subspace of $\mathcal{H}_\mathcal{P}$. This can either be seen as a consistency requirement [21] or as a consequence of superselection [22], which, in any case, can be traced back to the fact that the LQC quantum Hamiltonian constraint is a difference operator of constant step.
5. See, nevertheless, the variations on the LQG representation introduced by Koslowski and Sahlmann [26–28], and the related developments by M. Varadarajan and Campiglia [29,30].
6. Alternatively, the system can be considered as an axially symmetric field propagating in a static (2+1)-dimensional spacetime with the spatial topology of a two-torus.
7. Only very mild technical conditions on $s(t)$ are required, see [50].
8. Part of the techniques employed for fermions were already explored in the case of the scalar field in Bianchi I, in order to deal with the lack of conformal symmetry. A review of the range of different methods and improvements required to address the increasing degree of generalization encountered in the treatment of the scalar field can be found in [58].
9. In order to provide a unitary representation of translations, measures are required to satisfy the technical condition of quasi-invariance, which is satisfied, e.g., by any Gaussian measure.
10. With the obvious adaptations, taking into account that the spatial manifold is now S^1 instead of $\mathbb{R}^3$.

References

1. Dirac, P.A.M. *The Principles of Quantum Mechanics*, 4th ed.; Oxford University Press: Oxford, UK, 1958.
2. Reed, M.; Simon, B. *Methods of Modern Mathematical Physics I: Functional Analysis, Revised and Enlarged Edition*; Academic Press: San Diego, CA, USA, 1980.
3. Groenewold, H.J. On the principles of elementary quantum mechanics. *Physica* **1946**, *12*, 405. [CrossRef]
4. van Hove, L. Sur certaines représentations unitaires d'un groupe infini de transformations. *Proc. R. Acad. Sci. Belgium* **1951**, *26*, 1.
5. Gotay, M.J. Obstructions to Quantization. In *Journal of Nonlinear Science (eds) Mechanics: From Theory to Computation*; Springer: New York, NY, USA, 2000; pp. 171–216.
6. Reed, M.; Simon, B. *Methods of Modern Mathematical Physics II: Fourier Analysis, Self-Adjointness*; Academic Press: San Diego, CA, USA, 1975.
7. Gowdy, R.H. Vacuum spacetimes with two-parameter spacelike isometry groups and compact invariant hypersurfaces: Topologies and boundary conditions. *Ann. Phys.* **1974**, *83*, 203. [CrossRef]

8. Ashtekar A.; Lewandowski, J. Background independent quantum gravity: A status report. *Class. Quantum Grav.* **2004**, *21*, R53. [CrossRef]
9. Thiemann, T. *Modern Canonical Quantum General Relativity*; Cambridge University Press: Cambridge, UK, 2007.
10. Engle, J.; Hanusch, M.; Thiemann, T. Uniqueness of the representation in homogenic isotropic LQC. *Commun. Math. Phys.* **2017**, *354*, 231.; Erratum in **2018**, *362*, 759. [CrossRef]
11. Ashtekar, A.; Campiglia, M. On the uniqueness of kinematics of loop quantum cosmology. *Class. Quantum Grav.* **2012**, *29*, 242001. [CrossRef]
12. Bratteli, O.; Robinson, D.W. *Operator Algebras and Quantum Statistical Mechanics 1*; Springer: New York, NY, USA, 1987.
13. Shale, D. Linear symmetries of free boson fields. *Trans. Am. Math. Soc.* **1962**, *103*, 149. [CrossRef]
14. Kay, B.S. Linear spin-zero quantum fields in external gravitational and scalar fields I. A one particle structure for the stationary case. *Commun. Math. Phys.* **1978**, *62*, 55. [CrossRef]
15. Baez, J.C.; Segal, I.V.; Zhou, Z. *Introduction to Algebraic and Constructive Quantum Field Theory*; Princeton University Press: Princeton, NJ, USA, 1992.
16. Ashtekar, A.; Magnon, A. Quantum fields in curved space-times. *Proc. R. Soc. A* **1975**, *346*, 375.
17. Wald, R.M. *Quantum Field Theory in Curved Spacetime and Black Hole Thermodynamics*; Chicago University Press: Chicago, IL, USA, 1994.
18. Immirzi, G. Quantum gravity and Regee calculus. *Nucl. Phys. Proc. Suppl.* **1997**, *57*, 65. [CrossRef]
19. Ashtekar, A.; Bojowald, M.; Lewandowski, J. Mathematical structure of loop quantum cosmology. *Adv. Theor. Math. Phys.* **2003**, *7*, 233. [CrossRef]
20. Ashtekar, A.; Singh, P. Loop quantum cosmology: A status report. *Class. Quantum Grav.* **2011**, *28*, 213001. [CrossRef]
21. Velhinho, J.M. Comments on the kinematical structure of loop quantum cosmology. *Class. Quantum Grav.* **2004**, *21*, L109. [CrossRef]
22. Ashtekar, A.; Pawlowski, T.; Singh, P. Quantum nature of the big bang: Improved dynamics. *Phys. Rev. D* **2006**, *74*, 084003. [CrossRef]
23. Velhinho, J.M. The quantum configuration space of loop quantum cosmology. *Class. Quantum Grav.* **2007**, *24*, 3745. [CrossRef]
24. Lewandowski, J.; Okołów, A.; Sahlmann, H.; Thiemann, T. Uniqueness of diffeomorphism invariant states on holonomy–flux algebras. *Commun. Math. Phys.* **2006**, *267*, 703. [CrossRef]
25. Fleischhack, C. Representations of the Weyl algebra in quantum geometry. *Commun. Math. Phys.* **2009**, *285*, 67. [CrossRef]
26. Koslowski, T. Dynamical quantum geometry (DQG programme). *arXiv* **2007**, arXiv:0709.3465.
27. Sahlmann, H. On loop quantum gravity kinematics with a non-degenerate spatial background. *Class. Quantum Grav.* **2010**, *27*, 225007. [CrossRef]
28. Koslowski, T.; Sahlmann, H. Loop quantum gravity vacuum with nondegenerate geometry. *SIGMA* **2012**, *8*, 026. [CrossRef]
29. Campiglia, M.; Varadarajan, M. The Koslowski–Sahlmann representation: Gauge and diffeomorphism invariance. *Class. Quantum Grav.* **2014**, *31*, 075002. [CrossRef]
30. Campiglia, M.; Varadarajan, M. The Koslowski–Sahlmann representation: Quantum configuration space. *Class. Quantum Grav.* **2014**, *31*, 175009. [CrossRef]
31. Corichi, A.; Cortez, J.; Mena Marugán, G.A. Quantum Gowdy T^3 model: A unitary description. *Phys. Rev. D* **2006**, *73*, 084020. [CrossRef]
32. Corichi, A.; Cortez, J.; Mena Marugán, G.A. Unitary evolution in Gowdy cosmology. *Phys. Rev. D* **2006**, *73*, 041502. [CrossRef]
33. Cortez, J.; Mena Marugán, G.A. Feasibility of a unitary quantum dynamics in the Gowdy T^3 cosmological model. *Phys. Rev. D* **2005**, *72*, 064020. [CrossRef]
34. Pierri, M. Probing quantum general relativity through exactly soluble midi-superspaces II: Polarized Gowdy models. *Int. J. Mod. Phys. D* **2002**, *11*, 135. [CrossRef]
35. Corichi, A.; Cortez, J.; Quevedo, H. On Unitary Time Evolution in Gowdy T^3 Cosmologies. *Int. J. Mod. Phys. D* **2002**, *11*, 1451. [CrossRef]
36. Corichi, A.; Cortez, J.; Mena Marugán, G.A.; Velhinho, J.M. Quantum Gowdy T^3 model: A uniqueness result. *Class. Quantum Grav.* **2006**, *23*, 6301. [CrossRef]
37. Cortez, J.; Mena Marugán, G.A.; Velhinho, J.M. Uniqueness of the Fock quantization of the Gowdy T^3 model. *Phys. Rev. D* **2007**, *75*, 084027. [CrossRef]
38. Barbero, J.F.; Vergel, D.; Villaseñor, E. Hamiltonian dynamics of linearly polarized Gowdy models coupled to massless scalar fields. *Class. Quantum Grav.* **2007**, *24*, 5945. [CrossRef]
39. Barbero, J.F.; Vergel, D.; Villaseñor, E. Quantum unitary evolution of linearly polarized $S^1 \times S^2$ and S^3 models Gowdy models coupled to massless scalar fields. *Class. Quantum Grav.* **2008**, *25*, 085002. [CrossRef]
40. Cortez, J.; Mena Marugán, G.A.; Velhinho, J.M. Uniqueness of the Fock representation of the Gowdy $S^1 \times S^2$ and S^3 models. *Class. Quantum Grav.* **2008**, *25*, 105005. [CrossRef]
41. Mukhanov, V. *Physical Foundations of Cosmology*; Cambridge University Press: Cambridge, UK, 2005.
42. Mukhanov, V.F.; Feldman, H.A.; Bradenberger, R.H. Theory of cosmological perturbations. *Phys. Rep.* **1992**, *215*, 203. [CrossRef]
43. Bardeen, J.M. Gauge-invariant cosmological perturbations. *Phys. Rev. D* **1980**, *22*, 1882. [CrossRef]
44. Halliwell, J.J.; Hawking, S.W. Origin of structure in the Universe. *Phys. Rev. D* **1985**, *31*, 1777. [CrossRef]

45. Fernández-Méndez, M.; Mena Marugán, G.A.; Olmedo, J.; Velhinho, J.M. Unique Fock quantization of scalar cosmological perturbations. *Phys. Rev. D* **2012**, *85*, 103525. [CrossRef]
46. Birrell, N.D.; Davies, P.C.W. *Quantum Fields in Curved Space*; Cambridge University Press: Cambridge, UK, 1982.
47. Fulling, S.A. *Aspects of Quantum Field Theory in Curved Spacetime*; Cambridge University Press: Cambridge, UK, 1989.
48. Elizaga Navascués, B.; Mena Marugán, G.A. Perturbations in quantum cosmology: The continuum limit in Fourier space. *Phys. Rev. D* **2018**, *98*, 103522. [CrossRef]
49. Cortez, J.; Mena Marugán, G.A.; Serôdio, R.; Velhinho, J.M. Uniqueness of the Fock quantization of a free scalar field on S^1 with time dependent mass. *Phys. Rev. D* **2009**, *79*, 084040. [CrossRef]
50. Cortez, J.; Mena Marugán, G.A.; Velhinho, J.M. Fock quantization of a scalar field with time dependent mass on the three-sphere: Unitarity and uniqueness. *Phys. Rev. D* **2010**, *81*, 044037. [CrossRef]
51. Cortez, J.; Mena Marugán, G.A.; Olmedo, J.; Velhinho, J.M. Uniqueness of the Fock quantization of fields with unitary dynamics in nonstationary spacetimes. *Phys. Rev. D* **2011**, *83*, 025002. [CrossRef]
52. Cortez, J.; Mena Marugán, G.A.; Olmedo, J.; Velhinho, J.M. Criteria for the determination of time dependent scalings in the Fock quantization of scalar fields with a time dependent mass in ultrastatic spacetimes. *Phys. Rev. D* **2012**, *86*, 104003. [CrossRef]
53. Castelló Gomar, L.; Cortez, J.; Martín-de Blas, D.; Mena Marugán, G.A.; Velhinho, J.M. Uniqueness of the Fock quantization of scalar fields in spatially flat cosmological spacetimes. *J. Cosmol. Astropart. Phys.* **2012**, *1211*, 001. [CrossRef]
54. Cortez, J.; Elizaga Navascués, B.; Martín-Benito, M.; Mena Marugán, G.A.; Olmedo, J.; Velhinho, J.M. Uniqueness of the Fock quantization of scalar fields in a Bianchi I cosmology with unitary dynamics. *Phys. Rev. D* **2016**, *94*, 105019. [CrossRef]
55. Cortez, J.; Elizaga Navascués, B.; Martín-Benito, M.; Mena Marugán, G.A.; Velhinho, J.M. Unitary evolution and uniqueness of the Fock representation of Dirac fields in cosmological spacetimes. *Phys. Rev. D* **2015**, *92*, 105013. [CrossRef]
56. Cortez, J.; Elizaga Navascués, B.; Martín-Benito, M.; Mena Marugán, G.A.; Velhinho, J.M. Unique Fock quantization of a massive fermion field in a cosmological scenario. *Phys. Rev. D* **2016**, *93*, 084053. [CrossRef]
57. Cortez, J.; Elizaga Navascués, B.; Martín-Benito, M.; Mena Marugán, G.A.; Velhinho, J.M. Dirac fields in flat FLRW cosmology: Uniqueness of the Fock quantization. *Ann. Phys.* **2017**, *376*, 76. [CrossRef]
58. Cortez, J.; Mena Marugán, G.A.; Velhinho, J. Quantum linear scalar fields with time dependent potentials: Overview and applications to cosmology. *Mathematics* **2020**, *8*, 115. [CrossRef]
59. Cortez, J.; Elizaga Navascués, B.; Mena Marugán, G.A.; Prado, S.; Velhinho, J.M. Uniqueness criteria for the Fock quantization of Dirac fields and applications in hybrid loop quantum cosmology. *Universe* **2020**, *6*, 241. [CrossRef]

Review

Hawking Radiation and Black Hole Gravitational Back Reaction—A Quantum Geometrodynamical Simplified Model

João Marto [1,2]

[1] Departamento de Física, Universidade da Beira Interior, Rua Marquês D'Ávila e Bolama, 6200-001 Covilhã, Portugal; jmarto@ubi.pt
[2] Centro de Matemática e Aplicações da Universidade da Beira Interior, Rua Marquês D'Ávila e Bolama, 6200-001 Covilhã, Portugal

Abstract: The purpose of this paper is to analyse the back reaction problem, between Hawking radiation and the black hole, in a simplified model for the black hole evaporation in the quantum geometrodynamics context. The idea is to transcribe the most important characteristics of the Wheeler-DeWitt equation into a Schrödinger's type of equation. Subsequently, we consider Hawking radiation and black hole quantum states evolution under the influence of a potential that includes back reaction. Finally, entropy is estimated as a measure of the entanglement between the black hole and Hawking radiation states in this model.

Keywords: quantum gravity; Hawking radiation; entanglement entropy

1. Introduction

Since the discovery that black holes would have to emit radiation, there have been proposals to explain the loss of information associated with the apparent conversion of pure to mixed quantum states. From the beginning, this information loss was proposed to be fundamental and, the non unitary evolution of pure to mixed quantum states constituted a hypothesis to solve the problem associated with this loss. For example, Steven Hawking own proposal of the non unitary evolution is represented by the "dollar matrix" $\$$ [1]

$$\rho^{\text{final}} = \$\rho^{\text{initial}},$$

which allows the evolution of pure quantum states, characterised by the density matrix ρ^{initial}, into mixed states ρ^{final}.

The black hole evaporation mechanism and the problem of information loss, collected behind the event horizon, constituted a privileged arena for quantum gravity theories candidates (namely, quantum geometrodynamics [2], string theory [3–7] and loop quantum gravity [8–11]) to establish themselves beyond General Relativity. However, the scientific community was reluctant to give up unitarity, a crucial feature of Quantum Mechanics, and the hypothesis of a new principle of complementarity, between the points of views of an infinitely distant observer and a free falling observer near the event horizon, was raised [12]. Following a similar approach, it has been emphasised over the time the role of the gravitational back reaction effect [13,14] of Hawking radiation on the event horizon as a way to allow the information accumulated within the black hole to be encoded in the outgoing radiation. In this way, the emergence of a mechanism in which all the black hole information (a four dimensional object in General Relativity) would be accessible at the event horizon (which can be described as a membrane with one dimension less than the black hole), is somehow similar to what happens with a hologram [15]. This new holographic principle was simultaneously proposed and clarified [16,17] in order to incorporate the aforementioned principle of complementarity. The next step happened when it was conjectured the correspondence between classes of quantum gravity theories

(5-dimensional anti-De Sitter solutions in string theories) and conformal field theory (CFT-conformal field theory-4-dimensional boundary of the 5-dimensional solutions), the so called AdS/CFT conjecture [18–21]. This discovery was extremely important to ensure the possibility of a correspondence between the physics that describes the interior of the black hole (supposedly quantum gravity), and the existence of a quantum field theory at its boundary (the surface that defines the event horizon) that would allow to save the unitarity.

In 2012, in an effort to analyse important assumptions, such as: (1) the principle of complementarity proposed by Susskind and its colaborators, (2) the AdS/CFT correspondence and, (3) the equivalence principle of General Relativity, in the way that Hawking radiation could encode the information stored in the black hole, a new paradox was discovered [22]. In simple terms, the impossibility of having the particle, which leaves the black hole, in an maximally entangled state (or non factored state) with two systems simultaneously (the pair disappearing beyond the horizon and all the Hawking radiation emitted in the past, a problem related to the so-called monogamy of entanglement), leads to postulate the existence of a firewall that would destroy any free falling observer trying to cross the event horizon. The firewall existence is incompatible with Einstein's equivalence principle. However, if there is no firewall, and the principle of equivalence is respected, according to these authors, unitarity is lost and information loss is inevitable. Apparently, the situation is such that either General Relativity principles or Quantum Mechanics principles need to be reviewed [23,24]. This is an open problem and the role of gravitational back reaction, between Hawking radiation and the black hole, persists as an unknown and potentially enlightening mechanism on how to correctly formulate a quantum theory of the gravitational field.

In an attempt to study the possible gravitational back reaction, between Hawking radiation and the black hole, from the quantum geometrodynamics point of view, a toy model was proposed [25]. It was shown and discussed the conditions under which the Wheeler-DeWitt equation could be used to describe a quantum black hole. In particular, a simple model for the black hole evaporation was studied using a Schrödinger type of equation and, the cases for initial squeezed ground states and coherent states were taken to represent the initial black hole quantum state. One can ask, how can a complex equation such as the Wheeler-DeWitt be approximated by a Schrödinger type of equation? In the cosmological context, several formal derivations were carried [2,26–29] with the purpose of enabling to use the limit of a quantum field theory in an external space-time for the full quantum gravity theory. Such approaches usually involve procedures like the Born-Oppenheimer or Wentzel-Kramers-Brillouin (WKB) approximations.

In this work we review this toy model. It is important to notice that, even though, a full study of the time evolution of the Hawking radiation and black hole quantum states was performed when a simple back reaction term is introduced, an important part of the discussion about the time evolution of the resulting entangled state was left incomplete. In fact, it is exactly the motivation of this paper to address the problem of explicitly describe the time evolution of the degree of entanglement of this quantum system. In addition, another important goal is to get an approximate estimate of the Von Neumann entropy and check is the back reaction can induce a release of the quantum information in the Hawking radiation. These results can be interesting, in the quantum geometrodynamics context, as a simple starting point to more robustly address the black hole information paradox in a canonical quantization of gravity program.

This paper is organized as follow. In Sections 2 and 3 we present an introduction to quantum geometrodynamics and consider a semiclassical approximation of the Wheeler-DeWitt equation. In Sections 4 and 5 we derive a simple model of the back reaction between the Hawking radiation and the black hole quantum states, where its dynamics is governed by a Schrödinger type of equation. Finally, in Sections 6 and 7, we obtain and discuss the main result of this paper, namely the time evolution of the entanglement entropy and the behaviour of the quantum information of the Hawking radiation state.

2. Quantum Geometrodynamics and the Semiclassical Approximation

In the following, we mention a brief description of the bases of J.A. Wheeler's geometrodynamics, which consists in a 3 + 1 spacetime decomposition (ADM decomposition-R. Arnowitt, S. Deser and C.W. Misner [30]), and obtain General Relativity field equations in that context. The field equations, obtained in this procedure, will exhibit the evolution of a pair of dynamical variables (h_{ab}, K_{ab})-the 3-dimensional metric h_{ab} (induced metric) and the extrinsic curvature K_{ab}-on a Cauchy hypersurface Σ_t (three-dimensional surface).

General Relativity, defined by the Einstein-Hilbert action, here without the cosmological constant,

$$S_{EH} = \frac{c^4}{16\pi G} \int d^4x \sqrt{-g}\, R, \qquad (1)$$

can be expressed under the hamiltonian formalism. For this purpose, a 3 + 1 decomposition of spacetime ($\mathcal{M}$, g)[1] may be considered, where $\mathcal{M}$ is a smooth manifold and g lorentzian metric in $\mathcal{M}$. Moreover, this decomposition consists of the 4-dimensional spacetime foliation in a continuous sequence of Cauchy hypersurfaces Σ_t, parameterised by a global time variable t[2]. General Relativity covariance is maintained, in this procedure, by considering all possible ways of carrying this foliation. When we consider the hamiltonian formalism we need to define a pair of canonical variables, however, we can initially identify a pair of dynamical variables constituted, on the one hand, by the 3-dimensional metric induced in Σ_t by the spacetime metric

$$\mathbf{h} = \mathbf{g} + \mathbf{n} \otimes \mathbf{n} \quad (h_{\mu\nu} = g_{\mu\nu} + n_\mu n_\nu), \qquad (2)$$

where n_μ is an ortogonal vector to Σ_t. In this way we can separate the metric g, in its temporal e spatial components, according to the following expressions,

$$g_{\mu\nu} = \begin{pmatrix} N_a N^a - N^2 & N_b \\ N_c & h_{ab} \end{pmatrix} \text{ and } g^{\mu\nu} = \begin{pmatrix} -\frac{1}{N^2} & \frac{N^b}{N^2} \\ \frac{N^c}{N^2} & h_{ab} - \frac{N^a N^b}{N^2} \end{pmatrix}, \qquad (3)$$

or, in a more suitable compact form,

$$g_{\mu\nu}dx^\mu dx^\nu = -N^2 dt^2 + h_{ab}(dx^a + N^a dt)(dx^b + N^b dt). \qquad (4)$$

In the previous equation N is called the lapse function whereas N^a is the shift vector. The other canonical variable, on the other hand, is the extrinsic curvature

$$K_{\mu\nu} = h_\mu^\sigma \nabla_\sigma n_\nu. \qquad (5)$$

Hence, the dynamical variables pair (h_{ab}, K_{ab}) (with Latin letter indexes, defining 3-dimensional tensor fields) enable us to rewrite Einstein-Hilbert action (1) as

$$S_{EH} = \frac{c^4}{16\pi G} \int_\mathcal{M} dt d^3x\, \mathcal{L} = \frac{c^4}{16\pi G} \int_\mathcal{M} dt d^3x\, N\sqrt{h}\left(K_{ab}K^{ab} - K^2 + {}^{(3)}R\right). \qquad (6)$$

We can notice that the lapse function and the shift vector are Lagrange multipliers (since $\partial \mathcal{L}/\partial \dot{N} = 0$ e $\partial \mathcal{L}/\partial \dot{N}_a = 0$) and, according to Dirac [31] we can establish the existence of primary constraints which allow to write the action (6) as

$$S_{EH} = \frac{c^4}{16\pi G} \int_\mathcal{M} dt d^3x \left(p^{ab}\dot{h}_{ab} - N\mathcal{H}^g_\perp - N^a \mathcal{H}^g_a\right), \qquad (7)$$

---

[1] We assume that this spacetime is globally hyperbolic, such that we ensure that it can be foliated in Cauchy hypersurfaces.
[2] For which a flow of 'time' can be perceived when a observer world line crosses a sequence of Cauchy hypersurfaces.

with $p^{ab} = \partial \mathcal{L}/\partial \dot{h}_{ab}$ (conjugate momentum of the dynamical variable h_{ab}) and where

$$\begin{cases} \mathcal{H}_\perp^g = \dfrac{16\pi G}{c^4} G_{abcd} p^{ab} p^{cd} - \dfrac{c^4 \sqrt{h}}{16\pi G} {}^{(3)}R \\ \mathcal{H}_a^g = -2D_b\left(h_{ac} p^{bc}\right) \end{cases}, \qquad (8)$$

with $G_{abcd} = \dfrac{1}{2\sqrt{h}}(h_{ac}h_{bd} + h_{ad}h_{bc} - h_{ab}h_{cd})$ being the DeWitt metric and D_b is the covariant derivative. We can define the hamiltonian constraint

$$\mathcal{H}_\perp^g \approx 0, \qquad (9)$$

and the diffeomorphism constraint

$$\mathcal{H}_a^g \approx 0, \qquad (10)$$

through the variation of the action (7) with respect to N and N^a. Physically, constraints (9)–(10) express the freedom to choose any coordinate system in General Relativity. More precisely, the choice of the particular foliation Σ_t is equivalent to choose the lapse function N and, the spatial coordinates (x^i) choice is equivalent to choose a particular shift vector N^a. It is important to emphasise that, related to the DeWitt metric G_{abcd} definition, the kinetic term in Equation (9) is indefinite, since not all kinetic functional operators in Equation (8) share the same sign. This property will persist beyond the quantisation procedure and will play a crucial role in the semiclassical (where it will give rise to a negative kinetic term) approach to the black hole evaporation process.

3. Canonical Variables Quantisation and Wheeler-DeWitt Equation

The canonical quantisation programme, according to P.M. Dirac prescription, demands the transition of classical to quantum canonical variables $(h_{ab}, p_{ab}) \to \left(\hat{h}_{ab}, -i\hbar \delta/\delta h^{ab}\right)$, and also promotes Poisson brackets to commutators. We have to define a wave state functional $\Psi(h_{ab})$ belonging to the space of all 3-dimensional metrics Riem Σ. Nevertheless, there are important issues related with:

1. the correct factor ordering in building quantum observables from the fundamental variables $\left(\hat{h}_{ab}, -i\hbar \delta/\delta h^{ab}\right)$,
2. the interpretation of quantum observables as operators acting on the wave functional $\Psi(h_{ab})$ and the adequate definition of a Hilbert space,
3. the classical constraints (9)–(10) conversion to their quantum counterpart

$$\begin{cases} \mathcal{H}_\perp^g \Psi(h_{ab}) = 0 \\ \mathcal{H}_a^g \Psi(h_{ab}) = 0 \end{cases}, \qquad (11)$$

and their quantum interpretation,

4. the lack of time evolution in the previous quantum constraints.

These questions are thoroughly discussed in [2,32], as well as possible solutions and open problems till the present day. Among the previous mentioned issues, the problem related to the lack of time evolution seems to stand as an essential feature in the formulation of a quantum theory of the gravitational field. If we assume that the wave functional evolution over time depends on a time concept defined after the canonical quantisation, then, the time parameter t will be an emergent quantity [33].

In order to address the black hole evaporation problem and, to explore how information is eventually encoded in Hawking radiation, it would be important to obtain the entropy time evolution as a measure of the degree of quantum entanglement between radiation and black hole states. Since the quantum version of the hamiltonian constraint (9),

$$\mathcal{H}_\perp^g \Psi(h_{ab}) = \left(\frac{16\pi G \hbar^2}{c^4} G_{abcd} \frac{\delta^2}{\delta h_{ab} \delta h_{cd}} - \frac{c^4 \sqrt{h}}{16\pi G} {}^{(3)}R \right) \Psi(h_{ab}) = 0, \quad (12)$$

known as Wheeler-DeWitt equation, and the quantum diffeomorphism constraint

$$\mathcal{H}_a^g \Psi(h_{ab}) = D_b \left(h_{ac} \frac{\delta}{\delta h_{bc}} \right) \Psi(h_{ab}) = 0 \quad (13)$$

are both time independent, the wave functional is connected to a purely quantum and closed gravitational system. In the case involving the study of a black hole evaporation phase, Equations (12) and (13) describe a quantum black hole in the context of a purely quantum universe. This situation is not suitable if we consider that we must have several classical observers measuring and depicting the time evolution of the black hole outgoing radiation. These classical observers, experience and describe physical phenomena in a classical language that needs a time parameter. Hence, we need to consider a quantum black hole in a semiclassical universe where time appears as an emergent quantity.

Time is the product of an approximation which aims to extract, from the Wheeler-DeWitt equation, an external, semiclassical stage, in which black hole and Hawking radiation quantum states evolve.

In reference [2] (Section 5.4) we can find a derivation, from Equations (12) and (13), of a Schrödinger functional equation. In the following, we highlight some important details of this derivation. Let us start by writing the wave functional as

$$|\Psi(h_{ab})\rangle = e^{i m_{\text{Pl}}^2 S[h_{ab}]} |\psi(h_{ab})\rangle, \quad (14)$$

where $S[h_{ab}]$ is a solution of the vacuum Einstein-Hamilton-Jacobi function [34], since its WKB approximation enable us to extract, at higher orders, a Hamilton-Jacobi equation. In addition, $S[h_{ab}]$ is also solution to the Hamilton-Jacobi version of (12)–(13), namely

$$\frac{m_{\text{Pl}}^2}{2} G_{abcd} \frac{\delta S}{\delta h_{ab}} \frac{\delta S}{\delta h_{cd}} - 2 m_{\text{Pl}}^2 \sqrt{h}\, {}^{(3)}R + \langle \psi | \hat{\mathcal{H}}_\perp^m | \psi \rangle = 0, \quad (15)$$

$$-2 m_{\text{Pl}}^2 h_{ab} D_c \frac{\delta S}{\delta h_{bc}} + \langle \psi | \hat{\mathcal{H}}_a^m | \psi \rangle = 0, \quad (16)$$

with the definitions $m_{\text{Pl}}^2 = (32\pi G)^{-1}$, $\hbar = c = 1$ and $\hat{\mathcal{H}}_\perp^m$ and $\hat{\mathcal{H}}_a^m$ are assumed to be contributions from the non-gravitational fields. Having the solution $S[h_{ab}]$, we can now evaluate $|\psi(h_{ab})\rangle$ along a solution of the classical Einstein equations $h_{ab}(\mathbf{x}, t)$. In fact this solution is obtained from

$$\dot{h}_{ab} = N G_{abcd} \frac{\delta S}{\delta h_{cd}} + 2 D_{(a} N_{b)}, \quad (17)$$

after a choice of the lapse and shift function has been made. At this point, we can define the evolutionary equation for the quantum state $|\psi(h_{ab})\rangle$ as

$$\frac{\partial}{\partial t} |\psi(h_{ab})\rangle = \int d^3 x\, \dot{h}_{ab} \frac{\delta}{\delta h_{ab}} |\psi(h_{ab})\rangle, \quad (18)$$

which, since $\dot{h}_{ab}$ depends on the DeWitt metric G_{abcd}, will have differential operators with the wrong sign in its right hand side. Finally, we are in the position of defining a functional

Schrödinger equation for quantized matter fields in an external classical gravitational field as

$$i\hbar \frac{\partial}{\partial t} |\psi[h_{ab}(\mathbf{x},t)]\rangle = \hat{H}^m |\psi[h_{ab}(\mathbf{x},t)]\rangle ,$$

$$\hat{H}^m \equiv \int d^3x \{ N(\mathbf{x})\hat{\mathcal{H}}^m_\perp(\mathbf{x}) + N^a(\mathbf{x})\hat{\mathcal{H}}^m_a(\mathbf{x}) \} . \qquad (19)$$

Notice that the matter hamiltonian $\hat{H}^m$, is parametrically depending on metric coefficients of the curved space-time background and contain indefinite kinetic terms.

This derivation assumes a separation of the complete system (which state obeys the Wheeler-DeWitt equation and the quantum diffeomorphism invariance) in two parts, in total correspondence with the way a Born-Oppenheimer approximation is implemented. The physical system separation into two parts, one purely quantum and the other semiclassical, is essentially achieved by separating the gravitational from the non gravitational degrees of freedom through an expansion, with respect to the Planck mass m_{Pl}, of constraints (12)–(13). However, we notice that there are gravitational degrees of freedom that can be included in the purely quantum part (quantum density fluctuations whose origin is gravitational, for example). Equation (19), formally similar to Schrödinger equation, is an equation with functional derivatives, in which variable $\mathbf{x}$ is related to the 3-dimensional metric h_{ab}. As previously mentioned, we recall that due to the DeWitt G_{abcd} metric definition, a negative kinetic term emerges from the conjugated momentum p_{ab}.

In the following section, let us develop a simple model of the black hole evaporation stage [25], which incorporates one interesting feature of the Wheeler-DeWitt equation, namely the indefinite kinetic term, and study some of its consequences. The main objective, here, is to estimate the degree of entanglement between Hawking radiation and black hole quantum states, when we take into account a simple form of back reaction between the two.

4. Simplified Model with a Schrödinger Type of Equation

Equation (19) is a functional differential equation, its wave functional solution depends on the 3-metric h_{ab} describing the black hole and matter fields. It is obviously an almost impossible task to solve and find solutions to that equation. However, we can consider a simpler model, assuming a Schrödinger type of equation, which was first considered in [2]. In that work it was argued that in order to study the effect of the indefinite kinetic term in (19), as a first approach, and since we are dealing with an equation which is formally a Schrödinger equation, we could restrict our attention to finite amount of degrees of freedom. This first approach as been successful in cosmology, allowing to solve the Wheeler-DeWitt equation in minisuperspace, which brings a functional differential equation to a regular differential equation. We do not claim that we are doing the exact same process, but instead that a reduction of the physical system to a finite number of degrees of freedom could retain some aspects of quantum gravity that could be studied using much simpler equations. It is an acceptable concern if approximating a functional differential equation to a Schrödinger type of regular differential equation becomes an oversimplification. Nevertheless, it can also be acceptable to think that some physical insight can be obtained by assuming that the indefinite character of the functional equation is mimicked in the simpler model. Let us consider some assumptions in order to obtain the simpler equation.

1. Assuming that the hamiltonian $\hat{H}^m$ includes black hole and Hawking radiation parts, and ignoring other degrees of freedom, the simpler equation can take the form,

$$i\hbar \frac{\partial}{\partial t} \Psi(x,y,t) = \left(\frac{\hbar^2}{2m_{Pl}} \frac{\partial^2}{\partial x^2} - \frac{\hbar^2}{2m_y} \frac{\partial^2}{\partial y^2} + \frac{m_{Pl}\omega_x^2}{2}x^2 + \frac{m_y \omega_y^2}{2} y^2 \right) \Psi(x,y,t) . \quad (20)$$

This last equation, where the emergence of a negative kinetic term which plays the role of the functional derivative in the metric h_{ab} in (12), contrasts with an exact

Schrödinger equation. Therefore, because variable x is related to metric h_{ab}, we propose to identify it with the variation of the black hole radius[3] $2GM/c^2$, which turn out to be also a variation in the black hole mass or energy. Variable y will correspond to Hawking radiation with energy m_y.

2. Notice that the kinetic term of the gravitational part of the hamiltonian operator is suppressed by the Planck mass. As long as the black hole mass is large, this kinetic term is irrelevant. One would have in that case, only the Hawking radiation contribution. If, instead we consider the last stages of the evaporation process, when the black hole mass approaches the Planck mass, then the kinetic term associated with the black hole state becomes relevant.

3. The time parameter t in Equation (20) was obtained by means of a Born-Oppenheimer approximation and embodies all the semiclassical degrees of freedom of the universe.

4. In Equation (20) we consider harmonic oscillator potentials. Beside being simpler potentials, they allow for analytical solutions and, in the Hawking radiation case this regime is realistic [35,36]. For the black hole, this potential is an oversimplification, which can be far from realistic. However, it can help to disclose behaviours also present among more complex potentials, with respect to the entanglement between black hole and Hawking radiation quantum states, during the evaporation process. Furthermore, before dealing with the full problem, simpler models can identify physical phenomena that will reasonably manifest independently of the problem complexity (for example, the infinite square well helps to understand energy quantisation in the more complex Coulomb potential).

Let us assume that Equation (20) is solved by the variables separation method,

$$\Psi(x,y,t) = \psi_x(x,t)\psi_y(y,t), \tag{21}$$

so that we can obtain the two following equations[4],

$$i\hbar\dot{\psi}_x^*(x,t) = \left(-\frac{\hbar^2}{2m_{Pl}}\frac{\partial^2}{\partial x^2} - \frac{m_{Pl}\omega_x^2}{2}x^2\right)\psi_x^*(x,t)$$
$$i\hbar\dot{\psi}_y(y,t) = \left(-\frac{\hbar^2}{2m_y}\frac{\partial^2}{\partial y^2} + \frac{m_y\omega_y^2}{2}y^2\right)\psi_y(y,t) \tag{22}$$

Equations (22) describe an uncoupled system comprising a harmonic oscillator and an inverted one. In Figure 1 we illustrate the fact that having regular harmonic potential with a negative (indefinite) kinetic term is equivalent, in the quantum point of view, to the situation where an inverted oscillator potential has a positive kinetic term. In both situations we have to deal with an unstable system, which would correspond of having variable x varying uncontrollably. A wave function $\psi_0(x',0)$ that initially has a Gaussian profile, will evolve over time according to,

$$\int dx'\, G_{\text{inv.}}(x,x';t,0)\,\psi_0(x',0) = \psi(x,t), \tag{23}$$

where $G_{\text{inv.}}(x,x';t,0)$ is the inverted oscillator Green function [37,38],

$$G_{\text{inv.}}(x,x';t,0) = \sqrt{\frac{m_{Pl}\omega_x}{2\pi i\hbar \sinh(\omega_x t)}} \exp\left(im_{Pl}\omega_x \frac{((x^2+x'^2)\cosh(\omega_x t) - 2xx')}{2\hbar \sinh(\omega_x t)}\right), \tag{24}$$

---

[3] A black hole without rotation and charge which is simply described by the Schwarzschild static solution.
[4] Where ψ_x^* is the complex conjugate of ψ_x.

which can be obtained from the harmonic oscillator Green function

$$G_{\text{osc.}}(x, x'; t, 0) = \sqrt{\frac{m_y \omega_y}{2\pi i \hbar \sin(\omega_y t)}} \exp\left(i m_y \omega_y \frac{((x^2 + x'^2)\cos(\omega_y t) - 2xx')}{2\hbar \sin(\omega_y t)}\right), \quad (25)$$

by redefining $\omega \to (i\omega)$. The wave function obtained from the computation of Equation (23) shows a progressive squeezing of the state in phase space, which means an increasing uncertainty in the value of x. Physically, in this simplified model, that would correspond to an unstable variation of the Schwarzschild radius or mass of the black hole. Even though, conceivably, a strong squeezing of the black hole state would occur [39], driving its disappearance.

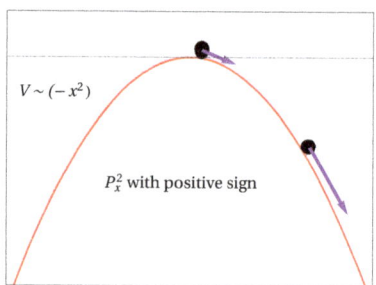 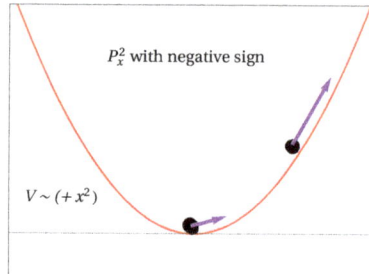

Figure 1. The behaviour of a particle in an inverted oscillator potential, with a positive kinetic term, is equivalent to the behaviour of a particle, with a negative kinetic term, in a regular harmonic oscillator potential.

In the next section we will introduce the effect of a back reaction, coupling effectively black hole and Hawking radiation quantum states, and see that, under particular circumstances, the system becomes stable and strongly entangled.

5. Back Reaction and Schrödinger Equation

In this section we review and reproduce some results obtained in reference [25]. Notice that in this work a slight change in some definitions will be carried. In addition some new aspects of the model will be discussed. In order to investigate the effects of a back reaction between Hawking radiation and black hole states, let us consider a linear coupling μxy between the variables, where μ is a constant, in equation

$$i\hbar \frac{\partial}{\partial t} \Psi(x, y, t) = \left(\frac{P_y^2}{2m_y} - \frac{P_x^2}{2m_{\text{Pl}}} + \frac{m_{\text{Pl}} \omega_x^2}{2} x^2 + \frac{m_y \omega_y^2}{2} y^2 + \mu xy\right) \Psi(x, y, t). \quad (26)$$

We should emphasize that, following the Born-Oppenheimer and WKB approximation used to obtain Equation (19), any phenomenological back reaction effect, here parametrized by μ must be suppressed by the Planck mass [2]. Therefore we can consider that this back reaction coupling constant, as the kinetic term of the gravitational part of the hamiltonian operator, only becomes relevant when the black hole approaches the Planck mass. Consequently we can assume that the constant $\mu \sim \mu'/m_{\text{Pl}}$. Suppose the initial state, describing the black hole, is the coherent state

$$\psi_{x_0}^\alpha(x, 0) = \left(\frac{m_{\text{Pl}} \omega_x}{\pi \hbar}\right)^{1/4} \exp\left(-\frac{m_{\text{Pl}} \omega_x}{2\hbar} x^2 + \alpha \sqrt{\frac{2 m_{\text{Pl}} \omega_x}{\hbar}} x - \frac{|\alpha|^2}{2} - \frac{\alpha^2}{2}\right), \quad (27)$$

where

$$\alpha = \sqrt{\frac{m_{\text{Pl}} \omega_x}{2\hbar}} x_0 + i \frac{p_0}{\sqrt{2 m_{\text{Pl}} \omega_x}}, \quad (28)$$

which represents a black hole whose Schwarzschild radius oscillates around the value $2GM/c^2$. A coherent state represents a displacement of the harmonic oscillator ground state $|0\rangle$

$$|\alpha\rangle = \hat{D}(\alpha)|0\rangle = e^{-\alpha \hat{a}^\dagger - \alpha^* \hat{a}}|0\rangle, \quad (29)$$

in order to get a finite excitation amplitude α. For the Hawking radiation initial state, let us consider the Gaussian distribution

$$\psi_{y_0}^H(y, 0) \propto \exp\left(-\frac{m_y \omega_y}{2\hbar} \coth\left(\frac{2\pi \omega_y GM}{c^3}\right) y^2\right), \quad (30)$$

which describes the radiation state [35,36,39] for a black hole with Schwarzschild radius $2GM/c^2$. Under these conditions, we can expect that, after the product state (27)–(30) evolves in time

$$\hat{U}|\Psi_0\rangle = \hat{U}\left(|\psi_{x_0}^\alpha\rangle \otimes |\psi_{y_0}^H\rangle\right) = |\Psi\rangle, \quad (31)$$

the emerging final state $|\Psi\rangle$ will be entangled, because the hamiltonian in Equation (26) includes a coupling such that $\hat{\mathcal{H}} \neq \hat{\mathcal{H}}_x \otimes \mathbb{1} + \mathbb{1} \otimes \hat{\mathcal{H}}_y$.

Determining the initial state $|\psi_{x_0}^\alpha\rangle \otimes |\psi_{y_0}^H\rangle$ evolution over time would be solved if we had the propagator related to the hamiltonian of Equation (26). Since this propagator is not available, we can instead redefine variables

$$\begin{pmatrix} P_x \\ P_y \\ x \\ y \end{pmatrix} = \begin{pmatrix} \sqrt{\frac{2m_{\text{Pl}}}{\cos 2\theta}} \cos\theta & \sqrt{\frac{2m_{\text{Pl}}}{\cos 2\theta}} \sin\theta & 0 & 0 \\ \sqrt{\frac{2m_y}{\cos 2\theta}} \sin\theta & \sqrt{\frac{2m_y}{\cos 2\theta}} \cos\theta & 0 & 0 \\ 0 & 0 & \frac{\cos\theta}{\sqrt{m_{\text{Pl}} \cos 2\theta}} & \frac{\sin\theta}{\sqrt{m_{\text{Pl}} \cos 2\theta}} \\ 0 & 0 & \frac{\sin\theta}{\sqrt{m_y \cos 2\theta}} & \frac{\cos\theta}{\sqrt{m_y \cos 2\theta}} \end{pmatrix} \begin{pmatrix} P_1 \\ P_2 \\ Q_1 \\ Q_2 \end{pmatrix}, \quad (32)$$

such that we can rewrite Equation (26) in the following way

$$i\hbar \frac{\partial}{\partial t}\Psi(Q_1, Q_2, t) = \left(\frac{1}{2}\left(P_2^2 - P_1^2\right) + \frac{1}{2}\left(\Omega_1^2 Q_1^2 + \Omega_2^2 Q_2^2\right) + \mathcal{K} Q_1 Q_2\right)\Psi(Q_1, Q_2, t). \quad (33)$$

In the previous equation, coordinates redefinition (32) implies that

$$\begin{aligned}
\Omega_1^2 \cos^2 2\theta &= \omega_x^2 \cos^2\theta + \omega_y^2 \sin^2\theta + \frac{\mu \sin 2\theta}{\sqrt{m_{\text{Pl}} m_y}} \\
\Omega_2^2 \cos^2 2\theta &= \omega_x^2 \sin^2\theta + \omega_y^2 \cos^2\theta + \frac{\mu \sin 2\theta}{\sqrt{m_{\text{Pl}} m_y}}
\end{aligned}, \quad (34)$$

and the coupling is

$$\mathcal{K} = \frac{1}{\cos^2 2\theta}\left(\frac{1}{2}\left(\omega_x^2 + \omega_y^2\right) \sin 2\theta + \frac{\mu}{\sqrt{m_{\text{Pl}} m_y}}\right). \quad (35)$$

If we impose that, in the new variables (Q_1, Q_2), the coupling is $\mathcal{K} = 0$, it follows that the coupling in the original variables (x, y) is given by,

$$\mu = -\frac{1}{2}\sqrt{m_{\text{Pl}} m_y}\left(\omega_x^2 + \omega_y^2\right) \sin 2\theta, \quad (36)$$

with $\theta \in]-\frac{\pi}{4}, \frac{\pi}{4}[$. We can check that $\mu = 0$ for $\theta = 0$. In the numerical simulations, to calculate the relevant physical quantities, we will assume that $m_y = 10^{-5} m_{\text{Pl}}$ and $\omega_y^2 = 10^5 \omega_x^2$ such that the potentials, in Equation (26), are of the same order, i.e., $m_y \omega_y^2 \sim m_{\text{Pl}} \omega_x^2$. This corresponds to assume that the Hawking radiation energy is well below Planck scale and, the fluctuations of the Schwarzschild radius have significantly smaller frequency than the Hawking radiation energy fluctuations. The numerical factor choice of 10^5 is arbitrary and does not influence the conclusions to be drawn from the results presented in subsequent sections. However, we can establish that the coupling is defined in the interval

$$-10^2 \leq \mu \leq 10^2, \tag{37}$$

which is sufficiently broad to explore the more relevant cases. If we substitute the coupling Equation (36) in the frequencies definition (34), we will obtain, in the new coordinates,

$$\Omega_1^2 = \frac{1}{\cos^2 2\theta} \left[\omega_x^2 \left(\cos^2 \theta - \frac{1}{2} \sin^2 2\theta \right) + \omega_y^2 \left(\sin^2 \theta - \frac{1}{2} \sin^2 2\theta \right) \right],$$

$$\Omega_2^2 = \frac{1}{\cos^2 2\theta} \left[\omega_x^2 \left(\sin^2 \theta - \frac{1}{2} \sin^2 2\theta \right) + \omega_y^2 \left(\cos^2 \theta - \frac{1}{2} \sin^2 2\theta \right) \right]. \tag{38}$$

We can notice an important observation related to Equation (38). Since, we assume that $\omega_y^2 \gg \omega_x^2$, it implies that Ω_1^2 remains strictly positive[5] in the significantly reduced sub interval of the possible angles $\theta \in]-\frac{\pi}{4}, \frac{\pi}{4}[$. We can verify that Ω_1^2 is only positive when

$$-\arctan\left(\sqrt{\frac{\omega_x^2}{\omega_y^2}}\right) < \theta < \arctan\left(\sqrt{\frac{\omega_x^2}{\omega_y^2}}\right). \tag{39}$$

This observation means that, for values outside the mentioned interval (39), Ω_1^2 is negative and Equation (33) turns out to be a Schrödinger equation describing two uncoupled harmonic oscillators in the coordinates (Q_1, Q_2), i.e.,

$$i\hbar \frac{\partial}{\partial t} \Psi(Q_1, Q_2, t) = \left(\frac{1}{2}\left(P_2^2 + \Omega_2^2 Q_2^2\right) - \frac{1}{2}\left(P_1^2 + \left|\Omega_1^2\right| Q_1^2\right) \right) \Psi(Q_1, Q_2, t). \tag{40}$$

In addition, we also have

$$\arctan\left(\sqrt{\frac{\omega_x^2}{\omega_y^2}}\right) < |\theta| < \frac{\pi}{4} \quad \Rightarrow \quad |\mu| > 1 \tag{41}$$

which implies that, in Equation (26), when the coupling is $|\mu| > 1$ the system becomes stable and this will restraint the influence of the inverted potential.

The calculation of the initial state $\left| \psi_{x_0}^\alpha \right\rangle \otimes \left| \psi_{y_0}^H \right\rangle$ time evolution, in coordinates (Q_1, Q_2), with the help of the harmonic (25) and inverted oscillator propagators,

$$\int \int dQ_1' dQ_2' \; G_{\text{inv.}}(Q_1, Q_1'; t) \cdot G_{\text{osc.}}(Q_2, Q_2'; t) \cdot \psi_{x_0}^\alpha(Q_1', Q_2'; 0) \cdot \psi_{y_0}^H(Q_1', Q_2'; 0), \tag{42}$$

enable us to obtain $\Psi(Q_1, Q_2, t)$, for which an explicit analytical expression is given in Appendix A (Equation (A1)). Subsequently, we can use the inverse transformation

$$\begin{cases} Q_1 = x\sqrt{m_{\text{Pl}}} \cos\theta - y\sqrt{m_y} \sin\theta \\ Q_2 = -x\sqrt{m_{\text{Pl}}} \sin\theta + y\sqrt{m_y} \cos\theta \end{cases}, \tag{43}$$

in order to retrieve the wave function in the original coordinates. This wave function has the generic form

$$\Psi(x, y, t) = F(t) \exp\left(-A(t)x^2 + B(t)x - C(t)y^2 + D(t)y + E(t)xy\right), \tag{44}$$

---

[5] Whereas Ω_2^2 is always positive, since the definition of Ω_2^2 can be further simplified to
$\Omega_2^2 = \frac{1}{2}\left(1 - \frac{1}{\cos 2\theta}\right) + \frac{m_{\text{Pl}}}{2m_y}\left(1 + \frac{1}{\cos 2\theta}\right) \sim \frac{10^5}{2}\left(1 + \frac{1}{\cos 2\theta}\right)$
which for $\theta \in]-\frac{\pi}{4}, \frac{\pi}{4}[$ is always positive.

where the time dependent functions can be found in Appendix A, more precisely in Equation (A4).

One of the main objectives in this paper is to quantify the entanglement degree between black hole and Hawking radiation quantum states. In order to proceed with that idea we have to define the system matrix density

$$\rho_{xy} = |\Psi\rangle\langle\Psi|. \tag{45}$$

Wave function (44) cannot be factored, hence, the initial density matrix $|\Psi_0\rangle\langle\Psi_0|$, corresponding to the factored pure state $|\psi_{x_0}^\alpha\rangle \otimes |\psi_{y_0}^H\rangle$, has evolved to a pure entangled state described by ρ_{xy}. Recalling the status of the classical observers outside the black hole, they can only access the state of the outgoing radiation, i.e., they can only experiment part of the system. Therefore, it is important to consider the reduced density matrix ρ_y obtained by taking the partial trace of the system density matrix ρ_{xy}, i.e., computing $\rho_y = \text{tr}_x(\rho_{xy})$. The reduced density matrix elements, for black hole and Hawking radiation, are respectively

$$\begin{aligned}\rho_{Bh}(x,x') &= \text{tr}_y\rho_{xy} = \int dy \, |\langle x', y|x, y\rangle|^2 \\ \rho_{Hr}(y,y') &= \text{tr}_x\rho_{xy} = \int dx \, |\langle x, y'|x, y\rangle|^2 \end{aligned} \tag{46}$$

where $|x,y\rangle \equiv \Psi(x,y,t)$, and with the generic form

$$\begin{aligned}\rho_{Bh}(x,x') &= \mathcal{N}_1 \exp\left(-\mathcal{A}_1 x^2 + \mathcal{B}_1 x - \mathcal{A}_1^* x'^2 + \mathcal{B}_1^* x' + |\mathcal{C}_1|xx'\right) \\ \rho_{Hr}(y,y') &= \mathcal{N}_2 \exp\left(-\mathcal{A}_2 y^2 + \mathcal{B}_2 y - \mathcal{A}_2^* y'^2 + \mathcal{B}_2^* y' + |\mathcal{C}_2|yy'\right)\end{aligned}. \tag{47}$$

The coefficients defined in the last equation are given in Appendix B (Equations (A9)–(A11)), and also depend directly on Equation (A4).

The diagonal reduced density matrix elements are

$$\rho_{Bh}(x,x) = |F|^2 \sqrt{\frac{1}{2\text{Re}(C)}} \exp\left(-2\text{Re}(A)x^2 + \frac{(\text{Re}(E)x + \text{Re}(D))^2}{2\text{Re}(C)} + 2\text{Re}(B)x\right) \tag{48}$$

$$\rho_{Hr}(y,y) = |F|^2 \sqrt{\frac{1}{2\text{Re}(A)}} \exp\left(-2\text{Re}(C)y^2 + \frac{(\text{Re}(E)y + \text{Re}(B))^2}{2\text{Re}(A)} + 2\text{Re}(D)y\right) \tag{49}$$

and, for illustration purposes, in Figure 2 we can observe their evolution over time. In that case, we have taken $\mu = 1.01$ for the back reaction coupling value. As we emphasised before, for this value, the system is stable and we can notice that the observed behaviours corresponds closely to squeezed coherent states[6], with an evident correlation between them.

---

[6] This observation will be corroborated by inspecting the behaviour of the Wigner functions in Appendix B. Squeezed coherent states are obtained through the action of two different operators over the ground state of the harmonic oscillator $|\alpha, \xi\rangle = \hat{D}(\alpha)\hat{S}(\xi)|0\rangle = \left(e^{-\alpha \hat{a}^\dagger - \alpha^* \hat{a}}\right)\left(e^{-\frac{1}{2}\xi^* \hat{a}^2 - \frac{1}{2}\xi \hat{a}^{\dagger 2}}\right)|0\rangle$, where $\hat{D}(\alpha)$ is the displacement operator and $\hat{S}(\xi)$ is the squeeze operator.

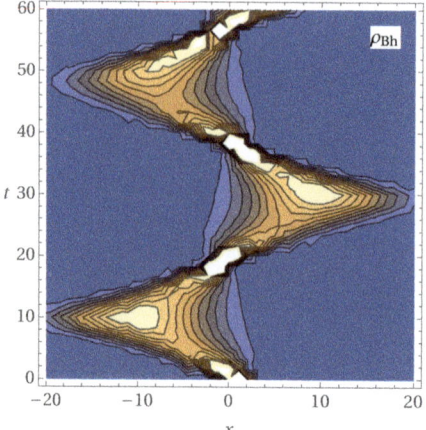
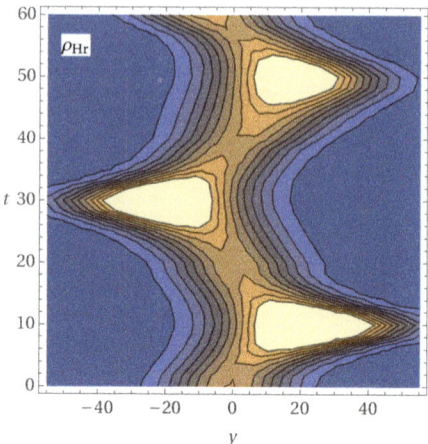

Figure 2. Diagonal reduced density matrix elements ρ_{Bh} and ρ_{Hr} (defined in Equations (48) and (49)) evolution over time, with $m_{Pl} = \hbar = \omega_x = x_0 = 1$; $p_0 = -1$. The back reaction coupling value is $\mu \approx 1.01$, where $\omega_y = \omega_x \times 10^{5/2}$ and $m_y = m_{Pl} \times 10^{-5}$. Light areas, in the density plots, correspond to higher values of the density matrix.

6. Entropy, Entanglement and Information

Theoretically, black holes emit radiation, when measured by an infinitely distant observer, approximately with a black body spectrum with an emission rate in a mode of frequency ω

$$\Gamma(\omega) = \frac{\gamma(\omega)}{\exp\left(\frac{\hbar\omega}{k_B T_H}\right) \pm 1} \frac{d^3k}{(2\pi)^3}, \tag{50}$$

where the Hawking temperature is,

$$T_H = \frac{\hbar c^3}{8\pi k_B G M} \tag{51}$$

and the factor $\gamma(\omega)$ embodies the effect of the non trivial geometry surrounding the black hole. Soon after this discovery, D. N. Page made important numerical estimates [40–42], of various particle emission rates, for black holes with and without rotation, and the evaporation average time for a black hole with mass M. Later, he made important conjectures [43] about the Von Neumann entropy

$$S_{VN} = -\text{tr}\left(\rho \log(\rho)\right) \tag{52}$$

of a quantum subsystem described by the reduced density matrix $\rho_A = \text{tr}_B \rho_{AB}$. If the Hilbert space of a quantum system, in a pure initial random state, has dimension mn, the average entropy of the subsystem of smaller dimension $m < n$ is conjectured to be given by

$$S_{m,n} \simeq \log m - \frac{m}{2n}. \tag{53}$$

Therefore, the given subsystem will be near its maximum entropy $\log m$ whenever $m < n$. Afterwards, he applied this new conjecture to the case of the black hole evaporation process [44]. Assuming that initially Hawking radiation and black hole are in a pure quantum state, described by the density matrix ρ_{AB}, he showed that the Von Neumann entropies related to the reduced density matrices (radiation - Hr - and black hole - Bh -),

$$S_{Hr} = -\text{tr}\left(\rho_{Hr} \log(\rho_{Hr})\right) \tag{54}$$

$$S_{Bh} = -\text{tr}\left(\rho_{Bh}\log(\rho_{Bh})\right) \tag{55}$$

display an information (defined as a measure of the departure of the actual entropy from its maximum value),

$$I_{Hr} = \log m - S_{Hr} \quad I_{Bh} = \log n - S_{Bh}. \tag{56}$$

In addition, he also described, through what is today known as the Page curve (a recent nice review can be found in [45]), the way entropy will evolve[7] (see Figure 3) while the black hole evaporates. More recently, he has numerically estimated, based on his previous works, about emission rates of several types of particles, the way Hawking radiation entropy should evolve in time [46]. It is believed that a correct quantum gravity theory should be able to show how Page curve emerges from the assumption of the outgoing radiation and black hole quantum states unitary evolution.

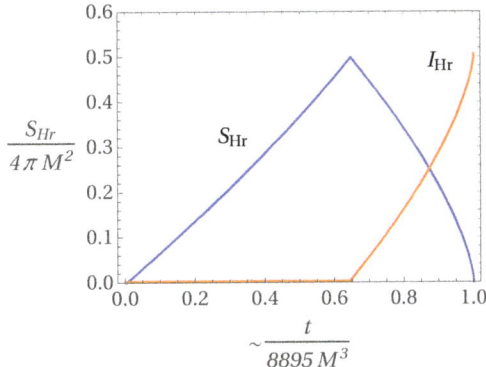

Figure 3. Time evolution of, the Von Neumann entropy (S_{Hr}) (according to Page curve [44,46]), and the information (I_{Hr}) for the Hawking radiation.

It seems pertinent to explore what the simplified model, under analysis, allow us to say about entropy and information. More precisely, we want to estimate how Hawking radiation entropy and information evolve over time according to Equation (56). Considering the reduced density matrices (47), we see that to properly calculate Von Neumann entropy (52) we have to diagonalize the matrices, i.e., compute their eigenvalues

$$\int_{-\infty}^{+\infty} dy' \, \rho_{Hr}(y, y') f_n(y') = \lambda_n f_n(y). \tag{57}$$

This particular calculation is only known for a few specific cases, as for example, for a system of two coupled harmonic oscillators [47], unfortunately a distinct situation from the case studied here, namely the coupling between harmonic and inverted oscillators. Solving the eigenvalues problem allows a great simplification and the evaluation of Von Neumann entropy becomes simply

$$S_{VN} = -\sum_n \left(\lambda_n \log(\lambda_n)\right). \tag{58}$$

However, considering the eigenvalues problem technical difficulty, instead of computing the Von Neumann entropy we can estimate the Wehrl entropy [48,49],

$$\begin{aligned} S_W &= -\text{tr}(H_{Bh}(x,p) \log(H_{Bh}(x,p))) \\ &= \iint \frac{dxdp}{\pi\hbar} H_{Bh}(x,p) \log(H_{Bh}(x,p))' \end{aligned} \tag{59}$$

---

[7] Also, the way information included in the correlations between black hole and Hawking radiation quantum states will evolve.

where $H_{Bh}(x, p)$ is the Husimi function [50]

$$H_{Bh}(x, p) = \int \frac{dx'dp'}{\pi\hbar} \exp\left(-\frac{\sigma(x-x')^2}{\hbar} - \frac{(p-p')^2}{\sigma\hbar}\right) W_{Bh}(x', p') \,. \tag{60}$$

The Husimi function is defined to access the classical phase space (x, p) representation of a quantum state, and it is obtained from a Gaussian average of the Wigner function

$$\begin{aligned} W_{Bh}(x, p) &= \frac{1}{\pi\hbar} \int d\eta \, e^{-ip\eta/\hbar} \left\langle x + \frac{\eta}{2} \middle| \rho_{Bh} \middle| x - \frac{\eta}{2} \right\rangle \\ &= \frac{1}{\pi\hbar} \int d\eta \, e^{-ip\eta/\hbar} \rho_{Bh}\left(x + \frac{\eta}{2}, x - \frac{\eta}{2}\right) \end{aligned} \tag{61}$$

The Wigner function give us an rough criterion on how much a quantum state is distant from its classical limit but, unfortunately, it is not a strictly positive function and cannot be taken as a probability distribution in phase space[8]. The Husimi representation (which is a Weierstrass transformation of Wigner function) enable us to define a strictly positive function and corresponds to the trace of the density matrix over the coherent states basis $|\alpha\rangle$, i.e.,

$$H_\alpha = \frac{1}{\pi} \langle \alpha | \rho | \alpha \rangle \,. \tag{62}$$

If we compare this last definition with Equations (58) and (59), we can understand Wehrl entropy as a classical estimate of the Von Neumann entropy, through the analogy

$$S_{VN} = -\sum_n (\lambda_n \log(\lambda_n)) \to S_W \approx -\sum (H_\alpha \log(H_\alpha)) \,. \tag{63}$$

Hence, Wehrl's entropy can be considered a measure of the classical entropy of a quantum system, and has already been used [51] in the contexts of cosmology and black holes. We should notice that Wehrl entropy gives an upper bound to Von Neumann entropy, i.e., $S_W(\rho) \geq S_{VN}(\rho)$.

The time has come to obtain, in this simplified model, Wehrl entropy and information evolution over time for Hawking radiation. In Figure 4 we can find the numerical estimates of Hawking radiation Wehrl entropy and information. These were obtained based on the calculation of Wigner (Appendix B) and Husimi functions, using the reduced density matrix (47). We can observe that the entropy start with lower values, this corresponds to a stage where entanglement and correlation between the states are weak. According to Figure 2 this happens in a phase where the quantum states become increasingly squeezed and displaced, under the influence of the inverted potential. However when the back reaction begins to grow, correlations and degree of entanglement between the two states increase, and consequently so does the entropy, and both subsystems are forced to oscillate (counteracting the inverted potential). Finally, both states return to their initial configurations, which brings a reduction of their entropies. It is in this last phase that, with a decreasing entropy, the information contained in the state describing Hawking radiation increases as expected from the Page curve.

At this point, we can ask ourselves: how much the estimate of the $S_W(\rho)$ give us an accurate description of the real behaviour of the Von Neumann entropy $S_{VN}(\rho)$? Since Wehrl entropy satisfies $S_W(\rho) \geq S_{VN}(\rho)$, inspection of Figure 2 tell us that the variation from a lower values of the entropy (initial stage of the time evolution) to higher (intermediate stage of the time evolution) and again to lower values (final stage of the time evolution) seems to indicate, with reasonable chance, that Von Neumann entropy can present a behaviour relatively close to Wehrl entropy. In addition, the fact that we have considered the unitary evolution of the pure state $\left|\psi_{x_0}^\alpha\right\rangle \otimes \left|\psi_{y_0}^H\right\rangle$, implies that the system

---

[8] In fact, Wigner function is considered a quasiprobability distribution.

matrix density remains a pure state ($S_{VN}(\rho_{AB}) = 0$), while the reduced density matrices, for the two subsystems, correspond to mixed states ($S_{VN}(\rho_A) \neq 0$).

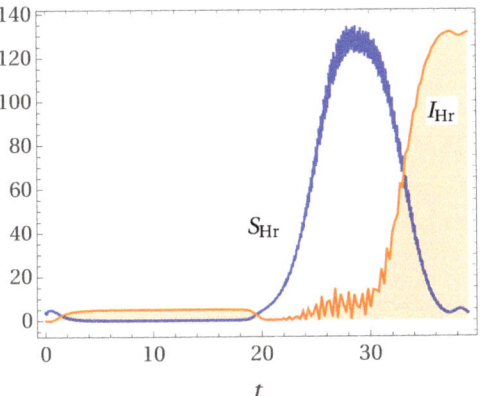

Figure 4. Hawking radiation's Wehrl entropy and information evolution over time, with $m_{Pl} = \hbar = \omega_x = x_0 = 1$; $p_0 = -1$. The back reaction coupling parameter is $\mu = 1.01$, where $\omega_y = \omega_x \times 10^{5/2}$ and $m_y = m_{Pl} \times 10^{-5}$.

7. Conclusions

Even though the simplified model, discussed in this paper, was based on modest assumptions (namely about the initial black hole quantum state, among others), it provides a simple mechanism where one can appreciate the temporal evolution of entropy and the behaviour of information (in a classical approach with Wehrl entropy being evaluated). The model has the advantage that it can be treated analytically and show how the coupling of a harmonic and inverted oscillators can produce results suggesting how the Page curve can emerge.

There are certainly many ways in which this model can become more realistic. However, it would also certainly no longer be able to be treated analytically, which would inevitably deprive it of its pedagogical appeal. In one hand, questions such as,

- how the squeezing parameter evolves in this model?
- what is the exact behaviour, in this model, of Von Neumann entropy $S_{VN}(\rho)$?
- which aspects of the discussed estimates would benefit by considering a more realistic model?
- how to apply the same procedure to the functional Schrödinger type of Equation (19)?

can be pursued as possible future topics of investigation. On the other hand, one can also try to understand to which extent entropy, and Hawking radiation information, estimates can be made in gravitational back reaction scenarios such as those proposed in [52,53]. In that proposal, it is assumed that particles moving at high speeds to and from the event horizon cause a drag [14] which has gravitational effects that can be described by the Aechelburg-Sexl metric [54,55]. It is worth to mention that the discussion of back reaction effects of the Hawking radiation and the correct way to derive the Page curve has been an active field of research in connection with the black hole information paradox. The reader can find complete reviews of the problem and recent progresses in that direction in [56–59] Finally, the black hole evaporation subject and the fate of the information enclosed inside it, are crucial aspects that any quantum gravity theory candidate will have to unveil. At a time when the first direct evidences of objects that in everything resemble what in General Relativity is described as a black hole are emerging, our scepticism about their real existence starts to fade away. However, it has been a long time since the conceptual problems associated with these hypothetical strange objects have challenged the limits of theoretical physics.

Funding: This research was funded by Fundação para a Ciência e a Tecnologia grant number UIDB/MAT/00212/2020.

Acknowledgments: In this section you can acknowledge any support given which is not covered by the author contribution or funding sections. This may include administrative and technical support, or donations in kind (e.g., materials used for experiments).

Conflicts of Interest: The authors declare no conflict of interest. The funders had no role in the design of the study; in the collection, analyses, or interpretation of data; in the writing of the manuscript, or in the decision to publish the results.

Appendix A. Wave Function Time Evolution

In this appendix, we explicitly write the analytical expressions, for the computation of Equation (42), and the various time functions which help to define state (44). Although some of the following expressions were originally presented in [25], a re-organisation, and introduction of new time functions, used to write Equation (44) justify the necessity to provide the reader with their accurate modifications.

Wave Function in the New Coordinates

When the initial state $|\psi_{x_0}^\alpha\rangle \otimes |\psi_{y_0}^H\rangle$ evolves in time, it defines a wave function that in variables (Q_1, Q_2) is,

$$\Psi(Q_1, Q_2, t) = \left(\frac{m_{\text{Pl}}\omega_x}{\pi\hbar}\right)^{1/4} \left(\frac{m_y \omega_y}{\hbar} \coth\left(\frac{2\pi\omega_y \text{GM}}{c^3} + i\omega_y t_0\right)\right)^{1/4} \left(-\frac{\Omega_1 \Omega_2}{\mathcal{F}_1 \mathcal{F}_3}\right)^{1/2}$$

$$\exp\left[-\left(\frac{Q_1^2}{2\hbar}\frac{\mathcal{F}_2}{\mathcal{F}_1} + \frac{Q_2^2}{2\hbar}\frac{\mathcal{F}_4}{\mathcal{F}_3}\right)\right]$$

$$\exp\left[-i\alpha^*\sqrt{\frac{2\widetilde{\omega}_x}{\hbar}}\left(\frac{\Omega_1 Q_1 \cos\theta}{\mathcal{F}_1} + \frac{\Omega_2 Q_2 \sin\theta}{\mathcal{F}_3}\right)\right]$$

$$\exp\left[\frac{\Omega_2 Q_2 \sin 2\theta}{2\mathcal{F}_1 \mathcal{F}_3}\left(\frac{\Omega_1 Q_1}{\hbar} - i\alpha^*\sqrt{\frac{2\widetilde{\omega}_x}{\hbar}}\sinh\Omega_1 t \cos\theta\right)(\widetilde{\omega}_y + \widetilde{\omega}_x)\right] \quad (\text{A1})$$

$$\exp\left[-\frac{\hbar \sin^2 2\theta \, \sin\Omega_2 t}{8\mathcal{F}_1^2 \mathcal{F}_3}\left(i\frac{\Omega_1 Q_1}{\hbar} + \alpha^*\sqrt{\frac{2\widetilde{\omega}_x}{\hbar}}\sinh\Omega_1 t \cos\theta\right)^2 (\widetilde{\omega}_y + \widetilde{\omega}_x)^2\right]$$

$$\exp\left[\alpha^{*2}\widetilde{\omega}_x\left(\frac{\cos^2\theta \, \sinh\Omega_1 t}{\mathcal{F}_1} + \frac{\sin^2\theta \, \sin\Omega_2 t}{\mathcal{F}_3}\right) - \frac{\alpha^{*2}}{2} - \frac{|\alpha|^2}{2}\right],$$

where

$$\widetilde{\omega}_x = \frac{\omega_x}{\cos^2 2\theta} \qquad \widetilde{\omega}_y = \frac{\omega_y}{\cos^2 2\theta} \coth\left(\frac{2\pi\omega_y \text{GM}}{c^3} + i\omega_y t_0\right) \quad (\text{A2})$$

$$\mathcal{F}_1 = -i\Omega_1 \cosh \Omega_1 t + \widetilde{\omega}_x \cos^2\theta \sinh \Omega_1 t + \widetilde{\omega}_y \sin^2\theta \sinh \Omega_1 t,$$
$$\mathcal{F}_2 = -\Omega_1^2 \sinh \Omega_1 t - i\Omega_1 \widetilde{\omega}_x \cos^2\theta \cosh \Omega_1 t - i\Omega_1 \widetilde{\omega}_y \sin^2\theta \cosh \Omega_1 t,$$
$$\mathcal{F}_3 = -i\Omega_2 \cos \Omega_2 t + \widetilde{\omega}_x \sin^2\theta \sin \Omega_2 t + \widetilde{\omega}_y \cos^2\theta \sin \Omega_2 t,$$
$$- (\widetilde{\omega}_y + \widetilde{\omega}_x)^2 \sin^2 2\theta \frac{\sinh \Omega_1 t \sin \Omega_2 t}{4\mathcal{F}_1} \quad\quad\quad\quad (A3)$$
$$\mathcal{F}_4 = \Omega_2^2 \sin \Omega_2 t - i\Omega_2 \widetilde{\omega}_x \sin^2\theta \cos \Omega_2 t - i\Omega_2 \widetilde{\omega}_y \cos^2\theta \cos \Omega_2 t$$
$$+ i\Omega_2 (\widetilde{\omega}_y + \widetilde{\omega}_x)^2 \sin^2 2\theta \frac{\sinh \Omega_1 t \cos \Omega_2 t}{4\mathcal{F}_1}.$$

When we reverse the coordinate transformation $\Psi(Q_1, Q_2, t) \to \Psi(x, y, t)$, applying transformations (43), we obtain the state defined in Equation (44), where,

$$F(t) = \left(\frac{m_{\text{Pl}}\omega_x}{\pi\hbar}\right)^{1/4} \left(\frac{m_y\omega_y}{\hbar} \coth\left(\frac{2\pi\omega_y GM}{c^3} + i\omega_y t_0\right)\right)^{1/4} \left(-\frac{\Omega_1\Omega_2}{\mathcal{F}_1\mathcal{F}_3}\right)^{1/2} \times$$
$$\exp\left[\alpha^{*2}\widetilde{\omega}_x\left(\frac{\cos^2\theta \sinh\Omega_1 t}{\mathcal{F}_1} + \frac{\sin^2\theta \sin\Omega_2 t}{\mathcal{F}_3}\right) - \frac{\alpha^{*2}}{2} - \frac{|\alpha|^2}{2}\right] \times$$
$$\exp\left[\frac{\alpha^{*2}\widetilde{\omega}_x(\widetilde{\omega}_y + \widetilde{\omega}_x)^2 \cos^2\theta \sin^2 2\theta \sinh^2\Omega_1 t \sin\Omega_2 t}{4\mathcal{F}_1^2 \mathcal{F}_3}\right],$$

$$A(t) = \frac{m_{\text{Pl}}}{2\hbar}\left[\frac{\mathcal{F}_2}{\mathcal{F}_1}\cos^2\theta + \frac{\mathcal{F}_4}{\mathcal{F}_3}\sin^2\theta + \frac{\Omega_1\Omega_2}{2\mathcal{F}_1\mathcal{F}_3}(\widetilde{\omega}_y + \widetilde{\omega}_x)\sin^2 2\theta\right.$$
$$\left.+ \frac{\Omega_1^2(\widetilde{\omega}_y + \widetilde{\omega}_x)^2 \sin^2 2\theta \cos^2\theta \sin\Omega_2 t}{4\mathcal{F}_1^2 \mathcal{F}_3}\right],$$

$$B(t) = -i\alpha^*\sqrt{\frac{2m_{\text{Pl}}\widetilde{\omega}_x}{\hbar}}\left[\frac{\Omega_1\cos^2\theta}{\mathcal{F}_1} - \frac{\Omega_2\sin 2\theta}{2\mathcal{F}_3} - \frac{\Omega_2(\widetilde{\omega}_y + \widetilde{\omega}_x)\sin^2 2\theta \sinh\Omega_1 t}{4\mathcal{F}_1\mathcal{F}_3}\right.\quad\quad (A4)$$
$$\left.- \frac{\Omega_1(\widetilde{\omega}_y + \widetilde{\omega}_x)^2 \sin^2 2\theta \cos^2\theta \sinh\Omega_1 t \sin\Omega_2 t}{4\mathcal{F}_1^2 \mathcal{F}_3}\right],$$

$$C(t) = \frac{m_y}{2\hbar}\left[\frac{\mathcal{F}_2}{\mathcal{F}_1}\sin^2\theta + \frac{\mathcal{F}_4}{\mathcal{F}_3}\cos^2\theta + \frac{\Omega_1\Omega_2}{2\mathcal{F}_1\mathcal{F}_3}(\widetilde{\omega}_y + \widetilde{\omega}_x)\sin^2 2\theta\right.$$
$$\left.+ \frac{\Omega_1^2(\widetilde{\omega}_y + \widetilde{\omega}_x)^2 \sin^2 2\theta \sin^2\theta \sin\Omega_2 t}{4\mathcal{F}_1^2 \mathcal{F}_3}\right],$$

$$D(t) = \frac{i\alpha^*}{2}\sqrt{\frac{2m_y\widetilde{\omega}_x}{\hbar}}\sin 2\theta\left(-\frac{\Omega_2}{\mathcal{F}_1\mathcal{F}_3}(\widetilde{\omega}_y + \widetilde{\omega}_x)\cos^2\theta \sinh\Omega_1 t - \frac{\Omega_2}{\mathcal{F}_3} + \frac{\Omega_1}{\mathcal{F}_1}\right.$$
$$\left.- \frac{\Omega_1(\widetilde{\omega}_y + \widetilde{\omega}_x)^2 \sin^2 2\theta \sinh\Omega_1 t \sin\Omega_2 t}{4\mathcal{F}_1^2 \mathcal{F}_3}\right),$$

$$E(t) = \frac{\sin 2\theta}{\hbar}\sqrt{m_{\text{Pl}}m_y}\left(\frac{\Omega_1\Omega_2}{2\mathcal{F}_1\mathcal{F}_3}(\widetilde{\omega}_y + \widetilde{\omega}_x)\cos 2\theta + \frac{\mathcal{F}_2}{\mathcal{F}_1} + \frac{\mathcal{F}_4}{\mathcal{F}_3}\right.$$
$$\left.+ \frac{\Omega_1^2(\widetilde{\omega}_y + \widetilde{\omega}_x)^2 \sin^2 2\theta \sin\Omega_2 t}{8\mathcal{F}_1^2 \mathcal{F}_3}\right).$$

Appendix B. Wigner Functions

In this appendix, we obtain the Wigner functions analytic expressions for the reduced density matrices (47). In addition, we present the numerical simulations for the Wigner function time evolution related to the black hole subsystem (moreover, for the Hawking radiation subsystem the simulations display some similarity). The main idea is also to illustrate the effects of the displacement $\hat{D}(\alpha)$ and squeeze $\hat{S}(\xi)$ operators, while Wigner function evolves in time in phase space.

Black Hole and Hawking Radiation Wigner Functions

The Wigner function can be defined as,

$$W_{Bh}(x,p) = \frac{1}{\pi\hbar}\int d\eta \frac{1}{\pi\hbar}\int d\eta \, e^{-ip\eta/\hbar}\rho_{Bh}\left(x+\frac{\eta}{2},x-\frac{\eta}{2}\right), \quad (A5)$$

where, upon the substitution of ρ_{Bh} by Equation (47), we get

$$\begin{aligned} W_{Bh}(x,p) &= \frac{1}{\pi\hbar}\int d\eta \, e^{-ip\eta/\hbar}\rho_{Bh}\left(x+\frac{\eta}{2},x-\frac{\eta}{2}\right) \\ &= \frac{\mathcal{N}_1}{\pi\hbar}\int d\eta \, e^{-ip\eta/\hbar}\exp\left(-\mathcal{A}_1\left(x-\frac{\eta}{2}\right)^2+\mathcal{B}_1\left(x-\frac{\eta}{2}\right)\right. \\ &\quad \left. -\mathcal{A}_1^*\left(x+\frac{\eta}{2}\right)^2+\mathcal{B}_1^*\left(x+\frac{\eta}{2}\right)+|\mathcal{C}_1|\left(x^2-\frac{\eta^2}{4}\right)\right) \end{aligned} \quad (A6)$$

After some algebraic manipulation, we obtain

$$\begin{aligned} W_{Bh}(x,p) =& \frac{\mathcal{N}_1}{\pi\hbar}\exp\left(-(2\text{Re}(\mathcal{A}_1)-|\mathcal{C}_1|)x^2+2\text{Re}(\mathcal{B}_1)x\right)\cdot \\ & \cdot \int d\eta \, \exp\left(-(2\text{Re}(\mathcal{A}_1)-|\mathcal{C}_1|)\frac{\eta^2}{4}+i\left(2\text{Im}(\mathcal{A}_1)x-\text{Im}(\mathcal{B}_1)-\frac{p}{\hbar}\right)\eta\right) \end{aligned} \quad (A7)$$

$$\begin{aligned} W_{Bh}(x,p) =& \frac{2\mathcal{N}_1}{\sqrt{\pi}\hbar\sqrt{2\text{Re}(\mathcal{A}_1)-|\mathcal{C}_1|}}\exp\left(-(2\text{Re}(\mathcal{A}_1)-|\mathcal{C}_1|)x^2+2\text{Re}(\mathcal{B}_1)x\right)\cdot \\ & \cdot \exp\left(-\frac{(2\text{Im}(\mathcal{A}_1)x-\text{Im}(\mathcal{B}_1)-p/\hbar)^2}{(2\text{Re}(\mathcal{A}_1)-|\mathcal{C}_1|)}\right) \end{aligned} \quad (A8)$$

where

$$\begin{cases} \mathcal{A}_1 = & A(t) - \dfrac{E^2(t)}{8\text{Re}(C(t))} \\ \mathcal{B}_1 = & B(t) + \dfrac{\text{Re}(D(t))}{\text{Re}(C(t))}\dfrac{E(t)}{2} \\ |\mathcal{C}_1| = & \dfrac{|E(t)|^2}{4\text{Re}(C(t))} \\ \mathcal{N}_1 = & |F(t)|^2\exp\left(\dfrac{\text{Re}(D(t))}{2\text{Re}(C(t))}\right)\sqrt{\dfrac{\pi\hbar}{m_y\text{Re}(C(t))}} \end{cases} \quad (A9)$$

Concerning the Hawking radiation, a similar procedure enable us to obtain the following Wigner function,

$$\begin{aligned} W_{Hr}(y,p) =& \frac{2\mathcal{N}_2}{\sqrt{\pi}\hbar\sqrt{2\text{Re}(\mathcal{A}_2)-|\mathcal{C}_2|}}\exp\left(-(2\text{Re}(\mathcal{A}_2)-|\mathcal{C}_2|)y^2+2\text{Re}(\mathcal{B}_2)y\right)\cdot \\ & \cdot \exp\left(-\frac{(2\text{Im}(\mathcal{A}_2)y-\text{Im}(\mathcal{B}_2)-p/\hbar)^2}{(2\text{Re}(\mathcal{A}_2)-|\mathcal{C}_2|)}\right) \end{aligned} \quad (A10)$$

where

$$\begin{cases} \mathcal{A}_2 = & C(t) - \dfrac{E^2(t)}{8\text{Re}(A(t))} \\ \mathcal{B}_2 = & D(t) + \dfrac{\text{Re}(B(t))}{\text{Re}(A(t))}\dfrac{E(t)}{2} \\ |\mathcal{C}_2| = & \dfrac{|E(t)|^2}{4\text{Re}(A(t))} \\ \mathcal{N}_2 = & |F(t)|^2 \exp\left(\dfrac{\text{Re}(B(t))}{2\text{Re}(A(t))}\right)\sqrt{\dfrac{\pi\hbar}{m_y\text{Re}(A(t))}} \end{cases} \qquad (A11)$$

In Figure A1 we display the time evolution for function $W_{Bh}(x,p)$ in the interval $t \sim [0, 40]$, which is related to Figure 2. This time interval can approximately be taken as measuring one full cycle of 'oscillation' for the black hole state, i.e., the average time required for the state to return to its initial configuration. Inspecting the aforementioned figure, we can notice that the initial state Wigner function (first left panel of the figure) describes a coherent state $|\psi_{x_0}^\alpha\rangle$, which is displaced from the origin of phase space, since

$$|\psi_{x_0}^\alpha\rangle = \hat{D}(\alpha)|0\rangle, \qquad (A12)$$

in agreement with Equation (29). After some time has elapsed (top right panel of the figure), the Wigner function starts to squeeze, in the density plot, deforming its initial circular shape to an elliptical one. This illustrates the action of the squeeze operator $\hat{S}(\xi)$, besides the displacement around the origin of phase space. Finally, we can observe that a full rotation of the displacement center point occurs around the origin of phase space, while various degrees of squeezing affect the shape of the state.

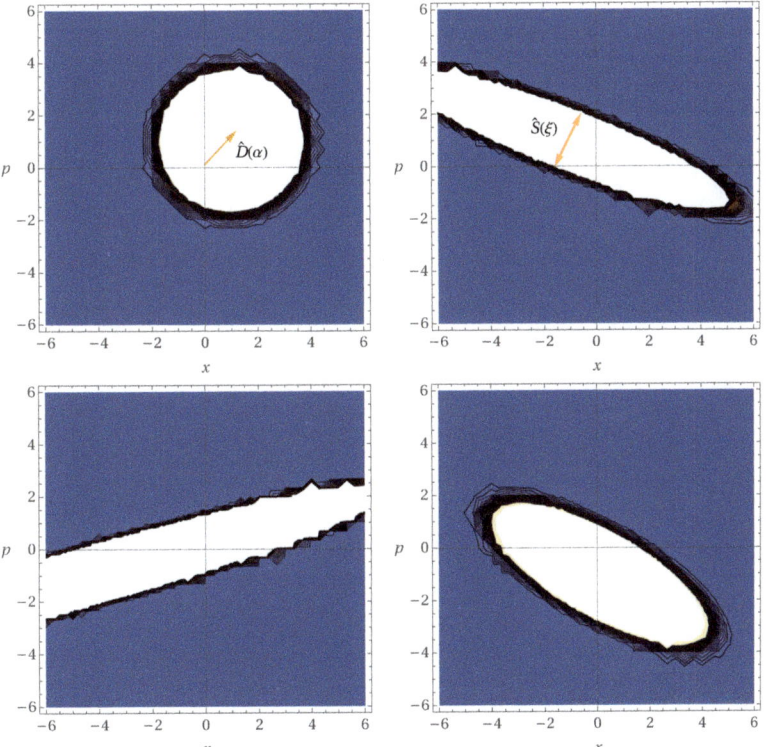

Figure A1. *Cont.*

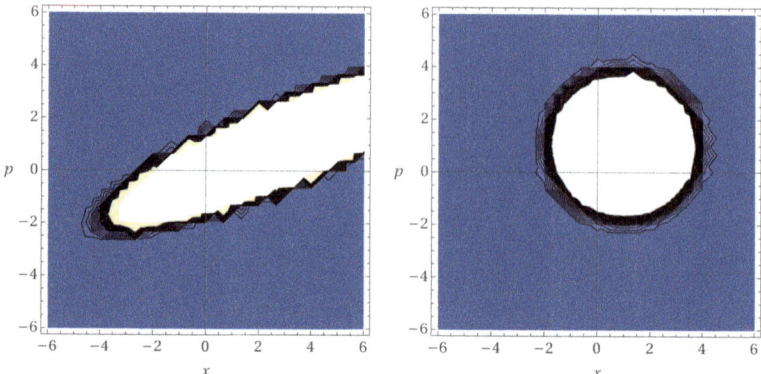

Figure A1. Time evolution of the black hole state Wigner function, with $m_{\text{Pl}} = \hbar = \omega_x = x_0 = 1$; $p_0 = -1$. The coupling parameter which defines the back reaction is $\mu = 1.01$, with $\omega_y = \omega_x \times 10^{5/2}$ and $m_y = m_{\text{Pl}} \times 10^{-5}$. We verify that, throughout the various stages of the evolution (corresponding to the various panels), the action of the operators $\hat{D}(\alpha)$ (displacement operator) and $\hat{S}(\xi)$ (squeeze operator), produces a full rotation of the displacement center point of the initial Wigner function around the origin of phase space, while various degrees of squeezing affect the shape of the state.

References

1. Hawking, S.W. Breakdown of predictability in gravitational collapse. *Phys. Rev. D* **1976**, *14*, 2460–2473. [CrossRef]
2. Kiefer, C. *Quantum Gravity*, 3rd ed.; International Series of Monographs on Physics; Oxford University Press: Oxford, UK, 2012. [CrossRef]
3. Ramond, P. Dual Theory for Free Fermions. *Phys. Rev. D* **1971**, *3*, 2415–2418. [CrossRef]
4. Neveu, A.; Schwarz, J.H. Tachyon-free dual model with a positive-intercept trajectory. *Phys. Lett. B* **1971**, *34*, 517–518. [CrossRef]
5. Scherk, J.; Schwarz, J.H. Dual models for non-hadrons. *Nucl. Phys. B* **1974**, *81*, 118–144. [CrossRef]
6. Becker, K.; Becker, M.; Schwarz, J.H. *String Theory and M-Theory: A Modern Introduction*; Cambridge University Press: Cambridge, UK, 2006. [CrossRef]
7. Zwiebach, B. *A First Course in String Theory*; Cambridge University Press: Cambridge, UK, 2004. [CrossRef]
8. Ashtekar, A. New Variables for Classical and Quantum Gravity. *Phys. Rev. Lett.* **1986**, *57*, 2244–2247. [CrossRef] [PubMed]
9. Rovelli, C. *Quantum Gravity*; Cambridge Monographs on Mathematical Physics; Cambridge University Press: Cambridge, UK, 2004. [CrossRef]
10. Thiemann, T. *Modern Canonical Quantum General Relativity*; Cambridge Monographs on Mathematical Physics; Cambridge University Press: Cambridge, UK, 2007. [CrossRef]
11. Bojowald, M. *Canonical Gravity and Applications: Cosmology, Black Holes, and Quantum Gravity*; Cambridge University Press: Cambridge, UK, 2010. [CrossRef]
12. Susskind, L.; Thorlacius, L.; Uglum, J. The Stretched horizon and black hole complementarity. *Phys. Rev. D* **1993**, *48*, 3743–3761. [CrossRef]
13. 't Hooft, G. On the quantum structure of a black hole. *Nucl. Phys. B* **1985**, *256*, 727–745. [CrossRef]
14. Dray, T.; 't Hooft, G. The gravitational shock wave of a massless particle. *Nucl. Phys. B* **1985**, *253*, 173–188. [CrossRef]
15. 't Hooft, G. Dimensional reduction in quantum gravity. *Conf. Proc.* **1993**, *C930308*, 284–296.
16. Susskind, L. The World as a hologram. *J. Math. Phys.* **1995**, *36*, 6377–6396. [CrossRef]
17. Susskind, L.; Lindesay, J. *An Introduction to Black Holes, Information and the String Theory Revolution*; World Scientific: Singapore, 2004. [CrossRef]
18. Maldacena, J.M. The Large N limit of superconformal field theories and supergravity. *Int. J. Theor. Phys.* **1999**, *38*, 1113–1133. [CrossRef]
19. Aharony, O.; Gubser, S.S.; Maldacena, J.M.; Ooguri, H.; Oz, Y. Large N field theories, string theory and gravity. *Phys. Rept.* **2000**, *323*, 183–386. [CrossRef]
20. Natsuume, M. AdS/CFT Duality User Guide. *Lect. Notes Phys.* **2015**, *903*, 1–294. [CrossRef]
21. Năstase, H. *Introduction to the AdS/CFT Correspondence*; Cambridge University Press: Cambridge, UK, 2015. [CrossRef]
22. Almheiri, A.; Marolf, D.; Polchinski, J.; Sully, J. Black holes: Complementarity or firewalls? *J. High Energy Phys.* **2013**, *2013*, 62. [CrossRef]
23. Polchinski, J. The Black Hole Information Problem. In Proceedings of the Theoretical Advanced Study Institute in Elementary Particle Physics: New Frontiers in Fields and Strings (TASI 2015), Boulder, CO, USA, 1–26 June 2015; pp. 353–397. [CrossRef]

24. Harlow, D. Jerusalem Lectures on Black Holes and Quantum Information. *Rev. Mod. Phys.* **2016**, *88*, 015002. [CrossRef]
25. Kiefer, C.; Marto, J.; Vargas Moniz, P. Indefinite oscillators and black-hole evaporation. *Annalen Phys.* **2009**, *18*, 722–735. [CrossRef]
26. Lapchinsky, V.G.; Rubakov, V.A. Canonical Quantization of Gravity and Quantum Field Theory in Curved Space-Time. *Acta Phys. Polon. B* **1979**, *10*, 1041–1048.
27. Halliwell, J.J.; Hawking, S.W. Origin of structure in the Universe. *Phys. Rev. D* **1985**, *31*, 1777–1791. [CrossRef]
28. Banks, T. TCP, quantum gravity, the cosmological constant and all that... *Nucl. Phys. B* **1985**, *249*, 332–360. [CrossRef]
29. Barvinsky, A. Perturbative quantum cosmology: The probability measure on superspace and semiclassical expansion. *Nucl. Phys. B* **1989**, *325*, 705–723. [CrossRef]
30. Arnowitt, R.; Deser, S.; Misner, C.W. Dynamical Structure and Definition of Energy in General Relativity. *Phys. Rev.* **1959**, *116*, 1322–1330. [CrossRef]
31. Dirac, P.A.M. Lectures on quantum mechanics. In *Belfer Graduate School of Science Monographs Series*; Belfer Graduate School of Science: New York, NY, USA, 1964; Volume 2, pp. v + 87.
32. Kiefer, C.; Sandhoefer, B. *Quantum Cosmology*; NNSA: Washington, DC, USA, 2008.
33. Kiefer, C. Conceptual Problems in Quantum Gravity and Quantum Cosmology. *ISRN Math. Phys.* **2013**, *2013*, 509316. [CrossRef]
34. Barvinsky, A.O.; Kiefer, C. Wheeler-DeWitt equation and Feynman diagrams. *Nucl. Phys. B* **1998**, *526*, 509–539. [CrossRef]
35. Demers, J.G.; Kiefer, C. Decoherence of black holes by Hawking radiation. *Phys. Rev. D* **1996**, *53*, 7050–7061. [CrossRef] [PubMed]
36. Kiefer, C. Hawking radiation from decoherence. *Class. Quantum Gravit.* **2001**, *18*, L151–L154. [CrossRef]
37. Müller-Kirsten, H.J.W. *Introduction to Quantum Mechanics*, 2nd ed.; World Scientific: Singapore, 2012. [CrossRef]
38. Kramer, T.; Moshinsky, M. Tunnelling out of a time-dependent well. *J. Phys. A Math. Gen.* **2005**, *38*, 5993–6003. [CrossRef]
39. Grishchuk, L.P.; Sidorov, Y.V. Squeezed quantum states of relic gravitons and primordial density fluctuations. *Phys. Rev. D* **1990**, *42*, 3413–3421. [CrossRef]
40. Page, D.N. Particle Emission Rates from a Black Hole: Massless Particles from an Uncharged, Nonrotating Hole. *Phys. Rev. D* **1976**, *13*, 198–206. [CrossRef]
41. Page, D.N. Particle Emission Rates from a Black Hole. 2. Massless Particles from a Rotating Hole. *Phys. Rev.* **1976**, *D14*, 3260–3273. [CrossRef]
42. Page, D.N. Particle Emission Rates from a Black Hole. 3. Charged Leptons from a Nonrotating Hole. *Phys. Rev.* **1977**, *D16*, 2402–2411. [CrossRef]
43. Page, D.N. Average entropy of a subsystem. *Phys. Rev. Lett.* **1993**, *71*, 1291–1294. [CrossRef]
44. Page, D.N. Information in black hole radiation. *Phys. Rev. Lett.* **1993**, *71*, 3743–3746. [CrossRef]
45. Alonso-Serrano, A.; Visser, M. Entropy/information flux in Hawking radiation. *Phys. Lett. B* **2018**, *776*, 10–16. [CrossRef]
46. Page, D.N. Time Dependence of Hawking Radiation Entropy. *JCAP* **2013**, *1309*, 028. [CrossRef]
47. Srednicki, M. Entropy and area. *Phys. Rev. Lett.* **1993**, *71*, 666–669. [CrossRef] [PubMed]
48. Wehrl, A. General properties of entropy. *Rev. Mod. Phys.* **1978**, *50*, 221–260. [CrossRef]
49. Wehrl, A. On the relation between classical and quantum-mechanical entropy. *Rep. Math. Phys.* **1979**, *16*, 353–358. [CrossRef]
50. Husimi, K. Some Formal Properties of the Density Matrix. *Proc. Phys. Math. Soc. Jpn.* **1940**, *22*, 264–314. [CrossRef]
51. Rosu, H.; Reyes, M. Shannon-Wehrl entropy for cosmological and black hole squeezing. *Int. J. Mod. Phys.* **1995**, *D4*, 327–332. [CrossRef]
52. 't Hooft, G. The Scattering matrix approach for the quantum black hole: An Overview. *Int. J. Mod. Phys.* **1996**, *A11*, 4623–4688. [CrossRef]
53. 't Hooft, G. Virtual Black Holes and Space-Time Structure. *Found. Phys.* **2018**, *48*, 1134–1149. [CrossRef]
54. Bonnor, W.B. The gravitational field of light. *Commun. Math. Phys.* **1969**, *13*, 163–174. [CrossRef]
55. Aichelburg, P.C.; Sexl, R.U. On the gravitational field of a massless particle. *Gen. Relativ. Gravit.* **1971**, *2*, 303–312. [CrossRef]
56. Wang, H.; Wang, J. The nonequilibrium back-reaction of Hawking radiation to a Schwarzschild black hole. *Adv. High Energy Phys.* **2020**, *2020*, 9102461. [CrossRef]
57. Marolf, D.; Maxfield, H. Observations of Hawking radiation: the Page curve and baby universes. *J. High Energy Phys.* **2021**, 272. [CrossRef]
58. Gautason, F.F.; Schneiderbauer, L.; Sybesma, W.; Thorlacius, L. Page curve for an evaporating black hole. *J. High Energy Phys.* **2020**, 91. [CrossRef]
59. Almheiri, A.; Mahajan, R.; Maldacena, J.; Zhao, Y. The Page curve of Hawking radiation from semiclassical geometry. *J. High Energy Phys.* **2020**, 149. [CrossRef]

Review

Classical and Quantum $f(R)$ Cosmology: The Big Rip, the Little Rip and the Little Sibling of the Big Rip

Teodor Borislavov Vasilev [1,*], Mariam Bouhmadi-López [2,3] and Prado Martín-Moruno [1]

1. Departamento de Física Teórica and IPARCOS, Universidad Complutense de Madrid, 28040 Madrid, Spain; pradomm@ucm.es
2. IKERBASQUE, Basque Foundation for Science, 48011 Bilbao, Spain; mariam.bouhmadi@ehu.eus
3. Department of Theoretical Physics, University of the Basque Country, UPV/EHU, P.O. Box 644, 48080 Bilbao, Spain
* Correspondence: teodorbo@ucm.es

Citation: Borislavov Vasilev, T.; Bouhmadi-López, M.; Martín-Moruno, P. Classical and Quantum $f(R)$ Cosmology: The Big Rip, the Little Rip and the Little Sibling of the Big Rip. *Universe* **2021**, *7*, 288. https://doi.org/10.3390/universe7080288

Academic Editor: Antonino Del Popolo

Received: 29 June 2021
Accepted: 29 July 2021
Published: 6 August 2021

Publisher's Note: MDPI stays neutral with regard to jurisdictional claims in published maps and institutional affiliations.

Copyright: © 2021 by the authors. Licensee MDPI, Basel, Switzerland. This article is an open access article distributed under the terms and conditions of the Creative Commons Attribution (CC BY) license (https://creativecommons.org/licenses/by/4.0/).

Abstract: The big rip, the little rip and the little sibling of the big rip are cosmological doomsdays predicted by some phantom dark-energy models that could describe the future evolution of our universe. When the universe evolves towards either of these future cosmic events, all bounded structures and, ultimately, space–time itself are ripped apart. Nevertheless, it is commonly believed that quantum gravity effects may smooth or even avoid these classically predicted singularities. In this review, we discuss the classical and quantum occurrence of these riplike events in the scheme of metric $f(R)$ theories of gravity. The quantum analysis is performed in the framework of $f(R)$ quantum geometrodynamics. In this context, we analyze the fulfilment of the DeWitt criterion for the avoidance of these singular fates. This review contains as well new unpublished work (the analysis of the equation of state for the phantom fluid and a new quantum treatment of the big rip and the little sibling of the big rip events).

Keywords: quantum cosmology; quantum geometrodynamics; Wheeler-DeWitt equation; extended theories of gravity; dark energy singularities

1. Introduction

The discovery that our universe is currently undergoing an accelerated expansion phase has supposed an inflexion point in our understanding of the physics of the cosmos. Even though this acceleration was first noticed more than 20 years ago [1,2], its mysteries are still not yet fully unraveled. On the contrary, understanding the mechanism involved in this phase represents one of the greatest milestones in modern cosmology. This behavior is so challenging because no form of matter we know from our ordinary experience can actually produce this phenomenon. In general relativity (GR), this phase is attributed to the existence of an exotic form of energy with a negative pressure that causes the repulsion of the matter content and, thus, pushes the universe into expansion. This exotic content is dubbed "dark energy" (DE). The simplest DE model, and still the one which best fits the observational data, is the standard ΛCDM theory, where DE plays the role of a cosmological constant. Nevertheless, this model contains various theoretical and observational unresolved puzzles. To mention some of them: the nature of dark matter (DM) and DE [3–7], the coincidence problem [8], the cosmological constant problem [9] and the new tensions in certain cosmological parameters. (For a review of the topic see, for instances, references [10,11] and references therein.) Therefore, one may expect the ΛCMD paradigm to be just a useful effective model. Hence, a great number of alternative models for DE describing the acceleration phase have been proposed. Some models are phantom DE [12,13], tachyonic matter [14,15], Chaplygin gas [16,17], holographic DE [18] and scalar fields in the form of quintessence [19,20] and k-essence [21], among other examples. Alternatively, the current cosmological expansion could be described not by the inclusion of new exotic content but with suitable modifications to the underlying theories of gravity. In that sense, modified theories of gravity provide an interesting framework for cosmologists. Some examples of this approach are Horndeski theories [22] (see also, e.g., reference [23]), Gauss–Bonnet gravity [24,25], $f(R)$ theories of gravity [26] (see also references [27–29]), $f(R, \mathcal{T})$ gravity [30], where $\mathcal{T}$ stands for the trace of the energy momentum tensor, $f(T)$ modified teleparallel gravity [31], being T the torsion scalar, and modified symmetric teleparallel $f(Q)$ theories of gravity [32], where Q denotes the non-metricity scalar. (See, for example, references [33–36] and references therein for a review of the state of the art.)

From a practical perspective, whatever the origin of DE may be, it can be described effectively by its equation of state (EoS) parameter. That is the ratio w between the pressure and energy density of the DE, or the effective pressure and energy density in the case of modified theories of gravity. Hence, observational constraints on w are crucial to unveil possible hints on the true nature of this exotic content. In this fashion, the cosmological data currently available constrain w to a very narrow band around -1 [37,38], see also [39–45]. Accordingly, the possibility of the expansion of the universe being fueled by phantom-type DE ($w < -1$) is not observationally excluded. Furthermore, it is even suggested by some data [46] and could help to alleviate the H_0 tension (see, for instance, references [47,48]). Nevertheless, considering phantom DE may entail some future cosmological singularities. (See reference [49] for counterexample.) All bounded structures in the universe and, ultimately, the fabric of space–time itself might be torn apart at a final big rip (BR) singularity [50]. Moreover, the universe could reach this singularity in a finite time from present epoch, resulting in a moment at which the size of the observable universe, the Hubble rate and its cosmic time derivative would diverge. On the other hand, for a phantom DE model with a w parameter converging sufficiently rapidly to the cosmological constant value ($w = -1$), the occurrence of a future singularity may be infinitely delayed in time [51]. In that case, the singularity is called an abrupt cosmic event. This is precisely the case of the little rip (LR) abrupt event [52,53] (see also [51,54]), where the scale factor, the Hubble rate and its cosmic time derivative explode at the infinite asymptotic future. Therefore, this abrupt event can be understood as a BR singularity that has been postponed indefinitely. Another characteristic abrupt event of phantom DE models is the little sibling of the big rip (LSBR). In this scenario, the size of the observable universe and the expansion rate grow infinitely but the cosmic time derivative of the Hubble parameter converges to a constant value [55]. It should be noted, however, that even though these abrupt events take place at the infinite distant future, bounded structures are destroyed in a finite time [51,55]. See a brief summary of the phenomenology of these riplike doomsdays in Table 1. (See also references [56,57] for observational constraints on some phantom DE models leading the expansion towards these events.)

The BR singularity, and the LR and the LSBR abrupt events are, in fact, intrinsic to phantom DE models. Nevertheless, a dark fluid could also induce the appearance of other cosmic singularities. For example, a phantom-dominated universe could also reach a big freeze singularity (BF) [58,59], where the scale factor reaches a maximum size at which the Hubble rate and its cosmic time derivative diverge. (See also references [60–62] for other examples of cosmological singularities and references [63–65] for a detailed classification of DE singularities in cosmology.) However, it is commonly believed that a consistent quantum description of the universe may prevent the appearance of classical singularities, see references [66,67] (see also [64,65,68–71]).

The quantum fate of classical singularities can be addressed in the setup of quantum cosmology: the application of quantum theory to the universe as a whole. Previous works in this framework have shown that the aforementioned phantom riplike cosmological doomsdays, among other singularities such as the big bang, can be avoided as a result of quantum effects emerging as the universe approaches the classical singularity [66,72,73]. However, since the same classical background evolution can be equivalently described in the context of GR or by alternative theories of gravity, it is natural to wonder whether these singularities are still avoided in the quantum realm for a different underlying theory of gravity. In this work, we gather and review the so far published results about the classical and quantum occurrence of these three riplike events when the description of gravity is that provided by $f(R)$ metric theories of gravity. For that purpose, we shall focus on the quantum cosmology scheme given by the Wheeler-DeWitt (WDW) equation [74] being adapted for the case of $f(R)$ gravity [75], namely $f(R)$ quantum geometrodynamics. Consequently, we explore here the possibility of avoiding the BR singularity, and the LR and the LSBR abrupt events in $f(R)$ quantum cosmology.

This manuscript is organized as follows. In Section 2, we review some phantom DE models predicting the classical appearance of future singularities and/or abrupt events. We also provide a new classification of the cosmic singularities and abrupt events found there. Moreover, we discuss the viability of those models from the point of view of cosmological observations. In Section 3, we consider that the classical background evolution found in the previous section can be equivalently described in the framework of metric $f(R)$ theories of gravity, where in the latter case the expansion of the universe will have a purely geometrical origin. For that aim, we briefly discuss a reconstruction method for metric $f(R)$ theories of gravity. Thereafter, in Sections 3.1–3.3, we apply this background reconstruction technique to the specific phantom DE models reviewed in Section 2. Thus, we present the group of metric $f(R)$ theories of gravity predicting a classical fate à la BR, LR or LSBR. Section 4 is entirely devoted to the revision of the fate of classical singularities in $f(R)$ quantum cosmology. Therefore, we summarize the $f(R)$ quantum geometrodynamics approach in Section 4.1. Then, the resolution of the modified Wheeler-DeWitt equation corresponding to the BR, LR and LSBR doomsdays is addressed in Sections 4.2–4.4, respectively. We also discuss there the avoidance of these cosmological catastrophes by means of the DeWitt criterion. Finally, we check the validity of the approximations performed to solve the modified Wheeler–DeWitt equations in Appendices A and B.

Table 1. Classification of the riplike cosmic events by means of the time of occurrence of the rip t_{rip}, the scale factor a, the Hubble parameter H and its cosmic time derivative $\dot{H}$. Please note that the pseudorip has been included for the sake of completeness, as this corresponds to a mild event (before which all bounded structures are disintegrated) rather than to a curvature singularity.

	t_{rip}	a	H	$\dot{H}$
Big rip	Finite	∞	∞	∞
Little rip	∞	∞	∞	∞
LSBR	∞	∞	∞	Finite
Pseudorip	∞	∞	Finite	Finite

2. A Phantom Dark-Energy Universe

Throughout this review we consider homogeneous and isotropic cosmological scenarios, which are described by the Friedmann-Lemaître-Robertson-Walker (FLRW) metric given by

$$ds^2 = -dt^2 + a(t)^2 ds_3^2, \tag{1}$$

where t stands for the cosmic time, a represents the scale factor, ds_3^2 denotes the three-dimensional spatial metric and we have used the geometric unit system $8\pi G = c = 1$. Assuming the content of the universe to be described by a perfect fluid with energy density ρ and pressure p, the Friedmann and Raychaudhuri equations are

$$H^2 = \frac{1}{3}\rho - \frac{k}{a^2}, \tag{2}$$

$$\dot{H} + H^2 = -\frac{1}{2}\left(p + \frac{\rho}{3}\right), \tag{3}$$

respectively, where the dot denotes the derivative with respect to the cosmic time t, H stands for the Hubble rate and k represents the spatial curvature (not fixed at this point). Within the standard interpretation of the observational data, the cosmic fluid at present is constituted by three species: radiation, matter (baryonic and dark) and dark energy. The present concentrations of matter and dark energy are roughly $\Omega_{M,0} = 0.315$ and $\Omega_{DE,0} = 0.685$, respectively, whereas the contribution of radiation to the total content of the universe is negligible at the present time [37,38]. Therefore, DE is the dominant ingredient today. Moreover, because we aim to study the asymptotic future expansion of the universe, we assume that the DE density increases, remains constant or decreases more slowly than the non-relativistic matter energy density and the spatial curvature term k/a^2. Thus, from a practical point of view, we can neglect the contribution of the other cosmic ingredients when studying the future asymptotic behavior of these models. Consequently, hereafter p and ρ denote the pressure and the energy density of the dark fluid. Then, the DE density evolves as

$$\dot{\rho} + 3H(p + \rho) = 0. \tag{4}$$

To evaluate the future fate of a DE dominated universe in the perfect fluid representation, an equation of state (EoS) for the dark fluid must be provided. We emphasize that presently observational data are compatible with the $w = -1$ value, though there is a significantly overlap with both possibilities $w < -1$ and $w > -1$. On the other hand, from a theoretical point of view, the value -1 in the w-axis represents a qualitative change in the properties of DE, as well as in the future evolution of the universe. Therefore, the cosmological constant case ($w = -1$) seems to play a special role from both theoretical and observational perspectives. Then, let us begin with a brief review of the phenomenology of the following EoS for the DE content, which is reminiscent of an expansion around a cosmological constant,

$$p = -\rho - A\rho^\alpha, \tag{5}$$

being A a positive constant[1]. This equation of state was first introduced in references [64,68] and thoroughly discussed in terms of singularity occurrence in reference [52]. In this section, of the manuscript, we review the analysis performed in reference [52] and present a novel alternative classification of the singularities there found by means of the behavior of the scale factor a, the Hubble rate H and its cosmic time derivative $\dot{H}$. That is a metric classification such as the one presented in [65]. (See also the characterization given in [64] and the study on cosmic singularities presented in references [76,77].) We argue for the need for such a classification, instead of only addressing the evolution of the scale factor a and the DE density ρ, since different cosmic events such as the LR and LSBR cannot be differentiated in the latter picture.

From the continuity Equation (4) it follows that the DE density evolves as

$$\rho = \rho_0 \left[1 + \frac{3(1-\alpha)A}{\rho_0^{1-\alpha}} \ln\left(\frac{a}{a_0}\right) \right]^{\frac{1}{1-\alpha}}, \tag{6}$$

for $\alpha \neq 1$, and

$$\rho = \rho_0 \left(\frac{a}{a_0}\right)^{3A}, \tag{7}$$

for the case of $\alpha = 1$, where the subscript "0" denotes the current value of the corresponding quantity. In either case, the EoS parameter w reads

$$w = -1 - \frac{A}{\rho_0^{1-\alpha} + 3(1-\alpha)A \ln\left(\frac{a}{a_0}\right)}. \tag{8}$$

Note that for a non-negative parameter A, the denominator in the r.h.s. in Equation (8) is always positive[2], whatever the value of α. Thus, the DE content modelled by the EoS (5) exhibits a phantom-like behavior when the parameter A is positive. Please note that this can be also seen directly from Equation (4) for an expanding ($H > 0$) universe.

On the other hand, note that $\alpha = 1$ and $\alpha = \frac{1}{2}$ are special values on the α-line that delimited qualitatively alike cosmological behaviors [52]. This can be deduced from the time dependence of the scale factor obtained from Equations (2) and (6). This is [52]

$$\ln\left(\frac{a}{a_0}\right) = \frac{2\sqrt{\rho_0}}{3A}\left\{\left[1 + \frac{3A}{2\sqrt{\rho_0}}\ln\left(\frac{a_\star}{a_0}\right)\right]\exp\left(\frac{\sqrt{3}A}{2}(t - t_\star)\right) - 1\right\}, \quad (9)$$

for the case of $\alpha = \frac{1}{2}$, and [52]

$$\ln\left(\frac{a}{a_0}\right) = \frac{1}{3(\alpha - 1)\rho_0^{\alpha-1}A}\left\{1 - \left[B - \frac{\sqrt{3}}{2}(2\alpha - 1)\rho_0^{\alpha-\frac{1}{2}}A(t - t_\star)\right]^{\frac{2(\alpha-1)}{2\alpha-1}}\right\}, \quad (10)$$

for $\alpha \neq \frac{1}{2}$ and $\alpha \neq 1$, and

$$\frac{a}{a_0} = \left[\left(\frac{a_\star}{a_0}\right)^{-\frac{3A}{2}} - \frac{\sqrt{3\rho_0}}{2}A(t - t_\star)\right]^{-\frac{2}{3A}}, \quad (11)$$

for $\alpha = 1$. B is a constant defined as

$$B := \left[1 + 3(1 - \alpha)\rho_0^{\alpha-1}A\ln\left(\frac{a_\star}{a_0}\right)\right]^{\frac{2\alpha-1}{2(\alpha-1)}}, \quad (12)$$

where we have denoted by $t_\star$ some arbitrary (future) moment in the expansion history of the universe from which we can safely neglect the contribution of matter fields and, therefore, assume that DE is the only content of the cosmos. In addition, $a_\star$ represents the corresponding scale factor. Therefore, when $\alpha = \frac{1}{2}$ the scale factor asymptotically approaches a double exponential growth on the cosmic time. However, for $\alpha > \frac{1}{2}$, it evolves as some function of $t_s - t$, being t_s the time of occurrence of the corresponding singularity. Moreover, the expression for t_s depends on the value of α. Hence, for the sake of the discussion, we proceed to summarize the main results found in [52] for each case. Those are: $\alpha > 1$, $\alpha = 1$, $\frac{1}{2} < \alpha < 1$, $\alpha = \frac{1}{2}$ and $\alpha < \frac{1}{2}$. Additionally, to the conclusion presented in [52], we also compute the corresponding H and $\dot{H}$ variables in each case. This allows us to provide a complementary classification of the cosmic events found in [52] similar to that introduced in reference [65], see Table 2.

For the case of $\alpha > 1$, the universe reaches a maximum size given by

$$a_{\max} := a_0 \exp\left(\frac{1}{3(\alpha - 1)\rho_0^{\alpha-1}A}\right). \quad (13)$$

Furthermore, this size is reached in a finite time into the future

$$t_s := t_\star + \frac{2B}{\sqrt{3}(2\alpha - 1)\rho_0^{\alpha-\frac{1}{2}}A}. \quad (14)$$

At that moment, the Hubble rate and its comics time derivative diverge, since

$$H = \sqrt{\frac{\rho_0}{3}} \left[\frac{\sqrt{3}}{2} (2\alpha - 1) \rho_0^{\alpha - \frac{1}{2}} A(t_s - t) \right]^{-\frac{1}{2\alpha - 1}}, \tag{15}$$

$$\dot{H} = \frac{\rho_0^{\alpha} A}{2} \left[\frac{\sqrt{3}}{2} (2\alpha - 1) \rho_0^{\alpha - \frac{1}{2}} A(t_s - t) \right]^{-\frac{2\alpha}{2\alpha - 1}}. \tag{16}$$

This asymptotic behavior corresponds to the occurrence of a BF singularity [58,59] (see also reference [76]). That is a type III singularity in the notation of [64].

The case of $\alpha = 1$ corresponds to a constant EoS parameter $w = -1 - A$. Accordingly, the size of the observable universe becomes infinite at a finite time from present epoch, namely $t_{\rm rip}$. This is

$$t_{\rm rip} := t_\star + \frac{2}{\sqrt{3\rho_0} A} \left(\frac{a_\star}{a_0} \right)^{-\frac{3A}{2}}, \tag{17}$$

see Equation (11). Furthermore, given that

$$H = \frac{2}{3A(t_{\rm rip} - t)}, \tag{18}$$

$$\dot{H} = \frac{2}{3A(t_{\rm rip} - t)^2}, \tag{19}$$

the Hubble rate and its cosmic time derivative also diverge at $t = t_{\rm rip}$. Therefore, for this value of α, the universe evolves towards a classical BR singularity, alike to that first introduced in [12,13,50]. This corresponds to a type I singularity according to the notation in reference [64].

On the other hand, for $\frac{1}{2} < \alpha < 1$, the scale factor diverge at finite cosmic time, see Equation (10) where the ratio $2(\alpha - 1)/(2\alpha - 1)$ is now negative. This makes the $\ln a$ proportional to some power of $1/(t_{\rm rip} - t)$. The time at which the observable universe becomes infinite is

$$t_{\rm rip} := t_\star + \frac{2B}{\sqrt{3}(2\alpha - 1) \rho_0^{\alpha - \frac{1}{2}} A}. \tag{20}$$

The Hubble rate and its time derivative follows the same relations given in Equations (15) and (16), respectively. Hence, these quantities also diverge along with the scale factor. Therefore, in a finite time into the future, the scale factor, the Hubble rate and $\dot{H}$ explode, whereas the EoS parameter w converges to -1 from below. Of course, this implies that the DE density and pressure blow up, as was found in [52]. This event is qualitatively equivalent to a BR singularity (see the singularities classified as type I in, for example, references [63–65]). Please note that this behavior was also found in reference [76], where it was dubbed "grand rip" (see type -1 singularities in reference [77]).

Another possible choice for α corresponds to the interesting case of $\alpha = \frac{1}{2}$. In this scenario, the scale factor asymptotically evolves as an exponential function of an exponential function on the cosmic time, i.e., $a \approx e^{e^t}$, see Equation (9). Accordingly, the Hubble rate and its cosmic time derivative are

$$H(t) = \sqrt{\frac{\rho_0}{3}} \left(1 + \frac{3A}{2\sqrt{\rho_0}} \ln \frac{a_\star}{a_0} \right) \exp \left[\frac{\sqrt{3}}{2} A(t - t_\star) \right], \tag{21}$$

$$\dot{H}(t) = \frac{A}{2} \sqrt{\rho_0} \left(1 + \frac{3A}{2\sqrt{\rho_0}} \ln \frac{a_\star}{a_0} \right) \exp \left[\frac{\sqrt{3}}{2} A(t - t_\star) \right]. \tag{22}$$

Thus, the scale factor, H, and $\dot{H}$ diverge at the infinite asymptotic future. This drives the universe towards a LR abrupt event, see classification in [63,65]. (See also reference [53].) In fact, the EoS (5) with $\alpha = \frac{1}{2}$ corresponds to the DE model for which the name "little rip" was first given [51], even though this cosmological behavior was already known from before [52] (see also [53] where the LR was found in brane cosmology and before that in some modified theories of gravity [78]).

Finally, for the case of $\alpha < \frac{1}{2}$, the scale factor obeys the relation given in Equation (10). However, since now the ratio $2(\alpha - 1)/(2\alpha - 1)$ is positive, then the equation for $\ln a$ reduces to a certain polynomial on the cosmic time. This makes the expansion of the observable universe to last indefinitely. Hence, no finite time singularities are present in this case. The Hubble rate and its cosmic time derivative read

$$H = \sqrt{\frac{\rho_0}{3}} \left[B + \frac{\sqrt{3}(1 - 2\alpha)A}{2\rho_0^{\frac{1}{2}-\alpha}} (t - t_\star) \right]^{\frac{1}{1-2\alpha}}, \qquad (23)$$

$$\dot{H} = \frac{A}{2} \rho_0^\alpha \left[B + \frac{\sqrt{3}(1 - 2\alpha)A}{2\rho_0^{\frac{1}{2}-\alpha}} (t - t_\star) \right]^{\frac{2\alpha}{1-2\alpha}}. \qquad (24)$$

For $0 < \alpha < \frac{1}{2}$, both quantities tend to infinity at the infinite asymptotic future, thus, leading to the occurrence of a LR abrupt event. On the contrary, $\dot{H}$ remains constant for $\alpha = 0$, or shrinks to zero when $\alpha < 0$. This behavior corresponds to a final fate à la LSBR, see reference [55]. Please note that since the DE density grows unbounded in both abrupt events, the LR and the LSBR, see Equation (6), this distinction could not be done analyzing only the asymptotic behavior of a and ρ. Additionally note that given some entire number n, all the higher order derivatives of H up to the n-th order diverge when $\alpha = (n-1)/2n$.

Therefore, we conclude that the phantom DE model described by the EoS (5) with a positive parameter A entails a great variety of cosmological singularities and abrupt events. We summarized the results from reference [52] and our new findings here in Table 2. We will present a more detailed analysis in another work.

Table 2. Classification of the singularities and abrupt events found for the phantom DE model given by the general EoS (5), see reference [52], accordingly to the value of the parameter α, the time of occurrence of the singular behavior t_s, the Hubble rate H and its cosmic time derivative $\dot{H}$. Please note that since the BR and grand rip [76] singularities have qualitatively the same behavior in terms of those quantities, we do not address the possible differences between both events in the following classification. Hence, we keep the term "big rip" for both of them.

α	t_s	a	H	$\dot{H}$	Event
$1 < \alpha$	Finite	Finite	∞	∞	Big freeze
$\frac{1}{2} < \alpha \leq 1$	Finite	∞	∞	∞	Big rip
$0 < \alpha \leq \frac{1}{2}$	∞	∞	∞	∞	Little rip
$\alpha \leq 0$	∞	∞	∞	Finite	LSBR

Cosmological Constraints

Since this review is mainly devoted to the analysis of the classical and quantum fate of the BR, the LR and the LSBR events in metric $f(R)$ theories of gravity, we should now restrict our attention to some particular phantom DE models on which to apply the reconstruction techniques in the next sections. Therefore, hereafter we shall consider only the following phantom DE models when addressing the occurrence of these cosmic events:

- For the BR singularity we consider the phantom DE model with a constant EoS parameter $w < -1$ [50]. This model corresponds to the choice of $\alpha = 1$ in the more general EoS (5) studied in the previous section.

- For a DE model with a future LR abrupt event, we select the EoS for DE described in reference [51], which corresponds to the case $\alpha = \frac{1}{2}$ in (5).
- For a universe doomed to evolve towards a LSBR abrupt event, we consider the DE content to be described by the EoS (5) with $\alpha = 0$. This LSBR was first introduced in reference [55].

Note that the BR model here considered is just a subcase of the more general wCDM scenario, which has been thoroughly analyzed in the literature (see, for example, references [37–45], among others). More importantly, these specific phantom DE models have been shown to be compatible with the current observational data; see, for instance, references [37–39,51,56,57,79]. Therefore, our own universe may evolve towards some of these singular fates.

In the following, for the sake of concreteness, we shall consider the cosmological constraints obtained in reference [57] when binding those DE models with observational data. Thus, the results presented in the incoming sections are subjected to the observational constraints on the parameter A presented in reference [57]. These constraints are summarized in Table 3. Please note that the small values for A there obtained suggest that tiny deviations from the ΛCDM scenario are, in fact, the observationally preferred situation today [57]. Nevertheless, we recall that the asymptotic evolution of these DE models is not that of a de Sitter universe, since the corresponding DE pressure and energy density do not converge to a constant value but diverge. In the next section, we will obtain the group of metric $f(R)$ theories of gravity able to reproduce the same asymptotic expansion history, which in GR corresponds to these particular phantom DE models.

Table 3. Best fit found in reference [57] for the phantom DE models discussed in Section 2, where A is dimensionless for the case of the BR, and has units of inverse of meter and inverse of square meter for the LR and LSBR models, respectively.

	α	A
BR	1	0.0276 ± 0.0240
LR	$\frac{1}{2}$	$(2.75 \pm 1.30) \times 10^{-28}$
LSBR	0	$(2.83 \pm 4.17) \times 10^{-54}$

3. Phantom Dark-Energy Models in $f(R)$ Cosmology

For a given cosmological background evolution it is possible to find a family of alternative theories of gravity that leads to the same expansion history. The group of techniques used to perform such a background "reconstruction" task are dubbed as "reconstruction methods" (for a review of the topic see, for instance, reference [80] and references therein). In this part of the review, we shall focus our attention on reconstruction methods within the scheme of metric $f(R)$ theories of gravity. Hence, we look for a metric $f(R)$ theory of gravity able to reproduce the same superaccelerated expansion profile to that of the general relativistic model filled with phantom DE with the EoS (5), for the values of α selected in Section 2. It is worth noting that whereas in GR the accelerated phase is attributed to the existence of an exotic form of energy with a negative pressure (DE), in the setup of $f(R)$ theories of gravity the same expansion has a rather geometrical origin. Furthermore, as the general relativistic model with the EoS (5) expands the cosmos towards some future doomsdays, then the resulting metric $f(R)$ theory of gravity will lead to the same classical fate. For previous works on reconstruction techniques in $f(R)$ gravity see, for instance, references [80–86]. See also references [87–91] for successful reconstruction of phantom DE-driven riplike events in metric $f(R)$ theories of gravity.

Henceforth, we shall refer to two cosmological evolutions as equivalent at the background level if the corresponding geometrical variables H, $\dot{H}$, R and $\dot{R}$ are identical [87]. In

GR, the expansion of a homogeneous and isotropic universe is ruled by the Friedmann and Raychaudhuri Equations (2) and (3), respectively. Accordingly, the scalar curvature reads

$$R = 6\left(\dot{H} + 2H^2 + \frac{k}{a^2}\right) = \rho - 3p, \tag{25}$$

where ρ and p denotes the total energy density and pressure, respectively, these are $\rho = \rho_{DE} + \rho_M$ and $p = p_{DE} + p_M$. Later we will assume that they just correspond to the DE component when the cosmic matter is diluted. Additionally, from the continuity equation for the perfect fluid (4) and the Friedmann Equation (2), it follows

$$\dot{\rho} = -3(p+\rho)\left(\frac{1}{3}\rho - \frac{k}{a^2}\right)^{\frac{1}{2}}, \tag{26}$$

$$\dot{p} = -3(p+\rho)\left(\frac{1}{3}\rho - \frac{k}{a^2}\right)^{\frac{1}{2}}\frac{dp}{d\rho}, \tag{27}$$

where we have assumed $p = p(\rho)$. Consequently, the cosmic time derivative of the scalar curvature R reads

$$\dot{R} = -3(p+\rho)\left(\frac{1}{3}\rho - \frac{k}{a^2}\right)^{\frac{1}{2}}\left(1 - 3\frac{dp}{d\rho}\right). \tag{28}$$

On the other hand, for the gravitational interaction being that provided by metric $f(R)$ theories of gravity, the evolution of the universe is described by the action

$$S = \frac{1}{2}\int d^4x \sqrt{-g} f(R) + S_m, \tag{29}$$

where S_m stands for the minimally coupled matter fields. In this framework, the field equations no longer coincide with (2) and (3). In fact, the modified Friedmann equation reads

$$3H^2 \frac{df}{dR} = \frac{1}{2}\left(R\frac{df}{dR} - f\right) - 3H\dot{R}\frac{d^2f}{dR^2} - 3\frac{k}{a^2} + \rho_m, \tag{30}$$

being ρ_m the energy density of the minimally coupled matter fields. Since we are interested in a metric $f(R)$ theory able to reproduce the same background cosmological expansion as a certain general relativistic model, then the preceding expression can be considered to be a differential equation for some, a priori unknown, function $f(R)$, where the coefficients are already fixed, i.e., when the geometrical quantities involved in Equation (30) are set to be equal to those of the GR model that we want to reproduce. Thus, the background cosmological expansion of the resulting metric $f(R)$ theory of gravity will be equivalent to that provided by the general relativistic model. Furthermore, as we are interested on the asymptotic future behavior of the universe, we can neglect the matter and spatial curvature contribution in Equations (25) and (30), which will be (more) quickly redshifted with the superaccelerated expansion. Therefore, we drop again the subindex DE for the energy density and pressure. Hence, in the following sections we present the most general solution to this reconstruction procedure when considering the EoS (5) for the different values of α discussed in Section 2. Nevertheless, the obtained $f(R)$ theories must be understood as useful asymptotic models for the study of the behavior of the universe near classical singularities, since a more realistic background reconstruction would imply the non-cancellation of the matter and spatial curvature contribution in Equations (25) and (30).

3.1. BR Singularity in $f(R)$ Gravity

For the BR singularity we considered the case of $\alpha = 1$ in the DE general EoS (5). This is

$$p = -(1+A)\rho. \tag{31}$$

For that specific value of α, the cosmic time derivative of the Hubble rate can be re-expressed as

$$\dot{H} = \frac{3}{2}AH^2, \tag{32}$$

compare Equations (18) and (19). Consequently, the scalar curvature and its cosmic time derivative reduces to

$$R = 3(4+3A)H^2, \tag{33}$$
$$\dot{R} = 9(4+3A)AH^3. \tag{34}$$

Therefore, the modified Friedmann Equation (30) reads

$$R^2 f_{RR} - \frac{2+3A}{6A} R f_R + \frac{4+3A}{6A} f = 0, \tag{35}$$

where we have used the notation $f_R := df/dR$ and $f_{RR} := d^2f/dR^2$. The most general solution for the function $f(R)$ was already obtained in reference [87]. That is

$$f(R) = c_+ R^{\gamma_+} + c_- R^{\gamma_-}, \tag{36}$$

being c_+ and c_- arbitrary integration constants and

$$\gamma_\pm := \frac{1}{2} \left\{ 1 + \frac{2+3A}{6A} \pm \sqrt{\left[1 + \frac{2+3A}{6A}\right]^2 - \frac{2(4+3A)}{3A}} \right\}. \tag{37}$$

In general, the parameter $\gamma_\pm$ may take complex values. However, for $\{A \leq (10 - 4\sqrt{6})/3\} \cup \{A \geq (10 + 4\sqrt{6})/3\}$, both branches are real valued. Please note that this is precisely the case for the values of A showed in Table 3. Hence, when the observational constraints [57] on the model are taken into account, both branches of the parameter $\gamma_\pm$ are real valued.

3.2. The LR in $f(R)$ Gravity

For the LR abrupt event, we consider here the EoS (5) for DE with $\alpha = \frac{1}{2}$ in (5). This is

$$p = -\rho - A\sqrt{\rho}. \tag{38}$$

Please note that for this choice of α, the Hubble rate in Equation (21) is an exponential function on the cosmic time. Subsequently, its cosmic time derivative is proportional to itself. Thus, for the sake of simplicity, we denoted by β that proportionality constant. That is

$$\dot{H} = \beta H, \tag{39}$$

where we have defined

$$\beta := \frac{\sqrt{3}}{2} A. \tag{40}$$

The curvature scalar and its cosmic time derivative in terms of the Hubble rate are

$$R = 6H(\beta + 2H), \quad (41)$$
$$\dot{R} = 6\beta H(\beta + 4H). \quad (42)$$

As noted in reference [88] (se also [91]), the modified Friedmann Equation (30) simplifies when rewritten in terms of H. Thus, we need to solve

$$\beta H^2(\beta + 4H)f_{HH} - \left[4\beta H^2 + H(\beta + H)(\beta + 4H)\right]f_H + (\beta + 4H)^2 f = 0, \quad (43)$$

where $f_H := df/dH$, $f_{HH} := d^2f/dH^2$, and ρ_m and k has been neglected, as discussed before. The most general solution to the above differential equation is [88]

$$f(H) = C_1\left(H^4 - 5\beta H^3 + 2\beta^2 H^2 + 2\beta^3 H\right) + C_2\left[\beta H\left(\beta^2 + 4\beta H - H^2\right)e^{\frac{H}{\beta}}\right. $$
$$\left. + \left(H^4 - 5\beta H^3 + 2\beta^2 H^2 + 2\beta^3 H\right)\text{Ei}\left(\frac{H}{\beta}\right)\right], \quad (44)$$

being C_1 and C_2 integration constants and Ei the exponential integral function (see definition, e.g., in 5.1.2 of reference [92]). To obtain the final $f(R)$ expression, the relation (41) must be inverted. This is

$$H = \frac{1}{12}\left(-3\beta \pm \sqrt{9\beta^2 + 12R}\right). \quad (45)$$

At a first sight, it seems that this transformation may be non-univocally defined. However, since the phantom DE density (6) increases with the expansion of the universe towards the LR abrupt event, whereas the matter content is quickly diluted, then the corresponding Hubble rate (21) and scalar curvature (41) also grow. Hence, only the positive branch in the preceding expression applies. Therefore,

$$f(R) = c_1\left[27\beta^4 + 150\beta^2 R - \beta\left(9\beta^2 + 12R\right)^{\frac{3}{2}} + 2R^2\right]$$
$$+ c_2\left\{\beta\left(-3\beta^2 - 2R + 9\beta\sqrt{9\beta^2 + 12R}\right)\right.$$
$$\times \left(-3\beta + \sqrt{9\beta + 12R}\right)\exp\left(-\frac{1}{4} + \frac{1}{4}\sqrt{1 + \frac{4R}{3\beta}}\right)$$
$$\left. + \left[27\beta^4 + 150\beta^2 R - \beta\left(9\beta^2 + 12R\right)^{\frac{3}{2}} + 2R^2\right]\text{Ei}\left(-\frac{1}{4} + \frac{1}{4}\sqrt{1 + \frac{4R}{3\beta}}\right)\right\} \quad (46)$$

represents the most general family of metric $f(R)$ theories of gravity predicting the occurrence of that specific LR abrupt event [88] (see also [91]). We recall that β, which depends on the parameter A, is observationally constrained in Table 3.

3.3. The LSBR in $f(R)$ Gravity

The selected DE model in this case corresponds to $\alpha = 0$ in the EoS (5). That is

$$p = -\rho - A, \quad (47)$$

where A takes the value showed in Table 3. In this scenario, see Equation (24), the cosmic time derivative of the Hubble rate remains constant. Its value is given by

$$\dot{H} = \frac{A}{2}. \quad (48)$$

Hence, the scalar curvature and its cosmic time derivative read

$$R = 12H^2 + 3A, \qquad (49)$$

$$\dot{R} = 12AH. \qquad (50)$$

Consequently, the modified Friedmann Equation (30) becomes

$$3A(R-3A)f_{RR} - \frac{1}{4}(R+3A)f_R + \frac{1}{2}f = 0, \qquad (51)$$

whose most general solution reads [90]

$$f(R) = c_1\left(9A^2 - 18AR + R^2\right) + c_2\left(\frac{R-3A}{12A}\right)^{\frac{3}{2}} {}_1F_1\left(-\frac{1}{2};\frac{5}{2};\frac{R-3A}{12A}\right), \qquad (52)$$

where c_1 and c_2 are integration constants and ${}_1F_1$ is the confluent hypergeometric functions or Kummer's function, see definition in chapter 13 of reference [92]. See Table 3 for the possible values of A we shall consider here.

3.4. Viability and Local System Tests

It is a well-known fact that local tests pose rather tight constraints on the metric formulation of $f(R)$ theories of gravity (see, for example, references [93–99]). Thus, any candidate for a reliable alternative to GR should pass or, somehow, evade these low-curvature-regime tests (see also [100–105] for an interesting discussion). However, the metric $f(R)$ models presented here were expressly built to reproduce a high curvature regime very different from that of (an effective) ΛCDM. Please note that the contributions of the matter (dark and baryonic) and the spatial curvature k to Equations (25) and (30) have been de facto ignored. Therefore, the resulting $f(R)$ theories obtained here must be seen as useful asymptotic models for the theoretical evaluation of the quantum fate of classical singularities rather than complete proposals for viable alternatives to GR at all scales.

4. $f(R)$ Quantum Geometrodynamics

Even though there is a lack of consensus on what is the correct quantum theory of gravity, the application of ordinary quantum mechanics to the universe as a whole leads to an interesting framework, known as quantum cosmology (for a review of the topic see, e.g., references [106,107]). Currently there are multiple proposals to quantize the cosmological background. A non-exhaustive listing of different approaches to quantum cosmology is string theory, loop quantum cosmology [108,109], causal dynamical triangulation [110–112] and quantum geometrodynamics [74], among other examples (see also reference [106]). Nonetheless, in this review we shall focus only on the latter approach, which corresponds, in fact, to one of the first attempts to quantize cosmological backgrounds [74]. In the De-Witt's pioneering work [74], a quantization procedure for a closed Friedmann universe was provided, leading to the first minisuperspace model in quantum cosmology. This quantum cosmology is based on a canonical quantization with the Wheeler-DeWitt equation for the wave function Ψ of the universe playing a central role [74,113,114]. The expression minisuperspace stands for a cosmological model truncated to a finite number of degrees of freedom. This nomenclature is derived from the usage of "superspace" to denote the full infinite-dimensional configuration space of GR and the prefix "mini" for the truncated versions. Moreover, DeWitt also proposed a criterion for the avoidance of classical singularities within the quantum regime, namely the DW criterion. This is, the classical singularity is potentially avoided if the wave function of the universe vanishes in the nearby configuration space. This criterion is based on a generalization of the interpretation of the wave function in ordinary quantum theory, where the wave function is the fundamental building block for any observable. Consequently, regions of the configuration space that lay outside of the support of Ψ are, therefore, irrelevant in practice[3]. It should be noted, however, that the non-vanishing of the wave function does not necessary entail a singularity. Therefore, the DW criterion can only be a sufficient but not necessary criterion for the avoidance of singularities. This criterion has been successfully applied in several cosmological scenarios, see, e.g., references [65,66,71–73,89,90,115–117] among others.

4.1. Modified Wheeler-DeWitt

Following the ideas presented in reference [75], the Wheeler-DeWitt equation must be adapted to the framework of $f(R)$ theories of gravity. It is worthy to note that when investigating the canonical quantization of an $f(R)$ theory we are interpreting that theory as a fundamental theory of gravity, rather than an effective framework coming from a quantum gravity proposal. As this classical theory of gravity implies the occurrence of singularities, one concludes that it must be quantized. In this section, we shall show how the formalism of quantum geometrodynamics can be adopted to perform such quantization. The resulting scheme is known as $f(R)$ quantum geometrodynamics.

For the gravitational interaction being that provided by metric $f(R)$ theories of gravity, cosmological models can be described by the action

$$S = \frac{1}{2} \int d^4x \sqrt{-g} f(R), \tag{53}$$

in the so-called Jordan frame. For a FLRW background metric, the above action reduces to

$$S = \frac{1}{2} \int dt \, \mathcal{L}(a, \dot{a}, \ddot{a}), \tag{54}$$

where the point-like Lagrangian reads

$$\mathcal{L}(a, \dot{a}, \ddot{a}) = \mathcal{V}_{(3)} a^3 f(R), \tag{55}$$

denoting by $\mathcal{V}_{(3)}$ the spatial three-dimensional volume. Please note that metric $f(R)$ theories of gravity have an additional degree of freedom when compared with GR (see Einstein's and Jordan's frame formulation in [118–121] and references therein). Consequently, a new variable can be introduced for the canonical quantization of these alternative theories of gravity. Furthermore, this new variable can be selected in such a way to remove the dependence of the action (54) on the second derivatives of the scale factor. Hence, following the line of reasoning presented in reference [75], we select the scalar curvature, R, to be the new variable. Then, the action (54) becomes

$$S = \frac{1}{2} \int dt \, \mathcal{L}(a, \dot{a}, R, \dot{R}). \tag{56}$$

However, since the scalar curvature and the scale factor are not independent (at the classical level), their relation needs to be properly introduced in the theory via a Lagrange multiplier, μ, for the classical constraint $R = R(a, \dot{a}, \ddot{a})$ given in Equation (25). Thence,

$$\mathcal{L} = \mathcal{V}_{(3)} a^3 \left\{ f(R) - \mu \left[R - 6 \left(\frac{\ddot{a}}{a} + \frac{\dot{a}^2}{a^2} + \frac{k}{a^2} \right) \right] \right\}. \tag{57}$$

The Lagrange multiplier can be determined by varying the action with respect to the scalar curvature. This leads to

$$\mu = f_R(R). \tag{58}$$

Accordingly, the point-like Lagrangian (57) can be reformulated as

$$\mathcal{L}(a,\dot{a},R,\dot{R}) = \mathcal{V}_{(3)}\left\{a^3\left[f(R) - Rf_R(R)\right] - 6a^2 f_{RR}(R)\dot{a}\dot{R} + 6af_R(R)(k-\dot{a}^2)\right\}. \quad (59)$$

For the sake of the quantization procedure, the derivative part of the above point-like Lagrangian can be diagonalized by the introduction of a new set of variables [75]. These are

$$q := a\sqrt{R_\star}\left(\frac{f_R}{f_{R_\star}}\right)^{\frac{1}{2}}, \quad (60a)$$

$$x := \frac{1}{2}\ln\left(\frac{f_R}{f_{R_\star}}\right), \quad (60b)$$

being $R_\star$ a constant needed for the change of variables to be well-defined. There are different proposals among the existing literature for the value of this constant, see, for instance, references [75,89–91]. In this work we adopt the convention discussed in reference [91], where $R_\star$ is defined as the value of the curvature scalar evaluated at some future scale factor $a = a_\star$ on which the description of the universe by means of DE only becomes appropriate. That is

$$R_\star := R(a,\dot{a},\ddot{a})\Big|_{a=a_\star}, \quad (61)$$

where $R(a,\dot{a},\ddot{a})$ obeys the classical relation (25). Furthermore, for the sake of concreteness, we can safely assume $a_\star = 100 a_0$ hereafter. Since at that moment in the expansion the matter content will be diluted by a factor of 10^{-6} with respect to the present concentration and, therefore, $\Omega_{DE} \approx 1$. Please note that from this definition it follows $R > R_\star$, since the scalar curvature asymptotically approaches an increasing function with the expansion of the universe. For different definitions of $R_\star$ see references [75,89,90]. In these new variables, Equation (59) transforms into

$$\mathcal{L}(q,\dot{q},x,\dot{x}) = \mathcal{V}_{(3)}\left(\frac{R_\star f_R}{f_{R_\star}}\right)^{-\frac{3}{2}} q^3\left[f - 6f_R\frac{\dot{q}^2}{q^2} - Rf_R + 6f_R\dot{x}^2 + 6k\frac{R_\star}{f_{R_\star}}\frac{f_R^2}{q^2}\right], \quad (62)$$

where f and f_R are now understood as functions of x. This form of the Lagrangian is already suitable for the quantization procedure.

Since the kinetic part of the point-like Lagrangian has been diagonalized, the derivation of the corresponding Hamiltonian is straightforward. The conjugate momenta are

$$P_q = \frac{\partial \mathcal{L}}{\partial \dot{q}} = -12\mathcal{V}_{(3)}R_\star^{-\frac{3}{2}}f_{R_\star}^{\frac{3}{2}}f_R^{-\frac{1}{2}}q\dot{q}, \quad (63)$$

$$P_x = \frac{\partial \mathcal{L}}{\partial \dot{x}} = 12\mathcal{V}_{(3)}R_\star^{-\frac{3}{2}}f_{R_\star}^{\frac{3}{2}}f_R^{-\frac{1}{2}}q^3\dot{x}. \quad (64)$$

Then, the corresponding Hamiltonian reads

$$\mathcal{H} = -\mathcal{V}_{(3)}q^3\left(\frac{R_\star f_R}{f_{R_\star}}\right)^{-3/2}\left\{f + 6k\frac{R_\star}{f_{R_\star}}\frac{f_R^2}{q^2} - Rf_R + \frac{6R_\star^3}{(12)^2 \mathcal{V}_{(3)}^2 f_{R_\star}^3}\frac{f_R^2}{q^4}\left[P_q^2 - \frac{P_x^2}{q^2}\right]\right\}. \quad (65)$$

For the quantization of the theory, we assume the procedure

$$P_q \to -i\hbar\frac{\partial}{\partial q}, \quad (66)$$

$$P_x \to -i\hbar\frac{\partial}{\partial x}. \quad (67)$$

Therefore, the Hamiltonian (65) of the classical system is promoted to a quantum operator. Consequently, the classical Hamiltonian constraint becomes the modified Wheeler-DeWitt (mWDW) equation for the wave function Ψ of the universe [74,75,106]. That is

$$\hat{\mathcal{H}}\Psi = 0. \tag{68}$$

After simple manipulations, the preceding expression can be cast in the form of the hyperbolic differential equation [75]

$$\left[\hbar^2 q^2 \frac{\partial^2}{\partial q^2} - \hbar^2 \frac{\partial^2}{\partial x^2} - V(q,x)\right]\Psi(q,x) = 0, \tag{69}$$

where the factor-ordering parameter has been set equal to zero. The effective potential entering the preceding equation is given by

$$V(q,x) = \frac{q^4}{\lambda^2}\left(k + \frac{q^2}{6R_\star f_{R_\star}}(f - Rf_R)e^{-4x}\right), \tag{70}$$

being $\lambda := R_\star/(12\mathcal{V}_{(3)}f_{R_\star})$. Please note that for a given $f(R)$ expression, the variables q and x are univocally fixed. Then, the relation in (60b) must be reversed to express f and Rf_R in terms of x. However, this may not always be possible analytically, limiting the non-numerical evaluation of the mWDW equation (69) to some suitable group of $f(R)$ theories. In the next sections, we will review the results for the wave function Ψ presented in the literature when considering the $f(R)$ expressions previously found by means of the reconstruction techniques (see Sections 3.1–3.3). We recall that those metric $f(R)$ theories of gravity lead to the same asymptotic background expansion as their respective general relativistic phantom DE models and, therefore, they predict (at the classical level) singular cosmological behaviors. Accordingly, we will neglect the contribution of the spatial curvature k in the effective potential (70) when studying the asymptotic fate of those models, since it will be quickly redshifted with the superacceletared expansion.

Before proceeding further, we want to address some comments on the structure of the mWDW equation (69). First, we want to emphasize the well-known ambiguity in the factor ordering in Equation (69), i.e., we could have chosen a different factor ordering when applying the quantization procedure on the Hamiltonian of the classical theory, which could have led to a different wave function of the universe. According to reference [75], a variation of the factor ordering affects only the pre-exponential factor of the semiclassical wave function. Thus, the actual value of this parameter can be considered to be unimportant for the evaluation of the DW criterion, which is the main goal of this review. In fact, the DW criterion was found to be satisfied independently of the chosen factor ordering for some particular models related to the topic of this review, see references [72,73,122,123]. On the other hand, note that the mWDW equation (69) is a globally hyperbolic differential equation, i.e., the signature of the minisuperspace DeWitt metric,

$$G^{AB} = \begin{pmatrix} q^2 & 0 \\ 0 & -1 \end{pmatrix}, \tag{71}$$

is $(+,-)$. This is quite different to what happens in GR when the DE content is described by a minimally coupled phantom scalar field. In that case, the DeWitt metric has a positive signature and, therefore, the WDW equation is of an elliptic type. Examples of these models can be found, for instance, in references [66,71–73]. Additionally, a change of signature in the WDW equation has also been noticed in the presence of non-minimally coupled scalar fields [124]. Moreover, the change in the signature has implications in the imposition of boundary conditions. Whereas the hyperbolic equation has a wave-like solution and, therefore, a well-posed initial value problem, a perturbation in the initial or boundary condition for an elliptic (or parabolic) equation will spread instantly over all the point in that domain.

4.2. The BR in $f(R)$ Quantum Cosmology

In sight of the form of expression (37), only the negative branch gives the expected limit to a de Sitter universe when the parameter A shrinks to zero. (On the contrary, the exponent γ_+ diverge when A vanishes) Since small deviations from the EoS of a cosmological constant are the observationally preferred situation today [57] (see also references [37–45], among others), hereafter we consider only the negative branch of the solution presented in Equation (36). This is, we assume $c_+ = 0$ in expression (36). Therefore, in this section, we quantize a subclass of the more general family of metric $f(R)$ theories of gravity predicting the classical occurrence of a BR singularity. That is

$$f(R) = c_- R^{\gamma_-}. \tag{72}$$

In the following of this section, we drop the subindex "-" for the sake of the notation but keeping in mind that the preceding expression corresponds only to one of the two independent solutions obtained from the reconstruction procedure. Additionally, note that cosmological constraints on the model [57] (see also references [37,38]) satisfy the condition $A \leq (10 - 4\sqrt{6})/3 \approx 0.067$. Therefore, those constraints favor a real valued exponent in Equation (72) [see discussion below Equation (37)]. Hence, hereafter we shall consider that the aforementioned exponent is real for values of A of physical interest. Additionally, note that the quantum fate of the BR singularity in $f(R)$ gravity has already been studied in the literature for some particular values for A, finding that the DW criterion can be satisfied, see reference [89].

For the choice of the $f(R)$ gravity in (72), the change of variables in Equation (60) reads

$$q = a\sqrt{R_\star}\left(\frac{R}{R_\star}\right)^{\frac{\gamma-1}{2}}, \tag{73a}$$

$$x = \frac{1}{2}\ln\left(\frac{R}{R_\star}\right)^{\gamma-1}. \tag{73b}$$

Moreover, from the evolution equations for the Hubble rate and its cosmic time derivative, and the definition adopted in expression (61), it follows that the constant $R_\star$ is

$$R_\star = (4 + 3A)100^{3A}\rho_0. \tag{74}$$

Accordingly, the effective potential (70) entering the mWDW equation (69) becomes

$$V(q,x) = -\frac{\gamma-1}{6\lambda^2\gamma}e^{-Cx}q^6, \tag{75}$$

where, for the sake of the calculation, we have adopted the notation

$$C := 2\frac{\gamma-2}{\gamma-1}. \tag{76}$$

Thus, the mWDW equation (69) reads

$$\left[\hbar^2 q^2 \frac{\partial^2}{\partial q^2} - \hbar^2 \frac{\partial^2}{\partial x^2} + \frac{\gamma-1}{6\lambda^2\gamma}e^{-Cx}q^6\right]\Psi(q,x) = 0. \tag{77}$$

However, the dependence of the effective potential on the minisuperspace variables can be simplified when considering the following change of variables [89] (see also reference [73])

$$q = r(z)\theta, \tag{78a}$$

$$x = z. \tag{78b}$$

Please note that this implies [73]

$$\frac{\partial^2}{\partial q^2} = r^{-2}\frac{\partial^2}{\partial\theta^2}, \tag{79}$$

$$\frac{\partial^2}{\partial x^2} = \left(\frac{r'}{r}\right)^2\left[\theta^2\frac{\partial^2}{\partial\theta^2} + \theta\frac{\partial}{\partial\theta}\right] - 2\frac{r'}{r}\theta\frac{\partial}{\partial\theta}\frac{\partial}{\partial z} + \left[\left(\frac{r'}{r}\right)^2 - \frac{r''}{r}\right]\theta\frac{\partial}{\partial\theta} + \frac{\partial^2}{\partial z^2}, \tag{80}$$

Then, for the choice of $r(z) = e^{Cz/6}$, the potential term will only depend on θ [89]. Accordingly, the mWdW equation (77) simplifies to

$$\left[\left(1 - \frac{C^2}{36}\right)\hbar^2\theta^2\frac{\partial^2}{\partial\theta^2} - \hbar^2\frac{\partial^2}{\partial z^2} + \frac{C}{3}\hbar^2\theta\frac{\partial}{\partial\theta}\frac{\partial}{\partial z} - \frac{C^2}{36}\hbar^2\theta\frac{\partial}{\partial\theta} + \frac{\gamma-1}{6\lambda^2\gamma}\theta^6\right]\Psi(\theta,z) = 0, \tag{81}$$

where the factor $1 - C^2/36$ is different from zero at least for the range of values for A of our interest, i.e., $A \leq 0.067$ [see discussion below Equation (72)]. Therefore, to analyze whether the BR singularity is avoided in the quantum realm by means of the DW criterion, we focus on solving the preceding equation for the wave function Ψ in the configuration space near the singularity. As we do not expect the wave function to be peaked along the classical trajectory in this regime, R and a may take completely independent values. Therefore, to consider a region close to the BR singularity, we should assume either $a \to \infty$ or $R \to \infty$. Both choices imply $\theta \to \infty$, but in the former case z is arbitrary whereas in the latter one $z \to \infty$. Please note that the divergence of the scalar curvature can be argued to be the dominant condition, from a geometric point of view, for the occurrence of the BR singularity. Thence, we shall consider both θ and z going to infinity as the main parametrization of the BR singularity in the configuration space. Nevertheless, the results and conclusions presented in this section are independent of this choice and still hold for $\theta \to \infty$ and z arbitrary.

Additionally, further simplifications can be made when solving (81) close to the BR. By considering the third term containing the cross partial derivatives to be unimportant when $\theta \to \infty$, the above equation can be solved via the separation ansatz

$$\Psi(\theta,z) = \sum_{\tilde{k}} b_{\tilde{k}}\chi_{\tilde{k}}(\theta)\varphi_{\tilde{k}}(z), \tag{82}$$

where $b_{\tilde{k}}$ gives the amplitude of each solution and $\tilde{k}$ is related to the associated energy. Please do not confuse $\tilde{k}$ with the spatial curvature k, which has been set to zero. (The validity of this approximation is analyzed in Appendix A.) Under these approximations, Equation (81) reduces to the following system of equations[4]

$$\hbar^2\frac{\partial^2\varphi_{\tilde{k}}}{\partial z^2} - \tilde{k}^2\varphi_{\tilde{k}} = 0, \tag{83}$$

$$\left(1 - \frac{C^2}{36}\right)\hbar^2\theta^2\frac{\partial^2\chi_{\tilde{k}}}{\partial\theta^2} - \frac{C^2}{36}\hbar^2\theta\frac{\partial\chi_{\tilde{k}}}{\partial\theta} + \left(\frac{\gamma-1}{6\lambda^2\gamma}\theta^6 - \tilde{k}^2\right)\chi_{\tilde{k}} = 0. \tag{84}$$

The former equation can be straightforwardly worked out. The solutions are

$$\varphi_{\tilde{k}}(z) = d_1\exp\left(\frac{\sqrt{\tilde{k}^2}}{\hbar}z\right) + d_2\exp\left(-\frac{\sqrt{\tilde{k}^2}}{\hbar}z\right), \tag{85}$$

being d_1 and d_2 arbitrary constants. This corresponds to either exponential or trigonometric functions on z, depending whether $\tilde{k}^2$ is positive or negative, respectively. On the other hand, Equation (84) can be solved in an exact way by means of Bessel functions, cf. 9.1.53 of reference [92]. This solution can be expressed as

$$\chi_{\tilde{k}}(\theta) = \theta^{\frac{18}{36-C^2}}\left[u_1J_\nu\left(\frac{\tilde{\lambda}}{3\hbar}\theta^3\right) + u_2Y_\nu\left(\frac{\tilde{\lambda}}{3\hbar}\theta^3\right)\right], \tag{86}$$

295

where u_1 and u_2 are integration constants, J$_\nu$ and Y$_\nu$ are the Bessel functions of first and second order, respectively, and

$$\nu^2 := \frac{36}{(36-C^2)^2}\left[1 + 4\frac{\tilde{k}^2}{\hbar^2}\left(1 - \frac{C^2}{36}\right)\right], \tag{87}$$

$$\tilde{\lambda}^2 := \frac{6(\gamma-1)}{\lambda^2 \gamma (36-C^2)}. \tag{88}$$

When near the BR singularity, i.e., for large θ, the $\chi_{\tilde{k}}$ part of the wave function Ψ behaves asymptotically as

$$\chi_{\tilde{k}}(\theta) \approx \sqrt{\frac{6\hbar}{\pi\tilde{\lambda}}}\, \theta^{-\frac{3(24-C^2)}{2(36-C^2)}} \left[\tilde{u}_1 \exp\left(i\frac{\tilde{\lambda}}{3\hbar}\theta^3\right) + \tilde{u}_2 \exp\left(-i\frac{\tilde{\lambda}}{3\hbar}\theta^3\right)\right], \tag{89}$$

where $\tilde{u}_1$ and $\tilde{u}_2$ now depend on $\tilde{k}$, cf. 9.2.1-2 in reference [92]. Thence, the total wave function of the universe asymptotically reads

$$\Psi(\theta, z) \approx \sqrt{\frac{6\hbar}{\pi\tilde{\lambda}}}\, \theta^{-\frac{3(24-C^2)}{2(36-C^2)}} \sum_{\tilde{k}} b_{\tilde{k}} \left[\tilde{u}_1 \exp\left(i\frac{\tilde{\lambda}}{3\hbar}\theta^3\right) + \tilde{u}_2 \exp\left(-i\frac{\tilde{\lambda}}{3\hbar}\theta^3\right)\right]$$
$$\times \left[d_1 \exp\left(\frac{\sqrt{\tilde{k}^2}}{\hbar}z\right) + d_2 \exp\left(-\frac{\sqrt{\tilde{k}^2}}{\hbar}z\right)\right]. \tag{90}$$

As a result, for the condition of $d_1 = 0$ when $\tilde{k}^2$ positive, the wave function vanishes at the BR singularity. Therefore, for this boundary condition, the DW criterion is satisfied. This result points towards the avoidance of this cosmological singularity in the quantum real of $f(R)$ cosmology. Nevertheless, it should be noted that by setting $d_1 = 0$ we have dismissed a subgroup of solutions to the mWDW equation as unphysical. If future investigations show the importance of these ignored solutions, then it would be concluded that the DW criterion may not always be satisfied.

4.3. The LR in $f(R)$ Quantum Cosmology

In this section, we address the quantum fate of the LR abrupt event in the framework of $f(R)$ quantum geometrodynamics. Previously, in Section 3.2, it was shown that the alternative theory of gravity given by the function (46) is the most general expression for a metric $f(R)$ theory of gravity which gives the same asymptotic expansion history as GR filled with phantom DE governed by Equation (38). Moreover, since the presence of the Ei function in the term multiplying c_2 spoils the analytical inversion of $x(R)$ in Equation (60b), which is crucial for the exact derivation of the mWDW equation (69), hereafter we consider $c_2 = 0$ [91]. Thus, we focus on the simple, still general, group of alternative metric $f(R)$ theories of gravity predicting the classical occurrence of a LR abrupt event given by [91]

$$f(R) = c_1\left[27\beta^4 + 150\beta^2 R - \beta\left(9\beta^2 + 12R\right)^{\frac{3}{2}} + 2R^2\right], \tag{91}$$

where we recall that $\beta = \sqrt{3}A/2$ and the parameter A is observationally constrained to approximately 2.75×10^{-28} m^{-1}, see Table 3. Given the expression for $f(R)$, the change of variables (60) reads

$$q = a\sqrt{\frac{2c_1 R_\star}{f_{R_\star}}}\left(75\beta^2 - 9\beta\sqrt{9\beta^2 + 12R} + 2R\right)^{\frac{1}{2}}, \tag{92a}$$

$$x = \frac{1}{2}\ln\left[\frac{2c_1}{f_{R_\star}}\left(75\beta^2 - 9\beta\sqrt{9\beta^2 + 12R} + 2R\right)\right], \tag{92b}$$

being $f_{R_\star} = 2c_1 \left(75\beta^2 - 9\beta\sqrt{9\beta^2 + 12R_\star} + 2R_\star \right)$. The definition of the constant $R_\star$ needed for the above change of variables to be well-defined was previously discussed for the most general case, see Equation (61). Consequently, from Equations (21) and (41) it follows [91]

$$R_\star = 4\rho_0 \left(1 + \beta\sqrt{\frac{3}{\rho_0}} \ln 100\right)^2 + 6\beta\sqrt{\frac{\rho_0}{3}} \left(1 + \beta\sqrt{\frac{3}{\rho_0}} \ln 100\right). \tag{93}$$

On the other hand, for the purpose of computing the effective potential entering the mWDW equation, the inverse of the relation (92b) applies. This is

$$R(x) = 84\beta^2 + \frac{1}{4c_1} f_{R_\star} e^{2x} \pm 54\beta \sqrt{\frac{1}{48c_1} f_{R_\star} e^{2x} + 2\beta^2}. \tag{94}$$

However, only the positive branch is compatible with R being an increasing function of x and $R > R_\star$. Thus, we choose the positive sign in the preceding expression. Thereafter, the effective potential in the mWDW equation becomes [91]

$$V(q,x) = -U(x)q^6, \tag{95}$$

where

$$U(x) = \tilde{U} \left\{ 1 + 1644 \frac{c_1\beta^2}{f_{R_\star}} e^{-2x} + 205992 \frac{c_1^2\beta^4}{f_{R_\star}^2} e^{-4x} \right.$$

$$+ 36\beta e^{-x} \left(1 + 336 \frac{c_1\beta^2}{f_{R_\star}} e^{-2x}\right) \sqrt{\frac{3c_1}{f_{R_\star}} \left(1 + 96 \frac{c_1\beta^2}{f_{R_\star}} e^{-2x}\right)}$$

$$- 12\beta \sqrt{\frac{3c_1}{f_{R_\star}}} e^{-x} \left[1 + 330 \frac{c_1\beta^2}{f_{R_\star}} e^{-2x} + 18\beta e^{-x} \sqrt{\frac{3c_1}{f_{R_\star}} \left(1 + 96 \frac{c_1\beta^2}{f_{R_\star}} e^{-2x}\right)}\right]$$

$$\left. \times \left[1 + 339 \frac{c_1\beta^2}{f_{R_\star}} e^{-2x} + 18\beta e^{-x} \sqrt{\frac{3c_1}{f_{R_\star}} \left(1 + 96 \frac{c_1\beta^2}{f_{R_\star}} e^{-2x}\right)}\right]^{\frac{1}{2}} \right\}, \tag{96}$$

being $\tilde{U} := f_{R_\star}/(48c_1\lambda^2 R_\star)$ a constant. Contrary to the previous section, a change of variables such as (78) will no longer be able to make the potential one-variable-dependent and, therefore, a separation ansatz such as (82) is not appropriate in this case. Instead, note that the $U(x)$ part of the effective potential converges very quickly to a constant value when the model is observationally constrained. This feature suggests an adiabatic semiseparability type ansatz for the wave function of the universe. This is based on the so-called Born-Oppenheimer (BO) ansatz originally formulated in the context of molecular physics [125] and first introduced in the framework of quantum cosmology in references [126–128]. Furthermore, in reference [91] we have argued that the scalar curvature can be considered more fundamental from a geometrical point of view than the scale factor, justifying that the following ansatz à la Born-Oppenheimer should apply

$$\Psi(q,x) = \sum_{\tilde{k}} b_{\tilde{k}} \chi_{\tilde{k}}(q,x) \varphi_{\tilde{k}}(x), \tag{97}$$

where we recall that x depends only on R, whereas q contains both R and a, see the definitions in (60). Additionally, $b_{\tilde{k}}$ represents the amplitude of each solution and $\tilde{k}$ is related to the associated energy. (Please do not confuse $\tilde{k}$ with the spatial curvature k which has been set to zero). As a result of this ansatz, the mWDW equation (69) becomes

$$\hbar^2 q^2 \varphi_{\tilde{k}} \frac{\partial^2 \chi_{\tilde{k}}}{\partial q^2} - \hbar^2 \varphi_{\tilde{k}} \frac{\partial^2 \chi_{\tilde{k}}}{\partial x^2} - 2\hbar^2 \frac{\partial \chi_{\tilde{k}}}{\partial x} \frac{d\varphi_{\tilde{k}}}{dx} - \hbar^2 \chi_{\tilde{k}} \frac{d^2 \varphi_{\tilde{k}}}{dx^2} + U(x) q^6 \chi_{\tilde{k}} \varphi_{\tilde{k}} = 0. \tag{98}$$

The contribution of the second and third terms in the above expression can be neglected by virtue of the adiabatic assumption[5]. Therefore, Equation (98) implies

$$\hbar^2 \frac{d^2 \varphi_{\tilde{k}}}{dx^2} - \tilde{k}^2 \varphi_{\tilde{k}} = 0, \qquad (99)$$

$$\hbar^2 q^2 \frac{\partial^2 \chi_{\tilde{k}}}{\partial q^2} + \left[U(x) q^6 - \tilde{k}^2 \right] \chi_{\tilde{k}} = 0. \qquad (100)$$

The former equation can be directly solved. The solutions are exponential and trigonometric functions on x, depending on the sign of $\tilde{k}^2$. These are [91]

$$\varphi_{\tilde{k}}(x) = d_1 \exp\left(\frac{\sqrt{\tilde{k}^2}}{\hbar} x \right) + d_2 \exp\left(-\frac{\sqrt{\tilde{k}^2}}{\hbar} x \right), \qquad (101)$$

where d_1 and d_2 are integration constants. The solutions for $\chi_{\tilde{k}}$, on the other hand, are obtained assuming that the potential $U(x)$ behaves as a quasiconstant. Please note that this approximation is based on the fact that $U(x)$ converges very quickly to a constant value when observational constraints on the parameter A are taken into account. Hence, the solutions are [91]

$$\chi_{\tilde{k}}(q,x) = \sqrt{q} \left[u_1 J_{\frac{1}{6}\sqrt{1+\frac{4\tilde{k}^2}{\hbar^2}}} \left(\frac{\sqrt{U(x)}}{3\hbar} q^3 \right) + u_2 Y_{\frac{1}{6}\sqrt{1+\frac{4\tilde{k}^2}{\hbar^2}}} \left(\frac{\sqrt{U(x)}}{3\hbar} q^3 \right) \right], \qquad (102)$$

being u_1 and u_2 integration constants, cf. 9.1.53 of [92]. Thus, near the LR abrupt event, when both q and x diverge[6], the resulting wave function of the universe becomes

$$\Psi(q,x) \approx \sqrt{\frac{6\hbar}{\pi}} \frac{1}{U(x)^{\frac{1}{4}} q} \sum_{\tilde{k}} b_{\tilde{k}} \left[\tilde{u}_1 \exp\left(i \frac{\sqrt{U(x)}}{3\hbar} q^3 \right) + \tilde{u}_2 \exp\left(-i \frac{\sqrt{U(x)}}{3\hbar} q^3 \right) \right]$$

$$\times \left[d_1 \exp\left(\frac{\sqrt{\tilde{k}^2}}{\hbar} x \right) + d_2 \exp\left(-\frac{\sqrt{\tilde{k}^2}}{\hbar} x \right) \right], \qquad (103)$$

where the integration constants $\tilde{u}_1$ and $\tilde{u}_2$ now depend on $\tilde{k}^2$, see discussion in reference [91]. Since $U(x) > 0$ and tends to a constant value when x grows, the wave function cancels at the LR abrupt event when one of the integrations constants is set to zero. This is $d_1 = 0$ when $\tilde{k}^2$ is positive. Hence, the DeWitt criterion can be, indeed, satisfied. This points towards the avoidance of the LR abrupt event in the quantum realm of metric $f(R)$ theories of gravity. Nevertheless, as discussed for the case of the BR in the previous section, the fulfilment of the DW criterion is conditioned to the cancellation of one of the integration constants in Ψ. Accordingly, if future investigations claim for the physical importance of the dismissed solution, then it would have to be concluded that the DW criterion may not always be satisfied.

4.4. The LSBR in $f(R)$ Quantum Cosmology

Finally, we present the quantum fate of the LSBR abrupt event predicted in the metric $f(R)$ theories of gravity (52) obtained in Section 3.3. Moreover, following a line of reasoning similar to that presented in the previous sections, we consider $c_2 = 0$ in (52) as a necessary assumption to analytically obtain the corresponding mWDW equation. Therefore, we focus on the simple, still general, $f(R)$ cosmological model exhibiting a future LSBR abrupt event given by

$$f(R) = c_1 \left(9A^2 - 18AR + R^2 \right). \qquad (104)$$

This model has been already studied in the $f(R)$ quantum geometrodynamics approach, see reference [90]. In the next part of this section, we not only review the conclusion there presented but also provide a different, less restrictive and more general, approximation when solving for the wave function.

For the $f(R)$ gravity model (104), the change of variables (60) reads

$$q = a\sqrt{\frac{2c_1 R_\star}{f_{R_\star}}}(R-9A)^{\frac{1}{2}}, \tag{105a}$$

$$x = \frac{1}{2}\ln\left[\frac{2c_1}{f_{R_\star}}(R-9A)\right], \tag{105b}$$

being $f_{R_\star} = 2c_1(R_\star - 9A)$. Following the spirit for a physically meaningful definition of the constant $R_\star$ given in Equation (61), and in view of Equations (23) and (49), we use

$$R_\star = 4\rho_0 + 3A[1+4\ln(100)], \tag{106}$$

where we recall that $a_\star = 100a_0$ has been set, for the sake of concreteness, as the moment in the expansion history from which we can safely assume that DE is the only relevant component of the universe [see discussion below Equation (61)].

Subsequently, the mWDW equation for the particular $f(R)$ expression considered in Equation (104) reads

$$\left[\hbar^2 q^2 \frac{\partial^2}{\partial q^2} - \hbar^2 \frac{\partial^2}{\partial x^2} + W(x)q^6\right]\Psi(q,x) = 0, \tag{107}$$

where $W(x)$ is

$$W(x) = \tilde{W}\left(1 + 36\frac{c_1 A}{f_{R_\star}}e^{-2x} + 288\frac{c_1^2 A^2}{f_{R_\star}^2}e^{-4x}\right), \tag{108}$$

with $\tilde{W} := f_{R_\star}/(24c_1\lambda^2 R_\star)$ a constant. It should be noted that (107) resembles the form of the corresponding mWDW equation for the LR case presented in the previous section. In fact, the $W(x)$ part of the effective potential also converges quickly to a constant value when A is observationally constrained. Therefore, we consider here the very same BO approximation discussed in the previous section[7]. That is ansatz (97),

$$\Psi(q,x) = \sum_{\tilde{k}} b_{\tilde{k}} \chi_{\tilde{k}}(q,x) \varphi_{\tilde{k}}(x), \tag{109}$$

where $b_{\tilde{k}}$ represents the amplitude of each solution and $\tilde{k}$ is related to the associated energy, do not confuse it with the spatial curvature k which has been neglected.

Accordingly to the results presented in the previous section, *mutatis mutandis*, the total wave function of the universe near the LSBR abrupt event reads

$$\Psi(q,x) \approx \sqrt{\frac{6\hbar}{\pi}} \frac{1}{W(x)^{\frac{1}{4}} q} \sum_{\tilde{k}} b_{\tilde{k}} \left[\tilde{u}_1 \exp\left(i\frac{\sqrt{W(x)}}{3\hbar}q^3\right) + \tilde{u}_2 \exp\left(-i\frac{\sqrt{W(x)}}{3\hbar}q^3\right)\right]$$
$$\times \left[d_1 \exp\left(\frac{\sqrt{\tilde{k}^2}}{\hbar}x\right) + d_2 \exp\left(-\frac{\sqrt{\tilde{k}^2}}{\hbar}x\right)\right], \tag{110}$$

where $\tilde{u}_1$ and $\tilde{u}_2$ now depend on $\tilde{k}^2$. The validity of the BO approximation in this case is verified in Appendix B. Since $W(x) > 0$ and tends to a positive constant value when x grows, the wave function cancels at the LSBR abrupt event when one of the integrations constants is set to zero. This is $d_1 = 0$ when $\tilde{k}^2$ is positive. Hence, the fulfilment of the De-Witt criterion points towards the avoidance of the LSBR abrupt event in the quantum realm of metric $f(R)$ theories of gravity. Nevertheless, as stated before, if future investigations find out the importance of the dismissed solutions, then it would be concluded that the DW criterion may not always be satisfied.

5. Conclusions

The BR, LR and LSBR are cosmic curvature doomsdays predicted in some cosmological models where the superaccelerated expansion of the universe leads to the disintegration of all bounded structures and, ultimately, to the tear down of space–time itself. Within the context of GR, these cosmological catastrophes are intrinsic to phantom DE models, although a phantom fluid may also induce other cosmic singularities. For a FLRW background, these events can be characterized by the behavior of the scale factor a, the Hubble rate H and its cosmic time derivative $\dot{H}$ near the singular fate, see Table 1. More importantly, some of these models have been shown to be compatible with the current observational data, see, for instance, reference [57]. Therefore, our own universe may evolve towards some of these (classical) doomsdays. Nevertheless, quantum gravity effects can ultimately become significant and smooth out, or even avoid, the occurrence of these classically predicted singularities. In that sense, quantum cosmology is the natural framework for addressing the quantum fate of cosmological singularities. Among the different approaches to quantum cosmology, we have focused on the canonical quantization of cosmological background due to DeWitt's pioneering paper [74]. In this scheme, the DW criterion can be understood as a sufficient but not necessary condition for the avoidance of a classical singularity. Thus, the classical singularity is potentially avoided if the wave function of the universe vanishes in the nearby configuration space. This criterion has been successfully applied in GR for the phantom DE models considered in Section 2, which predict a classical fate à la BR, LR and LSBR, see references [66,72,73]. Consequently, this hints towards the avoidance of these cosmological doomsdays for those specific phantom DE models.

On the other hand, since the background late-time classical evolution can be equivalently described in the context of GR or within the framework of alternative theories of gravity, it is interesting to wonder whether the DW criterion is still fulfilled in the quantum realm of a different underlying theory of gravity. To find that out, reconstruction methods can be applied to obtain the general group of alternative theories of gravity able to reproduce the same expansion history as that of a given general relativistic model. Thus, in Section 3, we have collected the different $f(R)$ theories of gravity found among the literature that produce the same asymptotic background expansion as the phantom DE models summarized in the preceding Section 2. Consequently, these $f(R)$ models predict the classical occurrence of a BR, LR or a LSBR abrupt event. For the evaluation of the quantum fate of these classical models, we have followed the $f(R)$ quantum geometrodynamics approach introduced by Vilenkin [75], where the Wheeler-DeWitt equation is adapted for the case of $f(R)$ theories of gravity. Within this framework, the application of the DW criterion has been discussed, showing that it is possible to find solutions to the mWDW equation with a vanishing wave function for the aforementioned cosmic events. Therefore, as it happens when the gravitational interaction is that provided by GR, this result hints towards the avoidance of these cosmological doomsdays within the scheme of metric $f(R)$ theories of gravity [89–91].

It should be noted, however, that the validity of the wave functions obtained in this review is subject to the fulfilment of the conditions discussed in the appendices. Thus, for the avoidance of the BR singularity, for instance, we found our approximations to the wave function of the universe to be legit only for very tiny values of the parameter A, indeed smaller even than those found in reference [57] where a strong ansatz was imposed on the possible values of w. Of course, by relaxing such an ansatz tinier values of A are compatible with cosmological observations; see, for example, references [37–39] among others. On the other hand, it should be also noted that we have applied the quantization procedure only to some particular solutions to the reconstruction method, i.e., we have quantized just some specific $f(R)$ functions from the more general family of metric $f(R)$ gravity found to classically predict a future fate à la BR, LR or LSBR. Furthermore, certain boundary conditions have also been imposed when solving for Ψ to obtain vanishing solutions at the singular points. These conditions typically involve the cancellation of one of the integration constants. Therefore, a whole subgroup of solutions to the mWDW equation has been disregarded as unphysical. If future investigations show the importance of the dismissed solutions, then it would be concluded that the DW criterion may not always be fulfilled for solutions of physical interest.

The pioneering investigations presented in this paper have raised additional questions that should be addressed. From a classical point of view, the metric $f(R)$ theories under consideration should be further analyzed, investigating their compatibility with different observational constraints, as those coming from Solar System tests. On the other hand, it would be also interesting to consider whether semiclassical effects may be important when approaching the quantum realm and, in that case, how they will affect the disintegration of bounded structures. Furthermore, we should highlight that strictly speaking, the DW criterion is not a sufficient condition to guarantee the avoidance of classical singularities. To have a complete analysis, one should obtain the expectation values of the relevant operators. Nonetheless, the calculation of expectation values and probability distributions is related to various open questions in quantum cosmology that still must be properly addressed in the framework of quantum geometrodynamics. Specifically, the correct boundary or initial conditions, the Hilbert space structure and the classical-quantum correspondence, see reference [106]. Finally, we hope that the topics summarized in this work and the presented open questions will motivated further investigations.

Author Contributions: All authors contributed equally to this review. All authors have read and agreed to the published version of the manuscript.

Funding: The research of T.B.V. and P.M.-M. is supported by MINECO (Spain) Project No. PID2019-107394GB-I00 (AEI/FEDER, UE). T.B.V. also acknowledge financial support from Project No. FIS2016-78859-P (AEI/FEDER, UE) through Grant No. PAII46/20-08/2020-03, and from Universidad Complutense de Madrid and Banco de Santander through Grant No. CT63/19-CT64/19. The research of M.B.-L. is supported by the Basque Foundation of Science Ikerbasque. She also would like to acknowledge the partial support from the Basque government through Project No. IT956-16 (Spain) and MINECO through Project No. FIS2017-85076-P (AEI/FEDER, UE).

Conflicts of Interest: The authors declare no conflict of interest.

Abbreviations

The following abbreviations are used in this manuscript:

ΛCDM	Lambda cold dark matter
BF	Big freeze
BR	Big rip
DE	Dark energy
DM	Dark matter
DW	DeWitt (criterion)
EoS	Equation of state
FLRW	Friedmann-Lemaître-Robertson-Walker
GR	General relativity
LR	Little rip
LSBR	Little sibling of the big rip
WDW	Wheeler-DeWitt (equation)
mWDW	Modified Wheeler-DeWitt (equation)
wCDM	w cold dark matter

Appendix A. Validity of the Wave Function Used to Describe the BR

When solving the mWDW equation for the BR singularity, we have neglected the contribution of the third term in (81). This approach is valid as long as the corresponding solutions satisfy

$$\frac{C}{3}\hbar^2\theta\frac{\partial\chi_{\tilde{k}}}{\partial\theta}\frac{\partial\varphi_{\tilde{k}}}{\partial z} \ll \left(1-\frac{C^2}{36}\right)\hbar^2\theta^2\varphi_{\tilde{k}}\frac{\partial^2\chi_{\tilde{k}}}{\partial\theta^2},\ \hbar^2\chi_{\tilde{k}}\frac{\partial^2\varphi_{\tilde{k}}}{\partial z^2},\ \frac{C^2}{36}\hbar^2\theta\varphi_{\tilde{k}}\frac{\partial\chi_{\tilde{k}}}{\partial\theta},\ \frac{\gamma-1}{6\lambda^2\gamma}\theta^6\chi_{\tilde{k}}\varphi_{\tilde{k}}. \quad (A1)$$

According to that approximation, the solutions for $\chi_{\tilde{k}}$ and $\varphi_{\tilde{k}}$ are presented in Equations (85) and (86). Therefore, the terms we kept in Equation (81) are

$$\left(1-\frac{C^2}{36}\right)\hbar^2\theta^2\varphi_{\tilde{k}}\frac{\partial^2\chi_{\tilde{k}}}{\partial\theta^2} \approx -\sqrt{\frac{6\tilde{\lambda}^3\hbar}{\pi}}\left(1-\frac{C^2}{36}\right)\theta^{6-\frac{3(24-C^2)}{2(36-C^2)}}\sum_{\tilde{k}}b_{\tilde{k}}\left[\bar{u}_1\exp\left(i\frac{\tilde{\lambda}}{3\hbar}\theta^3\right)\right.$$
$$\left.+\bar{u}_2\exp\left(-i\frac{\tilde{\lambda}}{3\hbar}\theta^3\right)\right]\left[d_1\exp\left(\frac{\sqrt{\tilde{k}^2}}{\hbar}z\right)+d_2\exp\left(-\frac{\sqrt{\tilde{k}^2}}{\hbar}z\right)\right], \quad (A2)$$

$$\hbar^2\chi_{\tilde{k}}\frac{\partial^2\varphi_{\tilde{k}}}{\partial z^2} \approx \sqrt{\frac{6\hbar}{\pi\tilde{\lambda}}}\tilde{k}^2\theta^{-\frac{3(24-C^2)}{2(36-C^2)}}\sum_{\tilde{k}}b_{\tilde{k}}\left[\bar{u}_1\exp\left(i\frac{\tilde{\lambda}}{3\hbar}\theta^3\right)\right.$$
$$\left.+\bar{u}_2\exp\left(-i\frac{\tilde{\lambda}}{3\hbar}\theta^3\right)\right]\left[d_1\exp\left(\frac{\sqrt{\tilde{k}^2}}{\hbar}z\right)+d_2\exp\left(-\frac{\sqrt{\tilde{k}^2}}{\hbar}z\right)\right], \quad (A3)$$

$$\frac{C^2}{36}\hbar^2\theta\varphi_{\tilde{k}}\frac{\partial\chi_{\tilde{k}}}{\partial\theta} \approx \frac{C^2}{36}\sqrt{\frac{6\tilde{\lambda}\hbar^3}{\pi}}\theta^{3-\frac{3(24-C^2)}{2(36-C^2)}}\sum_{\tilde{k}}b_{\tilde{k}}\left[\bar{u}_1\exp\left(i\frac{\tilde{\lambda}}{3\hbar}\theta^3\right)\right.$$
$$\left.-\bar{u}_2\exp\left(-i\frac{\tilde{\lambda}}{3\hbar}\theta^3\right)\right]\left[d_1\exp\left(\frac{\sqrt{\tilde{k}^2}}{\hbar}z\right)+d_2\exp\left(-\frac{\sqrt{\tilde{k}^2}}{\hbar}z\right)\right], \quad (A4)$$

$$\frac{\gamma-1}{6\lambda^2\gamma}\theta^6\chi_{\tilde{k}}\varphi_{\tilde{k}} \approx \frac{\gamma-1}{6\lambda^2\gamma}\sqrt{\frac{6\hbar}{\pi\tilde{\lambda}}}\theta^{6-\frac{3(24-C^2)}{2(36-C^2)}}\sum_{\tilde{k}}b_{\tilde{k}}\left[\bar{u}_1\exp\left(i\frac{\tilde{\lambda}}{3\hbar}\theta^3\right)\right.$$
$$\left.+\bar{u}_2\exp\left(-i\frac{\tilde{\lambda}}{3\hbar}\theta^3\right)\right]\left[d_1\exp\left(\frac{\sqrt{\tilde{k}^2}}{\hbar}z\right)+d_2\exp\left(-\frac{\sqrt{\tilde{k}^2}}{\hbar}z\right)\right]. \quad (A5)$$

However, the terms that we have neglected behave asymptotically as

$$\frac{C}{3}\hbar^2\theta\frac{\partial\chi_{\tilde{k}}}{\partial\theta}\frac{\partial\varphi_{\tilde{k}}}{\partial z} \approx i\sqrt{\frac{6\tilde{\lambda}\tilde{k}^2}{\pi\hbar}}\frac{C\hbar}{3}\theta^{3-\frac{3(24-C^2)}{2(36-C^2)}}\sum_{\tilde{k}}b_{\tilde{k}}\left[\bar{u}_1\exp\left(i\frac{\tilde{\lambda}}{3\hbar}\theta^3\right)-\bar{u}_2\exp\left(-i\frac{\tilde{\lambda}}{3\hbar}\theta^3\right)\right]$$
$$\times\left[d_1\exp\left(\frac{\sqrt{\tilde{k}^2}}{\hbar}z\right)-d_2\exp\left(-\frac{\sqrt{\tilde{k}^2}}{\hbar}z\right)\right]. \quad (A6)$$

We recall that the integration constants d_1 must be set to zero to fulfil the DW criterion at the BR when $\tilde{k}^2$ positive. Thus, to obtain the compliance region of the performed approximation we compare the largest of the neglected terms with the smallest of the saved ones. This is the ratio

$$\epsilon = \left|\frac{\frac{B}{3}\hbar^2\theta\partial_\theta\chi_{\tilde{k}}\partial_z\varphi_{\tilde{k}}}{\hbar^2\chi_{\tilde{k}}\partial_z^2\varphi_{\tilde{k}}}\right| \approx \frac{2}{3\tilde{k}}\frac{\gamma-2}{\gamma-1}\tilde{\lambda}\theta^3. \quad (A7)$$

This ratio keeps below one if $\gamma - 2$ is sufficiently small to compensate the increase of the variable θ towards the BR singularity. Therefore, for $\gamma \approx 2$, this approximation is valid throughout the semiclassical regime towards the BR singularity, where θ increases but not sufficiently rapidly to compensate the small value of $\gamma - 2$. In the expression (36) for the metric $f(R)$ theory of gravity predicting the BR singularity, this would correspond to having a small parameter A and $c_+ = 0$, since γ_+ diverge at $A \to 0$. Consequently, this argument would favor small deviations from the ΛCDM model. It is worth noting that the values estimated for A in reference [57] are not small enough to ensure $\epsilon \ll 1$;

however smaller values for A are compatible with the fits claimed in other references (see, for example, references [37–39]). Thus, we consider this approximation to Ψ valid for the appropriate γ value.

Additionally note that in this section, we have enhanced the discussion on the viability of the performed approximations for Ψ originally presented in reference [89], where the authors have focused only on the particular case of $\tilde{k} = 0$ when addressing the validity of the wave function there found.

Appendix B. Validity of the BO Approximation for Ψ

During the application of the BO-type ansatz (97) performed in Sections 4.3 and 4.4, we have considered that $\chi_{\tilde{k}}(q, x)$ depends adiabatically on x. Therefore, we have neglected the contribution of some parts in their corresponding mWDW equations. This approach is valid as long as the corresponding solutions satisfy

$$\hbar^2 \varphi_{\tilde{k}} \frac{\partial^2 \chi_{\tilde{k}}}{\partial x^2}, \ 2\hbar^2 \frac{\partial \chi_{\tilde{k}}}{\partial x} \frac{d\varphi_{\tilde{k}}}{dx} \ll \hbar^2 q^2 \varphi_{\tilde{k}} \frac{\partial^2 \chi_{\tilde{k}}}{\partial q^2}, \ \hbar^2 \chi_{\tilde{k}} \frac{d^2 \varphi_{\tilde{k}}}{dx^2}, \ U(x) q^6 \chi_{\tilde{k}} \varphi_{\tilde{k}}. \tag{A8}$$

As a result of this approximation, the solutions for $\varphi_{\tilde{k}}$ and $\chi_{\tilde{k}}$ obtained for the case of the LR abrupt event are presented in Equations (99) and (100), respectively (see also reference [91]). Then, the terms we keep in (98) read

$$\hbar^2 q^2 \varphi_{\tilde{k}} \frac{\partial^2 \chi_{\tilde{k}}}{\partial q^2} \approx -U(x) q^6 \chi_{\tilde{k}} \varphi_{\tilde{k}} \approx -\sqrt{\frac{6\hbar}{\pi}} U(x)^{\frac{3}{4}} q^5 \left[\tilde{u}_1 \exp\left(i \frac{\sqrt{U(x)}}{3\hbar} q^3\right) \right.$$
$$\left. + \tilde{u}_2 \exp\left(-i \frac{\sqrt{U(x)}}{3\hbar} q^3\right) \right] \left[d_1 \exp\left(\frac{\sqrt{\tilde{k}^2}}{\hbar} x\right) + d_2 \exp\left(-\frac{\sqrt{\tilde{k}^2}}{\hbar} x\right) \right], \tag{A9}$$

$$\hbar^2 \chi_{\tilde{k}} \frac{d^2 \varphi_{\tilde{k}}}{dx^2} \approx \sqrt{\frac{6\hbar}{\pi}} \frac{\tilde{k}^2}{U(x)^{\frac{1}{4}} q} \left[\tilde{u}_1 \exp\left(i \frac{\sqrt{U(x)}}{3\hbar} q^3\right) + \tilde{u}_2 \exp\left(-i \frac{\sqrt{U(x)}}{3\hbar} q^3\right) \right]$$
$$\times \left[d_1 \exp\left(\frac{\sqrt{\tilde{k}^2}}{\hbar} x\right) + d_2 \exp\left(-\frac{\sqrt{\tilde{k}^2}}{\hbar} x\right) \right]. \tag{A10}$$

However, the neglected terms behave asymptotically as

$$\hbar^2 \varphi_{\tilde{k}} \frac{\partial^2 \chi_{\tilde{k}}}{\partial x^2} \approx -\frac{1}{36} \sqrt{\frac{6\hbar}{\pi}} \frac{U'(x)^2}{U(x)^{\frac{5}{4}}} q^5 \left[\tilde{u}_1 \exp\left(i \frac{\sqrt{U(x)}}{3\hbar} q^3\right) + \tilde{u}_2 \exp\left(-i \frac{\sqrt{U(x)}}{3\hbar} q^3\right) \right]$$
$$\times \left[d_1 \exp\left(\frac{\sqrt{\tilde{k}^2}}{\hbar} x\right) + d_2 \exp\left(-\frac{\sqrt{\tilde{k}^2}}{\hbar} x\right) \right], \tag{A11}$$

$$2\hbar^2 \frac{\partial \chi_{\tilde{k}}}{\partial x} \frac{d\varphi_{\tilde{k}}}{dx} \approx \frac{i}{3} \sqrt{\frac{6\hbar \tilde{k}^2}{\pi}} \frac{U'(x)}{U(x)^{\frac{3}{4}}} q^2 \left[\tilde{u}_1 \exp\left(i \frac{\sqrt{U(x)}}{3\hbar} q^3\right) - \tilde{u}_2 \exp\left(-i \frac{\sqrt{U(x)}}{3\hbar} q^3\right) \right]$$
$$\times \left[d_1 \exp\left(\frac{\sqrt{\tilde{k}^2}}{\hbar} x\right) - d_2 \exp\left(-\frac{\sqrt{\tilde{k}^2}}{\hbar} x\right) \right]. \tag{A12}$$

Please note that for $\tilde{k}^2$ positive, the constants d_1 must be zero to have a vanishing wave function at the LR. Thus, the validity of the BO approximation can be verified comparing the largest of the neglected terms with the smallest of the saved ones. This is the ratio ε,

$$\varepsilon = \left| \frac{\hbar^2 \varphi_{\tilde{k}} \partial_x^2 \chi_{\tilde{k}}}{\hbar^2 \chi_{\tilde{k}} \partial_x^2 \varphi_{\tilde{k}}} \right|. \tag{A13}$$

For the case of the LR abrupt event analyzed in Section 4.3 that ratio reads

$$\varepsilon \approx \frac{U'(x)^2}{U(x)} \frac{q^6}{36|\tilde{k}^2|}. \tag{A14}$$

Consequently, the approximation is valid as long as $\varepsilon \ll 1$. To evaluate this condition, note that

$$\frac{U'(x)^2}{U(x)} \approx 36 \frac{\beta^2}{\lambda^2 R_\star} e^{-2x} \left[1 + \frac{40}{3} \sqrt{\frac{3c_1}{f_{R_\star}}} \beta e^{-x} + \frac{832}{3} \frac{c_1 \beta^2}{f_{R_\star}} e^{-2x} + \mathcal{O}\left(e^{-3x}\right)\right], \tag{A15}$$

when β is observationally constrained, see Table 3. Thus, in the configuration space near the LR cosmic event, we have

$$\varepsilon \approx \frac{\beta^2}{\lambda^2 R_\star |\tilde{k}^2|} e^{-2x} q^6, \tag{A16}$$

where both q and x diverge. Finally, $\varepsilon \ll 1$ near the LR if β is sufficiently small, i.e., for small value of A. Please note that this corresponds, in fact, to the observationally preferred situation [57]. (We recall that A is of order 10^{-28} m^{-1} when observationally constrained, see Table 3). Therefore, when the parameters of the theory are observationally constrained, the approximation is valid throughout the semiclassical regime towards the LR abrupt event; where the variables q and x increase but not sufficiently rapidly to compensate the small value of β^2. Hence, for the purpose of the present work, i.e., to analyze the fulfilment of the DW criterion in the configuration space close to the LR, this approximation is valid.

On the other hand, for the verification of the validity of the BO approximation performed in Section 4.4, $U(x)$ must be exchanged for $W(x)$ in the preceding formulas. Therefore, considering the expression for $W(x)$ given in (108), the ratio ε becomes

$$\varepsilon \approx \frac{W'(x)^2}{W(x)} \frac{q^6}{36|\tilde{k}^2|} \approx \frac{6c_1 A^2}{\lambda^2 R_\star f_{R_\star} |\tilde{k}^2|} e^{-4x} q^6. \tag{A17}$$

Following the same line of reasoning as that presented before, this ratio keeps below 1 since the parameter A for the LSBR is of order 10^{-54} m^{-2}, see Table 3. Therefore, the BO approximation applied in Section 4.4 is valid throughout the semiclassical regime towards the LSBR abrupt event.

Notes

1. Since the BR, the LR and the LSBR doomsdays are intrinsic to phantom-like DE models, we limit our discussions to a positive parameter A, which leads to $w < -1$. For an analysis of the most general case, we refer the reader to the reference [52].
2. This is trivial to check when $\alpha \leq 1$. For the case of $\alpha > 1$, on the other hand, the expansion of the universe stops at a finite value of the scale factor, namely $a_{\max}$, such as the denominator in Equation (8) never becomes negative [see Equation (13)].
3. Alternatively to this interpretation, the wave function Ψ can be linked in a heuristic way with the probability distribution. In that sense, having a vanishing wave function could be interpreted as having zero probability of reaching that point in the configuration space. Nevertheless, this interpretation is based on the existence of squared integral functions and a consistent probability interpretation of the wave function. The problem is that these assumptions would require a minisuperspace with a proper Hilbert space nature, and that is not obvious to be always doable for a quantum cosmology based on the WDW equation.
4. For a different approach to the asymptotic form of the wave function Ψ see reference [89].
5. The validity of this approximation is checked in Appendix B.
6. This corresponds to $R \to \infty$ and a arbitrary, which was argued to be the main condition for the appearance of a curvature singularity. Nevertheless, the results in this section would not change if we had considered both $a, R \to \infty$ instead.
7. Please note that in reference [90] we have used a different ansatz for solving the mWDW equation (107). Since the exponential terms appearing in (108) are strongly suppressed not only by the observational value of the parameter A, but also by the divergence of x, we have considered only the asymptotic value of the function $W(x)$, i.e., we have approached the effective potential in (107) to asymptotically depend only on the variable q. However, as argued in [91], subdominant contributions to $W(x)$ have imprints in the shape of Ψ near the abrupt event. Those subdominant contributions are, in fact, important when comparing different wave functions that share a common asymptotic regime.

References

1. Riess, A.G.; Filippenko, A.V.; Challis, P.; Clocchiatti, A.; Diercks, A.; Garnavich, P.M.; Gilliland, R.L.; Hogan, C.J.; Jha, S.; Kirshner, R.P.; et al. Observational evidence from supernovae for an accelerating universe and a cosmological constant. *Astron. J.* **1998**, *116*, 1009. [CrossRef]
2. Perlmutter, S.; Aldering, G.; Goldhaber, G.; Knop, R.A.; Nugent, P.; Castro, P.G.; Deustua1, S.; Fabbro, S.; Goobar, A.; Groom, D.E.; et al. Measurements of Omega and Lambda from 42 high redshift supernovae. *Astrophys. J.* **1999**, *517*, 565. [CrossRef]
3. Weinberg, S. The Cosmological constant problems. *arXiv* **2000**, arXiv:astro-ph/0005265.
4. Peebles, P.J.E.; Ratra, B. The Cosmological constant and dark energy. *Rev. Mod. Phys.* **2003**, *75*, 559. [CrossRef]
5. Padmanabhan, T. Cosmological constant: The Weight of the vacuum. *Phys. Rept.* **2003**, *380*, 235. [CrossRef]
6. Sahni, V. Dark matter and dark energy. *Lect. Notes Phys.* **2004**, *653*, 141–180.
7. Copeland, E.J.; Sami, M.; Tsujikawa, S. Dynamics of dark energy. *Int. J. Mod. Phys. D* **2006**, *15*, 1753. [CrossRef]
8. Velten, H.E.S.; vom Marttens, R.F.; Zimdahl, W. Aspects of the cosmological "coincidence problem". *Eur. Phys. J. C* **2014**, *74*, 3160. [CrossRef]
9. Adler, R.J.; Casey, B.; Jacob, O.C. Vacuum catastrophe: An elementary exposition of the cosmological constant problem. *Am. J. Phys.* **1995**, *63*, 620–626. [CrossRef]
10. Bull, P.; Akrami, Y.; Adamek, J.; Baker, T.; Bellini, E.; Jimenez, J.B.; Bentivegna, E.; Camera, S.; Clesse, S.; Davis, J.H.; et al. Beyond ΛCDM: Problems, solutions, and the road ahead. *Phys. Dark Univ.* **2016**, *12*, 56–99. [CrossRef]
11. Perivolaropoulos, L.; Skara, F. Challenges for ΛCDM: An update. *arXiv* **2021**, arXiv:2105.05208.
12. Caldwell, R.R. A Phantom menace? *Phys. Lett. B* **2002**, *545*, 23–29. [CrossRef]
13. Starobinsky, A.A. Future and origin of our universe: Modern view. *Gravit. Cosmol.* **2000**, *6*, 157–163.
14. Gibbons, G.W. Cosmological evolution of the rolling tachyon. *Phys. Lett. B* **2002**, *537*, 1–4. [CrossRef]
15. Padmanabhan, T. Accelerated expansion of the universe driven by tachyonic matter. *Phys. Rev. D* **2002**, *66*, 021301. [CrossRef]
16. Kamenshchik, A.Y.; Moschella, U.; Pasquier, V. An Alternative to quintessence. *Phys. Lett. B* **2001**, *511*, 265–268. [CrossRef]
17. Bento, M.C.; Bertolami, O.; Sen, A.A. Generalized Chaplygin gas, accelerated expansion and dark energy matter unification. *Phys. Rev. D* **2002**, *66*, 043507. [CrossRef]
18. Li, M. A Model of holographic dark energy. *Phys. Lett. B* **2004**, *603*, 1. [CrossRef]
19. Caldwell, R.R.; Dave, R.; Steinhardt, P.J. Cosmological imprint of an energy component with general equation of state. *Phys. Rev. Lett.* **1998**, *80*, 1582. [CrossRef]
20. Tsujikawa, S. Quintessence: A Review. *Class. Quant. Grav.* **2013**, *30*, 214003. [CrossRef]
21. Chiba, T.; Okabe, T.; Yamaguchi, M. Kinetically driven quintessence. *Phys. Rev. D* **2000**, *62*, 023511. [CrossRef]
22. Horndeski, G.W. Second-order scalar-tensor field equations in a four-dimensional space. *Int. J. Mod. Phys.* **1974**, *10*, 363–384. [CrossRef]
23. Kase, R.; Tsujikawa, S. Dark energy in Horndeski theories after GW170817: A review. *Int. J. Mod. Phys. D* **2019**, *28*, 1942005. [CrossRef]
24. Dehghani, M.H. Accelerated expansion of the Universe in Gauss-Bonnet gravity. *Phys. Rev. D* **2004**, *70*, 064009. [CrossRef]
25. Nojiri, S.; Odintsov, S.D. Modified Gauss-Bonnet theory as gravitational alternative for dark energy. *Phys. Lett. B* **2005**, *631*, 1–6. [CrossRef]
26. Nojiri, S.; Odintsov, S.D. Modified $f(R)$ gravity consistent with realistic cosmology: From matter dominated epoch to dark energy universe. *Phys. Rev. D* **2006**, *74*, 086005. [CrossRef]
27. Nojiri, S.; Odintsov, S.D. Gravity assisted dark energy dominance and cosmic acceleration. *Phys. Lett. B* **2004**, *599*, 137. [CrossRef]
28. Allemandi, G.; Borowiec, A.; Francaviglia, M.; Odintsov, S.D. Dark energy dominance and cosmic acceleration in first order formalism. *Phys. Rev. D* **2005**, *72*, 063505. [CrossRef]
29. Capozziello, S.; De Laurentis, M. Extended Theories of Gravity. *Phys. Rept.* **2011**, *509*, 167-321. [CrossRef]
30. Harko, T.; Lobo, F.S.N.; Nojiri, S.; Odintsov, S.D. $f(R, T)$ gravity. *Phys. Rev. D* **2011**, *84*, 024020. [CrossRef]
31. Bengochea, G.R.; Ferraro, R. Dark torsion as the cosmic speed-up. *Phys. Rev. D* **2009**, *79*, 124019. [CrossRef]
32. Jiménez, J.B.; Heisenberg, L.; Koivisto, T.S.; Pekar, S. Cosmology in $f(Q)$ geometry. *Phys. Rev. D* **2020**, *101*, 103507. [CrossRef]
33. Saridakis, E.N.; Lazkoz, R.; Salzano, V.; Moniz, P.V.; Capozziello, S.; Jiménez, J.B.; Laurentis, M.D.; Olmo, G.J.; Akrami, Y.; Bahamonde, S.; et al. Modified Gravity and Cosmology: An Update by the CANTATA Network. *arXiv* **2021**, arXiv:2105.12582.
34. Huterer, D.; Shafer, D.L. Dark energy two decades after: Observables, probes, consistency tests. *Rept. Prog. Phys.* **2018**, *81*, 016901. [CrossRef]
35. Bamba, K.; Capozziello, S.; Nojiri, S.; Odintsov, S.D. Dark energy cosmology: the equivalent description via different theoretical models and cosmography tests. *Astrophys. Space Sci.* **2012**, *342*, 155–228. [CrossRef]
36. Motta, V.; García-Aspeitia, M.A.; Hernández-Almada, A.; na, J.M.; Verdugo, T. Taxonomy of Dark Energy Models. *Universe* **2021**, *7*, 163. [CrossRef]
37. Aghanim, N.; Akrami, Y.; Ashdown, M.; Aumont, J.; Baccigalupi, C.; Ballardini, M.; Banday, A.J.; Barreiro, R.B.; Bartolo, N.; Basak, S.; et al. Planck 2018 results. VI. Cosmological parameters. *Astron. Astrophys.* **2020**, *641*, A6.
38. Ade, P.A.R.; Aghanim, N.; Arnaud, M.; Ashdown, M.; Aumont, J.; Baccigalupi, C.; Banday, A.J.; Barreiro, R.B.; Bartlett, J.G.; Bartolo, N.; et al. Planck 2015 results. XIII. Cosmological parameters. *Astron. Astrophys.* **2016**, *594*, A13.

39. Abbott, T.M.C.; Allam, S.; Andersen, P.; Angus, C.; Asorey, J.; Avelino, A.; Avila, S.; Bassett, B.A.; Bechtol11, K.; Bernstein, G.M.; et al. First Cosmology Results using Type Ia Supernovae from the Dark Energy Survey: Constraints on Cosmological Parameters. *Astrophys. J. Lett.* **2019**, *872*, L30. [CrossRef]
40. Lopez-Corredoira, M.; Melia, F.; Lusso, E.; Risaliti, G. Cosmological test with the QSO Hubble diagram. *Int. J. Mod. Phys. D* **2016**, *25*, 1650060. [CrossRef]
41. Vanderlinde, K.; Crawford, T.M.; de Haan, T.; Dudley, J.P.; Shaw, L.; Ade, P.A.R.; Aird, K.A.; Benson, B.A.; Bleem, L.E.; Brodwin, M.; et al. Galaxy Clusters Selected with the Sunyaev-Zel'dovich Effect from 2008 South Pole Telescope Observations. *Astrophys. J.* **2010**, *722*, 1180. [CrossRef]
42. Sehgal, N.; Trac, H.; Acquaviva, V.; Ade, P.A.R.; Aguirre, P.; Amiri, M.; Appel, J.W.; Barrientos, L.F.; Battistelli, E.S.; Bond, J.R.; et al. The Atacama Cosmology Telescope: Cosmology from Galaxy Clusters Detected via the Sunyaev-Zel'dovich Effect. *Astrophys. J.* **2011**, *732*, 44–45. [CrossRef]
43. Addison, G.E.; Hinshaw, G.; Halpern, M. Cosmological constraints from baryon acoustic oscillations and clustering of large-scale structure. *Mon. Not. R. Astron. Soc.* **2013**, *436*, 1674–1683. [CrossRef]
44. MSantos, V.D.; Reis, R.R.R.; Waga, I. Constraining the cosmic deceleration-acceleration transition with type Ia supernova, BAO/CMB and H(z) data. *J. Cosmol. Astropart. Phys.* **2016**, *2*, 066. [CrossRef]
45. Bonilla, A.; Castillo, J.E. Constraints On Dark Energy Models From Galaxy Clusters and Gravitational Lensing Data. *Universe* **2018**, *4*, 21. [CrossRef]
46. Risaliti, G.; Lusso, E. Cosmological constraints from the Hubble diagram of quasars at high redshifts. *Nat. Astron.* **2019**, *3*, 272–277. [CrossRef]
47. Di Valentino, E.; Melchiorri, A.; Linder, E.V.; Silk, J. Constraining Dark Energy Dynamics in Extended Parameter Space. *Phys. Rev. D* **2017**, *96*, 023523. [CrossRef]
48. Di Valentino, E.; Linder, E.V.; Melchiorri, A. Vacuum phase transition solves the H_0 tension. *Phys. Rev. D* **2018**, *97*, 043528. [CrossRef]
49. Bouhmadi-López, M.; Madrid, J.A.J. Escaping the big rip? *J. Cosmol. Astropart. Phys.* **2005**, *5*, 005. [CrossRef]
50. Caldwell, R.R.; Kamionkowski, M.; Weinberg, N.N. Phantom Energy: Dark Energy with $w < -1$ Causes a Cosmic Doomsday. *Phys. Rev. Lett.* **2003**, *91*, 071301. [PubMed]
51. Frampton, P.H.; Ludwick, K.J.; Scherrer, R.J. The Little Rip. *Phys. Rev. D* **2011**, *84*, 063003. [CrossRef]
52. Štefančić, H. Expansion around the vacuum equation of state - Sudden future singularities and asymptotic behavior. *Phys. Rev. D* **2005**, *71*, 084024. [CrossRef]
53. Bouhmadi-López, M. Phantom-like behaviour in dilatonic brane-world scenario with induced gravity. *Nucl. Phys. B* **2008**, *797*, 78–92. [CrossRef]
54. PFrampton, H.; Ludwick, K.J.; Nojiri, S.; Odintsov, S.D.; Scherrer, R.J. Models for Little Rip Dark Energy. *Phys. Lett. B* **2012**, *708*, 204. [CrossRef]
55. Bouhmadi-López, M.; Errahmani, A.; Martín-Moruno, P.; Ouali, T.; Tavakoli, Y. The little sibling of the big rip singularity. *Int. J. Mod. Phys. D* **2015**, *24*, 1550078. [CrossRef]
56. Albarran, I.; Bouhmadi-López, M.; Morais, J. Cosmological perturbations in an effective and genuinely phantom dark energy Universe. *Phys. Dark Univ.* **2017**, *16*, 94–108. [CrossRef]
57. Bouali, A.; Albarran, I.; Bouhmadi-López, M.; Ouali, T. Cosmological constraints of phantom dark energy models. *Phys. Dark Univ.* **2019**, *26*, 100391. [CrossRef]
58. Bouhmadi-López, M.; González-Díaz, P.F.; Martín-Moruno, P. Worse than a big rip? *Phys. Lett. B* **2008**, *659*, 1–5. [CrossRef]
59. Bouhmadi-López, M.; González-Díaz, P.F.; Martín-Moruno, P. On the generalised Chaplygin gas: Worse than a big rip or quieter than a sudden singularity? *Int. J. Mod. Phys. D* **2008**, *17*, 2269–2290. [CrossRef]
60. Barrow, J.D. Sudden future singularities. *Class. Quantum Gravity* **2004**, *21*, L79–L82. [CrossRef]
61. Lake, K. Sudden future singularities in FLRW cosmologies. *Class. Quantum Gravity* **2004**, *21*, L129. [CrossRef]
62. Barrow, J.D. More general sudden singularities. *Class. Quantum Gravity* **2004**, *21*, 5619–5622. [CrossRef]
63. Dąbrowski, M.P. Are singularities the limits of cosmology? *arXiv* **2014**, arXiv:1407.4851.
64. Nojiri, S.; Odintsov, S.D.; Tsujikawa, S. Properties of singularities in (phantom) dark energy universe. *Phys. Rev. D* **2005**, *71*, 063004. [CrossRef]
65. Bouhmadi-López, M.; Kiefer, C.; Martín-Moruno, P. Phantom singularities and their quantum fate: general relativity and beyond—A CANTATA COST action topic. *Gen. Relativ. Gravit.* **2019**, *51*, 135. [CrossRef]
66. Dąbrowski, M.P.; Kiefer, C.; Sandhöfer, B. Quantum phantom cosmology. *Phys. Rev. D* **2006**, *74*, 044022. [CrossRef]
67. Kamenshchik, A.; Kiefer, C.; Sandhöfer, B. Quantum cosmology with big-brake singularity. *Phys. Rev. D* **2007**, *76*, 064032. [CrossRef]
68. Nojiri, S.; Odintsov, S.D. The Final state and thermodynamics of dark energy universe. *Phys. Rev. D* **2004**, *70*, 103522. [CrossRef]
69. Elizalde, E.; Nojiri, S.; Odintsov, S.D. Late-time cosmology in (phantom) scalar-tensor theory: Dark energy and the cosmic speed-up. *Phys. Rev. D* **2004**, *70*, 043539. [CrossRef]
70. Nojiri, S.; Odintsov, S.D. Quantum escape of sudden future singularity. *Phys. Lett. B* **2004**, *595*, 1–8. [CrossRef]
71. Bouhmadi-López, M.; Kiefer, C.; Sandhöfer, B.; Moniz, P.V. On the quantum fate of singularities in a dark-energy dominated universe. *Phys. Rev. D* **2009**, *79*, 124035. [CrossRef]

72. Albarran, I.; Bouhmadi-López, M.; Cabral, F.; Martín-Moruno, P. The quantum realm of the "Little Sibling" of the Big Rip singularity. *J. Cosmol. Astropart. Phys.* **2015**, *11*, 044. [CrossRef]
73. Albarran, I.; Bouhmadi-López, M.; Kiefer, C.; Marto, J.; Moniz, P.V. Classical and quantum cosmology of the little rip abrupt event. *Phys. Rev. D* **2016**, *94*, 063536. [CrossRef]
74. DeWitt, B.S. Quantum Theory of Gravity. I. The Canonical Theory. *Phys. Rev.* **1967**, *160*, 1113. [CrossRef]
75. Vilenkin, A. Classical and Quantum Cosmology of the Starobinsky Inflationary Model. *Phys. Rev. D* **1985**, *32*, 2511. [CrossRef] [PubMed]
76. Fernández-Jambrina, L. Grand Rip and Grand Bang/Crunch cosmological singularities. *Phys. Rev. D* **2014**, *90*, 064014. [CrossRef]
77. Fernández-Jambrina, L. Initial directional singularity in inflationary models. *Phys. Rev. D* **2016**, *94*, 024049. [CrossRef]
78. Ruzmaĭkina, T.V.; Ruzmaĭkin, A.A. Quadratic corrections to the Lagrangian density of the gravitational field and the singularity. *Sov. Phys. JETP* **1970**, *30*, 372.
79. Bouali, A.; Albarran, I.; Bouhmadi-López, M.; Errahmani, A.; Ouali, T. Cosmological constraints of interacting phantom dark energy models. *arXiv* **2021**, arXiv:2103.13432.
80. Nojiri, S.; Odintsov, S.D. Unified cosmic history in modified gravity: From $F(R)$ theory to Lorentz non-invariant models. *Phys. Rep.* **2011**, *505*, 59–144. [CrossRef]
81. Capozziello, S.; Cardone, V.F.; Troisi, A. Reconciling dark energy models with $f(R)$ theories. *Phys. Rev. D* **2005**, *71*, 043503. [CrossRef]
82. de la Cruz-Dombriz, Á.; Dobado, A. A $f(R)$ gravity without cosmological constant. *Phys. Rev. D* **2006**, *74*, 087501. [CrossRef]
83. Nojiri, S.; Odintsov, S.D. Modified gravity and its reconstruction from the universe expansion history. *J. Phys. Conf. Ser.* **2007**, *66*, 012005. [CrossRef]
84. Nojiri, S.; Odintsov, S.D.; Toporensky, A.; Tretyakov, P. Reconstruction and deceleration-acceleration transitions in modified gravity. *Gen. Rel. Grav.* **2010**, *42*, 1997–2008. [CrossRef]
85. Dunsby, P.K.S.; Elizalde, E.; Goswami, R.; Odintsov, S.; Gómez, D.S. On the LCDM Universe in $f(R)$ gravity. *Phys. Rev. D* **2010**, *82*, 023519. [CrossRef]
86. Carloni, S.; Goswami, R.; Dunsby, P.K.S. A new approach to reconstruction methods in $f(R)$ gravity. *Class. Quant. Grav.* **2012**, *29*, 135012. [CrossRef]
87. Morais, J.; Bouhmadi-López, M.; Capozziello, S. Can $f(R)$ gravity contribute to (dark) radiation? *J. Cosmol. Astropart. Phys.* **2015**, *1509*, 041. [CrossRef]
88. Makarenko, A.N.; Obukhov, V.V.; Kirnos, I.V. From Big to Little Rip in modified F(R,G) gravity. *Astrophys. Space Sci.* **2013**, *343*, 481. [CrossRef]
89. Alonso-Serrano, A.; Bouhmadi-López, M.; Martín-Moruno, P. $f(R)$ quantum cosmology: avoiding the Big Rip. *Phys. Rev. D* **2018**, *98*, 104004 [CrossRef]
90. Vasilev, T.B.; Bouhmadi-López, M.; Martín-Moruno, P. Classical and quantum fate of the little sibling of the big rip in $f(R)$ cosmology. *Phys. Rev. D* **2019**, *100*, 084016. [CrossRef]
91. Vasilev, T.B.; Bouhmadi-López, M.; Martín-Moruno, P. The little rip in classical and quantum $f(R)$ cosmology. *Phys. Rev. D* **2021**, *103*, 124049. [CrossRef]
92. Abramowitz, M.; Stegun, I.A. *Handbook of Mathematical Functions*; Dover Publications: New York, NY, USA, 1972; ISBN 978-0486612720.
93. Bertotti, B.; Iess, L.; Tortora, P. A test of general relativity using radio links with the Cassini spacecraft. *Nature* **2003**, *425*, 374–376. [CrossRef] [PubMed]
94. Capozziello, S.; Troisi, A. PPN-limit of fourth order gravity inspired by scalar-tensor gravity. *Phys. Rev. D* **2005**, *72*, 044022. [CrossRef]
95. Chiba, T.; Smith, T.L.; Erickcek, A.L. Solar System constraints to general $f(R)$ gravity. *Phys. Rev. D* **2007**, *75*, 124014. [CrossRef]
96. Capozziello, S.; Stabile, A.; Troisi, A. The Newtonian Limit of $f(R)$ gravity. *Phys. Rev. D* **2007**, *76*, 104019. [CrossRef]
97. Capozziello, S.; Piedipalumbo, E.; Rubano, C.; Scudellaro, P. Testing an exact f(R)-gravity model at Galactic and local scales. *Astron. Astrophys.* **2009**, *505*, 21–28. [CrossRef]
98. O'Dwyer, M.; Joras, S.E.; Waga, I. γ gravity: Steepness control. *Phys. Rev. D* **2013**, *88*, 063520. [CrossRef]
99. Wang, J.Y.; Feng, C.J.; Zhai, X.H.; Li, X.Z. Solar System Tests of a New Class of $f(z)$ Theory. *Int. J. Mod. Phys. D* **2020**, *29*, 2050060. [CrossRef]
100. Sawicki, I.; Hu, W. Stability of Cosmological Solution in $f(R)$ Models of Gravity. *Phys. Rev. D* **2007**, *75*, 127502. [CrossRef]
101. Hu, W.; Sawicki, I. Models of $f(R)$ Cosmic Acceleration that Evade Solar-System Tests. *Phys. Rev. D* **2007**, *76*, 064004. [CrossRef]
102. Roshan, M.; Shojai, F. Notes on the post-Newtonian limit of massive Brans-Dicke theory. *Class. Quant. Grav.* **2011**, *28*, 145012. [CrossRef]
103. Guo, J.Q. Solar system tests of $f(R)$ gravity. *Int. J. Mod. Phys. D* **2014**, *23*, 1450036. [CrossRef]
104. Naik, A.P.; Puchwein, E.; Davis, A.C.; Arnold, C. Imprints of Chameleon $f(R)$ Gravity on Galaxy Rotation Curves. *Mon. Not. R. Astron. Soc.* **2018**, *480*, 5211–5225. [CrossRef]
105. Negrelli, C.; Kraiselburd, L.; Landau, S.J.; Salgado, M. Solar System tests and chameleon effect in $f(R)$ gravity. *Phys. Rev. D* **2020**, *101*, 064005. [CrossRef]
106. Kiefer, C. *Quantum Gravity*, 3rd ed.; Oxford University Press: Oxford, UK, 2012; ISBN 978-0199212521.

107. Kiefer, C.; Sandhöefer, B. Quantum Cosmology. *arXiv* **2008**, arXiv:0804.0672.
108. Bojowald, M. Loop quantum cosmology. *Living Rev. Relativ.* **2005**, *8*, 11. [CrossRef] [PubMed]
109. Ashtekar, A.; Singh, P. Loop Quantum Cosmology: A Status Report. *Class. Quant. Grav.* **2011**, *28*, 213001. [CrossRef]
110. Ambjørn, J.; Loll, R. Nonperturbative Lorentzian quantum gravity, causality and topology change. *Nucl. Phys. B* **1998**, *536*, 407–434. [CrossRef]
111. Ambjørn, J.; Jurkiewicz, J.; Loll, R. A Nonperturbative Lorentzian path integral for gravity. *Phys. Rev. Lett.* **2000**, *85*, 924–927. [CrossRef]
112. Ambjørn, J.; Jurkiewicz, J.; Loll, R. Dynamically triangulating Lorentzian quantum gravity. *Nucl. Phys. B* **2001**, *610*, 347–382. [CrossRef]
113. Kuchar, K.V.; Ryan, M.P. Is minisuperspace quantization valid?: Taub in mixmaster. *Phys. Rev. D* **1989**, *40*, 3982–3996. [CrossRef] [PubMed]
114. Wheeler, J.A. On the Nature of quantum geometrodynamics. *Annals Phys.* **1957**, *2*, 604. [CrossRef]
115. Bouhmadi-López, M.; Chen, C.Y. Towards the Quantization of Eddington-inspired-Born-Infeld Theory. *J. Cosmol. Astropart. Phys.* **2016**, *11*, 023. [CrossRef]
116. Bouhmadi-López, M.; Chen, C.Y.; Chen, P. On the Consistency of the Wheeler-DeWitt Equation in the Quantized Eddington-inspired Born-Infeld Gravity. *J. Cosmol. Astropart. Phys.* **2018**, *12*, 032. [CrossRef]
117. Albarran, I.; Bouhmadi-López, M.; Chen, C.Y.; Chen, P. Quantum cosmology of Eddington-Born–Infeld gravity fed by a scalar field: The big rip case. *Phys. Dark Univ.* **2019**, *23*, 100255. [CrossRef]
118. Higgs, P.W. Quadratic lagrangians and general relativity. *Il Nuovo Cimento* **1959**, *11*, 816-820 . [CrossRef]
119. Bicknell, G.V. Non-viability of gravitational theory based on a quadratic lagrangian. *J. Phys. A Math. Nucl. Gen.* **1974**, *7*, 1061. [CrossRef]
120. Whitt, B. Fourth-order gravity as general relativity plus matter. *Phys. Lett. B* **1984**, *145*, 176. [CrossRef]
121. Barrow, J.D.; Cotsakis, S. Inflation and the Conformal Structure of Higher Order Gravity Theories. *Phys. Lett. B* **1988**, *214*, 515. [CrossRef]
122. Albarran, I.; Bouhmadi-López, M.; Cabral, F.; Martín-Moruno, P. The Avoidance of the Little Sibling of the Big Rip Abrupt Event by a Quantum Approach. *Galaxies* **2018**, *6*, 21. [CrossRef]
123. Albarran, I.; Bouhmadi-López, M.; Chen, C.Y.; Chen, P. Doomsdays in a modified theory of gravity: A classical and a quantum approach. *Phys. Lett. B* **2017**, *772*, 814–818. [CrossRef]
124. Kiefer, C. Non-minimally coupled scalar fields and the initial value problem in quantum gravity. *Phys. Lett. B* **1989**, *225*, 227. [CrossRef]
125. Born, M.; Oppenheimer, J.R. On quantum theory of molecules. *Ann. Phys.* **1927**, *389*, 457. [CrossRef]
126. Kiefer, C. Continuous measurement of mini-superspace variables by higher multipoles. *Class. Quantum Grav.* **1987**, *4*, 1369. [CrossRef]
127. Brout, R.; Horwitz, G.; Weil, D. On the onset of time and temperature in cosmology. *Phys. Lett. B* **1987**, *192*, 318. [CrossRef]
128. Kiefer, C. Wave packets in minisuperspace. *Phys. Rev. D* **1988**, *38*, 1761. [CrossRef] [PubMed]

Review

Quantum and Classical Cosmology in the Brans–Dicke Theory

Carla R. Almeida [1,*], Olesya Galkina [1] and Julio César Fabris [1,2]

1. Núcleo Cosmo-ufes & Departamento de Física, Universidade Federal do Espírito Santo, Vitória 29075-910, ES, Brazil; olesya.galkina@cosmo-ufes.org (O.G.); julio.fabris@cosmo-ufes.org (J.C.F.)
2. National Research Nuclear University MEPhI, Kashirskoe sh. 31, 115409 Moscow, Russia
* Correspondence: cralmeida00@gmail.com

Abstract: In this paper, we discuss classical and quantum aspects of cosmological models in the Brans–Dicke theory. First, we review cosmological bounce solutions in the Brans–Dicke theory that obeys energy conditions (without ghost) for a universe filled with radiative fluid. Then, we quantize this classical model in a canonical way, establishing the corresponding Wheeler–DeWitt equation in the minisuperspace, and analyze the quantum solutions. When the energy conditions are violated, corresponding to the case $\omega < -\frac{3}{2}$, the energy is bounded from below and singularity-free solutions are found. However, in the case $\omega > -\frac{3}{2}$, we cannot compute the evolution of the scale factor by evaluating the expectation values because the wave function is not finite (energy spectrum is not bounded from below). However, we can analyze this case using Bohmian mechanics and the de Broglie–Bohm interpretation of quantum mechanics. Using this approach, the classical and quantum results can be compared for any value of ω.

Keywords: Brans–Dicke theory; bounce models; de Broglie–Bohm interpretation

---

Citation: Almeida, C.R.; Galkina, O.; Fabris, J.C. Quantum and Classical Cosmology in the Brans–Dicke Theory. *Universe* **2021**, *7*, 286. https://doi.org/10.3390/universe7080286

Academic Editor: Jaime Haro Cases

Received: 28 June 2021
Accepted: 2 August 2021
Published: 5 August 2021

Publisher's Note: MDPI stays neutral with regard to jurisdictional claims in published maps and institutional affiliations.

Copyright: © 2021 by the authors. Licensee MDPI, Basel, Switzerland. This article is an open access article distributed under the terms and conditions of the Creative Commons Attribution (CC BY) license (https://creativecommons.org/licenses/by/4.0/).

1. Introduction

General Relativity (GR) theory is a highly successful theory that describes gravitational interactions of the universe. It has successfully survived observational tests in the solar system, and its predictions about the emission of gravitational waves by binary systems were confirmed. However, there are strong indications that the theory is incomplete. For example, it cannot account for the observed gravitational anomalies, mainly in the cosmological context, and thus it is necessary to introduce the hypothesis of dark matter and dark energy to explain cosmological observational data within GR's framework. Another reason is the initial singularity at the beginning of the universe and at the final stage of some class of stars, which are generally predicted by the GR theory. These drawbacks led to many possible extensions of the GR that can address those problems. The prototype of alternative theory of gravity is Brans–Dicke theory (BD). Historically, it is one of the most important modifications to the standard GR theory, which was introduced by Brans and Dicke [1] as a possible implementation of Mach's principle in a relativistic theory (eventually, it did not work out as we will discuss later).

It is expected that a quantum formulation of GR might solve some of its problems, especially those related to the existence of singularities. However, there are many obstacles to this quantization [2,3]. Several attempts have been made to overcome the difficulties that appear when combining the principles of the GR theory and Quantum Mechanics (QM), be it via canonical methods or other procedures, like loop quantization or string theory. A simplified approach, like the quantization of the Einstein–Hilbert action in the minisuperspace in the presence of matter fields, shows that it is possible to obtain cosmological models without singularities. The construction of a quantum cosmological model encounters many problems, even when the minisuperspace restriction is used. The first one is the absence of an explicit time coordinate due to the invariance by time reparametrizations in the classical theory [4,5]. There are different ways to solve this problem. One of them is to allow the matter fields to play the role of time, which can be achieved, for example, through Schutz's description of a fluid [6]: its corresponding canonical formulation results in a Schrödinger-type equation since the conjugate momentum associated with the matter variables appears linearly in the Hamiltonian. We will employ this approach in the analysis to be exposed in the present text.

Another point of discussion is a choice of the suitable formalism to interpret the quantum theory and thus obtain specific predictions. This is a very sensible question. The usual Copenhagen interpretation is based on a probabilistic formalism, using concepts such as decoherence and a measurement mechanism through the spectral theorem. It is not ideal for a system consisting of a unique realization as it is the universe. Nevertheless, many adaptations of the Copenhagen interpretation are possible, like the Many World [7] or the Consistent Histories [8]. One alternative is the de Broglie–Bohm (dBB) interpretation of quantum mechanics [9,10], one we will address in this work. The dBB approach keeps the concept of trajectories of a given system, and a probabilistic analysis is not fundamental in this scheme [11].

In this work, we will investigate the BD theory. The reason to analyze this theory, which has been thoroughly studied in many contexts, is that it reserves some interesting and even unexpected features which deserve to be discussed in more detail. Classically, cosmological models constructed from the BD theory may lead to non-singular scenarios if the energy conditions are violated, as it happens in the GR theory. Such violations of the energy conditions occur when the Brans–Dicke parameter varies in the domain of $\omega < -3/2$. Surprisingly, it is also possible to obtain a non-singular model in the BD theory even if the energy conditions are satisfied. We will discuss this possibility using radiation as the matter content. The situation becomes more complex when we turn to quantum models: a consistent quantum model, from the point of view of the Copenhagen interpretation, is possible when $\omega < -3/2$, since only in this case is the energy bounded from below; if the energy conditions are satisfied, the spectrum of energy is not bounded from below. However, in both cases, the analysis becomes possible even if we use the dBB interpretation of quantum mechanics. Moreover, the use of the dBB formalism allows a comparison between the classical and quantum models in the BD theory, at least to some simple configurations. This possibility will be explored in the present work, revealing many peculiarities at the classical and quantum levels.

The paper is organized as follows: in Section 2, we review the classical BD theory. In Section 3, we present a non-singular model with radiative fluid in BD theory that obeys the energy conditions and does not contain ghosts. Section 4 presents the quantization of the model with a Lagrangian with a non-minimally coupled scalar field and matter fluid, establishing the corresponding Wheeler–DeWitt equation in the minisuperspace. Finally, we analyze the quantum solution via dBB interpretation in Section 5. Our conclusions are revealed in Section 6.

2. A Short Review of the Brans–Dicke Theory

The Brans–Dicke theory was proposed as a modification of the relativistic theory of gravity to include two new features: the possibility of a dynamical coupling to gravity and Mach's principle. Dirac had suggested earlier that the gravitational coupling in Einstein's theory might not be a constant, implying it could vary with time, at least in the cosmological context [12]. This idea was further developed by Jordan [13], but the rigorous implementation was made by Brans and Dicke in their seminal 1961 article [1]. They replaced the gravitational coupling by the inverse of a scalar field ϕ, such that

$$G \propto \frac{1}{\phi}. \tag{1}$$

A simple but elegant solution to implement this idea in a relativistic context is to consider in the action a non-minimal coupling between the new scalar field ϕ and the geometry represented by the Ricci scalar. The introduction of a kinetic term coupled by an arbitrary parameter ω complemented the theory, preserving the minimal coupling between gravity and matter and assuring the invariance by the full diffeomorphism group and the consequent conservation of the energy–momentum tensor.

The presence of a long-range scalar field was connected initially with the intention to implement the Mach's principle. The most common of many formulations of Mach's principle states that the inertial property of a given body results from its interaction with all matter present in the universe. It is not easy to implement such an appealing idea. General Relativity was considered to be a Machian theory since it relates the geometry of space-time to the matter distribution, but it has non-Machian features, such as the locality of the relation between space and matter and initial conditions for the matter fields. Perhaps, these issues could be circumvented by introducing a long-range scalar field as it occurs in the BD theory. In a Machian relativistic theory, the only solution in the absence of matter should be the Minkowski one. This is not the case for GR or for BD theory. In other words, both theories failed as a proposal to implement the Mach principle. However, introducing the scalar field ϕ non-minimally coupled to gravity led to many phenomenological applications and intriguing results: it opens new possibilities in comparison with the GR original framework.

The BD action reads as

$$S = \int d^4x \sqrt{-g} \left\{ \phi R - \frac{\omega}{\phi} \phi_{;\mu} \phi^{;\mu} \right\} + \int d^4x \sqrt{-g} \mathcal{L}_m(g_{\mu\nu}, \Psi), \tag{2}$$

where ω is a coupling constant and $\mathcal{L}_m$ is the matter term [1], with Ψ indicating generically the matter fields.

It is commonly understood that, in the $\omega \to \infty$ limit, BD theory coincides with GR [14,15]. Although it is true in most situations, this statement is not valid in general. When $\omega \gg 1$, the field equations show that $\Box \phi = \mathcal{O}\left(\frac{1}{\omega}\right)$, so we have

$$\phi = \frac{1}{G_N} + \mathcal{O}\left(\frac{1}{\omega}\right), \tag{3}$$

$$G_{\mu\nu} = 8\pi G_N T_{\mu\nu} + \mathcal{O}\left(\frac{1}{\omega}\right), \tag{4}$$

where G_N is Newton's gravitational constant and $G_{\mu\nu}$ is the Einstein tensor. However, there are some examples [16–18] where exact solutions cannot be continuously deformed into the corresponding GR solutions by taking the $\omega \to \infty$ limit. In this case, the solutions decay as

$$\phi = \frac{1}{G_N} + \mathcal{O}\left(\frac{1}{\sqrt{\omega}}\right). \tag{5}$$

Moreover, there is a particle solution that admits the appropriate asymptotic behavior given by Equation (3) but no GR limit [19].

From the action (2), the whole theory is derived. The field equations are obtained through the variation of the action (2) with respect to the metric and scalar field,

$$R_{\mu\nu} - \frac{1}{2} g_{\mu\nu} R = 8\pi \frac{T_{\mu\nu}}{\phi} + \frac{\omega}{\phi^2} \left(\phi_{;\mu} \phi_{;\nu} - \frac{1}{2} g_{\mu\nu} \phi_{;\alpha} \phi^{;\alpha} \right) + \frac{1}{\phi} (\phi_{;\mu;\nu} - g_{\mu\nu} \Box \phi), \tag{6}$$

$$\Box \phi = 8\pi \frac{T}{3 + 2\omega}. \tag{7}$$

The coupling with the curvature in the action (2) can be avoided if we perform a conformal transformation on the metric $g_{\mu\nu}$ such that

$$g_{\mu\nu} = \phi^{-1} \tilde{g}_{\mu\nu}. \tag{8}$$

With this, we change our frame of reference. With the reference frame $\tilde{g}_{\mu\nu}$, one could interpret the scalar field as a matter field, and thus recover general relativity. We, however, want to remain closer to Brans and Dicke's original idea of the gravitation not being purely geometrical, and thus the scalar field does not account for a matter content. The original frame is known as Jordan's frame, and, after the conformal transformation (8), it is called Einstein's frame. In this latter, the action becomes

$$S = \int d^4x \sqrt{-\tilde{g}} \left\{ \tilde{R} - \epsilon \zeta_{;\mu} \zeta^{;\mu} \right\} + \int d^4x \sqrt{-g} \mathcal{L}_m \left(e^{-\kappa \zeta} \tilde{g}_{\mu\nu}, \Psi \right), \quad (9)$$

where we have defined

$$\phi = e^{\kappa \zeta}, \quad (10)$$

$$\kappa = \frac{1}{\sqrt{\left|\omega + \frac{3}{2}\right|}}, \quad (11)$$

$$\epsilon = \text{sign}\left(\omega + \frac{3}{2}\right). \quad (12)$$

The resulting field equations are (ignoring the tildes in the redefined geometric terms),

$$R_{\mu\nu} - \frac{1}{2} g_{\mu\nu} R = 8\pi e^{-2\kappa \zeta} T_{\mu\nu} + \epsilon \left(\zeta_{;\mu} \zeta_{;\nu} - \frac{1}{2} g_{\mu\nu} \zeta_{;\rho} \zeta^{;\rho} \right), \quad (13)$$

$$\Box \zeta = \epsilon 4\pi T e^{-2\kappa \zeta}. \quad (14)$$

Please note that this set of equations implies a non-usual expression for the conservation laws, which is due to the non-minimal coupling between the scalar field and matter.

Thus, in the Einstein frame, the case $\epsilon = 1$ ($\omega > -\frac{3}{2}$) corresponds to an ordinary scalar field with positive energy density, while, for $\epsilon = -1$ ($\omega < -\frac{3}{2}$), the kinetic term of the scalar field changes sign, and it becomes a phantom field with negative energy density. In the special case $\omega = -3/2$ with a redefinition of the matter fields, the GR theory is recovered. When the matter field is given by radiation, the non-minimal coupling between the scalar field and the matter component is also broken since $T = 0$ for a radiative fluid. We will use this fact later.

Currently, the limits on the parameter ω are very stringent [20]. Cosmological constraints lead to values for ω of the order of some hundreds. Binary pulsars push these bounds to dozen of thousands. Hence, the BD theory essentially becomes GR in most of the cases, as mentioned before. Despite this, BD theory continues to be relevant for several reasons. For example, for a particular value of $\omega = -1$, the action (9) coincides with the effective string action for the dilatonic sector. In this case, the action (9) acquires some duality properties that have been explored in the Pre-Big Bang scenarios, in which there is a contracting phase in the evolution of the universe before the actual expanding phase. In fact, for $\omega = -1$, the action (2) in the cosmological context is invariant by the transformation,

$$a \to \frac{1}{a}, \quad \phi \to \phi - 2\ln a, \quad (15)$$

where a is the scale factor. Therefore, an expanding universe can be mapped into a contracting universe. For an excellent recent review of Pre-Big Bang scenarios, we refer to [21]. Even if there is a clear discrepancy with the observational constraint on the value of ω, it is possible to take into account a possible dependence of ω on energy scales that may reconcile observations with theoretical considerations. This may be achieved, for example, by supposing that ω is a function of the scalar field ϕ, $\omega = \omega(\phi)$.

Perhaps, one of the most delicate aspects concerning the pre-big bang scenario deals with the evolution of the perturbations in the primordial universe, which becomes highly discrepant with GR and, at the same time, strictly constrained by observations. In general, predictions for the evolution of perturbations depend on the transition from the contracting phase to an expanding phase. A possible solution to this problem may be found by studying perturbations behavior, taking into account the string configuration underlining the pre-big bang model since the transition may occur at the large curvature regime. Such analysis is not an easy task, technically and conceptually. Similar proposals, like the ekpyrotic scenario based on brane-world structures coming from string theories, have been developed in the literature [22].

In what concerns the primordial universe, we may also mention the extended inflationary model using the original proposal based on a cosmological constant [23] but implemented in the context of the BD theory. It is possible to obtain the transition to the radiative phase, but only if the value of the parameter ω is relatively small, in contradiction with the constraints mentioned above. In such a case, it is necessary to find a mechanism to change the value of ω during the universe's evolution since the energy scale may depend on this parameter, as already evoked above.

Multidimensional theories, when compactified to four dimensions, leads to action similar to the BD, with

$$\omega = -\frac{d-1}{d}. \tag{16}$$

In this case, gauge fields can appear, depending on the original configuration, with non-trivial coupling [24,25].

The proposal of modified gravity theories spiked a revival in the interest in the BD theory. Modified gravity theories were conceived mainly to address the problem of the acceleration of the universe without introducing a new exotic component in the universe called the dark energy. There are many types of the modified gravity theories. For example, in the Horndesky theories, the most general Lagrangian includes the scalar field in a non-trivial way and leads to second-order equations of motion. In fact, the BD theory can be considered the first of the modified theories, and it was formulated long before the Horndesky classification. Another class of modified gravity theories is the $f(R)$ theories. They are based on a nonlinear generalization of the Einstein–Hilbert action. Interestingly, the $f(R)$ theories can be recast as the BD theory, with $\omega = 0$ and a potential term that depends on the form of the $f(R)$ function. In general, all modified gravity theories must introduce a screening mechanism to reconcile the large-scale and small-scale constraints. According to these screening mechanisms, the supplementary degree of freedom associated with the scalar field does not propagate in a dense region (compared to the averaged cosmological density). The BD theory, when reformulated in the Einstein frame, gives the prototype of the chameleon mechanism, one of the most important screening mechanisms: the conformal transformation used to reformulate the BD theory in the Einstein frame introduces a non-minimal coupling between matter and scalar field, making the mass of the scalar field depend on the density of the medium, as it can be verified introducing a potential term in the set of Equations (13) and (14).

Intriguing results on the quantization of the BD theory adds to the relevance of the theory. Analysis on quantum cosmological scenarios using the BD theory was made in Refs. [26,27]. In Ref. [26], the authors considered the Einstein frame and used the WKB formalism to recover the notion of time. The result reveals a curious behavior: the quantum effects become relevant in a late phase of the expansion of the universe, contrary to expectation. Due to this property, the initial singularity existing in the FLRW models can not be avoided. In [27], the study was extended to the Jordan frame through the Bohm–de Broglie formalism to compute the quantum evolution of the universe. In this case, the opposite scenario was found: the quantum effects become important mainly in the early universe, and the initial singularity can be avoided.

We must also mention the study of the resulting quantum Schrödinger-like equation from the point of view of its self-adjointness properties [28], which are essential to employ the spectral theorem. The results indicated that, for the case $\omega < -3/2$, which corresponds to a phantom scalar field, it is possible to recover the self-adjointness of the quantum operator. We will discuss this in more detail in the forthcoming sections.

3. Bouncing Solutions and the Energy Conditions

The most common solution to avoiding a singularity in cosmological models is the introduction of exotic types of matter fields, for example, a scalar field with negative energy density. On the other hand, to obtain a bouncing solution in classical General Relativity, violation of the energy conditions is required. In this section, after briefly reviewing the bouncing scenarios, we present a non-singular model with radiative fluid in BD theory that obeys the energy conditions and does not contain ghosts [29]. To do so, we first briefly analyze the solutions determined by Gurevich et al. [30] for the cosmological isotropic and homogeneous flat universe with a perfect fluid with an equation of state $p = n\rho$, where the parameter n is given by $0 \leq n \leq 1$. Then, we focus on the analysis of these solutions in the case of radiative fluid.

The study of bouncing models is motivated by the search for cosmological solutions without singularities. Bouncing models have been widely studied to solve the initial-singularity problem and as an addition or alternative to inflation to describe the primordial universe because they can explain, in their own way, the horizon and flatness problems and justify the power spectrum of primordial cosmological perturbations inferred by observations [31–33].

In such models, an initial singularity is replaced with a bounce—a smooth transition from contraction to expansion. To obtain a bounce, one needs to change the value of the Hubble parameter $H \equiv \frac{\dot{a}(t)}{a(t)}$, which appears to be negative during the contracting phase, to a positive value for the following expanding phase. One of the options to switch the sign of the Hubble function lies within GR. It usually requires the violation of the null energy condition (NEC) [34]

$$\rho + p \geq 0, \tag{17}$$

where ρ is energy density, and p is pressure as usual. In most cases, the energy conditions reflect the nature of matter fields, "ordinary" (attractive effects) or "exotic" (repulsive effects). However, as we will see later, the bounce can be achieved through some non-standard coupling between the fields existing in a given theory.

In the early stages of the expansion of the universe, the curvature is not essential since the matter components are more relevant for the dynamic than the curvature term due to their dependence on the scale factor, and we can restrict our further analysis to the quasi-Euclidean variant of the isotropic model. In this case, for a flat FLRW metric,

$$ds^2 = N^2 dt^2 - a^2(t)\left(dx^2 + dy^2 + dz^2\right), \tag{18}$$

with $N^2 = 1$, the classical field Equations (6) and (7) reduce to

$$3\left(\frac{\dot{a}}{a}\right)^2 = 8\pi\frac{\rho}{\phi} + \frac{\omega}{2}\left(\frac{\dot{\phi}}{\phi}\right)^2 - 3\frac{\dot{a}}{a}\frac{\dot{\phi}}{\phi}, \tag{19}$$

$$2\frac{\ddot{a}}{a} + \left(\frac{\dot{a}}{a}\right)^2 = -8\pi\frac{p}{\phi} - \frac{\omega}{2}\left(\frac{\dot{\phi}}{\phi}\right)^2 - \frac{\ddot{\phi}}{\phi} - 2\frac{\dot{a}}{a}\frac{\dot{\phi}}{\phi}, \tag{20}$$

$$\ddot{\phi} + 3\frac{\dot{a}}{a}\dot{\phi} = \frac{8\pi}{3+2\omega}(\rho - 3p), \tag{21}$$

$$\dot{\rho} + 3\frac{\dot{a}}{a}(\rho + p) = 0. \tag{22}$$

The general solutions in [30] are obtained for $\omega < -\frac{3}{2}$ and $\omega > -\frac{3}{2}$. In the first case, there is violation of the energy conditions for the scalar field in Einstein's frame. In the latter case, the energy conditions for the scalar field are satisfied as we will discuss in this section. The solutions for the scale factor and scalar field for $\omega < -\frac{3}{2}$ are

$$a = a_0 \left[(\theta + \theta_-)^2 + \theta_+^2 \right]^{\frac{\sigma}{2A}} e^{\pm \frac{\sqrt{\frac{2}{3}|\omega|-1}}{A} \arctan \frac{\theta+\theta_-}{\theta_+}}, \tag{23}$$

$$\phi = \phi_0 \left[(\theta + \theta_-)^2 + \theta_+^2 \right]^{(1-3n)/2A} e^{\mp 3(1-n) \frac{\sqrt{\frac{2}{3}|\omega|-1}}{A} \arctan \frac{\theta+\theta_-}{\theta_+}}, \tag{24}$$

where $\sigma = 1 + \omega(1-n)$, $2A = (1-3n) + 3\sigma(1-n)$, a_0 is an arbitrary constant, and $\theta_- > \theta_+$ are integration constants. The time coordinate θ is connected with the cosmic time t by definition,

$$dt = a^{3n} d\theta. \tag{25}$$

When $\theta \to \infty$, the scale factor a does not vanish. The infinite contraction has a minimum a_{min}, and it is followed by the expansion. Thus, this model admits a cosmological bounce.

For $\omega > -\frac{3}{2}$, the general solutions are

$$a(\theta) = a_0(\theta - \theta_+)^{\omega/3(\sigma \mp \zeta)}(\theta - \theta_-)^{\omega/3(\sigma \pm \zeta)}, \tag{26}$$

$$\phi(\theta) = \phi_0(\theta - \theta_+)^{(1 \mp \zeta)/(\sigma \mp \zeta)}(\theta - \theta_-)^{(1 \pm \zeta)/(\sigma \pm \zeta)}, \tag{27}$$

where $\zeta = \sqrt{1 + \frac{2}{3}\omega}$. In this model, a regular bounce can be obtained for $\frac{1}{4} < n < 1$ and $-\frac{3}{2} < \omega \leq -\frac{4}{3}$. The case $n = 1$ is unusual and does not admit bounce solution [19].

Henceforth, we shall focus on the universe filled with radiative fluid, represented by the equation of state $p = \frac{1}{3}\rho$. In this case, the solutions for $\omega > -\frac{3}{2}$ are given by the following expressions:

$$a(\eta) = a_0(\eta - \eta_+)^{\frac{1}{2} \pm \frac{1}{2\sqrt{1+\frac{2}{3}\omega}}} (\eta - \eta_-)^{\frac{1}{2} \mp \frac{1}{2\sqrt{1+\frac{2}{3}\omega}}}, \tag{28}$$

$$\phi(\eta) = \phi_0(\eta - \eta_+)^{\mp \frac{1}{\sqrt{1+\frac{2}{3}\omega}}} (\eta - \eta_-)^{\pm \frac{1}{\sqrt{1+\frac{2}{3}\omega}}}, \tag{29}$$

where η is the conformal time and $\eta_\pm$ are constants such that $\eta_+ > \eta_-$. In Figure 1, we plot the scale factor and scalar field for the lower sign. The solutions for $\omega < -\frac{3}{2}$ are

$$a(\eta) = a_0 [(\eta + \eta_-)^2 + \eta_+^2]^{\frac{1}{4}} e^{\pm \frac{1}{\sqrt{\frac{2}{3}|\omega|-1}} \arctan \frac{\eta+\eta_-}{\eta_+}}, \tag{30}$$

$$\phi(\eta) = \phi_0 e^{\mp \frac{2}{\sqrt{\frac{2}{3}|\omega|-1}} \arctan \frac{\eta+\eta_-}{\eta_+}}. \tag{31}$$

In the case of the lower sign in Equations (28) and (29), we obtain bounce solutions for $-\frac{3}{2} < \omega < 0$. Nevertheless, there is a curvature singularity at $\eta = \eta_+$ for $-\frac{4}{3} < \omega < 0$, even if the scale factor diverges at this point. However, for $-\frac{3}{2} < \omega \leq -\frac{4}{3}$, bounce solutions are always regular without curvature singularity. In this case, there are two possible scenarios of the evolution of the universe due to the time reversal invariance. In the first one, the universe begins at $\eta = \eta_+$, with $a \to \infty$ and an infinite value for the gravitational coupling ($\phi = 0$). It evolves to the other asymptotic limit with $a \to \infty$, although with ϕ constant and finite. The second option is the reversal behavior of the first one for $-\infty < \eta < -\eta_+$. In both cases, the cosmic times varies as $-\infty < t < \infty$.

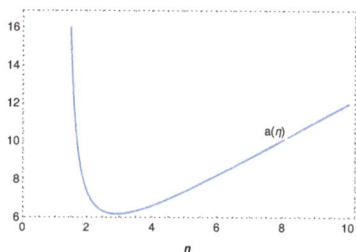
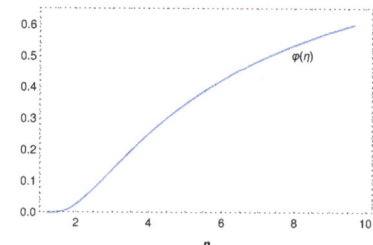

Figure 1. Behavior of the scale factor and scalar field in the case of the radiative fluid (Equations (28) and (29), lower sign) for $\omega = -1.43$.

The dual solution in the Einstein frame for $-\frac{3}{2} < \omega \leq -\frac{4}{3}$ is given by

$$b(\eta) = b_0(\eta - \eta_+)^{1/2}(\eta - \eta_-)^{1/2} \tag{32}$$

with $b = \phi^{1/2}a$ and contains an initial singularity. This can be considered as a specific case of "conformal continuation" in the scalar-tensor gravity proposed in [35].

Let us show that the energy conditions for the scalar field are satisfied for $\omega > -\frac{3}{2}$. In general, in order to have a bounce solution, violation of the energy conditions is required. Using the Friedmann and Raychaudhuri equations, we represent the strong and null energy conditions in GR as

$$\frac{\ddot{a}}{a} = -\frac{4\pi G}{3}(\rho + 3p) > 0 \tag{33}$$

$$-2\frac{\ddot{a}}{a} + 2\left(\frac{\dot{a}}{a}\right)^2 = 8\pi G(\rho + p) > 0. \tag{34}$$

We reformulate the BD theory in the Einstein frame so that it would be possible to use the energy condition in this form. One can see that both energy conditions are satisfied as long as $\omega > -\frac{3}{2}$. This is in agreement with the fact that, in the Einstein frame, the cosmological models are singular unless $\omega < -\frac{3}{2}$. However, in the original Jordan frame, there are non-singular models for $-\frac{3}{2} < \omega < -\frac{4}{3}$. However, in this range, the scalar field and the matter component obey the energy condition. The effects that lead to the absence of the singularity come from the non-minimal coupling. In Figure 2, we present the effective energy condition, defined in the left-hand side of Equations (33) and (34), considering the effects of the non-minimal coupling. If we analyze only the left-hand side in Equations (33) and (34), the effects of the interaction due to the non-minimal coupling are included, and the energy conditions can be violated even if the matter terms do not violate them. For more details, see Ref. [29].

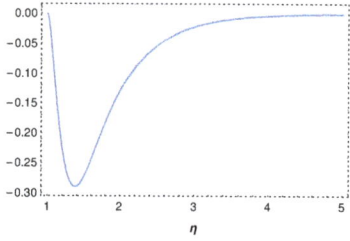
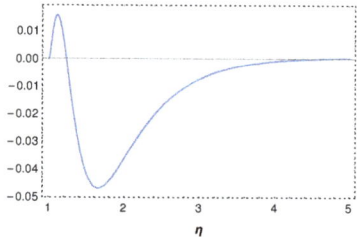

Figure 2. Behavior of the "effective" strong energy condition (**left**) and "effective" null energy (**right**) condition for $\omega = -1.43$ represented in the left-hand side of Equations (33) and (34), taking into account the effects of the non-minimal coupling.

It is important to observe that only in the case of radiative fluid is it possible to obtain a model without singularity preserving the energy conditions, at least in the BD theory. It is true also for the model with flat spatial sections. For a non-flat universe, one can obtain a singularity-free scenario even in General Relativity if the strong energy condition—but not necessarily the null energy condition—is violated.

4. Canonical Quantization of the BD Theory

In the early Universe, when the cosmo is compressed in a Planck scale, quantum effects become relevant, and the quantization of the classical models may be a necessity. Quantum cosmology stands on this principle, considering a wave description of the universe that satisfies the Hamiltonian constraint of the theory. In this section, we will investigate the canonical quantization of the BD theory in the minisuperspace, considering the FLRW metric (18). We shall quantize the Hamiltonian constraint $\mathcal{H}_{tot} \approx 0$ (where "$\approx$" means weak equality. See, for example, [36]) to obtain the Wheeler–DeWitt Equation $\hat{H}_{tot}\Psi = 0$, with Ψ being the wave function of the universe, and using the canonical operators ($\hbar = 1$),

$$x \to \hat{x} : \psi(x) \to x\psi(x) \quad ; \quad \pi_x \to \hat{\pi}_x : \psi(x) \to -i\partial_x \psi(x). \tag{35}$$

The total Hamiltonian $\mathcal{H} = \mathcal{H}_{tot}$ is formed by the gravitational part and the Hamiltonian of the matter content, $\mathcal{H} = \mathcal{H}_G + \mathcal{H}_M$. To find $\mathcal{H}_M$, let us consider a model of early-universe filled with a radiative fluid. Using Schutz formalism [6], the super-Hamiltonian of the fluid is given by

$$\mathcal{H}_M = \frac{N}{a}\pi_T, \tag{36}$$

where T is directly related to the entropy of the fluid [37]. With this, the canonical quantization of the total Hamiltonian in the original Jordan's frame is

$$\partial_a^2 \Psi + \frac{p}{a}\partial_a \Psi + \frac{6}{\omega}\frac{\phi^2}{a^2}\left\{\frac{a}{\phi}\partial_a\partial_\phi \Psi - \left(\partial_\phi^2 \Psi + \frac{q}{\phi}\partial_\phi \Psi\right)\right\} = -12i\frac{(3+2\omega)}{\omega}\phi\partial_T \Psi, \tag{37}$$

where p, q are ordering factors for the quantization of the momenta squared. For more details on the computation of (37), see Ref. [37]. Notice, however, that here we use a different sign convention.

The calculation is similar in Einstein's frame, considering the transformation (8). In this case, the canonical quantization results in

$$\partial_b^2 \Psi + \frac{\bar{p}}{b}\partial_b \Psi - \bar{\omega}\frac{\phi^2}{b^2}\left\{\partial_\phi^2 \Psi + \frac{\bar{q}}{\phi}\partial_\phi \Psi\right\} = -i\partial_T \Psi. \tag{38}$$

Here, again, we have that $\bar{p}$ and $\bar{q}$ are ordering factors.

Notice that Equations (37) and (38) are Schrödinger-like, that is,

$$\hat{H}\Psi = i\frac{\partial}{\partial t}\Psi, \tag{39}$$

if we consider the matter field playing the role of time. We can, therefore, treat it as a quantum system to analyze this cosmological scenario in the early-universe. The first step is to verify the conditions for the effective Hamiltonian operators of these models to be self-adjoint. It is known [37] that the Hamiltonian operator of this quantized BD model with a radiative fluid can be self-adjoint only if $q = \bar{q} = 1$. On the other hand, we have the quantum equivalence between Jordan's and Einstein's frames [38] if, and only if, $p = \bar{p} = 1$, and thus we can use Einstein's frame.

Now, in Einstein's frame, we choose the coordinate

$$\phi = e^{\sqrt{\frac{12}{|3+2\omega|}}\sigma} \tag{40}$$

instead of ϕ. With this, the relation between the scale factor in Jordan's and Einstein's frames is

$$a = e^{-\frac{\sigma}{\sqrt{|1+\frac{2}{3}\omega|}}} b. \tag{41}$$

In terms of the new variable σ, the Schrödinger Equation (39) is

$$\partial_b^2 \Psi + \frac{1}{b}\partial_b \Psi - \epsilon \frac{1}{b^2}\partial_\sigma^2 \Psi = -i\partial_T \Psi. \tag{42}$$

The measure of the Hilbert space is such that

$$\langle \psi | \psi \rangle = \int_{-\infty}^{\infty} \int_0^{\infty} \psi \psi^* b \, db d\sigma. \tag{43}$$

The regular solution for the Equation (42) is

$$\Psi(b,\sigma) = A(k,E) J_\nu(\sqrt{E}b) e^{i(k\sigma - ET)}, \quad \nu = \sqrt{-\epsilon}|k|, \tag{44}$$

where k is a separation constant. There is also another solution written in terms of the Bessel function $J_{-\nu}(x)$, which is not regular at the origin, at least when $\epsilon = -1$. For this reason, we will disregard it at the moment. We will choose $\epsilon = -1$; for this case, the Hamiltonian operator is bounded from below, and it is essentially self-adjoint for $-1 \leq \bar{p} \leq 3$ (see [37]), which is our case. Thus, $\nu = |k|$. Let us choose the coefficient $A(k,E)$ such that the wave packet becomes

$$\Psi(b,\sigma,T) = \frac{1}{N} \int_0^{\infty} \int_{-\infty}^{+\infty} e^{-k^2} x^{|k|+1} e^{-\alpha x^2} J_{|k|}(xb) e^{ik\sigma} dk dx, \tag{45}$$

where **N** is a normalization factor, and

$$\alpha = \gamma + iT, \quad x = \sqrt{E}, \tag{46}$$

with γ a positive real parameter. The integration in x gives us [39],

$$\Psi(b,\sigma,T) = \frac{1}{N} \int_{-\infty}^{+\infty} e^{-k^2+ik\sigma} \frac{b^{|k|}}{(2\alpha)^{|k|+1}} e^{-\frac{b^2}{4\alpha}} dk. \tag{47}$$

To find the normalization factor **N**, we should remember that

$$\langle \Psi, \Psi \rangle = \int_0^{\infty} \int_{-\infty}^{\infty} \Psi \Psi^* b \, db d\sigma = 1, \tag{48}$$

and thus

$$N^2 = \int_0^{\infty} \int_{-\infty}^{\infty} \left[\int_{-\infty}^{\infty} e^{-k^2} \frac{b^{|k|}}{(2\alpha)^{|k|+1}} e^{-\frac{b^2}{4\alpha}} e^{ik\sigma} dk \right]$$
$$\cdot \left[\int_{-\infty}^{+\infty} e^{-k'^2} \frac{b^{|k'|}}{(2\alpha^*)^{|k'|+1}} e^{-\frac{b^2}{4\alpha^*}} e^{-ik'\sigma} dk' \right] b db d\sigma.$$

Integrating over σ, k', and k, and defining $u = b/2|\alpha|$, we obtain

$$N^2 = 4\pi \sqrt{\frac{\pi}{8}} \int_0^{\infty} u^2 e^{-\gamma u^2} \left[1 - \Phi\left(\frac{\ln u}{\sqrt{2}}\right) \right] du$$
$$= 4\pi \sqrt{\frac{\pi}{8}} \left[\frac{\sqrt{\pi}}{4\gamma^{\frac{3}{2}}} - g_{(1)}(\gamma) \right],$$

where $\Phi(x)$ is the error function, and

$$g_{(1)}(\gamma) = \int_0^\infty u^2 e^{-\gamma u^2} \Phi\left(\frac{\ln u}{\sqrt{2}}\right) du. \tag{49}$$

Since $-1 < \Phi(x) < 1$ and $\int_0^\infty u^2 e^{-\gamma u^2}$ is strictly positive, we have

$$|g_{(1)}(\gamma)| < \frac{\sqrt{\pi}}{4\gamma^{\frac{3}{2}}}. \tag{50}$$

Therefore, $\mathbf{N}^2$ is positive, as expected.

We can now calculate the expected values of the scale factor b and scalar field σ. For the scale factor,

$$\langle b \rangle = \langle \Psi | b | \Psi \rangle. \tag{51}$$

Similarly to the calculation of the normalization factor, integrating in σ and k', we have

$$\begin{aligned}
\langle b \rangle &= \frac{8\pi |\alpha|}{\mathbf{N}^2} \sqrt{\frac{\pi}{8}} \int_0^\infty u^3 e^{-\gamma u^2} \left[1 - \Phi\left(\frac{\ln u}{\sqrt{2}}\right)\right] du \\
&= \frac{8\pi |\alpha|}{\mathbf{N}^2} \sqrt{\frac{\pi}{8}} \left[\frac{1}{2\gamma^2} - g_{(2)}(\gamma)\right],
\end{aligned}$$

where

$$|g_{(2)}(\gamma)| = \left| \int_0^\infty u^3 e^{-\gamma u^2} \Phi\left(\frac{\ln u}{\sqrt{2}}\right) du \right| < \frac{1}{2\gamma^2}. \tag{52}$$

Since $|\alpha| = \sqrt{\gamma^2 + T^2}$, and defining

$$\Omega(\gamma) = \frac{8\pi}{\mathbf{N}^2} \sqrt{\frac{\pi}{8}} \left[\frac{1}{2\gamma^2} - g_{(2)}(\gamma)\right] = 2 \left[\frac{\gamma^{-2} - 2g_{(2)}(\gamma)}{\frac{\sqrt{\pi}}{2\gamma^{\frac{3}{2}}} - 2g_{(1)}(\gamma)}\right], \tag{53}$$

which is strictly positive, the expected value of b is

$$\langle b \rangle = \Omega(\gamma) \sqrt{\gamma^2 + T^2}. \tag{54}$$

By definition, $\gamma > 0$. Therefore, $\langle b \rangle > 0$, that is, there is no singularity at $T = 0$. Moreover, for $T \gg \gamma$, at late times, $\langle b \rangle \to T$.

Using the fact that $\int f(x) \delta'(x - x_0) dx = -\int f'(x) \delta(x - x_0) dx$, a similar calculation for σ results in

$$\langle \sigma \rangle = -\frac{2\pi i}{\mathbf{N}^2} \int_0^\infty \int_{-\infty}^{+\infty} e^{-k^2} \frac{b^{|k|+1}}{(2\alpha)^{|k|+1}(2\alpha^*)} e^{-\left(\frac{\gamma b^2}{4|\alpha|^2}\right)} \partial_k \left[e^{-k^2} \frac{b^{|k|}}{(2\alpha^*)^{|k|}}\right] dk db.$$

The integral over k is zero because the function is odd with respect to this parameter. Thus,

$$\langle \sigma \rangle = 0. \tag{55}$$

This does not mean, however, that the scalar field ϕ is a constant, since the expectation value allows fluctuations. Notice that we have a symmetrical bouncing in both frames because of Equation (41).

5. Analysis of the Solution via the de Broglie–Bohm Approach

In the previous section, we introduced a quantum model of the BD theory, identifying the universe with a wave function that obeys a Schrödinger-like equation. Thus, we proceeded with the usual methods of quantum mechanics to analyze the behavior of the scale factor, which is directly connected with the volume of the universe. However, in the case of quantum cosmology, we must give up on the usual Copenhagen interpretation of QM, since it requires an external observer to cause the wave function to collapse into one state. An alternative is the many-world interpretation, which considers every state of the wave function as real, existing in parallel with each other. Our universe, therefore, is one of many. This interpretation does not require the collapse of the wave function, and so we can investigate different states separately from the wave packet.

The de Broglie–Bohm (dBB) interpretation, on the other hand, does not consider a wave description of the universe at all. Instead, it is a dynamical theory in which real trajectories can be obtained in the configuration space of the quantum system. Those trajectories are observer-independent, and, therefore, the de Broglie–Bohm interpretation does not rely on any collapse mechanism. This interpretation was formulated as a causal alternative to the probabilistic Copenhagen interpretation still in the early years of the quantum era [40,41], and it has replicated most of the classic results of quantum mechanics. In the dBB interpretation, observers are also described by quantum operators to be applied on a wave function that satisfies the Schrödinger equation. The wave function dictates the equations of motion to find the trajectories. In Bohmian mechanics, $\Psi(x_i, t)$ is decomposed as

$$\Psi(x_i, t) = R(x_i, t)e^{iS(x_i, t)} \qquad (56)$$

and the probability density of the trajectory is given by $\rho(x_i, t) = |\Psi(x_i, t)| = R^2(x_i, t)$. The Bohmian trajectories are realized through the momenta defined as

$$p_j = \partial_j S = \left(\frac{i}{2}\right)\frac{\Psi\Psi^*_{,j} - \Psi^*\Psi_{,j}}{|\Psi|^2}. \qquad (57)$$

Since it is not probabilistic, the observers in the dBB do not necessarily need to be represented by self-adjoint operators.

This ontological interpretation of the quantum theory has its critics, who argue that other interpretations may be better-suited [42], or take issue with the so-called hidden variables which determine the behavior of the trajectories in a many-body configuration [43]. Still, because of its distinct characteristic, the dBB interpretation is a suitable candidate to be applied in the quantization of cosmological scenarios [44,45], where the wave function becomes a guide to the possible observer-independent evolution of the universe. Notice that, contrary to the Copenhagen interpretation, the dBB interpretation of quantum mechanics does not necessarily need to work in the Hilbert space.

Let us compare these two approaches in particular cases. In the previous section, we have already made the computation using the expectation values finding, quite generally, that the expectation value for the scalar field is zero and the expectation value for the scale factor is the same as found in the corresponding case in GR. We will now analyze the predictions for the evolution of the universe using the dBB formulation. We will work initially in the Einstein frame. Remember the usual solution to the Equation (42) given in Equation (44):

$$\Psi(b, \sigma, T) = A(k, E)J_\nu(\sqrt{E}b)e^{i(k\sigma + ET)}, \quad \nu = \sqrt{-\epsilon}|k|. \qquad (58)$$

The construction of the wave packet exposed in the previous section is not convenient here. The reason is that the integration is made in the interval $-\infty < k < \infty$, but the terms coming from the order of the Bessel functions depend on $|k|$. Hence, the integration on the whole interval does not allow for obtaining a tractable form for the wave packet in view of using the de Broglie–Bohm approach. Let us consider the general wave packet given by

$$\Psi(b, \sigma, T) = \int_0^\infty \int_{-\infty}^{+\infty} A(k)x^{\nu+1}e^{-(\gamma+iT)x^2}J_\nu(xb)e^{ik\sigma}dkdx, \qquad (59)$$

with the definition $x = \sqrt{E}$ and $\nu = |k|$. The integration in x leads to

$$\Psi(b,\sigma,T) = \int_{-\infty}^{+\infty} A(k) \frac{b^{\nu}}{(\alpha)^{\nu+1}} e^{-\frac{b^2}{4\alpha}} e^{ik\sigma} dk. \tag{60}$$

With this, we investigate the behaviors of the scale factor and the scalar field in some special cases, and compare them with the results of the previous section.

5.1. The Scalar Field Is Absent

Let us consider the case where the factor A is given by a delta function. If we choose $A(k) = \delta(k)$, which means the world where $\nu = 0$, the contribution of the scalar field vanishes and the wave function is such that

$$\Psi(b,\sigma,T) = \frac{e^{-\frac{b^2}{4\alpha}}}{\alpha}. \tag{61}$$

Notice that this does not mean we recover General Relativity. This would be one world of the many in a BD theory of gravitation. Considering Equation (48), it is straightforward to verify that the norm of the wave function (61) is finite and time-independent. Similarly to the calculation in the last section, we can compute the expected value of the scale factor:

$$<b> = \frac{1}{\mathbf{N}^2} \int_0^\infty \frac{e^{-\gamma \frac{b^2}{2|\alpha|^2}}}{|\alpha|^2} b\, db = \frac{1}{\gamma \mathbf{N}^2} \sqrt{\gamma^2 + T^2}, \tag{62}$$

where $\mathbf{N}$ here is the normalization factor of the wave function (61), naturally. Therefore, in this world, a bounce occurs, and, for late times, that is, when $T \gg \gamma$, $\langle b \rangle \to T$.

For the Bohmian trajectories (57), we use the phase of the wavefunction (61), which is, in this case,

$$S = T \frac{b^2}{4|\alpha|^2} - \arctan\left(\frac{T}{\gamma}\right). \tag{63}$$

Remembering that the conjugate momentum is $p_b = \dot{b}/2$, we obtain for the Bohmian trajectories,

$$\dot{b} = \frac{bT}{|\alpha|^2}. \tag{64}$$

The solution of this differential equation is $b = b_0 \sqrt{\gamma^2 + T^2}$, which is the same result as before using expectation values. In a Bohmian analysis, this universe also has a bounce.

The results exposed in the previous section show quite generically that wave packets with constant finite norm lead to a zero expectation value for σ. From now on, we explore possibilities where we circumvent this restriction but at the price of having wave packets that are not finite, what is not an obstacle when dBB formulation is used, see Ref. [11].

5.2. A Single Scalar Mode

Let us consider now the superposition function such that $A(k) = \delta(k - k_0)$, where k_0 is a positive constant. This implies to consider a single scalar mode behaving like a plane wave. The wave function reads as

$$\Psi(b,\sigma,T) = \frac{b^{\nu_0}}{(\alpha)^{\nu_0+1}} e^{-\frac{b^2}{4\alpha}} e^{ik_0\sigma}, \tag{65}$$

with $\nu_0 = \sqrt{-\epsilon}k_0$. We will analyze the Bohmian scenario for both $\epsilon = 1$ and $\epsilon = -1$.

Let us start with $\epsilon = -1$. In this case, the phase of the wave function becomes

$$S = \frac{b^2}{4|\alpha|^2} T - (k+1)\arctan\left(\frac{T}{\gamma}\right) + k_0 \sigma. \tag{66}$$

The presence of the term $k_0\sigma$ changes the previous analysis, since now the variable σ is featured in the wave function. The guidance equations become:

$$\dot{b} = \frac{bT}{|\alpha|^2}, \tag{67}$$

$$\dot{\sigma}b^2 = 12k_0, \tag{68}$$

and their solutions are,

$$b = b_0\sqrt{\gamma^2 + T^2}, \tag{69}$$

$$\sigma = \sigma_0 \arctan\left(\frac{T}{\gamma}\right). \tag{70}$$

The scalar field is no longer a constant. In the original Jordan frame, Equation (69) yields

$$a \propto \exp\left[-\frac{\arctan\left(\frac{T}{\gamma}\right)}{\sqrt{|1 + \frac{2}{3}\omega|}}\right]\sqrt{\gamma^2 + T^2}. \tag{71}$$

Again, we have a non-singular solution recovering the classical solution asymptotically, but now the bounce in this case is asymmetric. Asymmetric bouncing models have been studied in Ref. [46].

Now, for the case $\epsilon = +1$, in an observer-dependent interpretation, this becomes a problematic situation since the energy E is not bounded from below. The signature of the Schrödinger equation is hyperbolic instead of elliptic. However, we can follow the same step as before to obtain the phase of the wave function and to compute the Bohmian trajectories. The main new feature now is that the order of the Bessel function is imaginary, $\nu = ik$. Hence, the phase of the wave function (60) for this case is

$$S = k_0 \ln\frac{b}{|\alpha|^2} + \frac{b^2}{4|\alpha|^2}T - \arctan\left(\frac{T}{\gamma}\right) + k_0\sigma. \tag{72}$$

The equations for the Bohmian trajectories become

$$\dot{b} = \frac{2k_0|\alpha|^2}{b} + \frac{T}{|\alpha|^2}b, \tag{73}$$

$$\dot{\sigma} = -12\frac{k_0}{b^2}. \tag{74}$$

The solutions are

$$b = b_0\sqrt{\gamma^2 + T^2}\sqrt{\sigma_0 - \bar{k}_0 \arctan\left(\frac{T}{\gamma}\right)}, \tag{75}$$

$$\sigma = 3\ln\left\{\sigma_0 - \bar{k}_0 \arctan\left(\frac{T}{\gamma}\right)\right\} \tag{76}$$

where b_0 and σ_0 are constants and $\bar{k}_0 = 36k_0/(b_0^2\gamma)$.

To return back to the scale factor in the Jordan frame, we use

$$a = \phi^{-1/2}b = e^{-\frac{\sigma}{\sqrt{1+\frac{2}{3}\omega}}}. \tag{77}$$

Hence,

$$a = a_0\sqrt{\gamma^2 + T^2}\left\{\sigma_0 + \bar{k}_0 \arctan\left(\frac{T}{\gamma}\right)\right\}^r, \tag{78}$$

with

$$r = \frac{1}{2}\frac{\sqrt{1+\frac{2}{3}\omega+6}}{\sqrt{1+\frac{2}{3}\omega}}. \qquad (79)$$

Notice that the solutions (75) and (76) are non-singular only if $\sigma_0 > \frac{\pi}{2}\bar{k}_0$ and $\sigma_0 > 0$. The classical solution is recovered asymptotically. This fact reveals again the very peculiar features when the energy conditions are satisfied.

We can compute the quantum potential that results from the modified Hamilton–Jacobi equation in the dBB formulation of quantum mechanics [11], which is given by

$$V_Q = \frac{\nabla^2 R}{R}, \quad R = \sqrt{\Psi^*\Psi}. \qquad (80)$$

The Laplacian operator is defined in the minisuperspace with variables b and σ. For the cases studied in this subsection, the result is the expected: the quantum potential reaches its maximum value at the bounce, and decreases to zero asymptotically where the classical solutions are recovered.

In performing the study above, we have used the Bessel function given in a (44). It can be explicitly verified that, if we have used the Bessel function of negative order, which is not regular at the origin, the results would be the same with a reversal of the time, $T \to -T$.

5.3. Multiple Scalar Modes

We can combine different scalar modes. One example can be achieved by the combination,

$$A(k) = \delta(k - k_0) + \eta\delta(k + k_0), \qquad (81)$$

with $\eta = \pm 1$. With a similar computation, we obtain, for the positive sign, $\cos k_0\sigma$ instead of e^{ik_0} and for a negative sign $\sin k_0\sigma$. In both of these cases, the scalar field is not present in the phase of the wave function, and we recover the same solutions already given in the case that the scalar field is absent.

6. Conclusions

The Brans–Dicke theory is one of the oldest proposals of a modification to the theory of general relativity. Although it has been studied for over sixty years, the Brans–Dicke theory continues to reveal intriguing aspects. Some of these were discussed in the present work concerning classical and quantum scenarios for the early universe. First of all, it is possible to obtain a singularity-free cosmological solution if the Brans–Dicke parameter ω varies as $-3/2 < \omega < -4/3$. In this range, the energy conditions are satisfied in the Einstein frame: the avoidance of the singularity is driven by the non-minimal coupling.

When we turn to the quantum scenario in the minisuperspace, other curious features appear. The energy is bounded only for $\omega < -3/2$, that is, when the energy condition is violated in the Einstein frame. On the other hand, in the case of $\omega > -3/2$, the energy conditions are satisfied, but the energy is not bounded from below, so it becomes problematic to employ the usual interpretation scheme based on the Copenhagen formulation of quantum mechanics: the wave function is not finite anymore.

This issue motivated us to consider the de Broglie–Bohm interpretation of quantum mechanics. The previous results are found again if $\omega < -3/2$. However, when $\omega > -3/2$, we have either a singular or non-singular solution. This implies that, in the interval $-\frac{3}{2} < \omega < -\frac{4}{3}$, the classical model displays singularity-free scenarios, while the quantum models may display either singular or non-singular solutions. This result raises the following question: can such results occur in more general modified gravity theories belonging to the Horndesky class?

We add to the previous discussion some additional remarks. First of all, it must be verified to what extent the results reported here depend on the wave packet construction and the choice of the time variable. On the other hand, the problems of convergence of the wave function, when the energy conditions are obeyed, is somehow general in the presence of scalar fields since it leads to a hyperbolic signature in the Hamiltonian and, consequently, in the Schrödinger-type equation. However, such a well-known fact acquires new features that were briefly described above. The analysis displayed in the present work shows that the well-studied Brans–Dicke cosmological models present interesting properties at classical and quantum levels as well, no matter which interpretation is chosen.

Author Contributions: All authors contributed equally to the present work. All authors have read and agreed to the published version of the manuscript.

Funding: This research was supported by the National Scientific and Technological Research Council (CNPq, Brazil) and the State Scientific and Innovation Funding Agency of Espírito Santo (FAPES, Brazil).

Conflicts of Interest: The authors declare no conflict of interest.

References

1. Brans, C.; Dicke, R.H. Mach's Principle and a Relativistic Theory of Gravitation. *Phys. Rev.* **1961**, *124*, 925–935. [CrossRef]
2. DeWitt, B.S. Quantum Theory of Gravity. I. The Canonical Theory. *Phys. Rev.* **1967**, *160*, 1113. [CrossRef]
3. Kiefer, C. *Quantum Gravity*; Oxford University Press: Oxford, UK , 2007.
4. Kuchar, K. Time and interpretation of quantum gravity. *Int. J. Mod. Phys. D* **2011**, *20*, 3–86. [CrossRef]
5. Isham, C.J. Canonical quantum gravity and the problem of time. *Sci. Ser. C* **1993**, *409*, 157–287.
6. Schutz, B.F. Perfect Fluids in General Relativity: Velocity Potentials and a Variational Principle. *Phys. Rev. D* **1970**, *2*, 2762. [CrossRef]
7. Everett, H. Relative state formulation of quantum mechanics. *Rev. Mod. Phys.* **1957**, *29*, 454–462. [CrossRef]
8. Omnès, R. *The Interpretation of Quantum Mechanics*; Princeton University Press: Princeton, NJ, USA, 1994.
9. Bohm, D.; Hiley, B.J. *The Undivided Universe: An Ontological Interpretation of Quantum Theory*; Routledge: London, UK, 1993.
10. Holland, P.R. *The Quantum Theory of Motion: An Account of the de Broglie–Bohm Causal Interpretation of Quantum Mechanics*; Cambridge University Press: Cambridge, UK, 1993.
11. Pinto-Neto, N.; Fabris, J.C. Quantum cosmology from the de Broglie–Bohm perspective. *Class. Quantum Gravity* **2013**, *30*, 143001. [CrossRef]
12. Dirac, P.A.M. The Cosmological Constants. *Nature* **1937**, *139*, 323. [CrossRef]
13. Jordan, P. The present state of Dirac's cosmological hypothesis. *Z. Phys.* **1959**, *157*, 112. [CrossRef]
14. Will, C.M. *Theory and Experiment in Gravitational Physics*; Cambridge University Press: Cambridge, UK, 2018.
15. Weinberg, S. *Gravitation and Cosmology*; John Wiley and Sons: Hoboken, NJ, USA, 1972.
16. Banerjee, N.; Sen, S. Does Brans–Dicke theory always yield general relativity in the infinite limit? *Phys. Rev. D* **1997**, *56*, 1334. [CrossRef]
17. Faraoni, V. Illusions of general relativity in Brans–Dicke gravity. *Phys. Rev. D* **1999**, *59*, 084021. [CrossRef]
18. Chauvineau, B. On the limit of Brans–Dicke theory when $\omega \to \infty$. *Class. Quantum Gravity* **2003**, *20*, 2617. [CrossRef]
19. Brando, G.; Fabris, J.C.; Falciano, F.T.; Galkina, O. Stiff matter solution in Brans–Dicke theory and the general relativity limit. *Int. J. Mod. Phys. D* **2019**, *28*, 1950156. [CrossRef]
20. Will, C.M. The Confrontation between General Relativity and Experiment. *Living Rev. Relativ.* **2014**, *17*, 4. [CrossRef]
21. Gasperini, M. From Pre- to Post-Big Bang: An (almost) self-dual cosmological history. *arXiv* **2021**, arXiv:2106.12865.
22. Khoury, J.; Ovrut, B.A.; Steinhardt, P.J.; Turok, N. Ekpyrotic universe: Colliding branes and the origin of the hot big bang. *Phys. Rev. D* **2001**, *64*, 123522. [CrossRef]
23. La, D.; Steinhardt, P.J. Extended Inflationary Cosmology. *Phys. Rev. Lett.* **1989**, *62*, 376. [CrossRef]
24. Bailin, D.; Love, A. Kaluza-Klein theories. *Rep. Prog. Phys.* **1987**, *50*, 1087. [CrossRef]
25. Lidsey, J.E.; Wands, D.; Copeland, E.J. Superstring cosmology. *Phys. Rep.* **2000**, *337*, 343. [CrossRef]
26. Colistete, R., Jr.; Fabris, J.C.; Pinto-Neto, N. Singularities and classical limit in quantum cosmology with scalar fields. *Phys. Rev. D* **1998**, *57*, 4707. [CrossRef]
27. Colistete, R., Jr.; Fabris, J.C.; Pinto-Neto, N. Gaussian superpositions in scalar tensor quantum cosmological models. *Phys. Rev. D* **2000**, *62*, 083507. [CrossRef]
28. Almeida, C.R.; Batista, A.B.; Fabris, J.C.; Moniz, P.R.V. Quantum cosmology with scalar fields: Self-adjointness and cosmological scenarios. *Gravit. Cosmol.* **2015**, *21*, 191. [CrossRef]
29. Galkina, O.; Fabris, J.C.; Falciano, F.T.; Pinto-Neto, N. Regular bouncing solutions, energy conditions, and the Brans–Dicke theory. *JETP Lett.* **2019**, *110*, 523–528. [CrossRef]

30. Gurevich, L.E.; Finkelstein, A.M.; Ruban, V.A. On the problem of the initial state in the isotropic scalar-tensor cosmology of Brans–Dicke. *Astrophys. Space Sci.* **1973**, *22*, 231. [CrossRef]
31. Battefeld, D.; Peter, P. A Critical Review of Classical Bouncing Cosmologies. *Phys. Rep.* **2015**, *571*, 1–66. [CrossRef]
32. Novello, M.; Bergliaffa, S.E.P. Bouncing Cosmologies. *Phys. Rep.* **2008**, *463*, 127–213. [CrossRef]
33. Ijjas, A.; Steinhardt, P.J.; Bouncing Cosmology made simple. *Class. Quantum Gravity* **2018**, *35*, 135004. [CrossRef]
34. Peter, P.; Pinto-Neto, N. Primordial perturbations in a non singular bouncing universe model. *Phys. Rev. D* **2002**, *65*, 023513. [CrossRef]
35. Bronnikov, K.A. Scalar-tensor gravity and conformal continuations. *J. Math. Phys.* **2002**, *43*, 6096–6115. [CrossRef]
36. Pinto-Neto, N. *Hamiltonian Formulation of General Relativity and Application*; PPGCosmo series; Livraria da Física: São Paulo, Brazil, 2020.
37. Almeida, C.R.; Batista, A.B.; Fabris, J.C.; Moniz, P.V. Quantum Cosmology os Scalar-tensor Theories and Self-adjointness. *J. Math. Phys.* **2017**, *58*, 042301. [CrossRef]
38. Almeida, C.R.; Batista, A.B.; Fabris, J.C.; Pinto-Neto, N. Quantum Cosmological Scenarios of Brans–Dicke Gravity in Einstein and Jordan Frames. *Gravit. Cosmol.* **2018**, *24*, 245–253. [CrossRef]
39. Gradshteyn, I.S.; Ryzhik, I.M. *Table of Integrals, Series, and Products*; Academic Press: Cambridge, MA, USA, 2007.
40. de Broglie, L. *An Introduction to the Study of Wave Machanics*; E.P. Dutton and Company: New York, NY, USA, 1930.
41. Bohm, D. A Suggested Interpretation of the Quantum Theory in Terms of "Hidden" Variables I. *Phys. Rev.* **1952**, *85*, 166–179. [CrossRef]
42. Brown, H.R.; Wallace, D. Solving the Measurement Problem: De Broglie–Bohm Loses Out to Everett. *Found. Phys.* **2005**, *35*, 517–540. [CrossRef]
43. Holland, P. What's Wrong with Einstein's 1927 Hidden-Variable Interpretation of Quantum Mechanics? *Found. Phys.* **2005**, *35*, 177–196. [CrossRef]
44. Pinto-Neto, N. The Bohm Interpretation of Quantum Cosmology. *Found. Phys.* **2005**, *35*, 577–603. [CrossRef]
45. Marto, J.; Moniz, P.V. de Broglie–Bohm FRW universes in quantum string cosmology. *Phys. Rev.* **2001**, *65*, 023516. [CrossRef]
46. Delgado, P.C.M.; Pinto-Neto, N. Cosmological models with asymmetric quantum bounces. *Class. Quantum Gravity* **2020**, *37*, 125002. [CrossRef]

Review

Cosmological Particle Production in Quantum Gravity

Yaser Tavakoli [1,2]

1 Department of Physics, University of Guilan, Namjoo Blv., Rasht 41335-1914, Iran; yaser.tavakoli@guilan.ac.ir
2 School of Astronomy, Institute for Research in Fundamental Sciences (IPM), Tehran 19395-5531, Iran

Abstract: Quantum theory of a test field on a quantum cosmological spacetime may be viewed as a theory of the test field on an emergent classical background. In such a case, the resulting dressed metric for the field propagation is a function of the quantum fluctuations of the original geometry. When the backreaction is negligible, massive modes can experience an anisotropic Bianchi type I background. The field modes propagating on such a quantum-gravity-induced spacetime can then unveil interesting phenomenological consequences of the super-Planckian scales, such as gravitational particle production. The aim of this paper is to address the issue of gravitational particle production associated with the massive modes in such an anisotropic dressed spacetime. By imposing a suitable adiabatic condition on the vacuum state and computing the energy density of the created particles, the significance of the particle production on the dynamics of the universe in Planck era is discussed.

Keywords: loop quantum cosmology; quantum fields in curved spacetime

Citation: Tavakoli, Y. Cosmological Particle Production in Quantum Gravity. *Universe* **2021**, *7*, 258. https://doi.org/10.3390/universe7080258

Academic Editor: Jaime Haro Cases

Received: 16 June 2021
Accepted: 20 July 2021
Published: 22 July 2021

Publisher's Note: MDPI stays neutral with regard to jurisdictional claims in published maps and institutional affiliations.

Copyright: © 2021 by the authors. Licensee MDPI, Basel, Switzerland. This article is an open access article distributed under the terms and conditions of the Creative Commons Attribution (CC BY) license (https://creativecommons.org/licenses/by/4.0/).

1. Introduction

Observation of cosmic microwave background (CMB) implies that the classical universe is homogeneous and isotropic on scales larger than 250 million light years. Based on the standard ΛCDM model of cosmology, the Friedmann–Lemaître–Robertson–Walker (FLRW) solution of Einstein's field equations provides a suitable explanation of such near perfect isotropy of the CMB and other astrophysical observables. Nevertheless, there is not yet a decisive answer to the question of whether or not the quantum structure of the universe in the super-Planckian regime has the same symmetry as observed in the CMB and how such structure can be traced in the observational data.

According to the standard model of cosmology, the structure formation at large scales are described via (inhomogeneous) perturbations at smaller scales in the early universe. Quantum field theory in classical, curved spacetime provides a good approximate description of such phenomena in a regime where the quantum effects of gravity are negligible [1,2]. Within such framework, the issue of the gravitational particle production induced by the time-dependent background in an expanding universe and its backreaction effect is of significant importance [3,4]. Nevertheless, when tracing further back in time, where the curvature of the universe reaches the Planck scales, the quantum effects of gravity become important. This might lead to additional phenomena that are expected to be important when exploring the dynamics of the early universe.

Loop quantum cosmology (LQC) is a promising candidate to investigate the quantum gravity effects in Planckian regime [5]. It follows the quantization scheme of loop quantum gravity (LQG), which is a background independent, non-perturbative approach to the quantization of general relativity [6–8]. This approach has provided a number of concrete results: the classical big bang singularity is resolved and is replaced by a quantum bounce [9–12]; the standard theory of cosmological perturbations has been extended to a self-consistent theory from the bounce in the super-Planckian regime to the onset of slow-roll inflation [13–17]. Further phenomenological consequences and observational predictions of LQC can be found, e.g., in Refs. [17–29], respectively.

The cosmological quantum backgrounds in LQC establish a fruitful ground to explore the quantum theory of inhomogeneous fields propagating on it. In a dressed metric approach, when the full quantum Hamiltonian constraint of the gravity-matter system is solved, an effectively dressed geometry emerges for the field modes. Depending on whether the field is massless or massive, various spacetime metrics can raise for the dressed background geometry. The emergent metric components, if the backreaction between the fields and the geometry is discarded, depend, generically, on the fluctuations of the original quantum metric operator only. For massive modes, quantum gravity effects may induce a small deviation from the initial isotropy [19,24]. If the backreaction is considered, other properties, such as emerging a mode-dependent background, can raise, which leads to the violation of the local Lorentz symmetry [18,20]. Even in the absence of the backreaction between the states of the gravity and matter fields, a backreaction effect may arise due to the gravitational particle production, which can challenge the validity of the test field approximations employed in the dressed metric scenario [22].

Motivated from the above paragraphs, our purpose in this article is to address some phenomenological issues associated with the massive test field propagating on a quantum cosmological spacetime. In particular, we will consider a scenario in which an anisotropic dressed geometry emerges for the massive modes due to quantum gravity effects and then revisit the occurrence of the gravitational particle production on such an anisotropic dressed background. Such effects will have important consequences in the super-Planckian regime and at the onset of the classical inflationary epoch. We thus organize this paper as follows. In Section 2, we will consider the propagation of a massive test field on a quantized FLRW spacetime. We will derive an effective evolution equation for the field and obtain a suitable anisotropic dressed background for the field propagation. In Section 3, we will study the theory of the quantum field by considering the many infinite modes on the emergent anisotropic spacetime. We will then explore the problem of the gravitational particle production by choosing a convenient, adiabatic vacuum state in the super-Planckian regime. Finally, in Section 5, we will present the conclusion and discussion of our work.

2. Quantum Fields on a Quantum FLRW Spacetime

In this section, we study the quantum theory of a massive scalar field on a quantum FLRW background. Firstly, we will analyze the quantum theory of a scalar field in the classical Friedmann universe. Next, we will quantize the background and obtain a quantum evolution equation for the composite state of the gravity-perturbation system. Finally, in the last step, we will extract an effective evolution equation for the field mode degrees of freedom, due to which we can extract an effective anisotropic background for propagation of the massive modes.

2.1. Quantum Field on a Cosmological Classical Spacetime

We start with considering a *massive* scalar field propagating on a more general classical, *anisotropic* background spacetime. In particular, we take a Bianchi type I model whose line element is represented by

$$g_{ab}dx^a dx^b = -N_{x_0}^2(x_0)(dx_0)^2 + \sum_i^3 a_i^2(x_0)\left(dx^i\right)^2, \qquad (1)$$

where N_{x_0} is the lapse function and a_i the scale factor in the x^i-direction. The coordinates $(x_0, \mathbf{x})$ are chosen such that $x_0 \in \mathbb{R}$ is a generic time coordinate, and $\mathbf{x} \in \mathbb{T}^3$ (i.e., a 3-torus with the coordinates $x^j \in (0, \ell_j)$).

We consider a real, minimally coupled scalar field, $\varphi(x_0, \mathbf{x})$, with a mass m, propagating on the background (Equation (1)). Having the Lagrangian density for $\varphi(x_0, \mathbf{x})$, with a quadratic potential,

$$\mathcal{L}_\varphi = -\frac{1}{2}\left[g^{ab}\nabla_a \varphi \nabla_b \varphi + m^2 \varphi^2\right], \qquad (2)$$

the equation of motion is obtained as

$$(g^{ab}\nabla_a\nabla_b - m^2)\varphi = 0. \tag{3}$$

On the $x_0 = const$ slice, by performing the Legendre transformation, we take a canonically conjugate momentum π_φ to the field φ. Then, the classical solutions of Equation (3) for the pair (φ, π_φ) can be Fourier expanded as[1]

$$\varphi(x_0, \mathbf{x}) = \frac{1}{\ell^{3/2}} \sum_{\mathbf{k} \in \mathscr{L}} \varphi_\mathbf{k}(x_0) \, e^{i\mathbf{k}\cdot\mathbf{x}}, \tag{4a}$$

$$\pi_\varphi(x_0, \mathbf{x}) = \frac{1}{\ell^{3/2}} \sum_{\mathbf{k} \in \mathscr{L}} \pi_\mathbf{k}(x_0) \, e^{i\mathbf{k}\cdot\mathbf{x}}, \tag{4b}$$

with $\mathscr{L} = \mathscr{L}_+ \cup \mathscr{L}_-$ being a 3-dimensional lattice:

$$\mathscr{L}_+ = \{\mathbf{k} : k_3 > 0\} \cup \{\mathbf{k} : k_3 = 0, k_2 > 0\} \cup \{\mathbf{k} : k_3 = k_2 = 0, k_1 > 0\}, \tag{5a}$$

and

$$\mathscr{L}_- = \{\mathbf{k} : k_3 < 0\} \cup \{\mathbf{k} : k_3 = 0, k_2 < 0\} \cup \{\mathbf{k} : k_3 = k_2 = 0, k_1 < 0\}, \tag{5b}$$

spanned by $\mathbf{k} = (k_1, k_2, k_3) \in (2\pi\mathbb{Z}/\ell)^3$, where $\mathbb{Z}$ is the set of integers [13,18].
Let us decompose the field modes into the real and imaginary parts as

$$\varphi_\mathbf{k}(x_0) = \frac{1}{\sqrt{2}}\left[\varphi_\mathbf{k}^{(1)}(x_0) + i\varphi_\mathbf{k}^{(2)}(x_0)\right]. \tag{6}$$

Then, the reality condition implies that $\varphi_\mathbf{k}^{(1)} = \varphi_{-\mathbf{k}}^{(1)}$ and $\varphi_{-\mathbf{k}}^{(2)} = -\varphi_\mathbf{k}^{(2)}$. By introducing a new variable $Q_\mathbf{k}$,

$$Q_\mathbf{k}(x_0) = \begin{cases} \varphi_\mathbf{k}^{(1)}(x_0), & \text{if } \mathbf{k} \in \mathscr{L}_+, \\ \varphi_{-\mathbf{k}}^{(2)}(x_0), & \text{if } \mathbf{k} \in \mathscr{L}_-, \end{cases} \tag{7}$$

associated to the real variables $\varphi_\mathbf{k}^{(1)}$ and $\varphi_\mathbf{k}^{(2)}$, we can rewrite the field modes $\varphi_\mathbf{k}$ as

$$\varphi_\mathbf{k}(x_0) = \frac{1}{\sqrt{2}}[Q_\mathbf{k}(x_0) + iQ_{-\mathbf{k}}(x_0)]. \tag{8}$$

Likewise, the decomposition of the momentum $\pi_\mathbf{k}$ as

$$\pi_\mathbf{k}(x_0) = \frac{1}{\sqrt{2}}\left[\pi_\mathbf{k}^{(1)}(x_0) + i\pi_\mathbf{k}^{(2)}(x_0)\right], \tag{9}$$

in terms of the real variables $\pi_\mathbf{k}^{(1)}$ and $\pi_\mathbf{k}^{(2)}$, and introducing the new variable $P_\mathbf{k}$ (conjugate to $Q_\mathbf{k}$ above), as

$$P_\mathbf{k}(x_0) = \begin{cases} \pi_\mathbf{k}^{(1)}(x_0), & \text{if } \mathbf{k} \in \mathscr{L}_+, \\ \pi_{-\mathbf{k}}^{(2)}(x_0), & \text{if } \mathbf{k} \in \mathscr{L}_-, \end{cases} \tag{10}$$

yields

$$\pi_\mathbf{k}(x_0) = \frac{1}{\sqrt{2}}[P_\mathbf{k}(x_0) + iP_{-\mathbf{k}}(x_0)]. \tag{11}$$

The conjugate variables $(Q_\mathbf{k}, P_\mathbf{k})$ satisfy the Poisson bracket, $\{Q_\mathbf{k}, P_{\mathbf{k}'}\} = \delta_{\mathbf{k},\mathbf{k}'}$.

Using the Equations (7) and (10), the Hamiltonian of the scalar test field, φ, can be written as the sum of the Hamiltonians, $H_{\mathbf{k}}(x_0)$, of the decoupled harmonic oscillators, each given in terms of $(Q_{\mathbf{k}}, P_{\mathbf{k}})$:

$$H_\varphi(x_0) := \sum_{\mathbf{k}\in\mathscr{L}} H_{\mathbf{k}}(x_0) = \frac{N_{x_0}(x_0)}{2|a_1 a_2 a_3|} \sum_{\mathbf{k}\in\mathscr{L}} \left(P_{\mathbf{k}}^2 + \omega_k^2(x_0) Q_{\mathbf{k}}^2 \right), \qquad (12)$$

where $\omega_k(x_0)$ is a time-dependent frequency, defined by

$$\omega_k^2(x_0) := |a_1 a_2 a_3|^2 \left[\sum_{i=1}^{3} \left(\frac{k_i}{a_i} \right)^2 + m^2 \right]. \qquad (13)$$

Note that $Q_{\mathbf{k}}$ is the field amplitude for the mode characterized by $\mathbf{k}$.

To quantize the field, we follow the Schrödinger representation. While the background spacetime is left as the classical, the quantization of $Q_{\mathbf{k}}$ for a fixed mode $\mathbf{k}$ resembles that of a quantum harmonic oscillator with the Hilbert space $\mathcal{H}_Q^{(\mathbf{k})} = L^2(\mathbb{R}, dQ_{\mathbf{k}})$, where $(Q_{\mathbf{k}}, P_{\mathbf{k}})$ are promoted to operators on $\mathcal{H}_Q^{(\mathbf{k})}$ as

$$\hat{Q}_{\mathbf{k}} \psi(Q_{\mathbf{k}}) = Q_{\mathbf{k}} \psi(Q_{\mathbf{k}}) \quad \text{and} \quad \hat{P}_{\mathbf{k}} \psi(Q_{\mathbf{k}}) = -i\hbar (\partial/\partial Q_{\mathbf{k}}) \psi(Q_{\mathbf{k}}). \qquad (14)$$

Then, the Hamiltonian operator $\hat{H}_{\mathbf{k}}$ generates the time evolution of the state $\psi(Q_{\mathbf{k}})$ via the Schrödinger equation

$$i\hbar \partial_{x_0} \psi(x_0, Q_{\mathbf{k}}) = \frac{\ell^3 N_{x_0}}{2V} \left(\hat{P}_{\mathbf{k}}^2 + \omega_k^2 \hat{Q}_{\mathbf{k}}^2 \right) \psi(x_0, Q_{\mathbf{k}}), \qquad (15)$$

where $V = \ell^3 |a_1 a_2 a_3|$ denotes the physical volume of the universe.

By setting $x_0 = \phi$ in Equation (15) as an *internal time* parameter, the evolution of the state $\psi(Q_{\mathbf{k}})$ with respect to ϕ on a Bianchi-I background with components $(\tilde{N}_\phi, \tilde{a}_i(\phi))$ reads

$$i\hbar \partial_\phi \psi(\phi, Q_{\mathbf{k}}) = \frac{\tilde{N}_\phi}{2|\tilde{a}_1 \tilde{a}_2 \tilde{a}_3|} \left[\hat{P}_{\mathbf{k}}^2 + \tilde{\omega}_k^2(\phi) \hat{Q}_{\mathbf{k}}^2 \right] \psi(\phi, Q_{\mathbf{k}}), \qquad (16)$$

where

$$\tilde{\omega}_k^2(\phi) = \left(\sum_i^3 \frac{\tilde{k}_i^2}{\tilde{a}_i^2} + \tilde{m}^2 \right) (\tilde{a}_1 \tilde{a}_2 \tilde{a}_3)^2. \qquad (17)$$

Clearly, one gets an isotropic background for the field by $\tilde{a}_1(\phi) = \tilde{a}_2(\phi) = \tilde{a}_3(\phi) \equiv \tilde{a}(\phi)$.

2.2. Quantization of the Background

In our model herein this paper, we will assume that the field, φ, propagates on an *isotropic FLRW spacetime* in a super-Planckian regime so that this isotropic background has to be quantized. However, the reason for constructing a general formalism of the field on an anisotropic background in the previous subsection is for the purpose of comparison, when an effectively dressed spacetime emerges from the isotropic quantum background. We will show that the emergent effective spacetime can have the same structure of an anisotropic Bianchi-I geometry for the field propagation.

Let us assume a harmonic time gauge, $x_0 = \tau$, and set the isotropic components as $a_1 = a_2 = a_3 = a(\tau)$ in Equation (1). Then, $N_\tau = a^3(\tau)$, and the Hamiltonian (12) becomes

$$H_\varphi^{(\text{iso})} = \sum_{\mathbf{k}} H_{\tau, \mathbf{k}} := \frac{1}{2} \sum_{\mathbf{k}} \left(P_{\mathbf{k}}^2 + \omega_{\tau, k}^2 Q_{\mathbf{k}}^2 \right), \qquad (18)$$

where

$$\omega_{\tau, k}^2(\tau) = k^2 a^4(\tau) + m^2 a^6(\tau). \qquad (19)$$

We note that the massless scalar field, $\phi(\tau)$, still serves as an internal time parameter.

In quantum theory, we will quantize not only the test field but also the background geometry. We will assume that the backreaction of the quantum field on the background quantized spacetime is negligible. This yields an evolution for the wave function, $\psi(Q_\mathbf{k})$, of the test field with respect to the internal time ϕ. Let $\mathcal{H}_{\text{kin}}^o = \mathcal{H}_{\text{grav}} \otimes \mathcal{H}_\phi$ denote the background Hilbert space, which consists of the Hilbert space of the gravity sector and that of the scalar clock variable ϕ; the matter sector is quantized according to the Schrödinger representation, $\mathcal{H}_\phi = L^2(\mathbb{R}, d\phi)$. Likewise, the Hilbert space of the field modes reads $\mathcal{H}_\varphi^{(\mathbf{k})} = L^2(\mathbb{R}, dQ_\mathbf{k})$, as before. Subsequently, the full kinematical Hilbert space of the system for a single mode $\mathbf{k}$ is given by $\mathcal{H}_{\text{kin}}^{(\mathbf{k})} = \mathcal{H}_{\text{kin}}^o \otimes \mathcal{H}_\varphi^{(\mathbf{k})}$.

In LQC, the background Hamiltonian constraint operator,

$$\hat{\mathcal{C}}_o = \hat{\mathcal{C}}_{\text{grav}} + \hat{\mathcal{C}}_\phi, \tag{20}$$

is well-defined on $\mathcal{H}_{\text{kin}}^o$; the physical states $\Psi_o(\phi, \nu) \in \mathcal{H}_{\text{kin}}^o$ are those lying on the kernel of $\hat{\mathcal{C}}_o$ and are solutions to the self-adjoint constraint Equation: [30]

$$N_\tau \hat{\mathcal{C}}_o \Psi_o(\nu, \phi) = -\frac{\hbar^2}{2\ell^3}(\partial_\phi^2 + \Theta)\Psi_o(\nu, \phi) = 0. \tag{21}$$

The quantum number ν is the eigenvalue of the background volume operator, $\hat{V}_o = \widehat{\ell^3 a^3}$, which acts on the states $\Psi_o(\phi, \nu) \in \mathcal{H}_{\text{kin}}^o$ as

$$\hat{V}_o \Psi_o(\nu, \phi) = 2\pi\gamma\ell_{\text{Pl}}|\nu|\Psi_o(\nu, \phi). \tag{22}$$

Moreover, Θ is a difference operator that acts on $\Psi_o(\nu)$, involving only the volume sector ν. Taking only the positive frequency solutions to Equation (21), we get a Schrödinger equation for the background as

$$-i\hbar\partial_\phi \Psi_o(\nu, \phi) = \hbar\sqrt{\Theta}\Psi_o(\nu, \phi) =: \hat{H}_o \Psi_o(\nu, \phi). \tag{23}$$

The solutions yield a physical Hilbert space, $\mathcal{H}_{\text{phys}}^o$, equipped by the inner product

$$\langle \Psi_o | \Psi_o' \rangle = \sum_\nu \Psi_o^*(\nu, \phi_0)\Psi_o'(\nu, \phi_0), \tag{24}$$

for an "instant" ϕ_0 of the internal time.

For a composite state $\Psi(\nu, Q_\mathbf{k}, \phi) \in \mathcal{H}_{\text{kin}}^{(\mathbf{k})}$ of the geometry-test field system, the action of the total quantum Hamiltonian constraint, $\hat{\mathcal{C}}_{\tau,\mathbf{k}}$, is written as [13]

$$\hat{\mathcal{C}}_{\tau,\mathbf{k}} \Psi(\nu, Q_\mathbf{k}, \phi) = (N_\tau \hat{\mathcal{C}}_o + \hat{H}_{\tau,\mathbf{k}})\Psi(\nu, Q_\mathbf{k}, \phi) = 0, \tag{25}$$

where $\hat{H}_{\tau,\mathbf{k}}$ is the Hamiltonian operator of the $\mathbf{k}$th field mode,

$$\hat{H}_{\tau,\mathbf{k}} = \frac{1}{2}\left[\hat{P}_\mathbf{k}^2 + \left(k^2 \hat{a}^4 + m^2 \hat{a}^6\right)\hat{Q}_\mathbf{k}^2\right]. \tag{26}$$

By replacing Equation (21) into the constraint Equation (25), we obtain

$$-i\hbar\partial_\phi \Psi(\nu, Q_\mathbf{k}, \phi) = \left[\hat{H}_o^2 - 2\ell^3 \hat{H}_{\tau,\mathbf{k}}\right]^{\frac{1}{2}} \Psi(\nu, Q_\mathbf{k}, \phi), \tag{27}$$

which represents a quantum evolution of $\Psi(\nu, Q_\mathbf{k}, \phi)$ with respect to the internal time ϕ. In a test field approximation, when the backreaction effect is omitted, the expression under the square root can be expanded up to the first-order terms, as [13]

$$-i\hbar\partial_\phi \Psi(\nu, Q_\mathbf{k}, \phi) \approx (\hat{H}_o - \hat{H}_{\phi,\mathbf{k}})\Psi(\nu, Q_\mathbf{k}, \phi). \tag{28}$$

In expanding the right-hand-side of Equation (27) to derive the equation above, we regarded $\hat{H}_o$ as the Hamiltonian of the heavy degree of freedom, whereas $\hat{H}_{\phi,\mathbf{k}}$, defined by

$$\hat{H}_{\phi,\mathbf{k}} := \ell^3 \hat{H}_o^{-\frac{1}{2}} \hat{H}_{\tau,\mathbf{k}} \hat{H}_o^{-\frac{1}{2}}, \qquad (29)$$

was considered as the Hamiltonian of the light degree of freedom (i.e, a perturbation term). In this approximation, it is suitable to separate the total state of the system as

$$\Psi(\nu, Q_{\mathbf{k}}, \phi) = \Psi_o(\nu, \phi) \otimes \psi(Q_{\mathbf{k}}, \phi). \qquad (30)$$

To explore the quantum evolution of a pure test field state, $\psi(Q_{\mathbf{k}}, \phi)$, on a time-dependent background, it is more convenient to employ an interaction picture. Thus, we introduce

$$\Psi_{\text{int}}(\nu, Q_{\mathbf{k}}, \phi) = e^{-(i\hat{H}_o/\hbar)(\phi - \phi_0)} \Psi(\nu, Q_{\mathbf{k}}, \phi), \qquad (31)$$

In this picture, the geometry evolves by $\hat{H}_o$ through Equation (23) for any $\Psi_o \in \mathcal{H}_{\text{kin}}^o$ in the Heisenberg picture,

$$\Psi_o(\nu, \phi) = e^{(i\hat{H}_o/\hbar)(\phi - \phi_0)} \Psi_o(\nu, \phi_0). \qquad (32)$$

Plugging this into Equation (31), we get

$$\Psi_{\text{int}}(\nu, Q_{\mathbf{k}}, \phi) = \Psi_o(\nu, \phi_0) \otimes \psi(Q_{\mathbf{k}}, \phi). \qquad (33)$$

Thereby, the geometrical sector in the composite state $\Psi_{\text{int}}(\nu, Q_{\mathbf{k}}, \phi)$ becomes frozen in the instant of time ϕ_0 so that Ψ_{int} represents a time-dependent test field state, ψ, solely.

By replacing Equation (33) into the evolution Equation (28) and tracing out the geometrical state, $\Psi_o(\nu, \phi_0)$, a quantum evolution for $\psi(Q_{\mathbf{k}}, \phi)$ is obtained as

$$i\hbar \partial_\phi \psi = \frac{1}{2}\left[\langle \hat{H}_o^{-1} \rangle \hat{P}_{\mathbf{k}}^2 + \left(k^2 \langle \hat{H}_o^{-\frac{1}{2}} \hat{a}^4(\phi) \hat{H}_o^{-\frac{1}{2}} \rangle + m^2 \langle \hat{H}_o^{-\frac{1}{2}} \hat{a}^6(\phi) \hat{H}_o^{-\frac{1}{2}} \rangle \right) \hat{Q}_{\mathbf{k}}^2 \right] \psi, \qquad (34)$$

where $\langle \cdot \rangle$ denotes the expectation value with respect to $\Psi_o(\nu, \phi_0)$. It is clear that the use of the interaction picture in Equation (34) provided the Heisenberg description for the quantum geometrical elements; that is, the geometry state, $\Psi_o(\nu, \phi_0)$, is fixed at time $\phi = \phi_0$, while the geometrical operator, $\hat{a}(\phi) = \hat{V}_o^{1/3}(\phi)/\ell$, evolves in time as

$$\hat{a}(\phi) = e^{-(i\hat{H}_o/\hbar)(\phi - \phi_0)} \hat{a} \, e^{(i\hat{H}_o/\hbar)(\phi - \phi_0)}. \qquad (35)$$

Therefore, Equation (34) represents a ϕ-evolution of the field, $\psi(Q_{\mathbf{k}}, \phi)$, on a *time-dependent* (classical) background, which is similar to the one we had in classical spacetime (cf. Equation (16)).

2.3. Emergence of Anisotropic Dressed Spacetimes

Equation (34) can be interpreted as an evolution equation for the field modes on an (effective) classical spacetime, whose components are generated by the expectation values of the original isotropic quantum geometry operators with respect to the unperturbed state Ψ_o. To explore the properties of such effective spacetime, we can compare Equation (34), for the evolution of the state ψ, with the corresponding Equation (16), for the same state ψ, on an anisotropic classical background. This comparison yields a set of relations between parameters of the Bianchi I geometry, $(\tilde{N}_\phi, \tilde{a}_i, \tilde{k}_i, \tilde{m})$, and those of the isotropic quantum geometry, $(\hat{a}, \hat{H}_o, k, m)$, as

$$\tilde{N}_\phi = \ell^3 |\tilde{a}_1 \tilde{a}_2 \tilde{a}_3| \langle \hat{H}_o^{-1} \rangle, \tag{36a}$$

$$\sum_{i=1}^{3} \frac{\tilde{k}_i^2}{\tilde{a}_i^2} \tilde{N}_\phi |\tilde{a}_1 \tilde{a}_2 \tilde{a}_3| = \sum_{i=1}^{3} k_i^2 \ell^3 \langle \hat{H}_o^{-1/2} \hat{a}^4 \hat{H}_o^{-1/2} \rangle, \tag{36b}$$

$$\tilde{N}_\phi \tilde{m}^2 |\tilde{a}_1 \tilde{a}_2 \tilde{a}_3| = \ell^3 m^2 \langle \hat{H}_o^{-1/2} \hat{a}^6 \hat{H}_o^{-1/2} \rangle. \tag{36c}$$

Note that $\hat{a} \equiv \hat{a}(\phi)$ and $\langle \hat{H}_o^{-1} \rangle = (\tilde{p}_\phi)^{-1}$. Equation (36a–c) provide an underdetermined system of five equations with eight unknowns $(\tilde{N}_\phi, \tilde{a}_i, \tilde{k}_i, \tilde{m})$. Thus, to be able to solve this system, we need to impose some arbitrary conditions on these parameters to reduce the number of unknowns to five. Different classes of solutions by imposing various conditions on the variables and their physical consequences were discussed in [24]. As an example, we will present two classes of such solutions; one is produced by a massless test field, $m = \tilde{m} = 0$, and the other is provided by the *dressed* massive modes, $\tilde{m}, m \neq 0$.

For a massless test field, $m = 0$, we will immediately obtain $\tilde{m} = 0$. In this case, we will have four equations for the seven unknown parameters $(\tilde{N}_\phi, \tilde{a}_i, \tilde{k}_i)$. Therefore, we will still need three more conditions to be able to solve the system in Equation (36a,b). The simplest choice is $k_i^2 = \tilde{k}_i^2$ (for each i) so that $\tilde{a}_1^2 = \tilde{a}_2^2 = \tilde{a}_3^2 = \tilde{a}^2$. Then, we obtain

$$\tilde{a}^4 = \frac{\langle \hat{H}_o^{-1/2} \hat{a}^4(\phi) \hat{H}_o^{-1/2} \rangle}{\langle \hat{H}_o^{-1} \rangle}, \tag{37a}$$

$$\tilde{N}_\phi = \ell^3 \langle \hat{H}_o^{-1} \rangle^{\frac{1}{4}} \langle \hat{H}_o^{-1/2} \hat{a}^4(\phi) \hat{H}_o^{-1/2} \rangle^{\frac{3}{4}} = \ell^3 \tilde{p}_\phi^{-1} \tilde{a}^3 \equiv \bar{N}_\phi. \tag{37b}$$

If $m, \tilde{m} \neq 0$ and $\tilde{m} \neq m$, we will have different ranges of solutions (cf. [24]). However, for our purpose in this paper, we will consider only a specific solution by imposing the condition $\tilde{k}_i = \alpha_{ii} k_i$ (with $i = 1, 2, 3$). This yields

$$\tilde{N}_\phi = \frac{\bar{N}_\phi}{\lambda}, \tag{38a}$$

$$\tilde{a}_i = \frac{\alpha_{ii}}{\lambda} \tilde{a}, \tag{38b}$$

$$\tilde{m}^4 = m^4 \lambda^4 \frac{\langle \hat{H}_o^{-1/2} \hat{a}^6(\phi) \hat{H}_o^{-1/2} \rangle^2 \langle \hat{H}_o^{-1} \rangle}{\langle \hat{H}_o^{-1/2} \hat{a}^4(\phi) \hat{H}_o^{-1/2} \rangle^3}, \tag{38c}$$

where $\lambda \equiv (\alpha_{11} \alpha_{22} \alpha_{33})^{1/2}$ is a constant. Note that, as a special subcase, when $\alpha_{ii} = 1$ (so $\lambda = 1$), the wave vector becomes undressed, $\tilde{k}_i = k_i$, and an isotropic dressed scale factor, $\tilde{a}_1 = \tilde{a}_2 = \tilde{a}_3 = \tilde{a}$, identical to Equation (37a), is obtained. However, here, different from the isotropic case above for $m = 0$, a nonzero dressed mass is obtained as

$$\tilde{m}^2 = m^2 \frac{\langle \hat{H}_o^{-1/2} \hat{a}^6(\phi) \hat{H}_o^{-1/2} \rangle \langle \hat{H}_o^{-1} \rangle^{\frac{1}{2}}}{\langle \hat{H}_o^{-1/2} \hat{a}^4(\phi) \hat{H}_o^{-1/2} \rangle^{\frac{3}{2}}} = \frac{\tilde{m}^2}{\lambda^2}. \tag{39}$$

This solution represents an isotropic dressed spacetime with the scale factor $\bar{a}(\phi)$, over which a massive mode with the mass $\tilde{m}$ and an undressed wave-vector $\mathbf{k} = (k_1, k_2, k_3)$ propagate.

3. QFT on the Dressed Spacetime

To date, we have seen that the massive field modes propagating on an isotropic quantum geometry can explore a classical anisotropic dressed background, $\tilde{g}_{ab}$ (see, e.g., Equation (38a–c), which is the solution of Equation (36a–c). In this section, we will study the QFT and the issue of the gravitational particle productions on such an emergent anisotropic spacetime.

3.1. Field Equation on the Dressed Anisotropic Background

We suppose that the scalar field, φ, propagates on the dressed Bianchi I background

$$\tilde{g}_{ab}dx^a dx^b = -\tilde{N}_\phi^2(\phi)d\phi^2 + \sum_{i=1}^{3} \tilde{a}_i^2(\phi)\left(dx^i\right)^2, \tag{40}$$

whose components are given as the solutions to Equation (36). For convenience, we introduce the new variables

$$\tilde{c}(\phi) := (\tilde{a}_1\tilde{a}_2\tilde{a}_3)^{\frac{2}{3}} = (\tilde{c}_1\tilde{c}_2\tilde{c}_3)^{\frac{1}{3}} \quad \text{and} \quad \tilde{c}_i(\phi) := \tilde{a}_i^2(\phi), \tag{41}$$

and consider a conformal time parameter, $\tilde{\eta}$, defined by

$$\tilde{N}_\phi d\phi = \tilde{N}_{\tilde{\eta}} d\tilde{\eta} = \tilde{c}^{1/2} d\tilde{\eta} \quad \Rightarrow \quad d\tilde{\eta} = \ell^3 \langle \hat{H}_o^{-1} \rangle \tilde{c}(\phi) d\phi. \tag{42}$$

Let us now take an auxiliary field, $\chi_\mathbf{k} \equiv \sqrt{\tilde{c}(\tilde{\eta})}\, \varphi_\mathbf{k}$. Then, in terms of the above variables, the equation of motion reads

$$\chi_\mathbf{k}'' + \left[\tilde{\omega}_{\tilde{\eta},k}^2(\tilde{\eta}) - \frac{\tilde{c}''}{2\tilde{c}} + \frac{\tilde{c}'^2}{4\tilde{c}^2}\right]\chi_\mathbf{k} = 0, \tag{43}$$

where a prime stands for a differentiation with respect to $\tilde{\eta}$, and $\tilde{\omega}_{\tilde{\eta},k}$ is given by

$$\tilde{\omega}_{\tilde{\eta},k}^2(\tilde{\eta}) = \tilde{c}^{-2}\left[k^2 \frac{\langle \hat{H}_o^{-\frac{1}{2}} \hat{a}^4 \hat{H}_o^{-\frac{1}{2}} \rangle}{\langle \hat{H}_o^{-1} \rangle} + m^2 \frac{\langle \hat{H}_o^{-\frac{1}{2}} \hat{a}^6 \hat{H}_o^{-\frac{1}{2}} \rangle}{\langle \hat{H}_o^{-1} \rangle}\right]. \tag{44}$$

Notice that in simplifying equations above, we have used the relations between components according to Equation (36).

The frequency from Equation (44) is specified only when the solutions for $\tilde{c}(\tilde{\eta})$ are determined. It turns out that only two classes of solutions for $\tilde{c}$ exist, which depend on the conditions on the field mass:

(i) For massive field with *undressed* mass, $\tilde{m} = m \neq 0$, the relation for $\tilde{c}(\tilde{\eta})$ has the form

$$\tilde{c}(\tilde{\eta}) = \frac{\langle \hat{H}_o^{-\frac{1}{2}} \hat{a}^6 \hat{H}_o^{-\frac{1}{2}} \rangle^{\frac{1}{3}}}{\langle \hat{H}_o^{-1} \rangle^{\frac{1}{3}}}. \tag{45}$$

(ii) For a massive field with the *dressed* mass, $\tilde{m} \neq m$, or a *massless* field, $\tilde{c}(\tilde{\eta})$ has the solutions of the form

$$\tilde{c}(\tilde{\eta}) = \zeta^2 \frac{\langle \hat{H}_o^{-\frac{1}{2}} \hat{a}^4 \hat{H}_o^{-\frac{1}{2}} \rangle^{\frac{1}{2}}}{\langle \hat{H}_o^{-1} \rangle^{\frac{1}{2}}} = \zeta^2 \tilde{a}^2(\tilde{\eta}), \tag{46}$$

where ζ is a parameter that distinguishes the anisotropic solutions from the isotropic one and depends on the conditions imposed on the additional unknown variables $\tilde{a}_i$'s and $\tilde{k}_i$'s of the system in Equation (36). The case $\zeta^2 = 1$ associates with the *isotropic* solution, whereas the case $\zeta \neq 1$ denotes the *anisotropic* solutions (cf. Equations (37) and (39) for different classes of the solutions).

3.2. The Adiabatic Condition and the Vacuum State

Any solution $\chi_\mathbf{k}(\tilde{\eta})$ to the Klein–Gordon equation (Equation (43)) can be expanded as

$$\chi_\mathbf{k}(\tilde{\eta}) = \frac{1}{\sqrt{2}}\left[a_\mathbf{k} u_k^*(\tilde{\eta}) + a_{-\mathbf{k}}^* u_k(\tilde{\eta})\right], \tag{47}$$

where $a_\mathbf{k}$ and $a_\mathbf{k}^*$ are constants of integration, and $u_k(\tilde{\eta})$ are the general mode solutions of Equation (43),

$$u_k'' + \left(\tilde{\omega}_{\tilde{\eta},k}^2(\tilde{\eta}) - \mathcal{Q}(\tilde{\eta})\right)u_k = 0, \tag{48}$$

with $\mathcal{Q} \equiv \tilde{c}''/2\tilde{c} - \tilde{c}'^2/4\tilde{c}^2$. These modes, together with their complex conjugates, $u_k^*(\tilde{\eta})$, form a complete set of orthonormal basis under the scalar product

$$W(u_k^*, u_k) := u_k u_k^{*\prime} - u_k^* u_k' = 2i. \tag{49}$$

In quantum theory, $a_\mathbf{k}$ and $a_\mathbf{k}^*$ become annihilation and creation operators, $\hat{a}_\mathbf{k}$ and $\hat{a}_\mathbf{k}^\dagger$, satisfying the commutation relation

$$[\hat{a}_\mathbf{k}, \hat{a}_{\mathbf{k}'}^\dagger] = \hbar \ell^3 \delta_{\mathbf{k},\mathbf{k}'}, \tag{50}$$

where all other commutation relations vanish. For the *positive frequency* solutions, a choice of the basis, $u_k(\tilde{\eta})$, determines a vacuum state, $|0\rangle$, which is defined as the eigenstate of the annihilation operators obeying the equation $\hat{a}_\mathbf{k}|0\rangle = 0$. Different families of the mode solutions distinguish different vacuum states. Accordingly, a Fock space is generated by repeatedly acting the operator $\hat{a}_\mathbf{k}^\dagger$ on a chosen vacuum state. Thereby, the evolution of the quantum fields on the quantum-gravity-induced dressed spacetime is determined. Now, the general (auxiliary) field solutions, $\chi(\tilde{\eta}, \mathbf{x})$, can be expanded as

$$\chi(\tilde{\eta}, \mathbf{x}) = \frac{1}{\ell^3} \sum_\mathbf{k} \chi_\mathbf{k}(\tilde{\eta}, \mathbf{x}), \tag{51}$$

where

$$\chi_\mathbf{k} = \frac{1}{\sqrt{2}}\left[a_\mathbf{k} u_k^*(\tilde{\eta}) e^{i\mathbf{k}\cdot\mathbf{x}} + a_\mathbf{k}^* u_k(\tilde{\eta}) e^{-i\mathbf{k}\cdot\mathbf{x}}\right]. \tag{52}$$

To have a well-defined vacuum state for the mode solutions, the (effective) frequency

$$\Omega_k^2(\tilde{\eta}) = \tilde{\omega}_{\tilde{\eta},k}^2(\tilde{\eta}) - \mathcal{Q}(\tilde{\eta}), \tag{53}$$

should be positive. Thus, the wave-number must satisfy $k^2 \geq k_*^2$, where

$$k_*^2(\tilde{\eta}) := \frac{\mathcal{Q}(\tilde{\eta})\tilde{c}^2(\tilde{\eta})\langle \hat{H}_o^{-1}\rangle}{\langle \hat{H}_o^{-\frac{1}{2}} \hat{a}^4 \hat{H}_o^{-\frac{1}{2}}\rangle} - m^2 \frac{\langle \hat{H}_o^{-\frac{1}{2}} \hat{a}^6 \hat{H}_o^{-\frac{1}{2}}\rangle}{\langle \hat{H}_o^{-\frac{1}{2}} \hat{a}^4 \hat{H}_o^{-\frac{1}{2}}\rangle}, \tag{54}$$

introduces a physical wavelength, $\lambda_*(\tilde{\eta}) \equiv 2\pi\sqrt{\tilde{c}(\tilde{\eta})}/k_*$, which provides an upper bound for the wavelengths of the physical mode functions, i.e., $\lambda \leq \lambda_*$. Therefore, modes with short wavelengths (i.e., large momenta), $\lambda \ll \lambda_*$, characterize the vacuum states in short distances, i.e., the *ultra-violet* (UV) regimes. This regime contains a limit of arbitrary slow time, $\tilde{\eta}$, variation of the metric functions called the *adiabatic* regime. We will, henceforth, explore the mode functions describing the physical vacuum and particles associated with this adiabatic regime.

A positive-frequency, adiabatic vacuum mode can be defined by a generalized WKB approximate solution to the Klein–Gordon equation, as in [31],

$$\underline{u}_k(\tilde{\eta}) = \frac{1}{\sqrt{W_k(\tilde{\eta})}} \exp\left(-i \int^{\tilde{\eta}} W_k(\eta) d\eta\right). \tag{55}$$

Substituting this relation in Equation (48), we obtain a relation for $W_k(\eta)$ as

$$\frac{W_k''}{W_k} - \frac{W_k'}{W_k^2} - \frac{1}{2}\frac{W_k'^2}{W_k^2} + 2\left(W_k^2 - \Omega_k^2\right) = 0. \tag{56}$$

An appropriate function can be chosen for $W_k(\tilde{\eta})$, such as [32]

$$W_k(\tilde{\eta}) = \left[Y(1+\underline{\epsilon_2})(1+\underline{\epsilon_4})\right]^{\frac{1}{2}}, \tag{57}$$

where $Y \equiv \Omega_k^2 = \tilde{\omega}_{\tilde{\eta},k}^2 - Q$ and

$$\epsilon_2 := -Y^{-\frac{3}{4}} \partial_{\tilde{\eta}} \left(Y^{-\frac{1}{2}} \partial_{\tilde{\eta}} Y^{\frac{1}{4}}\right), \tag{58}$$

$$\epsilon_4 := -Y^{-\frac{1}{2}}(1+\underline{\epsilon_2})^{-\frac{3}{4}} \partial_{\tilde{\eta}} \left\{[Y(1+\underline{\epsilon_2})]^{-\frac{1}{2}} \partial_{\tilde{\eta}}(1+\underline{\epsilon_2})^{\frac{1}{4}}\right\}. \tag{59}$$

To get the solutions for $\underline{u}_k(\tilde{\eta})$, one needs to solve Equation (56) for $W(\tilde{\eta})$. However, instead of solving (56) exactly, one way is to generate asymptotic series in orders of time $\tilde{\eta}$ derivatives of the background dressed metric. Terminating this series at a given order will specify an adiabatic mode $\underline{u}_k(\tilde{\eta})$ to that order. Up to an order n (being the power of $\tilde{\eta}$-derivatives of $\tilde{a}_i$), denoted $\mathcal{O}(\sqrt{\tilde{c}}/k\lambda_{n+\varepsilon})^{n+\varepsilon}$ (with positive real number ε), the asymptotic adiabatic expansion of W_k is defined to match

$$W_k(\tilde{\eta}) = W_k^{(0)} + W_k^{(2)} + \cdots + W_k^{(n)}, \tag{60}$$

and is obtained for higher order estimates by iteration.

When computing the expectation value of the energy-momentum operator of the scalar field, $\langle \tilde{0}|\hat{T}_{ab}|\tilde{0}\rangle$, with respect to the adiabatic vacuum, $|\tilde{0}\rangle$, associated to the mode function (Equation (55)), one encounters the UV divergences issues. All of these UV divergences are contained within the terms of adiabatic order equal to and smaller than four (cf. [33] and the appendix of Ref. [24]). Thus, the (adiabatic) regularization of the energy density and pressure of the scalar field are obtained from subtractions of the divergences contained within the adiabatic terms up to the fourth order; for example, the renormalized energy density of the field is given by

$$\langle \tilde{0}|\hat{\rho}_\varphi|\tilde{0}\rangle_{\text{ren}} = \frac{\hbar}{\ell^3 \tilde{c}^2} \sum_k \left(\rho_k[u_k(\tilde{\eta})] - \underline{\rho_k}[\underline{u_k}(\tilde{\eta})]\right), \tag{61}$$

where ρ_k is the energy density of each mode, $u_k(\tilde{\eta})$, and $\underline{\rho_k}$ is the energy density associated with the mode, $\underline{u_k}$, up to the fourth adiabatic order. We will, thus, restrict ourselves to the fourth-order adiabatic states only[2]. Then, by setting $n = 4$, we define $W_k(\tilde{\eta})$ to match the terms in Equation (60) that fall slowly in $\tilde{\omega}_{\tilde{\eta},k}$, rather than considering the exact mode solutions of Equation (48). Accordingly, we consider an appropriate asymptotic condition (up to the order four) at which the field's exact mode functions, $u_k(\tilde{\eta})$, match the adiabatic functions, $\underline{u_k}(\tilde{\eta})$, as

$$|u_k(\tilde{\eta}_b)| = |\underline{u_k}(\tilde{\eta}_b)|\left(1 + \mathcal{O}(\sqrt{\tilde{c}}/k\lambda_{4+\varepsilon})^{4+\varepsilon}\right), \tag{62a}$$

$$|u'_k(\tilde{\eta}_b)| = |\underline{u'_k}(\tilde{\eta}_b)|\left(1 + \mathcal{O}(\sqrt{\tilde{c}}/k\lambda_{4+\varepsilon})^{4+\varepsilon}\right), \tag{62b}$$

where $\tilde{\eta} = \tilde{\eta}_b$ is a natural choice for the preferred instant of time given at the quantum bounce in LQC. When a mode function, $u_k(\tilde{\eta})$, satisfies the conditions (62) at some initial time $\tilde{\eta}_b$, it will satisfy it for all times $\tilde{\eta}$. Therefore, an observable vacuum, $|0\rangle$, associated to $u_k(\tilde{\eta})$ will be of the fourth order for all times.

Following the above discussion, we thus discard the adiabatic order terms higher than four in Equation (57) and write

$$W_k(\tilde{\eta}) = \tilde{\omega}_{\tilde{\eta},k}(1+\epsilon_2+\epsilon_4)^{\frac{1}{2}}, \tag{63}$$

where ϵ_2, ϵ_4 are defined by

$$\epsilon_2 = \underline{\epsilon_2} - \tilde{\omega}_{\tilde{\eta},k}^{-2} \mathcal{Q}$$
$$= -\frac{1}{4}Y^{-2}Y'' + \frac{5}{16}Y^{-3}(Y')^2 - \tilde{\omega}_{\tilde{\eta},k}^{-2} \mathcal{Q}, \tag{64}$$

$$\epsilon_4 = \underline{\epsilon_4} - \underline{\epsilon_2}\tilde{\omega}_{\tilde{\eta},k}^{-2} \mathcal{Q}$$
$$= -\frac{1}{4}Y^{-1}(1+\underline{\epsilon_2})^{-2}\left[\underline{\epsilon_2}'' - \frac{1}{2}Y^{-1}Y'\underline{\epsilon_2}' - \frac{5}{4}(1+\underline{\epsilon_2})^{-1}(\underline{\epsilon_2}')^2\right] - \underline{\epsilon_2}\tilde{\omega}_{\tilde{\eta},k}^{-2}\mathcal{Q}, \tag{65}$$

and $\underline{\epsilon_2}$ is given by

$$\underline{\epsilon_2} = -\frac{1}{4}Y^{-2}Y'' + \frac{5}{16}Y^{-3}(Y')^2. \tag{66}$$

The leading order term of $\underline{\epsilon_2}$ is of the second order, whereas the leading order in $\underline{\epsilon_4}$ is four. Thus, ϵ_2 contains terms of orders two, four and higher, i.e,

$$\epsilon_2 = \epsilon_2^{(2)} + \epsilon_2^{(4)} + \text{Higher order terms}, \tag{67}$$

and ϵ_4 contains the leading order term four, i.e.,

$$\epsilon_4 = \epsilon_4^{(4)} + \text{Higher order terms}. \tag{68}$$

It should be noticed that we have considered only terms until the fourth order in their expressions (for more details, see [24]).

4. Gravitational Particle Production

Our aim in this section is to study the gravitational particle production in the herein quantum gravity regime due to quantum field on the anisotropic dressed background.

Once a vacuum state, $|\tilde{0}\rangle$, is specified due to the positive frequency solutions, $\underline{v}_k(\tilde{\eta})$, of Equation (48), a Fock space, $\mathcal{H}_F$, for the quantum field φ is generated. Let $\{v_k\}$ and $\{\underline{v}_k\}$ be two sets of WKB solutions given by Equation (55) in the herein adiabatic regime. These mode functions form two *normalized* bases[3] for $\mathcal{H}_F$, so they can be related to each other through the time-independent Bogolyubov coefficients, α_k and β_k, as

$$\underline{v}_k(\tilde{\eta}) = \alpha_k v_k(\tilde{\eta}) + \beta_k v_k^*(\tilde{\eta}). \tag{69}$$

The Bogolyubov coefficients satisfy the relation $|\alpha_k|^2 - |\beta_k|^2 = 1$ via the condition in Equation (49). Comparing Equations (69) and (52), it follows that the creation and annihilation operators associated with two families of mode functions (i.e, those with and without 'underline') are related as

$$\hat{a}_{\mathbf{k}} = \alpha_k \underline{\hat{a}}_{\mathbf{k}} + \beta_k^* \underline{\hat{a}}_{\mathbf{k}}^\dagger. \tag{70}$$

Working in the Heisenberg picture, an initial vacuum state of the system, say $|\tilde{0}\rangle$ connected to the 'underlined' modes, $\underline{v}_k(\tilde{\eta})$, is the vacuum state of the system for all times. Then, the number operator, $\hat{N}_{\mathbf{k}} = (\hbar \ell^3)^{-1}\hat{a}_{\mathbf{k}}^\dagger \hat{a}_{\mathbf{k}}$, associated to the particles in v_k mode, gives the average number of particles in the $|\tilde{0}\rangle$ vacuum. Thus, the v_k-mode-related vacuum state contains

$$\mathcal{N}_k := \langle \tilde{0}|\hat{N}_{\mathbf{k}}|\tilde{0}\rangle = |\beta_k|^2, \tag{71}$$

particles in the $\underline{v}_k$-mode vacuum.

Let us rewrite the mode solution, $\underline{v}_k$, in the WKB approximation from Equation (55), as[4]

$$\underline{v}_k(\tilde{\eta}) = \frac{1}{\sqrt{W_k(\tilde{\eta})}}\left[\alpha_k e_k(\tilde{\eta}) + \beta_k e_k^*(\tilde{\eta})\right], \tag{72}$$

where
$$e_k(\tilde{\eta}) := \exp\left(-i\int^{\tilde{\eta}} d\eta\, W_k(\eta)\right). \tag{73}$$

Taking the time ($\tilde{\eta}$) derivative of $\underline{v}_k(\tilde{\eta})$, we obtain
$$\underline{v}'_k(\tilde{\eta}) = -i\sqrt{W_k}\left[\alpha_k e_k(\tilde{\eta}) - \beta_k e_k^*(\tilde{\eta})\right]. \tag{74}$$

Inverting Equations (72) and (74) yields a relation for β_k as
$$\beta_k = \frac{\sqrt{W_k}}{2}\left(\underline{v}_k - \frac{i}{W_k}\underline{v}'_k\right)e_k. \tag{75}$$

Since the initial condition in LQC is fixed at the quantum bounce, $\tilde{\eta} = \tilde{\eta}_b$, by setting $\alpha_k = 1$ and $\beta_k = 0$ at $\tilde{\eta} = \tilde{\eta}_b$, we assume that no particle is created at the bounce. This yields the following conditions on $\underline{v}_k$:
$$\underline{v}_k(\tilde{\eta}_b) = 1/W_k(\tilde{\eta}_b) \quad \text{and} \quad \underline{v}'_k(\tilde{\eta}_b) = -iW_k(\tilde{\eta}_b)\,\underline{v}_k(\tilde{\eta}_b), \tag{76}$$

where $e_k(\tilde{\eta}_b) = 1$. Then, for any time $\tilde{\eta} > \tilde{\eta}_b$, the number of particles produced becomes
$$\mathcal{N}_k(\tilde{\eta}) = \frac{1}{4}\left(W_k\,|v_k|^2 + W_k^{-1}|v'_k|^2 - 2\right). \tag{77}$$

Here, for convenience, we have dropped the 'underline' for the mode functions after the bounce ($\tilde{\eta} > \tilde{\eta}_b$), i.e., we assume $v_k \equiv \underline{v}_k(\tilde{\eta}_b)$ and $v_k(\tilde{\eta}) \equiv \underline{v}_k(\tilde{\eta})$ for $\tilde{\eta} > \tilde{\eta}_b$.

Having the number of particles produced per mode, $\mathcal{N}_k(\tilde{\eta})$, at a given time $\tilde{\eta} > \tilde{\eta}_b$, we can compute the total number density of production, $\mathcal{N}(\tilde{\eta})$, as the limit of $\sum_k \mathcal{N}_k(\tilde{\eta})$ in a box of volume $\ell^3 \to \infty$, divided by the volume of the universe, $V = \ell^3|\tilde{a}_1\tilde{a}_2\tilde{a}_3| = \ell^3\tilde{c}^{3/2}$, as
$$\mathcal{N}(\tilde{\eta}) = \frac{\ell^3}{V}\sum_k \mathcal{N}_k(\tilde{\eta}) = \frac{1}{4\tilde{c}^{3/2}}\sum_k\left(W_k\,|v_k|^2 + W_k^{-1}|v'_k|^2 - 2\right). \tag{78}$$

The energy density of the created particles reads $\varrho_k(\tilde{\eta}) \equiv W_k(\tilde{\eta})\,\mathcal{N}_k(\tilde{\eta})$ for each mode. Thus, the total energy density is obtained by summing the overall modes as
$$\rho_{\text{par}} = \frac{1}{\tilde{c}^2}\sum_k \varrho_k(\tilde{\eta}), \tag{79}$$

where
$$\varrho_k(\tilde{\eta}) = \frac{1}{4}\left(|v'_k|^2 + W_k^2\,|v_k|^2 - 2W_k\right). \tag{80}$$

At the bounce, $\tilde{\eta} = \tilde{\eta}_b$, the energy density of production is zero, $\varrho_k(\tilde{\eta}_b) = 0$, as expected. However, for $\tilde{\eta} > \tilde{\eta}_b$, particles will be produced as the universe expands. In the following, we will analyze the energy density of production in the assumed adiabatic regime.

Following the (adiabatic) regularization scheme, the energy density of created particles is obtained as
$$\rho_{\text{par}}^{(\text{ren})} = \frac{1}{\tilde{c}^2}\sum_k\left(\varrho_k[v_k(\tilde{\eta})] - \underline{\varrho}_k(\tilde{\eta})\right), \tag{81}$$

where
$$\underline{\varrho}_k(\tilde{\eta}) = \frac{1}{16}\left[\frac{(\tilde{\omega}'_{k,\tilde{\eta}})^2}{\tilde{\omega}_{k,\tilde{\eta}}^3} + \tilde{\omega}_{k,\tilde{\eta}}(\epsilon_2^{(2)})^2 - \frac{(\tilde{\omega}'_{k,\tilde{\eta}})^2}{2\tilde{\omega}_{k,\tilde{\eta}}^3}\epsilon_2^{(2)} + \frac{\tilde{\omega}'_{k,\tilde{\eta}}}{\tilde{\omega}_{k,\tilde{\eta}}^2}\epsilon_2^{\prime(3)}\right]$$
$$= \varrho_k^{(0)} + \varrho_k^{(2)} + \varrho_k^{(4)}. \tag{82}$$

It turns out that ϱ_k does not fall off faster than k^{-4} when $k \to \infty$. The zeroth adiabatic order term in Equation (82) is zero, $\varrho_k^{(0)} = 0$, and the divergences are included in the second and fourth-order terms (for massive modes). For the massless modes, the divergences are included only in the fourth-order term. Therefore, the renormalized energy density of created particles, Equation (81), can be obtained by subtracting the adiabatic vacuum energy of the particle productions up to the fourth order. When considering the higher order terms (more than the fourth order) in the adiabatic mode functions, $v_k(\tilde{\eta})$, associated to the vacuum state $|0\rangle$, particles are produced and the total energy density of the created particles is proportional to $1/\tilde{c}^2(\tilde{\eta})$. For an isotropic case (either massive or massless), where $\tilde{c} = \tilde{a}^2$, we get $\rho_{\text{par}}^{(\text{ren})} \propto 1/\tilde{a}^4(\tilde{\eta})$, which scales as radiation. This is a similar situation derived in [22]. Therefore, unlike the classical FLRW cosmology, even massless modes contain nonzero particle production due to quantum gravity effects.

In the standard classical cosmology, the WKB ansatz yields a divergent asymptotic series in the adiabatic parameter so that the particle production phenomenon is associated with the violation of the WKB approximation or the region where the WKB approximation is not fulfilled very well [34]. However, in the present setting, since the spacetime region transitions between the super-Planckian and sub-Planckian regimes, where the nature of the background geometry differs from the original classical isotropic spacetime to an anisotropic dressed geometry due to quantum gravity effects, the changes in the vacuum and creation of particles are inevitable.

In the pre-inflationary scenario considered in [22], by assuming a massless field in an isotropic quantum cosmological background, $\tilde{a}(\tilde{\eta}) = \tilde{c}^{1/2}(\tilde{\eta})$, an integration range was chosen in Equation (79) by the window of observable modes of the CMB. Therein, it has been argued that when taking a UV cutoff at the characteristic momentum $k_b = k_*(\tilde{\eta}_b) = \sqrt{\tilde{a}''(\tilde{\eta})/\tilde{a}(\tilde{\eta})}|_{\tilde{\eta}_b} \approx 3.21\, m_{\text{Pl}}$ and an *infra-red* (IR) cutoff at $k_*(\tilde{\eta}) = \sqrt{\tilde{a}''(\tilde{\eta})/\tilde{a}(\tilde{\eta})} = \lambda_*^{-1}$ for all $\tilde{\eta} > \tilde{\eta}_b$, being the physical energy of particles after they reenter the effective horizon, the energy density of created particles becomes

$$\rho_{\text{par}} = 0.012\, m_{\text{Pl}}^4 / \tilde{a}^4(\tilde{\eta}). \tag{83}$$

This gives the energy density of particles produced in the region $k_* \leq k \leq k_b$. Thus, the main contribution to the production of particles is the modes whose wavelengths, λ, hold the range $\lambda_b \leq \lambda \leq \lambda_*$, i.e., the modes that reenter λ_* after the bounce during the pre-inflationary phase and will only reexit λ_* again in the slow-roll inflationary phase.

This implies that the energy density of particles produced in the quantum gravity regime, which scales as relativistic matter, is significant comparing to the background energy density. The background energy density during the slow-roll inflation is dominant over the backreaction of particle productions. However, it should be guaranteed that the production density is still dominant in the pre-inflationary phase before beginning the slow-roll phase. By assuming that the elapsed e-foldings between the bounce and the onset of inflation is about 4–5 e-folds, the energy density of the backreaction is smaller than the background energy density as $\rho_{\text{par}} \lesssim \langle \hat{\rho}_\varphi \rangle \approx 2 \times 10^{-5}\, m_{\text{Pl}}^4$. However, the analysis in [22] indicates that the energy density (Equation (83)) is two orders of magnitude larger than the required upper bound, $2 \times 10^{-5}\, m_{\text{Pl}}^4$, as estimated for the density of particle creation during the pre-inflationary phase. It turns out that the backreaction of produced particles cannot be neglected so that a more careful analysis of the backreacted wave function for the background quantum geometry is needed.

The above argument indicates that when starting from Equation (27), the backreaction of the field modes on the quantum background is not negligible; therefore, the total wave function, $\Psi(\nu, Q_{\mathbf{k}}, \phi)$, cannot be decomposed as $\Psi = \Psi_o \otimes \psi$. It may even make a further constraint on expanding the right-hand-side of Equation (27) to derive the evolution Equation (28). Nevertheless, if we suppose that the approximation in Equation (28) is valid, the presence of the backreaction would lead to a modification of the total state as $\Psi = \Psi_o \otimes \psi + \delta \Psi$ (cf. [18,20]). Taking into account such modification, the dispersion relation of the field, propagating on the emergent effective geometry, will be modified such that the local Lorentz symmetry becomes violated. In such a scenario, each mode feels a distinctive background geometry, which depends on that mode; a rainbow dressed background emerges. On such backgrounds, the standard approach for studying the infinite number of field modes fails, and an alternative procedure should be employed (e.g., see [35]). For other approaches in canonical quantum gravity, where the quantum theory of cosmological perturbations and their backreactions are implemented, see, e.g., [36–42].

5. Conclusions and Discussion

In this paper, the quantum theory of an (inhomogeneous) *massive* test field, φ, propagating on a quantized FLRW geometry is addressed. The background geometry constitutes of a homogeneous massless scalar field, ϕ, as matter source, which plays the role of internal time in quantum theory. From an effective point of view, due to quantum gravity effects on the background geometry, quantum modes of the field can experience a dressed spacetime whose geometry differs from the original FLRW metric. If the backreaction of the field modes, $Q_\mathbf{k}$, on the background is discarded, within a *test field* approximation, the full quantum state of the system can be decomposed as $\Psi(\nu, Q_\mathbf{k}, \phi) = \Psi_o(\nu, \phi) \otimes \psi(Q_\mathbf{k}, \phi)$; Ψ_o and ψ are the quantum states of the (unperturbed) background and the field modes, respectively.

In the interaction picture, an evolution equation emerges for $\psi(Q_\mathbf{k}, \phi)$, which resembles the Schrödinger equation for the same field modes propagating on a time-dependent dressed spacetime, whose metric components are functions of the quantum fluctuations of the FLRW geometry. For massive and massless modes, there exists a wide class of solutions for the effective dressed background metric. The *massless* modes can only experience an isotropic and homogeneous dressed background with a dressed scale factor, $\tilde{a}(\phi)$ (cf. see [13]). The *massive* modes, however, yield a general class of solutions for the emergent dressed geometries that resembles the anisotropic Biachi I spacetimes. Likewise, the scale factors $(\tilde{a}_1, \tilde{a}_2, \tilde{a}_3)$ of the consequent dressed Bianchi-I metric are functions of fluctuations of the isotropic quantum geometry.

Given a dressed anisotropic spacetime, as a solution discussed above, we reviewed the standard quantum field theory on such a background. More precisely, we investigated the issue of gravitational particle production associated with the field modes on the dressed Bianchi I geometry in a suitably chosen adiabatic regime. This led to some backreaction issues in the super-Planckian regime, which may affect the dynamics of the early universe. To have a regularized energy-momentum operator of the test field, the adiabatic vacuum state was chosen up to fourth-order terms. We computed the energy density of the particle production within this adiabatic limit. The divergences in the energy density of the produced particles were regularized within the fourth-order adiabatic terms and the remaining terms change as $\rho_{\text{par}}^{(\text{ren})} \propto 1/(\tilde{a}_1 \tilde{a}_2 \tilde{a}_3)^{4/3}$. Some phenomenological issues related to such particle production were discussed. It was demonstrated that the backreaction due to the particle production in the super-Planckina regime may have significant effects on the evolution of the universe and may subsequently modify the existing pre-inflationary scenario of LQC [16] (cf. [22]).

Funding: This paper is based upon work from the European Cooperation in Science and Technology (COST) action CA18108—Quantum gravity phenomenology in the multi-messenger approach supported by COST.

Acknowledgments: The results of the present review article are mainly based on my past collaborations with J. C. Fabris and S. Rastgoo.

Conflicts of Interest: The authors declare no conflict of interest.

Notes

1. We consider an elementary cell $\mathcal{V}$ by fixing its edge to lie along the coordinates (ℓ_1, ℓ_2, ℓ_3). We then denote the volume of $\mathcal{V}$ by $\mathring{V} = \ell_1 \ell_2 \ell_3 \equiv \ell^3$, and restrict all integrations in the Fourier integral to this volume.
2. Note that our aim at the moment is just to avoid the UV divergence terms when regularizing the energy-momentum of the field. However, to have a more complete mode solution, in particular, when exploring the issues of the particle productions or backreaction effects, as we will see later, higher order terms in $W_k(\bar{\eta})$ should be taken into account.
3. They satisfy the normalization condition (Equation (49)) for the mode functions.
4. It should be noticed that the standard WKB ansatz may yield a divergent asymptotic series in the adiabatic parameter. Thus, when investigating the particle production in a time-dependent background, an optimal number of terms in that series should be chosen so that the resulting truncated WKB series becomes exponentially small. Such precision would still be insufficient to describe particle production from vacuum. Therefore, an adequately precise approximation should be employed by improving the WKB ansatz [34].

References

1. Birrell, N.D.; Davies, P.C.W. *Quantum Fields in Curved Space*; Cambridge University Press: Cambridge, UK, 1984.
2. Parker, L.E.; Toms, D. *Quantum Field Theory in Curved Spacetime*; Cambridge Monographs on Mathematical Physics; Cambridge University Press: Cambridge, UK, 2009. [CrossRef]
3. Hu, B.L.; Parker, L. Anisotropy Damping Through Quantum Effects in the Early Universe. *Phys. Rev.* **1978**, *D17*, 933–945; Erratum: *Phys. Rev. D* **1978**, *17*, 3292. [CrossRef]
4. Zeldovich, Y.B. Particle production in cosmology. *Pisma Zh. Eksp. Teor. Fiz.* **1970**, *12*, 443–447.
5. Ashtekar, A.; Bojowald, M.; Lewandowski, J. Mathematical structure of loop quantum cosmology. *Adv. Theor. Math. Phys.* **2003**, *7*, 233–268. [CrossRef]
6. Ashtekar, A.; Lewandowski, J. Background independent quantum gravity: A Status report. *Class. Quant. Grav.* **2004**, *21*, R53, [CrossRef]
7. Rovelli, C. *Quantum Gravity*; Cambridge Monographs on Mathematical Physics; Cambridge University Press: Cambridge, UK, 2004. [CrossRef]
8. Thiemann, T. *Modern Canonical Quantum General Relativity*; Cambridge Monographs on Mathematical Physics; Cambridge University Press: Cambridge, UK, 2007. [CrossRef]
9. Ashtekar, A. Loop Quantum Cosmology: An Overview. *Gen. Rel. Grav.* **2009**, *41*, 707–741. [CrossRef]
10. Ashtekar, A.; Pawlowski, T.; Singh, P. Quantum Nature of the Big Bang: An Analytical and Numerical Investigation. I. *Phys. Rev.* **2006**, *D73*, 124038, [CrossRef]
11. Ashtekar, A.; Wilson-Ewing, E. Loop quantum cosmology of Bianchi I models. *Phys. Rev.* **2009**, *D79*, 083535, [CrossRef]
12. Ashtekar, A.; Singh, P. Loop Quantum Cosmology: A Status Report. *Class. Quant. Grav.* **2011**, *28*, 213001, [CrossRef]
13. Ashtekar, A.; Kaminski, W.; Lewandowski, J. Quantum field theory on a cosmological, quantum space-time. *Phys. Rev.* **2009**, *D79*, 064030, [CrossRef]
14. Agullo, I.; Ashtekar, A.; Nelson, W. Extension of the quantum theory of cosmological perturbations to the Planck era. *Phys. Rev.* **2013**, *D87*, 043507, [CrossRef]
15. Agullo, I.; Ashtekar, A.; Nelson, W. The pre-inflationary dynamics of loop quantum cosmology: Confronting quantum gravity with observations. *Class. Quant. Grav.* **2013**, *30*, 085014, [CrossRef]
16. Agullo, I.; Ashtekar, A.; Nelson, W. A Quantum Gravity Extension of the Inflationary Scenario. *Phys. Rev. Lett.* **2012**, *109*, 251301, [CrossRef] [PubMed]
17. Ashtekar, A.; Barrau, A. Loop quantum cosmology: From pre-inflationary dynamics to observations. *Class. Quant. Grav.* **2015**, *32*, 234001, [CrossRef]
18. Dapor, A.; Lewandowski, J.; Tavakoli, Y. Lorentz Symmetry in QFT on Quantum Bianchi I Space-Time. *Phys. Rev.* **2012**, *D86*, 064013, [CrossRef]
19. Dapor, A.; Lewandowski, J. Metric emerging to massive modes in quantum cosmological space-times. *Phys. Rev.* **2013**, *D87*, 063512, [CrossRef]
20. Lewandowski, J.; Nouri-Zonoz, M.; Parvizi, A.; Tavakoli, Y. Quantum theory of electromagnetic fields in a cosmological quantum spacetime. *Phys. Rev. D* **2017**, *96*, 106007, [CrossRef]
21. Tavakoli, Y.; Fabris, J.C. Creation of particles in a cyclic universe driven by loop quantum cosmology. *Int. J. Mod. Phys.* **2015**, *D24*, 1550062, [CrossRef]
22. Graef, L.L.; Ramos, R.O.; Vicente, G.S. Gravitational particle production in loop quantum cosmology. *Phys. Rev. D* **2020**, *102*, 043518, [CrossRef]
23. Garcia-Chung, A.; Mertens, J.B.; Rastgoo, S.; Tavakoli, Y.; Vargas Moniz, P. Propagation of quantum gravity-modified gravitational waves on a classical FLRW spacetime. *Phys. Rev. D* **2021**, *103*, 084053, [CrossRef]
24. Rastgoo, S.; Tavakoli, Y.; Fabris, J.C. Phenomenology of a massive quantum field in a cosmological quantum spacetime. *Ann. Phys.* **2020**, *415*, 168110, [CrossRef]
25. Lankinen, J.; Vilja, I. Gravitational Particle Creation in a Stiff Matter Dominated Universe. *JCAP* **2017**, *08*, 025, [CrossRef]

26. Scardua, A.; Guimarães, L.F.; Pinto-Neto, N.; Vicente, G.S. Fermion Production in Bouncing Cosmologies. *Phys. Rev. D* **2018**, *98*, 083505, [CrossRef]
27. Assanioussi, M.; Dapor, A.; Lewandowski, J. Rainbow metric from quantum gravity. *Phys. Lett.* **2015**, *B751*, 302–305. [CrossRef]
28. Celani, D.C.F.; Pinto-Neto, N.; Vitenti, S.D.P. Particle Creation in Bouncing Cosmologies. *Phys. Rev. D* **2017**, *95*, 023523, [CrossRef]
29. Bolliet, B.; Barrau, A.; Grain, J.; Schander, S. Observational exclusion of a consistent loop quantum cosmology scenario. *Phys. Rev. D* **2016**, *93*, 124011, [CrossRef]
30. Ashtekar, A.; Pawlowski, T.; Singh, P. Quantum nature of the big bang. *Phys. Rev. Lett.* **2006**, *96*, 141301, [CrossRef]
31. Parker, L.; Fulling, S.A. Adiabatic regularization of the energy momentum tensor of a quantized field in homogeneous spaces. *Phys. Rev.* **1974**, *D9*, 341–354. [CrossRef]
32. Chakraborty, B. The mathematical problem of reflection solved by an extension of the WKB method. *J. Math. Phys.* **1973**, *14*, 188. [CrossRef]
33. Fulling, S.A.; Parker, L.; Hu, B.L. Conformal energy-momentum tensor in curved spacetime: Adiabatic regularization and renormalization. *Phys. Rev.* **1974**, *D10*, 3905–3924. [CrossRef]
34. Winitzki, S. Cosmological particle production and the precision of the WKB approximation. *Phys. Rev. D* **2005**, *72*, 104011, [CrossRef]
35. Martin, J.; Brandenberger, R.H. The TransPlanckian problem of inflationary cosmology. *Phys. Rev. D* **2001**, *63*, 123501, [CrossRef]
36. Bojowald, M.; Ding, D. Canonical description of cosmological backreaction. *J. Cosmol. Astropart. Phys.* **2021**, *03*, 083, [CrossRef]
37. Fernández-Méndez, M.; Mena Marugán, G.A.; Olmedo, J. Hybrid quantization of an inflationary model: The flat case. *Phys. Rev. D* **2013**, *88*, 044013, [CrossRef]
38. Gomar, L.C.; Fernández-Méndez, M.; Marugán, G.A.M.; Olmedo, J. Cosmological perturbations in Hybrid Loop Quantum Cosmology: Mukhanov-Sasaki variables. *Phys. Rev. D* **2014**, *90*, 064015, [CrossRef]
39. Gomar, L.C.; Martín-Benito, M.; Marugán, G.A.M. Gauge-Invariant Perturbations in Hybrid Quantum Cosmology. *J. Cosmol. Astropart. Phys.* **2015**, *06*, 045, [CrossRef]
40. Bojowald, M.; Hossain, G.M.; Kagan, M.; Shankaranarayanan, S. Anomaly freedom in perturbative loop quantum gravity. *Phys. Rev. D* **2008**, *78*, 063547, [CrossRef]
41. Cailleteau, T.; Barrau, A.; Grain, J.; Vidotto, F. Consistency of holonomy-corrected scalar, vector and tensor perturbations in Loop Quantum Cosmology. *Phys. Rev. D* **2012**, *86*, 087301, [CrossRef]
42. Cailleteau, T.; Mielczarek, J.; Barrau, A.; Grain, J. Anomaly-free scalar perturbations with holonomy corrections in loop quantum cosmology. *Class. Quant. Grav.* **2012**, *29*, 095010, [CrossRef]

Review

Time and Evolution in Quantum and Classical Cosmology

Alexander Yu Kamenshchik [1,2,*,†], Jeinny Nallely Pérez Rodríguez [3,†] and Tereza Vardanyan [4,5,†]

1. Dipartimento di Fisica e Astronomia, Università di Bologna and INFN, Via Irnerio 46, 40126 Bologna, Italy
2. L.D. Landau Institute for Theoretical Physics, Russian Academy of Sciences, Kosygin Street 2, 119334 Moscow, Russia
3. Dipartimento di Fisica e Astronomia, Università di Bologna, Via Irnerio 46, 40126 Bologna, Italy; jeinnynallely.perez@studio.unibo.it
4. Dipartimento di Fisica e Chimica, Università di L'Aquila, 67100 Coppito, Italy; tereza.vardanyan@aquila.infn.it
5. INFN, Laboratori Nazionali del Gran Sasso, 67010 Assergi, Italy
* Correspondence: kamenshchik@bo.infn.it
† These authors contributed equally to this work.

Abstract: We analyze the issue of dynamical evolution and time in quantum cosmology. We emphasize the problem of choice of phase space variables that can play the role of a time parameter in such a way that for expectation values of quantum operators the classical evolution is reproduced. We show that it is neither necessary nor sufficient for the Poisson bracket between the time variable and the super-Hamiltonian to be equal to unity in all of the phase space. We also discuss the question of switching between different internal times as well as the Montevideo interpretation of quantum theory.

Keywords: classical and quantum cosmology; Wheeler-DeWitt equation; time

1. Introduction

The problem of "disappearance of time" in quantum gravity and cosmology is well known and has a rather long history (see, e.g., References [1–3] and references therein). Let us briefly recall the essence of the problem. Not all of the ten Einstein equations in the four-dimensional spacetime contain the second time derivatives: four of them include only the first time derivatives and are treated as constraints. The presence of such constraints is connected with the fact that the action is invariant under the spacetime diffeomorphisms group. To treat this situation, it is rather convenient to use the Arnowitt-Deser-Misner formalism [4]. The Hilbert-Einstein Lagrangian of General Relativity contains the Lagrange multipliers N and N^i, which are called lapse and shift functions. There are also the dynamical degrees of freedom connected with the spatial components of the metric g_{ij} and with the non-gravitational fields present in the universe. In order to use the canonical formalism, one introduces the conjugate momenta and makes a Legendre transformation. Then, one discovers that the Hamiltonian is proportional to the linear combination of constraints multiplied by the Lagrange multipliers [5]. Thus, the Hamiltonian vanishes if the constraints are satisfied. This can be interpreted as an impossibility of writing down a time-dependent Schrödinger equation. This problem can be seen from a somewhat different point of view. If one applies the Dirac quantization procedure [6], then the constraint in which the classical phase variables are substituted by the quantum operators should annihilate the quantum state of the system under consideration. Gravitational constraints contain momenta, which classically are time derivatives of fields, but their connection to the classical notion of time vanishes in the quantum theory, where momenta are simply operators satisfying commutation relations. The main constraint arising in General Relativity is quadratic in momenta and gives rise to the Wheeler-DeWitt equation [5,7]. It is time-independent, but we know that the universe lives in time. Where is it hidden?

Generally, all the explanations of the reappearance of time are based on the identification of some combination of the phase space variables of the system under consideration (in our case, the whole universe) with the time parameter. Then, the Wheeler-DeWitt equation is transformed into the Schrödinger-type equation with respect to this time parameter. As is well known, in gauge theories or, in other words, in theories with first class constraints, one should choose the gauge-fixing conditions (see, e.g., References [8,9]). If we choose a time-dependent classical gauge-fixing condition, we can define a classical time parameter expressed by means of a certain combination of phase variables. Some other variables are excluded from the game by resolving the constraints. As a result, one obtains a non-vanishing effective physical Hamiltonian that depends on the remaining quantized phase space variables. This Hamiltonian governs the evolution of the wave function des! cribing the degrees of freedom not included in the definition of time. Then, one can reconstruct a particular solution of the initial Wheeler-DeWitt equation, corresponding to the solution of the physical Schrödinger equation. The above procedure was elaborated in detail in Reference [10], and, in Reference [11], it was explicitly applied to some relatively simple cosmological models. It is important to note that, in this approach, the time parameter and the Hamiltonian are introduced before the quantization of the physical degrees of freedom.

Another approach is based on the Born-Oppenheimer method [12–15], which was very effective in the treatment of atoms and molecules [16]. This method relies on the existence of two time or energy scales. In the case of cosmology, one of these scales is related to the Planck mass and to the slow evolution of the gravitational degrees of freedom. The other one is characterized by a much smaller energy and by the fast motion of the matter degrees of freedom. The solution to the Wheeler-DeWitt equation is factorized into a product of two wave functions. One of them has a semiclassical structure with the action satisfying the Einstein-Hamilton-Jacobi equation and the expectation value of the energy-momentum tensor serving as the source for gravity; the semiclassical time arises from this equation. The second wave function describes quantum degrees of freedom of matter and satisfies the effective Schrödinger equation. However, the principles of separation between these two kinds of degrees of freedom in the two approaches described above are somewhat different. The comparison of these approaches and attempts to establish correspondence between them were undertaken in References [17–20].

We have already said that the very concept of time is in some sense classical. However, the time parameter, which arises when we work with the Wheeler-DeWitt equation, is a function of the phase space variables, which are quantum operators. Thus, we should "freeze" the quantum nature of these variables when using them to introduce time, and the time parameter should be such that expectation values of operators evolve classically. Here, we arrive to the main question of this review, which was stimulated by the relatively unknown paper by Asher Peres [21] with a rather significant title "Critique of the Wheeler-DeWitt equation". While we do not agree with the conclusions of this paper, we find it thought-provoking and worth analyzing. The author of Reference [21] assumes that the time parameter should be chosen in such a way that the classical equations of motion are satisfied not only on the constraint surface but in all of the phase space. We will show that this condition is neither necessary nor sufficient.

In Reference [21], just like in our papers [11,17], a very simple model was considered—a flat Friedmann universe filled with a minimally coupled scalar field. Despite its simplicity, this model appears to be rather instructive. In the recent paper, Reference [22], it is studied in detail in the framework of the relational approach to quantum theory and quantum gravity developed in the series of preceding papers [23–26]. In References [27,28], the ideas of the relational or quantum reference frame approach are further developed. These papers attracted attention of the authors of Reference [29], where their connection with the Montevideo interpretation of quantum theory is studied [30–34].

The structure of this paper is the following: In the next section, we present for completeness some basic formulae of the Arnowitt-Deser-Misner formalism and the Dirac quantization of gravity, arriving to the general form of the Wheeler-DeWitt equation. In the third section, we present a simple toy model, which was analyzed by us in References [11,17], as well as by Peres in Reference [21]. Section 3.1 contains the description of the classical dynamics of the model—a flat Friedmann universe filled with a minimally coupled massless scalar field. In Section 3.2, we present the Hamiltonian formalism, Wheeler-DeWitt equation and the gauge fixing procedure for this model in the spirit of References [10,11,17]. As is done in Reference [21], in Section 3.3, we introduce the time parameter by using direct analysis of the Poisson brackets, without the super-Hamiltonian constraint. We show that, in Reference [21], an essential additional free parameter was overlooked and that our approach in References [10,11,17] corresponds to one choice of this parameter, whereas, in Reference [21], a different choice was made. In Section 3.4, we reproduce the results of 3.3 by using a canonical transformation and constructing the corresponding generating function. In Section 3.5, we show that only our choice of the free parameter and the explicit implementation of the super-Hamiltonian constraint permits to obtain the quantum evolution that is in agreement with the classical one. Section 4 contains the comparison of our results with those obtained in Reference [22] and the discussion of some related questions. In the last section, we discuss the obtained results and possible future directions of research.

2. The Arnowitt-Deser-Misner Formalism, Dirac Quantization and the Wheeler-DeWitt Equation

To make the transition from the Lagrangian formalism to the Hamiltonian formalism, one should break the relativistic covariance and introduce a variable that will play the role of time. The 3+1 foliation of the pseudo-Riemannian manifold serves exactly this purpose (see, e.g., Reference [35] for a very clear and detailed exposition of the 3+1 formalism). Constructing 3+1 foliation of the spacetime begins with an embedding of a three-dimensional hypersurface into a four-dimensional spacetime. Such an embedding can be realized by defining a scalar function $t(p)$ on a four-dimensional manifold (here, p denotes a point on the manifold). A hypersurface can then be defined as a level surface of this function. The gradient of $t(p)$ is a one-form that annihilates the vectors tangent to this hypersurface. The vector field obtained from the contraction of this gradient with the contravariant spacetime metric (we can call it $\vec{\nabla} t$) is orthogonal to the tangent spaces of the hypersurface. If this hypersurface is spacelike, then the orthogonal vector is timelike and can be normalized. The normalized vector field is called the unit normal vector field and is denoted by $\vec{n}$.

We can now define the three-dimensional induced metric γ_{ij} on the hypersurface by using the pullback operation of the spacetime metric $g_{\mu\nu}$ to the three-dimensional hypersurface. The induced metric will play the role of phase space variables in the Hamiltonian formalism of General Relativity. On the hypersurface, we define a covariant derivative D associated with the induced three-metric γ_{ij}. Then, one can introduce the corresponding three-dimensional Riemann–Christoffel curvature tensor, which describes the internal geometry of the hypersurface. To describe the position of a three-dimensional hypersurface inside a four-dimensional spacetime, we also define an extrinsic curvature tensor. To this end, one introduces the Weingarten transformation that maps a tangent vector $\vec{u}$ into the covariant derivative $\nabla_{\vec{u}} \vec{n}$ (associated with the metric $g_{\mu\nu}$) of $\vec{n}$ along $\vec{u}$. To show that the vector $\nabla_{\vec{u}} \vec{n}$ is also tangent to the hypersurface, it is enough to prove that it is orthogonal to the normal vector $\vec{n}$. In this section, it is convenient for us to use the spacetime signature $(-,+,+,+)$. Then, the unit normal vector satisfies the relation $\vec{n}^2 = -1$; hence,

$$\vec{n} \cdot \nabla_{\vec{u}} \vec{n} = \frac{1}{2} \nabla_{\vec{u}}(\vec{n}^2) = 0.$$

By contracting the Weingarten operator with the induced metric, one obtains a symmetric tensor, which we shall call the extrinsic curvature K_{ij}:

$$K_{ij} v^i u^j \equiv -\gamma_{ij} v^i (\nabla_{\vec{u}} \vec{n})^j, \tag{1}$$

where $\vec{v}$ and $\vec{u}$ are tangent vectors, and i, j run over the spatial coordinates. In what follows, Greek indices run over all spacetime coordinates, whereas Latin indices run only over the spatial coordinates. As will be shown, the extrinsic curvature is related to the time derivatives of the induced metric and the corresponding conjugate momenta and, hence, plays a very important role in the Hamiltonian formalism.

The next step is the construction of the projector from the spacetime to the hypersurface. This projector transforms any vector $\vec{u}$ into the vector $\vec{u} + (\vec{u} \cdot \vec{n})\vec{n}$. In the index notation, this projector can be represented as

$$\gamma_\beta^\alpha = \delta_\beta^\alpha + n^\alpha n_\beta. \tag{2}$$

One can easily see that (2) acts as an identity operator for vectors tangent to the hypersurface, whereas it annihilates vectors orthogonal to the hypersurface. Obviously, the projector can act not only on vectors, but also on arbitrary tensors. We can also introduce an extension of tensors defined on the hypersurface by treating them as a tensors defined in the spacetime. The principle is the following: the contraction of the extended tensor with vectors should give the same result as the contraction of the original one if the vectors are tangent to the hypersurface and give zero if at least one of the vectors is orthogonal to the hypersurface.

Interestingly, the extended version of the induced metric can be obtained from (2) by lowering its index with the spacetime metric,

$$\gamma_{\alpha\beta} = g_{\alpha\beta} + n_\alpha n_\beta. \tag{3}$$

For the extrinsic curvature, the extended expression has the following form:

$$K_{\alpha\beta} = -\nabla_\beta n_\alpha - a_\alpha n_\beta, \tag{4}$$

where the acceleration vector $\vec{a}$ is defined as $\vec{a} \equiv \nabla_{\vec{n}} \vec{n}$.

Using the projector (2), one can find the difference between the covariant derivative ∇ and the covariant derivative D acting on the vector fields tangent to the hypersurface:

$$D_{\vec{u}} \vec{v} = \nabla_{\vec{u}} \vec{v} + K_{ij} v^i u^j \vec{n}. \tag{5}$$

This difference is directed along the vector normal to the hypersurface and depends on the extrinsic curvature.

Now, we can try to express the four-dimensional curvature tensor in terms of the three-dimensional Ricci tensor and the extrinsic curvature. The corresponding equations are called Gauss-Codazzi relations. Using the Ricci identity for the covariant derivative D,

$$D_\mu D_\nu u^\alpha - D_\nu D_\mu u^\alpha = {}^{(3)}R^\alpha_{\beta\mu\nu} u^\beta, \tag{6}$$

the analogous identity for the covariant derivative ∇ and the projector (2), one can show that

$$\gamma_\alpha^\mu \gamma_\beta^\nu \gamma_\rho^\gamma \gamma_\delta^\sigma R^\rho_{\sigma\mu\nu} = {}^{(3)}R^\gamma_{\delta\alpha\beta} + K_\alpha^\gamma K_{\delta\beta} - K_\beta^\gamma K_{\alpha\delta}, \tag{7}$$

where ${}^{(3)}R^\gamma_{\delta\alpha\beta}$ is the Riemann curvature tensor (also called intrinsic curvature) associated with the induced metric $\gamma_{\alpha\beta}$. By contracting this identity with respect to the indices α and γ, we obtain the following relation:

$$\gamma_\alpha^\mu \gamma_\beta^\nu R_{\mu\nu} + \gamma_{\alpha\mu} \gamma_\beta^\rho n^\nu n^\sigma R^\mu_{\nu\rho\sigma} = {}^{(3)}R_{\alpha\beta} + K K_{\alpha\beta} - K_{\alpha\mu} K_\beta^\mu, \tag{8}$$

where $K \equiv K_\mu^\mu$. Now, we can contract this relation with the contravariant induced metric $\gamma^{\alpha\beta}$ and obtain

$$R + 2n^\mu n^\nu R_{\mu\nu} = {}^{(3)}R + K^2 - K_i^j K_j^i, \tag{9}$$

where we used that $K_\mu^\nu K_\nu^\mu = K_i^j K_j^i$.

We can also apply the Ricci identity to the vector n^α and project the result on the hypersurface; this results in

$$\gamma_\rho^\gamma \gamma_\alpha^\mu \gamma_\beta^\nu n^\sigma R^\rho_{\sigma\mu\nu} = D_\beta K_\alpha^\gamma - D_\alpha K_\beta^\gamma. \tag{10}$$

By contracting this relation with respect to the indices α and γ, we obtain

$$\gamma_\beta^\nu n^\sigma R_{\sigma\nu} = D_\beta K - D_\nu K_\beta^\nu. \tag{11}$$

An embedding of a three-dimensional spacelike hypersurface into a four-dimensional spacetime is not sufficient for the construction of the Hamiltonian formalism: one should also know the time evolution of the embedded hypersurface. In other words, we need the complete 3+1 foliation of the spacetime. Thus, we need to introduce a family of spacelike hypersurfaces that cover the whole spacetime and are parametrized by the parameter t, which we shall later identify with time. We start by defining the famous lapse function N (this name was invented by J.A. Wheeler in 1964 [36]):

$$\vec{n} = -N\vec{\nabla}t. \tag{12}$$

The lapse function plays a very important role in both the 3+1 foliation of a four-dimensional spacetime and in the Hamiltonian formalism of General Relativity. It is also convenient to introduce another vector field orthogonal to the spacelike hypersurfaces of the foliation, the normal evolution vector $\vec{m}$:

$$\vec{m} = N\vec{n}. \tag{13}$$

If we start at the hypersurface parameterized by t and move along an integral curve of the vector field $\vec{m}$, denoting the change of the curve's parameter by δt, we will find ourselves at the hypersurface parameterized by $t + \delta t$. One can obtain different useful relations involving the vector fields $\vec{n}$ and $\vec{m}$ and the lapse function N. The acceleration a can be expressed as

$$a_\alpha = D_\alpha \ln N. \tag{14}$$

The Lie derivative of the induced metric along the normal evolution field is

$$\mathcal{L}_{\vec{m}} \gamma_{\alpha\beta} = -2NK_{\alpha\beta}, \tag{15}$$

while the Lie derivative of the induced metric along the unit normal field is

$$\mathcal{L}_{\vec{n}} \gamma_{\alpha\beta} = -2K_{\alpha\beta}. \tag{16}$$

Now, we can complete the expression for R in terms of $^{(3)}R$, the extrinsic curvature and the lapse function. We apply the Ricci identity to the normal unit vector and project the result two times on the hypersurface and one time on the normal direction. As a result, we obtain

$$\gamma_{\alpha\mu} \gamma_\beta^\nu n^\rho n^\sigma R^\mu_{\rho\nu\sigma} = \frac{1}{N} \mathcal{L}_{\vec{m}} K_{\alpha\beta} + \frac{1}{N} D_\alpha D_\beta N + K_{\alpha\mu} K_\beta^\mu. \tag{17}$$

Combining this equation with Equation (8), we obtain

$$\gamma_\alpha^\mu \gamma_\beta^\nu R_{\mu\nu} = -\frac{1}{N} \mathcal{L}_{\vec{m}} K_{\alpha\beta} - \frac{1}{N} D_\alpha D_\beta N + {}^{(3)}R_{\alpha\beta} + K K_{\alpha\beta} - 2 K_{\alpha\mu} K_\beta^\mu. \tag{18}$$

Taking the trace of this equation with respect to the metric γ gives

$$R + n^\mu n^\nu R_{\mu\nu} = {}^{(3)}R + K^2 - \frac{1}{N} \mathcal{L}_{\vec{m}} K - \frac{1}{N} D_i D^i N. \tag{19}$$

Now, we can combine this equation with Equation (9) and arrive at the following expression:

$$R = {}^{(3)}R + K^2 + K_{ik}K^{ij} - \frac{2}{N}L_{\vec{m}}K - \frac{2}{N}D_iD^iN. \quad (20)$$

However, to transition to the Hamiltonian formalism, one should use the explicit time derivatives of field variables. Thus, we shall identify the parameter t, defining the hypersurfaces of the 3+1 foliation, with time. Then, we can introduce the vector field $\frac{\partial}{\partial t}$. This vector field is tangent to the integral curves, which are the curves along which the three spatial coordinates are constant. The action of this vector field on the gradient of the function defining 3+1 foliation is equal to unity, but, at the same time, it is not necessarily orthogonal to the spacelike hypersurfaces. Thus, we can write the following identity

$$\frac{\partial}{\partial t} = N\vec{n} + \vec{N}, \quad (21)$$

where $\vec{N}$ is the shift vector field. This vector field is orthogonal to the normal and the components of the metric in terms of the coordinates t and x_i, where $i = 1, 2, 3$, are given by the expression

$$ds^2 = -(N^2 - N_iN^i)dt^2 + 2N_i dt dx^i + \gamma_{ij}dx^i dx^j. \quad (22)$$

The components of the contravariant metric are

$$g^{00} = -\frac{1}{N^2}, \quad g^{i0} = \frac{N^i}{N^2}, \quad g^{ij} = \gamma^{ij} - \frac{N^i N^j}{N^2}. \quad (23)$$

The determinant of the four-dimensional metric is

$$g = -N^2\gamma. \quad (24)$$

From Equation (21), it follows that the Lie derivative of a tensor field along the normal evolution vector $\vec{m}$ is

$$L_{\vec{m}}T = L_{\frac{\partial}{\partial t}}T - L_{\vec{N}}T. \quad (25)$$

Using (25), it is easy to show that

$$L_{\vec{m}}K_{ij} = \left(\frac{\partial}{\partial t} - L_{\vec{N}}\right)K_{ij}, \quad (26)$$

and

$$K_{ij} = -\frac{1}{2N}\left(\frac{\partial \gamma_{ij}}{\partial t} - D_iN_j - D_jN_i\right). \quad (27)$$

From Equation (27), we see that the time derivative of the spatial metric is related to the extrinsic curvature. Now, the Hilbert-Einstein action

$$\int d^4x\sqrt{-g}R \quad (28)$$

can be expressed as

$$S = \int dt \int d^3x \sqrt{\gamma}N({}^{(3)}R + K_{ij}K^{ij} - K^2). \quad (29)$$

We can finally make the Legendre transformation to transition to the Hamiltonian formalism. The time derivatives of the lapse and shift functions are not present in the action; thus, they will play the role of Lagrange multipliers. The corresponding conjugate momenta are equal to zero and should remain equal to zero during the evolution. The momenta conjugate to the components of the three-metric γ_{ij} are defined as usual,

$$\pi^{ij} = \frac{\partial \mathcal{L}}{\partial \gamma_{ij}} = \sqrt{\gamma}(K\gamma^{ij} - K^{ij}). \tag{30}$$

We can resolve Equation (30) with respect to the extrinsic curvature:

$$K_{ij} = \frac{1}{\sqrt{\gamma}}\left(\frac{1}{2}\gamma_{ij}\gamma_{kl}\pi^{kl} - \gamma_{ik}\gamma_{jl}\pi^{kl}\right). \tag{31}$$

Using (27) and (31), we find

$$\frac{\partial \gamma_{ij}}{\partial t} = D_i N_j + D_j N_i + \frac{N}{\sqrt{\gamma}}(2\gamma_{ik}\gamma_{jl}\pi^{kl} - \gamma_{kl}\gamma_{ij}\pi^{kl}). \tag{32}$$

Now, using (29)–(32), we can calculate the Hamiltonian density

$$\mathcal{H} = \pi^{ij}\frac{\partial \gamma_{ij}}{\partial t} - \mathcal{L} = NH_\perp + N^i H_i, \tag{33}$$

where

$$H_\perp = \frac{1}{\sqrt{\gamma}}\left(\gamma_{ik}\gamma_{jl} - \frac{1}{2}\gamma_{ij}\gamma_{kl}\right)\pi^{ij}\pi^{kl} - \sqrt{\gamma}\,{}^{(3)}R, \tag{34}$$

$$H_i = -\gamma_{ik}D_j\pi^{jk} - \gamma_{ij}D_k\pi^{jk}. \tag{35}$$

The expression (34) represents the super-Hamiltonian constraint, which has a rather complicated structure. The expressions (35) are the supermomenta constraints and are responsible for the independence of the geometry of the three-dimensional spatial sections of the spacetime on the particular choice of spatial coordinates. These constraints constitute a system of first class constrains, i.e., the Poisson brackets between them are again proportional to the constraints. Namely, they have the following form:

$$\{H_\perp(\vec{x}), H_\perp(\vec{y})\} = \frac{\partial \delta(\vec{x},\vec{x})}{\partial x^i}(\gamma^{ij}(\vec{x})H_j(\vec{x}) + \gamma^{ij}(\vec{y})H_j(\vec{y})),$$

$$\{H_i(\vec{x}), H_\perp(\vec{y})\} = H_\perp(\vec{y})\frac{\partial \delta(\vec{x},\vec{y})}{\partial x^i},$$

$$\{H_i(\vec{x}), H_j(\vec{y})\} = H_i(\vec{x})\frac{\partial \delta(\vec{x},\vec{y})}{\partial x^i} + H_i(\vec{y})\frac{\partial \delta(\vec{x},\vec{y})}{\partial x^j}. \tag{36}$$

Remarkably, if we add an action of matter fields to the action (28), the structure of the Hamiltonian density will not changed and will still be proportional to the linear combination of constraints. The super-Hamiltonian constraint will acquire an additional term equal to the energy density of the corresponding field, while the supermomenta will acquire terms equal to the mixed temporal-spatial components of the corresponding energy-momentum tensor [37–39].

To quantize the theory, we should substitute all the phase space variables with the corresponding operators and the Poisson brackets (36) with the corresponding commutators. Naturally, such a substitution raises the question of consistency of the quantum version of the relations (36). In string and superstring theories, analyzing this question resulted in the discovery of the dimensionalities of spacetimes where these theories can be consistently realized. The corresponding analysis in General Relativity is more complicated and the number of works concerning this topic is rather limited. Here, we can mention Reference [40], where the Hamiltonian BFV-BRST method of the quantization of the constrained systems [41–43] was applied to analyze the system (36) in relatively simple models in quantum cosmology.

If one considers simple models where it is possible to choose a coordinate system in such a way that the shift functions and the corresponding supermomenta constraints are absent, then the problem of the consistency of the quantized relations (36) vanishes, and it is necessary to consider only the super-Hamiltonian constraint. As we have already explained in the Introduction, applying the operator version of this constraint to the wave function of the universe gives the Wheeler-DeWitt equation. However, even in this case, we stumble upon some technical and conceptual problems. Thus, the rest of this paper will be devoted to the analysis of these problems based on the consideration of a very simple model—a flat Friedmann universe filled with a massless scalar field.

3. Flat Friedmann Universe with a Massless Scalar Field
3.1. Classical Dynamics

Let us consider the simplest toy model—a flat Friedmann universe with the metric

$$ds^2 = N^2(t)dt^2 - a^2(t)dl^2, \tag{37}$$

where N is, as usual, the lapse function, and $a(t)$ is the scale factor. (In this section, we shall use a different convention for the signature of the spacetime to simplify the comparison with the papers [17,21].) This universe is filled with a massless spatially homogeneous scalar field $\phi(t)$ minimally coupled to gravity. For this minisuperspace model, the Lagrangian can be written as

$$\mathcal{L} = -\frac{L^3 M^2 \dot{a}^2 a}{2N} + \frac{L^3 \dot{\phi}^2 a^3}{2N}, \tag{38}$$

where M is a conveniently rescaled Planck mass, and L is a length scale, introduced to provide the correct dimensionality for the Lagrangian. It will be convenient to use another parametrization of the scale factor

$$a(t) = e^{\alpha(t)}. \tag{39}$$

Then,

$$\mathcal{L} = -\frac{L^3 M^2 \dot{\alpha}^2 e^{3\alpha}}{2N} + \frac{L^3 \dot{\phi}^2 e^{3\alpha}}{2N}. \tag{40}$$

The variation of the Lagrangian (40) with respect to the lapse function N gives the first Friedmann equation

$$M^2 \dot{\alpha}^2 = \dot{\phi}^2, \tag{41}$$

while its variation with respect to ϕ gives the first integral of the Klein-Gordon equation

$$\frac{L^3 \dot{\phi} e^{3\alpha}}{N} = p_\phi = \text{const.} \tag{42}$$

Here, p_ϕ is the conjugate momentum, which is conserved during the classical time evolution of our universe. It will be convenient to choose the cosmic time t as a time parameter, which is equivalent to fixing $N = 1$. Upon substituting Equation (42) into Equation (41), we find that, for the expanding universe and $0 \leq t \leq \infty$,

$$e^{3\alpha} = 3\frac{|p_\phi|}{ML^3}t, \tag{43}$$

and for the contracting universe, when $-\infty < t \leq 0$,

$$e^{3\alpha} = -3\frac{|p_\phi|}{ML^3}t. \tag{44}$$

In what follows, we will consider the expanding universe and choose the positive sign for p_ϕ without losing generality.

3.2. Hamiltonian formalism, Wheeler-DeWitt Equation, and the Gauge Fixing Procedure

On introducing the conjugate momenta p_ϕ (see Equation (42)) and

$$p_\alpha = -\frac{L^3 M^2 \dot{\alpha} e^{3\alpha}}{N}, \qquad (45)$$

and making the Legendre transformation, we see that the Hamiltonian is

$$\mathcal{H} = N\left(-\frac{p_\alpha^2 e^{-3\alpha}}{2M^2 L^3} + \frac{p_\phi^2 e^{-3\alpha}}{2L^3}\right) = NH, \qquad (46)$$

where H is nothing but the super-Hamiltonian constraint. From Equations (41), (42), and (45), it is obvious that the Hamiltonian is constrained to vanish:

$$H = 0. \qquad (47)$$

The action in the Hamiltonian form is

$$S = \int dt(p_\alpha \dot{\alpha} + p_\phi \dot{\phi} - NH). \qquad (48)$$

On performing the procedure of the Dirac quantization of the system with constraints [6], we obtain the Wheeler-DeWitt equation:

$$\hat{H}|\Psi\rangle = 0. \qquad (49)$$

Here, the operator $\hat{H}$ arises when we substitute the phase space variables by the corresponding operators and fix some particular operator ordering, and $|\Psi\rangle$ is the quantum state of the universe. Now, we shall choose the simplest operator ordering, such that

$$\hat{H} = e^{-3\hat{\alpha}}\left(-\frac{\hat{p}_\alpha^2}{2M^2 L^3} + \frac{\hat{p}_\phi^2}{2L^3}\right). \qquad (50)$$

It will be convenient to consider the quantum state $|\Psi\rangle$ in the (α, p_ϕ) representation. Thus, the Wheeler-DeWitt equation will have the following form:

$$\left(\frac{\partial^2}{\partial \alpha^2} + M^2 p_\phi^2\right)\Psi(\alpha, p_\phi) = 0. \qquad (51)$$

The general solution of this equation is

$$\Psi(\alpha, p_\phi) = \psi_1(p_{\phi_1})e^{iM|p_\phi|\alpha} + \psi_1(p_{\phi_2})e^{-iM|p_\phi|\alpha}. \qquad (52)$$

We are going to derive the effective Schrödiger equation for the physical wave function and the physical Hamiltonian following the recipe described in detail in References [10,11]. First of all, we have to introduce a time-dependent gauge-fixing condition. Let us try to use the following one:

$$\xi(\alpha, p_\alpha, t) = \frac{L^3 M^2 e^{3\alpha}}{3p_\alpha} - t = 0. \qquad (53)$$

This gauge condition coincides with the classical solution of the Friedmann equation, giving the dependence of the scale factor a on the cosmic time t. Then, on requiring the conservation of the gauge condition in time and using the equation

$$\frac{d\xi}{dt} = \frac{\partial \xi}{\partial t} + N\{\xi, H\} = 0, \qquad (54)$$

where the curly braces mean the Poisson brackets, one can easily see that the lapse function is equal to unity as it should be. If we solve the constraint (47) and use the gauge-fixing condition (53) to define the time, the action (48) can be written in the following form

$$S = \int dt (p_\phi \dot\phi - H_{\text{phys}}),\tag{55}$$

where the physical Hamiltonian is

$$H_{\text{phys}} = \frac{M p_\phi}{3t}.\tag{56}$$

The corresponding Schrödiger equation is

$$i \frac{\partial \psi_{\text{phys}}(p_\phi, t)}{\partial t} = H_{\text{phys}} \psi_{\text{phys}}(p_\phi, t) = \frac{M p_\phi}{3t} \psi_{\text{phys}}(p_\phi, t).\tag{57}$$

The solution of this equation is

$$\psi_{\text{phys}}(p_\phi, t) = \tilde\psi(p_\phi) e^{-\frac{iM p_\phi}{3} \ln t}.\tag{58}$$

Using Equation (53), one can express the time t as a function of the variables α and p_ϕ; hence, we come back to one of the two branches of the general solution of the Wheeler-DeWitt equation (52). Let us note that the probability density $\tilde\psi^* \tilde\psi$ corresponding to the function (58) does not depend on the cosmic time t, and this function can be normalized.

Interestingly, we can take one of the branches of the general solution of the Wheeler-DeWitt equation and express the variable α as a function of time and of the variable p_ϕ. On considering this function as the physical wave function satisfying the Schrödinger equation, we can calculate its partial time derivative to find the physical Hamiltonian. The results will coincide with those of Equation (56). Note that, to obtain these results, we have used the super-Hamiltonian constraint.

3.3. Introducing Time without Using the Super-Hamiltonian Constraint

In Reference [21], the author considers the same model in slightly different notations. We will analyze his approach using our notations. The main feature of the approach developed in Reference [21] is the requirement that the time parameter, as a combination of phase space variables, should be such that its Poisson bracket with the super-Hamiltonian is equal to 1 not only on the constraint surface but on all of the phase space. Note that this is not the case for the time parameter chosen in Reference [17] and described in the preceding subsection. Let us now consider our arguments from this point of view and compare them with those in Reference [21].

We would like to introduce a time parameter by identifying it with some function of phase space variables. It would be convenient to choose it in such a way that the time parameter coincided with the cosmic time, which is equivalent to the fixing of the lapse function $N = 1$. This fixation of the time parameter is equivalent to the gauge fixing of the type

$$\chi(t, \alpha, p_\alpha, \phi, p_\phi) = t - \tilde\chi(\alpha, p_\alpha, \phi, p_\phi) = 0.\tag{59}$$

By taking the time derivative of Equation (59), we obtain the condition of the gauge conservation

$$1 = N\{\tilde\chi, \mathcal{H}\},\tag{60}$$

where $\{,\}$ is the classical Poisson bracket.

Thus, to have $N = 1$, we should choose the function $\tilde\chi$ in such a way that

$$1 = \{\tilde\chi, \mathcal{H}\}.\tag{61}$$

Obviously the choice of the function $\tilde{\chi}$ is not unique. We can choose a function that depends only on geometrical variables α and p_α. Let us try the following:

$$\tilde{\chi}(\alpha, p_\alpha) = -\frac{L^3 M^2 e^{3\alpha}}{3 p_\alpha}. \tag{62}$$

As we have seen in Section 3.1, the classical solution of the Friedmann equation of the model under consideration is

$$a(t) = a_0 t^{1/3}, \tag{63}$$

or

$$\alpha(t) = \ln a_0 + \frac{1}{3}\ln t, \quad \dot{\alpha} = \frac{1}{3t}. \tag{64}$$

Substituting Equation (64) into Equation (45) with $N = 1$ gives

$$p_\alpha = -\frac{L^3 M^2 a_0^3}{3}. \tag{65}$$

From (64), (65), and (62), we can confirm that

$$\tilde{\chi}(\alpha, p_\alpha) = t. \tag{66}$$

Calculating the Poisson bracket of the function $\tilde{\chi}(\alpha, p_\alpha)$ with the super-Hamiltonian (46), we obtain

$$\{\tilde{\chi}, \mathcal{H}\} = \frac{1}{2}\left(1 + \frac{M^2 p_\phi^2}{p_\alpha^2}\right), \tag{67}$$

which is different from 1. However, if we use the constraint $\mathcal{H} = 0$, we see that on the constraint surface our Poisson bracket is equal to 1. Now, we can consider the function $\tilde{\chi}(\alpha, p_\alpha)$ as a new phase space variable T and find its conjugate momentum. The momentum

$$p_T = -\frac{p_\alpha^2 e^{-3\alpha}}{L^3 M^2} \tag{68}$$

satisfies the relation

$$\{T, p_T\} = 1, \tag{69}$$

as it should be. Besides,

$$\{T, \phi\} = \{T, p_\phi\} = \{p_T, \phi\} = \{p_T, p_\phi\} = 0.$$

Now, we can make the canonical transformation from α, p_α to T, p_T, without involving the variables p_ϕ and ϕ. This gives us an opportunity to write the reduced physical Hamiltonian as

$$H_{\text{phys}} = -p_T = -\frac{p_\alpha^2 e^{-3\alpha}}{L^3 M^2}. \tag{70}$$

Using Equations (62), (66), and (46), we can rewrite the expression (70) as follows (up to a sign)

$$H_{\text{phys}} = -p_T = \frac{M p_\phi}{3t}. \tag{71}$$

Thus, we have reproduced the formula obtained in Section 3.2.

In Reference [21], Peres considered the same model (in slightly different notations and with different normalizations and variables). He wanted to introduce the time parameter that would coincide with the cosmic time and looked for the function of the phase space variables whose Poisson bracket with the Hamiltonian is equal to 1. This function would play the role of the cosmic time. To find this function, he solved the second-order equations of motion of the model, without using the super-Hamiltonian constraint (first-order equation of motion). Thus, his time parameter should be represented by the phase space variables everywhere and not only on the constraint surface. Otherwise, he says, the method would not be consistent.

In practice, one can find the corresponding function without solving equations of motion, but by simply looking for the function of the phase space variables whose Poisson bracket with the super-Hamiltonian is equal to 1 everywhere. Obviously, such a function cannot depend only on the geometrical phase variables α and p_α, but should also depend on p_ϕ. After some calculations, we find the following function

$$T_P = \frac{2M^2 L^3 e^{3\alpha}}{3(Mp_\phi - p_\alpha)}. \tag{72}$$

One can easily check that its Poisson bracket with the super-Hamiltonian is equal to unity in all of the phase space and not only on the constraint surface. If we resolve the constraint and choose the appropriate sign, i.e., $Mp_\phi = -p_\alpha$, (72) transforms into the function T introduced earlier (see (62)). There is also another function

$$T_{P1} = -\frac{2M^2 L^3 e^{3\alpha}}{3(Mp_\phi + p_\alpha)}. \tag{73}$$

Its Poisson bracket with the super-Hamiltonian is identically equal to one. It coincides with our function T if we resolve the constraint as $Mp_\phi = p_\alpha$.

Now, let us look for the momentum conjugate to the new phase space coordinate T_P: by requiring that $\{T_P, P_{T_P}\} = 1$, we obtain

$$P_{T_P} = \frac{e^{-3\alpha}}{M^2 L^3}\left(-\frac{1}{2}p_\alpha^2 + Ap_\phi^2 + \left(\frac{M}{2} - \frac{A}{M}\right)p_\phi p_\alpha\right), \tag{74}$$

where A is a free real parameter. Note that, in Reference [21], the existence of this additional freedom of choice is overlooked, and the conjugate momentum has a definite expression without free parameters.

In Reference [21], the super-Hamiltonian is represented as a sum of two terms. One of these terms coincides with the momentum P_{T_P} given by the expression (74), whereas the remaining term does not depend on P_{T_P}. It is this second term that plays the role of the physical Hamiltonian. Such an interpretation is based on the following reasoning. The action of the super-Hamiltonian on the quantum state is equal to zero. The conjugate momentum P_{T_P} can be represented as

$$P_{T_P} = -i\frac{\partial}{\partial T_P}. \tag{75}$$

Then, from

$$\mathcal{H}|\psi\rangle = (P_{T_P} + H_{\text{phys-P}})|\psi\rangle = \left(-i\frac{\partial}{\partial T_P} + H_{\text{phys-P}}\right)|\psi\rangle = 0, \tag{76}$$

it follows that

$$i\frac{\partial}{\partial T_P}|\psi\rangle = H_{\text{phys-P}}|\psi\rangle. \tag{77}$$

Thus, we have obtained the Schrödinger equation where T_P plays the role of time and $H_{\text{phys-P}}$ is the physical Hamiltonian. Explicitly,

$$H_{\text{phys-P}} = \mathcal{H} - P_{T_P} = \frac{e^{-3\alpha}(M^2 - 2A)p_\phi(Mp_\phi - p_\alpha)}{2L^3 M^3}. \tag{78}$$

We can get rid of the factor $e^{-3\alpha}$ using our new variable T_P:

$$e^{-3\alpha} = \frac{2M^2L^3}{3T_P(Mp_\phi - p_\alpha)}. \tag{79}$$

Substituting the expression (79) into Equation (78) gives the following expression for the physical Hamiltonian

$$H_{\text{phys}-P} = \frac{(M^2 - 2A)p_\phi}{3Mt}, \tag{80}$$

where $t \equiv T_P$.

The expression (80) has some interesting features.

1. It has the same structure as the expression (71); namely, it is proportional to the momentum p_ϕ and inversely proportional to the cosmic time parameter t.
2. The expression for the time parameter as a function of phase space variables was obtained without using the constraint $\mathcal{H} = 0$.
3. The combination of the phase variables representing the cosmic time parameter was chosen in such a way that its Poisson bracket with the super-Hamiltonian is equal to 1 in all of the phase space and not only on the constraint surface.
4. There is a free parameter A in the expressions (74) and (80). The existence of such a freedom was not noticed in Reference [21]. Indeed, one can see that the expression for P_{T_P} and, hence, for $H_{\text{phys}-P}$ in Reference [21], corresponds to the choice

$$A = -\frac{M^2}{2}.$$

Our physical Hamiltonian (71) arises if we choose

$$A = 0.$$

5. Although we did not use the classical constraint to introduce the time parameter as a function of the phase space variables, when defining the physical Hamiltonian, we used the fact that the quantized super-Hamiltonian eliminates the physical state.

Let us stress that it is our physical Hamiltonian (71) that is compatible with the corresponding classical evolution of the model.

So far, we have demonstrated that the whole family of reduced physical Hamiltonians, which depend on a free parameter A, corresponds to the same choice of the classical cosmic time parameter. How can this parameter be fixed? Let us write the Schrödinger equation for the Hamiltonian (80):

$$i\frac{\partial \psi(p_\phi, t)}{\partial t} = \frac{(M^2 - 2A)p_\phi}{3Mt} \psi(p_\phi, t). \tag{81}$$

Its solution is

$$\psi(p_\phi, t) = \psi(p_\phi) \exp\left(-i\frac{(M^2 - 2A)p_\phi}{3M} \ln t\right), \tag{82}$$

where the function $\psi(p_\phi)$ should satisfy the normalizability condition

$$\int dp_\phi \psi^*(p_\phi) \psi(p_\phi) = 1. \tag{83}$$

The expectation value of the momentum

$$\langle p_\phi \rangle \equiv \langle \psi(p_\phi, t) | p_\phi | \psi(p_\phi, t) \rangle = \int dp_\phi p_\phi \psi^*(p_\phi) \psi(p_\phi) \tag{84}$$

does not depend on time, just like its classical analogue. What happens with the expectation value of the scalar field ϕ?

$$\langle \phi \rangle \equiv \langle \psi(p_\phi,t)|\phi|\psi(p_\phi,t)\rangle = i\int dp_\phi \psi^*(p_\phi,t)\frac{\partial \psi(p_\phi,t)}{\partial p_\phi}$$
$$= \langle \phi \rangle_{t=1} + \frac{M^2-2A}{3M}\ln t, \qquad (85)$$

where

$$\langle \phi \rangle_{t=1} \equiv i\int dp_\phi \psi^*(p_\phi,t)\frac{d\psi(p_\phi,t)}{dp_\phi}$$

is the expectation value at $t=1$.

Let us now find the classical evolution of the scalar field ϕ. Using the Lagrangian (40) and making variation with respect to the lapse function N, we obtain the following constraint in the Lagrangian formalism

$$M^2 \dot{\alpha}^2 = \dot{\phi}^2. \qquad (86)$$

Note that this constraint does not depend on the choice of the time parametrization. Let us choose the solution as follows

$$\dot{\phi} = M\dot{\alpha}. \qquad (87)$$

We have already mentioned that $\dot{\alpha} = \frac{1}{3t}$ (see Equation (64)). Thus,

$$\dot{\phi} = \frac{M}{3t}. \qquad (88)$$

We see that (88) and (85) are compatible if $A=0$, i.e., if the reduced Hamiltonian is equal to our Hamiltonian (71).

3.4. Hamilton-Jacobi Type Equation and the Choice of the Time Parameter and of the Hamiltonian

As a matter of fact, the author of Reference [21] gives his definition of time and of the Hamiltonian using the solution of the Hamiltonian-Jacobi type equation and not the method of direct construction of new canonical variables, which we used in the preceding subsection. One can try to find out if the solution of the corresponding partial differential equation can fix these variables in an unambiguous way. We will show in this subsection that this not the case and that the arbitrariness, which was not noticed in Reference [21], is still present.

Let us introduce the volume variable

$$v \equiv a^3 \equiv e^{3\alpha}. \qquad (89)$$

In terms of this variable, the Lagrangian (38) is as follows:

$$\mathcal{L} = -\frac{L^3 M^2 \dot{v}^2}{18Nv} + \frac{L^3 \phi^2 v}{2N}. \qquad (90)$$

We can put $N=1$ because we will be looking for the time parameter that coincides with the cosmic time (the author of Reference [21] calls it "auxiliary time τ"). After this, we can write the Euler-Lagrange equation with respect to v, which is nothing but the second Friedmann equation:

$$2\ddot{v}v - \dot{v}^2 + \frac{9p_\phi^2}{L^3 M^2} = 0, \qquad (91)$$

where we have used the relation

$$\dot{\phi} = \frac{p_\phi}{L^3 v}. \qquad (92)$$

Now, we can find the solution of the second Friedmann Equation (91) without using the first Friedmann equation (the super-Hamiltonian constraint). With the initial condition $v(0) = 0$, the solution is

$$v(t) = Kt^2 + 3\frac{p_\phi}{L^3 M}t, \qquad (93)$$

where K is an arbitrary constant. Using the expression (93), we can calculate explicitly the expression for the super-Hamiltonian. Indeed,

$$\dot{v} = 2Kt + 3\frac{p_\phi}{L^3 M}. \qquad (94)$$

Then,

$$p_v = -\frac{L^3 M^2}{9}\frac{\dot{v}}{v}, \qquad (95)$$

and

$$H = -\frac{9}{2M^2 L^3}v p_v^2 + \frac{1}{2L^3}\frac{p_\phi^2}{v} = -\frac{2}{9}KL^3 M^2. \qquad (96)$$

Thus, we see that the super-Hamiltonian does not depend on time and is proportional to the constant K. If we take into account the first Friedmann equation, or, in other words the super-Hamiltonian constraint, the constant K should be equal to zero. Note that the dependence of the volume on time given by Equation (93) arises naturally in the universe filled with dust and a massless scalar field, which behaves as a stiff matter (see, e.g., Reference [44]). On the other hand, this behavior arises in the generalized unimodular theory of gravity [45], where the lapse function is treated not as a Lagrange multiplier but as a certain function of the determinant of the spatial metric. If one chooses the lapse function to be a constant (or in other words, chooses a gauge fixing condition corresponding to the cosmic time), an effective dust arises in the universe. If the lapse function is inversely proportional to the determinant of the spatial metric, one has the unimodular theory of gravity, and an effective cosmological constant appears [46,47].

Now, using Equations (93)–(95), we can easily express the time parameter t as a function of the phase variables v, p_v and p_ϕ:

$$t = \frac{2M^2 L^3 v}{3(Mp_\phi - 3v p_v)}. \qquad (97)$$

It is easy to check that this expression coincides with the Formula (72), which we obtained by simply looking for the combination of the phase variables whose Poisson bracket with the super-Hamiltonian equals to 1 in all of the phase space. Note that the constant K does not enter into this expression.

The next step is to find the canonical momentum conjugate to the new coordinate given by the expression (97). In the preceding subsection, we did it using the definition of the Poisson bracket. In Reference [21], the author constructed the corresponding canonical transformation. For this purpose the generating function of the type F_1 (see, e.g., Reference [48]) was used. The generating function can be written as $S(q, Q)$, where q are the old coordinates, while Q are the new coordinates. Then,

$$p_k = \frac{\partial S}{\partial q^k}, \qquad (98)$$

$$P_\mu = \frac{\partial S}{\partial Q^\mu}. \qquad (99)$$

Now, as a first new coordinate we choose the coordinate T, which coincides with the expression of the right-hand side of Equation (97). It will be convenient for us to go back to our old variable α. Thus,

$$T = \frac{2M^2 L^3 e^{3\alpha}}{3(Mp_\phi - p_\alpha)}. \tag{100}$$

As a second new coordinate we choose

$$Q = p_\phi. \tag{101}$$

Obviously, the Poisson bracket between the expressions (100) and (101) is equal to zero. From Equation (100), we have

$$p_\alpha = MQ - \frac{2}{3} \frac{M^2 L^3 e^{3\alpha}}{T}. \tag{102}$$

The generating function for which we are looking depends on four variables:

$$S = S(\alpha, \phi, T, Q). \tag{103}$$

Thus,

$$p_\alpha = \frac{\partial S}{\partial \alpha} = MQ - \frac{2}{3} \frac{M^2 L^3 e^{3\alpha}}{T}. \tag{104}$$

The general solution of this equation is

$$S(\alpha, \phi, T, Q) = MQ\alpha - \frac{2}{9} \frac{M^2 L^3 e^{3\alpha}}{T} + \tilde{S}(\phi, T, Q). \tag{105}$$

We can impose an additional constraint on the function $\tilde{S}$. The partial derivative of the generating function with respect to ϕ should give us $p_\phi = Q$. Hence,

$$\frac{\partial S}{\partial \phi} = \frac{\partial \tilde{S}}{\partial \phi} = p_\phi = Q. \tag{106}$$

Then,

$$\tilde{S} = \phi Q + \bar{S}(T, Q). \tag{107}$$

We still have some freedom of choice for the function $\bar{S}(T, Q)$. To make the comparison with the results of the preceding two subsections simpler, let us choose it as follows:

$$\bar{S}(T, Q) = \frac{2}{3}\left(\frac{M}{2} + \frac{A}{M}\right) Q \ln T, \tag{108}$$

where A is an arbitrary constant.

Now, we can find the momentum P_T using the Formula (99):

$$P_T = -\frac{\partial S}{\partial T} = -\frac{2}{9} \frac{M^2 L^3 e^{3\alpha}}{T^2} - \frac{2}{3}\left(\frac{M}{2} + \frac{A}{M}\right)\frac{Q}{T}. \tag{109}$$

It is easy to check that, if we go back to the old phase space variables $\alpha, \phi, p_\alpha, p_\phi$, the Formula (109) coincides with the Formula (74). Thus, using the canonical transformation and the generating function we came to the same result as was obtained by the direct study of the corresponding Poisson brackets. Naturally, the effective physical Hamiltonian is given by the Formula (80).

3.5. Can the Choice $A \neq 0$ Correspond to Some Classical Evolution?

At the end of Section 3.3, we saw that, if A is different from zero (for example, if $A = -\frac{M^2}{2}$, as was implicitly chosen in Reference [21]), then the quantum evolution is not compatible with the classical: the time derivative of the expectation value of the field is given by

$$\frac{d\langle\phi\rangle}{dt} = \frac{M^2 - 2A}{3Mt}, \tag{110}$$

instead of

$$\frac{d\langle\phi\rangle}{dt} = \frac{M}{3t}. \tag{111}$$

We can ask if the classical counterpart of (110) exists in the theory where the super-Hamiltonian constraint is not imposed. To answer this question, let us recall that the relation between the momentum p_ϕ, the time derivative of the scalar field $\dot\phi$ and the volume $v = a^3 = e^{3\alpha}$,

$$p_\phi = L^3 v \dot\phi, \tag{112}$$

does not depend on imposing the super-Hamiltonian constraint and follows simply from the Klein-Gordon equation for the scalar field. By combining Equations (110) and (112), we obtain

$$v(t) = \frac{3 p_\phi M}{L^3 (M^2 - 2A)} t. \tag{113}$$

On the other hand, from the second Friedmann equation and the Klein-Gordon equation, we obtain the expression (93). We see that it cannot coincide with (113) unless $A = 0$.

4. Switching Internal Times and the Montevideo Interpretation of Quantum Theory

In the recent paper by Reference [22], the author considers the same simple model that we discussed in the preceding section. His approach to the problem of time and related problems in quantum gravity was laid out in References [23–26]. The author states that the question of the quantum notion of general-relativistic covariance received little attention; he addresses this question by using quantum reference frames. One of the main points of Reference [22] is the explicit and detailed study of switching between different internal times. The toy model is very convenient from this point of view. Indeed, all the calculations can be done analytically, and the obtained formulae are rather simple. There is also another advantage: the symmetry between the logarithm of the scale factor and the scalar field. This symmetry is especially clear if one looks at the super-Hamiltonian constraint (46)–(47), which we can conveniently rewrite as follows:

$$p_\alpha^2 - M^2 p_\phi^2 = 0. \tag{114}$$

Generally, we find an essential correspondence between the approaches of References [10,11,17,22] and Section 3.2 of the present paper. Indeed, Ref. [22] uses a method that "identifies a consistent reduction procedure that maps the Dirac quantized theory to the various reduced quantized versions of it to different choices of quantum reference systems. It identifies the physical Hilbert space of the Dirac quantization as a reference-system-neutral-quantum superstructure and the various reduced quantum theories as the physics described relative to the corresponding choice of reference system." Hence, the quantum state annihilated by the Dirac constraint (or, in other words, satisfying the Wheeler-DeWitt equation) does not have a direct physical interpretation. This opinion coincides with ours. We also stressed that the physical (probabilistic) interpretation arises when we construct the time parameter from a part of the phase space variables and this parameter, as well as its conjugate momentum, are excluded from the effective Hamiltonian by using the constraint. Let us note that the author of Reference [22] also uses the constraint at all the stages of his procedure, in contrast to the author of Reference [21].

Because of the symmetry of our model, there are four natural ways to introduce a time parameter or, in other words, a quantum clock [22]. One can choose a function of the scale factor as a time variable, and, in this case, the wave function will depend on the scalar field, or one can consider the scalar field as a quantum clock. Two other choices arise when we use Equation (114) to exclude the momentum conjugate to the time variable from the effective Hamiltonian. In the language of Reference [22], it is called "trivialize the constraint to the internal time to render it redundant".

In Section 3.2, the function of the scale factor was chosen to be the internal time. Let us try to describe the switching between different internal times in our terms. We assign the role of time to a suitable function of the scalar field ϕ. Solving jointly the Friedmann Equation (41) and the Klein-Gordon Equation (42) gives

$$\phi(t) = \frac{M}{3} \ln \frac{t}{t_0}, \tag{115}$$

where t_0 is a constant. Now, we can introduce a new canonical variable

$$T = T_0 \exp\left(\frac{3\phi}{M}\right); \tag{116}$$

its conjugate momentum is

$$p_T = \frac{M p_\phi}{3 T_0} \exp\left(-\frac{3\phi}{M}\right) = \frac{M p_\phi}{3T}. \tag{117}$$

Using the algorithm described in Section 3.2 and the constraint (114), we obtain the following effective physical Hamiltonian:

$$H_{\text{phys}-p_\alpha} = \pm \frac{p_\alpha}{3t}, \tag{118}$$

where the sign "$\pm$" corresponds to the two options existing in the quadratic form of the constraint (114). It is easy to see that the form of the Hamiltonian (118) is quite similar to that of (56). Likewise, the solution of the corresponding Schrödinger equation $\psi(p_\alpha, t)$ has the structure quite similar to that of (58). The switching between two choices of time gives similar results due to the symmetry and simplicity of the model as was noted in Reference [22]. The two options—the time related to purely geometric characteristic of the universe and the time connected with the scalar field—look quite natural in this model. However, other choices also exist. For example, instead of α and ϕ, we can introduce a different pair of canonical variables:

$$T = T_0 \exp\left(\frac{3}{2M}(\phi + M\alpha)\right),$$
$$\beta = M\alpha. \tag{119}$$

The conjugate momenta are

$$p_T = \frac{2M p_\phi}{3 T_0} \exp\left(-\frac{3}{2M}(\phi + M\alpha)\right),$$
$$p_\beta = -p_\phi + \frac{p_\alpha}{M}. \tag{120}$$

Implementing the procedure described above, we arrive at the effective Hamiltonian

$$H_{\text{phys}-p_\beta} = \frac{M p_\beta}{3t}, \tag{121}$$

which again has the same form as Hamiltonians considered before.

We should mention that the internal time parameters that do not mix the geometrical variable α and the matter variable ϕ look more natural and are more convenient for work with. However, it is not always so. When one considerers the models where the scalar field is non-minimally coupled to gravity, it is more convenient to introduce an internal time parameter that mixes the geometrical and the matter variables; see, e.g., References [49,50].

Further elaboration of the procedure of switching between internal times [27,28] has some implications for the Montevideo interpretation of quantum mechanics [29]. To discuss this topic, let us first recapitulate the main features of this interpretation [30–34].

As with all interpretations of quantum mechanics, the main target of the Montevideo interpretation is the problem of measurement. This problem can be formulated as the necessity to understand how the principle of superposition and the linearity of the Schrödinger equation can be reconciled with the fact that, in every measurement, we see only one outcome. The Copenhagen interpretation suggests that there are two realms—the quantum and the classical—and there is a complementarity between them. The Montevideo interpretation is based on the principle that the quantum description of reality is the only fundamental one. This is akin to the Many-Worlds interpretation of quantum mechanics [51,52]. The distinguishing feature of the Montevideo interpretation is its attention to the notion of time in quantum theory. The clear distinction is made between the quantum clock, which is connected with some quantum variable and is an operator, and the coordinate time. It is important that there are quantum limits on the precision of the time measurement by quantum clocks. First of all, there is the well-known time-energy uncertainty relation or Mandelstam-Tamm relation [53]. Besides, the presence of gravity implies that there are additional constraints on the resolution of quantum clocks. Thus, according to the Montevideo interpretation, while the quantum system has a unitary evolution with respect to the coordinate time, its evolution with respect to the quantum clock is non-unitary, and a pure state transforms into a mixed one. This means that the corresponding density matrix evolves not according to the quantum Liouville (von Neumann) equation but according to the equation with some additional terms of the Lindblad type [54,55]. This equation has the following form

$$\frac{\partial \rho(T)}{\partial T} = i[\rho(T), H] + \sigma(T)[H, [H, \rho(T)]] + \cdots, \tag{122}$$

where T is a quantum clock time. It is written in Reference [29] that the Inclusion of the of quantum gravitational notion of time in the spirit of the papers [27,28] makes the construction based on the Montevideo interpretation more solid. In this connection we see an interesting technical problem. Can we starting with different definitions of internal time, like those considered in Reference [22] and in the present paper, to obtain the forms of the function $\sigma(T)$ responsible for the non-unitary evolution of the density matrix? It would be curios to see how it works at least for simple toy models.

Before finishing this section, we would like to say once again that from the point of view of classification of different interpretations of quantum mechanics, the Montevideo interpretation is rather close to the Everett interpretation but with a special emphasis on the role of time and gravitational effects. Besides, the branching that arises in the Everett interpretation as a result of a measurement-like process becomes approximate. On the other hand, the Montevideo interpretation can be considered as one of the approaches to the synthesis of gravity and quantum theory. Investigations of a possible influence of gravitational effects on quantum physics have a rather long history [56–61]. The authors proposing the Montevideo interpretation stress that they do not modify quantum mechanics. They simply consistently take into account the gravitational effects, which results in resolution of some long-standing problems of quantum theory. An interesting example of quite a different attitude to the problem of synthesis between quantum theory and gravity is the recently proposed Correlated Worldline Theory [62–65]. In the framework of this theory, the generalized equivalence principle is implemented and new path integrals representing the generating functionals are obtained. The linearity of the quantum theory disappears, but is restored in the limit when the gravitational interaction is switched off. Thus, we see another interesting problem here: what happens with the notion of time in this new theory?

5. Discussion

In this review various ideas concerning the problem of time in quantum gravity and cosmology were studied by using a simple model of a flat Friedmann universe filled with a massless minimally coupled scalar field. This model was studied in a relatively old paper, Reference [21], and in some new works [11,17,22]. Why does this model attract the attention of researchers? It is very simple, and all the calculations can be done analytically for classical and quantum dynamics. At the same time, most of the problems can be seen by looking at this simple example. What would change if we included a non-vanishing mass? As we can see from the pioneering paper of Reference [66], even classical dynamics becomes very rich and there are no exact solutions. One can also consider scalar fields with other potentials. For example, for a constant potential one can find the exact solutions of the Wheeler-DeWitt equation [11]; the same is true for an exponential potential [67]. However, these more complicated minisuperspace models give almost no additional insight. The same can be said about more realistic models, which contain anisotropies or/and perturbations. To introduce an internal time or a quantum clock, one usually uses macroscopic variables, such as the scale factor and the homogenous mode of a scalar field. Thus, one can try to understand the basic features of the internal time by playing with relatively simple models, which we tried to do in the present review.

Upon analyzing the method of introduction of time without using the super-Hamiltonian constraint, we found a flaw in the calculations of Reference [21] and demonstrated that some freedom of choice of variables was overlooked. One can represent (at least partially, but it was enough for our analysis) this freedom by a single parameter A. Choosing this parameter as $A = 0$ gives the evolution of the quantum average of the scalar field that coincides with its classical counterpart; with any other choice (including the one implicitly made in the Reference [21]) there is not a sensible classical counterpart. On the other hand, the correct expression for the effective Hamiltonian was obtained in the framework of the formalism where the super-Hamiltonian constraint was used at all the stages of the problem [10,11,17]. Thus, it can be concluded that the introduction of new variables with the desired value of the Poisson brackets in all of the phase space is ne! ither necessary nor sufficient for a reasonable treatment of the Wheeler-DeWitt equation.

Nevertheless, the paper written by A. Peres is, certainly, thought-provoking because it makes the reader think about such complicated questions as the problem of time in quantum cosmology, the correct treatment of the Wheeler-DeWitt equation, the transition to the Hamiltonian formalism in theories that possess reparametrization invariance, and, last but not least, the complicated interrelations between the classical and quantum theories. The author of Reference [21] advocates the Copenhagen interpretation of quantum mechanics. He writes: "On the other hand, quantum theory, unlike general relativity, is not a "theory of everything". Its mathematical formalism can be given a consistent physical interpretation only by arbitrarily dividing the physical world into two parts: the system under study, represented by vectors and operators in a Hilbert space, and the observer (and the rest of the world), for which a classical description is used". The reference to the famous paper by Niels Bohr [68] is given in the Introduction of Reference [21].

In all the calculations and considerations of Section 3 of the present paper, the treatment of time as a classical variable and the comparison between the classical and quantum evolutions are used. At the same time, a different point of view becomes more and more popular among researchers working in quantum cosmology. It is related to the many-worlds interpretation of quantum mechanics [51,52] and to the idea that the quantum theory is the only fundamental theory. The classical behavior emerges only in some limits and contexts (see, e.g., Reference [69] for a recent review). If we accept the fundamental character of the quantum theory, we should expect the appearance of some quantum properties of time, considering that the time parameter in quantum gravity and cosmology is a combination of quantum operators. Here, this "quantumness" was in a way frozen, but this phenomenon can hardly have a universal character. Is it possible to see some flashes revealing the true quantum nature of time?

Let us note that ideas concerning this quantum nature of time were already discussed in some works. For example, in Reference [70], a simple cosmological model, similar to that in the present paper but with the massive scalar field, was considered. Using the Born-Oppenheimer approach, the authors factorized the solution of the Wheeler-DeWitt equation into two parts: one depending only on the geometrical variables (scale factor), and the other depending also on the matter variables and the perturbations of the homogeneous background. The purely geometrical part of the wave function is responsible for the emergence of time, while the other part describes the quantum evolution of the matter degrees of freedom with respect to this time parameter. It was discovered that, to make this definition of time reasonable, it is necessary to use some kind of course-graining procedure. Without this procedure the time variable behaves in a strange, oscillatory way.

The idea of Reference [70] was further developed in Reference [71], where the attempt to construct a "non-semiclassical" wave function of the universe was undertaken, and the possible observable effects, such as the influence of this non-classicality on the spectrum of the cosmic microwave background, were predicted.

Some interesting features of time and causality in quantum gravity were studied in References [72–74]. As is known, when introducing the higher-derivative terms in the gravitational action to make it renormalizable, one encounters the problem of unitarity. To resolve this problem, the author of References [72–74] introduduces particles which he calls "fakeons". The propagators of these particles are not of Feynman type, and the standard notion of causality is not valid for them. Thus, one can say they live in a "different time". This is even more evident in References [75,76]. To study the structure of the propagator of the graviton, the authors introduce the notion of the so-called "Merlin modes", which live in a time that runs backwards. These states are highly unstable, their appearance does not break the causality on the observable distances.

In our opinion, the question of genuine quantum nature of time and its relation to the classical evolution is one of the most interesting and challenging problems of quantum cosmology.

Author Contributions: All authors contributed equally to the present work. All authors have read and agreed to the published version of the manuscript.

Funding: This research was supported by the Russian Foundation for Basic Research grant No 20-02-00411 and by the research grant "The Dark universe: A Synergic Multimessenger Approach" No. 2017X7X85K under the program PRIN 2017 funded by the Ministero dell'Istruzione, delle Università e della Ricerca (MIUR), Italy.

Acknowledgments: We are grateful to A.O. Barvinsky, A. Tronconi and G. Venturi for fruitful discussions.

Conflicts of Interest: The authors declare no conflict of interest.

References

1. Kuchar, K.V. Time and interpretations of quantum gravity. *Int. J. Mod. Phys. D* **2011**, *20*, 3. [CrossRef]
2. Kiefer, C. *Quantum Gravity*, 3rd ed.; Oxford University Press: Oxford, UK, 2012.
3. Ryan, M.P.; Shepley, L.C. *Homogeneous Relativistic Cosmologies*; Princeton University Press: Princeton, NJ, USA, 1975.
4. Arnowitt, R.; Deser, S.; Misner, C. Dynamical structure and definition of energy in general relativity. *Phys. Rev.* **1959**, *116*, 1322–1330. [CrossRef]
5. DeWitt, B.S. Quantum Theory of Gravity. I. The Canonical Theory. *Phys. Rev.* **1967**, *160*, 1113. [CrossRef]
6. Dirac, P.A.M. *Lectures on Quantum Mechanics*; Yeshiva University Press: New York, NY, USA, 1964.
7. Wheeler, J.A. *Einstein's Vision*; Springer: Berlin, Germany, 1968.
8. Sundermeyer, K. *Constrained Dynamics*; Springer: Berlin, Germany, 1982.
9. Henneaux, M.; Teitelboim, C. *Quantisation of Gauge Systems*; Princeton University Press: Princeton, NJ, USA, 1992.
10. Barvinsky, A.O. Unitarity approach to quantum cosmology. *Phys. Rept.* **1993**, *230*, 237. [CrossRef]
11. Barvinsky, A.O.; Kamenshchik, A.Y. Selection rules for the Wheeler-DeWitt equation in quantum cosmology. *Phys. Rev. D* **2014**, *89*, 043526. [CrossRef]
12. Brout, R. On the Concept of Time and the Origin of the Cosmological Temperature. *Found. Phys.* **1987**, *17*, 603. [CrossRef]
13. Brout, R.; Venturi, G. Time in Semiclassical Gravity. *Phys. Rev. D* **1989**, *39*, 2436. [CrossRef] [PubMed]
14. Venturi, G. Minisuperspace, matter and time. *Class. Quant. Grav.* **1990**, *7*, 1075. [CrossRef]

15. Kamenshchik, A.Y.; Tronconi, A.; Venturi, G. The Born—Oppenheimer method, quantum gravity and matter. *Class. Quant. Grav.* **2018**, *35*, 015012. [CrossRef]
16. Born, M.; Oppenheimer, J.R. Zur Quantentheorie der Molekeln. *Ann. Phys.* **1927**, *84*, 457. [CrossRef]
17. Kamenshchik, A.Y.; Tronconi, A.; Vardanyan, T.; Venturi, G. Time in quantum theory, the Wheeler-DeWitt equation and the Born–Oppenheimer approximation. *Int. J. Mod. Phys. D* **2019**, *28*, 1950073. [CrossRef]
18. Chataignier, L. Gauge Fixing and the Semiclassical Interpretation of Quantum Cosmology. *Z. Naturforsch. A* **2019**, *74*, 1069. [CrossRef]
19. Chataignier, L. Construction of quantum Dirac observables and the emergence of WKB time. *Phys. Rev. D* **2020**, *101*, 086001. [CrossRef]
20. Chataignier, L.; Kraemer, M. Unitarity of quantum-gravitational corrections to primordial fluctuations in the Born-Oppenheimer approach. *Phys. Rev. D* **2021**, *103*, 066005. [CrossRef]
21. Peres, A. Critique of the Wheeler-DeWitt equation. In *On Einstein's Path—Essays in Honor of Engelbert Schucking*; Harvey, A., Ed.; Springer: Berlin, Germany, 1999; pp. 367–379.
22. Höhn, P.A. Switching Internal Times and a New Perspective on the Wave Function of the Universe. *Universe* **2019**, *5*, 116. [CrossRef]
23. Bojowald, M.; Höhn, P.A.; Tsobanjan, A. An Effective approach to the problem of time. *Class. Quant. Grav.* **2011**, *28*, 035006. [CrossRef]
24. Bojowald, M.; Höhn, P.A.; Tsobanjan, A. Effective approach to the problem of time: General features and examples. *Phys. Rev. D* **2011**, *83*, 125023. [CrossRef]
25. Vanrietvelde, A.; Höhn, P.A.; Giacomini, F.; Castro-Ruiz, E. A change of perspective: Switching quantum reference frames via a perspective-neutral framework. *Quantum* **2020**, *4*, 225. [CrossRef]
26. Höhn, P.A.; Vanrietvelde, A. How to switch between relational quantum clocks. *New. J. Phys.* **2020**, *22*, 123048. [CrossRef]
27. Höhn, P.E.; Smith, A.R.H.; Lock, M.P.E. The trinity of Relational Quantum Dynamics. *arXiv* 2019, arXiv:1912.00033.
28. Höhn, P.E.; Smith, A.R.H.; Lock, M.P.E. Equivalence of approaches to relational quantum dynamics in relativistic settings. *arXiv* 2020, arXiv:2007.00580.
29. Gambini, R.; Pullin, J. The Montevideo Interpretation: How the inclusion of a Quantum Gravitational Notion of Time Solves the Measurement Problem. *Universe* **2020**, *6*, 236. [CrossRef]
30. Gambini, R.; Porto, R.A.; Pullin, J. Fundamental decoherence from quantum gravity: Apedagogical review. *Gen. Relativ. Gravit.* **2007**, *39*, 1143. [CrossRef]
31. Gambini, R.; Pullin, J. The Montevideo interpretation of quantum mechanics: Frequently asked questions. *J. Phys. Conf. Ser.* **2009**, *174*, 012003. [CrossRef]
32. Gambini, R.; Garcia-Pintos, L.P.; Pullin, J. An axiomatic formulation of the Montevideo interpretation of quantum mechanics. *Stud. Hist. Philos. Sci. B* **2011**, *42*, 256. [CrossRef]
33. Gambini, R.; Pullin, J. The Montevideo Interpretation of Quantum Mechanics: A Short Review. *Entropy* **2018**, *20*, 413. [CrossRef] [PubMed]
34. Butterfield, J. Assessing the Montevideo Interpretation of Quantum Mechanics. *Stud. Hist. Philos. Sci. B* **2015**, *52*, 75. [CrossRef]
35. Gourgoulhon, É. 3 + 1 Formalism in General Relativity. *Bases of Numerical Relativity*; Lecture Notes in Physics; Springer: Berlin/Heidelberg, Germany, 2012.
36. Wheeler, J.A. Geometrodynamics and the issue of the final state. In *Relativity, Groups and Topology*; DeWitt, C., DeWitt, B.S., Eds.; Gordon and Breach: New York, NY, USA, 1964; p. 316.
37. Kuchar, K. Geometry of hyperspace. I. *J. Math. Phys.* **1976**, *17*, 777. [CrossRef]
38. Kuchar, K. Kinematics of tensor fields in hyperspace. II. *J. Math. Phys.* **1976**, *17*, 792. [CrossRef]
39. Kuchar, K. Dynamics of tensor fields in hyperspace. III. *J. Math. Phys.* **1976**, *17*, 801. [CrossRef]
40. Kamenshchik, A.Y.; Lyakhovich, S.L. Hamiltonian BFV-BRST theory of closed quantum cosmological models. *Nucl. Phys. B* **1997**, *495*, 309. [CrossRef]
41. Fradkin, E.S.; Vilkovisky, G.A. Quantization of relativistic systems with constraints. *Phys. Lett. B* **1975**, *55*, 224. [CrossRef]
42. Batalin, I.A.; Vilkovisky, G.A. Relativistic S-matrix of dynamical systems with boson and fermion constraints. *Phys. Lett. B* **1977**, *69*, 309. [CrossRef]
43. Batalin, I.A.; Fradkin, E.S. Operatorial quantizaion of dynamical systems subject to constraints. A Further study of the construction. *Ann. l'IHP Phys. Théorique* **1988**, *49*, 145.
44. Khalatnikov, I.M.; Kamenshchik, A.Y. A Generalization of the Heckmann-Schucking cosmological solution. *Phys. Lett. B* **2003**, *553*, 119. [CrossRef]
45. Barvinsky, A.O.; Kamenshchik, A.Y. Darkness without dark matter and energy—Generalized unimodular gravity. *Phys. Lett. B* **2017**, *774*, 59. [CrossRef]
46. Henneaux, M.; Teitelboim, C. The Cosmological Constant and General Covariance. *Phys. Lett. B* **1989**, *222*, 195. [CrossRef]
47. Unruh, W.G. A Unimodular Theory of Canonical Quantum Gravity. *Phys. Rev. D* **1989**, *40*, 1048. [CrossRef]
48. Goldstein, H.; Poole, C.P.; Safko, J. Classical Mechanics, 3rd ed. Available online: https://dokumen.tips/documents/classical-mechanics-3rd-edition-goldstein-pool-safko.html (accessed on 30 June 2021).

49. Kamenshchik, A.Y.; Tronconi, A.; Venturi, G. Induced Gravity and Quantum Cosmology. *Phys. Rev. D* **2019**, *100*, 023521. [CrossRef]
50. Kamenshchik, A.Y.; Tronconi, A.; Venturi, G. Quantum cosmology and the inflationary spectra from a nonminimally coupled inflaton. *Phys. Rev. D* **2020**, *101*, 023534. [CrossRef]
51. Everett, H. Relative state formulation of quantum mechanics. *Rev. Mod. Phys.* **1957**, *29*, 454. [CrossRef]
52. *The Many-Worlds Interpretation of Quantum Mechanics*; DeWitt, B.S.; Graham, N., Eds.; Princeton University Press: Princeton, NJ, USA, 1973.
53. Mandelstam, L.; Tamm, I. The Uncertainty Relation Between Energy and Time in Non-relativistic Quantum Mechanics. *J. Phys. USSR* **1945**, *9*, 249.
54. Lindblad, G. On the generators of quantum dynamical semigroups. *Commun. Math. Phys.* **1976**, *48*, 119. [CrossRef]
55. Gorini, V.; Kossakowski, A.; Sudarshan, E.C.G. Completely Positive Dynamical Semigroups of N Level Systems. *J. Math. Phys.* **1976**, *17*, 821. [CrossRef]
56. Salecker, H.; Wigner, E.P. Quantum Limitations of the Measurement of Space-Time Distances. *Phys. Rev.* **1958**, *109*, 571. [CrossRef]
57. Karolyhazy, F. Gravitation and quantum mechanics of macroscopic. *Il Nuovo C. A* **1966**, *42*, 390. [CrossRef]
58. Diosi, L. Gravitation and quantum-mechanical localization of macro-objects. *Phys. Lett. A* **1984**, *105*, 199. [CrossRef]
59. Penrose, R. On Gravity Role in Quantum State Reduction. *Gen. Relativ. Grav.* **1996**, *28*, 581. [CrossRef]
60. Kibble, T.W.B. Relativistic Models of Nonlinear Quantum Mechanics. *Commun. Math. Phys.* **1978**, *64*, 73. [CrossRef]
61. Kibble, T.W.B.; Randjbar-Daemi, S. Nonlinear Coupling of Quantum Theory and Classical Gravity. *J. Phys. A* **1980**, *13*, 141. [CrossRef]
62. Stamp, P.C.E. Environmental Decoherence versus Intrinsic Decoherence. *Phil. Trans. R. Soc. Lond. A* **2012**, *370*, 4429. [CrossRef]
63. Stamp, P.C.E. Rationale for a Correlated Worldline Theory of Quantum Gravity. *New J. Phys.* **2015**, *17*, 065017. [CrossRef]
64. Barvinsky, A.O.; Carney, D.; Stamp, P.C.E. Structure of Correlated Worldline Theories of Quantum Gravity. *Phys. Rev. D* **2018**, *98*, 084052. [CrossRef]
65. Barvinsky, A.O.; Wilson-Gerow, J.; Stamp, P.C.E. Correlated Worldline theory: Structure and Consistency. *Phys. Rev. D* **2021**, *103*, 064028. [CrossRef]
66. Belinskii, V.A.; Grishchuk, L.P.; Zeldovich, Y.B.; Khalatnikov, I.M. Inflationary stages in cosmological models with a scalar field. *Sov. Phys. JETP* **1985**, *62*, 195. [CrossRef]
67. Andrianov, A.A.; Novikov, O.O.; Lan, C. Quantum cosmology of the multi-field scalar matter: Some exact solutions. *Theor. Math. Phys.* **2015**, *184*, 1224. [CrossRef]
68. Bohr, N. Can Quantum-Mechanical Description of Physical Reality be Considered Complete? *Phys. Rev.* **1935**, *48*, 696. [CrossRef]
69. Barvinsky, A.O.; Kamenshchik, A.Y. Preferred basis, decoherence and a quantum state of the universe. *arXiv* **2020**, arXiv:2006.16812.
70. Tronconi, A.; Vacca, G.P.; Venturi, G. The Inflaton and time in the matter gravity system. *Phys. Rev. D* **2003**, *67*, 063517. [CrossRef]
71. Kamenshchik, A.Y.; Tronconi, A.; Vardanyan, T.; Venturi, G. Quantum Gravity, Time, Bounces and Matter. *Phys. Rev. D* **2018**, *97*, 123517. [CrossRef]
72. Anselmi, D. On the quantum field theory of the gravitational interactions. *JHEP* **2017**, *1706*, 86. [CrossRef]
73. Anselmi, D.; Piva, M. Quantum Gravity, Fakeons And Microcausality. *JHEP* **2018**, *1811*, 21. [CrossRef]
74. Anselmi, D. The quest for purely virtual quanta: Fakeons versus Feynman-Wheeler particles. *JHEP* **2020**, *2003*, 142. [CrossRef]
75. Donoghue, J.F.; Menezes, G. Arrow of Causality and Quantum Gravity. *Phys. Rev. Lett.* **2019**, *123*, 171601. [CrossRef]
76. Donoghue, J.F.; Menezes, G. Quantum causality and the arrows of time and thermodynamics. *Prog. Part. Nucl. Phys.* **2020**, *115*, 103812. [CrossRef]

Article

Cosmic Tangle: Loop Quantum Cosmology and CMB Anomalies

Martin Bojowald

Institute for Gravitation and the Cosmos, The Pennsylvania State University, 104 Davey Lab, University Park, PA 16802, USA; bojowald@gravity.psu.edu

Abstract: Loop quantum cosmology is a conflicted field in which exuberant claims of observability coexist with serious objections against the conceptual and physical viability of its current formulations. This contribution presents a non-technical case study of the recent claim that loop quantum cosmology might alleviate anomalies in the observations of the cosmic microwave background.

Keywords: loop quantum cosmology; observations

"Speculation is one thing, and as long as it remains speculation, one can understand it; but as soon as speculation takes on the actual form of ritual, one experiences a proper shock of just how foreign and strange this world is." [1]

1. Introduction

Quantum cosmology is a largely uncontrolled and speculative attempt to explain the origin of structures that we see in the universe. It is uncontrolled because we do not have a complete and consistent theory of quantum gravity from which cosmological models could be obtained through meaningful restrictions or approximations. It is speculative because we do not have direct observational access to the Planck regime in which it is expected to be relevant.

Nevertheless, quantum cosmology is important because extrapolations of known physics and observations of the expanding universe indicate that matter once had a density as large as the Planck density. Speculation is necessary because it can suggest possible indirect effects that are implied by Planck-scale physics, but manifest themselves on more accessible scales. Speculation is therefore able to guide potential new observations.

Speculation becomes a problem when it is based on assumptions that are unmentioned, poorly justified, or, in some cases, already ruled out. When this happens, speculation is turned into a ritual followed by a group of practitioners who continue to believe in their assumptions and ignore outside criticism. The uncontrolled nature of quantum cosmology makes it particularly susceptible to this danger.

A recent example is the claim [2] that loop quantum cosmology may alleviate various anomalies in observations of the cosmic microwave background. Given the commonly accepted distance between quantum cosmology and observations available at present, this claim, culminating in the statement that "these results illustrate that LQC has matured sufficiently to lead to testable predictions," is surprising and deserves special scrutiny, all the more so because a large number of conceptual and physical shortcomings have been uncovered in loop quantum cosmology over the last few years (all citations from [2] refer to its preprint version).

Upon closer inspection, we encounter a strange world in the claims of [2] according to which, to mention just one obvious misjudgment, "many of the specific technical points [of a recent critique of the methods used in the analysis] were already addressed, e.g., in [42, 64, 65] and in the Appendix of [66]" even though [42, 64, 65, 66] were published between nine and twelve years before the recent critique that they were supposed to have addressed.

Clearly, the authors of [2] have not sufficiently engaged with relevant new criticism. There is therefore a danger that their work is based on rituals. Our analysis will, unfortunately, confirm this suspicion. A dedicated review of these claims is especially important because they have been widely advertized, for instance in a press release. [1]

The present paper provides a close reading of these claims and highlights various shortcomings. It therefore serves as a case study of the complicated interplay between quantum cosmological modeling and observations. A detailed technical discussion of some of the underlying problems of current versions of loop quantum cosmology was already given in [3]. The analysis here is held at a non-technical level and is more broadly accessible. It highlights conceptual problems that are relevant for [2], but were not considered in [3], for instance concerning suitable justifications of initial states in a bouncing cosmological model.

Since the focus of our exposition was on quantum cosmology, other questions that may well be relevant for an assessment of the claims of [2] are not discussed. These questions include a proper analysis of all early-universe observables in addition to a select number of anomalies and the possible role of sub-Planckian unknowns such as further modifications of gravity not considered in [2] or uncertainties about the energy contributions in the early universe; see for instance [4–6] for related examples and reviews.

2. Inconsistencies

The model constructed in [2] was based on several key assumptions made elsewhere in the literature on loop quantum cosmology:

Assumption 1. *In loop quantum cosmology, there are quantum-geometry corrections that imply a bounce of the universe at about the Planck density;*

Assumption 2. *During quantum evolution through the bounce and long before and after, the quantum state of the universe remains sharply peaked as a function of the scale factor;*

Assumption 3. *Although the dynamics is modified and geometry is quantum, general covariance in its usual form is preserved on all scales, including the Planck scale;*

Assumption 4. *At the bounce, the spatial geometry and matter distribution are as homogeneous and isotropic as possible, restricted only by uncertainty relations for modes of perturbative inhomogeneity.*

These assumptions are necessary for the model to work as it does. For instance, the bounce (Assumption 1) and spatial symmetries (Assumption 4) imply a certain cut-off in the primordial power spectrum that quickly turns out to be useful in the context of anomalies in the cosmic microwave background. A sharply peaked state (Assumption 2) and general covariance (Assumption 3) make it possible to analyze the model by standard methods of spacetime physics, using line elements and well-known results from cosmological perturbation theory [7].

Upon closer inspection, however, it is hard to reconcile these assumptions with established physics. Some of them are even mutually inconsistent. We first discuss these inconsistencies and then examine how the authors tried to justify their assumptions.

2.1. Bounce

In general relativity, there are well-known singularity theorems that prove the existence of singularities in the future or the past, under certain assumptions such as energy inequalities or the initial condition of a universe expanding at one time [8,9]. In their most general form, these theorems do not require the specific dynamics of general relativity, but only use the properties of Riemannian geometry, such as the geodesic deviation equation. Loop quantum cosmology can avoid the big bang singularity and replace it by a bounce only if it violates at least one of the assumptions of singularity theorems. Just having a theory with modified dynamics compared with general relativity is not sufficient.

Loop quantum cosmology does not question that the universe is currently expanding, thus obeying the same initial condition commonly used to infer the singular big bang (it is known that the specific modeling of an expanding universe can be weakened so as to evade singularity theorems [10], but loop quantum cosmology does not avail itself of this option). Moreover, as emphasized in [2], "These qualitatively new features arise without having to introduce matter that violates any of the standard energy conditions." According to Assumption 3, spacetime in loop quantum cosmology remains generally covariant on all scales (described by a suitable line element) and therefore obeys the general properties of Riemannian geometry. Loop quantum cosmology does modify the gravitational dynamics, but this is not a required ingredient of singularity theorems.

Since all the conditions of singularity theorems still hold, according to the explicit or implicit assumptions in [2], loop quantum cosmology should be singular just as classical general relativity. How can it exhibit a bounce at high density? For a detailed resolution of this conundrum, see [11].

2.2. Peakedness

Setting aside special dynamics such as the harmonic oscillator, quantum states generically spread out and change in complicated ways as they evolve. For a macroscopic object, such as a heavy free particle, it takes longer for such features to become significant than for a microscopic system, but they are nevertheless present.

In late-time cosmology, it is justified to assume that a simple homogeneous and isotropic spatial geometry of a large region describes the dynamics very well. At late times, a state in quantum cosmology may therefore be assumed to remain sharply peaked because it describes a large region that is conceptually analogous to a heavy free particle.

When such a state is extrapolated to the big bang, however, very long time scales are involved. Is it still justified to assume that the state does not change significantly and spread out during these times? Moreover, if the usual assumption of a homogeneous and isotropic geometry is maintained for a tractable description of quantum cosmology, such a region would have to be chosen smaller and smaller as we approach the big bang, not only because of the shrinking space of an expanding universe in time in reverse, but also, and more importantly, because attractive gravity implies structure formation within co-moving volumes. As the big bang is approached, a valid homogeneous approximation of quantum cosmology more and more resembles quantum mechanics of a microscopic object.

Because we do not know what a realistic geometry of our universe at the Planck scale should be, we do not know how macroscopic it may still be considered to be in a homogeneous approximation. The only indication is the Belinskii–Khalatnikov–Lifshitz (BKL) scenario [12] of generic spacetime properties near a spacelike singularity, which suggests that there is no lower limit to the size of homogeneous regions in classical general relativity. Quantum gravity is expected to modify the dynamics of general relativity, but it is not known whether and at what scale it would be able to prevent the BKL-type behavior.

This result suggests that a microscopic description should be assumed in the Planck regime. However, according to [2], "One can now start with a quantum state $\Psi(a,\phi)$ that is peaked on the classical dynamical trajectory at a suitably late time when curvature is low, and evolve it back in time towards the big bang using either the WDW equation or the LQC evolution equation. Interestingly the wave function continues to remain sharply peaked in both cases." Why should quantum cosmological dynamics be such that it maintains a sharply peaked state even over long time scales that include a phase in which the evolved object should be considered microscopic?

2.3. Covariance

According to Assumption 1, the bounce in loop quantum cosmology is supposed to happen because quantum geometry modifies the classical dynamics: "Of particular interest is the area gap—the first non-zero eigenvalue Δ of the area operator. It is a fundamental microscopic parameter of the theory that then governs important macroscopic phenomena in LQC that lead, e.g., to finite upper bounds for curvature." There is supposed to be a certain quantum structure of space that leads, among other things, to a discrete area spectrum.

However, the description of perturbative inhomogeneity in [2] or the underlyingin [13] assumes that this quantum geometry can be described by a line element. To be sure, coefficients of the "dressed" metric in this line element contain quantum corrections, but it has not been shown that a quantum geometry with some discrete area spectrum can be described by any line element at all, even at the Planck scale, where discreteness is supposed to be significant enough to change the dynamics of the classical theory. While specific corrections in metric coefficients have been derived, the claim that they should appear in a line element of some Riemannian geometry has not been justified.

There are now several no-go results [14,15] that demonstrate violations of covariance in regimes envisioned by the authors of [2]. A proof that Riemannian geometry and line elements can nevertheless be used in their context would therefore require a detailed discussion of how these no-go results can be circumvented. However, there is no hint of such an attempt. It is worth noting that problems with covariance have also occurred in a different application of the same formalism to black holes, in which several other physical problems were quickly found [16–19]. The spacetime description assumed in [2] is therefore unreliable.

Given the uncertain status of what structure spacetime geometry should have in the presence of modifications from loop quantum cosmology, the meaning of "finite upper bounds for curvature" that are supposed to be implied by the discrete area spectrum is unclear.

2.4. Symmetry

Spatial homogeneity and isotropy are assumed at various places in [2]. First, in older papers referred to for justifications of some claims, quantum evolution is numerically computed for states of an exactly homogeneous and isotropic geometry, using methods from quantum cosmology. Secondly, states for inhomogeneous matter perturbations on such a background are assumed to preserve the symmetry as much as possible while obeying uncertainty relations.

2.4.1. Background

An attempt was made in [2] to justify these assumptions: "On the issue of simplicity of the LQC description, we note that in the 1980s it was often assumed that the early universe is irregular at all scales and therefore quite far from being as simple as is currently assumed at the onset of inflation. Yet now observations support the premise that the early universe is exceedingly simple in that it is well modeled by a FLRW spacetime with first order cosmological perturbations. Therefore, although a priori one can envisage very complicated quantum geometries, it is far from being clear that they are in fact realized in the Planck regime."

However, this attempted justification is invalid because the scale probed by early-universe observations referred to in this statement is vastly different from the Planck scale in which the outcome is applied. Cosmic inflation was proposed precisely to address, among other things, the homogeneity problem even if the initial state may be much less regular, and inflation is still used in most models of loop quantum cosmology. In such scenarios, large-scale homogeneity at later times does not show in any way that the universe must have been homogeneous at the Planck scale.

The assumed homogeneity is also in conflict with the claimed discrete structure of space that, according to Assumption 1, might imply a bounce. The authors of [2] never addressed the relevant question of how their symmetry assumptions can be reconciled with a discrete geometry on ultraviolet scales that is supposed to imply all the claimed effects (in other words, the authors ignored the trans-Planckian problem of inflationary cosmology [20–22]). Of course, suitable superpositions of discrete states may lead to a continuum of expectation values even of operators that have a discrete spectrum. However, if this argument were used as a possible explanation of homogeneity, it would put in doubt the strong emphasis on a single eigenvalue Δ of the area spectrum that is claimed to govern the new dynamics.

2.4.2. Perturbations

The authors of [2] used symmetry assumptions not only for the background on which inhomogeneous modes evolve, but also for the modes themselves, described perturbatively. These modes cannot be exactly symmetric because they are subject to uncertainty relations and therefore have, at least, non-zero fluctuations even if their expectation values may be zero. It is claimed that these modes should be as symmetric as possible in the bounce phase, that is have zero expectation values and fluctuations such that they saturate uncertainty relations.

The homogeneity assumption for perturbations is motivated by Penrose's Weyl curvature hypothesis [23], introduced in [2]: "Finally, the principle that determines the quantum state $\Psi(Q,\phi)$ of scalar modes involves a quantum generalization of Penrose's Weyl curvature hypothesis in the Planck regime near the bounce, which physically corresponds to requiring that the state should be 'as isotropic and homogeneous in the Planck regime, as the Heisenberg uncertainty principle allows'." However, conceptually, there is a significant difference between these two proposals. Penrose's hypothesis was given in a big bang setting in which the initial state (close to the big bang singularity) was to be restricted by geometrical considerations. As always, one may question the specific motivation for a certain choice of initial states, but there are no physical objections provided general conditions (such as uncertainty relations) are obeyed.

The symmetry assumption employed for matter perturbations in [2] is of a very different nature. It is used to determine an initial state only of our current expanding phase of the universe, but it is set in a bounce model with a pre-history before the big bang. The symmetry assumption for matter perturbations is therefore a final condition for the collapse phase and violates determinism. Potential violations of deterministic behavior have indeed been derived in models of loop quantum cosmology in the form of signature change at high density [24–26]. However, Reference [2] did not use this option, which would in fact be in conflict with the line element they assumed to formulate a wave equation for cosmological perturbations. Moreover, the derived versions of signature change would set the beginning of the Lorentzian expanding branch of the universe later than assumed in [2], at the very end of the bounce phase in which modifications from loop quantum cosmology subside.

Setting aside the question of determinism, a restriction of the state during the bounce phase, a transitory stage, is not an initial condition, but rather an assumption about the state to which preceding collapse may have led. It is then questionable that gravitational collapse of a preceding inhomogeneous universe should lead to a bounce state that is as homogeneous as possible, respecting uncertainty relations. In this scenario, the collapse of a preceding universe is supposed to have led to a very homogeneous state at the Planck density of more than one trillion solar masses in a proton-sized region. Given the inherently unstable nature of gravitational collapse, one would rather expect that any slightly overdense region in a collapsing universe would quickly become denser and magnify initial inhomogeneities that had been present when collapse commenced.

It is hard to see how collapse could, instead, lead to a state that is as homogeneous as possible. Such an assumption would at least require a dedicated justification, in particular because it directly implies the crucial features of a new scale in loop quantum cosmology that is then used to alleviate anomalies. Unfortunately, no attempted justification can be found in [2].

Upon closer inspection, the arguments given in [2] were even circular. The very setup that led the authors to their formulation of a homogeneous initial condition already assumed the near homogeneity of a collapsing universe: The specific statement in [2] referred to "the Planck regime near the bounce" and implicitly assumed that (in the

Riemannian geometry, according to Assumption 3) there is a time coordinate such that "the bounce" happens everywhere within a large region at the same time. However, if the collapsing geometry is inhomogeneous, overdense regions will become denser during collapse and reach the Planck density earlier than their neighbors. Once they bounce and start expanding, it is not obvious that a simple near-homogeneous slicing with a uniform bounce time still exists (this process suggests a multiverse rather than a single nearly homogeneous universe [27]).

There is no unique bounce time in this picture, and therefore, an implicit assumption (a meaningful "near the bounce") used in the condition of initial homogeneity is unphysical. Crucial statements such as "In our LQC model, the physical principles used to select the background quantum geometry imply that the corresponding ΛCDM universe has undergone approximately 141 e-folds of expansion since the quantum bounce until today." therefore remain unjustified.

2.5. Attempted Justifications

The authors of [2] realized that some of these assumptions should be justified, while they were apparently unaware of additional implicit assumptions that they did not mention.

1. The bounce is justified by quantum-geometry corrections from loop quantum gravity, but the conflict with singularity theorems has apparently gone unnoticed;
2. The peakedness of states is justified by referring to detailed numerical studies of evolving states in loop quantum cosmology. However, the authors failed to notice that these studies implicitly assume that the universe is still macroscopic (large-scale homogeneity), even in the Planck regime, ignoring structure formation within co-moving regions in a collapsing universe, as well as BKL-type behavior;
3. Covariance is justified only vaguely by referring to wave equations for perturbations on a background, without asking the relevant question of whether modified perturbation and background equations can still be consistent with a single metric that is being perturbed;
4. In their discussion of initial conditions for perturbations, the authors seemed to be unaware of problems posed by the pre-history of a collapsing universe.

3. A Brief Engagement with the Previous Critique

It is instructive to analyze the brief, but rapid-fire response given in [2] to the previous criticism of certain claims in loop quantum cosmology [3]. As already mentioned in the Introduction, the authors stated that "Many of the specific technical points were already addressed, e.g., in [42, 64, 65] and in the Appendix of [66]." However, the references provided tried (but failed [28,29]) to address an older issue, cosmic forgetfulness [30,31], that did not play a major role in the recent discussion of [3] (since the publication of these older papers, cosmic forgetfulness has been strengthened to signature change).

3.1. Effective Descriptions

The authors of [2] went on and stated that "First, although 'effective equations' are often used in LQC, conceptually they are on a very different footing from those used in effective field theories: One does not integrate out the UV modes of cosmological perturbations. The term 'effective' is used in a different sense in LQC: these equations carry some of the leading-order information contained in sharply peaked quantum FLRW geometries $\Psi(a, \phi)$." (Note that the authors were hedging their statement by correctly saying that "these equations carry *some* of the leading-order information" (emphasis added). The fact that they carry only some, but not all of the leading-order information is a problem in itself that will not be discussed here; see [3,32] for details.).

The admission that "the term 'effective' is used in a different sense in LQC" does not make this formalism more meaningful. It is in fact one of the major problems in current realizations of the framework of loop quantum cosmology. Equations of loop quantum cosmology are used on vastly different scales, in the Planck regime to analyze the possibility of a bounce, and at low curvature to justify a peaked late-time state. A suitable effective theory (not in the sense used, according to [2], in loop quantum cosmology) would be needed to determine how parameters of the model may change by infrared or ultraviolet renormalization. The authors were simply assuming that a single effective theory without any running parameters can be used to describe the quantum universe on a vast range of scales. There is no justification for this assumption.

In addition, the authors referred not only to different scales in cosmology, but also to different geometries, including those of black hole horizons: When they justified certain parameter choices for the dynamics of loop quantum cosmology, they made statements such as "The eigenvalues of [the area operator] $\hat{A}_S$ are discrete in all γ-sectors. But their numerical values are proportional to [the Barbero–Immirzi parameter (a quantization ambiguity)] γ and vary from one γ sector to another.", "In LQG, a direct measurement of eigenvalues of geometric operators would determine γ. But of course such a measurement is far beyond the current technological limits.", and "Specifically, in LQG the number of microstates of a black hole horizon grows exponentially with the area, whence one knows that the entropy is proportional to the horizon area. But the proportionality factor depends on the value of γ. Therefore if one requires that the leading term in the statistical mechanical entropy of a spherical black hole should be given by the Bekenstein-Hawking formula $S = A/4\ell_{Pl}^2$, one determines γ and thus the LQG sector Nature prefers."

These statements refer to at least three different regimes of some underlying theory of loop quantum gravity: direct microscopic measurements of eigenvalues, macroscopic black hole horizons, and the entire universe at various densities. It is simply assumed that the same value of γ (as well as other parameters) may be used in all these situations without suitable renormalization. The only justification attempted for this assumption is the statement that "the term 'effective' is used in a different sense in LQC", which ignores the physical reasons for standard ingredients in effective theories.

The possibility of running is also ignored in statements such as "In IV C, we will show that the interplay between LQC and observations is a 2-way bridge, in that the CMB observations can also be used to constrain the value of the area gap Δ, the most important of fundamental microscopic parameters of LQG.", "As we saw in section III, the area gap Δ is the key microscopic parameter that determines values of important new, macroscopic observables such as the matter density and the curvature at the bounce. Its specific value, $\Delta = 5.17\ell_{Pl}^2$, is determined by the statistical mechanical calculation of the black hole entropy in loop quantum gravity (see, e.g., [55, 56, 59, 60]).", and "Clearly, the value $\Delta \approx 5.17\ell_{Pl}^2$ chosen in Sec. III C and used in this paper, is within 68% (1σ) confidence level of the constraint obtained from Planck 2018. This not only indicates a synergy between the fundamental theoretical considerations and observational data, but also provides internal consistency of the LQC model."

3.2. Covariance

In their reply to [3], the authors stated that "As we will see in Section III B, equations satisfied by the cosmological perturbations are indeed covariant." In Section III.B, however, less than half a sentence was devoted to this important question: "Note also that the equation is covariant w.r.t. [the dressed metric] $\tilde{g}_{ab}$ and $\tilde{g}_{ab}$ rapidly tends to the classical FLRW metric of GR outside the Planck regime." This attempted justification of covariance apparently refers to the equation $(\tilde{\Box} + \tilde{U}/\tilde{a}^2)\hat{Q} = 0$ for modes $\hat{Q}$, with certain functions $\tilde{U}$ and $\tilde{a}$ of the background scale factor. However, nowhere did the authors address the crucial question of whether background $\tilde{g}_{ab}$, which defines the d'Alembertian $\tilde{\Box}$, and perturbation $\hat{Q}$ can, after modifications, still be obtained from a single covariant metric (obeying the tensor-transformation law), as in the underlying classical theory.

The authors' statement about covariance refers only to transformations of the perturbative mode, $\hat{Q}$, and therefore to small inhomogeneous coordinate changes that preserve the perturbative nature. In addition, potentially large homogeneous transformations of the background time coordinate, such as transforming from proper time to conformal time, are relevant in cosmological models of perturbative inhomogeneity. The usual curvature perturbations are not invariant with respect to these transformations [33]. The authors' statements had nothing to say about the question of whether their model of modified perturbation equations is covariant with respect to large transformations of background time. The failure of covariance in the underlying construction was shown in [14].

3.3. Symmetry

We already addressed the authors' erroneous view that observations can be used to justify near homogeneity in the Planck regime. The authors finally stated that "Nonetheless, one should keep in mind that, as in other approaches to quantum cosmology, in LQC the starting point is the symmetry reduced, cosmological sector of GR. Difference from the Wheeler-DeWitt theory is that one follows the same systematic procedure in this sector as one does in full LQG. But the much more difficult and fundamental issue of systematically deriving LQC from full LQG is still open mainly because dynamics of full LQG itself is still a subject of active investigation." Here, they were attempting to construct a false binary choice between exactly isotropic models of symmetry-reduced geometries, on one hand, and calculations in full loop quantum gravity without any symmetry assumptions, on the other. If this choice were correct, one might as well give up because exactly isotropic models are unrealistic, and the full theory is intractable.

What the authors were missing is a proper effective theory that not only amends isotropic equations by certain leading-order corrections, but also tries to go beyond strictly isotropic models by parameterizing all the ignorance in the parameter choices implied by the intractable nature of full loop quantum gravity. Without such an effective description, which does not require direct calculations in the full theory and is therefore feasible, but would not be as simple as the authors assumed, no observational claims can be reliable.

4. Conclusions

How can we reconcile the claim that "these results illustrate that LQC has matured sufficiently to lead to testable predictions" [2] with the availability of several serious and independent concerns that have shown in recent years how loop quantum cosmology, as it is commonly practiced, has overlooked a large number of important conceptual and physical requirements? In the present paper, we provided detailed evidence to show that [2] merely ignored or insufficiently addressed relevant criticism. Moreover, the main new claims of [2] are implied rather directly by specific assumptions that remain unjustified. We summarize these observations in this concluding section.

In their quest to show that loop quantum cosmology naturally resolves anomalies in observations of the cosmic microwave background and therefore makes testable predictions, the authors of [2] used or introduced several conceptually distinct assumptions. Some of these assumptions, including those about bounces and covariance, are questionable within loop quantum cosmology. Others, for instance about free parameters, rely on an oversimplified presentation of loop quantum cosmology as some special version of an effective theory that could be used to describe the dynamics of quantum gravity on a vast range of scales, including the Planck regime, without any running parameters.

Yet another set of assumptions, referring to the state of perturbations and their spatial homogeneity, is independent of loop quantum cosmology, but has been packaged with the other assumptions in a way that gives the erroneous impression of a single coherent theory. Moreover, these assumptions are physically questionable because they implicitly make strong and unrealistic claims about the generic final state of a collapsing universe. We are observing a single universe that might perhaps have emerged from a special version of preceding collapse as envisioned by the authors. However, by simply formulating the desired behavior as an assumption without addressing a possible relationship with a pre-history, the authors hid its restrictive nature and ultimately failed to explain the initial state of cosmological perturbations and the observed microwave background.

The authors of [2] were aware of the previous criticism and gave a half-hearted attempt to address it. In Section 3, we saw how inadequate their response was. To summarize, the cited papers that supposedly addressed "many of the specific technical points", made in [3], had been published between nine and twelve years before this recent critique. The crucial question of general covariance in models of quantum gravity, discussed from different viewpoints for instance in [34,35], was acknowledged in [2] only by the misleading "as we will see in Section 3 B, equations satisfied by the cosmological perturbations are indeed covariant" to announce a brief statement "note also that the equation is covariant w.r.t. [the dressed metric] $\tilde{g}_{ab}$" that reflects a misunderstanding of the issue of covariance in the setting of modified perturbation equations (not only perturbations, but also the background must be included in a covariance analysis [14]).

Notably, there are also points of critique that the authors of [2] did not bother to address. An important example, in addition to the prevalence of quantization ambiguities, is the implicit assumption in current formulations of loop quantum cosmology that the universe remains large-scale homogeneous (over at least several hundred Planck lengths) even at the Planck density. As explained in [3], this assumption is related to the incorrect implementation of effective descriptions in loop quantum cosmology. Without this unjustified and unmentioned assumption, the authors of [2] would not even be able to formulate their restrictive condition that cosmological perturbations be as homogeneous as possible at the bounce. This assumption ultimately leads to a suppression of power on large scales of the cosmic microwave background and allows claims about resolved anomalies. However, this assumption is not only unjustified, but it is also based on another and implicit assumption that had already been ruled out.

Therefore, the results of [2] relied on several major assumptions, made explicitly or implicitly, that turned out to be unjustified. The authors' claims therefore went well beyond the usual level of speculation that is common (and unavoidable) in quantum cosmology. Their response to objections that have been published during the last few years was inadequate in some cases and non-existent in others. Ashtekar et al. created a cosmic tangle of rituals that they no longer wish to be questioned.

Funding: This research was funded by NSF Grant Number PHY-1912168.

Conflicts of Interest: The author declares no conflict of interest.

Note

1 https://news.psu.edu/story/626795/2020/07/29/research/cosmic-tango-between-very-small-and-very-large (accessed on 2 June 2021).

References

1. Arendt, H. Letter to Gershom Scholem, January 1, 1952. In *The Correspondence of Hannah Arendt and Gershom Scholem*; The University of Chicago Press: Chicago, IL, USA; London, UK, 2017.
2. Ashtekar, A.; Gubt, B.; Sreenath, V. Cosmic tango between the very small and the very large: Addressing CMB anomalies through Loop Quantum Cosmology. *arXiv* **2021**, arXiv:2103.14568.
3. Bojowald, M. Critical Evaluation of Common Claims in Loop Quantum Cosmology. *Universe* **2020**, *6*, 36. [CrossRef]
4. Sola, J.; Gomez-Valent, A.; de Cruz Perez, J.; Moreno-Pulido, C. Brans-Dicke cosmology with a Λ-term: A possible solution to ΛCDM tensions. *Class. Quantum Grav.* **2020**, *37*, 245003.
5. Di Valentino, E.E.A. Cosmology Intertwined II: The Hubble Constant Tension. *Astropart. Phys.* **2021**, *131*, 102605. [CrossRef]
6. Di Valentino, E.; Mena, O.; Pan, S.; Visinelli, L.; Yang, W.; Melchiorri, A.; Mota, D.F.; Riess, A.G.; Silk, J. In the Realm of the Hubble tension—A Review of Solutions. *arXiv* **2021**, arXiv:2103.01183.
7. Mukhanov, V.F.; Feldman, H.A.; Brandenberger, R.H. Theory of cosmological perturbations. *Phys. Rept.* **1992**, *215*, 203–333. [CrossRef]
8. Hawking, S.W.; Ellis, G.F.R. *The Large Scale Structure of Space-Time*; Cambridge University Press: Cambridge, UK, 1973.
9. Senovilla, J.M.M. Singularity Theorems and Their Consequences. *Gen. Rel. Grav.* **1998**, *30*, 701–848. [CrossRef]
10. Senovilla, J.M.M. New Class of Inhomogeneous Cosmological Perfect-Fluid Solutions without Big-Bang Singularity. *Phys. Rev. Lett.* **1990**, *64*, 2219–2221. [CrossRef]
11. Bojowald, M. Black-hole models in loop quantum gravity. *Universe* **2020**, *6*, 125. [CrossRef]

12. Belinskii, V.A.; Khalatnikov, I.M.; Lifschitz, E.M. A general solution of the Einstein equations with a time singularity. *Adv. Phys.* **1982**, *31*, 639–667. [CrossRef]
13. Agulló, I.; Ashtekar, A.; Nelson, W. An Extension of the Quantum Theory of Cosmological Perturbations to the Planck Era. *Phys. Rev. D* **2013**, *87*, 043507. [CrossRef]
14. Bojowald, M. Non-covariance of the dressed-metric approach in loop quantum cosmology. *Phys. Rev. D* **2020**, *102*, 023532. [CrossRef]
15. Bojowald, M. No-go result for covariance in models of loop quantum gravity. *Phys. Rev. D* **2020**, *102*, 046006. [CrossRef]
16. Bodendorfer, N.; Mele, F.M.; Münch, J. A note on the Hamiltonian as a polymerisation parameter. *Class. Quantum Grav.* **2019**, *36*, 187001. [CrossRef]
17. Bodendorfer, N.; Mele, F.M.; Münch, J. Effective Quantum Extended Spacetime of Polymer Schwarzschild Black Hole. *Class. Quantum Grav.* **2019**, *36*, 195015. [CrossRef]
18. Bouhmadi-López, M.; Brahma, S.; Chen, C.Y.; Chen, P.; Yeom, D.h. Asymptotic non-flatness of an effective black hole model based on loop quantum gravity. *Phys. Dark Univ.* **2020**, *30*, 100701. [CrossRef]
19. Faraoni, V.; Giusti, A. Unsettling physics in the quantum-corrected Schwarzschild black hole. *Symmetry* **2020**, *12*, 1264. [CrossRef]
20. Martin, J.; Brandenberger, R.H. The Trans-Planckian Problem of Inflationary Cosmology. *Phys. Rev. D* **2001**, *63*, 123501. [CrossRef]
21. Brandenberger, R.H.; Martin, J. The Robustness of Inflation to Changes in Super-Planck-Scale Physics. *Mod. Phys. Lett. A* **2001**, *16*, 999–1006. [CrossRef]
22. Niemeyer, J.C. Inflation with a Planck-scale frequency cutoff. *Phys. Rev. D* **2001**, *63*, 123502. [CrossRef]
23. Penrose, R. *General Relativity: An Einstein Centenary Survey*; Cambridge University Press: Cambridge, UK, 1979.
24. Bojowald, M.; Paily, G.M. Deformed General Relativity and Effective Actions from Loop Quantum Gravity. *Phys. Rev. D* **2012**, *86*, 104018. [CrossRef]
25. Mielczarek, J. Signature change in loop quantum cosmology. *Springer Proc. Phys.* **2014**, *157*, 555.
26. Bojowald, M.; Mielczarek, J. Some implications of signature-change in cosmological models of loop quantum gravity. *JCAP* **2015**, *8*, 52. [CrossRef]
27. Bojowald, M. A loop quantum multiverse? *AIP Conf. Proc.* **2013**, *1514*, 21–30.
28. Bojowald, M. Comment on "Quantum bounce and cosmic recall". *Phys. Rev. Lett.* **2008**, *101*, 209001. [CrossRef]
29. Bojowald, M.; Tsobanjan, A. Effective Casimir conditions and group coherent states. *Class. Quantum Grav.* **2014**, *31*, 115006. [CrossRef]
30. Bojowald, M. What happened before the big bang? *Nat. Phys.* **2007**, *3*, 523–525. [CrossRef]
31. Bojowald, M. Harmonic cosmology: How much can we know about a universe before the big bang? *Proc. Roy. Soc. A* **2008**, *464*, 2135–2150. [CrossRef]
32. Bojowald, M. Quantum cosmology: A review. *Rep. Prog. Phys.* **2015**, *78*, 023901. [CrossRef]
33. Stewart, J.M. Perturbations of Friedmann–Robertson–Walker cosmological models. *Class. Quantum Grav.* **1990**, *7*, 1169–1180. [CrossRef]
34. Collins, J.; Perez, A.; Sudarsky, D.; Urrutia, L.; Vucetich, H. Lorentz invariance and quantum gravity: An additional fine-tuning problem? *Phys. Rev. Lett.* **2004**, *93*, 191301. [CrossRef]
35. Polchinski, J. Comment on [arXiv:1106.1417] "Small Lorentz violations in quantum gravity: Do they lead to unacceptably large effects?". *arXiv* **2011**, arXiv:1106.6346.

Review

Quantum String Cosmology

Maurizio Gasperini [1,2]

1. Dipartimento di Fisica, Università di Bari, Via G. Amendola 173, 70126 Bari, Italy; maurizio.gasperini@ba.infn.it
2. Istituto Nazionale di Fisica Nucleare, Sezione di Bari, Via E. Orabona 4, 70125 Bari, Italy

Abstract: We present a short review of possible applications of the Wheeler-De Witt equation to cosmological models based on the low-energy string effective action, and characterised by an initial regime of asymptotically flat, low energy, weak coupling evolution. Considering in particular a class of duality-related (but classically disconnected) background solutions, we shall discuss the possibility of quantum transitions between the phases of pre-big bang and post-big bang evolution. We will show that it is possible, in such a context, to represent the birth of our Universe as a quantum process of tunneling or "anti-tunneling" from an initial state asymptotically approaching the string perturbative vacuum.

Keywords: string cosmology; quantum cosmology; Wheeler-DeWitt equation

Citation: Gasperini, M. Quantum String Cosmology. *Universe* **2021**, *7*, 14. https://doi.org/10.3390/universe7010014

Received: 26 November 2020
Accepted: 9 January 2021
Published: 12 January 2021

Publisher's Note: MDPI stays neutral with regard to jurisdictional claims in published maps and institutional affiliations.

Copyright: © 2021 by the author. Licensee MDPI, Basel, Switzerland. This article is an open access article distributed under the terms and conditions of the Creative Commons Attribution (CC BY) license (https://creativecommons.org/licenses/by/4.0/).

1. Introduction

In the standard cosmological context the Universe is expected to emerge from the big bang singularity and to evolve initially through a phase of very high curvature and density, well inside the quantum gravity regime. Quantum cosmology, in that context, turns out to be a quite appropriate formalism to describe the "birth of our Universe", possibly in a state approaching the de Sitter geometric configuration typical of inflation (see, e.g., [1] for a review).

In the context of string cosmology, in contrast, there are scenarios where the Universe emerges from a state satisfying the postulate of "asymptotic past triviality" [2] (see [3] for a recent discussion): in that case the initial phase is classical, with a curvature and a density very small in string (or Planck) units. Even in that case, however, the transition to the decelerated radiation-dominated evolution, typical of standard cosmology, is expected to occur after crossing a regime of very high-curvature and strong coupling. The birth of our Universe, regarded as the beginning of the standard cosmological state, corresponds in that case to the transition (or "bounce") from the phase of growing to decreasing curvature, and even in that case can be described by using quantum cosmology methods, like for a Universe emerging from an initial singularity.

There is, however, a crucial difference between a quantum description of the "big bang" and of the "big bounce": indeed, the bounce is preceded by a long period of low-energy, classical evolution, while the standard big bang picture implies that the space-time dynamics suddenly ends at the singularity, with no classical description at previous epochs (actually, there are no "previous" epochs, as the time coordinate itself ends at the singularity). In that context the initial state of the Universe is unknown, and has to be fixed through some ad hoc prescription: hence, different choices for the initial boundary conditions are in principle allowed [4–9], leading in general to different quantum pictures for the very early cosmological evolution. Such an approach, based in particular on "no-boundary" initial conditions, has been recently applied also to the ekpyrotic scenario [10], leading to the production of ekpyrotic instantons [11,12]. In the class of string cosmology models considered in this paper, in contrast, the initial state is uniquely determined by a fixed choice of pre-big bang (or pre-bounce) evolution (see, e.g., [13–17]), which starts asymptotically from the string perturbative vacuum and which, in this way, unambiguously determines the initial "wave function" of the Universe and the subsequent transition probabilities.

In this paper we report the results of previous works, based on the study of the Wheeler–De Witt (WDW) equation [18,19] in the "minisuperspace" associated with a class of cosmological backgrounds compatible with the dynamics the low-energy string effective action [20–26]. It is possible, in such a context, to obtain a non-vanishing transition probability between two different geometrical configurations—in particular, from a pre-big bang to a post-big bang state—even if they are classically disconnected by a space-time singularity. There is no need, to this purpose, of adding higher-order string-theory contributions (like α' and loop corrections) to the WDW equation, except those possibly encoded into an effective (non-local) dilaton potential (but see [27,28] for high-curvature contributions to the WDW equation). It will be shown, also, that there are no problems of operator ordering in the WDW equation, as the ordering is automatically fixed by the duality symmetry of the effective action. Other possible problems—of conceptual nature and typical of the WDW approach to quantum cosmology—however, remain, like the the validity of a probabilistic interpretation of the wave function [29], the existence and the possible meaning of a semiclassical limit [30], the unambiguous identification of a time-like coordinate in superspace (see however [23], and see [31] for a recent discussion).

Let us stress that this review is dedicated in particular to string cosmology backgrounds of the pre-big bang type, and limited to a class of spatially homogeneous geometries. It should be recalled, however, that there are other important works in a quantum cosmology context which are also directly (or indirectly) related to the string effective action, and which are applied to more general classes of background geometries not necessarily characterised by spatial Abelian isometries, and not necessarily emerging from the string vacuum.

We should mention, in particular, the quantum cosmology results for the bosonic sector of the heterotic string with Bianchi-type IX geometry [32] and Bianchi class A geometry [33]; solutions for the WDW wave function with quadratic and cubic curvature corrections [34] (typical of $f(R)$ models of gravity), describing a phase of conventional inflation; two-dimensional models of dilaton quantum cosmology and their supersymmetric extension [35]; WDW equation for a class of scalar-tensor theories of gravity with a generalised form of scale-factor duality invariance [36,37]. We think that discussing those (and related) works should deserve by itself a separate review paper.

This paper is organized as follows. In Section 2 we present the explicit form of the WDW equation following from the low-energy string effective action, for homogeneous backgrounds with d Abelian spatial isometries, and show that it is free from operator-ordering ambiguities thanks to its intrinsic $O(d,d)$ symmetry. In Section 3, working in the simple two-dimensional minisuperspace associated with a class of exact gravi-dilaton solutions of the string cosmology equations, we discuss the scattering of the WDW wave function induced by the presence of a generic dilaton potential. In Section 4 we show that an appropriate quantum reflection of the wave function can be physically interpreted as representing the birth of our Universe as a process of tunnelling from the string perturbative vacuum. Similarly, in Section 5, we show that the parametric amplification of the WDW wave function can describe the birth of our Universe as a process of "anti-tunnelling" from the string perturbative vacuum. Section 6 is finally devoted to a few conclusive remarks.

2. The Wheeler-De Witt Equation for the Low-Energy String Effective Action

In a quantum cosmology context the Universe is described by a wave function evolving in the so-called superspace and governed by the WDW equation [18,19], in much the same way as in ordinary quantum mechanics a particle is described by a wave function evolving in Hilbert space [38], governed by the Schrodinger equation. Each point of superspace corresponds to a possible geometric configuration of the space-like sections of our cosmological space-time, and the propagation of the WDW wave function through this manifold describes the quantum dynamics of the cosmological geometry (thus providing, in particular, the transition probabilities between different geometric states).

The WDW equation, which implements the Hamiltonian constraint $H = 0$ in the superspace of the chosen cosmological scenario, has to be obtained, in our context, from the appropriate string effective action. Let us consider, to this purpose, the low-energy, tree-level, $(d+1)$-dimensional (super)string effective action, which can be written as [39–41]

$$S = -\frac{1}{2\lambda_s^{d-1}} \int d^{d+1}x \sqrt{-g}\, e^{-\phi} \left(R + \partial_\mu \phi \partial^\mu \phi - \frac{1}{12} H_{\mu\nu\alpha} H^{\mu\nu\alpha} + V \right), \qquad (1)$$

where ϕ is the dilaton [42], $H_{\mu\nu\alpha} = \partial_\mu B_{\nu\alpha} + \partial_\nu B_{\alpha\mu} + \partial_\alpha B_{\mu\nu}$ is the field strength of the NS-NS two form $B_{\mu\nu} = -B_{\nu\mu}$ (also called Kalb-Ramond axion), and $\lambda_s \equiv (\alpha')^{1/2}$ is the fundamental string length parameter. We have also added a possible non-trivial dilaton potential $V(\phi)$.

For the purpose of this paper it will be enough to consider a class of homogeneous backgrounds with d Abelian spatial isometries and spatial sections of finite volume, i.e., $(\int d^d x \sqrt{-g})_{t=\text{const}} < \infty$. In the synchronous frame where $g_{00} = 1$, $g_{0i} = 0 = B_{0i}$, and where the fields are independent of all space-like coordinates x^i ($i, j = 1, .., d$), the above action can then be rewritten as follows [43,44]:

$$S = \int dt\, L(\overline{\phi}, M), \qquad L = -\frac{\lambda_s}{2} e^{-\overline{\phi}} \left[(\dot{\overline{\phi}})^2 + \frac{1}{8} \text{Tr}\, \dot{M}(M^{-1})\dot{} + V \right]. \qquad (2)$$

Here a dot denotes differentiation with respect to the cosmic time t, and $\overline{\phi}$ is the so-called "shifted" dilaton field,

$$\overline{\phi} = \phi - \ln\sqrt{-g}, \qquad (3)$$

where we have absorbed into ϕ the constant shift $-\ln(\lambda_s^{-d} \int d^d x)$. Finally, M is the $2d \times 2d$ matrix

$$M = \begin{pmatrix} G^{-1} & -G^{-1}B \\ BG^{-1} & G - BG^{-1}B \end{pmatrix}, \qquad (4)$$

where G and B are, respectively, $d \times d$ matrix representations of the spatial part of the metric (g_{ij}) and of the antisymmetric tensor (B_{ij}). For constant V, or for $V = V(\overline{\phi})$, the above action (2) is invariant under global $O(d,d)$ transformations [43,44] that leave the shifted dilaton invariant, and that are parametrized in general by a constant matrix Ω such that

$$\overline{\phi} \to \overline{\phi}, \qquad M \to \Omega^T M \Omega, \qquad (5)$$

where Ω satisfies

$$\Omega^T \eta \Omega = \eta, \qquad \eta = \begin{pmatrix} 0 & I \\ I & 0 \end{pmatrix}, \qquad (6)$$

and I is the d-dimensional identity matrix. It can be easily checked that, in the particular case in which $B = 0$ and Ω coincides with η, Equation (5) reproduces the well-known transformation of scale-factor duality symmetry [45,46].

From the effective Lagrangian (2) we can now obtain the (dimensionless) canonical momenta

$$\Pi_{\overline{\phi}} = \frac{\delta L}{\delta \dot{\overline{\phi}}} = -\lambda_s \dot{\overline{\phi}} e^{-\overline{\phi}}, \qquad \Pi_M = \frac{\delta L}{\delta \dot{M}} = \frac{\lambda_s}{8} e^{-\overline{\phi}} M^{-1} \dot{M} M^{-1}, \qquad (7)$$

and the associated classical Hamiltonian:

$$H = \frac{e^{\overline{\phi}}}{2\lambda_s} \left[-\Pi_{\overline{\phi}}^2 + 8\text{Tr}(M \Pi_M M \Pi_M) + \lambda_s^2 V e^{-2\overline{\phi}} \right]. \qquad (8)$$

The corresponding WDW equation, implementing in superspace the Hamiltonian constraint $H = 0$ through the differential operator representation $\Pi_{\overline{\phi}} = \pm i\delta/\delta\overline{\phi}$, $\Pi_M =$

$\pm i\delta/\delta M$, would seem thus to be affected by the usual problems of operator ordering, since $[M, \Pi_M] \neq 0$.

The problem disappears, however, if we use the $O(d,d)$ covariance of the action (2), and the symmetry properties of the axion-graviton field represented by the matrix (4), which satisfies the identity $M\eta = \eta M^{-1}$. Thanks to this property, in fact, we can identically rewrite the axion-graviton part of the kinetic term appearing in the Lagrangian (2) as follows:

$$\text{Tr}\,\dot{M}(M^{-1})\dot{} = \text{Tr}(\dot{M}\eta\dot{M}\eta). \qquad (9)$$

The corresponding canonical momentum becomes

$$\Pi_M = -\frac{\lambda_s}{8} e^{-\overline{\phi}} \eta \dot{M} \eta, \qquad (10)$$

and the associated Hamiltonian

$$H = \frac{e^{\overline{\phi}}}{2\lambda_s}\left[-\Pi_{\overline{\phi}}^2 - 8\text{Tr}(\eta\Pi_M\eta\Pi_M) + \lambda_s^2 V e^{-2\overline{\phi}}\right] \qquad (11)$$

has a flat metric in momentum space, and leads to a WDW equation

$$\left[\frac{\delta^2}{\delta\overline{\phi}^2} + 8\text{Tr}\left(\eta\frac{\delta}{\delta M}\eta\frac{\delta}{\delta M}\right) + \lambda_s^2 V e^{-2\overline{\phi}}\right]\Psi(\overline{\phi}, M) = 0 \qquad (12)$$

which is manifestly free from problems of operator ordering.

Finally, it may be interesting to note that the quantum ordering imposed by the duality symmetry of the effective action is exactly equivalent to the order fixed by the condition of reparametrization invariance in superspace.

To check this point let us consider a simple spatially isotropic background, with $B_{ij} = 0$ and scale factor $a(t)$, so that $G_{ij} = -a^2(t)\delta_{ij}$. The effective Lagrangian (2) becomes

$$L(\overline{\phi}, a) = -\frac{\lambda_s}{2} e^{-\overline{\phi}} \left(\dot{\overline{\phi}}^2 - d\frac{\dot{a}^2}{a^2} + V\right), \qquad (13)$$

with associated canonical momenta

$$\Pi_{\overline{\phi}} = \frac{\delta L}{\delta \dot{\overline{\phi}}} = -\lambda_s \dot{\overline{\phi}} e^{-\overline{\phi}}, \qquad \Pi_a = \frac{\delta L}{\delta \dot{a}} = \lambda_s d \frac{\dot{a}}{a^2} e^{-\overline{\phi}}, \qquad (14)$$

and Hamiltonian constraint:

$$2\lambda_s e^{-\overline{\phi}} H = -\Pi_{\overline{\phi}}^2 + \frac{a^2}{d}\Pi_a^2 + \lambda_s^2 V e^{-2\overline{\phi}} = 0. \qquad (15)$$

The differential implementation of this constraint in terms of the operators $\Pi_{\overline{\phi}} \to \pm i\partial/\partial\overline{\phi}$, $\Pi_a \to \pm i\partial/\partial a$ has to be ordered, because $[a, \Pi_a] \neq 0$. It follows that in general, for the kinetic part of the Hamiltonian $H_k = -\Pi_{\overline{\phi}}^2 + a^2\Pi_a^2/d$, we have the following differential representation

$$H_k = \frac{\partial}{\partial\overline{\phi}^2} - \frac{a^2}{d}\frac{\partial^2}{\partial a^2} - \epsilon\frac{a}{d}\frac{\partial}{\partial a}, \qquad (16)$$

where ϵ is a numerical parameter depending on the imposed ordering. However, if we perform a scale-factor duality transformation $\overline{\phi} \to \overline{\phi}$, $a \to \widetilde{a} = a^{-1}$ (which exactly corresponds to the class of transformations (5) for the particular class of backgrounds that we are considering), we find

$$H_k(a) = H_k(\widetilde{a}) + \frac{2}{d}(\epsilon - 1)\widetilde{a}\frac{\partial}{\partial\widetilde{a}}. \qquad (17)$$

The duality invariance of the Hamiltonian thus requires $\epsilon = 1$ (which, by the way, is also the value of ϵ that we have to insert into Equation (16) to be in agreement with the general result (12) if we consider the particular class of geometries with $B = 0$ and $G = -a^2 I$). See also [36,37] for the WDW equation with a generalised form of scale-factor duality symmetry.

Let us now consider the kinetic part of the Hamiltonian operator (15), which is given as a quadratic form in the canonical momenta $\Pi_A = (\Pi_{\overline{\phi}}, \Pi_a)$, written in a 2-dimensional minisuperspace with a non-trivial metric γ_{AB} and coordinates $x^A = (\overline{\phi}, a)$, such that:

$$H_k = -\Pi_{\overline{\phi}}^2 + \frac{a^2}{d}\Pi_a^2 \equiv \gamma^{AB}\Pi_A\Pi_B, \qquad \gamma_{AB}(\overline{\phi},a) = \text{diag}\left(-1, \frac{d}{a^2}\right). \tag{18}$$

If we impose on the differential representation of the Hamiltonian constraint the condition of general covariance with respect to the given minisuperspace geometry [47], we obtain

$$H_k = -\gamma^{AB}\nabla_A\nabla_B = -\frac{1}{\sqrt{-\gamma}}\partial_A(\sqrt{-\gamma}\gamma^{AB}\partial_B) \equiv \frac{\partial}{\partial\overline{\phi}^2} - \frac{a^2}{d}\frac{\partial^2}{\partial a^2} - \frac{a}{d}\frac{\partial}{\partial a}, \tag{19}$$

and this result exactly reproduces the differential operator (16) with $\epsilon = 1$. The duality symmetry of the action, and the requirement of reparametrisation invariance in superspace, are thus equivalent to select just the same ordering prescription, as previously anticipated.

3. Quantum Scattering of the Wheeler-De Witt Wave Function in Minisuperspace

For an elementary discussion of this topic, and for the particular applications we have in mind—namely, a quantum description of the "birth" of our present cosmological state from the string perturbative vacuum—we shall consider the homogeneous, isotropic and spatially flat class of $(d+1)$-dimensional backgrounds already introduced in the previous section, with $B_{\mu\nu} = 0$, $g_{00} = 1$ and scale factor $a(t)$. We shall thus work in a two-dimensional minisuperspace, spanned by the convenient coordinates $(\overline{\phi}, \beta)$ where $\beta = \sqrt{d}\ln a$. With such variables the effective Lagrangian (2) takes the form

$$L(\beta,\overline{\phi}) = -\lambda_s \frac{e^{-\overline{\phi}}}{2}\left[\dot{\overline{\phi}}^2 - \dot{\beta}^2 + V(\beta,\overline{\phi})\right], \tag{20}$$

and the momenta, canonically conjugate to the coordinates $\overline{\phi}, \beta$, are given by

$$\Pi_{\overline{\phi}} = \frac{\delta L}{\delta\dot{\overline{\phi}}} = -\lambda_s \dot{\overline{\phi}} e^{-\overline{\phi}}, \qquad \Pi_\beta = \frac{\delta L}{\delta\dot{\beta}} = \lambda_s \dot{\beta} e^{-\overline{\phi}}. \tag{21}$$

The Hamiltonian constraint (15) becomes

$$-\Pi_{\overline{\phi}}^2 + \Pi_\beta^2 + \lambda_s^2 V(\beta,\overline{\phi}) e^{-2\overline{\phi}} = 0, \tag{22}$$

corresponding to an effective WDW equation

$$\left[\partial_{\overline{\phi}}^2 - \partial_\beta^2 + \lambda_s^2 V(\beta,\overline{\phi}) e^{-2\overline{\phi}}\right]\Psi(\beta,\overline{\phi}) = 0. \tag{23}$$

For $V = 0$ we have the free D'Alembert equation, and the general solution can be written in terms of plane waves as

$$\Psi(\beta,\overline{\phi}) = \psi_\beta^\pm \psi_{\overline{\phi}}^\pm \sim e^{\mp ik\beta} e^{\mp ik\overline{\phi}}. \tag{24}$$

Here $k > 0$, and $\psi_\beta^\pm, \psi_{\overline{\phi}}^\pm$ are free momentum eigenstates, satisfying the eigenvalue equations

$$\Pi_\beta \psi_\beta^\pm = \pm k \psi_\beta^\pm, \qquad \Pi_{\overline{\phi}} \psi_{\overline{\phi}}^\pm = \pm k \psi_{\overline{\phi}}^\pm. \tag{25}$$

Let us now recall that, for $V = 0$, the equations following from the effective Lagrangian (20) admit a class of exact solution describing four (physically different) cosmological phases, two expanding and two contracting, parametrized by [13,15–17]:

$$a(t) \sim (\mp t)^{\mp 1/\sqrt{d}}, \qquad \overline{\phi}(t) \sim -\ln(\mp t), \qquad (26)$$

They are defined on the disconnected time ranges $[-\infty, 0]$ and $[0, +\infty]$, and are related by duality transformations $a \to a^{-1}$, $\overline{\phi} \to \overline{\phi}$ and time-reversal transformation, $t \to -t$. They may represent the four asymptotic branches of the low-energy string cosmology solutions even in the presence of a non-vanishing dilaton potential, provided the effective contribution of the potential is localized in a region of finite extension of the $(\overline{\phi}, \beta)$ plane, and goes (rapidly enough) to zero as $\overline{\phi}, \beta \to \pm \infty$.

The above solutions satisfy the condition

$$\dot{\overline{\phi}} = \pm \sqrt{d}\, \frac{\dot{a}}{a} = \pm \dot{\beta}, \qquad (27)$$

so that, according to the definitions (21), they correspond to configurations with canonical momenta related by $\Pi_\beta = \pm \Pi_{\overline{\phi}}$. By recalling that the phase of (expanding or contracting) pre-big bang evolution is characterized by growing curvature and growing dilaton [15,16] (namely, $\dot{\overline{\phi}} > 0$, $\Pi_{\overline{\phi}} < 0$), while the curvature and the dilaton are decreasing in the (expanding or contracting) post-big bang phase (where $\dot{\overline{\phi}} < 0$, $\Pi_{\overline{\phi}} > 0$), we can conclude, according to Equations (21), (25), that the classical solutions (26) of the string cosmology equations admit the following plane-wave representation in minisuperspace in terms of $\psi_\beta^\pm$, $\psi_{\overline{\phi}}^\pm$:

- expansion $\longrightarrow$ $\dot{\beta} > 0$ $\longrightarrow$ ψ_β^+,
- contraction $\longrightarrow$ $\dot{\beta} < 0$ $\longrightarrow$ ψ_β^-,
- pre-big bang (growing dilaton) $\longrightarrow$ $\dot{\overline{\phi}} > 0$ $\longrightarrow$ $\psi_{\overline{\phi}}^-$,
- post-big bang (decreasing dilaton) $\longrightarrow$ $\dot{\overline{\phi}} < 0$ $\longrightarrow$ $\psi_{\overline{\phi}}^+$.

Let us now impose, as our physical boundary condition, that the initial state of our Universe describes a phase of expanding pre-big bang evolution, asymptotically emerging from the string perturbative vacuum (identified with the limit $\beta \to -\infty$, $\phi \to -\infty$). It follows that the initial state Ψ_{in} must represent a configuration with $\dot{\beta} > 0$ and $\dot{\overline{\phi}} > 0$, namely a state with positive eigenvalue of Π_β and negative (opposite) eigenvalue of $\Pi_{\overline{\phi}}$, i.e., $\Psi_{\text{in}} \sim \psi_\beta^+ \psi_{\overline{\phi}}^-$.

In such a context, a quantum transition from the pre- to the post-big bang regime can be described as a process of scattering of the initial wave function induced by the presence of some appropriate dilaton potential, which we shall assume to have non-negligible dynamical effects only in a finite region localized around the origin of the minisuperspace spanned by the $(\overline{\phi}, \beta)$ coordinates. In other words, we shall assume that the contributions of $V(\phi)$ to the WDW equation tend to disappear not only in the initial but also in the final asymptotic regime where $\beta \to +\infty$, $\phi \to +\infty$. As a consequence, also the final asymptotic configuration Ψ_{out}, emerging from the scattering process, can be represented in terms of the free momentum eigenstates $\psi_\beta^\pm$ and $\psi_{\overline{\phi}}^\pm$.

However, unlike the initial state fixed by the chosen boundary conditions—and selected to represent a configuration with $\Pi_\beta > 0$ and $\Pi_{\overline{\phi}} < 0$—the final state is not constrained by such a restriction and can describe in general different configurations. In particular, the scattering process may lead to configurations asymptotically described by a wave function Ψ_{out} which is a superposition of different momentum eigenstates: for instance, waves with the same $\Pi_\beta > 0$ and opposite values of $\Pi_{\overline{\phi}}$, i.e., $\Psi_{\text{out}} \sim \psi_\beta^+ \psi_{\overline{\phi}}^\pm$ (see Figure 1, cases (a) and (b)); or waves with the same $\Pi_{\overline{\phi}} < 0$ and opposite values of Π_β, i.e., $\Psi_{\text{out}} \sim \psi_{\overline{\phi}}^- \psi_\beta^\pm$ (see Figure 1, cases (c) and (d)).

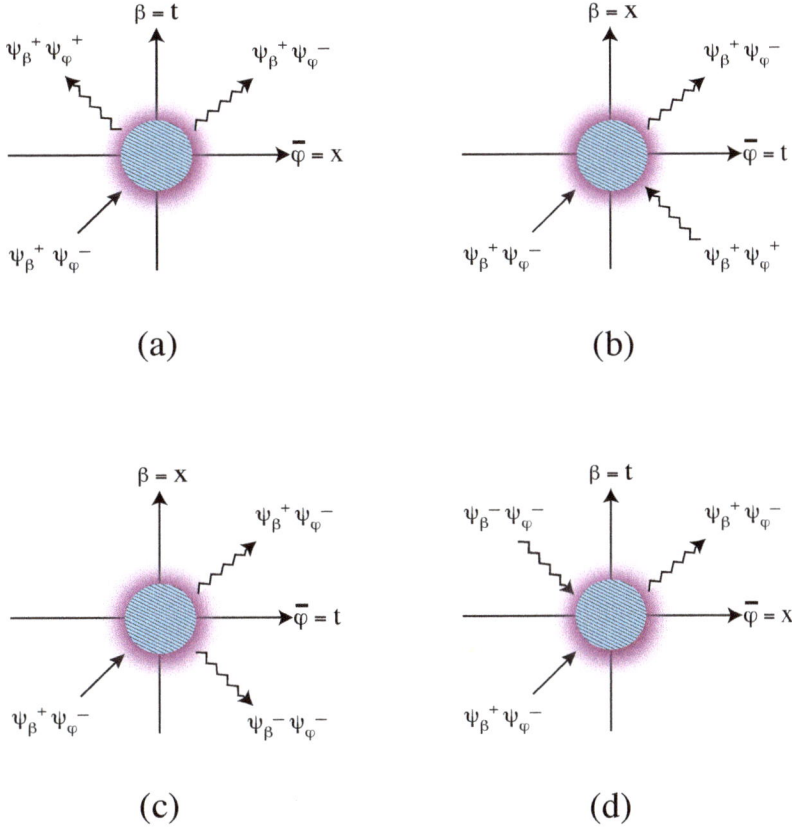

Figure 1. Four different classes of scattering processes for the incoming wave function describing a phase of expanding pre-big bang evolution, asymptotically emerging from the string perturbative vacuum (straight, solid line). The outgoing state is represented by a mixture of eigenfunctions of Π_β and $\Pi_{\bar\phi}$ with positive and negative eigenvalues. See the main text for a detailed explanation of the four different cases (**a**–**d**) illustrated in this figure.

It may be interesting to note that those different configurations may be interpreted as different possible "decay channels" of the string perturbative vacuum [24]. Also, it should be stressed (as clearly illustrated in Figure 1) that one of the two components of the outgoing wave function Ψ_{out} must always correspond to the "transmitted" part of the incident wave Ψ_{in}, namely must correspond to a state with $\Pi_\beta > 0$ and $\Pi_{\bar\phi} < 0$, represented by $\psi_\beta^+ \psi_{\bar\phi}^-$. However, the "reflected" part of the wave function may have different physical interpretations, also depending on the chosen identification of the time-like coordinate in minisuperspace [23,25]: the β axis for the cases (a) and (d), the $\bar\phi$ axis for the case (b) and (c).

It turns out that only the cases (a) and (c) of Figure 1 represent a true process of reflection of the incident wave along a spacelike coordinate (the axes $\bar\phi$ and β, respectively). In case (a), in particular, the evolution along β is monotonic, the Universe always keeps expanding, and the incident wave Ψ_{in} is partially transmitted towards the pre-big bang singularity (with unbounded growth of the curvature and of the dilaton, $\beta \to +\infty$, $\bar\phi \to +\infty$), and partially reflected back towards the expanding, low-energy, post-big bang regime ($\beta \to +\infty$, $\bar\phi \to -\infty$). As we shall show in Section 4, this type of quantum reflection can also be interpreted as a process of "tunnelling" from the string perturbative vacuum.

The cases (b) and (d) of Figure 1 are qualitatively different, as the final state is a superposition of modes of positive and negative frequency with respect to the chosen timelike coordinate (the axes $\overline{\phi}$ and β, respectively). Namely, Ψ_{out} is a superposition of positive and negative energy eigenstates, and this represents a quantum process of "parametric amplification" of the wave function [48,49] or, in the language of third quantization [50–54]—i.e., second quantization of the WDW wave function—a process of "Bogoliubov mixing" of the energy modes (see, e.g., [55,56]), associated with the production of "pairs of universes" from the vacuum. For that process, the mode "moving backwards" with respect to the chosen time coordinate has to be "reinterpreted": as an anti-particle in the usual quantum field theory context, as an "anti-universe" in a quantum cosmology context.

Such a re-interpretation principle produces, as usual, states of positive energy and opposite momentum. It turns out, in particular, that the case (d) of Figure 1 describes—after the correct re-interpretation—the production of universe/anti-universe pairs in which both members of the pair have positive energy and positive momentum along the β axis. Hence, they are both expanding: one falls inside the pre-big bang singularity ($\overline{\phi} \to +\infty$), but the other expands towards the low-energy post-big bang regime ($\overline{\phi} \to -\infty$). As we shall show in Section 5, this quantum effect of pair production can also be interpreted as a process of "anti-tunneling" from the string perturbative vacuum.

4. Birth of the Universe as a Tunnelling from the String Perturbative Vacuum

To illustrate the process of quantum transition from the pre- to the post-big bang regime as a wave reflection in superspace we shall consider here the simplest (almost trivial) case of constant dilaton potential, $V = V_0 = \text{const}$ (see also [37], and see, e.g., [20] for more general dynamical configurations). With this potential the classical background solutions for the cosmological equations of the effective Lagrangian (20) are well known [57], and can be written as

$$a(t) = a_0 \left[\tanh\left(\mp\sqrt{V_0}t/2\right)\right]^{\mp 1/\sqrt{d}}, \qquad \overline{\phi} = \phi_0 - \ln\left[\sinh\left(\mp\sqrt{V_0}t\right)\right], \qquad (28)$$

where a_0 and ϕ_0 are integration constants.

These solutions have two branches, of the pre-big bang type ($\dot{\overline{\phi}} > 0$) and post-big bang type ($\dot{\overline{\phi}} < 0$), defined respectively over the disconnected time ranges $t < 0$ and $t > 0$, and classically separated by a singularity of the curvature and of the effective string coupling ($\exp\overline{\phi}$) at $t = 0$. For $t \to \pm\infty$ they approach, asymptotically, the free vacuum solution (26) obtained for $V = 0$. It is important to note, also, that each branch of the above solution can describe either expanding or contracting geometric configurations, which are both characterized by a constant canonical momentum along the β axis, given (according to Equation (21)), by

$$\Pi_\beta = \lambda_s \dot{\beta} e^{-\overline{\phi}} = \pm k, \qquad k = \lambda_s \sqrt{V_0} e^{-\phi_0}. \qquad (29)$$

Let us now apply the WDW Equation (23) to compute the (classically forbidden) probability of transition from the pre- to the post-big bang branches of the solution (28). We are interested, in particular, in the transition between expanding configurations, and we shall thus consider the quantum process described by the case (a) of Figure 1, with a wave function monotonically evolving along the positive direction of the β axis (also in agreement with the role of time-like coordinate asssigned to β). In that case $\dot{\beta} > 0$, and the conserved canonical momentum (29) is positive, $\Pi_\beta > 0$. By imposing momentum conservation as a differential condition on the wave function,

$$\Pi_\beta \Psi_k(\beta,\overline{\phi}) = i\partial_\beta \Psi_k(\beta,\overline{\phi}) = k\,\Psi_k(\beta,\overline{\phi}), \qquad (30)$$

we can then separate the variables in the solution of the WDW Equation (23), and we obtain

$$\Psi(\beta,\overline{\phi}) = e^{-ik\beta}\psi_k(\overline{\phi}), \qquad \left(\partial_{\overline{\phi}}^2 + k^2 + \lambda_s^2 V_0 e^{-2\overline{\phi}}\right)\psi_k(\overline{\phi}) = 0. \qquad (31)$$

The general solution of the above equation can now be written as a linear combination of Bessel functions [58], $AJ_\nu(z) + BJ_{-\nu}(z)$, of index $\nu = ik$ and argument $z = \lambda_s \sqrt{V_0} \exp(-\overline{\phi})$. Consistently with the chosen boundary conditions for the process illustrated in case (a) of Figure 1 (namely, with the choice of an initial wave function asymptotically incoming from the string perturbative vacuum), we have now to impose that there are only right-moving waves (along $\overline{\phi}$) approaching the high-energy region and the final singularity in the limit $\beta \to +\infty$, $\overline{\phi} \to +\infty$. Namely, waves of the type $\psi_{\overline{\phi}}^-$—see Equations (24) and (25)—representing a state with $\dot{\overline{\phi}} > 0$ and $\Pi_{\overline{\phi}} < 0$. By using the small argument limit of the Bessel functions [58],

$$\lim_{\overline{\phi} \to +\infty} J_{\pm ik}\left(\lambda_s \sqrt{V_0}\, e^{-\overline{\phi}}\right) \sim e^{\mp ik\overline{\phi}}, \tag{32}$$

we can then eliminate the $J_\nu(z)$ component and uniquely fix the WDW solution (modulo an arbitrary normalization factor N_k) as follows:

$$\Psi_k(\beta, \overline{\phi}) = N_k J_{-ik}\left(\lambda_s \sqrt{V_0}\, e^{-\overline{\phi}}\right) e^{-ik\beta}. \tag{33}$$

Let us now consider the wave content of this solution in the opposite, low-energy limit $\overline{\phi} \to -\infty$, where the large argument limit of the Bessel functions gives [58]

$$\begin{aligned}\lim_{\overline{\phi} \to -\infty} \Psi_k(\beta, \overline{\phi}) &= \frac{N_k e^{-ik\beta}}{(2\pi z)^{1/2}} \left[e^{-i(z-\pi/4)} e^{k\pi/2} + e^{i(z-\pi/4)} e^{-k\pi/2} \right] \\ &\equiv \Psi_k^-(\beta, \overline{\phi}) + \Psi_k^+(\beta, \overline{\phi}), \end{aligned} \tag{34}$$

and where the two wave components Ψ_k^- and Ψ_k^+ are asymptotically eigenstates of $\Pi_{\overline{\phi}}$ with negative and positive eigenvalues, respectively. Hence, we find in this limit a superposition of right-moving and left-moving modes (along $\overline{\phi}$), representing, respectively, the initial, pre-big bang incoming state Ψ_k^- (with $\Pi_{\overline{\phi}} < 0$, i.e., growing dilaton), and the final, post-big bang reflected component Ψ_k^+ (with $\Pi_{\overline{\phi}} > 0$, i.e., decreasing dilaton). Starting from an initial pre-big bang configuration, we can then obtain a finite probability for the transition to the "dual" post-big bang regime, represented as a reflection of the wave function in minisuperspace, with reflection coefficient

$$R_k = \frac{|\Psi_k^+(\beta, \overline{\phi})|^2}{|\Psi_k^-(\beta, \overline{\phi})|^2} = e^{-2\pi k}. \tag{35}$$

The probability for this quantum process is in general nonzero, even if the corresponding transition is classically forbidden.

It may be interesting to evaluate R_k in terms of the string-scale variables, for a region of d-dimensional space of given proper volume Ω_s. By computing the constant momentum k of Equation (29) at the string epoch t_s, when $\dot{\beta}(t_s) = \sqrt{d}(\dot{a}/a)(t_s) \simeq \sqrt{d}\lambda_s^{-1}$, and using the definition (3) of $\overline{\phi}$, we find

$$R_k \sim \exp\left\{ -\frac{2\pi\sqrt{d}\,\Omega_s}{g_s^2\,\lambda_s^d} \right\}, \tag{36}$$

where the proper spatial volume is given by $\Omega_s = a^d(t_s)\int d^d x$, and where $g_s = \exp(\phi_s/2)$ is the effective value of the string coupling when the dilaton has the value $\phi_s \equiv \phi(t_s)$. Note that, for values of the coupling $g_s \sim 1$, the above probability is of order one for the formation of spacelike "bubbles" of unit size (or smaller) in string units. In general, the probability has a typical "instanton-like" dependence on the coupling constant, $R_k \sim \exp(g_s^{-2})$.

It may be observed, finally, that an exponential dependence of the transition probability is also typical of tunnelling processes (induced by the presence of a cosmological constant Λ) occurring in the context of standard quantum cosmology, where the tunnelling probability can be estimated as [6,29,59,60]

$$P \sim \exp\left\{-\frac{4}{\lambda_P^2 \Lambda}\right\} \tag{37}$$

(λ_P is the Planck length). That scenario is different because, in that case, the Universe emerges from the quantum era in a classical inflationary configuration, while, in the string scenario, the Universe is expected *to exit* (and not to enter) the phase of inflation thanks to quantum cosmology effects.

In spite of such important differences, it turns out that the string cosmology result (36) is formally very similar to the result concerning the probability that the birth of our Universe may be described as a quantum process of "tunneling from nothing" [6,29,59,60]. The explanation of this formal coincidence is simple, and based on the fact that the choice of the string perturbative vacuum as initial boundary condition implies—as previously stressed—that the are only outgoing (right-moving) waves approaching the singularity at $\overline{\phi} \to +\infty$. This is exactly equivalent to imposing tunneling boundary conditions, that select "... *only outgoing modes at the singular space-time boundary*" [29,60]. In this sense, the process illustrated in this Section can also be interpreted as a tunneling process, not "from nothing" but "from the string perturbative vacuum".

5. Birth of the Universe as Anti-Tunnelling from the String Perturbative Vacuum

In this Section we shall illustrate the possible transition from the pre- to the post-big bang regime as a process of parametric amplification of the WDW wave function, also equivalent—as previously stressed—to a quantum process of pair production from the vacuum. We shall consider, in particular, an example in which both the initial and final configurations are expanding, like in the case (d) of Figure 1.

What we need, to this purpose, is a WDW equation with the appropriate dilaton potential, able to produce an outgoing configuration which is a superposition of states with positive and negative eigenvalues of the momentum Π_β (see Figure 1). Since we are starting from an initial expanding (pre-big bang) regime, it follows that the contribution of the potential has to break the traslational invariance of the effective Hamiltonian (22) along the β axis (i.e., $[\Pi_\beta, H] \neq 0$), otherwise the final configuration described in case (d) of Figure 1 would be forbidden by momentum conservation.

We shall work here with the simple two-loop dilaton potential already introduced in [26], possibly induced by an effective cosmological constant $\Lambda > 0$, appropriately suppressed in the low-energy, classical regime, and given explicitly by

$$V(\beta, \overline{\phi}) = \Lambda\,\theta(-\beta)\,e^{2\phi} = \Lambda\,\theta(-\beta)\,e^{2\overline{\phi} + 2\sqrt{d}\beta}. \tag{38}$$

The Heaviside step function θ has been inserted to mimic an efficient damping of the potential outside the interaction region (in particular, in the large radius limit $\beta \to +\infty$ of the expanding post-bb configuration). The explicit form and mechanism of the damping, however, is not at all a crucial ingredient of our discussion, and other, different forms of damping would be equally appropriate.

With the given potential (38) the effective Hamiltonian is no longer translational invariant along the β direction, but we still have $[\Pi_{\overline{\phi}}, H] = 0$, so that we can conveniently separate the variables in the WDW Equation (23) by factorizing the eigenstates (25) of the canonical momentum $\Pi_{\overline{\phi}}$, and we are lead to

$$\Psi(\beta, \overline{\phi}) = e^{ik\overline{\phi}}\psi_k(\beta), \qquad \left[\partial_\beta^2 + k^2 - \lambda_s^2\,\Lambda\,\theta(-\beta)\,e^{2\sqrt{d}\beta}\right]\psi_k(\beta) = 0. \tag{39}$$

The above WDW equation can now be exactly solved by separately considering the two ranges of the "temporal coordinate" β, namely $\beta < 0$ and $\beta > 0$.

For $\beta < 0$ the contribution of the potential is non vanishing, and the general solution is a linear combination of Bessel functions $J_\mu(\sigma)$ and $J_{-\mu}(\sigma)$, of index $\mu = ik/\sqrt{d}$ and argument $\sigma = i\lambda_s \sqrt{\Lambda/d}\, e^{\sqrt{d}\beta}$. As before, we have to impose our initial boundary conditions requiring that, in the limit $\beta \to -\infty$, the solution may asymptotically represent a low-energy pre-big bang configuration with $\Pi_\beta = -\Pi_{\overline{\phi}} = k$, namely (according to Equations (24) and (25)):

$$\lim_{\beta \to -\infty} \Psi(\beta, \overline{\phi}) \sim \psi_\beta^+ \psi_{\overline{\phi}}^- \sim e^{ik\overline{\phi} - ik\beta}. \tag{40}$$

By using the small argument limit (32), and imposing the above condition, we can then uniquely fix (modulo a normalization factor N_k) the solution of the WDW Equation (39), for $\beta < 0$, as follows:

$$\Psi_k(\beta, \overline{\phi}) = e^{ik\overline{\phi}} N_k J_{-ik/\sqrt{d}}\left(i\lambda_s \sqrt{\Lambda/d}\, e^{\sqrt{d}\beta}\right), \qquad \beta < 0. \tag{41}$$

In the complementary regime $\beta > 0$ the potential (38) is exactly vanishing, and the general outgoing solution is a linear superposition of eigenstates of Π_β with positive and negative eigenvalues, represented by the frequency modes $\psi_\beta^\pm$ of Equations (24) and (25). We can then write

$$\Psi_k(\beta, \overline{\phi}) = e^{ik\overline{\phi}}\left[A_+(k)e^{-ik\beta} + A_-(k)e^{ik\beta}\right], \qquad \beta > 0, \tag{42}$$

and the numerical coefficients $A_\pm(k)$ can be fixed by the two matching conditions imposing the continuity of Ψ_k and $\partial_\beta \Psi_k$ at $\beta = 0$.

Let us now recall that the so-called Bogoliubov coefficients $|c_\pm(k)|^2 = |A_\pm(k)|^2/|N_k|^2$, determining the mixing of positive and negative energy modes in the asymptotic outgoing solution [55,56], also play the role of destruction and creation operators in the context of the third quantization formalism [50–54], thus controlling the "number of universes" $n_k = |c_-(k)|^2$ produced from the vacuum, for each mode k. It turns out, in particular, that such a transition from the initial vacuum to the final standard regime, represented as a quantum scattering of the initial wave function, is an efficient process only when the final wave function is not damped but, on the contrary, turns out to be "parametrically amplified" by the interaction with the effective potential barrier [48,49]. This is indeed what happens for the solutions of our WDW Equation (39), provided the dilaton potential satisfies the condition $k < \lambda_s \sqrt{\Lambda}$ [26] (as can be checked by an explicit computation of our coefficients $A_\pm(k)$).

In order to illustrate this effect we have numerically integrated Equation (39), with the boundary conditions (40), for $d = 3$ spatial dimensions. The results are shown in Figure 2, where we have plotted the evolution in superspace of the real part of the WDW wave function, for different values of k (the behavior of the imaginary part is qualitatively similar). We have used units where $\lambda_s^2 \Lambda = 1$, so that the effective potential barrier of Equation (39) is non-negligible only for very small (negative) values of β (the grey shaded region of Figure 2). Also, we have imposed on all modes the same formal normalization $|\Psi_k|^2 = 1$ at $\beta \to -\infty$, to emphasize that the amplification is more effective at lower frequency.

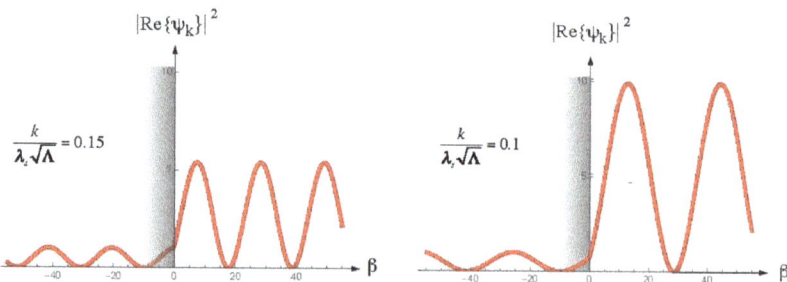

Figure 2. Evolution in superspace of the Wheeler–De Witt (WDW) solution which illustrates the anti-tunnelling effect produced by the effective potential barrier (grey shaded region) due to the dilaton potential (38). The wave function is not damped but parametrically amplified provided $k < \lambda_s \sqrt{\Lambda}$, and the effect is larger for smaller k.

Concerning this last point, we can find an interesting (and reasonable) interpretation of the condition $k < \lambda_s \sqrt{\Lambda}$ by considering the realistic case of a transition process occurring at the string scale, with $\dot{\beta} \sim \lambda_s$, with coupling constant g_s, and for a spatial region of proper volume Ω_s. In that case, by using the result (36) for the momentum k expressed in terms of string-scale variables, we can write the condition of efficient parametric amplification in the following form:

$$k \sim g_s^{-2}(\Omega_s/\lambda_s^d) \lesssim \lambda_s \sqrt{\Lambda}. \tag{43}$$

It implies that the birth of our present, expanding, post-big bang phase can be efficiently described as a process of anti-tunnelling—or, in other words, as a forced production of pairs of universes—from the string perturbative vacuum, in the following cases: initial configurations of small enough volume in string units, and/or large enough coupling g_s, and/or large enough cosmological constant in string units. Quite similar conclusions were obtained also in the case discussed in the previous section.

In view of the above results, we may conclude that, for an appropriate initial configuration, and if triggered by the appropriate dilaton potential, the decay of the initial string perturbative vacuum can efficiently proceed via parametric amplification of the WDW wave function in superspace, and can be described as a forced production of pairs of universes from the quantum fluctuations. One member of the pair disappears into the pre-big bang singularity, the other bounces back towards the low-energy regime. The resulting effect is a net flux of universes that may escape to infinity in the post-big bang regime (as qualitatively illustrated in Figure 3), with a process which can describe the birth of our Universe as "anti-tunnelling from the string perturbative vacuum".

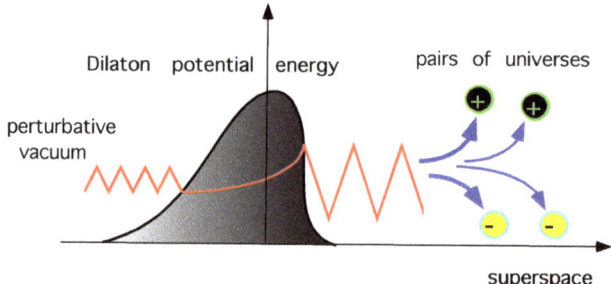

Figure 3. Birth of the universe represented as an anti-tunneling (parametric amplification) effect of the wave function in superspace, or—in the language of third quantization—as a process of pair production from the string perturbative vacuum.

6. Conclusions

The quantum cosmology scenarios reported in this review are based on the low-energy, tree-level string effective action, which is physically appropriate to describe early enough and late enough cosmological phases, approaching, respectively, the initial perturbative vacuum and the present cosmological epoch.

Such an action cannot used to classically describe the high-curvature, strong coupling regime without the inclusion of higher-order corrections. However, when at least some of these corrections and of possible non-perturbative effects are accounted for by an appropriate dilaton potential, the WDW equation obtained from the low-energy action action permits a quantum analysis of the background evolution, and points out new possible interesting ways for a Universe born from the string vacuum to reach more standard configurations, and evolve towards the present cosmological regime. In such a context, the possible (future) detection of a stochastic background of cosmic gravitons with the typical imprints of the pre-big bang dynamics (see, e.g., [61]) might thus represent also an "indirect" indication that some quantum cosmology mechanism has been effective to trigger the transition to the cosmological state in which we are living.

Funding: This research received no external funding.

Institutional Review Board Statement: Not applicable.

Informed Consent Statement: Not applicable.

Data Availability Statement: Not applicable.

Acknowledgments: It is a great pleasure to thank all colleagues and friends who collaborated and made many important contributions to the original research articles reported in this paper. Let me mention, in particular (and in alphabetical order): Alessandra Buonanno, Marco Cavaglià, Michele Maggiore, Jnan Maharana, Carlo Ungarelli, Gabriele Veneziano. This work is supported in part by INFN under the program TAsP (*Theoretical Astroparticle Physics*), and by the research grant number 2017W4HA7S (*NAT-NET: Neutrino and Astroparticle Theory Network*), under the program PRIN 2017, funded by MUR.

Conflicts of Interest: The author declares no conflict of interest.

Abbreviations

The following abbreviations are used in this manuscript:

WDW   Wheeler-De Witt equation

References

1. Vilenkin, A. Predictions from Quantum Cosmology. In *Contribution to the 4th Course of the International School of Astrophysics D, Proceedings of the String Gravity and Physics at the Planck Energy Scale, A NATO Advanced Study Institute, Erice, Italy, 8–19 September 1995*; NATO: Brussels, Belgium, 1996; Volume 476, pp. 345–367.
2. Buonanno, A.; Damour, T.; Veneziano, G. Pre–big bang bubbles from the gravitational instability of generic string vacua. *Nucl. Phys.* **1999**, *B543*, 275.
3. Gasperini, M. On the initial regime of pre-big bang cosmology. *J. Cosmol. Astropart. Phys.* **2017**, *9*, 001.
4. Hartle, J.B.; Hawking, S.W. Wave Function of the Universe. *Phys. Rev.* **1983**, *28*, 2960.
5. Hawking, S.W. The Quantum State of the Universe. *Nucl. Phys.* **1984**, *239*, 257.
6. Vilenkin, A. Quantum Creation of Universes. *Phys. Rev.* **1984**, *30*, 509.
7. Linde, A.D. Quantum creation of an inflationary universe. *Sov. Phys. JETP* **1984**, *60*, 211.
8. Zeldovich, Y.; Starobinski, A.A. Quantum creation of a universe in a nontrivial topology. *Sov. Astron. Lett.* **1984**, *10*, 135.
9. Rubakov, V.A. Quantum Mechanics in the Tunneling Universe. *Phys. Lett.* **1984**, *B148*, 280.
10. Khoury, J.; Ovrut, B.A.; Steinhardt, P.J.; Turok, N. The Ekpyrotic universe: Colliding branes and the origin of the hot big bang. *Phys. Rev.* **2001**, *64*, 123522.
11. Lehners, J.-L. Classical Inflationary and Ekpyrotic Universes in the No-Boundary Wavefunction. *Phys. Rev.* **2015**, *91*, 083525.
12. Lehners, J.-L. New ekpyrotic quantum cosmology. *Phys. Lett.* **2015**, *750*, 242.
13. Gasperini, M.; Veneziano, G. Pre-big bang in string cosmology. *Astropart. Phys.* **1993**, *1*, 317.

14. Gasperini, M. *Elementary Introduction to Pre-Big Bang Cosmology and to the Relic Graviton Background*; Contribution to the SIGRAV Graduate School in Contemporary Relativity and Gravitational Physics (Center "A. Volta", Como, April 1999); Gravitational Waves; Ciufolini, I., Gorini, V., Moschella, U., Fre', P., Eds.;IOP Publishing: Bristol, UK, 2001; pp. 280–337. ISBN 0-7503-0741-2.
15. Gasperini, M.; Veneziano, G. The Pre-big bang scenario in string cosmology. *Phys. Rep.* **2003**, *373*, 1.
16. Gasperini, M. *Elements of String Cosmology*; Cambridge University Press: Cambridge, UK, 2007. [CrossRef]
17. Gasperini, M.; Veneziano, G. String Theory and Pre-big bang Cosmology. *Nuovo Cim.* **2016**, *38*, 160.
18. De Witt, B.S. Quantum Theory of Gravity. I. The Canonical Theory. *Phys. Rev.* **1967**, *160*, 1113.
19. Wheeler, J.A. *Battelle Rencontres*; De Witt, C., Wheeler, J.A., Eds.; Benjamin: New York, NY, USA, 1968.
20. Gasperini, M.; Maharana, J.; Veneziano, G. Graceful exit in quantum string cosmology. *Nuc. Phys.* **1996**, *472*, 349.
21. Gasperini, M.; Veneziano, G. Birth of the universe as quantum scattering in string cosmology. *Gen. Rel. Grav.* **1996**, *28*, 1301.
22. Buonanno, A.; Gasperini, M.; Maggiore, M.; Ungarelli, C. Expanding and contracting universes in third quantized string cosmology. *Class. Quantum Grav.* **1997**, *14*, L97.
23. Cavaglià, M.; De Alfaro, V. Time gauge fixing and Hilbert space in quantum string cosmology. *Gen. Rel. Grav.* **1997**, *29*, 773.
24. Gasperini, M. Low-energy quantum string cosmology. *Int. J. Mod. Phys.* **1998**, *A13*, 4779.
25. Cavaglià, M.; Ungarelli, C. Canonical and path integral quantization of string cosmology models. *Class. Quantum Grav.* **1999**, *16*, 1401.
26. Gasperini, M. Birth of the universe as anti-tunneling from the string perturbative vacuum. *Int. J. Mod. Phys.* **2001**, *10*, 15.
27. Pollock, M.D. On the Quantum Cosmology of the Superstring Theory Including the Effects of Higher Derivative Terms. *Nucl. Phys.* **1989**, *324*, 187.
28. Pollock, M.D. On the derivation of the Wheeler-DeWitt equation in the heterotic superstring theory. *Int. J. Mod. Phys.* **1992**, *7*, 4149.
29. Vilenkin, A. Boundary Conditions in Quantum Cosmology. *Phys. Rev.* **1986**, *33*, 3650.
30. Bento, M.G.; Bertolami, O. Scale factor duality: A Quantum cosmological approach. *Class. Quantum Grav.* **1995**, *12*, 1919.
31. Kamenshchik, A.Y.; Tronconi, A.; Venturi, G. The Born-Oppenheimer approach to quantum cosmology. arXiv **2020**, arXiv:2010.15628.
32. Lidsey, J.E. Bianchi-IX Quantum Cosmology of the Heterotic String. *Phys. Rev.* **1994**, *49*, R599.
33. Lidsey, J.E. String quantum cosmology of the Bianchi class A. arXiv **1994**, arXiv:gr-qc/9404050.
34. van Elst, H.; Lidsey, J.E.; Tavakol, R. Quantum Cosmology and Higher-Order Lagrangian Theories. *Class. Quantum Grav.* **1994**, *11*, 2483.
35. Lidsey, J.E. Quantum cosmology of generalized two-dimensional dilaton-gravity models. *Phys. Rev.* **1995**, *51*, 6829.
36. Lidsey, J.E. Scale factor duality and hidden supersymmetry in scalar-tensor cosmology. *Phys. Rev.* **1995**, *52*, R5407.
37. Lidsey, J.E. Inflationary and deflationary branches in extended pre-big-bang cosmology. *Phys. Rev.* **1997**, *55*, 3303.
38. De Sabbata, V.; Gasperini, M. Neutrino Oscillations in the Presence of Torsion. *Nuovo Cim.* **1981**, *65*, 479–500.
39. Lovelace, C. Strings in Curved Space. *Phys. Lett.* **1984**, *135*, 75.
40. Fradkin, E.S.; Tseytlin, A.A. Quantum String Theory Effective Action. *Nucl. Phys.* **1985**, *261*, 1.
41. Callan, C.G.; Martinec, E.J.; Perry, M.J.; Friedan, D. Strings in Background Fields. *Nucl. Phys.* **1985**, *262*, 593.
42. Gasperini, M. Dilatonic interpretation of the quintessence? *Phys. Rev.* **2001**, *64*, 043510. [CrossRef]
43. Meissner, K.A.; Veneziano, G. Manifestly O(d,d) invariant approach to space-time dependent string vacua. *Mod. Phys. Lett.* **1991**, *6*, 3397.
44. Meissner, K.A.; Veneziano, G. Symmetries of cosmological superstring vacua. *Phys. Lett.* **1991**, *267*, 33. [CrossRef]
45. Veneziano, G. Scale factor duality for classical and quantum strings. *Phys. Lett.* **1991**, *265*, 287. [CrossRef]
46. Tseytlin, A.A. Duality and dilaton. *Mod. Phys. Lett.* **1991**, *6*, 1721
47. Ashtekar, A.; Geroch, R. Quantum theory of gravitation. *Rep. Prog. Phys.* **1974**, *37*, 1211.
48. Grishchuk, L.P. Amplification of gravitational waves in an istropic universe. *Sov. Phys. JETP* **1975**, *40*, 409.
49. Starobinski, A.A. Spectrum of relict gravitational radiation and the early state of the universe. *JETP Lett.* **1979**, *30*, 682.
50. Rubakov, V.A. On the Third Quantization and the Cosmological Constant. *Phys. Lett.* **1988**, *214*, 503.
51. Kozimirov, N.; Tkachev, I.I. Dimension of space-time in third quantized gravity. *Mod. Phys. Lett.* **1988**, *4*, 2377.
52. McGuigan, M. Third Quantization and the Wheeler-dewitt Equation. *Phys. Rev.* **1988**, *38*, 3031.
53. McGuigan, M. Universe Creation From the Third Quantized Vacuum. *Phys. Rev.* **1989**, *39*, 2229.
54. McGuigan, M. Universe Decay and Changing the Cosmological Constant. *Phys. Rev.* **1990**, *41*, 418.
55. Birrel, N.D.; Davies, P.C.W. *Quantum Fields in Curved Spaces*; University Press: Cambridge, UK, 1982.
56. Schumaker, B.L. Quantum mechanical pure states with gaussian wave functions. *Phys. Rep.* **1986**, *135*, 317.
57. Muller, M. Rolling radii and a time-dependent dilation. *Nucl. Phys.* **1990**, *337*, 37.
58. Abramowitz, M.; Stegun, I.A. *Handbook of Mathematical Functions*; Dover: New York, NY, USA, 1972.
59. Vilenkin, A. Creation of Universes from Nothing. *Phys. Lett.* **1982**, *117*, 25.
60. Vilenkin, A. Quantum Cosmology and the Initial State of the Universe. *Phys. Rev.* **1988**, *37*, 888.
61. Gasperini, M. Observable gravitational waves in pre-big bang cosmology: An update. *J. Cosmol. Astropart. Phys.* **2016**, *12*, 010. [CrossRef]

MDPI
St. Alban-Anlage 66
4052 Basel
Switzerland
Tel. +41 61 683 77 34
Fax +41 61 302 89 18
www.mdpi.com

Universe Editorial Office
E-mail: universe@mdpi.com
www.mdpi.com/journal/universe